D1653514

Edited by Valery V. Tuchin

Advanced Optical Flow Cytometry

Related Titles

Lakshmipathy, U., Chesnut, J. D., Thyagarajan, B.

Emerging Technology Platforms for Stem Cells

2009
506 pages
Hardcover
ISBN: 978-0-470-14693-4

Wolfbeis, O. S. (ed.)

Fluorescence Methods and Applications
Spectroscopy, Imaging, and Probes

2008
452 pages
Softcover
ISBN: 978-1-57331-716-0

Murphy, D. B.

Fundamentals of Light Microscopy and Electronic Imaging

2008
E-Book
ISBN: 978-0-470-24202-5

Prasad, P. N.

Introduction to Biophotonics

2004
E-Book
ISBN: 978-0-471-46539-3

Durack, G., Robinson, J. P. (eds.)

Emerging Tools for Single-Cell Analysis
Advances in Optical Measurement Technologies

2002
Online Book Wiley Interscience
ISBN: 978-0-471-22484-6

Givan, A. L.

Flow Cytometry
First Principles

296 pages
2001
Softcover
ISBN: 978-0-471-38224-9

Edited by Valery V. Tuchin

Advanced Optical Flow Cytometry

Methods and Disease Diagnoses

WILEY-VCH Verlag GmbH & Co. KGaA

The Editor

Prof. Valery V. Tuchin
Saratov State University
Optics and Biophotonics
Saratov, Russian Feeration
tuchinvv@mail.ru

Library of Congress Card No.: applied for

British Library Cataloguing-in-Publication Data
A catalogue record for this book is available from the British Library.

Bibliographic information published by the Deutsche Nationalbibliothek
The Deutsche Nationalbibliothek lists this publication in the Deutsche Nationalbibliografie; detailed bibliographic data are available on the Internet at <http://dnb.d-nb.de>.

Cover Design Grafik-Design Schulz, Fußgönheim
Typesetting Laserwords Private Limited, Chennai, India
Printing and Binding Strauss GmbH, Mörlenbach

Printed in the Federal Republic of Germany
Printed on acid-free paper

ISBN: 978-3-527-40934-1
ePDF ISBN: 978-3-527-63430-9
oBook ISBN: 978-3-527-63428-6
ePub ISBN: 978-3-527-63429-3
Mobi ISBN: 978-3-527-63431-6

Contents

Preface

Flow cytometry was invented in the late 1960s, and since then the flow cytometer has become an indispensable tool in modern research and clinical laboratories [1–7]. Beyond the routine usage, new trends can be observed in the development of flow cytometric techniques. The main technological improvements include high-speed sorting, phase-sensitive flow cytometry, multicolor flow cytometry, high-throughput multiplex bead assays, and spectral detection, and it provides the basis for extensive data collection.

Classical flow cytometry (FC) uses an instrument system for making, processing, and displaying one or more measurements on individual cells in flowing cell suspension [1–7]. Cells may be stained with one or more fluorescent dyes specific to cell components of interest, for example, DNA, and fluorescence of each cell is measured as cells one by one rapidly transverse the excitation beam (laser or mercury arc lamp). Fluorescence provides a quantitative measure of various biochemical and biophysical properties of the cell. Other measurable optical parameters, which are applicable to the measurement of cell size, shape, density, granularity, and stain uptake, include light absorption, light scattering, and polarization degree.

Numerous clinical and research applications, especially in anatomic pathology for detection and study of malignant lesions, use the so-called image cytometry. This technique encompasses morphometry and densitometry as measuring techniques, and neural networks and expert systems for processing of collected data.

Another cytometric technique is the microscope-based *laser scanning cytometry*, which allows one to make fluorescence measurements and topographic analysis on individual cells. Laser-induced fluorescence of labeled cellular specimens is detected using multiple discrete wavelengths, and the spatially resolved data are processed to quantify cell proliferation, apoptosis, gene expression, protein transport, and other cellular processes. For instance, confocal microscopy and two-photon imaging techniques are able to detect fluorescently labeled cells not only *in vitro* but also *in vivo* [8].

Improvements in image cytometric techniques speeded up in the last two decades, applying more and more sensitive detectors and introducing nonlinear optics (two photon excitation) for lifetime measurements (fluorescence lifetime imaging microscopy, FLIM). The latest developments were able to break the diffraction limit as in the scanning near-field optical microscopy (SNOM), total internal reflection microscopy (TIRFM), fluorescence resonance energy transfer (FRET), stimulated emission depletion (STED), and 4Pi and multiobjective microscopy.

Conventional FC is currently the method of choice for rapid quantification of cells, but it requires invasive extraction of cells from a living organism and associated procedures (e.g., fluorescence labeling and sorting), which may lead to unpredictable artifacts, and prevents long-term cell monitoring in the native biological environment. Among *in vivo* techniques, both nonoptical (e.g., PET and MRI) and optical (e.g., scattering, fluorescence, confocal, and multiphoton) techniques can be used for visualizing only single static or slowly migrating cells [8–10]. To detect fast moving cells in blood and lymph flows, a number of methods providing *in vivo* FC have been developed [11–21]. In particular, the principle of FC has been adapted to the *in vivo* monitoring of labeled cells in ear blood vessels, and a few modifications of *in vivo* flow cytometers that are capable of real-time confocal detection of fluorescently labeled cells in both the arterial and venous circulation of small animals, have been built [12–14].

The alternative photothermal (PT) and photoacoustic (PA) techniques for *in vivo* blood FC, which do not require cell labeling and are not sensitive to light scattering and autofluorescence background, have also been recently suggested [15–17]. These techniques have potential application in the study of normal and abnormal cells in their native condition in blood or lymph flows *in vivo*, including molecular imaging, studying the metabolism and pathogenesis of diseases at a cellular level, and monitoring and quantifying metastatic and apoptotic cells and/or their responses to therapeutic interventions (e.g., drug or radiation).

Video microscopy and particle tracking methods adapted and integrated with an ultrahigh-speed imaging camera were used to measure lymph velocities throughout the entire lymphatic contraction cycle in the rat mesentery [18–23]. *In vivo*, label-free, high-speed (up to 10 000 with the potential for 40 000 fps), high-resolution (up to 300 nm) optical imaging of circulating individual erythrocytes, leukocytes, and platelets in fast blood flow has been developed [22]. Different potential applications of *in vivo* digital video microscopy include visualization of circulating cells and their deformability in lymph and blood flows and the study of the kinetics of platelets and leukocyte rolling, with high sensitivity and resolution.

Multiphoton fluorescence flow cytometry and its confocal and fiber-optic modifications hold a great promise for *in vivo* monitoring of multiple circulating cell populations in blood and lymph flows by exciting and detecting the emission from multiple fluorophores, such as fluorescent proteins and exogenous chromophores, important for multilabeling of cells of interest [24, 25].

There are many books on cytometry published since the 1980s (see, for example, the list on the web site [1]). They could be classified as books on general flow cytometry and cell sorting, clinical cytometry, and microscopic and imaging cytometry. The most recent and comprehensive are Refs [2, 3, 5, 8–10, 26–30]. The book by Shapiro [5] is the fourth edition on classic flow cytometry. This is one of the best textbooks, and covers well the field of practical flow cytometry prior to 2003 well. Recently, two books on flow cytometry and cellular diagnostics have been published in German and French [26, 27]. Both books were written and edited by well-known experts in the field. Karger has published the English translation of the German book edited by Ulrich Sack, Attila Tárnok, and Gregor Rothe, which is a bestseller in German [28]. Practical cytometry protocols have been given in the third edition of the book edited by Michael G. Ormerod [29]. The recent second edition of the book by Wojciech Gorczyca is more clinically and practically oriented [30]. Michael G. Ormerod has also designed an introductory book "to give that knowledge, aiming at people coming to flow cytometry for the first time," in which all the major applications in mammalian biology are covered [31].

While the above-mentioned books describe clinical diagnostic methods and receipts of their applications, the current book is more research oriented and opens new perspectives in the development of flow cytometry for *in vivo* studies. It contains novel results of basic research on light scattering by different types of cells, which are very important for the improvement of already existing technologies and for designing new technologies in optical cytometry. The recently invented and fast-moving-to-practice methods of *in vivo* flow cytometry, based on ultrafast video and phase intra-vital microscopy and light scattering, diffraction, speckle, fluorescence, multiphoton, Raman, photothermal, and photoacoustic phenomena, are presented in the book.

In Chapter 1, *Perspectives in Cytometry* by Anja Mittag and Attila Tárnok, in addition to definitions, historical aspects, and the importance of cytometry in the development of biology and medicine, its prospective application as a science and diagnostic tool are discussed – in particular, for comprehensive analyses, on the basis of the simultaneous detection of several parameters, of up to millions of individual cells in one sample.

Chapter 2 by Herbert Schneckenburger *et al.*, *Novel Concepts and Requirements in Cytometry*, presents slide-based cytometry techniques and the concepts of high content screening (HCS) where detailed information is accumulated from a single cell or examination of multicellular spheroids where 3D detection methods are required. These techniques include microscopic setups, fluorescence reader systems, and microfluidic devices with micromanipulation, for example, cell sorting.

In Chapter 3 by Stoyan Tanev *et al.*, *Optical Imaging of Cells with Gold Nanoparticle Clusters as Light Scattering Contrast Agents: A Finite-Difference Time-Domain Approach to the Modeling of Flow Cytometry Configurations*, a brief summary of different formulations of the finite-difference time-domain (FDTD) approach is presented in the framework of its strengths, for cytometry in general and for potential applications in *in vivo* flow cytometry based on light scattering, including nanoscale targets. This chapter focuses on comparison of light scattering by a single biological cell alone under controlled refractive index matching conditions and by cells labeled using gold nanoparticle clusters. The optical phase contrast microscopy (OPCM) is analyzed as a prospective modality for *in vivo* flow cytometry.

In Chapter 4 by Valeri P. Maltsev *et al.*, *Optics of White Blood Cells: Optical Models, Simulations, Experiments*, a state-of-the-art summary of analytical and numerical simulating methods and experimental approaches for precise description and detection of elastic light scattering from white blood cells (WBCs) are presented. The discussion of the instrumental tools for measurement of light scattering lays emphasis on scanning flow cytometry. This chapter gives some basis for understanding the methods, techniques, and experimental results presented in the following chapters, as it presents solutions for the inverse light-scattering problem to obtain cellular characteristics from light scattering data.

The optical properties of blood are discussed in Chapter 5 by Martina Meinke *et al.*, *Optical Properties of Flowing Blood Cells.* Authors use the transport theory accounting for the multiplicity of light scattering events, where the optical properties of blood are described by the absorption and scattering coefficients and the anisotropy factor. The double integrating sphere measurement technique combined with inverse Monte Carlo simulation is applied for extraction of the optical parameters of undiluted blood. It is shown that the influence of the shear rate and osmolarity have to be taken into account when the blood is prepared *ex vivo* and the physiological environment cannot be ensured.

In Chapter 6, *Laser Diffraction on RBC and Deformability Measurement* by Alexander V. Priezzhev *et al.*, RBC shape variability and deformability as intrinsic properties, their strong relation to RBC aggregation and blood rheology, and the determination of the general hemorheologic status of human organism are considered. It is shown that laser diffraction can be efficiently applied to quantitatively assess the deformability properties of RBC in the blood of a particular individual. The theoretical basis of diffractometry and implementation of particular experimental techniques to experimental and clinical measurements, as well as potentialities and pitfalls of the technique are discussed.

The principles and fundamentals of flicker spectroscopy as a quantitative tool to measure static and dynamic mechanical properties of composite cell membrane are presented in Chapter 7, *Characterization of Red Blood Cells Rheological and Physiological State Using Optical Flicker Spectroscopy* by Vadim L. Kononenko. These mechanical properties are associated with cell membrane and cytoplasm molecular organization and composition and cell metabolic activity and could be characteristic not only for RBC, but for other blood cell types as well. Microscope-based flicker spectroscopy technique in combination with quantitative phase imaging and fluorescence microscopy can be easy integrated into the slide-based cytometry arrangement. As the author states, the approximate models developed are good, but not enough to reconstruct precisely the details of erythrocyte cell membrane mechanical properties; thus a more advanced theory of flicker spectroscopy is needed.

Chapter 8, *Digital Holographic Microscopy for Quantitative Live Cell Imaging and Cytometry*, by Björn Kemper and Jürgen Schnekenburger, demonstrates the principles and applications of quantitative cell imaging using digital holographic microscopy (DHM). The quantitative phase contrast imaging, cell thickness determination, multifocus imaging, and 2D cell tracking provided by DHM show that it is a suitable method for the label-free characterization of dynamic live cell processes involving morphological alterations and migration and for the analysis of cells in 3D environments. Examples and illustrations of applicability of the technique in tumor cell biology and for the development of improved systems for drug and toxicity testing are presented.

Chapter 9 by József Bocsi *et al.*, *Comparison of Immunophenotyping and Rare Cell Detection by Slide-Based Imaging Cytometry and by Flow Cytometry*, allows one to get answers on the following questions: are flow cytometry (FC) and slide-based cytometry (SBC) comparable? What is the type of cytometer and analysis technique that should be chosen for the given biological problem to be solved? Answers are illustrated by applying scanning fluorescence microscope (SFM) to determine of CD4/CD8 T cell ratio, laser scanning cytometer (LSC) to multiparametric leukocyte phenotyping and apoptosis analysis on the basis of DNA content measurements, and SFM and LSC to rare and frequent tumor cell detection.

A brief overview and discussion of recent progress in microfluidic flow cytometry, including the main components of full scale flow cytometers, containing systems for fluidic control, optical detection and cell sorting, each of which are being developed into on-chip microfluidic platforms, are given by Shawn O. Meade *et al.* in Chapter 10, *Microfluidic Flow Cytometry: Advancements Toward Compact, Integrated Systems.*

In Chapter 11 by Xin-Hua Hu and Jun Q. Lu, *Label-Free Cell Classification with a Diffraction Imaging Flow Cytometer*, aiming at the accurate modeling of light scattering from biological cells with realistic cell structures and the development of a high contrast diffraction imaging flow cytometer for experimental study, the authors are focused on the application of the FDTD method for modeling of coherent light scattering from cells. Numerical and experimental results are presented and their implications to future improvement of the flow cytometry are discussed.

In Chapter 12 by Rabindra Tirouvanziam *et al.*, *An Integrative Approach for Immune Monitoring of Human Health and Disease by Advanced Flow Cytometry Methods*, the authors show key steps to move past the current limitations and truly enable the use of advanced flow cytometry tools for human research, promoting simplified, low-cost, and better standardized methods for sample collection, highlighting the enormous opportunities for research on reagents available for advanced flow cytometry analysis of human samples, and novel insights into relations of human immunity with age, gender, ethnicity, environmental exposure, health conditions, and therapies.

R. Dasgupta and P.K. Gupta, in Chapter 13, *Optical Tweezers and Cytometry*, give a brief introduction to optical tweezers and an overview of their use in cytometric applications, including measurements of viscoelastic properties of cells, in particular RBC, and Raman spectroscopic studies at single cell level. A few examples illustrating the potential of this approach for cytometric applications are also presented.

Chapter 14 by Valery V. Tuchin *et al.*, *In vivo Image Flow Cytometry*, presents one of the novel approaches in flow cytometry – *in vivo* video imaging digital flow cytometry. The fundamentals and instrumentation of video imaging flow cytometry, as well as spatial and temporal resolution of the method, are discussed. Experimental animal models, data on imaging and detection of individual cells in lymph and blood flows, and cell velocity measurements in lymph and blood vessels are presented and discussed. Intravessel RBC deformability measurement, monitoring of intralymphatic cell aggregation, and many other cell interaction phenomena are demonstrated and quantified. Perspectives of the technique for disease diagnostics and monitoring and cell flow response on drugs, pollutions, and toxins are shown.

Chapter 15 by Stephen P. Morgan and Ian M. Stockford, *Instrumentation for In Vivo Flow Cytometry: A Sickle Cell Anemia Case Study*, discusses label-free monitoring of the properties of circulating blood cells for the *in vivo* monitoring of sickle cell anemia. For discriminating sickled RBCs in a background of normal cells, absorption measurements associated with sickle cell lower oxygen saturation and polarization measurements to identify their polymerization ability via cell adhesion to the vascular walls and, thus, more slow flow, are used. Illumination methods overcoming surface reflections, such as orthogonal polarization spectral imaging, dark field epi-illumination, and sidestream dark field illumination, are analyzed. In humans blood cell imaging has been performed either on the lower lip or under the tongue where the superficial mucosal tissue above the microcirculation is thinner than at other sites on the body. All steps of clinical instrumentation design, starting from discussion of the clinical needs for the measurements, the illumination and detection requirements, image processing methods for correction of image distortions, and a Monte Carlo model of the image formation process, up to engineering of the clinical prototype and presentation of clinical results are highlighted by the authors.

Accounting for the clinical importance of detection and quantification of circulating tumor cells (CTCs) for cancer diagnosis, staging, and treatment, in Chapter 16, *Advances in Fluorescence-Based In Vivo Flow Cytometry for Cancer Applications*, Cherry Greiner and Irene Georgakoudi, review the principles and instrumentation designs of fluorescence-based *in vivo* flow cytometry (IVFC) and present data on the *in vivo* quantification of CTCs. The confocal and multiphoton microscopic techniques and systems adapted for the detection of fluorescently labeled CTCs in blood vessels are described. The noninvasive nature of IVFC systems and their capability to provide sensitive, continuous and dynamic monitoring of CTCs in blood flow are proved.

Chapter 17, *In Vivo Photothermal and Photoacoustic Flow Cytometry* by Valery V. Tuchin *et al.*, is devoted to presentation of the prospective approaches of IVFC that use laser-induced photothermal (PT) and photoacoustic (PA) effects. The authors analyze in detail integrated IVFC techniques combining a few different methods, such as PT imaging conjugated with thermo-lens and PA imaging, transmittance digital microscopy, and phase-sensitive and fluorescence imaging. The unique capabilities of the PT/PAFC (photoacoustic flow cytometry) technique for IVFC are illustrated in many examples of *in vivo* and *ex vivo* studies within lymph and blood vessels of animal models. Data on cell velocity measurements, detection, and real-time monitoring of circulating blood and lymph cells, bacteria, CTCs, contrast agents, and nanoparticles, and on quantification of cell interactions are presented. Perspectives of PT/PAFC technique for early diagnostics of cancer are discussed.

Martin J. Leahy and Jim O'Doherty, in Chapter 18, *Optical Instrumentation for the Measurement of Blood Perfusion, Concentration, and Oxygenation in Living Microcirculation*, compare the operation of an established microcirculation imaging technique, such as laser Doppler perfusion imaging (LDPI), which, for example, has been shown to accurately assess burn depth, with laser speckle perfusion imaging (LSPI) and tissue viability imaging (TiVi) in human skin tissue using the occlusion and reactive hyperaemia response. On the basis of the presented experimental data they conclude that LSPI and TiVi are both welcome tools in the study of the microcirculation, but care must be taken in the interpretation of the images since blood flow velocity and blood concentration in tissue are essentially different parameters.

In Chapter 19, *Blood Flow Cytometry and Cell Aggregation Study with Laser Speckle* by Qingming Luo *et al.*, the fundamentals and instrumentation of laser speckle contrast imaging (LSCI) are analyzed with a discussion of important imaging parameters for the optimal imaging conditions. Some recent advancements, such as spatio-temporal algorithm for laser speckle contrast analysis and fast blood flow visualization using GPU (Graphics Processing Unit), are overviewed. The application of LSCI for investigation of RBC aggregation is discussed.

Chapter 20, *Modifications of Blood Optical Properties during Photodynamic Reactions In Vitro and In Vivo*, by Alexandre Douplik *et al.*, describes photodynamic reactions where blood cells are involved. Change in blood properties on light delivery and modification of photosensitizer (PS) optical properties on interaction with blood at photodynamic therapy (PDT) are considered. The blood cell uptake of PSs and modification of blood optical properties caused by PDT reactions *in vitro* and *in vivo* are analyzed. The authors of the chapter believe that IVFC can be applied to study PDT-induced processes related to blood and blood cells.

As it follows from the above, this book is focused on state-of-the-art research in a novel field, noninvasive *in vivo* cytometry, and its applications, with particular emphasis on the novel biophotonic methods, disease diagnosis, and monitoring of disease treatment at single cell level in stationary and flow conditions. However, discussions of advanced methods and techniques of classical flow cytometry are also presented. The use of photonic technologies in medicine is a rapidly emerging and potentially powerful approach for disease detection and treatment. This book seeks to advance scholarly research that spans from fundamental interactions between light, cells, vascular tissue, and cell labeling particles, to strategies and opportunities for preclinical and clinical research. General topics include light scattering by cells, fast video microscopy, polarization, laser scanning, fluorescence, Raman, multiphoton, photothermal, and photoacoustic methods for cellular diagnostics and monitoring of disease treatment in living organisms. Specific topics include optics of erythrocytes, leukocytes, platelets, and lymphocytes; novel photonic cytometry techniques; *in vivo* studies using animal models; *in vivo* cytometry techniques used in humans (mucosal, nail bed, and skin); detection of metastatic cancer cells and labeling of nanoparticles in blood and lymph microvessels; immune monitoring of human health; cytomics in regenerative/predictive medicine; comparison of laser scanning slide-based and flow cytometry, and so on.

The book is for research workers, practitioners, and professionals in the field of cytometry. Advanced students (MS and Ph.D.) as well as undergraduate students specialized in biomedical physics and engineering, biomedical optics and biophotonics, and medical science may use this book as a comprehensive tutorial helpful in the preparation of their research work and diploma. Scientists or professionals and students in other disciplines, such as laser and optical engineering and technology, spectroscopy, tomography, developmental and other directions of biology, tissue engineering, and different specialties in medicine, are also potential readership.

This book represents a valuable contribution by well-known experts in the field of photonic cytometry with their particular interest in a variety of advanced cytometry problems, including *in vivo* flow cytometry. The contributors are drawn from Canada, China, Denmark, Germany, India, Ireland, Russia, The Netherlands, UK, and the USA. I greatly appreciate the cooperation and contributions of all authors in the book, who have done great work in the preparation of their chapters.

It should be mentioned that this book presents results of international collaborations and exchanges of ideas among many research groups participating in the book project. This book project was supported by many international grants, which are credited in the particular chapters. Here, I would like to mention only a few, PHOTONICS 4 LIFE, which is a consortium of a well-balanced pan-European dimension. The inclusion of the members of this consortium in this book is of great significance, encompassing five chapters. My own work on the book was supported by grants 2.1.1/4989 and 2.2.1.1/2950 of RF Program on the Development of High School Potential and RF Governmental contracts 02.740.11.0484, 02.740.11.0770, and 02.740.11.0879 of The Program "Scientific and Pedagogical Personnel of Innovative Russia."

I am grateful to Valerie Moliere for her suggestion to publish this book and help on project activation and to Ulrike Werner for the technical editing of the book and communication with the contributors.

I would like to thank all those authors and publishers who freely granted permissions to reproduce their copyrighted works.

I greatly appreciate the cooperation, contributions, and support of all my colleagues from the Optics and Biophotonics Chair and Research-Educational Institute of Optics and Biophotonics of the Physics Department of Saratov State University and the Institute of Precise Mechanics and Control of the Russian Academy of Science.

I express my gratitude to my family, especially to my wife Natalia and grandchildren Dasha, Zhenya, Stepa, and Serafim for their indispensable support, understanding, and patience during my writing and editing of this book.

August 2010 *Saratov*

References

1. Robinson, J.P. (2011) *Reference Books in Cytometry*, Purdue University Cytometry Laboratories, *http://www.cyto.purdue.edu/flowcyt/books/bookindx.htm*, February 21, 2011.
2. Melamed, M.R., Lindmo, T., and Mendelsohn, M.L. (eds) (1990) *Flow Cytometry and Sorting*, 2nd edn, Wiley-Liss, New York.
3. Hudnall, S.D. (ed.) (2000) *Introduction to Diagnostic Flow Cytometry: An Integrated*

Case-Based Approach (Pathology and Laboratory Medicine)*, Humana Press.

4. Givan, A.L. (2001) Principles of flow cytometry: an overview. *Methods Cell. Biol.*, **63**, 19–50.
5. Shapiro, H.M. (2003) *Practical Flow Cytometry*, 4th edn, Wiley-Liss, New York.
6. De Rosa, S.C., Brenchley, J.M., and Roederer, M. (2003) Beyond six colors: a new era in flow cytometry. *Nat. Med.*, **9**, 112–117.
7. Hogan, H. (2006) Flow cytometry moves ahead. *Biophotonics Int.*, **13**, 38–42.
8. Pawley, J.P. (1995) *Handbook of Biological Confocal Microscopy*, 2nd edn, Plenum, New York.
9. Diaspro, A. (ed.) (2002) *Confocal and Two-Photon Microscopy: Foundations, Applications, and Advances*, Wiley-Liss, New York.
10. Masters, B.R. and So, T.P.C. (2004) *Handbook of Multiphoton Excitation Microscopy and other Nonlinear Microscopies*, Oxford University Press, New York.
11. Tuchin, V.V., Tárnok, A., and Zharov, V.P. (Guest Editors) (2009) Towards *in vivo* flow cytometry. *J. Biophoton.*, **2** (8–9), 457–547.
12. Georgakoudi, I., Solban, N., Novak, J., Rice, W., Wei, X., Hasan, T., and Lin, C.P. (2004) *In vivo* flow cytometry: a new method for enumerating circulating cancer cells. *Cancer Res.*, **64**, 5044–5047.
13. Sipkins, D.A., Wei, X., Wu, J.W., Runnels, J.M., Côté, D., Means, T.K., Luster, A.D., Scadden, D.T., and Lin, C.P. (2005) *In vivo* imaging of specialized bone marrow endothelial microdomains for tumour engraftment. *Nature (London)*, **435**, 969–973.
14. Boutrus, S., Greiner, C., Hwu, D., Chan, M., Kuperwasser, C., Lin, C.P., and Georgakoudia, I. (2007) Portable two-color *in vivo* flow cytometer for real-time detection of fluorescently-labeled circulating cells. *J. Biomed. Opt.*, **12** (2), 020507-1–020507-3.
15. Zharov, V.P., Galanzha, E.I., and Tuchin, V.V. (2005) Photothermal image flow cytometry *in vivo*. *Opt. Let.*, **30** (6), 628–630.
16. Zharov, V.P., Galanzha, E.I., and Tuchin, V.V. (2006) *In vivo* photothermal flow cytometry: imaging and detection of individual cells in blood and lymph flow. *J. Cell Biochem.*, **97** (5), 916–930.
17. Zharov, V.P., Galanzha, E.I., Shashkov, E.V., Khlebtsov, N.G., and Tuchin, V.V. (2006) *In vivo* photoacoustic flow cytometry for monitoring of circulating single cancer cells and contrast agents. *Opt. Lett.*, **31**, 3623–3625.
18. Galanzha, E.I., Brill, G.E., Aisu, Y., Ulyanov, S.S., and Tuchin, V.V. (2002) in *Handbook of Optical Biomedical Diagnostics*, vol. PM107, Chapter 16 (ed. V.V. Tuchin), SPIE Press, Bellingham, WA, pp. 881–937.
19. Galanzha, E.I., Tuchin, V.V., and Zharov, V.P. (2005) *In vivo* integrated flow image cytometry and lymph/blood vessels dynamic microscopy. *J. Biomed. Opt.*, **10**, 54018-1–54018-8.
20. Dixon, J.B., Zawieja, D.C., Gashev, A.A., and Coté, G.L. (2005) Measuring microlymphatic flow using fast video microscopy. *J. Biomed. Opt.*, **10**, 064016.
21. Dixon, B.J., Greiner, S.T., Gashev, A.A., Coté, G.L., Moore, J.E., and Zawieja, D.C. (2006) Lymph flow, shear stress, and lymphocyte velocity in rat mesenteric prenodal lymphatics. *Microcirculation*, **13**, 597–610.
22. Zharov, V.P., Galanzha, E.I., Menyaev, Y., and Tuchin, V.V. (2006) *In vivo* high-speed imaging of individual cells in fast blood flow. *J. Biomed. Opt.*, **11** (5), 054034-1–054034-4.
23. Galanzha, E.I., Tuchin, V.V., and Zharov, V.P. (2007) Advances in small animal mesentery models for *in vivo* flow cytometry, dynamic microscopy, and drug screening (invited review). *World J. Gastroenterol.*, **13** (2), 198–224.
24. Tkaczyk, E.R., Zhong, C.F., Ye, J.Y. *et al.* (2008) *In-vivo* monitoring of multiple circulating cell populations using two-photon flow cytometry. *Opt. Commun.*, **281**, 888–894.
25. Chang, Y.-C., Yong Ye, J., Thomas, T.P., Cao, Z., Kotlyar, A., Tkaczyk, E.R., Baker, J.R. Jr., and Norris, T.B. (2010)

Fiber-optic multiphoton flow cytometry in whole blood and *in vivo*. *J. Biomed. Opt.*, **15**, 047004.

26. Sack, U., Tárnok, A., and Rothe, G. (eds) (2006) *Zelluläre Diagnostik: Grundlagen, Methoden und klinische Anwendungen der Durchflusszytometrie*, Karger.
27. Ronot, X., Grunwald, D., Mayol, J.F., and Boutannat, J. (eds) (2006) *La Cytométrie en Flux*, TEC & DOC, EM Inter.
28. Sack, U, Tárnok, A, and Rothe, G (eds) (2008) *Cellular Diagnostics: Basic Principles, Methods and Clinical Applications of Flow Cytometry*, Karger.
29. Ormerod, M.G. (ed.) (2000) *Flow Cytometry. A Practical Approach*, 3rd edn, IRL Press at Oxford University Press, Oxford.
30. Gorczyca, W. (2010) *Flow Cytometry in Neoplastic Hematology: Morphologic-Immunophenotypic Correlation*, 2nd edn, Informa Healthcare. 184184702X.
31. Ormerod, M.G. (2008) *Flow Cytometry – A Basic Introduction*, De Novo software, the electronic version of the book (with colour figures), free for you to read, *http://flowbook.denovosoftware.com/*, page last modified 08:42, 12 Aug 2010 by Michael Ormerod, February 21, 2011.

List of Contributors

József Bocsi
University of Leipzig
Department of Paediatric
Cardiology
Heart Centre
Strümpellstr. 39
04275 Leipzig
Germany

and

University of Leipzig
LIFE - Leipzig Research Center
for Civilization Diseases
Faculty of Medicine
Philipp-Rosenthal-Str. 27
04103 Leipzig
Germany

Thomas Bruns
Hochschule Aalen
Institut für Angewandte
Forschung
Beethovenstr. 1
73431 Aalen
Germany

Chun-Hao Chen
University of California
San Diego
Department of Bioengineering
9500 Gilman Drive
MC 0407
La Jolla, CA 92093
USA

Sung Hwan Cho
University of California
San Diego
Department of Materials Science
and Engineering
9500 Gilman Drive
MC 0407
La Jolla, CA 92093
USA

Monique Crubezy
Stanford University School of
Medicine
Beckman Center
279 Campus Drive
Stanford, CA 94305-5318
USA

Raktim Dasgupta
Laser Biomedical Applications
and Instrumentation Division
Raja Ramanna Centre for
Advanced Technology
P.O: CAT
Indore 452013
India

Daisy Diaz
Stanford University
School of Medicine
Beckman Center
279 Campus Drive
Stanford, CA 94305-5318
USA

Alexandre Douplik
Clinical Photonics Lab at
Erlangen Graduate School in
Advanced Optical
Technologies (SAOT)
Friedrich-Alexander Universität
Erlangen-Nürnberg
Erlangen
Germany

and

Medical Photonics Engineering at
Chair of Photonics Technologies
Friedrich-Alexander Universität
Erlangen-Nürnberg
Erlangen
Germany

Moritz Friebel
Laser- und Medizin-Technologie
GmbH
Fabeckstrasse 60–62
14195 Berlin
Germany

Ekaterina I. Galanzha
Saratov State University
Optics and Biophotonics
Department and Research-
Educational Institute of Optics
and Biophotonics
83 Astrakhanskaya St.
Saratov 410012
Russia

and

University of Arkansas
for Medical Sciences
Philips Classic Laser Laboratories
4301 W. Markham St.
Little Rock, AR 72205 7199
USA

Irene Georgakoudi
Tufts University
Biomedical Engineering
Department
4 Colby Street
Medford, MA 02155
USA

Yael Gernez
Stanford University
School of Medicine
Beckman Center
279 Campus Drive
Stanford, CA 94305-5318
USA

Jessica Godin
University of California
San Diego
Department of Electrical and
Computer Engineering
9500 Gilman Drive
MC 0407
La Jolla, CA 92093
USA

Cherry Greiner
Tufts University
Biomedical Engineering
Department
4 Colby Street
Medford, MA 02155
USA

Pradeep Kumar Gupta
Laser Biomedical Applications
and Instrumentation Division
Raja Ramanna Centre for
Advanced Technology
P.O: CAT
Indore 452013
India

Jürgen Helfmann
Laser- und Medizin-Technologie
GmbH
Fabeckstrasse 60–62
14195 Berlin
Germany

Alfons G. Hoekstra
University of Amsterdam
Faculty of Science
Computational Science
Science Park 904
1098 XH Amsterdam
The Netherlands

Xin-Hua Hu
East Carolina University
Department of Physics
MS 563
Greenville, North Carolina 27858
USA

Björn Kemper
University of Muenster
Center for Biomedical
Optics and Photonics
Robert-Koch-Str. 45
48149 Muenster
Germany

Vadim L. Kononenko
Russian Academy of Sciences
N.M. Emanuel Institute of
Biochemical Physics
Kosygin Str. 4
Moscow, 119334
Russia

Julie Laval
Stanford University
School of Medicine
Beckman Center
279 Campus Drive
Stanford, CA 94305-5318
USA

Martin J. Leahy
University of Limerick
Department of Physics
Castletroy
Plassey Park Rd.
Limerick
Ireland

and

Royal College of Surgeons in
Ireland
123. St., Stephen's Green
Dublin
Ireland

Pengcheng Li
Huazhong University of
Science and Technology
Britton Chance Center for
Biomedical Photonics
Wuhan National Laboratory for
Optoelectronics
1037 Luoyu Road
Wuhan 430074
China

Yu-Hwa Lo
University of California
San Diego
Department of Electrical and
Computer Engineering
9500 Gilman Drive
MC 0407
La Jolla, CA 92093
USA

Viktor Loshchenov
General Physics Institute
Russian Academy of Science
Moscow
Russia

Andrei E. Lugovtsov
Lomonosov Moscow
State University
Department of Physics
International Laser Center
Moscow 119991
Russia

Jun Q. Lu
East Carolina University
Department of Physics
MS 563
Greenville, North Carolina 27858
USA

Qingming Luo
Huazhong University of
Science and Technology
Britton Chance Center for
Biomedical Photonics
Wuhan National Laboratory for
Optoelectronics
1037 Luoyu Road
Wuhan 430074
China

Megha Makam
Stanford University
School of Medicine
Beckman Center
279 Campus Drive
Stanford, CA 94305-5318
USA

Valeri P. Maltsev
Institute of Chemical Kinetics
and Combustion
Laboratory of Cytometry and
Biokinetics
Institutskaya 3
Novosibirsk 630090
Russia

and

Novosibirsk State University
Physics Department
Pirogova 2
Novosibirsk 630090
Russia

Shawn O. Meade
University of California
San Diego
Department of Electrical and
Computer Engineering
9500 Gilman Drive
MC 0407
La Jolla, CA 92093
USA

Martina C. Meinke
Charité – Universitätsmedizin
Berlin
Klinik für Dermatologie
Venerologie und Allergologie
Charité Campus Mitte
Charitéplatz 1
10117 Berlin
Germany

Anja Mittag
University of Leipzig
Department of Paediatric
Cardiology
Heart Centre
Strümpellstr. 39
04275 Leipzig
Germany

and

University of Leipzig
Translational Centre for
Regenerative Medicine (TRM)
Leipzig
Philipp-Rosenthal-Str. 55
04103 Leipzig
Germany

Stephen P. Morgan
University of Nottingham
Electrical Systems and Optics
Research Division
Faculty of Engineering
University Park
Nottingham, NG7 2RD
UK

Sergei Yu. Nikitin
Lomonosov Moscow
State University
Department of Physics
International Laser Center
Moscow 119991
Russia

Jim O'Doherty
Royal Surrey County Hospital
Department of Medical Physics
Egerton Road
Guildford GU2 7XX
UK

James Pond
Lumerical Solutions
201-1290 Homer Street
Vancouver, BC V6B 2Y5
Canada

Alexander V. Priezzhev
Lomonosov Moscow
State University
Department of Physics
International Laser Center
Moscow 119991
Russia

Wen Qiao
University of California
San Diego
Department of Electrical and
Computer Engineering
9500 Gilman Drive
MC 0407
La Jolla, CA 92093
USA

Jianjun Qiu
Huazhong University of
Science and Technology
Britton Chance Center for
Biomedical Photonics
Wuhan National Laboratory for
Optoelectronics
1037 Luoyu Road
Wuhan 430074
China

Herbert Schneckenburger
Hochschule Aalen
Institut für Angewandte
Forschung
Beethovenstr. 1
73431 Aalen
Germany

and

Universität Ulm
Institut für Lasertechnologien in
der Medizin und Messtechnik
Helmholtzstr. 12
89081 Ulm
Germany

Jürgen Schnekenburger
University of Muenster
Department of Medicine B
Gastroenterological Molecular
Cell Biology
Domagkstr. 3A
48149 Muenster
Germany

Ian M. Stockford
University of Nottingham
Electrical Systems and Optics
Research Division
Faculty of Engineering
University Park
Nottingham NG7 2RD
UK

Alexander Stratonnikov
General Physics Institute
Russian Academy of Science
Moscow
Russia

Wenbo Sun
Science Systems and Applications
Inc.
10210 Greenbelt Road
Lanham, MD 20706
USA

Stoyan Tanev
University of Southern Denmark
Department of Technology and
Innovation
Integrative Innovation
Management Unit
Niels Bohrs Alle 1
DK-5230 Odense M
Denmark

Attila Tárnok
University of Leipzig
Department of Paediatric
Cardiology
Heart Centre
Strümpellstr. 39
04275 Leipzig
Germany

and

University of Leipzig
LIFE - Leipzig Research Center
for Civilization Diseases
Faculty of Medicine
Philipp-Rosenthal-Str. 27
04103 Leipzig
Germany

Rabindra Tirouvanziam
Stanford University School of Medicine
Beckman Center
279 Campus Drive
Stanford, CA 94305-5318
USA

Frank S. Tsai
University of California
San Diego
Department of Electrical and Computer Engineering
9500 Gilman Drive
MC 0407
La Jolla, CA 92093
USA

Valery V. Tuchin
Saratov State University
Optics and Biophotonics Department and Research-Educational Institute of Optics and Biophotonics
83 Astrakhanskaya St.
Saratov 410012
Russia

and

Laboratory on Laser Diagnostics of Technical and Living Systems
Institute of Precise Mechanics and Control of RAS
24, Rabochaya st.
Saratov 410028
Russia

Michael Wagner
Hochschule Aalen
Institut für Angewandte Forschung
Beethovenstr. 1
73431 Aalen
Germany

Petra Weber
Hochschule Aalen
Institut für Angewandte Forschung
Beethovenstr. 1
73431 Aalen
Germany

Maxim A. Yurkin
Institute of Chemical Kinetics and Combustion
Laboratory of Cytometry and Biokinetics
Institutskaya 3
Novosibirsk 630090
Russia

and

Novosibirsk State University
Physics Department
Pirogova 2
Novosibirsk 630090
Russia

Vladimir P. Zharov
University of Arkansas for Medical Sciences
Philips Classic Laser Laboratories
4301 W. Markham St.
Little Rock
AR 72205 7199
USA

Olga Zhernovaya
International Research-Educational Center of Optical Technologies for Industry and Medicine "Photonics"
Saratov State University
Saratov
Russia

1
Perspectives in Cytometry

Anja Mittag and Attila Tárnok

1.1
Background

Cytometry is the general term for quantitative single cell analyses. Without cytometric analyses, work in modern life sciences would be unthinkable. Since its introduction, cytometry has been influencing and promoting development in biology and medicine. A high number of molecular parameters are analyzable within heterogeneous cell systems by cytometry. If the normality of a heterogeneous cell system is known, changes can be identified. Hence, biological alterations induced by malignancies, infections, and so on, are diagnosable. Such phenotypic changes allow for understanding disease-related (or induced) alterations of molecule expression patterns and hence, the functionality of the whole biological system. This interest to unravel molecular properties of single cells of healthy and diseased organisms (and to compare them) led to the development of the first cytophotometric instruments in the middle of the last century [1].

Analyses in those days were usually based on different light absorption capabilities of cell constituents of cells fixed on microscopic slides, with or without staining (e.g., Feulgen). Since these analyses were very time consumptive (5–10 min per nucleus or cytoplasm region), measurements of high cell numbers were simply not possible [2]. This technology was followed by instruments for blood cell counting with a higher throughput where cell concentrations were enumerated by counting electrical voltage pulse during cell transit [3].

Application of fluorescence dyes opened the way for obtaining more information per cell. In 1961, the first use of fluorescence for quantitation was reported [4]. Since then, development of new instruments was focused toward fluorescence analysis. In 1969, the first impulse cytophotometer (ICP-11) (Phywe GmbH, Göttingen [5]) was commercially launched where the fluorescence (resulting from mercury arc lamp excitation) of several thousand cells per second was measured by photomultiplier tubes (PMTs). Later, lasers were employed as stable light sources for excitation of fluorescence dyes. The first flow cytometer equipped with two lasers was available in 1976 [6]. Several fluorescence dyes could now be measured simultaneously. The basic principle of this technology is still applied in modern flow cytometers: cells

Advanced Optical Flow Cytometry: Methods and Disease Diagnoses, First Edition. Edited by Valery V. Tuchin.

are separated by sheath fluid, (hydrodynamically) focused, and excited by (laser) light in flow. The scattered and emitted or absorbed light is measured.

The demand for comprehensive analyses and with it the simultaneous detection of several parameters on many (thousands to millions of) individual cells in one sample led to further developments in the field of cytometric analyses. More lasers as well as detectors were included to be able to perform three- or four-color measurements (plus information of scattered light) routinely.

This was sufficient for many applications, or at least, it had to be. Detailed cellular subtyping, coexpression of specific markers, cytokine analyses of certain cell types, cell–cell interaction, and so on, are, in the majority of cases, not possible by using only four fluorescence parameters. The list of applications is long where multiparametric analyses are essential.

1.2 Basics of Cytometry

The beauty of labeling specific markers or cellular functions with fluorescence dyes lies in its multicolor approach and therewith the feasibility of simultaneous analysis of many parameters. If cells are stained with different "colors," each single color can be distinguished from each other and multiple information can be obtained for single cells. Admittedly, differentiation of more than three colors by the eye is almost impossible, but light detectors in cytometers (PMTs or camera), in combination with appropriate bandpass filters, are able to detect wavelength ranges (i.e., specific "colors") of interest. Discrimination of fluorescence dyes is hereby possible by defining certain wavelength ranges, suitable for specific fluorescence dyes. However, the usually very broad emission spectra of fluorescence dyes make it sometimes hard to differentiate between dyes in one detection filter owing to spectral overlap. This problem is known for fluorescence dyes such as fluorescein isothiocyanate (FITC) and phycoerythrin (PE) but can be mathematically solved by compensation, that is, "purification" of specific fluorescence from unwanted signals. Another novel possibility to overcome the problem of spectral overlap is multispectral analyses (known from microscopy) although it is rarely used in flow cytometry (FCM) [7].

In general, there are two different types of cytometric analyses named by the analytical technique: FCM and slide-based cytometry (SBC). As previously mentioned, cytometry has its roots in the analysis of cells on a slide. Owing to the higher throughput, development moved to the FCM, although now, with higher computing and storage capacity of workstations, cytometry by microscopy has been revitalized. Nevertheless, both methods are quite similar and identical in many details.

1.2.1 Flow Cytometry

It is apparent from the name cytometry that cells are analyzed in flow. Generally, cells in suspension are sucked or pressed into the cytometer by overpressure or mechanical pumps. Covered with sheath fluid, cells are separated (like pearls on a string) and move actively to the place of analysis. Lasers (or also light emitting diodes nowadays) excite the cell (i.e., the fluorescence dye on it) and the emitted fluorescence is detected by PMTs. On the basis of specific characteristics (mainly fluorescence of a certain label but also light scattering properties), separation of wanted cell types and its concentration and purification can be accomplished. However, fluorescence-activated cell sorting (FACS) is necessary for that, which can be time consumptive. In "normal" FCM, the sample is usually lost after analysis. Up to 50 000 cells can be measured per second, although the normal throughput is usually around 1000–10 000 cells per second.

Fluorescence information of the cells can be displayed as histograms and dotplots. Clever experimental setup (marker selection, fluorescence combination) and smart gating strategies allow for extraction of multiple information out of a three- or four-color staining. Nevertheless, detailed subtyping or functional information (e.g., activation) of specific cellular subtypes can hardly be obtained by such low-color analyses [8].

Although the main principle in FCM has not dramatically changed since its beginning, there are of course some new developments besides increasing number of detectors and lasers. There is not only hydrodynamic focusing of cells (with the need for utilization of sheath fluid) but also focusing derived from acoustic radiation pressure forces [9] or the utilization of photodetectors for sensing the position of particles in the sample stream without sheath fluid [10]. Without sheath fluid (but with a unique flow cell design), even usage of FCM in space is conceivable [11]. Another development is the implementation of imaging in FCM. There are flow cytometers available that are able to capture images of analyzed cells in flow for morphological analysis [12].

Cellular analyses in FCM, however, are restricted to cell suspensions. Solid tissue or adherent cells cannot be analyzed, that is, not without prior trypsinization or disintegration of tissue. For these types of specimens, SBC was developed.

1.2.2 Slide-Based Cytometry

There are two major types of SBC systems: camera-based detection in combination with lamp illumination (e.g., [13]) and laser excitation and fluorescence detection via PMTs (e.g., [14]). However, mixed systems, for example, lasers and camera, are available, too. No matter which modality is used for excitation and detection, the core of these instruments is a fluorescence microscope. But that does not mean that every fluorescence microscope is capable of cytometric analysis. Cytometric analysis means quantitative analysis of the whole cell, that is, it requires optics with a relatively low numerical aperture. Analysis of single slices of a cell, as, for example, in confocal microscopy, is not cytometric. Moreover, analyses using microscopes with lamps or diodes as excitation source and no corrections (optical or software solutions) for light excitation intensity, that is, stability of the excitation light, are also not cytometric. Owing to unstable excitation intensity, one cannot be certain that the resulting fluorescence intensity of cells in different fields of view (or different samples, irrespective of the same acquisition setups) will provide the same fluorescence intensity. Qualitative statements about existing fluorescences are possible but no quantitative conclusion can be made about cell activation or other marker expression of cells. Prerequisites for cytometric analyses with microscopes are stable excitation power, even illumination of the sample, and a steady and sensitive detection of the emitted fluorescence.

Even though cytometric analyses were slide based at the beginning and the modern concept of SBC was presented in the 1980s, the first type of such instruments, the laser scanning cytometer (LSC), became commercially available a decade later [15]. The reason for this was probably the time needed for image analyses in the past. However, higher processing power and storage capacities of modern computers promoted development in this field.

Unlike FCM, samples in SBC analyses are fixed on a slide or plate. Although mainly developed for tissue analysis, LSC was used for many different applications, for example, cell cycle studies [16–18], apoptosis [19, 20], immunophenotyping [21–24], tumor analysis in solid tissue [25], fine-needle aspirate biopsies [26], circulating tumor cell analysis [27], stem cell analysis [28, 29], or study of the effects of drugs [30, 31].

The principle of LSC analysis is comparable to FCM. Fluorescence dyes on (or in) cells are excited by laser light and the emitted fluorescence is split into certain wavelength bands by optical filters and detected by PMTs. The deviation from FCM is that cells remain on the slide and can be further analyzed or even cultured.

LSC allows for studying growth and the variety of expression of specific markers during development of cells in their "natural" environment [31]. Possible effects of cell preparation, for example, stress or activation caused by detachment of cells from the surface (like for FCM analyses), can be avoided. Moreover, interactions between single cells can hardly be observed on detached cells. This applies for cell cultures as well as tissue sections. Another advantage is that cells can be traced and analyzed during culturing [29].

1.3 Cytomics

Since the complexity of biological systems is very high, a multiplicity of different information from cells, their interaction, and triggered reactions (e.g., by external stimuli or diseases) is necessary to understand such systems. For this purpose, different concepts were and still are under development to get a better insight into biological processes in organisms. Cytomics is one of these concepts. Its aim is to characterize single cells in cell systems and to unravel the interactions of cells within these systems [32]. Another concept is systems biology. The aim of systems biology is similar to that of cytomics, but it focuses more on understanding intracellular behavior, that is, the interaction of single cellular constituents such as genes, proteins, metabolites, and organelles and *in silico* modeling [33]. Interconnection of different analyses is very important for this purpose, that is, to obtain all needed information and combine them appropriately. In contrast to other concepts like genomics (characterization of genome [34]), proteomics (analysis of proteome [35]), lipidomics (cellular lipid constituents [36]), or other -omics, where only certain components of cells are in the focus of interest, cytomics and systems biology focus on interaction of cells and cellular constituents.

Always, biological conditions are the result of the interaction of all components of a complex system. Therefore, such a system must be analyzed as a unit to unravel its secrets. For example, different developmental stages of an organism have the same genome but are different (also phenotypically) in their protein composition [37]. Cytomics and systems biology start there and go even further. Not only single components are under investigation but also the relations and interactions between different components. Therewith, changes in cell systems can be understood – from work flows within the cell (systems biology) to interactions of the whole system (cytomics). If these actions are known, alterations (even before clinical manifestation) can be classified and can lead to predictive and preventive individualized medicine [38, 39]. Cells are the elementary building units of an organism and hence, their analysis is the easiest way to identify diseases or reasons for diseases. Alterations to healthy conditions can be found by differential screening, that is, examining a multiplicity of cell types for phenotype, activation, or cytokine production (multiparameter analyses), and extracting important and relevant cell types and marker combinations for a further diagnostic panel. However, it is clear that the mass of information obtained from multiparametric cytometry must be analyzed appropriately to find causal connections. Bioinformatics tools, that is, algorithms for cluster analyses, can be applied [40–42].

1.4 Cytometry – State of the Art

Routine are still fluorescence analyses with a relatively low amount of measured parameters (sample stained with three to four colors). However, developments of new fluorescence dyes with promising spectral characteristics, for example, UV excitable dyes for protein labeling (e.g., Quantum dots [43]) and, of course, instruments capable for multiparametric analyses, led to the increased usage of multiparametric, that is, multicolor, analyses in laboratories worldwide. Cellular analyses with 6–12 colors (polychromatic cytometry) simultaneously are no longer a rarity (e.g., [8, 22, 44]).

Multicolor analyses allow for detailed understanding of complex cellular structures, cell subtypes, and cell–cell interactions. All fluorescence information, that is, the tagged biological components, generate a network from where information of interest can be extracted. Admittedly, there are attempts to perform it with three- or four-parameter analyses. To this end, as an example, in FCM analyses many aliquots of the same sample (e.g., blood) are stained for different antibodies. All these single tubes are measured and afterward, information obtained from each tube is combined, for example, for cellular subtyping of the sample (but not on a single cell level). The presence of main cellular subtypes can certainly be detected but it is not possible to get details. For example, a staining for CD3, CD4, and CD8 in one tube yields information on the amount and distribution of T helper and cytotoxic T cells as well as double positive and double negative T cells. Conclusions on further subtypes, for example, existence of $\alpha\beta$-TCR (T cell receptor) or $\gamma\delta$-TCR or expression of CD16 or CD56 in NKT cells, are not possible. If cells are stained for CD3, CD19, and CD16/CD56 one can get an overview of lymphocyte subtypes (i.e., T cells, B cells, NK cells, and NKT cells) but merging this information with one of the other tubes and drawing a conclusion on, for example, the presence of $\alpha\beta$-TCR$^+$CD4$^+$CD8$^+$ NKT cells is impossible. Therefore, one needs this information on a single cell basis, for example, a sample with CD3, CD4, CD8, CD16/CD56 (or even better, CD16 and CD56), and $\alpha\beta$-TCR. Only then, the required information can be extracted from the analysis.

Admittedly, some recent developments promise to combine several (three to four color) FCM measurements of a sample into one metadata file (i.e., a virtual multiparameter analysis by data merging) on the basis of a leading antibody [45] but this might not be feasible with every desired cellular subtype and, of course, needs verification with a true polychromatic analysis. For answering complex questions, multiparametric analyses are indispensable [8]. Another advantage of multiparametric analyses is the fact that information density increases with each parameter added to the existing setting. The resulting network provides the opportunity to find interactions of cells, coexpression of markers, and so on, that were never be thought of before. Moreover, polychromatic cytometry reduces costs of analysis. Although a complex pattern of fluorescence dyes means also the use of "unusual" dyes or fluorescence marker combinations (mostly more expensive than the commonly used ones), the combination of markers in one analysis prevents the repetitive use of markers in several tubes for identifying main populations.

1.4.1 Multiparametric Analyses

As an example for the complexity of multicolor analyses, an eight-color measurement on LSC [22] is described here.

For precise analysis and correct interpretation of data, a careful selection of antibodies and their corresponding fluorescence dyes is essential [46]. Only optimal experiment settings allow for unequivocal identification of cell types, clear discrimination between positive and negative cells, and doubt-free analysis of data. This means that in most of the cases different antibody fluorescence combinations need to be tested before selecting the final experiment settings. For the eight-color experiment on LSC, the following panel was selected for a simultaneous staining of several surface markers in a sample of human peripheral blood leukocytes: CD14-FITC, CD4-PE, CD56-PECy5, CD16-PECy7, CD45-APC (Allophycocyanin), CD8-APCCy5.5, and CD3-APCCy7. This mixture allowed for identification of the required lymphocyte cell subtypes (focus on NK and NKT cells) by an appropriate gating strategy. In Figure 1.1, analysis including gating strategy is shown. Since LSC lacks side scatter, which is usually used in FCM to discriminate between leukocyte subsets (neutrophils, monocytes, lymphocytes), fluorescence signals were used for this purpose. Nevertheless, leukocytes could be subdivided because of their different expression of CD45 (pan-leukocyte marker) and CD14 (LPS-Receptor) into the subsets: monocytes, lymphocytes, granulocytes (neutrophilic and eosinophilic), and basophilic granulocytes after excluding cell debris, artifacts, and aggregates based on their forward scatter maximum pixel and area values (for details, see [22]). The lack of a CD14 signal on basophils enables their differentiation from neutrophils and eosinophils (both $CD14^{weak}$) and monocytes ($CD14^{bright}$). Within the monocytes there is a small but prominent population of $CD16^+$ cells with a slightly reduced CD14 expression: proinflammatory monocytes. $CD16^{weak}$ cells within the granulocyte population are eosinophils. Thereby, the main leukocyte populations were discriminated.

The focus of this LSC analysis though was lymphocytes. Lymphocytes were subtyped by their CD3 and CD16 expression: $CD3^+CD16^-$, $CD3^+CD16^+$, $CD3^-CD16^-$, and $CD3^-CD16^+$. CD19 expression allowed for unequivocal identification of B cells in the $CD3^-CD16^-$ population.

Since cytomics and multiparametric analyses provide complex networks of data (fluorescence, scatter, size, etc.) on a single cell level, expression of every marker can be verified for each cell, or more practically, for each identified cell population.

Without further investigation, one would expect all the $CD3^+CD16^-$ population to be T cells. But that is not the truth. CD56 expression revealed a small population (~7%) of $CD3^+CD16^-$ NKT cells. The presence or absence of CD56 (in combination with the previous step: CD16) allowed for discrimination of several NK and NKT cell subphenotypes also within the other lymphocyte subsets ($CD3^+CD16^+$, $CD3^-CD16^+$, $CD3^-CD16^-$) with varying combinations of CD16 and/or CD56. All of these newly identified populations of T cells, NK cells, and NKT cells were further investigated for their expression of CD4 and CD8. As can be seen in Figures 1.1 and 1.2, the diversification of all these populations goes deeper and deeper. One can imagine that with additional antibodies further subsets can be discriminated, although it must be clearly said that only at a certain concentration, that is, number of cells, a cell population can be identified as a "population" [47]. If only a very low number of cells can be expected at the end of such a gating cascade, a very high number of cells must be analyzed in the beginning.

1.5 Perspectives

Biological systems never were simple – but their analyses were, in the past. The reason for these "simple" analyses may be limitations in understanding such systems or limitations of techniques for their analyses. However, understanding is based on observations that are only possible with suitable instruments. Most important for unraveling complex systems is the ability to look at many parameters at a time

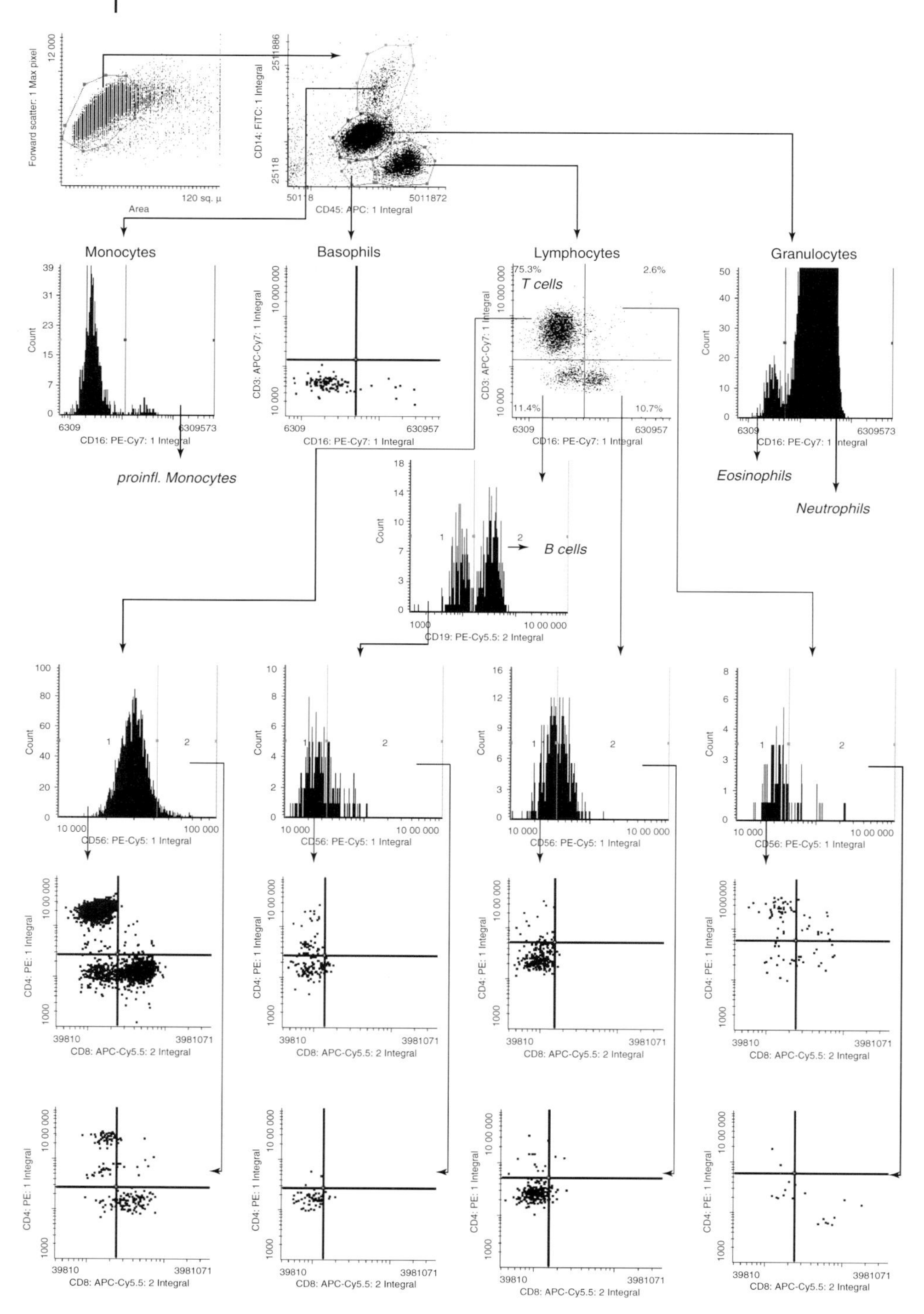
Forward scatter: 1 Max pixel
Area
120 sq. µ
CD14: FITC: 1 Integral
CD45: APC: 1 Integral
Monocytes
Basophils
Lymphocytes
Granulocytes
T cells
75.3%
2.6%
11.4%
10.7%
CD16: PE-Cy7: 1 Integral
CD3: APC-Cy7: 1 Integral
Count
proinfl. Monocytes
Eosinophils
Neutrophils
B cells
CD19: PE-Cy5.5: 2 Integral
CD56: PE-Cy5: 1 Integral
CD4: PE: 1 Integral
CD8: APC-Cy5.5: 2 Integral

Figure 1.1 Eight-color analysis by LSC. EDTA anticoagulated blood was stained for CD3, CD4, CD8, CD14, CD16, CD19, CD45, and CD56. Artifacts were excluded by FSC (forward light scatter) MaxPixel versus area for further analysis (top left) and events of nonleukocyte origin by the lack of CD45 expression (top center). Leukocytes were further subdivided by their different antigen expression. Figure was published earlier in: Mittag *et al.* [22].

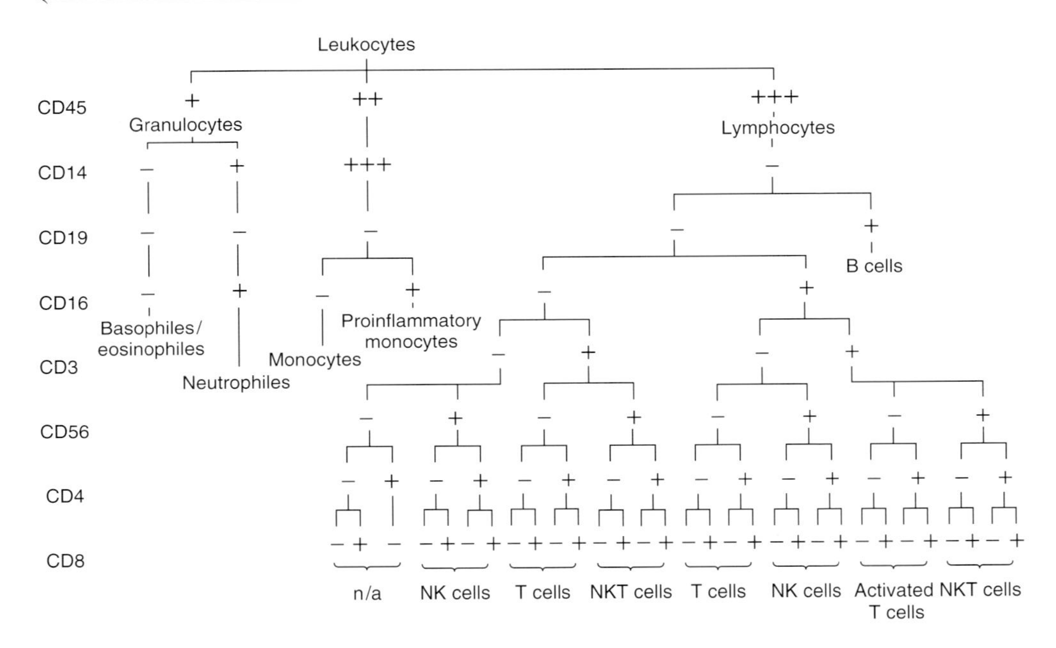

Figure 1.2 Polychromatic characterization of leukocytes. Detailed subtyping of cells, leukocytes in this example, is only possible with polychromatic cytometry, that is, the simultaneous use of several markers. Figure was published earlier in: Mittag [48].

to find interactions and causal connections. Thus, there is an increasing demand for multiparametric analyses in the field of biological and clinical investigations.

Perfetto *et al.* [8] point out that markers for identification of different developmental stages of T cells are expressed on several T cell types (e.g., CD45RA, CD27, CD28, CD62L, CCR7) and that in the case of four-color experiments one cannot make a statement about expression of (presumably analog) markers such as CD62L or CD27 on $CD45^{-}CCR7^{+}$ cells. Only if more markers are added to the commonly used "simple" protocols for immunophenotyping, questions about coexpression of antigens can be answered unequivocally without theoretical assumptions. This applies even more to the investigation of antigen specificities or cell functions [8]. The authors, therefore, emphasize the need for technologies for polychromatic cytometry with more than six colors. Only then can investigations of cellular interrelations be accomplished as well as a more precise analysis of rare cell populations feasible. It may also help explain immunopathogenesis of infectious or autoimmune diseases and will substantially improve currently available diagnostic tools. Hence, the discovery of sensitive and specific panels of biomarkers might be essential for future diagnostic and therapeutic monitoring of complex diseases [49].

One of the perspectives of cytometry therefore definitely is cytomics and multiparametric analyses, that is, their further development, whereby the information density of the analyzed material is exponentially increased with each additional analyzed parameter. Technical possibilities are exhausted and new methods for multiparametric approaches are under development.

1.5.1 New Technologies and Methods

Perfetto *et al.* [8] reported an FCM analysis of 17 fluorescence dyes simultaneously. This high number of fluorescence-labeled markers is necessary to get to the bottom of the immunological reasons for disease patterns. For example, even if the overall number of T cells is not significantly altered, alterations in the composition of the immune system can be detected by a closer look on T cell subsets of patients with diseases caused by HIV [8]. Alterations in T cell subsets, such as depletion of naïve or newly produced T cells or viral infection of memory and HIV-specific $CD4^+$ T cells, provide information on the progression of disease or are valuable for diagnosis or prognosis [50–54]. For studying functions of these cellular subtypes (e.g., cytokine pattern analysis) or other disease-related characteristics, multiparametric analysis is mandatory.

To the best of our knowledge, there is no cytometer available yet on the market which is capable of measuring such a high amount of parameters simultaneously as described by Perfetto *et al.* [8]. The authors built, that is, modified their own instrument in a way that it fulfilled the requested tasks. Nevertheless, manufacturers of cytometers recognized the signs of the need for multicolor analyses in laboratories. Several instruments capable of 10- to 12-color measurements were launched recently. This might be enough for cytometric investigations at the moment, at least for routine analyses, but researchers tend to look a little bit deeper. This means, to make it short, there can never be enough parameters in cytometric analyses. The only problem is that, at a certain point, hardware has its limitations. Software solutions for increasing measurable parameters might therefore be a good addition or even an alternative. Implementation of additional lasers for excitation and PMTs for fluorescence detection changed standard 3-4 colour FCM analysis to >8 colour polychromatic cytometry. However, for many technical reasons this development will not continue for ever. SBC seems to be a very interesting alternative in this field. Besides its ability to analyze not only cell suspensions but also cells in culture or tissue, the main advantage is that the sample is not lost after analysis. This opens the way for reanalysis and merging of information for one sample from different analyses – on a single cell level. Cellular reactions of added compounds can be tested [31] or a detailed phenotyping of cells by sequential analysis [22], for example, restaining [55], is possible, among other exciting possibilities.

1.5.1.1 **Sequential Analyses**

In the eight-color leukocyte analysis by LSC mentioned before, merging of two separate measurements was part of the analysis [22]. Although the sample was stained for all eight markers simultaneously, the cytometric analysis consisted of two parts, that is, two different measurements. This was necessary because of the limit of four PMTs in LSC. In combination with a 488 and a 633 nm laser for excitation, there is a limit for fluorescence detection of six parameters. The exchange of filters between the measurements enables an analysis of the fluorescence signals, the detection of which was excluded by the settings for the first measurement. This second measurement, that is, the obtained data, was merged with the first data set. In the resulting data file, all information, obtained from measurement 1 or 2, is available for each individual cell. In this example, only two scans were performed and only eight markers were analyzed. However, this method of sequential analysis can be extended; there is no limit to merging data files.

Another application of this approach is the sequential staining of samples. This means staining of samples with one set of antibodies (or only one marker) with a subsequent analysis, followed by staining with a second set (same fluorescence dyes can be applied). Between the two measurements a bleaching step can be applied, although it is not mandatory as Laffers *et al.* [55] could demonstrate. Bleaching, of course, allows "resetting" the sample and starting with a "blank one" but with the information from the previous analysis [56, 57]. The combination of staining, analysis, bleaching, and restaining could be extended and brought to perfection by Schubert *et al.* [56] for an analysis of more than 90 molecules in one tissue section. The multiepitope-ligand-cartography (MELC) technology, developed by this group, allows for mapping the topological position of many proteins simultaneously in a cell and can therewith unravel hierarchies of proteins. This, in turn, indicates cell function or dysfunction and allows identification of specific proteins in protein networks of cells and tissues [58]. Mapping of toponome, that is, identification of protein colocalization, could be demonstrated on peripheral blood leukocytes, on a rhabdomyosarcoma cell line as well as on skin biopsies and rat spinal cords by repetitive fluorescence labeling of proteins [56]. This technology might be relevant for pharmacology to find new target proteins for drugs.

1.5.1.2 **Spectral Analyses**

If bandpass filters are used for analysis, always a part of the fluorescence will not be detected since most of the fluorescence dyes have a very broad emission spectrum. The narrower the bandpass is in front of the detector the more specific is the detection of fluorescence signals (i.e., without substantial interferences of spillover signals from other dyes) of the dyes in multicolor experiments. However, a lesser amount of light is transmitted, resulting in lower fluorescence signals (i.e., signal-to-noise ratios). This can be of hindrance in analysis of weakly expressed antigens.

An alternative to splitting fluorescence emission of samples into several broad bands by filters is spectral analysis. Image stacks of 1 nm distance (up to broader image acquisition steps) can cover the whole range of visible and invisible light although main applications (based on fluorescence characteristics) are in the visible range. Wavelength changes in spectral imaging are usually done by acousto-optic tunable filters (AOTFs) or liquid crystal tunable filters (LCTFs) as used in some *in vivo* imaging systems (e.g., [59]). AOTFs, implemented in microscopic systems (e.g., spectrofluorometer, confocal laser scanning microscopy), allow for multi-parametric analyses [60]. In such analyses, the spectrum or at least the spectral range of interest is analyzed and recorded ("spectral fingerprinting"). All used fluorescence dyes (or cell type specific spectra – without staining) can be extracted (spectral deconvolution) and distinguished by reference values (e.g., measurement of "pure" fluorescence signals) or virtual filters [44].

However, spectral analysis is not applicable only for SBC. Spectral measurements of fluorescence and Raman labels are also feasible with a commercial (although modified) large particle analyzer using a spectrograph and charge coupled device (CCD) array detector [7].

1.5.1.3 **Fluorescence Modifications for Analyses**

When using SBC, one is confronted with photobleaching of fluorescence signals. Most fluorescence dyes are sensitive to prolonged light exposure. In SBC, the microscope-based cytometry technology, excitation times are by at least one order of magnitude longer than in FCM. Therefore, photobleaching will affect analytical results of SBC. On the one hand, bleaching effects are very annoying, but on the other hand, it is possible to use bleaching as a tool for analysis. The combination of stable and photosensitive fluorescence dyes for staining biological samples can be used to extend the number of measurable parameters per analysis [61, 62]. Bleaching of fluorescence dyes was also used as a tool for single-cell fluorescence spectroscopy. The electronic interaction of DNA base pairs leads to oxidation of bound fluorescence dye and this in turn to fluorescence bleaching, which enables discrimination of DNA information at the single-molecule level [63]. Moreover, fluorescence recovery after photobleaching (FRAP) can be used for biological investigations of molecular dynamics in living cells. Examples for FRAP analyses are studies on integrin dynamics [64] or RNA motion [65].

Fluorescence resonance energy transfer (FRET) is probably the main application of directed bleaching in cytometry. With FRET analyses, interactions of surface molecules on two interacting cells can be studied [66] or the intracellular trafficking and state of toxin subunits can be determined [67].

Alternatively, studies of protein interactions can also be performed with a new technology, called *in situ proximity ligation assay* [68]. This method utilizes antibodies coupled to oligonucleotides. If the target proteins are in close proximity, they can form a circular DNA molecule that can be amplified and detected by adding complementary oligonucleotides labeled with a fluorescence dye. This fluorescence can then be analyzed by FCM [68].

Another interesting technique for manipulation of fluorescence to gain information is photoconversion [69]. Fluorescent proteins (e.g., KAEDE) or fluorescence dyes (e.g., Lucifer Yellow) are transferred into stable reaction products by light excitation. These reaction products can be detected and analyzed by microscopic techniques [70, 71]. Similar to FRET analyses, there is a change in the "color" of emitted fluorescence. This method is usually used for the study of fluorescence-labeled cellular structures [70].

All these methods can provide additional information from already stained samples. Mechanical manipulation of cells is not necessary. However, for some applications no manipulation of the cells at all (i.e., also no staining) is desirable. A prominent example is the isolation of stem cells for therapeutic use.

1.5.1.4 Label-Free Analyses

Fluorescence analyses enable specific labeling and characterization of biomarkers; even analyses using combination of multiple markers are possible (polychromatic cytometry). Nevertheless, tagging cells with antibodies or the use of DNA dyes influences cells' behavior and can be toxic as well. For live cell analyses and the intention to further use these cells (e.g., for therapeutic use or cell differentiation studies), characterization of cells by a label-free approach would be preferable.

Some promising efforts were made for label-free analysis of cells. Impedance analysis is one of them. With this technology it is possible to analyze adherent cells, for example, for detection of ischemic effects on cardiomyocytes [72] or detection of tau hyperphosphorylation in a neuroblastoma cell line [73]. Impedance analysis is also feasible with flow cytometric techniques, that is, analysis of cells in flow [74]. Cells can be analyzed at different frequencies. The analysis over a wide frequency range provides information of cellular characteristics depending on the frequency, that is, cell size (0.1–20 MHz, but mainly determined in the range of 0.1–1 MHz), membrane capacitance (1–5 MHz), and cytoplasm conductivity (3–20 MHz). The amplitude, opacity, and phase information can be used for discrimination between different cell populations without the use of cell markers [75], viability and apoptosis analyses, or the visualization of microbiological life cycle phases [76].

Conventional FCM can also be used for label-free analysis. Scatter as a parameter provides opportunity for label-free detection since all particles scatter light when illuminated. Commonly used in FCM are forward light scatter (which is dependent on cell size and refraction index) and orthogonal side scatter (provides information of granularity). These two parameters are sufficient for rough leukocyte discrimination [77, 78] but cannot be used for separation of cells with similar size, shape, or structure. Therefore, an analysis of the complex scatter pattern of particles (e.g., cells) is necessary, that is, information about size, shape, refraction index, density, and morphology obtained by different light scattering properties. Rajwa *et al.* [79] demonstrated the classification of particles on the basis of their discrete scatter patterns with a modified flow cytometer capable of measuring five different angles of scattered light and axial light loss. Label-free detection of bacterial pathogens (colonies growing on agar plates) can be accomplished by their forward-scatter signatures at 635 nm and image analysis. Multiple bacterial pathogens can be identified although their visual morphology is similar [80].

Another approach is based on image analysis. Label-free analysis intends to replace evaluated markers by parameters formerly not used for identification of specific cells or characteristics. Newberg and Murphy [81] demonstrated an automated image analysis of subcellular patterns of proteins by immunohistochemistry for studying the functions of proteins. For this purpose, proteins were stained and analyzed using morphological parameters for further prediction. Ishikawa *et al.* [82] succeeded in identifying alga cells without using any labels. Fujita and Smith [83] outline the importance of label-free analysis for the examination of living cells. The authors describe the chemical and biochemical impact of markers on living cells and suggest using Raman scattering and coherent anti-Stokes Raman scattering (CARS) to image molecules in cells. Moreover, biomolecular composition of tissue can be determined by Raman spectroscopy, for example, applied for discrimination of normal liver, viable tumor, and fibrotic hepatoblastoma [84]. For Raman scattering analysis, a relatively high concentration of molecules (or high laser intensity) is necessary. Spectrum of single molecules, therefore, cannot be detected. Surface enhanced raman scattering (SERS) can solve this problem and make this method sensitive enough for FCM [85].

Another vibrational spectroscopic technique, near infrared spectroscopy (NIRS), allows for noninvasive monitoring of oxygenation and perfusion [86–89], functional brain imaging [90], or examination of normal and malignant tissue, for example, applied for breast cancer imaging [91]. Although this method is mainly used for imaging in clinical applications, it might also be interesting for tissomics and cytometry, for example, for distinguishing intact cartilage from the enzymatically digested one [92] or the characterization, discrimination, and identification of microorganisms [93].

Autofluorescence can also be used as a relevant discriminatory parameter in analysis. The combination of second-harmonic generation (SHG) and multiphoton excited autofluorescence (MAF) signals can be used for a label-free discrimination of cancerous from noncancerous lung tissue [94].

1.5.2
Automation

Particularly in cytometric analyses for diagnostics, high-throughput and standardized reliable data are mandatory. Therefore, instruments capable of automated measurement and analysis are necessary. This applies, of course, also for multiparametric analysis. Automated measurement might not be the problem; robot units are applied in many routine laboratories, but an automated analysis, that is, precise setting of gates for analysis, is still at its beginning.

Hence, modern instruments will be equipped with a high-precision hardware and sophisticated, reliable software suitable for high-speed data acquisition and computer-assisted fully automated data analysis. Currently available cytometers allow for detection of increasing numbers of cell surface markers. However, this development is challenging for "traditional" methods of identifying cell populations, that is, manual gating for positive and negative cells. Easy-to-use software tools are needed, allowing for fast and effective analysis of multiparametric datasets based on supervised or unsupervised data mining algorithms [95]. The routine generation of vast amounts of data makes cytometric analyses a logical target for the application of data mining techniques. Clustering algorithms help arrange the multidimensional datasets containing large numbers of parameters describing molecular features of individual cells in a small number of relatively homogeneous clusters on the basis of differences and similarities between analyzed objects [40–42]. The rapid advance in automatic cytometric analysis will facilitate the development of computer-assisted clinical decision-making systems and will provide additional assistance to medical professionals [96].

Automation in cellular analyses implies not only automatic gating of flow cytometric data but also automated image analysis. Image analysis can be simple if only the size of objects or fluorescence intensity is of interest. But image analysis provides more opportunities. A long list of numerical parameters can be extracted from images or more precisely from identified objects, describing shape, texture, pixel-intensity statistics, and so on. Classification algorithms can then be trained to discriminate between cell phenotypes [97]. Such software applications can be of high accuracy in classifying cell types but reach their limits in recognizing new phenotypes. Hence, the trend points to "intelligent" classification systems, which automatically learn and define new classes with similar characteristics [97]. However, such an automatic identification and classification system can also be used for protein analyses (proteomics). It is a valuable tool in location proteomics, that is, specifying the location of proteins within cells [98].

The main obstacle in live cell image analysis is the monitoring of single cells over time. Cell motility hinders direct retrieval of cells in single images. Cell tracking algorithms, however, allow for connecting objects in time. Even tracking of object splitting (cell division) or merging (cell fusion) is possible. Analysis of time-lapsed data sets can provide information of, for example, individual cell cycle progression [99], cell migration [100], or cell motility behavior [101].

1.5.3
Cytometry – the Other Side

Multiparametric cytometry has been the major trend for many years. However, more and more complex analyses cause more and more costs, of reagents, instrumentation, and so on. But not everywhere can money be spent for such purposes although cellular analyses are badly needed – simple analyses, easy to accomplish, but too expensive to be done in some areas of this world. CD4 counting in HIV patients in Africa is such an example. The procedure of counting cells of a single cell population is far from being tricky but the technology is too expensive as are the costs for each analysis. Howard Shapiro's appeal in 2000 on a Purdue Cytometry Discussion Forum for low-cost and effective CD4 testing tools was a starting point for many researchers to take a serious look at the situation in Africa. Since then, the development of low-cost tests for CD4 counting or for malaria for resource-limited countries is another trend in cytometry.

FCM, as well as SBC, is of interest for reducing costs and developing low-cost instruments. Since FCM is expensive because of the sophisticated technology and costs for maintenance, the cheaper alternative will be SBC. However, SBC instruments available today with sophisticated microscopes and complex optics (e.g., [14, 44, 102, 103]) are expensive. If they could be reduced to inexpensive light sources (e.g., diodes), elementary optics, and simple image analysis software, such instruments will become a valuable tool for low-cost cell analyses and adequate diagnostics in third-world countries [104]. An example of such an approach is the device developed by Moon *et al.* [105], which utilizes immobilized anti-CD4 antibodies, a CCD sensor, and an automatic cell-counting software.

Nevertheless, an FCM is also in the focus of low-cost analyzers. Zaragosa [106] developed a portable, low-cost flow cytometer that uses acoustic micromanipulating techniques to focus particles in a flow stream instead of the commonly used hydrodynamic focusing. This technique reduces maintenance costs because sheath fluid, a big cost factor in consumables, is no longer required. Additionally, inexpensive light sources and cheaper detector systems (less sensitive) can be used, which further reduce costs [106]. However, costs for preparation of blood samples should not be neglected. Greve *et al.* [107] presented a method for less expensive determination of CD4 T cells. With a no-lyse, no-wash flow-cytometric method in combination with a simplified gating strategy, it is possible to reduce the costs per sample from €30 to below €2 [107, 108].

It should be kept in mind that not only HIV patients need adequate diagnostics but also patients suffering from other global level diseases such as malaria and tuberculosis. In this context, Wolfgang Göhde and his great effort in developing and spreading affordable diagnostic tests cannot be forgotten. Göhde and the Partec Essential Healthcare program aim to bring innovative techniques directly to where they are needed most, including to those patients far from large cities, living in remote areas where the existing health services have not reached until now [109].

1.6 Conclusion

Cytometry offers many opportunities for cellular investigations. It varies from flow- to slide-based cytometry and from analysis of cell suspension to solid tissue. As broad the methodological spectrum is so broad is the range of applications. Nevertheless, two major trends can be identified: (i) very complex, multiparametric (mainly multicolor) analyses and (ii) "back to basics" – low-cost and simple analyses for diagnostics in resource-limited countries. Multiparametric cytometry is mainly for gaining deeper understanding of biological processes or identifying rare cell types (for diagnoses or drug target screening) in the western world, while simple (partially only monocolor or even label-free) analyses are definitely cheaper and can be utilized in a wider area, where they are desperately needed.

References

1. Sandritter, W., Schiemer, G., Muller, D., and Schroder, H. (1958) Construction and operation of a simple microspectrophotometer (cytophotometer) for visible light. *Z. Wiss. Mikrosk.*, **63** (8), 453–476.

2. Valet, G. (2007) *Zelluläre Diagnostik*, Karger, Basel, pp. 1–26.
3. Coulter, W.H. (1956) High speed automatic blood cell counter and cell size analyzer. *Proc. Natl. Electron. Conf.*, **12**, 1034–1040.
4. Rotman, B. (1961) Measurement of activity of single molecules of beta-d-galactosidase. *Proc. Natl. Acad. Sci. U.S.A.*, **47**, 1981–1991.
5. Dittrich, W. and Göhde, W. (1969) Impulszytometrie bei einzelzellen in suspension. *Z. Naturforsch.*, **B24**, 221–228.
6. Stöhr, M. (1976) *Pulse Cytophotometry II*, European Press, Ghent, pp. 39–45.
7. Watson, D.A., Gaskill, D.F., Brown, L.O., Doorn, S.K., and Nolan, J.P. (2009) Spectral measurements of large particles by flow cytometry. *Cytometry A*, **75** (5), 460–464.
8. Perfetto, S.P., Chattopadhyay, P.K., and Roederer, M. (2004) Seventeen-colour flow cytometry: unravelling the immune system. *Nat. Rev. Immunol.*, **4** (8), 648–655.
9. Ward, M., Turner, P., DeJohn, M., and Kaduchak, G. (2009) Fundamentals of acoustic cytometry. *Curr. Protoc. Cytom.*, **53**, 1.22.1–1.22.12.
10. Swalwell, J.E., Petersen, T.W., and van den Engh, G. (2009) Virtual-core flow cytometry. *Cytometry A*. **75** (11), 960–965.
11. Crucian, B. and Sams, C. (2005) Reduced gravity evaluation of potential spaceflight-compatible flow cytometer technology. *Cytometry B Clin. Cytom.*, **66** (1), 1–9.
12. Basiji, D.A., Ortyn, W.E., Liang, L., Venkatachalam, V., and Morrissey, P. (2007) Cellular image analysis and imaging by flow cytometry. *Clin. Lab. Med.*, **27** (3), 653–670, viii.
13. Varga, V.S., Ficsor, L., Kamarás, V., Jónás, V., Virág, T., Tulassay, Z., and Molnár, B. (2009) Automated multichannel fluorescent whole slide imaging and its application for cytometry. *Cytometry A*, **75** (12), 1020–1030.
14. Kamentsky, L.A., Burger, D.E., Gershman, R.J., Kamentsky, L.D., and Luther, E. (1997) Slide-based laser scanning cytometry. *Acta Cytol.*, **41** (1), 123–143.
15. Kamentsky, L.A. and Kamentsky, L.D. (1991) Microscope-based multiparameter laser scanning cytometer yielding data comparable to flow cytometry data. *Cytometry*, **12** (5), 381–387.
16. Mosch, B., Morawski, M., Mittag, A., Lenz, D., Tarnok, A., and Arendt, T. (2007) Aneuploidy and DNA replication in the normal human brain and alzheimer's disease. *J. Neurosci.*, **27** (26), 6859–6867.
17. Wojcik, E.M., Saraga, S.A., Jin, J.K., and Hendricks, J.B. (2001) Application of laser scanning cytometry for evaluation of DNA ploidy in routine cytologic specimens. *Diagn. Cytopathol.*, **24** (3), 200–205.
18. Sasaki, K., Kurose, A., Miura, Y., Sato, T., and Ikeda, E. (1996) DNA ploidy analysis by laser scanning cytometry (lsc) in colorectal cancers and comparison with flow cytometry. *Cytometry*, **23** (2), 106–109.
19. Darzynkiewicz, Z., Galkowski, D., and Zhao, H. (2008) Analysis of apoptosis by cytometry using tunel assay. *Methods*, **44** (3), 250–254.
20. Abdel-Moneim, I., Melamed, M.R., Darzynkiewicz, Z., and Gorczyca, W. (2000) Proliferation and apoptosis in solid tumors. Analysis by laser scanning cytometry. *Anal. Quant. Cytol. Histol.*, **22** (5), 393–397.
21. Al-Za'abi, A.M., Geddie, W.B., and Boerner, S.L. (2008) Equivalence of laser scanning cytometric and flow cytometric immunophenotyping of lymphoid lesions in cytologic samples. *Am. J. Clin. Pathol.*, **129** (5), 780–785.
22. Mittag, A., Lenz, D., Gerstner, A.O.H., Sack, U., Steinbrecher, M., Koksch, M., Raffael, A., Bocsi, J., and Tárnok, A. (2005) Polychromatic (eight-color) slide-based cytometry for the phenotyping of leukocyte, nk, and nkt subsets. *Cytometry A*, **65** (2), 103–115.
23. Harnett, M.M. (2007) Laser scanning cytometry: understanding the immune system in situ. *Nat. Rev. Immunol.*, **7** (11), 897–904.
24. Tárnok, A. and Gerstner, A.O.H. (2003) Immunophenotyping using a laser

scanning cytometer. *Curr. Protoc. Cytom.*, 6.13.1–6.13.15.

25. Gorczyca, W., Bedner, E., Burfeind, P., Darzynkiewicz, Z., and Melamed, M.R. (1998) Analysis of apoptosis in solid tumors by laser-scanning cytometry. *Mod. Pathol.*, **11** (11), 1052–1058.
26. Gerstner, A.O.H. and Tárnok, A. (2002) Analysis of fine-needle aspirate biopsies from solid tumors by laser scanning cytometry (lsc). *Curr. Protoc. Cytom.*, 7.20.1–7.20.10.
27. Zabaglo, L., Ormerod, M.G., Parton, M., Ring, A., Smith, I.E., and Dowsett, M. (2003) Cell filtration-laser scanning cytometry for the characterisation of circulating breast cancer cells. *Cytometry A*, **55** (2), 102–108.
28. Lenz, D., Lenk, K., Mittag, A., Adams, V., Kränkel, N., Boldt, A., Gerstner, A.O.H., Raida, M., Weiss, T., Hambrecht, R., and Tarnok, A. (2005) Detection and quantification of endothelial progenitor cells by flow and laser scanning cytometry. *J. Biol. Regul. Homeost. Agents J.*, **19** (3–4), 180–187.
29. Oswald, J., Jørgensen, B., Pompe, T., Kobe, F., Salchert, K., Bornhäuser, M., Ehninger, G., and Werner, C. (2004) Comparison of flow cytometry and laser scanning cytometry for the analysis of cd34+ hematopoietic stem cells. *Cytometry A*, **57** (2), 100–107.
30. Bedner, E., Ruan, Q., Chen, S., Kamentsky, L.A., and Darzynkiewicz, Z. (2000) Multiparameter analysis of progeny of individual cells by laser scanning cytometry. *Cytometry*, **40** (4), 271–279.
31. Pozarowski, P., Huang, X., Gong, R.W., Priebe, W., and Darzynkiewicz, Z. (2004) Simple, semiautomatic assay of cytostatic and cytotoxic effects of antitumor drugs by laser scanning cytometry: effects of the bis-intercalator wp631 on growth and cell cycle of t-24 cells. *Cytometry A*, **57** (2), 113–119.
32. Valet, G., Leary, J.F., and Tárnok, A. (2004) Cytomics--new technologies: towards a human cytome project. *Cytometry A*, **59** (2), 167–171.
33. Hwang, D., Rust, A.G., Ramsey, S., Smith, J.J., Leslie, D.M., Weston, A.D., de Atauri, P., Aitchison, J.D., Hood, L., Siegel, A.F., and Bolouri, H. (2005) A data integration methodology for systems biology. *Proc. Natl. Acad. Sci. U.S.A.*, **102** (48), 17296–17301.
34. Lander, E.S., Linton, L.M., Birren, B., Nusbaum, C., Zody, M.C., Baldwin, J., Devon, K., Dewar, K., Doyle, M., FitzHugh, W., Funke, R., Gage, D., Harris, K., Heaford, A., Howland, J., Kann, L., Lehoczky, J., LeVine, R., McEwan, P., McKernan, K., Meldrim, J., Mesirov, J.P., Miranda, C., Morris, W., Naylor, J., Raymond, C., Rosetti, M., Santos, R., Sheridan, A., Sougnez, C., Stange-Thomann, N., Stojanovic, N., Subramanian, A., Wyman, D., Rogers, J., Sulston, J., Ainscough, R., Beck, S., Bentley, D., Burton, J., Clee, C., Carter, N., Coulson, A., Deadman, R., Deloukas, P., Dunham, A., Dunham, I., Durbin, R., French, L., Grafham, D., Gregory, S., Hubbard, T., Humphray, S., Hunt, A., Jones, M., Lloyd, C., McMurray, A., Matthews, L., Mercer, S., Milne, S., Mullikin, J.C., Mungall, A., Plumb, R., Ross, M., Shownkeen, R., Sims, S., Waterston, R.H., Wilson, R.K., Hillier, L.W., McPherson, J.D., Marra, M.A., Mardis, E.R., Fulton, L.A., Chinwalla, A.T., Pepin, K.H., Gish, W.R., Chissoe, S.L., Wendl, M.C., Delehaunty, K.D., Miner, T.L., Delehaunty, A., Kramer, J.B., Cook, L.L., Fulton, R.S., Johnson, D.L., Minx, P.J., Clifton, S.W., Hawkins, T., Branscomb, E., Predki, P., Richardson, P., Wenning, S., Slezak, T., Doggett, N., Cheng, J.F., Olsen, A., Lucas, S., Elkin, C., Uberbacher, E., Frazier, M., Gibbs, R.A., Muzny, D.M., Scherer, S.E., Bouck, J.B., Sodergren, E.J., Worley, K.C., Rives, C.M., Gorrell, J.H., Metzker, M.L., Naylor, S.L., Kucherlapati, R.S., Nelson, D.L., Weinstock, G.M., Sakaki, Y., Fujiyama, A., Hattori, M., Yada, T., Toyoda, A., Itoh, T., Kawagoe, C., Watanabe, H., Totoki, Y., Taylor, T., Weissenbach, J., Heilig, R., Saurin, W., Artiguenave, F., Brottier, P., Bruls, T., Pelletier, E., Robert, C., Wincker, P., Smith, D.R., Doucette-Stamm, L., Rubenfield, M., Weinstock, K., Lee, H.M., Dubois, J., Rosenthal, A., Platzer, M., Nyakatura,

G., Taudien, S., Rump, A., Yang, H., Yu, J., Wang, J., Huang, G., Gu, J., Hood, L., Rowen, L., Madan, A., Qin, S., Davis, R.W., Federspiel, N.A., Abola, A.P., Proctor, M.J., Myers, R.M., Schmutz, J., Dickson, M., Grimwood, J., Cox, D.R., Olson, M.V., Kaul, R., Raymond, C., Shimizu, N., Kawasaki, K., Minoshima, S., Evans, G.A., Athanasiou, M., Schultz, R., Roe, B.A., Chen, F., Pan, H., Ramser, J., Lehrach, H., Reinhardt, R., McCombie, W.R., de la Bastide, M., Dedhia, N., Blöcker, H., Hornischer, K., Nordsiek, G., Agarwala, R., Aravind, L., Bailey, J.A., Bateman, A., Batzoglou, S., Birney, E., Bork, P., Brown, D.G., Burge, C.B., Cerutti, L., Chen, H.C., Church, D., Clamp, M., Copley, R.R., Doerks, T., Eddy, S.R., Eichler, E.E., Furey, T.S., Galagan, J., Gilbert, J.G., Harmon, C., Hayashizaki, Y., Haussler, D., Hermjakob, H., Hokamp, K., Jang, W., Johnson, L.S., Jones, T.A., Kasif, S., Kaspryzk, A., Kennedy, S., Kent, W.J., Kitts, P., Koonin, E.V., Korf, I., Kulp, D., Lancet, D., Lowe, T.M., McLysaght, A., Mikkelsen, T., Moran, J.V., Mulder, N., Pollara, V.J., Ponting, C.P., Schuler, G., Schultz, J., Slater, G., Smit, A.F., Stupka, E., Szustakowski, J., Thierry-Mieg, D., Thierry-Mieg, J., Wagner, L., Wallis, J., Wheeler, R., Williams, A., Wolf, Y.I., Wolfe, K.H., Yang, S.P., Yeh, R.F., Collins, F., Guyer, M.S., Peterson, J., Felsenfeld, A., Wetterstrand, K.A., Patrinos, A., Morgan, M.J., de Jong, P., Catanese, J.J., Osoegawa, K., Shizuya, H., Choi, S., Chen, Y.J., and Szustakowki, J. (2001) Initial sequencing and analysis of the human genome. *Nature*, **409** (6822), 860–921.

35. Geisow, M.J. (1998) Proteomics: one small step for a digital computer, one giant leap for humankind. *Nat. Biotechnol.*, **16** (2), 206.
36. Lagarde, M., Géloën, A., Record, M., Vance, D., and Spener, F. (2003) Lipidomics is emerging. *Biochim. Biophys. Acta*, **1634** (3), 61.
37. Rehm, H. (2006) *Der Experimentator: Proteinbiochemie/Proteomics*, Spektrum Akademischer Verlag.
38. Valet, G. (2002) Predictive medicine by cytomics: potential and challenges. *J. Biol. Regul. Homeost. Agents*, **16** (2), 164–167.
39. Hood, L., Heath, J.R., Phelps, M.E., and Lin, B. (2004) Systems biology and new technologies enable predictive and preventative medicine. *Science*, **306** (5696), 640–643.
40. Lugli, E., Pinti, M., Nasi, M., Troiano, L., Ferraresi, R., Mussi, C., Salvioli, G., Patsekin, V., Robinson, J.P., Durante, C., Cocchi, M., and Cossarizza, A. (2007) Subject classification obtained by cluster analysis and principal component analysis applied to flow cytometric data. *Cytometry A*, **71** (5), 334–344.
41. Zeng, Q.T., Pratt, J.P., Pak, J., Ravnic, D., Huss, H., and Mentzer, S.J. (2007) Feature-guided clustering of multi-dimensional flow cytometry datasets. *J. Biomed. Inform.*, **40** (3), 325–331.
42. Steinbrich-Zöllner, M., Grün, J.R., Kaiser, T., Biesen, R., Raba, K., Wu, P., Thiel, A., Rudwaleit, M., Sieper, J., Burmester, G., Radbruch, A., and Grützkau, A. (2008) From transcriptome to cytome: integrating cytometric profiling, multivariate cluster, and prediction analyses for a phenotypical classification of inflammatory diseases. *Cytometry A*, **73** (4), 333–340.
43. Chan, W.C.W., Maxwell, D.J., Gao, X., Bailey, R.E., Han, M., and Nie, S. (2002) Luminescent quantum dots for multiplexed biological detection and imaging. *Curr. Opin. Biotechnol.*, **13** (1), 40–46.
44. Ecker, R.C., de Martin, R., Steiner, G.E., and Schmid, J.A. (2004) Application of spectral imaging microscopy in cytomics and fluorescence resonance energy transfer (fret) analysis. *Cytometry A*, **59** (2), 172–181.
45. Pedreira, C.E., Costa, E.S., Barrena, S., Lecrevisse, Q., Almeida, J., van Dongen, J.J.M., and Orfao, A. (2008) Generation of flow cytometry data files with a potentially infinite number of dimensions. *Cytometry A*, **73** (9), 834–846.

46. McLaughlin, B.E., Baumgarth, N., Bigos, M., Roederer, M., De Rosa, S.C., Altman, J.D., Nixon, D.F., Ottinger, J., Oxford, C., Evans, T.G., and Asmuth, D.M. (2008) Nine-color flow cytometry for accurate measurement of t cell subsets and cytokine responses. part i: panel design by an empiric approach. *Cytometry A*, **73** (5), 400–410.
47. Roederer, M. (2008) How many events is enough? are you positive? *Cytometry A*, **73** (5), 384–385.
48. Mittag, A. (2009) *Objektträgerbasierte Zytometrie als Analyseverfahren in Diagnostik und Polychromatischer Zellphänotypisierung*, Göttingen, Sierke.
49. Apweiler, R., Aslanidis, C., Deufel, T., Gerstner, A., Hansen, J., Hochstrasser, D., Kellner, R., Kubicek, M., Lottspeich, F., Maser, E., Mewes, H., Meyer, H.E., Müllner, S., Mutter, W., Neumaier, M., Nollau, P., Nothwang, H.G., Ponten, F., Radbruch, A., Reinert, K., Rothe, G., Stockinger, H., Tárnok, A., Taussig, M.J., Thiel, A., Thiery, J., Ueffing, M., Valet, G., Vandekerckhove, J., Wagener, C., Wagner, O., and Schmitz, G. (2009) Approaching clinical proteomics: current state and future fields of application in cellular proteomics. *Cytometry A*, **75** (10), 816–832.
50. Roederer, M., Dubs, J.G., Anderson, M.T., Raju, P.A., Herzenberg, L.A., and Herzenberg, L.A. (1995) Cd8 naive t cell counts decrease progressively in hiv-infected adults. *J. Clin. Invest.*, **95** (5), 2061–2066.
51. Rabin, R.L., Roederer, M., Maldonado, Y., Petru, A., Herzenberg, L.A., and Herzenberg, L.A. (1995) Altered representation of naive and memory cd8 t cell subsets in hiv-infected children. *J. Clin. Invest.*, **95** (5), 2054–2060.
52. Hatzakis, A., Touloumi, G., Karanicolas, R., Karafoulidou, A., Mandalaki, T., Anastassopoulou, C., Zhang, L., Goedert, J.J., Ho, D.D., and Kostrikis, L.G. (2000) Effect of recent thymic emigrants on progression of hiv-1 disease. *Lancet*, **355** (9204), 599–604.
53. Schnittman, S.M., Lane, H.C., Greenhouse, J., Justement, J.S., Baseler, M., and Fauci, A.S. (1990) Preferential infection of cd4+ memory t cells by human immunodeficiency virus type 1: evidence for a role in the selective t-cell functional defects observed in infected individuals. *Proc. Natl. Acad. Sci. U.S.A.*, **87** (16), 6058–6062.
54. Douek, D.C., Brenchley, J.M., Betts, M.R., Ambrozak, D.R., Hill, B.J., Okamoto, Y., Casazza, J.P., Kuruppu, J., Kunstman, K., Wolinsky, S., Grossman, Z., Dybul, M., Oxenius, A., Price, D.A., Connors, M., and Koup, R.A. (2002) Hiv preferentially infects hiv-specific cd4+ t cells. *Nature*, **417** (6884), 95–98.
55. Laffers, W., Mittag, A., Lenz, D., Tárnok, A., and Gerstner, A.O.H. (2006) Iterative restaining as a pivotal tool for n-color immunophenotyping by slide-based cytometry. *Cytometry A*, **69** (3), 127–130.
56. Schubert, W., Bonnekoh, B., Pommer, A.J., Philipsen, L., Böckelmann, R., Malykh, Y., Gollnick, H., Friedenberger, M., Bode, M., and Dress, A.W.M. (2006) Analyzing proteome topology and function by automated multidimensional fluorescence microscopy. *Nat. Biotechnol.*, **24** (10), 1270–1278.
57. Bedner, E., Du, L., Traganos, F., and Darzynkiewicz, Z. (2001) Caffeine dissociates complexes between DNA and intercalating dyes: application for bleaching fluorochrome-stained cells for their subsequent restaining and analysis by laser scanning cytometry. *Cytometry*, **43** (1), 38–45.
58. Schubert, W. (2003) Topological proteomics, toponomics, melk-technology. *Adv. Biochem. Eng. Biotechnol.*, **83**, 189–209.
59. Levenson, R.M. and Mansfield, J.R. (2006) Multispectral imaging in biology and medicine: slices of life. *Cytometry A*, **69** (8), 748–758.
60. Tran, C.D. and Furlan, R.J. (1993) Spectrofluorometer based on acousto-optic tunable filters for rapid scanning and multicomponent sample analyses. *Anal. Chem.*, **65** (13), 1675–1681.

61. Mittag, A., Lenz, D., Bocsi, J., Sack, U., Gerstner, A.O.H., and Tárnok, A. (2006) Sequential photobleaching of fluorochromes for polychromatic slide-based cytometry. *Cytometry A*, **69** (3), 139–141.
62. Panchuk-Voloshina, N., Haugland, R.P., Bishop-Stewart, J., Bhalgat, M.K., Millard, P.J., Mao, F., Leung, W.Y., and Haugland, R.P. (1999) Alexa dyes, a series of new fluorescent dyes that yield exceptionally bright, photostable conjugates. *J. Histochem. Cytochem.*, **47** (9), 1179–1188.
63. Takada, T., Fujitsuka, M., and Majima, T. (2007) Single-molecule observation of DNA charge transfer. *Proc. Natl. Acad. Sci. U.S.A.*, **104** (27), 11179–11183.
64. Wehrle-Haller, B. (2007) Analysis of integrin dynamics by fluorescence recovery after photobleaching. *Methods Mol. Biol.*, **370**, 173–202.
65. Braga, J., McNally, J.G., and Carmo-Fonseca, M. (2007) A reaction-diffusion model to study rna motion by quantitative fluorescence recovery after photobleaching. *Biophys. J.*, **92** (8), 2694–2703.
66. Bacsó, Z., Bene, L., Bodnár, A., Matkó, J., and Damjanovich, S. (1996) A photobleaching energy transfer analysis of cd8/mhc-i and lfa-1/icam-1 interactions in ctl-target cell conjugates. *Immunol. Lett.*, **54** (2–3), 151–156.
67. Bastiaens, P.I., Majoul, I.V., Verveer, P.J., Söling, H.D., and Jovin, T.M. (1996) Imaging the intracellular trafficking and state of the ab5 quaternary structure of cholera toxin. *EMBO J.*, **15** (16), 4246–4253.
68. Leuchowius, K., Weibrecht, I., Landegren, U., Gedda, L., and Söderberg, O. (2009) Flow cytometric in situ proximity ligation analyses of protein interactions and post-translational modification of the epidermal growth factor receptor family. *Cytometry A*, **75** (10), 833–889.
69. Maranto, A.R. (1982) Neuronal mapping: a photooxidation reaction makes lucifer yellow useful for electron microscopy. *Science*, **217** (4563), 953–955.
70. Tozer, J.T., Henderson, S.C., Sun, D., and Colello, R.J. (2007) Photoconversion using confocal laser scanning microscopy: a new tool for the ultrastructural analysis of fluorescently labeled cellular elements. *J. Neurosci. Methods*, **164** (2), 240–246.
71. Hatta, K., Tsujii, H., and Omura, T. (2006) Cell tracking using a photoconvertible fluorescent protein. *Nat. Protoc.*, **1** (2), 960–967.
72. Krinke, D., Jahnke, H., Pänke, O., and Robitzki, A.A. (2009) A microelectrode-based sensor for label-free in vitro detection of ischemic effects on cardiomyocytes. *Biosens. Bioelectron.*, **24** (9), 2798–2803.
73. Jahnke, H., Rothermel, A., Sternberger, I., Mack, T.G.A., Kurz, R.G., Pänke, O., Striggow, F., and Robitzki, A.A. (2009) An impedimetric microelectrode-based array sensor for label-free detection of tau hyperphosphorylation in human cells. *Lab Chip*, **9** (10), 1422–1428.
74. Gawad, S., Cheung, K., Seger, U., Bertsch, A., and Renaud, P. (2004) Dielectric spectroscopy in a micromachined flow cytometer: theoretical and practical considerations. *Lab Chip*, **4** (3), 241–251.
75. Cheung, K., Gawad, S., and Renaud, P. (2005) Impedance spectroscopy flow cytometry: on-chip label-free cell differentiation. *Cytometry A*, **65** (2), 124–132.
76. Schade-Kampmann, G., Huwiler, A., Hebeisen, M., Hessler, T., and Di Berardino, M. (2008) On-chip non-invasive and label-free cell discrimination by impedance spectroscopy. *Cell Prolif.*, **41** (5), 830–840.
77. Loken, M.R., Sweet, R.G., and Herzenberg, L.A. (1976) Cell discrimination by multiangle light scattering. *J. Histochem. Cytochem.*, **24** (1), 284–291.
78. Salzman, G.C., Crowell, J.M., Martin, J.C., Trujillo, T.T., Romero, A., Mullaney, P.F., and LaBauve, P.M. (1975) Cell classification by laser light scattering: identification and separation of unstained leukocytes. *Acta Cytol.*, **19** (4), 374–377.
79. Rajwa, B., Venkatapathi, M., Ragheb, K., Banada, P.P., Hirleman, E.D., Lary, T., and Robinson, J.P. (2008)

Automated classification of bacterial particles in flow by multiangle scatter measurement and support vector machine classifier. *Cytometry A*, **73** (4), 369–379.
80. Banada, P.P., Huff, K., Bae, E., Rajwa, B., Aroonnual, A., Bayraktar, B., Adil, A., Robinson, J.P., Hirleman, E.D., and Bhunia, A.K. (2009) Label-free detection of multiple bacterial pathogens using light-scattering sensor. *Biosens. Bioelectron.*, **24** (6), 1685–1692.
81. Newberg, J. and Murphy, R.F. (2008) A framework for the automated analysis of subcellular patterns in human protein atlas images. *J. Proteome Res.*, **7** (6), 2300–2308.
82. Ishikawa, F.N., Stauffer, B., Caron, D.A., and Zhou, C. (2009) Rapid and label-free cell detection by metal-cluster-decorated carbon nanotube biosensors. *Biosens. Bioelectron.*, **24** (10), 2967–2972.
83. Fujita, K. and Smith, N.I. (2008) Label-free molecular imaging of living cells. *Mol. Cells*, **26** (6), 530–535.
84. Lorincz, A., Haddad, D., Naik, R., Naik, V., Fung, A., Cao, A., Manda, P., Pandya, A., Auner, G., Rabah, R., Langenburg, S.E., and Klein, M.D. (2004) Raman spectroscopy for neoplastic tissue differentiation: a pilot study. *J. Pediatr. Surg.*, **39** (6), 953–956; discussion 953–956.
85. Watson, D.A., Brown, L.O., Gaskill, D.F., Naivar, M., Graves, S.W., Doorn, S.K., and Nolan, J.P. (2008) A flow cytometer for the measurement of raman spectra. *Cytometry A*, **73** (2), 119–128.
86. du Plessis, A.J. (1995) Near-infrared spectroscopy for the in vivo study of cerebral hemodynamics and oxygenation. *Curr. Opin. Pediatr.*, **7** (6), 632–639.
87. Krohn, K.A., Link, J.M., and Mason, R.P. (2008) Molecular imaging of hypoxia. *J. Nucl. Med.*, **49** (Suppl. 2), 129S–148S.
88. Hamaoka, T., McCully, K.K., Quaresima, V., Yamamoto, K., and Chance, B. (2007) Near-infrared spectroscopy/imaging for monitoring muscle oxygenation and oxidative metabolism in healthy and diseased humans. *J. Biomed. Opt.*, **12** (6), 062105.
89. Lima, A. and Bakker, J. (2005) Noninvasive monitoring of peripheral perfusion. *Intensive Care Med.*, **31** (10), 1316–1326.
90. Irani, F., Platek, S.M., Bunce, S., Ruocco, A.C., and Chute, D. (2007) Functional near infrared spectroscopy (fnirs): an emerging neuroimaging technology with important applications for the study of brain disorders. *Clin. Neuropsychol.*, **21** (1), 9–37.
91. van de Ven, S.M.W.Y., Elias, S.G., van den Bosch, M.A.A.J., Luijten, P., and Mali, W.P.T.M. (2008) Optical imaging of the breast. *Cancer Imaging*, **8**, 206–215.
92. Brown, C.P., Bowden, J.C., Rintoul, L., Meder, R., Oloyede, A., and Crawford, R.W. (2009) Diffuse reflectance near infrared spectroscopy can distinguish normal from enzymatically digested cartilage. *Phys. Med. Biol.*, **54** (18), 5579–5594.
93. Harz, M., Rösch, P., and Popp, J. (2009) Vibrational spectroscopy – a powerful tool for the rapid identification of microbial cells at the single-cell level. *Cytometry A*, **75** (2), 104–113.
94. Wang, C., Li, F., Wu, R., Hovhannisyan, V.A., Lin, W., Lin, S., So, P.T.C., and Dong, C. (2009) Differentiation of normal and cancerous lung tissues by multiphoton imaging. *J. Biomed. Opt.*, **14** (4), 044034.
95. Pyne, S., Hu, X., Wang, K., Rossin, E., Lin, T., Maier, L.M., Baecher-Allan, C., McLachlan, G.J., Tamayo, P., Hafler, D.A., De Jager, P.L., and Mesirov, J.P. (2009) Automated high-dimensional flow cytometric data analysis. *Proc. Natl. Acad. Sci. U.S.A.*, **106** (21), 8519–8524.
96. Lakoumentas, J., Drakos, J., Karakantza, M., Nikiforidis, G.C., and Sakellaropoulos, G.C. (2009) Bayesian clustering of flow cytometry data for the diagnosis of b-chronic lymphocytic leukemia. *J. Biomed. Inform.*, **42** (2), 251–261.

97. Pepperkok, R. and Ellenberg, J. (2006) High-throughput fluorescence microscopy for systems biology. *Nat. Rev. Mol. Cell Biol.*, **7** (9), 690–696.
98. Newberg, J., Hua, J., and Murphy, R.F. (2009) Location proteomics: systematic determination of protein subcellular location. *Methods Mol. Biol.*, **500**, 313–332.
99. Chen, X., Zhou, X., and Wong, S.T.C. (2006) Automated segmentation, classification, and tracking of cancer cell nuclei in time-lapse microscopy. *IEEE Trans. Biomed. Eng.*, **53** (4), 762–766.
100. Degerman, J., Thorlin, T., Faijerson, J., Althoff, K., Eriksson, P.S., Put, R.V.D., and Gustavsson, T. (2009) An automatic system for in vitro cell migration studies. *J. Microsc.*, **233** (1), 178–191.
101. Fotos, J.S., Patel, V.P., Karin, N.J., Temburni, M.K., Koh, J.T., and Galileo, D.S. (2006) Automated time-lapse microscopy and high-resolution tracking of cell migration. *Cytotechnology*, **51** (1), 7–19.
102. Varga, V.S., Bocsi, J., Sipos, F., Csendes, G., Tulassay, Z., and Molnár, B. (2004) Scanning fluorescent microscopy is an alternative for quantitative fluorescent cell analysis. *Cytometry A*, **60** (1), 53–62.
103. Bajaj, S., Welsh, J.B., Leif, R.C., and Price, J.H. (2000) Ultra-rare-event detection performance of a custom scanning cytometer on a model preparation of fetal nrbcs. *Cytometry*, **39** (4), 285–294.
104. Shapiro, H.M. and Perlmutter, N.G. (2006) Personal cytometers: slow flow or no flow? *Cytometry A*, **69** (7), 620–630.
105. Moon, S., Keles, H.O., Ozcan, A., Khademhosseini, A., Haeggstrom, E., Kuritzkes, D., and Demirci, U. (2009) Integrating microfluidics and lensless imaging for point-of-care testing. *Biosens. Bioelectron.*, **24** (11), 3208–3214.
106. Zaragosa, K.A. (2006) Lanl's low-cost device. *Innov.: Am. J. Technol. Commer.*, **4** (4), *http://www.innovation-america.org/lanls-low-cost-device.*
107. Greve, B., Cassens, U., Westerberg, C., Jun, W., Sibrowski, W., Reichelt, D., and Gohde, W. (2003) A new no-lyse, no-wash flow-cytometric method for the determination of cd4 t cells in blood samples. *Transfus. Med. Hemother.*, **30** (1), 8–13.
108. Cassens, U., Göhde, W., Kuling, G., Gröning, A., Schlenke, P., Lehman, L.G., Traoré, Y., Servais, J., Henin, Y., Reichelt, D., and Greve, B. (2004) Simplified volumetric flow cytometry allows feasible and accurate determination of cd4 t lymphocytes in immunodeficient patients worldwide. *Antivir. Ther. (Lond.)*, **9** (3), 395–405.
109. *www.partec.com* (accessed September 2009).

2
Novel Concepts and Requirements in Cytometry

Herbert Schneckenburger, Michael Wagner, Petra Weber, and Thomas Bruns

2.1
Introduction

In addition to high-throughput flow cytometry, slide-based techniques [1] play an increasing role in modern cytomics. This holds, in particular, for high content screening (HCS), where detailed information is collected from individual samples, for example, single cells. Experimental systems include microscopic setups, fluorescence reader systems based on microtiter plates, and microfluidic devices. Examples of these systems are provided, and some basic concepts and requirements are outlined in this chapter.

2.2
Fluorescence Microscopy

2.2.1
Light Dose

Optical microscopy includes trans- and epi-illumination methods, among which fluorescence microscopy plays a predominant role. This is due to the fact that in addition to intrinsic fluorophores, a large number of organic [2, 3] and inorganic [4] fluorescent probes are now available for specific staining of cells and organelles. In addition, numerous cellular proteins fluorescing in blue, green, yellow, and red ranges of the spectrum have been generated genetically [5, 6].

In fluorescence microscopy, excitation light is commonly focused onto small areas of cells or tissues, often resulting in rather high light doses upon exposure. This holds, in particular, for laser scanning microscopy [7, 8] and related techniques, for example, 4-Pi and stimulated emission depletion (STED) microscopy [9], where very high lateral and axial resolutions have been obtained. Only recently, some attention has been paid to the phototoxic effects of irradiation [10, 11], as shown in Figure 2.1. This figure shows the percentage of colony formation (plating efficiency) after seeding of single U373-MG glioblastoma cells and illumination at various

Advanced Optical Flow Cytometry: Methods and Disease Diagnoses, First Edition. Edited by Valery V. Tuchin.

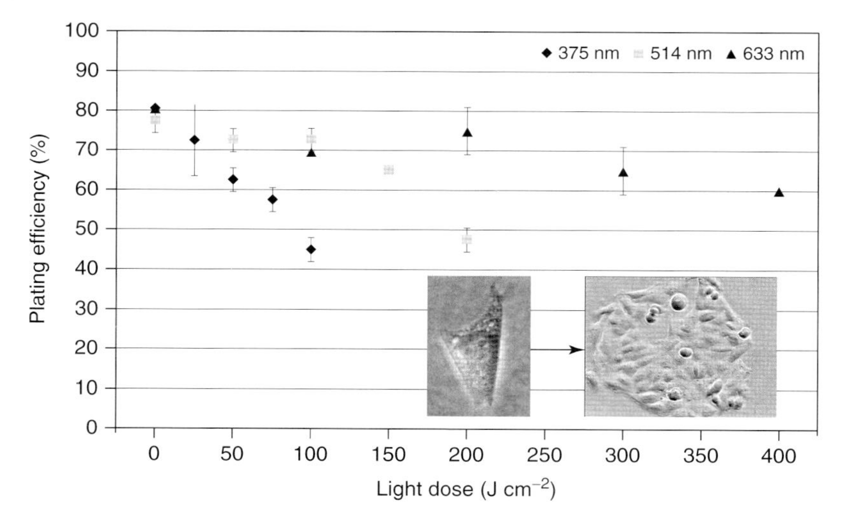

Figure 2.1 Percentage of colony formation ("plating efficiency") of single nonincubated U373-MG glioblastoma cells upon exposure to different excitation wavelengths and light doses. Values represent medians ± median absolute deviations (MADs) of four experiments with irradiation of 20 cells in each case. The inlay visualizes colony formation (within seven days) by phase contrast microscopy. Reproduced with permission from Ref. [11] with modifications.

wavelengths and light doses. Plating efficiency is around 80% for nonirradiated controls, but decreases upon light exposure. If cell viability is defined as less than 10% reduction of the plating efficiency, cells remain viable up to a light dose of about 25 J cm^{-2} at an irradiation wavelength of 375 nm, 100 J cm^{-2} at 514 nm, and about 200 J cm^{-2} at 633 nm. These doses correspond to 250, 1000, and 2000 s of solar irradiance, respectively.

Cells incubated with fluorescent dyes or transfected with genes encoding for a fluorescent protein are generally more sensitive to irradiation, and nonphototoxic light doses are often limited to about 10 J cm^{-2} [11]. This corresponds to 100 s of solar irradiation or 1 s of irradiation of a sample (100 μm × 100 μm in size) with a power of 1 mW.

2.2.2
Cell Systems

In fluorescence microscopy, cells from various organs are commonly cultivated as monolayers on glass or plastic dishes, for example, microscope objective slides, as shown in the inlay of Figure 2.1. Three-dimensional cell cultures, however, are more similar to real tissue with respect to its architecture and metabolism, and therefore a multicellular spheroid model has been developed recently [12]. This and similar models may play an increasing role in the near future, but require microscopic techniques with high axial resolution and low light exposure.

2.2.3
Methods

Owing to the limitation of light exposure, wide-field microscopy with simultaneous detection of whole samples appears to be more preferable to laser scanning microscopy. However, an image from the focal plane is commonly superposed by out-of-focus images from other parts of the sample. Therefore, techniques are required for optical sectioning. A well-established method uses structured illumination for this purpose [13]. By imaging an optical grid from three positions on the sample and by application of an appropriate algorithm, an image from the focal plane can be recovered, whereas out-of-focus images are eliminated. More recent techniques use a three-dimensional interference pattern, permitting doubling of optical resolution down to about 100 nm in lateral direction and 200 nm in axial direction [14]. Structured illumination, however, needs numerous light exposures for imaging various planes of a sample with subsequent 3D reconstruction. While the resulting light doses may affect cell viability, light doses are minimized in single plane illumination microscopy (SPIM) or light sheet-based fluorescence microscopy (LSFM), where only those planes of a sample that are examined simultaneously are illuminated [15]. This requires illumination from the perpendicular direction in a microscope using, for example, specific cylindrical optics.

Total internal reflection fluorescence microscopy (TIRFM) is a method for selective measurements of cell surfaces, for example, plasma membranes where a laser beam is totally reflected on a surface (e.g., cell–glass interface) with its evanescent electromagnetic field penetrating a small distance into the sample [16, 17]. Layers of around 100 nm thickness are selected, the resolution of which may be below 10 nm if the angle of incidence is varied [18]. Applications of this technique are numerous including studies of membrane dynamics or membrane-associated molecular interactions, as well as single molecule detection [19]. An experimental setup with easy switching between variable-angle TIRFM and conventional wide-field microscopy is shown in Figure 2.2.

2.3
Fluorescence Reader Systems

2.3.1
Cell-Based Fluorescence Screening

Since the 1990s, cell-based assay technologies have played an increasing role in high-throughput screening (HTS) and HCS [20, 21]. Main topics are the detection of cancer [22] and other diseases, drug discovery [23], as well as pharmaceutical or toxicology testing [24, 25]. Experimental equipments are manifold, ranging from laser scanning fluorometry [26] to fluorescence lifetime imaging [27]. However, the use of specific slides or microtiter plates is a common feature of many systems.

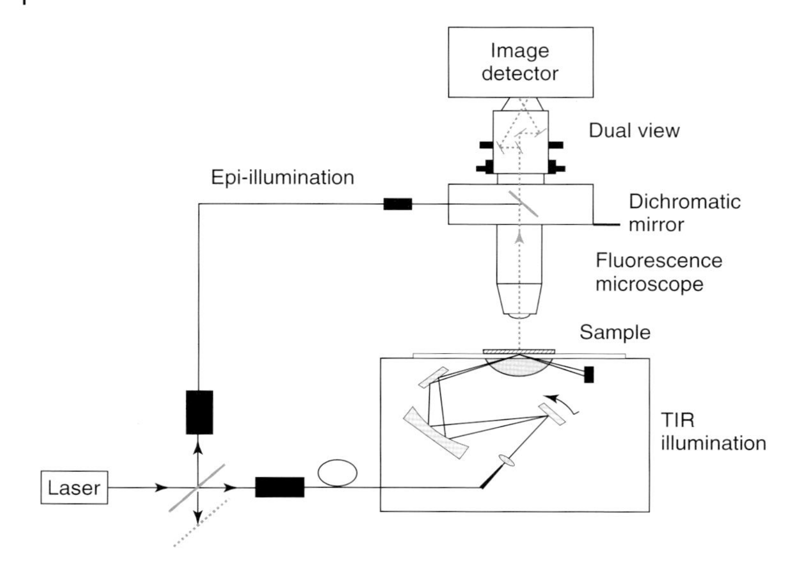

Figure 2.2 Fluorescence microscope permitting easy switching between total internal reflection (TIR) and epi-illumination as well as simultaneous detection of two images.

2.3.2
TIR Fluorescence Reader

Only in very few cases, the principle of total internal reflection (TIR) has been used for reader technology [28, 29]. TIR measurements, however, may give valuable information upon focal adhesion during cell growth, cell migration, or cell death, as well as upon signal transduction after stimulation or upon modification of membrane properties as a result of certain diseases or pharmacological agents. In addition, at a penetration depth of around 100 nm of the evanescent electromagnetic field very little of the surrounding medium or supernatant is illuminated, and the signal/background ratio may be very high.

As shown in Figure 2.3, a TIR fluorescence reader for 96 well microtiter plates is described with a laser beam being split into 8 individual beams, each of which is totally reflected at the bottom of 12 cavities. Currently, at a distance of 9 mm between individual samples, this is fulfilled by a glass bottom of 2 mm thickness permitting an angle of incidence of 66°, which is fairly larger than the critical angle $\Theta_C = 64^\circ$.

Low light absorption by the glass bottom and use of noncytotoxic adhesives are main prerequisites, as described in Ref. [28]. Fluorescence of the whole plate may be recorded simultaneously within a few seconds by an integrating camera [28], or the plate may be scanned by a point detector permitting acquisition of multiparametric data, for example, fluorescence spectra, lifetimes, or polarization [29, 30].

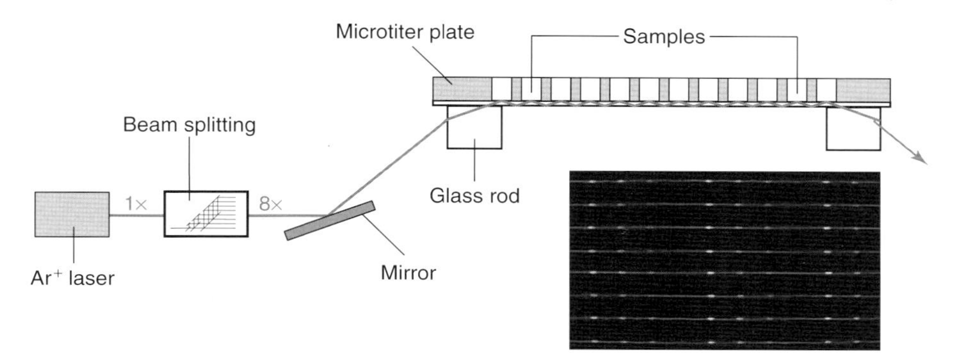

Figure 2.3 TIR fluorescence reader using 8-fold beam splitting and 12-fold total internal reflection of each laser beam to illuminate 96 samples of a microtiter plate simultaneously. The inlay represents an image with various concentrations of T47D-EYFP-Mem breast cancer cells upon seeding of 0, 125, 250, or 500 cells mm^{-2}. Reproduced with permission from Ref. [28] with modifications.

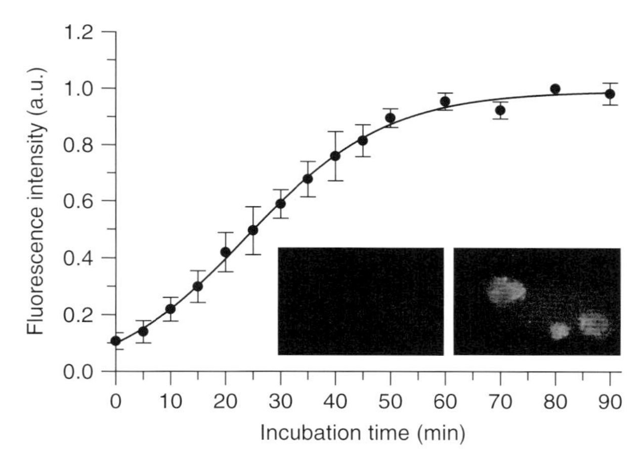

Figure 2.4 Time course of TIR fluorescence intensity of T47D-PKCα-EGFP breast cancer cells after incubation with 100 nM PMA in a microtiter plate (excitation wavelength: 488 nm; fluorescence measured at $\lambda \geq 515$ nm; recording time: 60 s; density of seeded cells: 500 mm^{-2}); mean values ± standard deviations of eight samples including fitting function $I_F = I_0 + (I_{max} - I_0) \times \left[1 + e^{-(t-t_0)/\tau}\right]^{-1}$. The inlays represent TIR fluorescence micrographs (140 μm × 100 μm) prior to activation and 60 min after activation with PMA. Reproduced with permission from Ref. [28] with modifications.

An application of the TIR fluorescence reader is depicted in Figure 2.4 showing the fluorescence intensity of T47D breast cancer cells stably transfected with a plasmid encoding for a green fluorescent protein kinase c fusion protein (T47D-PKCα-EGFP), as further described in Ref. [28]. PKCα was activated by phorbol-12-myristate-13-acetate (PMA) and, as proven by the TIR reader (Figure 2.3), dislocated from the cytoplasm toward the plasma membrane.

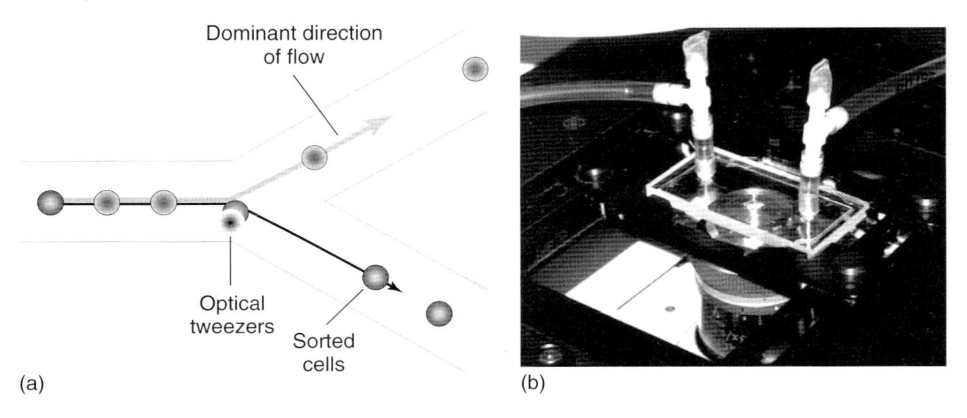

Figure 2.5 Principle of a microfluidic system for single-cell sorting with laser tweezers (a) and corresponding experimental setup (b).

2.4
Microfluidics Based on Optical Tweezers

Although the principle of optical tweezers was reported about 40 years ago [31], laser tweezers have been combined with microfluidic systems only recently [32]. Owing to the novel techniques of etching microchannels, those systems have now become rather promising, in particular, for single-cell sorting [32–34], when few cells have to be isolated from a larger cell collective. As shown in Figure 2.5, single cells may be moved out of a cellular flow either by application of single laser pulses or by deflection in a permanent laser beam. In both cases, they can be collected in a small reservoir for further (e.g., biochemical or PCR) measurements. Major problems could be sealing of the chamber and pumping of extremely low volumes (around 100 pl s^{-1}) through the tiny cross section of the microchannels [35]. In addition, cell viability must be maintained upon exposure to the focused laser beam. Previous experiments with a near-infrared laser (1064 nm) or a red laser diode (670 nm), however, have shown that viability is maintained up to a light dose of about 1 GW cm^{-2} [36], corresponding to 4 min of irradiation of 1 μm^2 with a laser power around 40 mW. This implies that upon illumination of very small areas by laser tweezers cell survival is considerably higher than upon light exposure of whole cells, as reported above.

2.5
Conclusion

Some slide-based techniques, which play an increasing role in modern cytometry, have been described. These techniques, including microscopic setups, fluorescence reader systems, and microfluidic devices, are of particular interest for multiparameter analysis in HCS (HTS), and in some cases also for micromanipulation, for example, cell sorting. While many experiments are performed with single cells, alternative systems like multicellular spheroids are becoming more popular and require novel 3D detection methods. Light dose upon irradiation is an important parameter for maintaining cell viability, which should be considered for high irradiance or long exposure times.

Acknowledgment

Current research is supported by Land Baden-Württemberg, Europäische Union, Europäischer Fonds für die regionale Entwicklung, and Bundesministerium für Bildung und Forschung (BMBF).

References

1. Tarnok, A. (2006) Slide-based cytometry for cytomics – a minireview. *Cytometry*, **69A**, 555–562.
2. Johnson, I. (1988) Fluorescent probes for living cells. *Histochem. J.*, **30**, 123–140.
3. Mullins, J.M. (1999) Overview of fluorochromes. *Methods Mol. Biol.*, **115**, 97–105.
4. Bruchez, M., Moronne, M., Gin, P., Weis, S., and Alivisatos, A.P. (1998) Semiconductor nanocrystals as fluorescence biological labels. *Science*, **281**, 2013–2016.
5. Rizzuto, R., Brini, M., Pizzo, P., Murgia, M., and Pozzan, T. (1995) Chimeric green fluorescent protein as a tool for visualizing subcellular organelles in living cells. *Curr. Biol.*, **5**, 635–642.
6. Ikawa, M., Yamada, S., Nakanishi, T., and Okabe, M. (1999) Green fluorescent protein (GFP) as a vital marker in mammals. *Curr. Top. Dev. Biol.*, **44**, 1–20.
7. Pawley, J. (1990) *Handbook of Biological Confocal Microscopy*, Plenum Press, New York.
8. Denk, W., Strickler, J.H., and Webb, W.W. (1990) Two-photon laser scanning microscope. *Science*, **248**, 73–76.
9. Willig, K.I., Harke, B., Medda, R., and Hell, S.W. (2007) STED microscopy with continuous wave beams. *Nat. Methods*, **4**, 915–918.
10. Hoebe, R.A., van Oven, C.H., Gadella, T.W.J. Jr., Dhonukshe, P.B., vanNoorden, C.J.F., and Manders, E.M.M. (2007) Controlled light-exposure microscopy reduces photobleaching and phototoxicity in fluorescence live-cell imaging. *Nat. Biotechnol.*, **25**, 249–253.
11. Wagner, M., Weber, P., Bruns, T., Strauss, W.S.L., Wittig, R., and Schneckenburger, H. (2010) Light dose is a limiting factor to maintain cell viability in fluorescence microscopy and single molecule detection. *Int. J. Mol.Sci.*, **11**, 956–966.
12. Kunz-Schughart, L.A., Freyer, J.P., Hofstädter, F., and Ebner, R. (2004) The use of 3-D cultures for high throughput screening: the multicellular spheroid model. *J. Biomol. Screen.*, **9**, 273–285.
13. Neil, M.A.A., Juskaitis, R., and Wilson, T. (1997) Method of obtaining optical sectioning by structured light in a conventional microscope. *Opt. Lett.*, **22**, 1905–1907.
14. Gustafsson, M.T.L., Shao, L., Carlton, P.M., Wang, C.J.R., Golubovskaya, I.N., Cande, W.Z., Agard, D.A., and Sedat, J.W. (2008) Three-dimensional resolution doubling in wide-field fluorescence

microscopy by structured illumination. *Biophys. J.*, **94**, 4957–4970.

15. Huisken, J., Swoger, J., del Bene, F., Wittbrodt, J., and Stelzer, E.H.K. (2004) Optical sectioning deep inside live embryos by SPIM. *Science*, **305**, 1007–1009.
16. Axelrod, D. (1981) Cell-substrate contacts illuminated by total internal reflection fluorescence. *J. Cell Biol.*, **89**, 141–145.
17. Schneckenburger, H. (2005) Total internal reflection fluorescence microscopy: Technical innovations and novel applications. *Curr. Opin. Biotechnol.*, **16**, 13–18.
18. Wagner, M., Weber, P., Strauss, W.S.L., Lassalle, H.-P., and Schneckenburger, H. (2008) Nanotomography of cell surfaces with evanescent fields. *Adv. Opt. Technol.*, **2008**, 254317:1–254317:7.
19. Sako, Y. and Uyemura, T. (2002) Total internal reflection fluorescence microscopy for single molecule imaging in living cells. *Cell. Struct. Funct.*, **27**, 357–365.
20. Kamentsky, L.A., Burger, D.E., Gershman, R.J., Kamentsky, L.-D., and Luther, E. (1997) Slide-based laser scanning cytometry. *Acta Cytol.*, **41**, 123–143.
21. Hertzberg, R.P., and Pope, A.J. (2000) High throughput screening: new technology for the 21st century. *Curr. Opin. Chem. Biol.*, **4**, 445–451.
22. Méhes, G., Lörch, T., and Ambros, P.F. (2000) Quantitative analysis of disseminated tumor cells in the bone marrow by automated fluorescence image analysis. *Cytometry*, **42**, 357–362.
23. Zheng, W., Spencer, R.H., and Kiss, L. (2004) High throughput assay technologies for ion channel drug discovery. *Assay Drug Dev. Technol.*, **2**, 543–552.
24. Wolff, M., Kredel, S., Wiedenmann, J., Nienhaus, G.U., and Heilker, R. (2008) Cell-based assays in practice: cell markers from autofluorescent proteins of the GFP-family. *Comb. Chem. High Throughput Screen.*, **11**, 602–609.
25. Fritzsche, M. and Mandenius, C.F. (2010) Fluorescent cell-based sensing approaches for toxicity testing. *Anal. Bioanal. Chem.*, **389**, 181–191.
26. Bowen, W.P. and Wylie, P.G. (2006) Application of laser-scanning fluorescence microplate cytometry in high content screening. *Assay Drug Dev. Technol.*, **4**, 209–221.
27. Ramanujan, V.K., Zhang, J.H., Biener, E., and Herman, B. (2005) Multiphoton fluorescence lifetime contrast in deep tissue imaging: prospects in redox imaging and disease diagnosis. *J. Biomed. Opt.*, **10**, 051407.
28. Bruns, T., Strauss, W.S.L., Sailer, R., Wagner, M., and Schneckenburger, H. (2006) Total internal reflectance fluorescence reader for selective investigations of cell membranes. *J. Biomed. Opt.*, **11**, 034011.
29. Bruns, T., Angres, B., Steuer, H., Weber, P., Wagner, M., and Schneckenburger, H. (2009) A FRET-based total internal reflection (TIR) fluorescence reader for apoptosis. *J. Biomed. Opt.*, **14**, 021003.
30. Bruns, T., Strauss, W.S.L., and Schneckenburger, H. (2008) Total internal reflection fluorescence lifetime and anisotropy screening of cell membrane dynamics. *J. Biomed. Opt.*, **13**, 041317.
31. Ashkin, A. (1969) Acceleration and trapping of particles by radiation pressure. *Phys. Rev. Lett.*, **24**, 156–159.
32. Enger, J., Goksör, M., Ramser, K., Hagberg, P., and Hanstorp, D. (2004) Optical tweezers applied to a microfluidic system. *Lab Chip*, **4**, 196–200.
33. Perroud, T.D., Kaiser, J.N., Sy, J.C., Lane, T.W., Branda, C.S., Singh, A.K., and Patel, K.D. (2008) Microfluidic-based cell sorting of Francisella tularensis infected macrophages using optical forces. *Anal. Chem.*, **80**, 6365–6372.
34. Eriksson, E., Sott, K., Lundqvist, F., Sveningsson, M., Scrimgeour, J., Hanstorp, D., Goksör, M., and Granéli, A. (2010) A microfluidic device for reversible environmental changes around single cells using optical tweezers for

cell selection and positioning. *Lab Chip*, **10**, 617–625.

35. Bruns, T., Becsi, L., Talkenberg, M., Wagner, M., Weber, P., Mescheder, U., and Schneckenburger, H. (2010) Microfluidic system for single cell sorting with optical tweezers. *Proc. SPIE*, Vol. 7376, 7360M.

36. Schneckenburger, H., Hendinger, A., Sailer, R., Gschwend, M.H., Strauss, W.S.L., Bauer, M., and Schütze, K. (2000) Cell viability in optical tweezers: high power red laser diode versus Nd:YAG laser. *J. Biomed. Opt.*, **5**, 40–44.

3
Optical Imaging of Cells with Gold Nanoparticle Clusters as Light Scattering Contrast Agents: A Finite-Difference Time-Domain Approach to the Modeling of Flow Cytometry Configurations

Stoyan Tanev, Wenbo Sun, James Pond, Valery V. Tuchin, and Vladimir P. Zharov

3.1
Introduction

Flow cytometry is one of the most widely used techniques for the characterization of biological cells with an increasing number of applications in modern biomedical practice. It is a well-established, powerful analytical tool that has led to many revolutionary discoveries in cell biology [1–3]. It is based on the analysis of single cells that are introduced into an artificial flow and hydrodynamically focused in such a way that laser-induced fluorescence and/or the light scattered by them could be detected by appropriately positioned photodetectors. This is a promising and highly accurate cell characterization and imaging technology providing high-speed (a few million cells in a minute) multiparameter quantification of the biological properties of individual cells at subcellular and molecular levels including cell size, shape, and intracellular heterogeneities. Conventional flow cytometry measures the intensity of scattered light in both, forward and orthogonal directions (Figure 3.1). For example, the analysis of forward and orthogonal scattering of light allows discrimination between lymphocytes, monocytes, and granulocytes [2].

The scattered light angular distribution can also be measured and the information contained in the scattered light can thereby be put to greater use. The angular distribution of the scattered light intensity is significantly more informative than the light scattered in the forward and orthogonal directions as it manifests a higher sensitivity to cell morphology and characteristics. This idea has already been implemented in more advanced flow cytometry instrumentation devices such as scanning flow cytometers [4]. The flow-cytometric light scattering angular patterns provided by such devices open entirely new possibilities for the characterization of normal and pathological cells.

It should be pointed out that the invasive extraction of cells from a living system may lead to changes in cell properties and eventually thwart the long-term study of cells in their native biological environment. *In vivo* flow cytometry using vessels with natural bioflows containing cells of interest is expected to overcome these problems. One of the potential challenges for *in vivo* flow cytometry could be the interference between the light scattered from the cells under examination and

Advanced Optical Flow Cytometry: Methods and Disease Diagnoses, First Edition. Edited by Valery V. Tuchin.

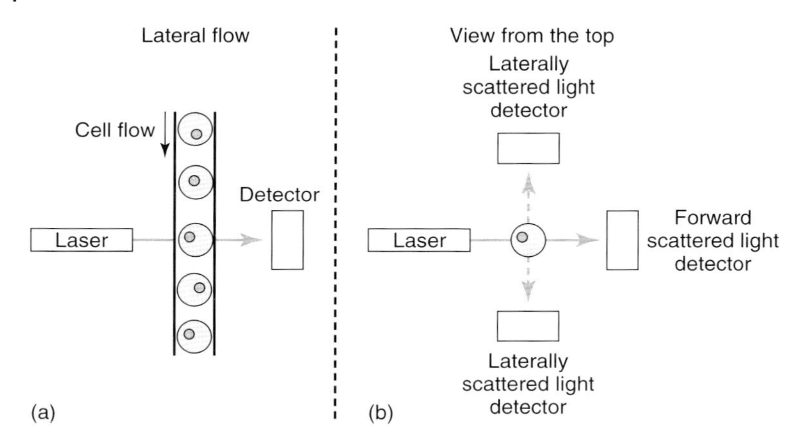

Figure 3.1 A simplified schematic representation of a conventional flow cytometer: (a) lateral view and (b) top view.

the light scattering background from cells in the surrounding tissue. To reduce the interference caused by light scattering background from surrounding tissue, gold nanoparticles (NPs) with strong plasmon scattering resonance properties can be applied as flow cytometry contrast agents. It is, therefore, important to study the light scattering properties of cells containing gold NP clusters as light scattering contrast agents within a flow cytometry configuration [5]. This requires an understanding of the light interaction phenomena at the micro- and nanoscales with cells alone as well as with cells in the presence of NPs. Information about the nature of the light scattering mechanisms of cell microstructures determines the sensitivity of the light scattering parameters to pathological changes in the cellular morphology but unfortunately, the biological origins of the differences in the light scattering patterns from normal and pathological (for example, precancerous and cancerous) cells are not fully understood [1, 6–9]. This makes the interpretation of the *in vivo* flow cytometry results difficult and, very often, quite inefficient.

In several studies of flow cytometry and cell imaging, optical software simulation and modeling tools are the only means to a deeper understanding of the underlying physical optics phenomena, and biological and biochemical processes [10]. The computational modeling of light scattering from single biological cells is of particular interest as it could provide information about the fundamental light–cell interaction phenomena that is highly relevant for the practical interpretation of cell scattering signatures and images by pathologists. The modeling of light interaction with cells is usually approached from a single particle electromagnetic wave scattering perspective. In particular, the single particle scattering approach is of particular relevance for experimental configurations based on flow cytometry and could be characterized by two specific features. First, the wavelength of light is larger than or comparable to the size of the scattering subcellular structures. Second, biological cells have irregular shapes and inhomogeneous refractive index distributions, which makes it impossible to use analytical modeling approaches.

Both features necessitate the use of numerical modeling approaches derived from rigorous electromagnetic theory, such as the method of separation of variables, finite element method, method of lines, point matching method, method of moments, discrete dipole approximation method, null-field (extended boundary condition) method, **T**-matrix electromagnetic scattering approach, surface Green's function electromagnetic scattering approach, and finite-difference time-domain (FDTD) method [11].

The FDTD simulation and modeling of light interaction with single and multiple, normal and pathological biological cells, and subcellular structures have attracted the attention of researchers since 1996 [10, 12–25] (Chapter 11). The FDTD approach was first considered a better alternative to Mie theory [26] allowing for the modeling of irregular cell shapes and inhomogeneous distributions of complex refractive index values. The emerging relevance of nanoscale cell imaging research has established the FDTD method as one of the powerful tools for studying the nature of light–cell interactions within the context of flow cytometry. A number of cytometry-related research directions based on the FDTD approach can be identified. The first one focuses on studying the lateral light scattering patterns for early detection of pathological changes in cancerous cells, such as increased nuclear size, higher degrees of nuclear pleomorphism and increased nuclear-to-cytoplasmic ratios [10, 12–18]. The second research direction explores the application of the FDTD method to modeling of forward light transmission and scattering from cells for advanced cell imaging based on optical phase contrast microscopy (OPCM) techniques [21–25].

This chapter has two main objectives. First, to review a number of examples illustrating the application of the FDTD approach to the modeling of some typical light scattering configurations that could be associated with flow cytometry [5]. Second, to provide a thorough discussion of these new developments in advanced cytometry research by pointing out potential new research directions. A brief description of the FDTD method focusing on the features associated with its application to modeling of light scattering and OPCM cell imaging experiments is provided. The examples include light scattering from OPCM imaging of single biological cells in conditions of controlled refractive index matching (RIM) and labeling by diffused and clustered gold NPs [5]. The chapter concludes with a discussion and suggestions for future research.

3.2 Fundamentals of the FDTD Method

3.2.1 The Basic FDTD Numerical Scheme

The FDTD technique is an explicit numerical method for solving Maxwell's equations. It was invented by Yee in 1966 [27]. The advances of the various FDTD approaches and applications have been periodically reviewed by Taflove *et al.* [28].

Details about the numerical aspects of the explicit finite-difference approximation of Maxwell's equations in space and time can be found in Refs [28–31].

In a source-free absorptive dielectric medium, Maxwell's equations have the form:

$$\nabla \times \mathbf{E} = -\mu_0 \frac{\partial \mathbf{H}}{\partial t} \tag{3.1a}$$

$$\nabla \times \mathbf{H} = \varepsilon_0 \varepsilon \frac{\partial \mathbf{E}}{\partial t} \tag{3.1b}$$

where **E** and **H** are the vectors of the electric and magnetic fields, respectively, μ_0 is the vacuum permeability, and $\varepsilon_0\varepsilon$ is the permittivity of the medium. Assuming a harmonic [$\propto \exp(-\mathrm{i}\omega t)$] time dependence of the electric and magnetic fields and a complex value of the relative permittivity $\varepsilon = \varepsilon_r + \mathrm{i}\varepsilon_i$ transforms Equation 3.1b as follows:

$$\begin{aligned}\nabla \times \mathbf{H} = \varepsilon_0 \varepsilon \frac{\partial \mathbf{E}}{\partial t} &\Leftrightarrow \nabla \times \mathbf{H} = \omega \varepsilon_0 \varepsilon_i \mathbf{E} + \varepsilon_0 \varepsilon_r \frac{\partial \mathbf{E}}{\partial t} \Leftrightarrow \frac{\partial \left(\exp(\tau t)\,\mathbf{E}\right)}{\partial t} \\ &= \frac{\exp(\tau t)}{\varepsilon_0 \varepsilon_r} \nabla \times \mathbf{H}\end{aligned} \tag{3.2}$$

where $\tau = \omega\varepsilon_r/\varepsilon_i$, and ω is the angular frequency of the light. The continuous coordinates $(x, y, \text{and } z)$ are replaced by discrete spatial and temporal points: $x_i = i\Delta s$, $y_j = j\Delta s$, $z_k = k\Delta s$, and $t_n = n\Delta t$, where $i = 0, 1, 2, \ldots, I$, $j = 0, 1, 2, \ldots, J$, $k = 0, 1, 2, \ldots, K$, and $n = 0, 1, 2, \ldots, N$. Δs and Δt denote the cubic cell size and time increment, respectively. Using central difference approximations for the temporal derivatives over the time interval $[n\Delta t, (n+1)\Delta t]$ leads to

$$E^{n+1} = \exp(-\tau\Delta t)\, E^n + \exp(-\tau\Delta t/2)\, \frac{\Delta t}{\varepsilon_0 \varepsilon_r} \nabla \times \mathrm{E}^{n+1/2} \tag{3.3}$$

where the electric and the magnetic fields are calculated at alternating half-time steps. The discretization of Equation 3.3 over the time interval $[(n-1/2)\Delta t, (n+1/2)\Delta t]$ (one half-time step earlier than the electric field) ensures second-order accuracy of the numerical scheme and takes the form

$$\begin{aligned}E_x^{n+1}\left(i+1/2, j, k\right) &= \exp\left[-\frac{\varepsilon_i\left(i+1/2, j, k\right)}{\varepsilon_r\left(i+1/2, j, k\right)}\omega\Delta t\right] E_x^n\left(i+1/2, j, k\right) \\ &+ \exp\left[-\frac{\varepsilon_i\left(i+1/2, j, k\right)}{\varepsilon_r\left(i+1/2, j, k\right)}\omega\Delta t/2\right] \frac{\Delta t}{\varepsilon_0\varepsilon_r\left(i+1/2, j, k\right)\Delta s} \\ &\times \Big[H_y^{n+1/2}\left(i+1/2, j, k-1/2\right) - H_y^{n+1/2}\left(i+1/2, j, k+1/2\right) \\ &+ H_z^{n+1/2}\left(i+1/2, j+1/2, k\right) - H_z^{n+1/2}\left(i+1/2, j-1/2, k\right)\Big]\end{aligned} \tag{3.4}$$

where E_x, and H_y and H_z denote the electric and magnetic field components, respectively. Equation 3.1b is discretized in a similar way [28–31]. The numerical stability of the FDTD scheme is ensured through the Courant–Friedrichs–Levy condition [26, 27]: $c\Delta t \leq \left(1/\Delta x^2 + 1/\Delta y^2 + 1/\Delta z^2\right)^{-1/2}$, where c is the speed of light in the host medium and Δx, Δy, and Δz are the spatial steps in x, y, and z directions, respectively. In our case, $\Delta x = \Delta y = \Delta z = \Delta s$ and $\Delta t = \Delta s/2c$.

3.2.2 Input Wave Excitation

The FDTD approach uses the so-called total-field/scattered-field (TFSF) formulation [28, 31] to excite the input magnetic and electric fields and simulate a linearly polarized plane wave propagating in a finite region of a homogeneous absorptive dielectric medium. In this formulation, an arbitrary (in most cases rectangular) shape closed surface is defined inside the computational domain and around the scattering object. Using the equivalence theorem [28] allows for demonstration that the input wave excitation within the closed surface can be replaced with the equivalent electric and magnetic currents located at the closed surface. Therefore, the input wave can be numerically excited at the closed surface by adding the electric and magnetic field incident sources as follows:

$$\mathbf{H} \Leftarrow \mathbf{H} - \frac{\Delta t}{\mu_0 \Delta s}(\mathbf{E}_{\text{inc}} \times \mathbf{n}) \tag{3.5a}$$

$$\mathbf{E} \Leftarrow \mathbf{E} - \frac{\Delta t}{\varepsilon_0 \varepsilon \Delta s}(\mathbf{n} \times \mathbf{H}_{\text{inc}}) \tag{3.5b}$$

where $\mathbf{E}_{\text{inc}}$ and $\mathbf{H}_{\text{inc}}$ are the incident fields and $\mathbf{n}$ is the inward normal vector of the closed surface [28]. If there is a scattering object inside the closed surface, the interior fields will be the total fields (incident plus scattered) and the exterior fields will be just the scattered fields. Details about the specific form of the components of Equation 3.5 in the case of a rectangular closed surface can be found in Refs [28, 31].

3.2.3 Uniaxial Perfectly Matched Layer Absorbing Boundary Conditions

The FDTD numerical scheme shown here uses the uniaxial perfectly matcher layer (UPML) suggested by Sacks *et al.* [32] to truncate the absorptive host medium in the FDTD computational domain. The UPML approach is based on the physical introduction of absorbing anisotropic, perfectly matched medium layers at all sides of the rectangular computational domain. The anisotropic medium of each of these layers is uniaxial and is composed of both electric permittivity and magnetic permeability tensors [33, 34]. The introduction of the UPMLs leads to a modified UPML FDTD numerical scheme, which is applied in the UPML parts of the computational domain. In the non-UPML region, the unmodified FDTD formulation is used (Equation 3.4). More details on the numerical implementation of the UPML boundary conditions can be found in Refs [28–34].

3.2.4
FDTD Formulation of the Light Scattering Properties from Single Cells

The calculation of the light scattering and extinction cross sections by cells in free space requires the far-field approximation for the electromagnetic fields [7, 28, 35]. The far-field approach has also been used [36] to study the scattering and absorption of spherical particles in an absorptive host medium. However, when the host medium is absorptive, the scattering and extinction rates depend on the distance from the cell center. This could lead to ambiguities in the presentation and interpretation of the light scattering results. Recently, an alternative approach based on Mie theory was suggested to calculate the single-scattering properties of a spherical object in an absorptive medium using the electromagnetic fields on the surface of the scattering object [37, 38]. This alternative approach was then extended and numerically implemented within the FDTD context to enable the calculation of the absorption and extinction rates of arbitrarily shaped scattering objects in an absorptive medium. The full details of this FDTD implementation can be found in Ref. [31]. The light scattering approach is used here. The absorption and extinction rates calculated in this manner depend on the size, shape, and optical properties of the scattering object and the surrounding medium, but do not depend on the distance from it. The possibility to use the FDTD method in studying the lateral light scattering patterns from single cells in absorptive media is of critical importance for the modeling of flow cytometry experimental configurations, where the absorption properties of the host (extracellular) medium could be an important scattering factor.

In order to define some of the key light-scattering parameters, the incident field is decomposed into two components along the unit vectors $\mathbf{e}_\alpha$ and $\mathbf{e}_\beta$, both lying in a plane perpendicular to the direction of propagation of the input wave and defined as parallel and perpendicular to the scattering plane, respectively [20]:

$$\mathbf{E}_0 = \mathbf{e}_\alpha E_{0,\alpha} + \mathbf{e}_\beta E_{0,\beta} \tag{3.6}$$

The numerically calculated scattered field is decomposed into two components along the unit vectors $\boldsymbol{\alpha}$ and $\boldsymbol{\beta}$, which are both perpendicular to the direction of propagation and parallel and perpendicular to the scattering plane, respectively:

$$\mathbf{E}_s = \boldsymbol{\alpha} E_{s,\alpha} + \boldsymbol{\beta} E_{s,\beta} \tag{3.7}$$

Thus, the incident and scattered fields are specified relative to different sets of base vectors. The relationship between the incident and scattered fields can be conveniently expressed in a matrix form

$$\mathbf{E}_s = \overline{\overline{S}} \times \mathbf{E}_0 \tag{3.8}$$

where $S_k = S_k(\theta, \phi)$, $k = 1, 2, 3,$ and 4 are the elements of the amplitude scattering matrix and, in general, depend on the scattering angle θ and azimuth angle ϕ [20]. In the case of the scattered wave, the four Stokes parameters which take the form [7, 20, 36]

$$
\begin{aligned}
I_s &= \langle E_{s,\alpha} E^*_{s,\alpha} + E_{s,\beta} E^*_{s,\beta} \rangle \\
Q_s &= \langle E_{s,\alpha} E^*_{s,\alpha} - E_{s,\beta} E^*_{s,\beta} \rangle \\
U_s &= \langle E_{s,\alpha} E^*_{s,\beta} + E_{s,\beta} E^*_{s,\alpha} \rangle \\
V_s &= \langle E_{s,\alpha} E^*_{s,\beta} - E_{s,\beta} E^*_{s,\alpha} \rangle
\end{aligned}
\tag{3.9}
$$

The relation between the incident and the scattered Stokes parameters is given by the Mueller scattering matrix as follows:

$$
\begin{pmatrix} I_s \\ Q_s \\ U_s \\ V_s \end{pmatrix} = \frac{1}{k_h^2 R^2} \begin{pmatrix} P_{11} & P_{12} & P_{13} & P_{14} \\ P_{21} & P_{22} & P_{23} & P_{24} \\ P_{31} & P_{32} & P_{33} & P_{34} \\ P_{41} & P_{42} & P_{43} & P_{44} \end{pmatrix} \begin{pmatrix} I_0 \\ Q_0 \\ U_0 \\ V_0 \end{pmatrix}
\tag{3.10}
$$

where $k_h = \omega\sqrt{\mu_0 \varepsilon_0 \varepsilon_h}$, ε_h is the complex relative permittivity of the host medium, I_0, Q_0, U_0, and V_0 are the Stokes parameters of the incident wave, R is the distance from the cell, and

$$
\begin{aligned}
P_{11} &= \frac{1}{2}\left(|S_1|^2 + |S_2|^2 + |S_3|^2 + |S_4|^2\right), P_{12} = \frac{1}{2}\left(|S_2|^2 - |S_1|^2 + |S_4|^2 - |S_3|^2\right) \\
P_{13} &= \mathrm{Re}\left(S_2 S_3^* + S_1 S_4^*\right), P_{14} = \mathrm{Im}\left(S_2 S_3^* - S_1 S_4^*\right) \\
P_{21} &= \frac{1}{2}\left(|S_2|^2 - |S_1|^2 - |S_4|^2 + |S_3|^2\right), P_{22} = \frac{1}{2}\left(|S_2|^2 + |S_1|^2 - |S_4|^2 - |S_3|^2\right) \\
P_{23} &= \mathrm{Re}\left(S_2 S_3^* - S_1 S_4^*\right), P_{24} = \mathrm{Im}\left(S_2 S_3^* + S_1 S_4^*\right) \\
P_{31} &= \mathrm{Re}\left(S_2 S_4^* + S_1 S_3^*\right), P_{32} = \mathrm{Re}\left(S_2 S_4^* - S_1 S_3^*\right) \\
P_{33} &= \mathrm{Re}\left(S_1 S_2^* + S_3 S_4^*\right), P_{34} = \mathrm{Im}\left(S_2 S_1^* + S_4 S_3^*\right) \\
P_{41} &= \mathrm{Im}\left(S_2 S_4^* + S_1 S_3^*\right), P_{42} = \mathrm{Im}\left(S_4 S_2^* - S_1 S_3^*\right) \\
P_{43} &= \mathrm{Im}\left(S_1 S_2^* - S_3 S_4^*\right), P_{44} = \mathrm{Re}\left(S_1 S_2^* - S_3 S_4^*\right)
\end{aligned}
\tag{3.11}
$$

The P matrix elements contain the full information about the scattering process. In nonabsorptive media, the elements of the Mueller matrix (Equation 3.10) can be used to define the scattering cross section and anisotropy. The scattering cross section σ_s is defined as the geometrical cross section of a scattering object that would produce an amount of light scattering equal to the total observed scattered power in all directions. It can be expressed by the elements of the scattering matrix P and the Stokes parameters I_0, Q_0, U_0, and V_0 of the incident light as follows [20]:

$$
\sigma_s = \frac{1}{k_h^2 I_0} \int_{4\pi} \left[I_0 P_{11} + Q_0 P_{12} + U_0 P_{13} + V_0 P_{14}\right] \mathrm{d}\Omega
\tag{3.12}
$$

In the case of a spherically symmetrical scattering object and nonpolarized light, the Mueller scattering matrix component $P_{11}(\theta)$ becomes the most relevant and the relationship (Equation 3.12) is reduced to the usual integral of the indicatrix with respect to the scattering angle:

$$\sigma_s = \frac{2\pi}{k_h^2} \int_0^{\pi} P_{11}(\theta) \sin(\theta)\, d\theta \tag{3.13}$$

Using $P_{11}(\theta)$, the anisotropy parameter g can be expressed as follows:

$$g = \langle \cos(\theta) \rangle = \frac{2\pi}{k_h^2} \int_0^{\pi} \cos(\theta) P_{11}(\theta) \sin(\theta)\, d\theta \tag{3.14}$$

In an absorptive medium, the elements of the Mueller scattering matrix (Equation 3.10) depend on the radial distance from the scattering object and cannot be directly related to the scattering cross section as described previously in Equation 3.12 In this case, the different elements of the matrix are used individually in the analysis of the scattering phenomena. In practice, their values are normalized by the total scattered rate in the radiation zone around the object, which is proportional to the integral of P_{11} for all scattering angles.

3.2.5
FDTD Formulation of Optical Phase Contrast Microscopic (OPCM) Imaging

The 3D FDTD formulation provided here is based on a modified version of the TFSF formulation [28, 31] that was described briefly in association with Equation 3.5a and b. It could be more appropriately called *total-field/reflected-field* (*TFRF*) formulation. The 3D TFRF formulation uses a TFSF region which contains the biological cell and extends beyond the limits of the simulation domain (Figure 3.2a and b). The extension of the transverse dimension of the input field beyond the limits of the computational domain through the UPML boundaries would lead to distortions of the ideal plane wave shape and eventually distort the simulation results. To avoid these distortions, one must use Bloch periodic boundary conditions (Figure 3.2a and b) in the lateral x- and y directions which are perpendicular to the direction of propagation – z [28].

Phase contrast microscopy is utilized to produce high-contrast images of transparent specimens such as microorganisms, thin tissue slices, and living cells and subcellular components such as nuclei and organelles, and therefore can be used as a basis for phase-sensitive slide or flow cytometry. In a conventional flow cytometry configuration, a beam of light of a single (or several) wavelength is directed onto a hydrodynamically focused stream of fluid driving a periodical array of cells to flow through it. The OPCM simulation model requires the explicit availability of the forward scattered transverse distribution of the fields. The phase of the scattered field accumulated by a plane wave propagating through a biological cell within a cytometric cell flow will be used in the FDTD model of the OPCM that is described as follows.

Figure 3.3 shows a possible flow cytometry configuration of a phase contrast microscope, where an image with a strong contrast ratio is created by coherently interfering a reference (R) with a beam (D) that is diffracted from one particular

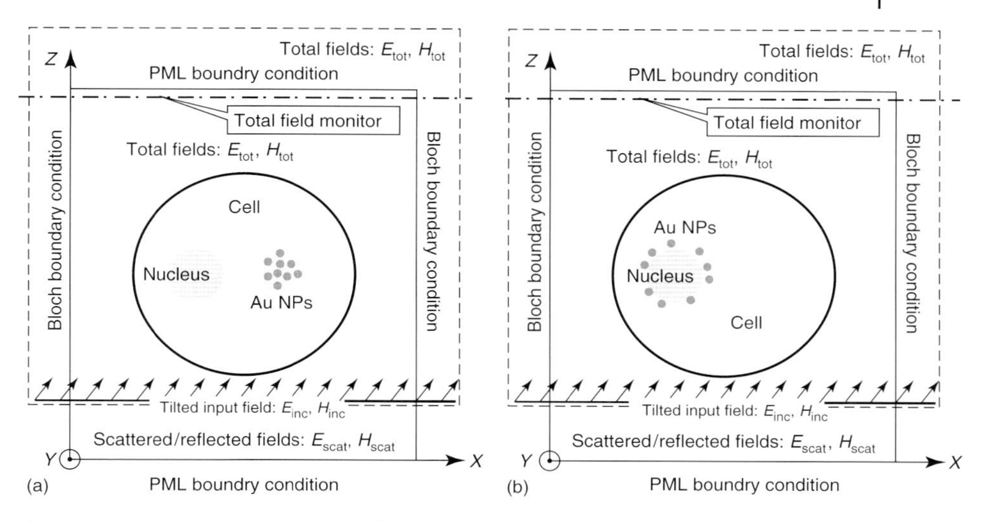

Figure 3.2 Schematic representation of the 3D FDTD formulation including: (a) a cell with a nucleus and a cluster of gold nanoparticles in the cytoplasm and (b) a cell with gold nanoparticles randomly distributed on the surface of the nucleus.

cell in the cell flow. Relative to the reference beam, the diffracted beam has lower amplitude and is retarded in phase by approximately $\pi/2$ through interaction with the cell. The main feature of the design of the phase contrast cytometer is the spatial separation of the R beam from the D wavefront emerging from the cell. In addition, the amplitude of the R beam light must be reduced and the phase must be advanced or retarded by another $\pm\pi/2$ in order to maximize the differences in the intensity between the cell and the background in the image plane. The mechanism for generating relative phase retardation has two steps: (i) the D beam is retarded in phase by a quarter wavelength (i.e., $\pi/2$) at the specimen and (ii) the R beam is advanced (or retarded) in phase by a phase plate positioned in or very near the objective rear focal plane (Figure 3.3). This two-step process is enabled by a specially designed annular diaphragm – the annulus. The condenser annulus, which is placed in the condenser front focal plane, is matched in diameter and optically conjugated to the phase plate residing in the objective rear focal plane. The resulting image, where the total phase difference is translated by interference at the image plane into an amplitude variation, can have a high-contrast ratio, particularly if both beams have the same amplitude.

Figure 3.3 also provides a visual representation illustrating the major steps in the FDTD OPCM model. The phase contrast microscope uses incoherent annular illumination that could be approximately modeled by adding up the results of eight different simulations using ideal input plane waves incident at a given polar angle (30°), an azimuthal angle (0, 90, 180, or 270°), and a specific light polarization (parallel or perpendicular to the plane of the graph). Every single FDTD simulation provides the near-field components in a transverse monitoring plane located behind the cell (Figure 3.1). The far-field transformations use the calculated near fields behind the cell and return the three complex components of the electromagnetic fields far enough from the location of the near fields, that is, in the far field [28]. The amplitudes and phases of the calculated far-field components can be used in Fourier optics with both the scattered and reference beams. An ideal optical lens system that could be characterized by a given magnification factor is assumed. This simple model can be easily extended to include the numerical equivalent of the two lenses together with an additional model to take into account any aberrational effects. The magnification factor was implemented by merely modifying the angle of light propagation – it was applied to the far fields before the interference of the diffracted (D) and reference (R) beams (Figure 3.3) at the image plane [28]. It was also possible to apply the effect of a numerical aperture (NA) which clips any light that has too steep an angle and would not be collected by the lens system. The OPCM images at the image plane are calculated by adding up the scattered and reference beams at any desired phase offset Ψ:

$$I = \left|E_{x-\mathrm{D}} + aE_{x-\mathrm{R}}\exp\left(\mathrm{i}\Psi\right)\right|^2 + \left|E_{y-\mathrm{D}} + aE_{y-\mathrm{R}}\exp\left(\mathrm{i}\Psi\right)\right|^2 + \left|E_{z-\mathrm{D}} + aE_{z-\mathrm{R}}\exp\left(\mathrm{i}\Psi\right)\right|^2 \tag{3.15}$$

The coefficient a and phase Ψ are simulation parameters corresponding to the ability of the OPCM to adjust the relative amplitudes and phase difference between the two beams.

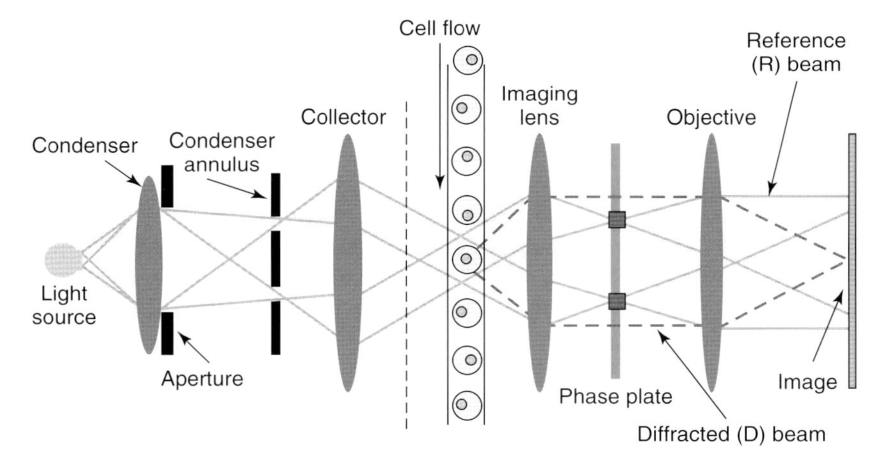

Figure 3.3 Schematics of an OPCM cytometer with a 2D visual representation of the FDTD OPCM model using incoherent illumination by two planes waves at a polar angle of 30°. For each of the two plane waves the propagation of light is modeled as a combination of two parallel wave phenomena: (i) propagation of the reference (R) beam without the cell flow and (ii) propagation of the diffracted (D) beam due to one cell from the cell flow.

3.3
FDTD Simulation Results of Light Scattering Patterns from Single Cells

3.3.1
Effect of Extracellular Medium Absorption on the Light Scattering Patterns

This section describes the FDTD simulation results for the effect of absorption in the extracellular medium on the light scattering patterns from a single cell [20]. Two different cell geometries are considered: the one considered in the previous section and another with a shape of a spheroid (both cytoplasm and nucleus). The light is incident in the x direction along the long axis of the spheroid. The size parameters of the cytoplasm are defined by $2\pi R_a/\lambda_0 = 40$ and $2\pi R_{b,c}/\lambda_0 = 20$, where $R_a = a/2$ and $R_{b,c} = b/2 = c/2$. The size parameters of the nucleus are defined by $2\pi r_a/\lambda_0 = 20$ and $2\pi r_{b,c}/\lambda_0 = 10$, where r_a and r_b are the large and small radii of the nucleus, respectively. The cell refractive indices are as follows: cytoplasm, 1.37; nucleus, 1.40; extracellular medium, 1.35 and $1.35 + 0.05i$. The FDTD cell size is $\Delta s = \lambda_0/20$. The number of mesh point is as follows: $N_x = 147$, $N_y = 147$, and $N_z = 275$. The number of simulation time steps is 14 000. Figure 3.4a shows FDTD simulation results for the phase function (the P_{11} element of the Mueller scattering matrix is described in Equation 3.11) of a simple spherical biological cell containing only a cytoplasm and a nucleus, both embedded in a nonabsorptive extracellular medium. Figure 3.4b shows similar results for the normalized scattering matrix element P_{44}. Figure 3.4 validates our simulation approach by comparing the numerical results to the exact analytical ones provided by the Mie theory.

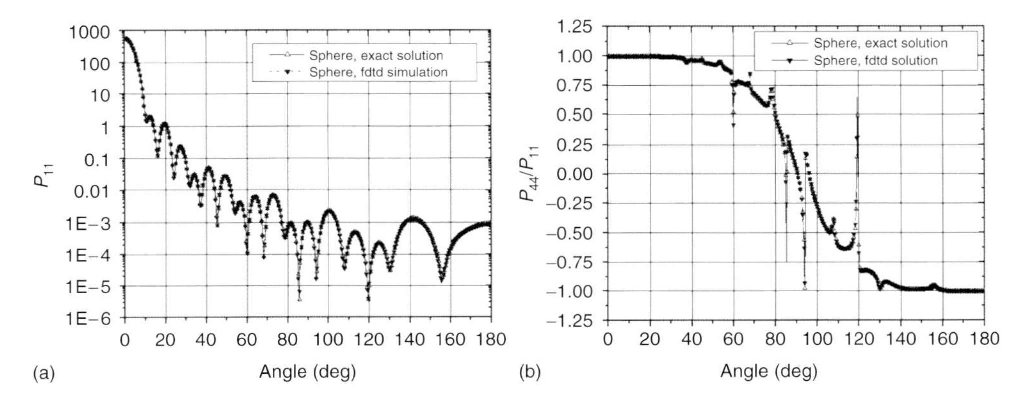

Figure 3.4 Comparison of the FDTD simulation results with exact analytical (Mie theory) results: (a) Normalized light scattering intensity (P_{11} element of the Mueller scattering matrix) distribution with scattering angle. (b) Angular distributions of the normalized scattering matrix element P_{44} as calculated by the FDTD method. There is a very good agreement between exact analytical and the numerical solutions with a relative error of approximately 5%.

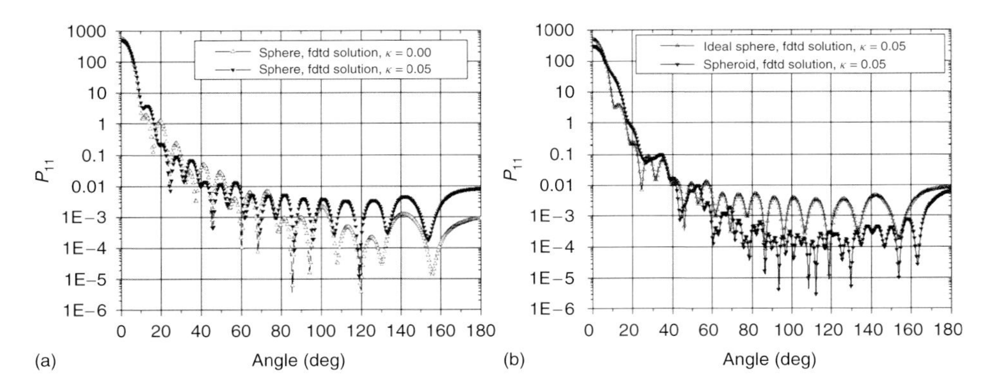

Figure 3.5 Normalized light scattering intensity distribution with scattering angle for two different cases: (a) absorptive and nonabsorptive extracellular medium for ideally spherical cells and (b) absorptive extracellular medium for an ideal sphere and a spheroid. The values of the imaginary part of the refractive index of the extracellular medium are indicated in the insight.

Figure 3.5 shows (i) the effect of absorption in the extracellular medium of spherical cells and (ii) the effect of cell shape (ideal sphere vs spheroid) in an absorptive extracellular medium [18, 19]. A more detailed analysis of the two graphs shown in Figure 3.5 leads to some interesting findings. First, absorption in the extracellular material of spherical cells (Figure 3.5a) increases the intensity of the light scattering up to one order in the angle range between 90° (transverse scattering) and $\Psi = 180^\circ$ (backward scattering). The same light scattering feature was also found in the case of spheroid cells (not shown here). Second, the influence of absorption in the extracellular material is relatively more pronounced in the case of spherical cells as compared with spheroid cells (Figure 3.5b).

Third, in the case of exact backward scattering and absorptive extracellular material, the light scattering intensity is approximately equal for both cell shapes. However, this is not true for nonabsorptive cell surroundings. This last finding could be very important for OCT and confocal imaging systems, especially for studies in the wavelength ranges within the hemoglobin, water, and lipid bands [5].

These findings show that the analysis of light scattering from isolated biological cells should necessarily account for the absorption effect of the surrounding medium. It could be particularly relevant in the case of optical immersion techniques using intratissue administration of appropriate chemical agents with absorptive optical properties [39–41]; however, this relevance does not appear to have been studied. It should be pointed out that whenever there is a matching of the refractive indices of a light scatterer and the background material, the scattering coefficient becomes zero and it is only the absorption in the scatterer or in the background material that is responsible for the light beam extinction. Former research studies [42, 43] have examined the light scattering role of the absorption within the particle-based tissue phantoms. The results shown here illustrate the light scattering role of absorption in the background material.

3.4 FDTD OPCM Nanobioimaging Simulation Results

3.4.1 Cell Structure

The 3D FDTD modeling of OPCM imaging of single biological cells in a number of different scenarios is based on the FDTD OPCM model described previously for the optical magnification factor $M = 10$ and the NA $= 0.8$. The cell is modeled as a dielectric sphere with a radius $R_c = 5\ \mu m$ (Figure 3.1). The cell membrane thickness is $d = 20$ nm, which corresponds to effective (numerical) thickness of approximately 10 nm. The cell nucleus is also spherical with a radius $R_n = 1.5\ \mu m$ centered at a position which is 2.0 μm shifted from the cell center in a direction perpendicular to the direction of light propagation. The refractive index of the cytoplasm is $n_{cyto} = 1.36$, that of the nucleus $n_{nuc} = 1.4$, that of the membrane $n_{mem} = 1.47$, and that of the extracellular material $n_{ext} = 1.33$ (no RIM) or 1.36 (RIM).

3.4.2 Optical Clearing Effect

The RIM between the cytoplasm and the extracellular medium leads to the optical clearing of the cell image. The optical clearing effect leads to increased light transmission through the cell due to matching of the refractive indices of some of their components to that of the extracellular medium. The refractive index of the extracellular fluid can be externally controlled by the administration of an appropriate chemical agent [39–41]. Figure 3.6 shows the cross sections of cell images for different values of the phase offset Ψ between the reference and diffracted beams of the OPCM: 90, 120, 150, and 180°. The images illustrate the nature of the optical clearing effect and the value of its potential applications for the early detection of cancerous cells by a careful examination of their nucleus size, eccentricity, morphology, and chromatin texture (refractive index fluctuations) [17]. For example, in all cases, at no RIM condition the images of the nuclei are represented by a dip in the cell image. At RIM conditions, the image contrast of the cells is drastically reduced to zero level and it is only the images of the nuclei that remain sharply visible. The images of the nuclei are represented by well-defined peaks associated with three-dimensional optical phase accumulation corresponding to their perfectly spherical shapes and homogeneous refractive index distributions. A finer analysis of two of the graphs shown in Figure 3.6 ($\Psi = 180^\circ$ and $\Psi = 90^\circ$) shows that the diameter of the nucleus (the full width at the half-height of the nucleus image contrast peak) depends on the phase delay Ψ. At $\Psi = 180^\circ$ and no RIM conditions, the diameter of the nucleus is estimated to be ~2.3 μm as compared with RIM conditions, where its value is 3.3 μm (the estimation accounts for the optical magnification factor ×10 of the system). At $\Psi = 90^\circ$ and no RIM conditions, the diameter of the nucleus is estimated to be ~3.05 μm as compared

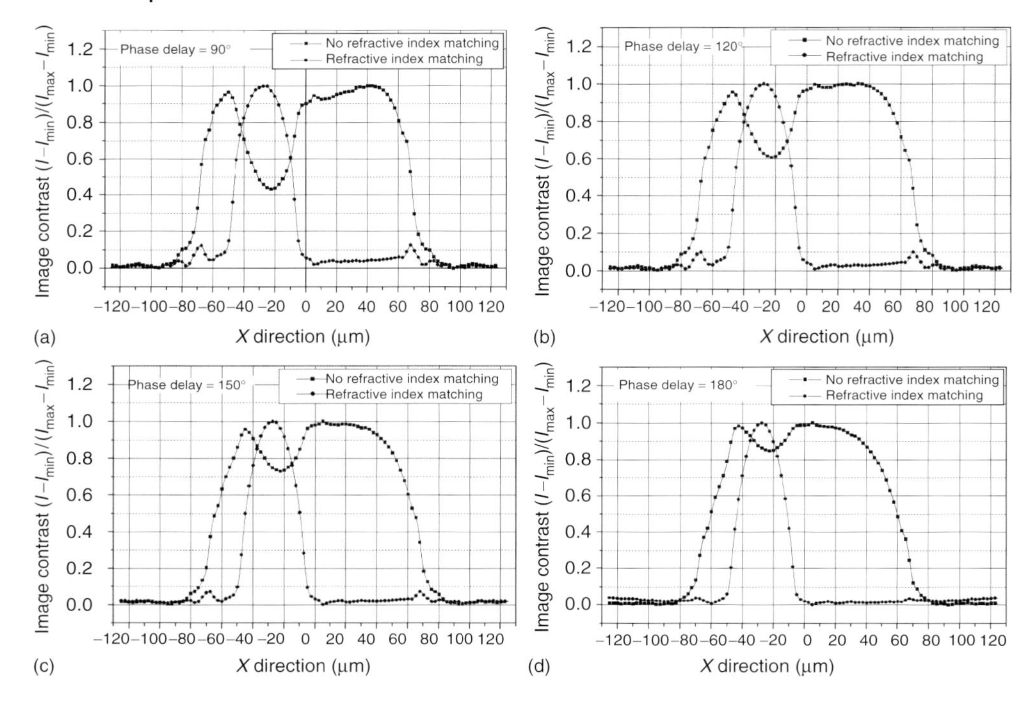

Figure 3.6 Cross sections of FDTD-generated OPCM images of a single cell illustrating the optical clearing effect for different values of the phase offset Ψ between the reference and diffracted beam of the OPCM: 90° (a), 120° (b), 150° (c), and 180° (d). Matching the refractive index value of the extracellular material with that of the cytoplasm enhances the optical contrast and leads to a finer view of the morphological structure of the nucleus.

with RIM conditions, where its value is 3.75 μm (the cell model used in the FDTD simulations has a nucleus with a diameter 3.0 μm). This shows that the OPCM should be a preliminary setup at a given optimum phase delay and the OPCM images should be used for relative measurements only after a correct calibration. The analysis of the graphs, however, shows an unprecedented opportunity for using the optical clearing effect for the analysis of any pathological change in the eccentricity and the chromatin texture of cell nuclei within the context of OPCM cytometry configurations. This new opportunity is associated with the fact that at RIM the cell image is practically transformed into a much finer image of the nucleus.

3.4.3
The Cell Imaging Effect of Gold Nanoparticles

Gold NPs have the ability to resonantly scatter visible and near infrared light. The scattering ability is due to the excitation of surface plasmon resonances (SPRs). It is extremely sensitive to their size, shape, and aggregation state, offering a great potential for optical cellular imaging and detection labeling studies [44–49]. The FDTD approach [21] uses the dispersion model for gold NPs derived from the experimental data provided by Johnson and Christy [50] where the total complex-valued permittivity is given as

$$\varepsilon(\omega) = \varepsilon_{\mathrm{REAL}} + \varepsilon_{\mathrm{L}}(\omega) + \varepsilon_{\mathrm{P}}(\omega) \tag{3.16}$$

The first term represents the contribution from the basic, background permittivity. The second and third terms represent Lorentz and plasma contributions:

$$\varepsilon_{\mathrm{L}}(\omega) = \varepsilon_{\mathrm{LORENTZ}}\omega_0^2/\left(\omega_0^2 - 2\mathrm{i}\delta_0\omega - \omega^2\right), \varepsilon_{\mathrm{P}}(\omega) = \omega_{\mathrm{P}}^2/\left(i\omega\nu_{\mathrm{C}} + \omega^2\right) \tag{3.17}$$

where all material constants are as follows: $\varepsilon_{\mathrm{REAL}} = 7.077$, $\varepsilon_{\mathrm{LORENTZ}} = 2.323$, $\omega_0 = 4.635 \times 10^{15}$ Hz, $\delta_0 = 9.267 \times 10^{14}$ Hz, $\omega_{\mathrm{P}} = 1.391 \times 10^{16}$ Hz, and $\nu_{\mathrm{C}} = 1.411 \times 10^7$ Hz [50].

Both the resonant and nonresonant cases were modeled. The ability to model these two different cases, together with the optical clearing effect, provides the opportunity to numerically study the possibility of imaging the uptake of NP clusters – a scenario which needs to be further studied [20]. The FDTD technique has also been used to calculate the extinction cross sections over a 400–900 nm wavelength range for a single 50 nm diameter gold NP immersed in a material having the properties of the cytoplasm $\left(n_{\mathrm{cyto}} = 1.36\right)$ and a resolution of $\mathrm{d}x = \mathrm{d}y = \mathrm{d}z = 10$ nm. The calculated extinction cross section has a maximum of 3.89 at around 543.0 nm corresponding to one of the radiation wavelengths of He–Ne lasers. The result for $\lambda = 676.4$ nm (a Krypton laser wavelength) corresponds to the nonresonant case (extinction cross section value 0.322, ~12 times smaller than 3.89). The FDTD results were validated by comparing them with the theoretical curve calculated by Mie theory.

The OPCM cell images are the result of simulations using nonuniform meshing, where the number of mesh points in space is automatically calculated to ensure a higher number of mesh points in materials with higher values of the refractive index.[1] Figure 3.7a shows the schematic positioning of a cluster of 42 NPs in the cytoplasm, which were used in simulations. The cell center is located in the middle $\left(x = y = z = 0\right)$ of the computational domain with dimensions of 15 μm × 12 μm × 15 μm (Figure 3.2a). The nucleus' center is located at $x = -2$ μm and $y = z = 0$ μm. The cluster of gold NPs is located at $x = 2$ μm and $y = z = 0$ μm. Figure 3.7b shows another simulation scenario where the cluster of 42 NPs is randomly distributed on the surface of the cell nucleus.

The realistic cell dimensions (including both cell radius and membrane) require a very fine numerical resolution, making the simulations computationally intensive. The numerical resolution of the NPs was hard-coded to $\mathrm{d}x = \mathrm{d}y = \mathrm{d}z = 10$ nm ensuring that their numerically manifested optical resonant properties will be the same as the ones that were discussed above. This leads to additional requirements for the CPU time and memory (~120 Gbs RAM) requiring high performance computing resources.

1) Nonuniform meshing is a standard feature of the FDTD Solutions™ software.

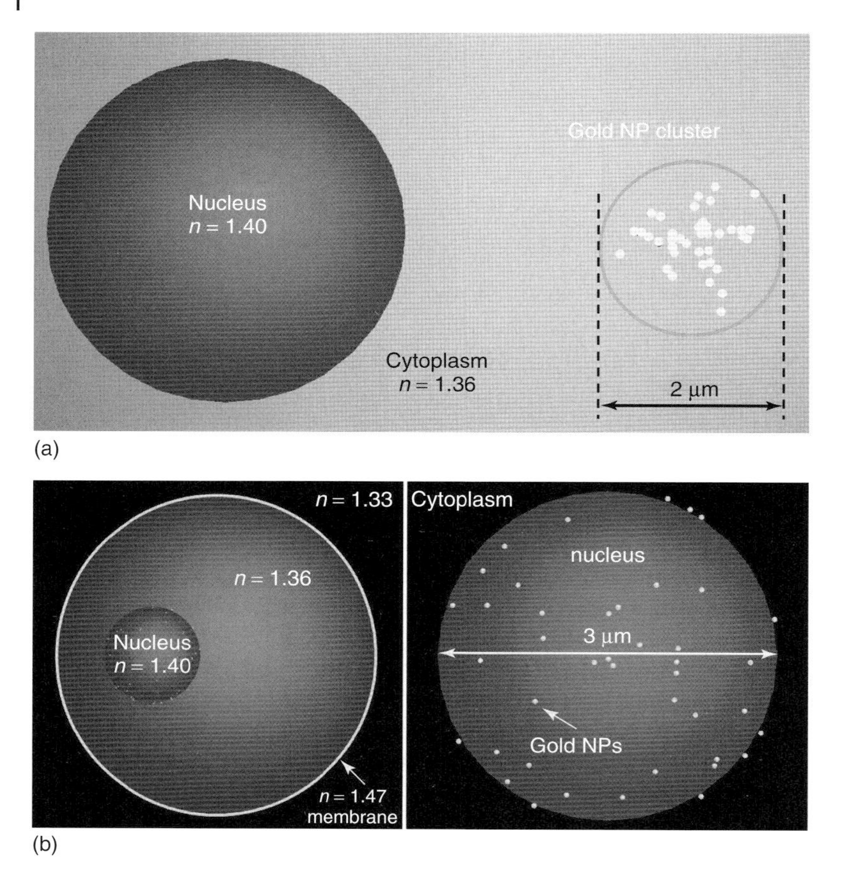

Figure 3.7 (a) A cluster of 42 gold NPs in the cytoplasm. (b) A group of 42 gold NPs randomly distributed on the surface of the cell nucleus. The NP size is slightly exaggerated.

3.4.3.1 A Cell with a Cluster of Gold Nanoparticles Located in the Cytoplasm

On the basis of the fact that RIM enhances significantly the imaging of the cell components, we have used the FDTD OPCM model to create OPCM images of the cell at optical immersion conditions with and without a cluster of 42 gold NPs located in the cytoplasm (Figure 3.8) and for four different values (−150, −90, +90, and 180°) of the phase offset Ψ between the reference and scattered beams (assuming $a = 1$, Equation 3.15). The presence of the NPs can be seen more clearly in the graph shown in Figure 3.9, representing a comparison of the cross sections of the images shown in Figure 3.8d.

The effect of the optical resonance of the gold NPs can be seen clearly in Figure 3.10. For $\Psi = 180^\circ$, the resonant optical contrast of the gold NP peak is ~2.24 times larger than the nonresonant one (Figure 3.11a). For $\Psi = -90^\circ$, the resonant enhancement factor is 1.71. Thus, the enhanced imaging of the gold NP

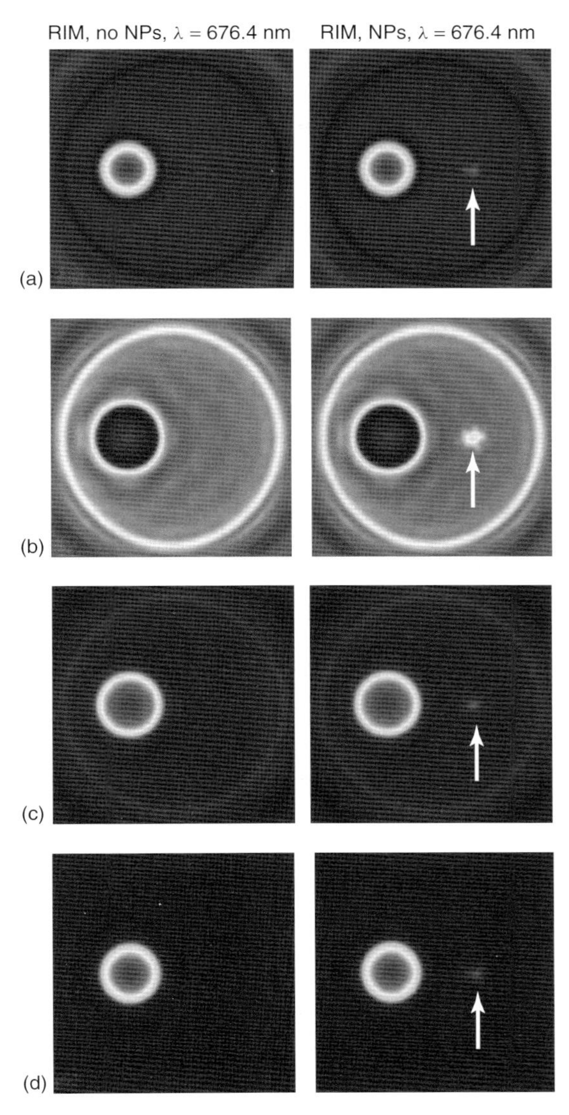

Figure 3.8 OPCM images of a single cell for different values ((a) -150°, (b) -90°, (c) $+90^\circ$, and (d) $+180^\circ$) of the phase offset Ψ between the reference and diffracted beam of the OPCM at RIM (optical immersion) conditions with and without a cluster of 42 gold NPs located in a position symmetrically opposite to the nucleus (nonresonant conditions $-\lambda = 676.4$ nm). The arrows indicate the position of the cluster.

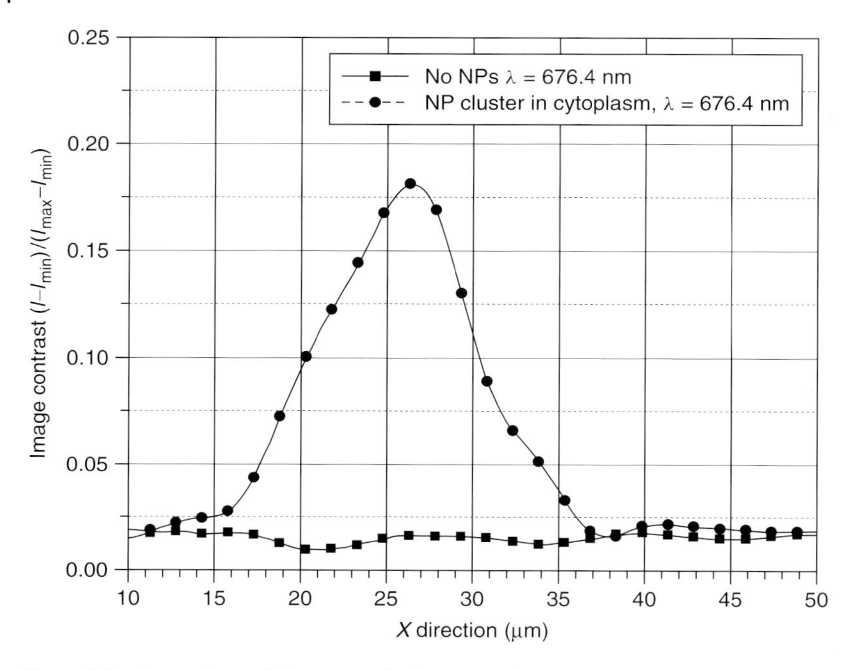

Figure 3.9 Comparison of the geometrical cross sections at $y = 0$ μm of two of the OPCM images shown in Figure 3.8d corresponding to a phase offset $\Psi = 180^\circ$ between the reference and the diffracted beam in terms of optical contrast. The effect of the gold NPs at no-resonance ($\lambda = 676.4$ nm) is clearly visible.

cluster at resonant conditions is clearly demonstrated. It, however, needs to be further studied as a function of particular phase offset Ψ between the reference beam and the scattered beam. A calculation of the optical contrast due to the gold NP cluster as a function of the phase offsets between the reference (R) and the diffracted (D) beams of the optical phase microscope (Figure 3.11b) shows that the enhancement of the optical contrast due to the NP resonance changes significantly from a minimum of 0.0 $(\Psi = 0^\circ)$ to a maximum of 3.60 $(\Psi = -150^\circ)$ [5, 21]. Taking into account this dependence could be very important in real life OPCM imaging flow cytometry experiments.

3.4.3.2 A Cell with a Cluster of Gold Nanoparticles Randomly Distributed on the Surface of its Nucleus

Figure 3.12 shows the OPCM images of a cell without NPs (left) and with a group of 42 gold NPs (right) randomly located on the surface of the cell nucleus [5, 24]. The images are for different values of the phase offset Ψ (−90, −30, +30, and +90°) at optical immersion and at nonresonance.

Figure 3.13 shows the OPCM images of a cell including a group of 42 gold NPs randomly located on the surface of the cell nucleus [5, 24] at optical immersion

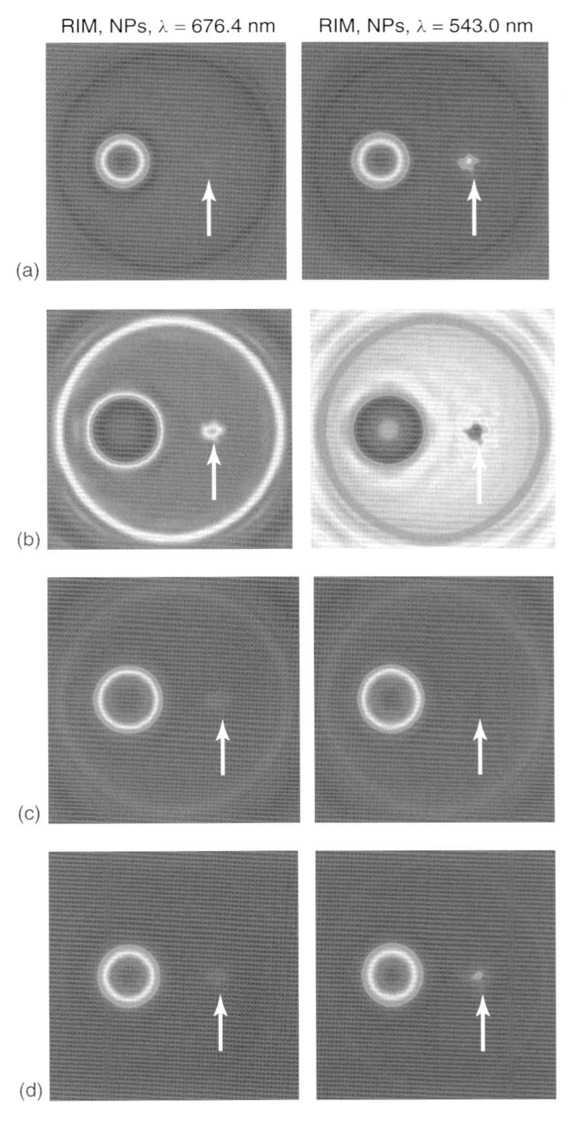

Figure 3.10 OPCM images of a single cell for different values ((a) -150°, (b) -90°, (c) $+90^\circ$, and (d) $+180^\circ$) of the phase offset Ψ between the reference and diffracted beam of the OPCM at RIM (optical immersion) conditions including a cluster of 42 gold NPs located in a position symmetrically opposite to the nucleus at resonance (right column, $\lambda = 543.0$ nm) and at no-resonance (left column, $\lambda = 676.4$ nm). The arrows indicate the position of the cluster.

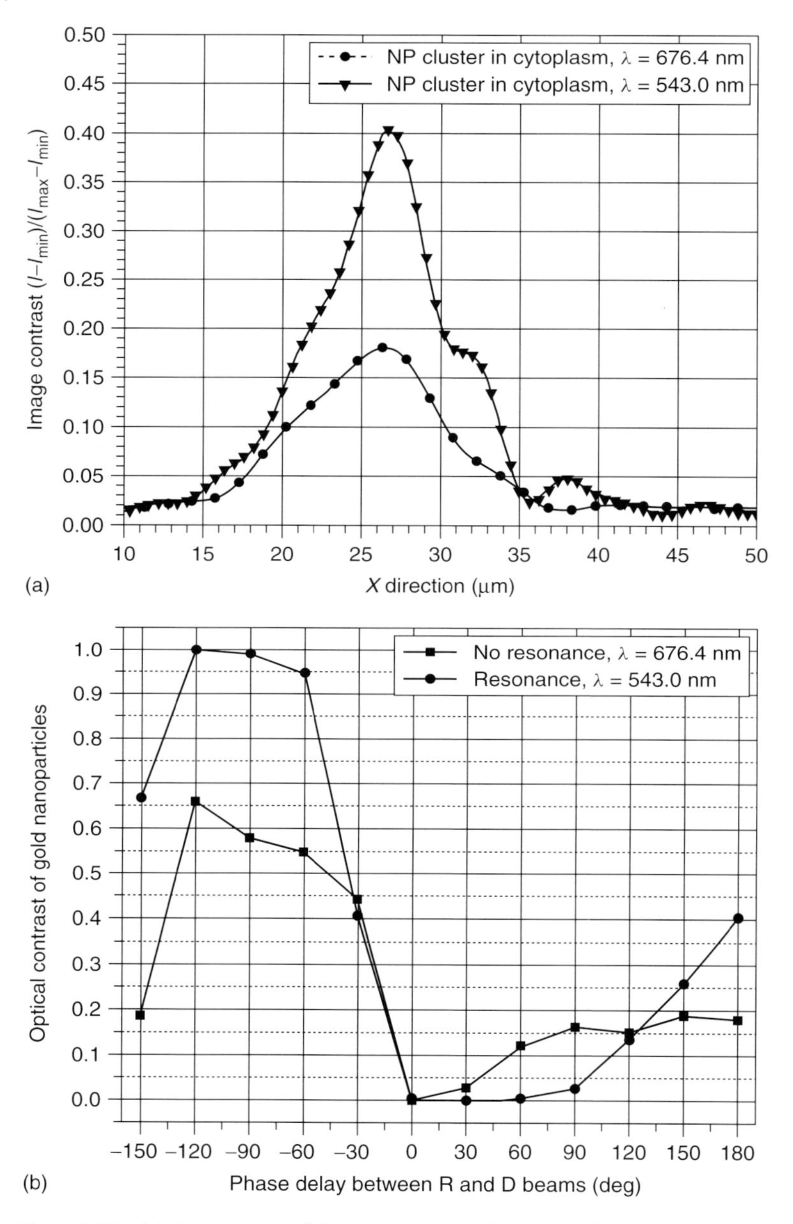

Figure 3.11 (a) Comparison of the geometrical cross sections at $y = 0$ μm of two of the OPCM images shown in Figure 3.10d corresponding to a phase offset $\Psi = 180^\circ$ between the reference and the diffracted beams. The effect of the optical resonance ($\lambda = 543.0$ nm as compared with $\lambda = 676.4$ nm when there is no resonance) on the imaging contrast of the gold NPs is clearly demonstrated. (b) Optical contrast due to the optical resonance of the gold NP cluster as a function of the phase offsets between the reference (R) and the diffracted (D) beams of the OPCM.

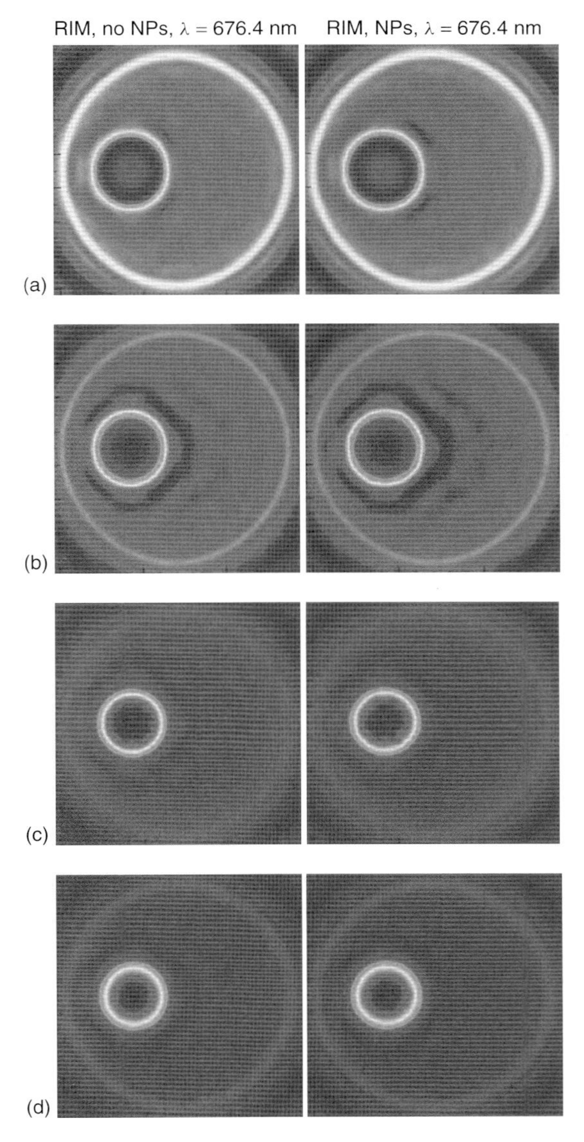

Figure 3.12 OPCM images of the cell for different values of the phase offset Ψ ((a) -90°, (b) -30°, (c) $+30^\circ$, (d) $+90^\circ$) between the reference and diffracted beams of the OPCM at optical immersion conditions without NPs (left) and including 42 gold NPs (right) randomly located at the surface of the cell nucleus (nonresonant case $-\lambda =$ 676.4 nm).

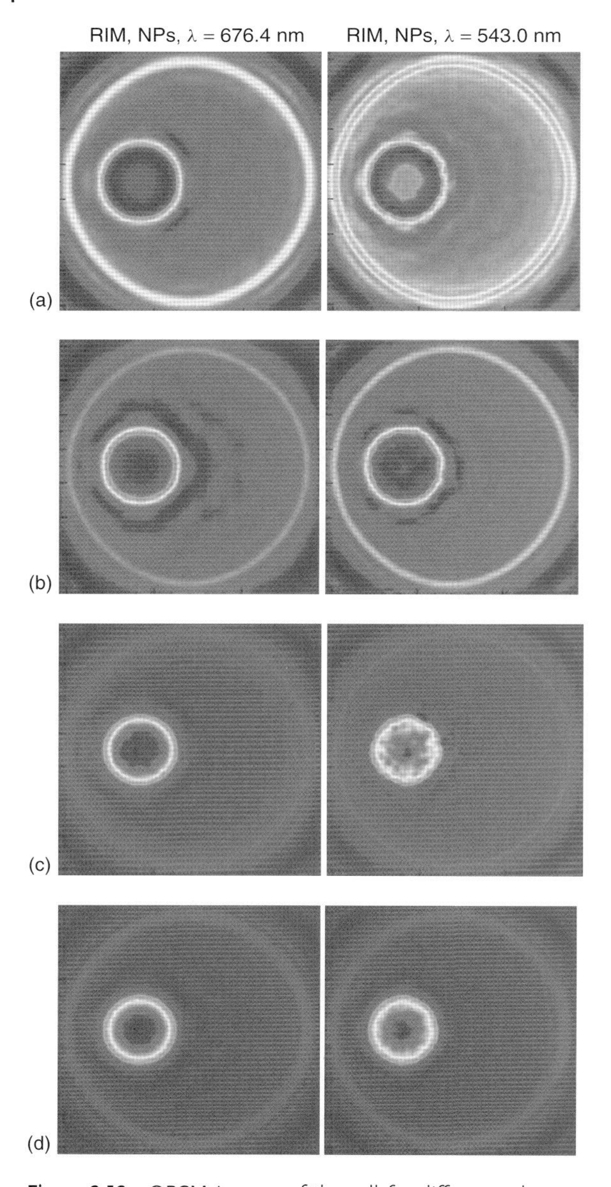

Figure 3.13 OPCM images of the cell for different values of the phase offset Ψ ((a) -90°, (b) -30°, (c) $+30^{\circ}$, (d) $+90^{\circ}$) between the reference and diffracted beams of the OPCM at optical immersion conditions including 42 gold NPs randomly located at the surface of the cell nucleus at no-resonance (left) and resonance (right).

and at both nonresonant (left) and resonant (right) conditions. The images are for different values of the phase offset Ψ: -90, -30, $+30$, and $+90^{\circ}$.

It should be pointed out that the optical wave phenomena involved in the simulation scenario considered here are fundamentally different from the ones considered earlier in this chapter, where the gold NPs are randomly distributed within the homogeneous material of the cytoplasm and their presence is manifested by means of their own absorption and scattering properties. Currently, the NPs are located at the interface of the nucleus and the cytoplasm, which is characterized by a relatively large refractive index difference ($\Delta n = 0.04$) and which is, therefore, expected to largely dominate and modify the visual effect of the NPs. A close examination of the OPCM images in Figure 3.8 illustrates this fact.

The analysis of Figures 3.12–3.14 leads to a number of interesting findings. First, the images of the cell without the gold NPs are hardly distinguishable from the images with the gold NPs at nonresonant conditions ($\lambda = 676.4$ nm). Second, the visual effect of the gold NP presence at resonant conditions ($\lambda = 543.0$ nm) depends significantly on the phase offset Ψ. Third, the presence of the gold NPs at resonant conditions can be identified by a specific fragmentation of the image of the nucleus for specific values of the offset Ψ. This shows the importance of the ability to adjust the offset Ψ between the reference and diffracted beams of the OPCM in practical circumstances. This ability to finely adjust the offset Ψ should be highly relevant for the development and calibration of phase-sensitive flow cytometers.

3.5 Conclusion

In this chapter, a brief summary of different formulations of the FDTD method for application in flow cytometry-related biophotonics problems was provided. The FDTD approach was then applied to three different modeling scenarios: (i) light scattering from single cells, (ii) OPCM imaging of realistic size cells, and (iii) OPCM imaging of gold NPs in singe cells. The FDTD ability to model OPCM microscopic imaging by, first, reproducing the effect of optical immersion on the OPCM images of a realistic size cell containing a cytoplasm, a nucleus, and a membrane; second, including the presence of a cluster of gold NPs in the cytoplasm at optical immersion conditions as well as the enhanced imaging effect of the optical resonance of the NPs; third, studying the imaging effect of gold NPs randomly distributed on the surface of the cell nucleus was demonstrated. The results do not allow analysis of the scaling of the NP imaging effect as a function of the number of the NPs. However, the validation of the model provides a basis for future research on cytometry-related OPCM nanobioimaging including the effects of NP cluster size, NP size, and number, as well as average distance between the NPs.

The shift from the modeling of the light scattering properties of single cells alone to the construction of images of cells containing gold NPs represents a major step forward in extending the application of the FDTD approach to *in vivo* flow cytometry (Chapters 14 and 17). Future development will include

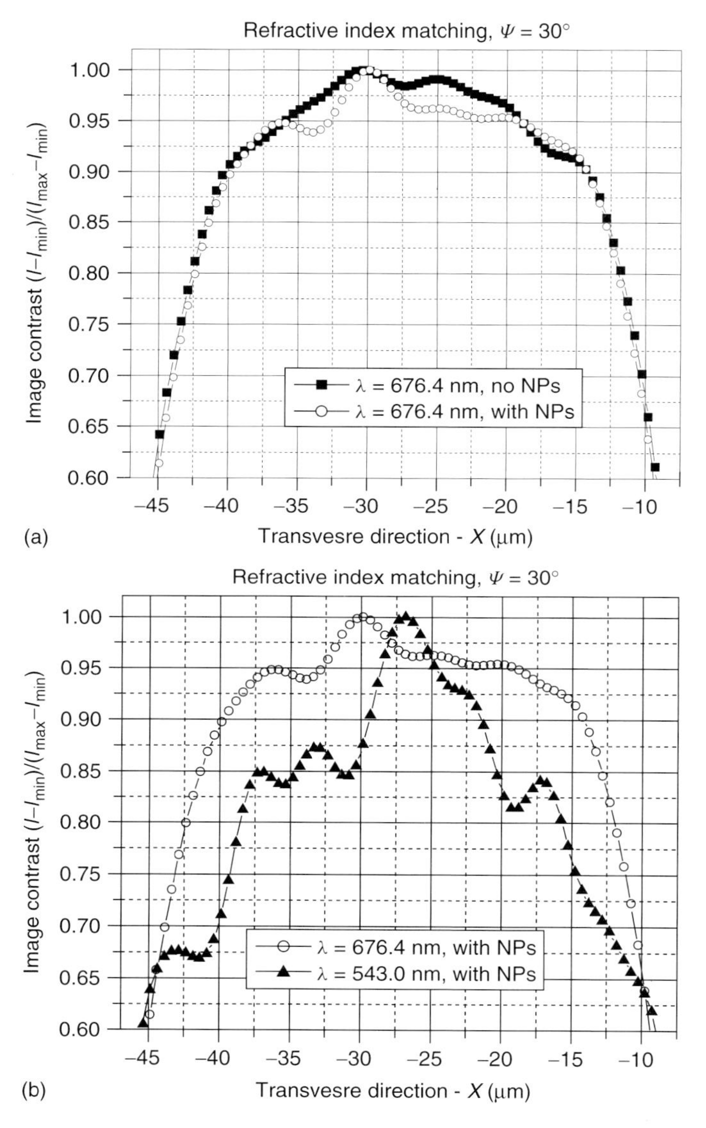

Figure 3.14 Cross sections of the images shown in (a) Figure 3.12c and (b) Figure 3.13c corresponding to phase offset $\Psi = +30^{\circ}$. The specific fragmentation of the nucleus' image is due to the presence of the Gold NPs at resonant condition.

detection of enhanced scattering effects from laser-induced microbubbles around the endogenous absorbing cellular zones and NPs with different methods including thermo-lens schematics as was discussed in Refs [47, 51, 52] and demonstrated *in vivo* for nonmoving cells in Ref. [53].

Acknowledgment

ST and JP acknowledge the use of the computing resources of WestGrid (Western Canada Research Grid). VVT was supported by grants: **224014 PHOTONICS4LIFE-FP7-ICT-2007-2** RF Ministry of Science and Education 2.1.1/4989 and 2.2.1.1/2950, RFBR-09-02-90487_Ukr_f_a, and the Special Program of RF "Scientific and Pedagogical Personnel of Innovative Russia," Governmental contracts 02.740.11.0484, 02.740.11.0770, and 02.740.11.0879. VPZ was supported by the National Institute of Health grants **EB005123, EB000873, CA131164.**

References

1. Pawley, J.P. (1995) *Handbook of Biological and Confocal Microscopy*, 2nd edn, Plenum, New York.
2. (a) Shapiro, H.M. (2003) *Practical Flow Cytometry*, 4th edn, Wiley-Liss, New York; (b) Carey, J.L., McCoy, J.P., and Keren, D.F. (2007) *Flow Cytometry*, 4th edn, ASCP Press, Chicago.
3. Sack, U., Tárnok, A., and Rothe, G. (eds) (2006) *Zelluläre Diagnostik: Grundlagen, Methoden und Klinische Anwendungen der Durchflusszytometrie*, Karger.
4. Maltsev, V. (2000) Scanning flow cytometry for individual particle analysis. *Rev. Sci. Instrum.*, **71**, 243–255.
5. Tanev, S., Sun, W., Pond, J., Tuchin, V.V., and Zharov, V.P. (2009) Flow cytometry with gold nanoparticles and their clusters as scattering contrast agents: FDTD simulation of light-cell interaction. *J. Biophoton.*, (Special issue "Towards *in vivo* flow cytometry"), pp. 505–520. doi: 10.1002/jbio.200910039
6. Prasad, P.N. (2003) *Introduction to Biophotonics*, Chapter 7, John Wiley & Sons, Inc., pp. 203–249.
7. Tuchin, V.V. (2007) *Tissue Optics: Light Scattering Methods and Instruments for Medical Diagnosis*, 2nd edn, SPIE Press, Bellingham, WA, PM 166.
8. Drexler, W. and Fujimoto, J.G. (eds) (2008) *Optical Coherence Tomography: Technology and Applications*, Springer, Berlin.
9. Wilson, T. (2003) in *Biomedical Photonics Handbook*, Chapter 10 (ed. T. Vo-Dinh), CRC Press, Boca Raton, FL, pp. 10.1–10.18.
10. Drezek, R., Guillaud, M., Collier, T., Boiko, I., Malpica, A., Macaulay, C., Follen, M., and Richards-Kortum, R. (2003) Light scattering from cervical cells throughout neoplastic progression: influence of nuclear morphology, DNA content, and chromatin texture. *J. Biomed. Opt.*, **8**, 7–16.
11. Kahnert, F.M. (2003) Numerical methods in electromagnetic scattering theory. *J. Quant. Spectr. Radiat. Trans.*, **73**, 775–824.
12. Dunn, A. and Richards-Kortum, R. (1996) Three-dimensional computation of light scattering from cells. *IEEE J. Selec. Top. Quant. Electr.*, **2**, 898–894.
13. Dunn, A. (1997) Light scattering properties of cells, PhD Dissertation, Biomedical Engineering, University of Texas at Austin, Austin TX.
14. Dunn, A., Smithpeter, C., Welch, A.J., and Richards-Kortum, R. (1997)

Finite-difference time-domain simulation of light scattering from single cells. *J. Biomed. Opt.*, **2**, 262–266.
15. Drezek, R., Dunn, A., and Richards-Kortum, R. (1999) Light scattering from cells: finite-difference time-domain simulations and goniometric measurements. *Appl. Opt.*, **38**, 3651–3661.
16. Drezek, R., Dunn, A., and Richards-Kortum, R. (2000) A pulsed finite-difference time-domain (FDTD) method for calculating light scattering from biological cells over broad wavelength ranges. *Opt. Exp.*, **6**, 147–157.
17. Arifler, D., Guillaud, M., Carraro, A., Malpica, A., Follen, M., and Richards-Kortum, R. (2003) Light scattering from normal and dysplastic cervical cells at different epithelial depths: finite-difference time domain modeling with a perfectly matched layer boundary condition. *J. Biomed. Opt.*, **8**, 484–494.
18. Tanev, S., Sun, W., Zhang, R., and Ridsdale, A. (2004) The FDTD approach applied to light scattering from single biological cells. *Proc. SPIE*, **5474**, 162–168.
19. Tanev, S., Sun, W., Zhang, R., and Ridsdale, A. (2004) Simulation tools solve light-scattering problems from biological cells. *Laser Focus World*, 67–70. *http://www.optoiq.com/index/photonics-technologies-applications/lfw-display/lfw-article-display/197494/articles/laser-focus-world/volume-40/issue-1/departments/software-and-computing/simulation-tools-solve-light-scattering-problems-from-biological-cells.html* (accessed 11 January 2011).
20. Tanev, S., Sun, W., Loeb, N., Paddon, P., and Tuchin, V.V. (2005) in *Advances in Biophotonics*, NATO Science Series I, Vol. 369 (eds B.C. Wilson, V.V. Tuchin, and S. Tanev), IOS Press, Amsterdam, pp. 45–78.
21. Tanev, S., Pond, J., Paddon, P., and Tuchin, V.V. (2008) A new 3D simulation method for the construction of optical phase contrast images of gold nanoparticle clusters in biological cells. *Adv. Opt. Techn.*, 9. Article ID 727418. doi: 10.1155/2008/727418. *http://www.hindawi.com/GetArticle.aspx?doi=10.1155/2008/727418* (accessed 11 January 2011).
22. Tanev, S., Tuchin, V., and Paddon, P. (2006) Cell membrane and gold nanoparticles effects on optical immersion experiments with non-cancerous and cancerous cells: FDTD modeling. *J. Biomed. Opt.*, **11**, 064037-1–064037-6.
23. Tanev, S., Tuchin, V.V., and Paddon, P. (2006) Light scattering effects of gold nanoparticles in cells: FDTD modeling. *Laser Phys. Lett.*, **3**, 594–598.
24. Tanev, S., Tuchin, V.V., and Pond, J. (2008) Simulation and modeling of optical phase contrast microscope cellular nanobioimaging. *Proc. SPIE*, **7027**, 16-1–16-8.
25. Li, X., Taflove, A., and Backman, V. (2005) Recent progress in exact and reduced-order modeling of light-scattering properties of complex structures. *IEEE J. Selec. Top. Quant. Electr.*, **11**, 759–765.
26. Sun, W., Loeb, N.G., and Fu, Q. (2004) Light scattering by coated sphere immersed in absorbing medium: a comparison between the FDTD and analytic solutions. *J. Quant. Spectr. Radiat. Trans.*, **83**, 483–492.
27. Yee, K.S. (1966) Numerical solution of initial boundary value problems involving Maxwell's equation in isotropic media. *IEEE Trans. Anten. Propag.*, **AP-14**, 302–307.
28. Taflove, A. and Hagness, S.C. (eds) (2005) *Computational Electrodynamics: The Finite-Difference Time-Domain Method*, 3rd edn, Artech House, Norwood, MA.
29. Sun, W., Fu, Q., and Chen, Z. (1999) Finite-difference time-domain solution of light scattering by dielectric particles with a perfectly matched layer absorbing boundary condition. *Appl. Opt.*, **38**, 3141–3151.
30. Sun, W. and Fu, Q. (2000) Finite-difference time-domain solution of light scattering by dielectric particles with large complex refractive indices. *Appl. Opt.*, **39**, 5569–5578.
31. Sun, W., Loeb, N.G., and Fu, Q. (2002) Finite-difference time-domain solution of light scattering and absorption by

particles in an absorbing medium. *Appl. Opt.*, **41**, 5728–5743.
32. Sacks, Z.S., Kingsland, D.M., Lee, R., and Lee, J.F. (1995) A perfectly matched anisotropic absorber for use as an absorbing boundary condition. *IEEE Trans. Anten. Propag.*, **43**, 1460–1463.
33. Gedney, S.D. (1996) An anisotropic perfectly matched layer absorbing media for the truncation of FDTD lattices. *IEEE Trans. Anten. Propag.*, **44**, 1630–1639.
34. Gedney, S.D. and Taflove, A. (2000) in *Computational Electrodynamics: The Finite-Difference Time Domain Method* (eds A. Taflove and S. Hagness), Artech House, Boston, pp. 285–348.
35. Yang, P. and Liou, K.N. (1996) Finite-difference time domain method for light scattering by small ice crystals in three-dimensional space. *J. Opt. Soc. Am. A*, **13**, 2072–2085.
36. Bohren, C.F. and Huffman, D.R. (1998) *Absorption and Scattering of Light by Small Particles*, John Wiley & Sons, Inc., New York.
37. Goedecke, G.H. and O'Brien, S.G. (1988) Scattering by irregular inhomogeneous particles via the digitized Green's function algorithm. *Appl. Opt.*, **27**, 2431–2438.
38. Fu, Q. and Sun, W. (2001) Mie theory for light scattering by a spherical particle in an absorbing medium. *Appl. Opt.*, **40**, 1354–1361.
39. Barer, R., Ross, K.F.A., and Tkaczyk, S. (1953) Refractometry of living cells. *Nature*, **171**, 720–724.
40. Fikhman, B.A. (1967) *Microbiological Refractometry*, Medicine, Moscow.
41. Tuchin, V.V. (2006) *Optical Clearing of Tissues and Blood*, SPIE Press, Bellingham, WA, PM 154.
42. Beck, G.C., Akgun, N., Rück, A., and Steiner, R. (1997) Developing optimized tissue phantom systems for optical biopsies. *Proc. SPIE*, **3197**, 76–85.
43. Beck, G.C., Akgun, N., Rück, A., and Steiner, R. (1998) Design and characterization of a tissue phantom system for optical diagnostics. *Lasers Med. Sci.*, **13**, 160–171.
44. Sokolov, K., Follen, M., Aaron, J., Pavlova, I., Malpica, A., Lotan, R., and Richards-Kortum, R. (2003) Real time vital imaging of pre-cancer using anti-EGFR antibodies conjugated to gold nanoparticles. *Cancer Res.*, **63**, 1999–2004. *http://cancerres.aacrjournals.org/cgi/content/full/63/9/1999#B9* (accessed 11 January 2011).
45. El-Sayed, I.H., Huang, X., and El-Sayed, M.A. (2005) Surface plasmon resonance scattering and absorption of anti-EGFR antibody conjugated gold nanoparticles in cancer diagnostics: applications in oral cancer. *Nano Lett.*, **5**, 829–834.
46. Khlebtsov, N.G., Melnikov, A.G., Dykman, L.A., and Bogatyrev, V.A. (2004) in *Photopolarimetry in Remote Sensing*, NATO Science Series II, Mathematics, Physics, and Chemistry, Vol. 161 (eds G. Videen, Ya.S. Yatskiv, and M.I. Mishchenko), Kluwer, Dordrecht, pp. 265–308.
47. Zharov, V.P., Kim, J.-W., Curiel, D.T., and Everts, M. (2005) Self-assembling nanoclusters in living systems: application for integrated photothermal nanodiagnostics and nanotherapy. *Nanomed. Nanotech., Biol., Med.*, **1**, 326–345.
48. Zharov, V.P., Mercer, K.E., Galitovskaya, E.N., and Smeltzer, M.S. (2006) Photothermal nanotherapeutics and nanodiagnostics for selective killing of bacteria targeted with gold nanoparticles. *Biophys. J.*, **90**, 619–627.
49. Tuchin, V.V., Drezek, R., Nie, S., and Zharov, V.P. (Guest Editors) (2009) Special section on nanophotonics for diagnostics, protection and treatment of cancer and inflammatory diseases. *J. Biomed. Opt.* **14** (2), 020901, 021001–021017.
50. Johnson, P.B. and Christy, R.W. (1972) Optical constants of the noble metals. *Phys. Rev. B*, **6**, 4370–4379.
51. Zharov, V.P., Galitovsky, V., and Chowdhury, P. (2005) Nanocluster model of photothermal assay: application for high-sensitive monitoring of nicotine – induced changes in metabolism, apoptosis and necrosis at a cellular level. *J. Biomed. Opt.*, **10**, 044011.
52. Zharov, V.P., Menyaev, Y.A., Shashkov, E.V., Galanzha, E.I., Khlebtsov, B.N.,

Scheludko, A.V., Zimnyakov, D.A., and Tuchin, V.V. (2006) Fluctuation of probe beam in thermolens schematics as potential indicator of cell metabolism, apoptosis, necrosis and laser impact. *Proc. SPIE*, **6085**, 608504-1–608504-12.

53. Lee, H., Alt, C., Pitsillides, C.M., and Lin, C.P. (2007) Optical detection of intracellular cavitation during selective laser targeting of the retinal pigment epithelium: dependence of cell death mechanism on pulse duration. *J. Biomed. Opt.*, **12**, 064034.

4 Optics of White Blood Cells: Optical Models, Simulations, and Experiments

Valeri P. Maltsev, Alfons G. Hoekstra, and Maxim A. Yurkin

4.1 Introduction

4.1.1 White Blood Cells

The role of white blood cells (WBCs), or leukocytes, is to control various disease conditions and eliminate invading microorganisms. Although these cells do most of their work outside the circulatory system, they use the blood for transportation to sites of infection. Five types of white blood cells are normally found in the circulating blood: two mononuclear types (lymphocytes and monocytes) and three types of granulocytes (eosinophils, basophils, and neutrophils). Their presence relative to all leukocytes in normal conditions is in the ranges 18–44%, 2.6–8.5%, 46–73%, 0–4.4%, and 0.2–1.2%, respectively. And their characteristic size is 6–10 μm [1].

Lymphocytes and monocytes are cells with clear, transparent cytoplasm and large round nucleii. The lymphocytes mediate highly specific immunity against microorganisms and other sources of foreign macromolecules. B lymphocytes confer immunity through the production of specific, soluble antibodies, while T lymphocytes direct a large variety of immunity functions, including killing cells that bear foreign molecules on their surface membranes. Granulocytes and monocytes can exit from blood vessels and migrate among the cells of many tissues. These cells play key roles in inflammation and phagocytosis. Contrary to mononuclear cells, the nucleus of granulocytes is separated into definite lobes with a very narrow filament or strand connecting the lobes. Usually, two to three lobes are observed.

4.1.2 Particle Identification and Characterization

There exist two levels of scientific description of single particles in a disperse media: identification and characterization. Identification (also known as classification) means attributing the particles to one of several (usually predefined) classes, which

Advanced Optical Flow Cytometry: Methods and Disease Diagnoses, First Edition. Edited by Valery V. Tuchin.

can be distinguished from each other due to their morphological or/and functional characteristics. Characterization means quantitative measurement of a particle by its morphological and functional characteristics.

With reference to WBC analysis, identification corresponds to differential WBC count, that is, measuring the number (or fraction) of each of the five WBC types in the samples. This problem has been largely solved 20 years ago (Sections 4.1.3 and 4.3.2.1), using empirical methods based on vast experimental data. To develop the characterization method, one has to devise an optical model of the particle, and solve direct and inverse light-scattering problems.

Existing optical models of WBCs based on literature data and recently obtained-confocal images are described in Section 4.2. The methods used to simulate light scattering by optical models of WBCs range from the Mie theory to the discrete dipole approximation (DDA), and depend on the complexity of the model. Progress in the development of these methods and increase in computer literacy opened avenues for the development of new methods to solve the inverse light-scattering problem. In particular, it has become possible to perform an extensive search of the parameter space of a coated sphere model to perform fit of experimental light-scattering patterns of mononuclear WBCs. Both direct and inverse light-scattering problems for WBCs are discussed in Section 4.3.

4.1.3
Experimental Techniques

Light scattering and fluorescence offer a number of advantages in the identification and characterization of particles. For this purpose, various fluorescence-based and light-scattering methods have been used in scientific research and industries. Instrumentally, these methods can be divided into following categories: (i) methods based on analysis of light scattered by a particle suspension, and (ii) methods based on analysis of light scattered by individual particles. The first can be realized by using instruments measuring light scattering and fluorescence of particle ensembles. Today, however, the parameters of a distribution of particle characteristics can only be evaluated for very rough models of disperse media. Particle ensembles can be effectively studied using a single-particle analysis when light scattering and fluorescence are measured from individual particles with a sufficient rate for statistical data processing. In this case, determination of particle characteristics from light scattering is simplified substantially, whereas such instrumental approach requires solutions of technical problems that are related to fast detection of single particles. Fortunately, these problems were solved by flow cytometry, which utilizes hydrofocusing as a means for measuring scattering and fluorescent signals of individual particles.

Today, all large hospitals have facilities based on flow cytometry for detailed screening and identification of bacterial and blood cells of patients. Also, many research laboratories specialized in cell biology or immunology, all over the world, routinely apply cell identification and sorting to facilitate their research. A flow cytometer can identify single cells (at speeds as high as 50 000 cells/s) using elastic light scattering from the cell, and fluorescence from fluorescent probes, which bind to specific molecules on the cell surface or inside the cell. Several fluorescence signals are measured, in combination with light-scattering signals. In this way, a number of independent measurements are obtained for every cell in the sample, and this information is used to identify different subsets in the original cell sample. However, at present, 95% of sorting facility of flow cytometers is based on an analysis of fluorescence, disregarding the advantages of light scattering for cell characterization.

During the last 15 years, researchers from the laboratory of Cytometry and Biokinetics, Institute of Chemical Kinetics and Combustion, Novosibirsk, Russia, have developed a revolutionary approach toward measuring light scattering in flow cytometry [2]. This scanning flow cytometer (SFC) allows detailed measurement of the entire angular light-scattering pattern (i.e., much more than in standard flow cytometry, in which light scattering is measured over only two different angles). The SFC offers exciting new ways to use flow cytometry, ranging from extending and improving routine measurements in hematology laboratories to applying it in new fields (such as microbiology).

Section 4.4 reviews application of the SFC to analysis of WBCs, including measuring absolute differential scattering cross section (DSCS) and characterization of lymphocytes from the measured light-scattering patterns.

4.2 Optical Models of White Blood Cells

4.2.1 Confocal Imaging of White Blood Cells

Confocal imaging of WBCs to construct optical models for light-scattering simulations was pioneered by Brock *et al.* [3] for cultured NALM-6 cells (a human B cell precursor). A Carl Zeiss LSM 510 META confocal microscope was used to develop optical models of lymphocytes and neutrophils. The parameters used for cell imaging are summarized in Table 4.1. The typical images of a lymphocyte and a neutrophil are shown in Figures 4.1 and 4.2, respectively.

4.2.2 Optical Models of Mononuclear Cells

The simplest model of mononuclear cells is a homogeneous sphere. It has been successfully used (explicitly or implicitly) to size lymphocytes using light-scattering signals by Neukammer *et al.* [4], Yang *et al.* [5], and Semyanov *et al.* [6]. To extract more morphological characteristics of mononuclear cells from experimental data, a more detailed model is required. Zharinov *et al.* [7] tested a multilayered sphere model, and showed that a five-layer model gave the best agreement between experimental and theoretical light-scattering profiles (LSPs).

Table 4.1 Parameters of the confocal microscopy.

Cell	Objective	Wavelengths (nm)	Filters (nm)	Scaling (nm)
Neutrophil	Plan-Apochromat 100 × /1.4 Oil DIC	405	BP 420–480	40 (x axis)
		488	505–530	40 (y axis)
		543	LP 560	310 (z axis)
Lymphocyte	Plan-Apochromat 100 × /1.4 Oil DIC	405	BP 560–615	20 (x axis)
		543	LP 420	20 (y axis)
		–	–	310 (z axis)

BP, bandpass; LP, landpass.

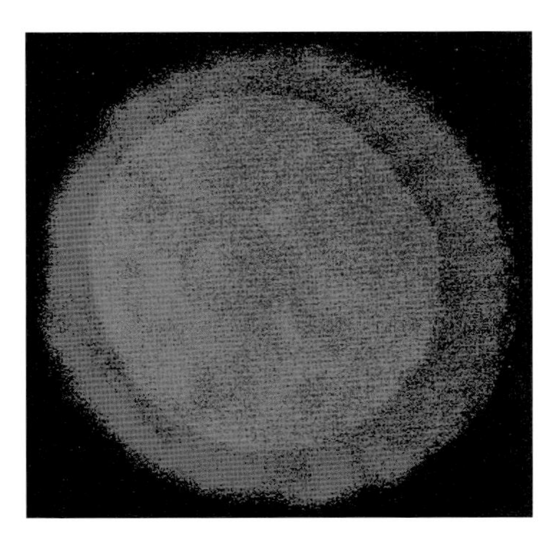

Figure 4.1 A slice from a confocal image of a lymphocyte.

However, this model is too complicated to provide reliable solution for the inverse light-scattering problems (Section 4.3.3), and its biological relevancy is not immediately clear.

For a simpler optical model, a more reliable procedure was recently proposed by Strokotov *et al.* [8]. From the confocal images (Figure 4.1), one can see that a coated sphere is an adequate first-order model, and this model has been used to solve the inverse light-scattering problem. However, it is also clear that the nucleus is certainly not perfectly homogeneous. To study the effect of such nuclear inhomogeneity on the light scattering, the optical model was enhanced by introducing an oblate spheroid into the nucleus. This model was used to mimic the effect that nuclear inhomogeneity may have on the LSPs and to verify the results of global optimization algorithms in the presence of model errors.

Another modification of coated sphere model was proposed by Hoekstra *et al.* [9] by shifting the nucleus from the cell center. Loiko *et al.* [10] proposed a

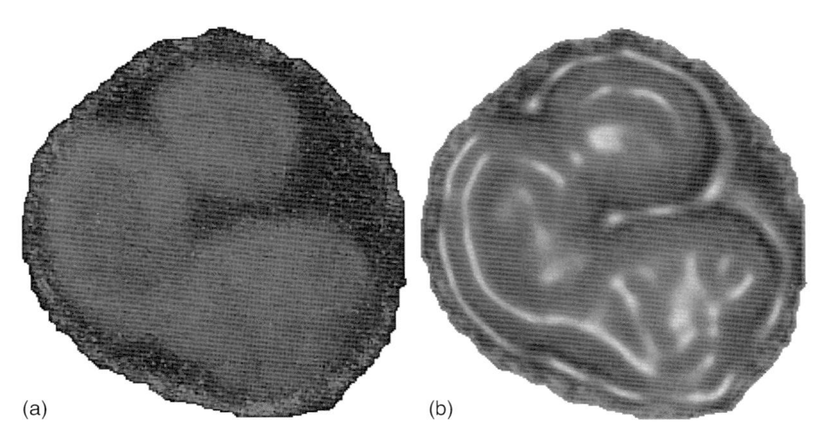

Figure 4.2 Slices from the confocal image of a neutrophil: original (a) and visually improved (b).

morphometric model of lymphocytes based on the images obtained by an optical microscope. This model consists of two nonconcentric spheroids and its main goal is to assist in the systematization of observational data. 3D models of precursor B cells have been constructed by Brock *et al.* [3] directly from confocal images. Unfortunately, such complicated models cannot be practically used to solve the inverse light-scattering problem.

4.2.3
Optical Models of Granular Cells

Granulocytes have the most complex morphology among all blood cells, featuring numerous granules and multilobed nucleus [1]. Although a simple homogeneous sphere model was used to determine the diameters of granulocytes [6], more detailed models are clearly required. Dunn [11] summarized two approaches to generate such models: (i) granules and nucleus (with certain random variations in size and shape) placed randomly inside the cytoplasm, and (ii) continuous random distribution of the refractive index within the cell or nucleus with certain spatial correlation lengths. In particular, Dunn [11] showed that both models produce LSPs, which were drastically different from that of homogeneous spheres, except for the forward-scattering direction.

A particulate model (i.e., of the type 1 above) of granulocytes was also used by Yurkin *et al.* [12]. It constitutes a sphere filled with identical spherical granules (Figure 4.3a) and its main goal was to study the influence of granules on LSPs. However, to match measured LSPs with simulations, a nucleus had to be included in the theoretical modeling. Orlova *et al.* [13] modeled a nucleus of neutrophils by four spheroids of various sizes, which were randomly placed and oriented inside the cytoplasm (Figure 4.3b). Geometrical parameters of this neutrophil model were based on literature data: the diameter of the cell $d_c = 9.6\ \mu\text{m}$ [6, 13]; granule

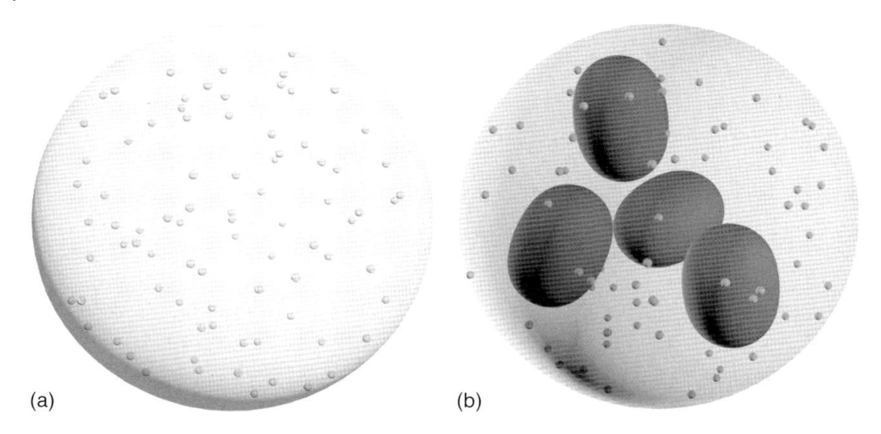

Figure 4.3 Optical models of (a) nuclei-free granulocyte and (b) neutrophil.

diameters $d_g = 0.1$, 0.15, and 0.2 μm [14, 15] and their volume fraction $f_g = 0.1$ [15, 16]; volume fraction of nuclei $f_n = 0.11$ [17, 18], resulting in total volume fraction of noncytoplasm material, granules and nucleus, equal to 0.2.

4.2.4
Refractive Indices of White Blood Cells and their Organelles

A necessary component of optical models is the refractive index of all its constituents. However, for WBCs, such data is hardly available in the literature. Dunn [11] collected information on refractive indices of the cell cytoplasm and constituents from several sources. However, none of them directly corresponds to human WBCs, while general ranges for cytoplasm and mitochondria are rather broad (1.35–1.38 and 1.40–1.42, respectively). Brunsting and Mullaney [19] reported a refractive index m_n of 1.39 for the nucleus of hamster ovary cells, which was used in simulations of granulocyte-type cells by Dunn [11].

Recently, refractive indices of mononuclear blood cells were determined by fitting LSP with five- or two-layer spherical models (Sections 4.3.3 and 4.4.4). Zharinov *et al.* [7] determined m_n of lymphocytes and monocytes in one sample to be 1.43 $\pm$ 0.05 (standard deviation, SD) and 1.43 $\pm$ 0.04, respectively. Strokotov *et al.* [8] studied samples of lymphocytes from seven donors and reported mean values between 1.44 and 1.45 with SD 0.01. Concerning the refractive index of cytoplasm m_c, Zharinov *et al.* [7] reported 1.356 $\pm$ 0.009 and 1.348 $\pm$ 0.004 for lymphocytes and monocytes, respectively. However, these results are arguable as they are based on the ambiguous correspondence between cytoplasm and the fourth layer of five-layer model. Strokotov *et al.* [8] reported significantly larger m_c: mean values between 1.376 and 1.377 with SD 0.003.

We are aware of only one other study where m_c of human lymphocytes was measured [20]. The index-matching method was used leading to mean value of m_c equal to 1.3572 $\pm$ 0.0002 (standard error of mean) for several normal donors, which is lower than values obtained by Strokotov *et al.* [8]. This is because the latter determines the integral (effective) refractive index of the whole cytoplasm, including all granules and other inclusions, while index matching yields the value for the ground substance of the cytoplasm.

Brock *et al.* [3] used 1.3675 and 1.4 for m_c and m_n, respectively, for their B cell optical model. However, these values were taken from paper by Mourant *et al.* [21], who derived them from the study of light scattering by rat fibroblast cells.

It is almost impossible to choose reliable values for the refractive index of granulocyte granules m_g, because there seems to be no literature data available and there exist at least several different types of granules with different composition. Hence, the choice of any value of m_g to be used in light-scattering simulations is, to some extent, arbitrary. Yurkin *et al.* [12] varied the m_g from 1.47 to 1.6, the latter being close to the refractive index of dried protein [22]. Orlova *et al.* [13] chose intermediate refractive index 1.538 for both nuclei and granules to eliminate one free parameter of the model. Melanin with refractive index 1.7 was also used as granule material in optical models of cells [23, 24].

It is important to note that light-scattering patterns are sensitive to refractive index, and hence potentially this parameter can be determined with good accuracy from the solution of the inverse light-scattering problem (see above). However, the appropriateness of the used optical model has to be assumed. Moreover, we believe that all existing methods to determine refractive index of cell constituents are in some respect indirect: they either assume a certain optical model or measure some effective value (as index-matching technique). The ultimate goal is to develop a model-free method to solve inverse light-scattering problem, which, for example, computes the distribution of refractive index in the volume of the particle. But this goal is still far from reaching satisfactory accuracy [25].

4.3 Direct and Inverse Light-Scattering Problems for White Blood Cells

4.3.1 Simulation of Light Scattering by Mononuclear Cells

Spherical models of mononuclear cells are trivial with respect to light-scattering simulation. The theory of light scattering by single- or multilayered spheres is well known [26, 27] and LSPs are easily and quickly computed. To simulate light scattering by models lacking spherical symmetry, more general and slower methods have to be used. Methods based on surface discretization, such as extended boundary condition method [28] and null-field method with discrete sources [29], are especially effective for axisymmetric particles. Volume discretization methods, such as the DDA [30] and the finite difference time domain (FDTD, [31]) can handle any scatterers including inhomogeneous ones.

Hoekstra *et al.* [9] applied the DDA to a coated sphere model with shifted nucleus. It should be noted, however, that such a model can also be simulated by a faster analytical algorithm [32]. Strokotov *et al.* [8] simulated light scattering by lymphocytes with inhomogeneous nucleus using the DDA. Brock *et al.* [3] and Ding *et al.* [33] used the FDTD to simulate light scattering by realistic 3D models of B cells precursors. Although both methods may now be used routinely for different cell models, the DDA has been proven to be more efficient than the FDTD for large biological cells (a factor of 10–100 faster with the same accuracy [34]).

4.3.2 Simulation of Light Scattering by Granular Cells

There were only a few studies that included simulation of light scattering by realistic models of granulocytes. The FDTD simulations of light scattering by a granulocyte-type model were performed by Dunn and coworkers (reviewed by Dunn [11]). Yurkin *et al.* [12] performed DDA simulations of light scattering by granulated sphere models, varying model parameters in the range of all granulocyte subtypes. Orlova *et al.* [13] simulated light scattering by a neutrophil model including four-lobed nucleus and compared it with experimental LSPs measured by an SFC. Below, we discuss the latter two papers in detail.

4.3.2.1 Granulocyte Model Without Nucleus

Yurkin *et al.* [12] employed a granulated sphere model for granulocytes (Figure 4.3a) with the goal to explain the difference in depolarization ratio between neutrophils and eosinophils reported by de Grooth *et al.* [35]. The default parameters of the model, based on literature data on morphology of granulocytes, were as follows: diameter of the cell $d_c = 8\,\mu m$, volume fraction of granules $f = 0.1$, $m_c = 1.357$, $m_g = 1.2$, $m_0 = 1.337$, $\lambda = 0.66\,\mu m$. The main variable was the granule diameter d_g, varying from 0.075 to $2\,\mu m$. Each DDA simulation was repeated 10 times for different random granule placements, resulting in mean value $\pm 2 \times$ SD for each simulated value. Additionally, the result for the limiting case of $d_g = 0$ was obtained by the Mie theory applied to a homogeneous sphere with a refractive index obtained by the Maxwell–Garnett effective medium theory [26].

Some of the model parameters varied from their default values; however, to limit the number of DDA simulations, each parameter was varied with others fixed. For each set of parameters, a whole range of d_g was calculated. Yurkin *et al.* [12] tried four additional values of f: 0.02, 0.05, 0.2, and 0.3; two values of d_c: 4 and $14\,\mu m$; two of m_g: 1.1 and 1.15; and one of m_c: 1; that is, no cytoplasm at all. All simulations were carried out using the ADDA computer code v. 0.76, which is capable of running on a cluster of computers parallelizing a single DDA computation [36].[1)]

The side-scattering signals were simulated exactly as described by de Grooth *et al.* [35]: total side-scattering intensity I_{SS}, total depolarized intensity $I_\perp$, and the depolarization ratio $D_{SS} = I_\perp / I_{SS}$. Aperture widths $\Delta\theta$ and $\Delta\phi$ were defined so

1) This code is freely available at *http://code.google.com/p/a-dda/*.

that scattering angles $\theta = 90^\circ \pm \Delta\theta$ and $\phi = 90^\circ \pm \Delta\phi$ are collected (the default values are $\Delta\theta = \Delta\phi = 25^\circ$). The experimental findings of de Grooth *et al.* [35] can be summarized as follows. The scatter plot of a sample containing eosinophils and neutrophils in $I_\perp$ versus I_{SS} coordinates shows a clear discrimination of two subtypes. The I_{SS} signals are almost the same, while $I_\perp$ are larger for eosinophils. Moreover, there is a significant correlation between $I_\perp$ and I_{SS} for both neutrophils and eosinophils. Discrimination between two subtypes is the most pronounced using D_{SS}, with mean values of 0.044 and 0.013 for eosinophils and neutrophils, respectively (averaged over a single sample). D_{SS} is almost constant when $\Delta\phi$ is decreased at fixed $\Delta\theta$ for both granulocyte subtypes.

The results of DDA simulations of side scattering by granulated spheres with the default set of parameters and varying granule diameter are shown in Figure 4.4. For each d_g, mean values of I_{SS} and $I_\perp$ and error bars corresponding to two SDs are shown. Labels near some of the points indicate the values of d_g. There are two ranges of d_g: from 0.1 to 0.25 μm and from 0.4 to 2 μm, corresponding to distinct regions. The range from 0.25 to 0.4 μm is intermediate, where I_{SS} and $I_\perp$ strongly, albeit differently, depend on d_g. The two regions in Figure 4.4 qualitatively correspond to the neutrophils and eosinophils on the plots by de Grooth *et al.* [35]. Considering the morphological characteristics of neutrophil and eosinophils granules, the difference in depolarized side scattering can be explained solely by the difference in d_g. Moreover, Yurkin *et al.* [12] showed that the qualitative two-region picture does not change when varying model parameters (as described above).

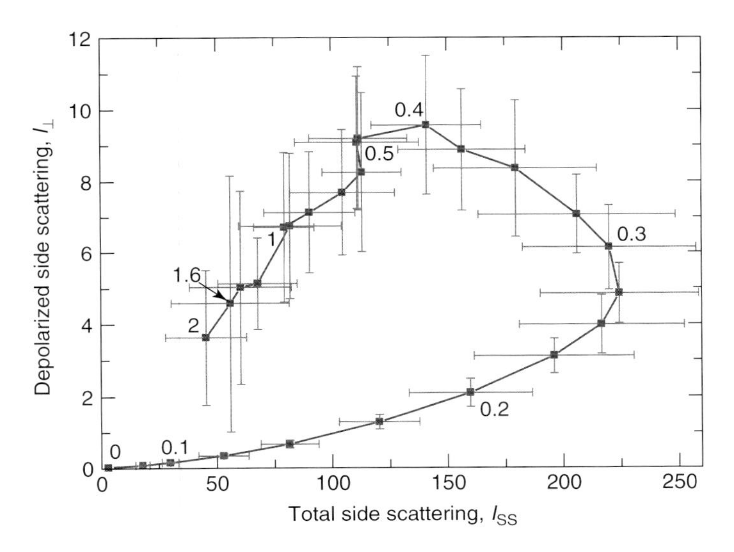

Figure 4.4 Depolarized versus total side-scattering intensities for several granule diameters from 0 to 2 μm, indicated for some points by labels, for the default set of parameters.

4.3.2.2 Approximate Theories

In addition to rigorous DDA simulations, Yurkin *et al.* [12] derived expressions for side-scattering signals based on approximate theories. The Rayleigh–Debye–Gans (RDG, [26]) approximation was used to calculate I_{SS}, and a two-order Born approximation (2-Born, e.g., [30]) was developed for $I_{\perp}$. Moreover, additional approximations were assumed during the derivation to obtain analytical expressions for values of I_{SS} and $I_{\perp}$, averaged over different granule placements. Accuracy of the theoretical expressions was assessed by comparison with DDA results for the default parameters of the granulated sphere model (Section 4.3.2.1). RDG results were calculated for three volume fractions $f = 0.02$, 0.05, and 0.1, while 2-Born was calculated only for the default one ($f = 0.1$). Yurkin *et al.* [12] concluded that the RDG is an accurate approximation for small f, especially for small x_g. However, it systematically underestimates I_{SS} for larger f owing to the ignored multiple scattering effects. The agreement between the 2-Born and the DDA is good, especially up to the first maximum. For larger x_g, the 2-Born approximation systematically underestimates the depolarized intensity, which is due to neglecting higher order scattering contributions and the fact that even a single large granule is treated inaccurately in the framework of the 2-Born approximation.

Although approximate theories are much faster than the DDA, their main added value consists in physical insight into the light-scattering problem, and the opportunity for an approximate solution for the inverse light-scattering problem. For instance, Yurkin *et al.* [12] derived the following scaling laws:

$$\langle I_{SS}\rangle \sim x_c^3 f \left|\frac{m_g - m_c}{m_0}\right|^2 \times \begin{cases} [1 + O(f)]x_g^3, x_g \ll 1 \\ O\left(x_g^{-1}\right), x_g \gg 1 \end{cases} \tag{4.1}$$

$$\langle I_{\perp}\rangle \sim x_c^3 f^2 \left|\frac{m_g - m_c}{m_0}\right|^4 \times \begin{cases} x_g^3[1 + O(f) + O(x_c x_g^3)], x_g \ll 1 \\ x_c O\left(x_g^{-2}\ln x_g\right), x_g \gg 1 \end{cases} \tag{4.2}$$

$$D_{SS} \sim f \left|\frac{m_g - m_c}{m_0}\right|^2 \times \begin{cases} 1 + O(f) + O\left(x_c x_g^3\right), x_g \ll 1 \\ x_c O\left(x_g^{-1}\ln x_g\right), x_g \gg 1 \end{cases} \tag{4.3}$$

where x_g and x_c are size parameters of granules and the whole cell, respectively. DDA simulations varying all model parameters (Section 4.3.2.1) showed that these scaling laws are more accurate, especially for D_{SS}, than approximate theories themselves.

4.3.2.3 Neutrophil Model with Nucleus

A limited set of light-scattering simulations for a neutrophil model with nucleus (Figure 4.3b) was performed by Orlova *et al.* [13], using ADDA v. 0.76 [36]. Geometrical parameters of the model are described in Section 4.2.3; a refractive index 1.357 was used for the cytoplasm, 1.458 for both the nuclei and the granules, and 1.337 for the outer medium. LSPs were evaluated in a form that corresponds to

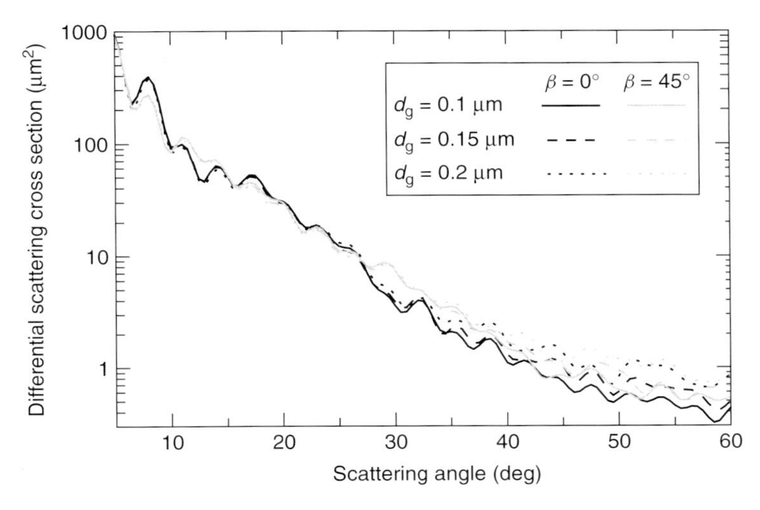

Figure 4.5 The LSPs calculated from DDA for the optical model of a neutrophil.

Eq. 4.17. Additional to varying d_g, two different orientations of the model with respect to the incident radiation were used (different by 45° rotation over Euler angle β). Six simulated LSPs are presented in Figure 4.5. They have different intensities for different granule sizes, but the differences are significant only for $\theta > 30°$, which agree with the results obtained by Yurkin *et al.* [12]. The overall adequacy of this optical model was proven by comparing it with experimental LSPs (Section 4.4.5).

4.3.3
Inverse Light-Scattering Problem for Mononuclear Cells

Unfortunately, the inverse light-scattering problem is much harder than the direct light-scattering problem. Previous studies, reviewed by Maltsev and Semyanov [2], concentrated on homogeneous spheres, prolate spheroids in a fixed orientation, and coated spheres (determining only the diameters). Recent advances include a spectral sizing method [37] – an empirical technique, whose applicability was demonstrated by extensive simulations for spheres [37], two-layer spheres [2], and erythrocytes [38]. This technique was applied to lymphocytes, monocytes, and granulocytes by Semyanov *et al.* [6] and produced meaningful results. Although the spectral method allows determination of refractive index of homogeneous spheres [37], it is at least hard to extend this property to a coated sphere and other models.

Zharinov *et al.* [7] proposed a solution for the inverse light-scattering problem for a five-layer spherical model of mononuclear blood cells, on the basis of an optimization procedure using multistart Levenberg–Marquardt algorithm. However, it has a few drawbacks. First, as the model has 10 free parameters (five radii and five refractive indices for the corresponding layers), the search space is quite complex and it is hard to guarantee finding the global minimum. Although it was empirically proven that as many as 2500 restarts of the optimization procedure resulted in a reliable global minimum, such testing of robustness consumes a lot of computational time (about 1 h/cell). Second, and more importantly, that method did not provide (statistically reliable) errors of estimates of characteristics of individual cells, which is crucial to assess the overall adequacy of the fitting procedure.

Strokotov *et al.* [8] developed a solution for the inverse-light-scattering problem for mononuclear cells, modeled as a coated sphere. It is based on a global optimization technique and a new advanced statistical method to estimate errors of the deduced cellular characteristics. They also proposed a method to characterize the sample on the basis of individual measurements with varying errors. Below, we describe this approach in more detail.

4.3.3.1 Global Optimization

The solution for the inverse problem is performed by a least-square method, that is, by minimizing the weighted sum of squares:

$$S(\boldsymbol{\beta}) = \sum_{i=1}^{N} z_i^2, z_i = w(\theta_i)(I_{\text{th}}(\theta_i, \boldsymbol{\beta}) - I_{\text{exp}}(\theta_i)) \tag{4.4}$$

where $\boldsymbol{\beta}$ is a vector of p model parameters, I_{th} and I_{exp} are theoretical and experimental LSP, respectively, $N = 160$ (in the range of θ from 12 to 50°), and $p = 4$ (described below). $w(\theta)$ is a weighting function given by

$$w(\theta) = \frac{1^\circ}{\theta} \exp\left(-2\ln^2(\theta/54^\circ)\right) \tag{4.5}$$

which goal is to reduce an effect of the noise on the fitting result [8].

To perform global minimization, DIRECT algorithm [39] is used, which performs an extensive search over the parameter space, described by a four-dimensional rectangular parallelepiped $\mathbf{B}$, requiring a lot of computational time (a few minutes per lymphocyte). However, in addition to finding the global minimum of $S(\boldsymbol{\beta})$, that is, the best estimate $\boldsymbol{\beta}_0$, it provides an approximate description of the whole least-square surface, which we further use to estimate the parameter errors.

The following cellular characteristics are considered: cell diameter D_c, ratio of nucleus and cell diameters D_n/D_c, refractive indices of cytoplasm m_c, and nucleus m_n. For a particular case of human blood lymphocytes, the region $\mathbf{B}$ is defined by the following boundary values: $D_c \in [4.5, 10]$ μm, $D_n/D_c \in [0.7, 0.95]$, $m_c \in [1.34, 1.41]$, $m_n \in [1.41, 1.58]$.

4.3.3.2 Errors of Parameter Estimates

Most of the theories on nonlinear regression and errors of parameter estimates (see, e.g., Seber and Wild [40] and Bates and Watts [41]) are worked out for the case of normally distributed and independent experimental errors. Unfortunately, there is significant dependence between residuals z_i in the case of lymphocytes, which can be characterized by a sample autocorrelation function ρ_k [40]. It is mainly caused by (i) model errors, that is, systematic difference between lymphocyte shape and a coated sphere, and (ii) imperfect alignment of SFC and noncentral position of the particle in the flow. Both types of these errors can be considered random for any particular lymphocyte.

Rigorous methods to deal with correlated residuals exist [40]; however, they are quite cumbersome. Strokotov *et al.* [8] proposed an approximate but much simpler approach, on the basis of the assumption that, for the purpose of statistical inference, N correlated residuals are equivalent to n independent residuals. The effective number of degrees of freedom n is determined by autocorrelation function

$$n = \frac{N^2}{N + 2\sum_{k=1}^{N-1} (N-k)\rho_k^2} \tag{4.6}$$

For lymphocytes that are studied, n is typically found to be from 10 to 25.

Next, a Bayesian inference method with standard noninformative (or homogeneous) prior $P(\sigma, \boldsymbol{\beta}) \sim \sigma^{-1}$ [40] is applied. The posterior probability density of $\boldsymbol{\beta}$ given a specific experimental LSP is then $P(\boldsymbol{\beta}|I_{\text{exp}}) = \kappa[S(\boldsymbol{\beta})]^{-n/2}$, where κ is a normalization constant obtained as

$$\kappa = \left(\int_{\mathbf{B}} [S(\boldsymbol{\beta})]^{-n/2} \, \mathrm{d}\boldsymbol{\beta} \right)^{-1} \tag{4.7}$$

To calculate integrals involving $[S(\boldsymbol{\beta})]^{-n/2}$ over $\mathbf{B}$, one can use the output of the DIRECT algorithm – a partition of $\mathbf{B}$ into $M(\sim 10^4)$ parts with volumes V_i and centers $\boldsymbol{\beta}_i$; with values $S_i = S(\boldsymbol{\beta}_i)$ also known. This is used to calculate an average of any quantity $f(\boldsymbol{\beta})$:

$$\langle f(\boldsymbol{\beta}) \rangle = \kappa \int_{\mathbf{B}} f(\boldsymbol{\beta}) \left[S(\boldsymbol{\beta}) \right]^{-n/2} \mathrm{d}\boldsymbol{\beta} = \kappa \sum_{i=1}^{M} f(\boldsymbol{\beta}_i) S_i^{-n/2} V_i, \kappa = \left(\sum_{i=1}^{M} S_i^{-n/2} V_i \right)^{-1} \tag{4.8}$$

Knowing the complete probability distribution of model parameters for any measured particle, one can infer any statistical characteristics of this distribution, for example, the expectation value $\boldsymbol{\mu} = <\boldsymbol{\beta}>$ and the covariance matrix $\mathbf{C} =<(\boldsymbol{\beta} - \boldsymbol{\mu})(\boldsymbol{\beta} - \boldsymbol{\mu})^{\mathrm{T}}>$. Other quantities of interest are the highest posterior density (HPD) confidence regions, defined as $\mathbf{B}_{\text{HPD}}(P_0) = \{\boldsymbol{\beta} | P(\boldsymbol{\beta}|I_{\text{exp}}) > P_0\}$ [40]. Its confidence level is

$$\alpha = \kappa \int_{\mathbf{B}_{\text{HPD}}(P_0)} \left[S(\boldsymbol{\beta}) \right]^{-n/2} \mathrm{d}\boldsymbol{\beta} \tag{4.9}$$

These confidence regions are used to assess the influence of parameter space boundaries on the results. Although $\mathbf{B}$ is quite large, its boundaries cut off tails of the distribution, causing bias in the determination of both $\boldsymbol{\mu}$ and $\mathbf{C}$. The bias increases with relative weight of the cut tail, so a confidence level α of the particle (more precisely, of associated probability distribution over $\boldsymbol{\beta}$) is defined as the confidence level of the HPD region touching the boundary of **B**. The main purpose of α is to assess biases of $\boldsymbol{\mu}$ and **C**. Relative goodness of fit is better assessed by the matrix **C**.

4.3.3.3 Theoretical Tests Based on More Complicated Model

Since real lymphocytes can only approximately be described by a coated sphere, a characterization method can be tested by a virtual experiment using a more complicated model. First, one can shift the nucleus inside the coated sphere model. However, Strokotov *et al.* [8] reported that, when nucleus is relatively large ($D_n/D_c = 0.9$), the nucleus shift showed no considerable effect on the structure of the LSP, except for scattering angles larger than 40°. Such LSPs were successfully processed with the global optimization, resulting in a confidence level above 0.95 and an accurate estimation of the model parameters.

Second, the model including nucleus inhomogeneity (Figure 4.6) was tested. The following parameters for the initial coated sphere model were used: $D_c = 6.38\ \mu\text{m}$, $D_n/D_c = 0.903$, $m_c = 1.3765$, average $m_n = 1.4476$. Dividing the nucleus into two domains, 1.3765 and 1.457 were chosen as the refractive indices for the inner spheroid and the rest of the nucleus, respectively. This was done to maximize the contrast and make the average (effective medium) refractive index of the nucleus equal to the one specified above. The spheroid had axis ratios 1 : 2 : 2, its volume was 0.108 of the whole nuclear volume, its center coincided with the center of the cell, the orientation of the symmetry axis with respect to the light incidence direction θ_0 was from 0 to 90° with a step of 30°. Results of processing the simulated LSPs with the global optimization are shown in Figure 4.7. The important conclusion is that the global optimization performed reliably on these distorted LSPs, that is, the real values of the initial model were well inside the confidence intervals (although the latter may be large).

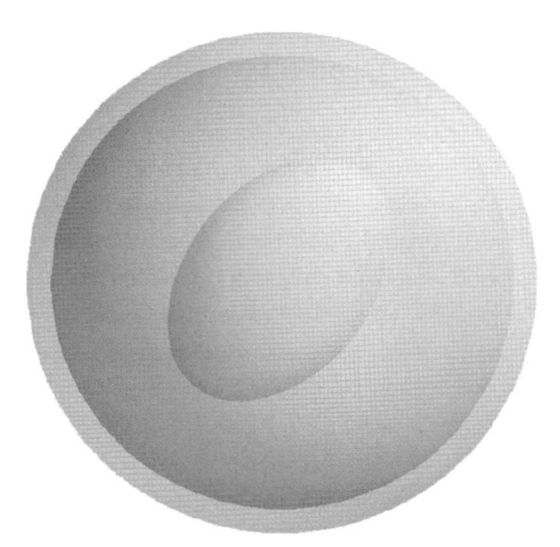

Figure 4.6 Optical model of a mononuclear cell with an inhomogeneous core.

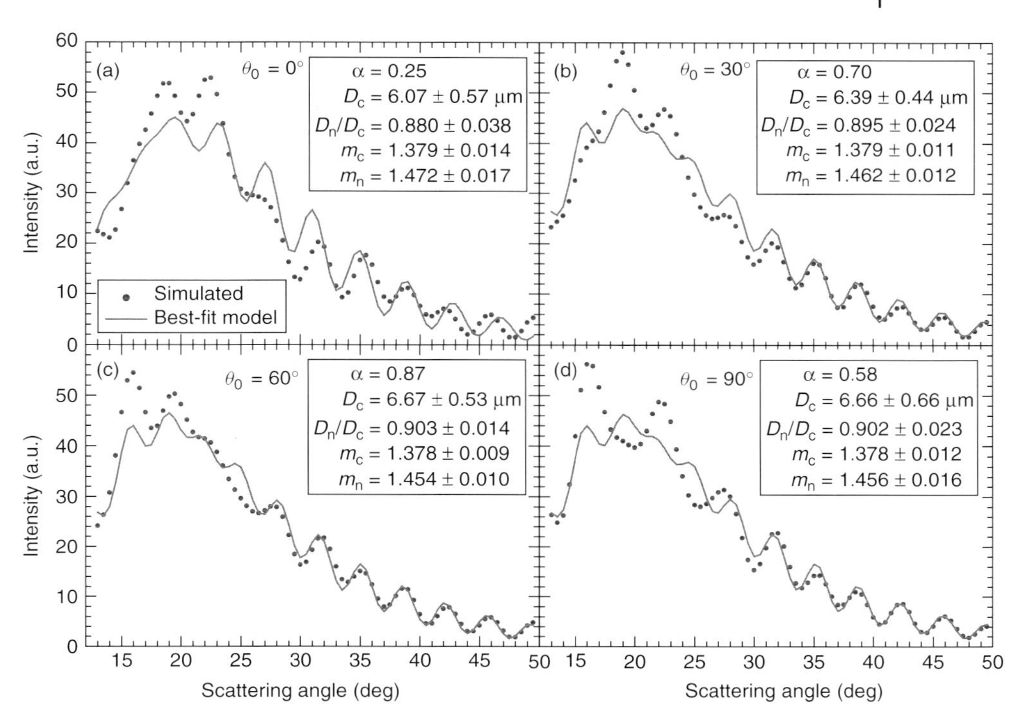

Figure 4.7 The LSPs calculated from DDA for the optical model of a lymphocyte with inhomogeneous nucleus.

4.3.3.4 Sample Characterization

Having determined the characteristics of each lymphocyte with certain errors (i.e., $\boldsymbol{\mu}_i$ and $\mathbf{C}_i$, $i = 1, \ldots, N_s$, N_s is the number of cells analyzed), Strokotov *et al.* [8] proposed the following method to estimate population mean and variances. General normal distributions are assumed for both the measurement results of individual particles and the true distribution of characteristics over a sample, the latter is described by a mean value **h** and covariance matrix **A**. An estimator proposed by Fuller [42] is used, whose first step is based on the mean of $\boldsymbol{\mu}_i$ and $\mathbf{C}_i$ and a sample covariance matrix based on $\boldsymbol{\mu}_i$:

$$\overline{\boldsymbol{\mu}} = \frac{1}{n}\sum_i \boldsymbol{\mu}_i, \overline{\mathbf{C}} = \frac{1}{n}\sum_i \mathbf{C}_i, \overline{\mathbf{M}} = \frac{1}{n-1}\sum_i (\boldsymbol{\mu}_i - \overline{\boldsymbol{\mu}})(\boldsymbol{\mu}_i - \overline{\boldsymbol{\mu}})^{\mathrm{T}} \tag{4.10}$$

$\overline{\mathbf{M}} - \overline{\mathbf{C}}$ is an unbiased estimator of **A**, but it should be modified to ensure positive semidefiniteness: $\mathbf{A}^{(0)} = \mathbf{F}(\overline{\mathbf{M}}, \overline{\mathbf{C}})$, where the transformation **F** is defined as

$$\mathbf{F}(\mathbf{X}, \mathbf{Y}) = \mathbf{X}^{1/2}\mathbf{Q}\mathbf{D}'\mathbf{Q}^{\mathrm{T}}\mathbf{X}^{1/2}, \text{where } \mathbf{I} - \mathbf{X}^{-1/2}\mathbf{Y}\mathbf{X}^{-1/2} = \mathbf{Q}\mathbf{D}\mathbf{Q}^{\mathrm{T}}, \mathbf{D}' = \max(0, \mathbf{D}) \tag{4.11}$$

The second step is the following:

$$\mathbf{X}_i = \left(\frac{n-1}{n}\left(\mathbf{A}^{(0)} + \mathbf{C}_i\right) + \frac{1}{n}\overline{\mathbf{M}}\right)^{-1}, \mathbf{Z} = \left(\sum_i \mathbf{X}_i\right)^{-1}, \boldsymbol{\mu}_0 = \mathbf{Z}\sum_i \mathbf{X}_i\boldsymbol{\mu}_i$$
$$\mathbf{M}_0 = \mathbf{G}\left(\left\{\mathbf{X}_i, (\boldsymbol{\mu}_i - \boldsymbol{\mu}_0)(\boldsymbol{\mu}_i - \boldsymbol{\mu}_0)^{\mathrm{T}}\right\}\right), \mathbf{C}_0 = \mathbf{G}\left(\{\mathbf{X}_i, \mathbf{C}_i\}\right), \mathbf{A}^{(1)} = \mathbf{F}(\mathbf{M}_0, \mathbf{C}_0) \quad (4.12)$$

where $\mathbf{G}(\{\mathbf{X}_i, \mathbf{Y}_i\})$ is a solution of the matrix equation

$$\sum_i \mathbf{X}_i\left(\mathbf{Y}_i - \mathbf{G}\right)\mathbf{X}_i = 0 \quad (4.13)$$

Equation 4.12 is iterated to convergence. The resulting $\boldsymbol{\mu}_0$ is an estimate for **h**, whose covariance matrix (error of mean) is **Z** [42], and $\mathbf{A}^{(k)}$ is an estimate of **A**.

The advantage of this procedure is that measurements which show large errors become insignificant. Therefore, one can seamlessly include all measurements into the estimation of cell characteristics, even those that seem unreliable due to large sum of squares and/or SDs of the characteristics. However, doing so is not the best option because of the bias due to the region boundaries, which are especially important for particles with small confidence level α. Therefore, only particles with $\alpha \geq \alpha_0$ are considered for characterization of the samples. Strokotov *et al.* [8] showed that $\alpha_0 = 0.8$ is a logical compromise between large number of processed cells (required to decrease error of mean) and smallness of systematic bias. Moreover, the moderate variation of α_0 does not significantly change the results.

It is important to note that the proposed method allows one to estimate a full covariance matrix of the sample of individual lymphocytes as well, including correlations between different characteristics. In Section 4.4.4, we show and analyze only the diagonal elements of the matrices **A** and **Z**, that is, variances of the characteristics over the sample and errors of mean values. However, correlations are significant, for example, the correlation between D_{c} and both m_{c} and m_{n} are generally between −0.6 and −0.8, which is an interesting topic for future study.

We also emphasize that the method described in this section, which was originally developed for lymphocytes by Strokotov *et al.* [8], can be easily adapted to any biological cell modeled by a concentric coated sphere, including monocytes and stem cells.

4.4 Experimental Measurement of Light Scattering by White Blood Cells

4.4.1 Scanning Flow Cytometer

SFC is an instrument designed to measure the angular LSP of an individual particle [43–46]. The LSP is measured by guiding a particle or a cell through the measurement cuvette of the SFC and illuminating it with laser light. At present, the

SFC allows measurement of LSPs of individual cells with a rate of 500 particles/s within the measurable range of scattering angles ranging from 5 to 100°. The basic principle of the SFC was patented [47].

Combining the SFC with a solution of the inverse light-scattering problem potentially allows characterization of individual particles from light scattering with a typical rate of a few hundred particles per second. Characterization means determination of the physical particle characteristics such as size, shape, density, and so on. This is made possible by information contained in the measured LSP contrary to the ordinary flow cytometry, in which only the forward- and side-scattering signals are used.

4.4.1.1 The Current State of the Art of the SFC

The current state of the art of SFC technology relates to the instrumental platform of the universal analyzer for biology and medicine BioUniScan™ (CytoNova Ltd., Novosibirsk, Russian Federation). The core of the BioUniScan is formed by an SFC joining the optical and hydrodynamic parts *within* the optical scanning system. Individual particles are illuminated by laser sources, and fluorescence and elastic light scattering are collected by dedicated optics. The instrumental platform allows one to encircle the SFC with lasers and detectors, which provide an effective solution in analyzing individual particles. Automatic sampler, stepper-motor-controlled flow system, and a powerful electronic unit allow the effective control of the operation of the BioUniScan. The optical benches of the BioUniScan are flexible in operational wavelengths. The photograph of the BioUniScan system is shown in Figure 4.8.

The schematic layout of the BioUniScan optical system is presented in Figure 4.9. Two lasers are used to illuminate the particles within the cuvette. The coaxial laser 1 is used to generate the scattering pattern and the orthogonal laser 2 is used to excite the fluorescence, to control flow speed and as a reference point in time to angle transformation. The beam of laser 1 is directed coaxially with the stream by lens 1 through a hole in mirror 3. The hydrodynamic focusing head (flow cell in Figure 4.9) produces two concentric fluid streams: a sheath stream without particles and a probe stream that carries the analyzed particles. A differential

Figure 4.8 The instrumental platform of universal analyzer for biology and medicine "BioUniScan."

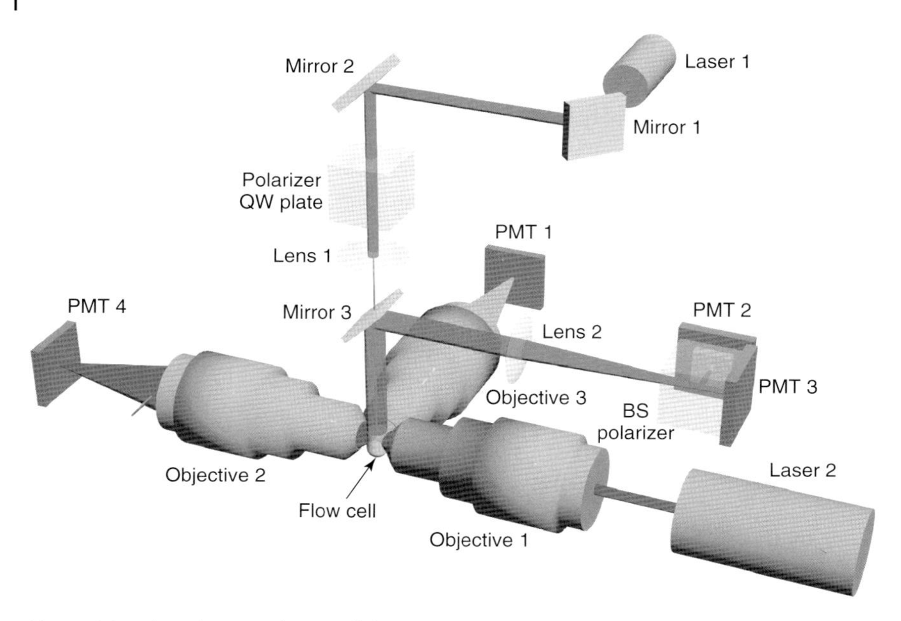

Figure 4.9 The schematic layout of the instrumental platform of universal analyzer for biology and medicine "BioUniScan."

pressure regulator is used for sample flow control. The fluidics system directs the streams into the capillary (diameter of 0.25 cm) of the flow cell (Figure 4.9). Finally fluidic system provides a probe stream with 12 μm diameter within the testing zone of the SFC. The hydrodynamically focused stream is fed into the cuvette at the spherical end through a cone where the hydrodynamic focusing head is connected by mechanical pressure. The flow channel is truncated in the flat end of the cuvette where it is connected to an orthogonal outlet. For this purpose, the cuvette is combined with a quartz plate. The 0.5 mm wide outlet is grooved at the flat end surface of the cuvette.

The LSP signal is obtained by measuring the signal intensity as a function of time by means of the light-scattering unit formed by mirror 3, lens 2, BS (beam splitter), polarizer, PMT 2, and PMT 3. Regular and polarized LSPs are measured by PMT 3 and PMT 2, respectively. The system is equipped with laser 2, which is orthogonal to the probe stream and provides the trigger signal by means of objective 2 and PMT 4. The latter, controlling the AD- converter, can also be taken from the fluorescence channel formed by objective 3 and PMT 1. In fact, the PMT 1 schematically represents the fluorescence unit of the BioUniScan formed by dichroic mirrors and band-pass filters requested. The ADC is operated continuously and the AD-conversion is interrupted for read out when a trigger signal is obtained.

4.4.1.2 Mueller Matrix of the SFC

The intensity and polarizing properties of a laser beam are described via a Stokes vector [48]. The transformation of the Stokes vector of incident light by interaction with a particle within the optical system can be described by a corresponding Mueller matrix. Hence, the complete information on the scattering from a particle can be presented in the form of a 4×4 Mueller matrix [26]

$$M = \begin{pmatrix} S_{11} & S_{12} & S_{13} & S_{14} \\ S_{21} & S_{22} & S_{23} & S_{24} \\ S_{31} & S_{32} & S_{33} & S_{34} \\ S_{41} & S_{42} & S_{43} & S_{44} \end{pmatrix} \tag{4.14}$$

In general, the S_{ij} depends on polar θ and azimuthal ϕ scattering angles, $S_{ij} = S_{ij}(\theta, \phi)$. Averaging over ϕ from 0 to 360° by the SFC optical cuvette results in the following scattering intensity for circularly polarized incident laser beam:

$$I_{\mathrm{r}} = \frac{1}{2\pi} \int_0^{2\pi} \left[S_{11}(\theta, \varphi) + S_{14}(\theta, \varphi)\right] \mathrm{d}\varphi \tag{4.15}$$

$$\begin{aligned} I_{\mathrm{p}} &= \frac{1}{2\pi} \int_0^{2\pi} \big[S_{11}(\theta, \varphi) + S_{14}(\theta, \varphi) + (S_{21}(\theta, \varphi) + S_{24}(\theta, \varphi)) \cos(2\varphi) \\ &\quad + (S_{31}(\theta, \varphi) + S_{34}(\theta, \varphi)) \sin(2\varphi)\big] \mathrm{d}\varphi \end{aligned} \tag{4.16}$$

for regular and polarized LSPs, respectively. For axisymmetric scatterers, Equation 4.15 becomes [49]

$$I(\theta) = \frac{1}{2\pi} \int_0^{2\pi} S_{11}(\theta, \varphi) \mathrm{d}\varphi \tag{4.17}$$

4.4.2 Differential Scattering Cross Section of White Blood Cells

The essential feature of the SFC is the ability to measure the absolute DSCS of single particles of any shape and structure. This feature is realized by measurement of a mixture of unknown particles and polymer microspheres [13]. The LSP of the polymer microsphere, measured with the SFC, gives a high-accuracy agreement with Mie theory [45]. This allows calibration in absolute light-scattering units. The DSCS was calculated from the following equation:

$$\frac{\mathrm{d}\sigma}{\mathrm{d}\theta} = I_{\mathrm{r}}(\theta) \left(\frac{\lambda}{2\pi m_0}\right)^2 \tag{4.18}$$

where I_{r} is the measured LSP (Equation 4.15), λ is the wavelength of the incident light, m_0 is the refractive index of the surrounding medium (1.337).

This method was applied to lymphocytes by Strokotov *et al.* [8] using $\lambda = 488$ nm. Fifty DSCSs of individual T lymphocytes and one polystyrene microsphere (diameter 1.8 μm) are presented in Figure 4.10. Variation in LSP structure and intensity among lymphocytes is discussed in Section 4.4.4. Orlova *et al.* [13] measured DCSC of neutrophils using $\lambda = 660$ nm. These results are presented in Figure 4.11, showing 25 LSPs of individual neutrophils and polystyrene microspheres (diameter 5 μm).

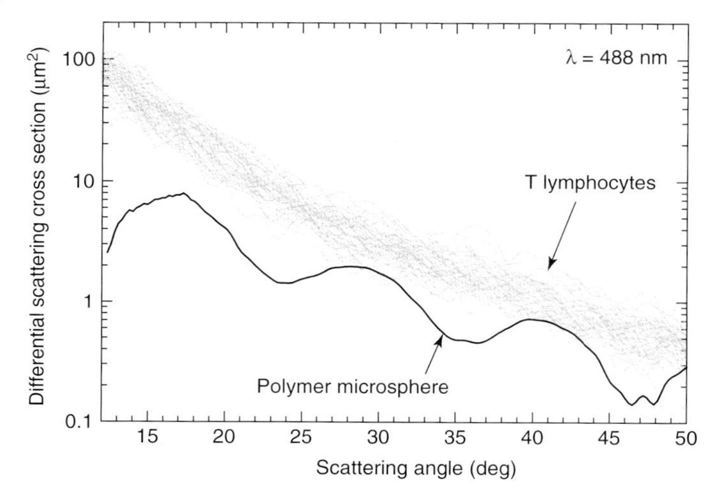

Figure 4.10 DSCS of T lymphocytes and a microsphere.

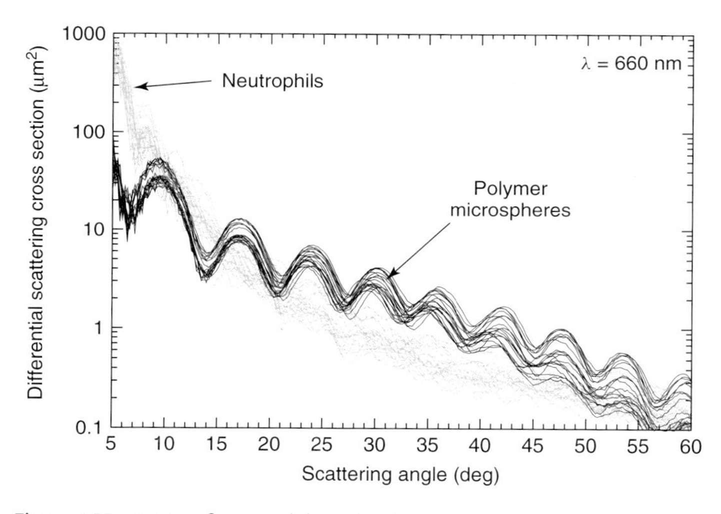

Figure 4.11 DSCS of neutrophils and polymer microspheres.

To directly compare lymphocytes and granulocytes, we have separated them from a single blood sample (Section 4.4.3) and measured their DSCS at the wavelength 488 nm. The results for representative number of individual cells as well as average of all lymphocytes and of all granulocytes in the sample are presented in Figure 4.12.

4.4.3
Measurement of Light Scattering of Mononuclear Cells

Ordinary flow cytometry allows one to measure the forward and side scattering from single cells, which can be used to discriminate lymphocytes, monocytes,

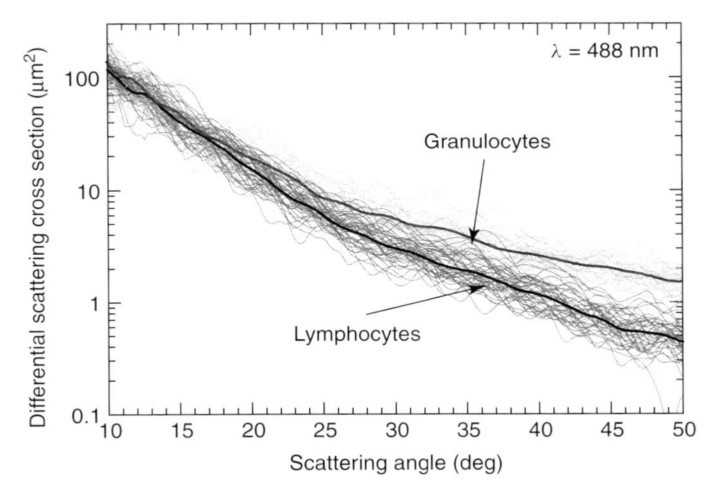

Figure 4.12 DSCS of lymphocytes and granulocytes. Bold lines are average DCSC over the whole sample for each cell type.

and granulocytes (Terstappen *et al.* [50]). Moreover, Terstappen *et al.* [51] reported the correlation between fraction of cells with large side scattering and fraction of cytotoxic lymphocytes, which can be potentially used to determine the latter fraction from light scattering only. Ding *et al.* [33] employed a goniometer system with a photoelastic modulation scheme to determine selected Mueller matrix elements of B cell hydrosol samples. The angular dependence of S_{11}, S_{12}, and S_{34} was determined from the scattered light signals between 10 and 160° at three wavelengths.

Angle-resolved light-scattering measurement of single mononuclear cells was pioneered by Doornbos *et al.* [52], who measured LSPs of optically trapped human lymphocytes in a range of scattering angles from 15 to 60°. Similar measurements for human monocyte in a full range of scattering angles were performed by Watson *et al.* [53]. LSPs of single cells were also measured by an SFC [6–8]. SFC can also be used in a two-parameter regime of ordinary flow cytometry to discriminate leukocytes subtypes [7]. Instead of side scattering, absolute values of the integrated whole light-scattering trace may be used. An example of such cytogram for a blood sample from a healthy donor is shown in Figure 4.13. Strokotov *et al.* [8] combined this discrimination with CD3 and CD19 immunofluorescence markers to identify T and B lymphocytes, respectively.

One possibility of looking at differences between T and B lymphocytes is by studying average LSPs. Averaged 100 experimental LSPs from the samples of T and B lymphocytes for a single donor are shown in Figure 4.14. There is difference in the intensity of scattering angles from 15 to 35°. However, this difference is much smaller than the natural variability of LSPs inside each sample (Figure 4.10). Strokotov *et al.* [8] showed that this difference can be partly explained

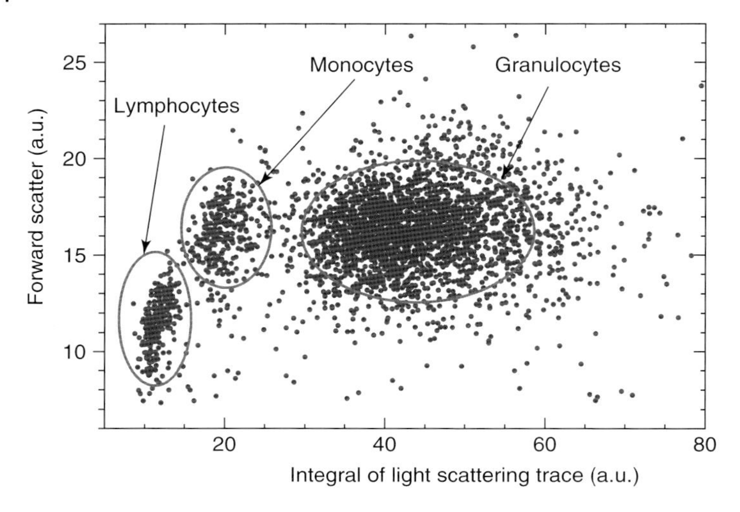

Figure 4.13 Light-scattering cytogram of leucocytes.

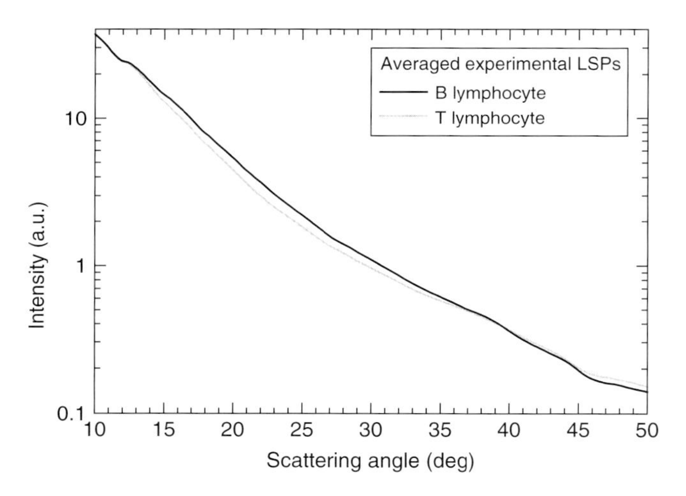

Figure 4.14 The experimental LSPs of T and B lymphocytes averaged over a sample.

by the difference in the mean diameter (Table 4.2); however, further study is required to clarify it.

4.4.4
Characterization of Mononuclear Cells from Light Scattering

Zharinov *et al.* [7] have analyzed a blood sample containing lymphocytes and monocytes using SFC and processed it with a five-layer spherical model. They have used multistart Levenberg–Marquardt algorithm (Section 4.3.3), which produced satisfactory results only for a small percentage of cells in a sample. The results

Table 4.2 Comparison of model characteristics of T and B lymphocytes of two donors.

	Donor 1		Donor 2	
	T	B	T	B
Sample size	225	146	330	86
$<D_c>$ (μm)	6.31	6.63	6.38	6.63
$\sigma(D_c)$(μm)	0.50	0.65	0.57	0.73
$\sigma(D_c)$(μm)	0.04	0.06	0.04	0.09
$<D_n/D_c>$	0.901	0.904	0.903	0.905
$\sigma(D_n/D_c)$	0.005	0.007	0.005	0.008
$\sigma(D_n/D_c)$	0.001	0.001	0.001	0.001
$<m_c>$	1.3759	1.3766	1.3765	1.3766
$\sigma(m_c)$	0.0026	0.0024	0.0026	0.0027
$\sigma(m_c)$	0.0005	0.0006	0.0004	0.0008
$<m_n>$	1.4479	1.4502	1.4476	1.4490
$\sigma(m_n)$	0.0080	0.0086	0.0080	0.0094
$\sigma(m_n)$	0.0007	0.0009	0.0006	0.0013

$<.>$, $\sigma(.)$, and $\sigma(<.>)$ denote estimated mean, standard deviation, and standard error of mean, respectively.

of this small fraction were used to characterize the whole sample. The following empirical correspondence between parameters of the five-layer model and studied cells was proposed: outer diameter of the third layer, that is the nucleus diameter; outer diameter of the fifth layer, that is, the cell diameter; refractive index of the fourth layer, that is, the cytoplasm refractive index; and the volumetric mean of refractive indices of first three layers, that is, the nucleus refractive index. Using this correspondence, it was shown that the obtained results fall within the range of literature data for lymphocytes and monocytes.

Strokotov *et al.* [8] measured leukocytes of seven donors identifying the T and B cells using the procedure described in Section 4.4.3. The global optimization algorithm described in Section 4.3.3 was applied to each cell. Results of this procedure for four typical cells from a sample of T cells of one donor are shown in Figure 4.15. Comparing this to Figure 4.7, one can see that the model errors for real lymphocytes are of the same magnitude as are the errors due to nuclear inhomogeneity with a volume fraction of only 10%. So, the latter may be responsible for a significant part of the former. However, more simulations are required to make any definite conclusions.

The characterization of the sample is performed as mentioned in Section 4.3.3.4. The ratio of LSPs that passed the confidence level threshold varied from 10% to 20% from sample to sample. On the basis of simulations described in Section 4.3.3.3, Strokotov *et al.* [8] suggested that using the threshold $\alpha_0 = 0.8$ leaves only cells with a relatively homogeneous nucleus (but possibly shifted from the center of the cell). These may be the cells that are in the G1-phase of a cell cycle [54].

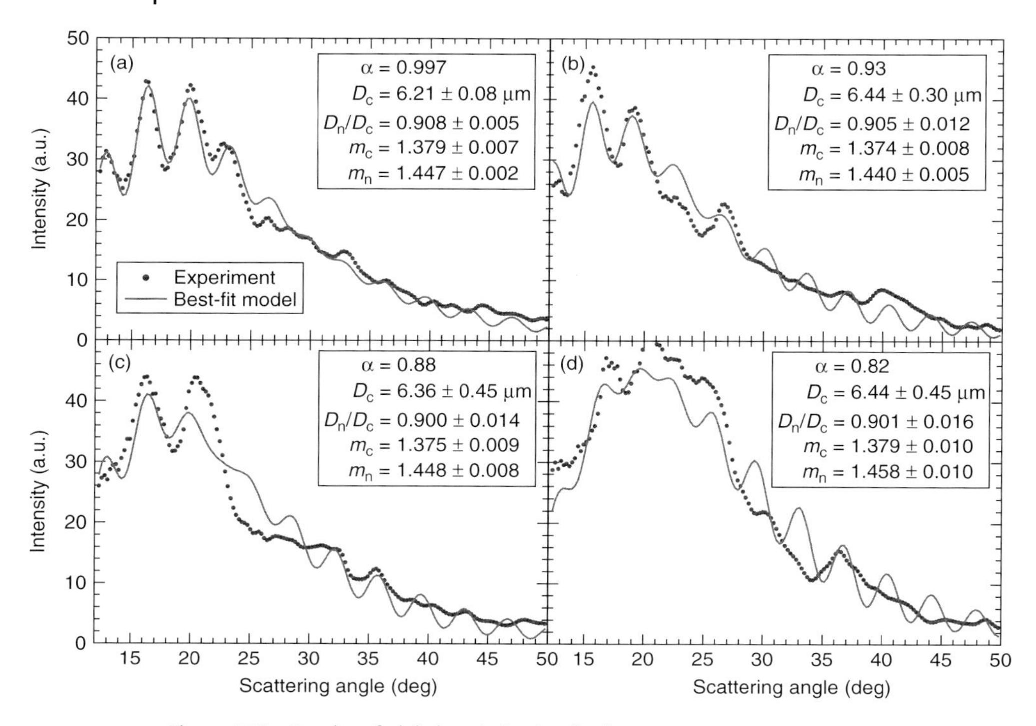

Figure 4.15 Results of global optimization for four experimental LSPs of T lymphocytes showing experimental and best-fit weighted LSPs. The graphs correspond to different reliability level of the fit α, decreasing from (a) to (d). Estimates of the cell characteristics (expectation value ± standard deviation) are also shown.

Unfortunately, the number of processed B lymphocytes that passed the threshold was statistically small for most of the samples except two. Results for the latter comparing B and T lymphocytes are shown in Table 4.2.

Analyzing literature data on cell and nucleus diameters obtained by the Coulter principle and electron and optical microscopy, Strokotov *et al.* [8] concluded that there is a disagreement between different methods, but current results fall within the broad range of existing data. However, there is an unusually small variation in the mean size of T cells and their nucleii among all seven donors (even accounting for error of mean). This fact disagrees with recent microscopic measurements [10, 17], which currently cannot be explained. A detailed study combining SFC-based and microscopic methods is required to resolve this issue.

Comparing the results of the global optimization for characterization of T and B lymphocytes (Table 4.2), one can see that the main difference in T and B cell morphology is the mean cell size, $<D_c>$. Although the difference in $<D_c>$ of about 5% is statistically significant, it is smaller than the estimated biological variability $\sigma(D_c)$ in each sample. Hence, at present, one cannot reliably classify a single lymphocyte as being T- or B-, using only its morphological characteristics that are determined from LSP. However, it may be possible to estimate the relative count of T to B lymphocytes in a sample, processing the distribution of all lymphocytes over morphological characteristics as a bimodal distribution (similar to Terstappen *et al.* [51]).

Strokotov *et al.* [8] also noted that B cells are generally better fitted by the coated sphere model than T cells, which may be caused by smaller deviation in the shape of B lymphocytes from the coated sphere; in particular, by a relatively smaller inhomogeneity of the nucleus.

4.4.5
Measurement of Light Scattering of Granular Cells

Light scattering from a suspension of granulocytes was employed in dynamical studies. Light-scattering intensity was used as a measure of neutrophil aggregation [55] and change in shape [56]. Neutrophil degranulation was shown to correlate with right-angle light-scattering intensity [57] and granulocyte aggregation – with light extinction [58]. Flow cytometrical studies of neutrophil biology were reviewed by Carulli [59].

In flow cytometers, granulocytes can be discriminated from other leukocytes by their higher forward and side scattering, except the basophils which are found in the light-scattering region of lymphocytes [60]. To perform further discrimination of granulocytes in order to classify them into subclasses, more measured parameters are needed. This can be achieved by using monoclonal antibodies to particular surface antigens or by measuring autofluorescence (reviewed by Semyanov *et al.* [6]). Depolarized side scattering, a parameter proposed by de Grooth *et al.* [35], can be used to discriminate eosinophils from neutrophils. This was actually employed in cytometrical protocol, applying higher value of this signal for eosinophils [50, 61]. Only recently, Yurkin *et al.* [12] showed that this experimental fact can be explained by the difference in granule sizes of neutrophils and eosinophils (Section 4.3.2.1).

LSPs of individual granulocytes were measured using an optical trap [53] and an SFC [6]. However, it was not specifically determined to which subtype of granulocytes the analyzed cells belonged.

Orlova *et al.* [13] used an SFC to measure LSPs of neutrophils that were identified using CD16b fluorescent markers. Samples from four donors were studied. Twenty-five randomly chosen experimental LSPs from the first sample are shown in Figure 4.16 together with the most similar LSP from theoretical DDA simulation. The simulation was performed for an optical model of the neutrophil (Figure 4.3b) with the following parameters: cell size of 9.6 μm, granule diameter of 0.2 μm with a volume fraction of 0.1, and nucleus volume fraction of 0.1. Experimental LSPs demonstrate a relatively large variability in their intensities and shapes, and theoretical LSP falls within the range of this variation, supporting the adequacy of the optical model. However, Orlova *et al.* [13] noted that simulated LSPs do not agree with experimental LSPs of other three samples as well, as they do with the first one. So, further theoretical study by varying the model parameters is required.

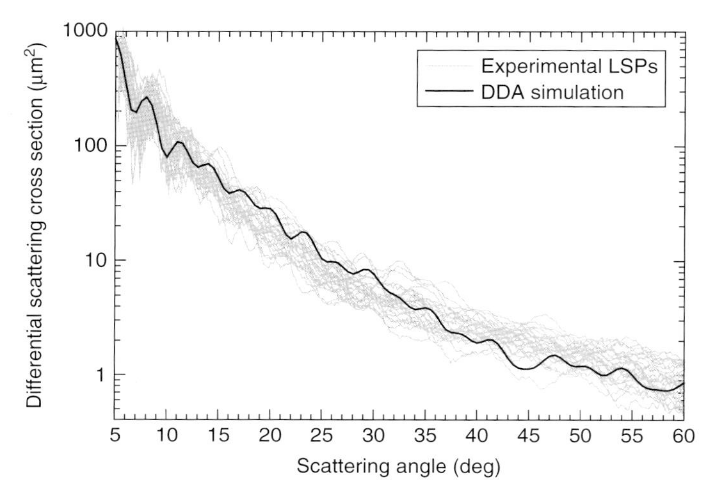

Figure 4.16 The experimental LSPs of single neutrophils (gray) and LSP calculated by DDA (black) from the optical model of the cell.

To illustrate the difference between LSPs of neutrophils from four donors, Figure 4.17 presents LSPs averaged over all neutrophils in each sample. These averaged LSPs are almost featureless except for a minimum and maximum between 7 and 10°. These extrema are present in all individual LSPs (Figure 4.16) as well as in the averaged LSP. The position of extrema in individual LSPs for small scattering angles is mostly determined by neutrophil diameter; that is, it can be described by diffraction. The simulated LSPs also have these two extrema at similar angles, which proves that more or less correct size was used for the neutrophil model.

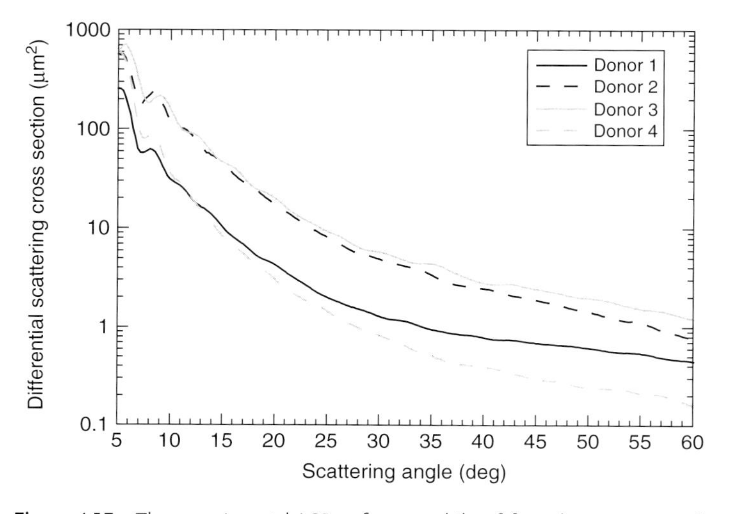

Figure 4.17 The experimental LSPs of neutrophils of four donors averaged over sample.

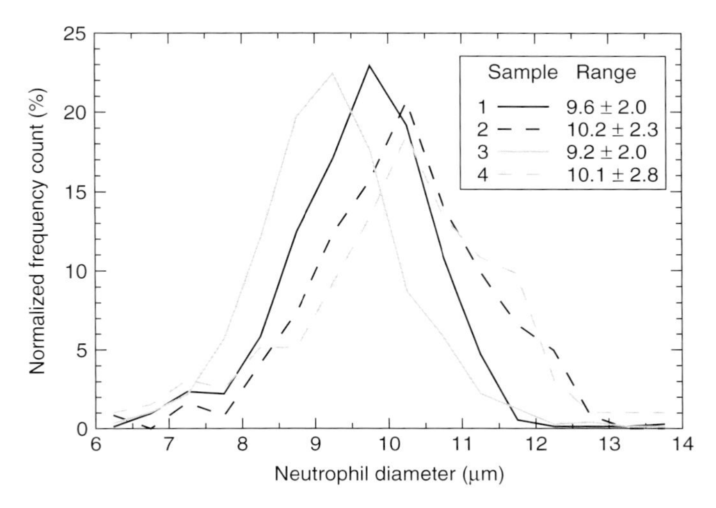

Figure 4.18 Size distributions of neutrophils from four samples determined by the spectral method.

The complete characterization of granulocytes, and neutrophils in particular, does not seem currently possible due to their complexity, as compared to the characterization of mononuclear cells (Section 4.4.4). That is why Orlova *et al.* [13] proposed to use averaged LSPs as a diagnostic parameter because these are essentially different between individuals; for example, they have different overall magnitudes and decay rates for larger scattering angles. These differences are caused by morphological differences in neutrophils; however, clinical study of both normal and abnormal samples is required to clarify the diagnostic value of the averaged LSP.

Experimental LSPs were also processed with a spectral sizing method (Section 4.3.3), although applicability of the latter for granulocytes has not been thoroughly studied. Size distributions of neutrophils from four samples are shown in Figure 4.18. On the basis of comparison with a recent microscopic study [17], Orlova *et al.* [13] concluded that spectral sizing method produces meaningful results for neutrophils (mean values within normal physiological ranges), but results for each individual cell may contain significant random errors leading to widening of distributions. Moreover, the difference in sample-averaged neutrophil sizes between the samples is less than 10%. This implies that significant changes in neutrophil internal structure (granularity, nucleus size, refractive indices, etc.) must be considered to explain the fivefold intersample variation of LSP magnitude (Figure 4.17).

4.5 Conclusion

In this chapter, we have summarized the state of the art of light-scattering studies of WBCs. The ultimate goal of these studies is to reliably measure light-scattering signals and use them to deduce morphological characteristics of the cells. The latter relies on a solution of inverse light-scattering problem based on an adequate optical model of a WBC. The complexity of this goal lies in the combination of different experimental and theoretical techniques; to name a few: confocal microscopy, DDA, SFC, and global optimization.

Recent progress in related techniques as well as a smart combination of them has led to a significant progress in the field of light scattering by WBCs. In particular, this chapter describes recent results for lymphocytes and neutrophils, obtained using SFC. A method to obtain detailed morphological characterization from the measured light-scattering pattern was developed for lymphocytes (as well as for other mononuclear cells), which allowed determination of size and refractive index of both the cell and the nucleus of individual cells. For neutrophils, a good agreement between experiments and theoretical simulations was obtained, which supports the overall adequacy of the proposed optical model.

New knowledge about optical properties of the WBCs should encourage researchers to develop new methods for enhanced characterization of blood cells with optical technologies, and to incorporate these methods into automatic cell analyzers. New cellular characteristics should provide medical doctors with additional indicators, which could play an important role in the analysis of abnormal cellular phenomenon, of patient pathologies.

Acknowledgments

We thank our current and former colleagues from the Laboratory of Cytometry and Biokinetics: Andrey Chernyshev, Konstantin Gilev, Vyacheslav Nekrasov, Konstantin Semyanov, and Dmitry Strokotov.

We also acknowledge the financial support from Siberian Branch of the Russian Academy of Sciences (integration grant N7-2009), the Presidium of the Russian Academy of Sciences (grant N4.5.10), the President of the Russian Federation programme for state support of the leading scientific schools (grant number: NSh-65387.2010.4.) and program of the Russian Government "Research and educational personnel of innovative Russia" (contracts P422, P2497, and P1039).

References

1. Greer, J.P., Foerster, J., and Lukens, J.N. (eds) (2003) *Wintrobe's Clinical Hematology*, 11th edn, Lippincott Williams & Wilkins Publishers, Baltimore.
2. Maltsev, V.P. and Semyanov, K.A. (2004) *Characterisation of Bio-Particles from Light Scattering*, VSP, Utrecht.
3. Brock, R.S., Hu, X., Weidner, D.A., Mourant, J.R., and Lu, J.Q. (2006) Effect of detailed cell structure on light scattering distribution: FDTD study of a B-cell with 3D structure constructed from confocal images. *J. Quant. Spectrosc. Radiat. Transfer*, **102**, 25–36.
4. Neukammer, J., Gohlke, C., Höpe, A., Wessel, T., and Rinneberg, H. (2003) Angular distribution of light scattered by single biological cells and oriented

particle agglomerates. *Appl. Opt.*, **42**, 6388–6397.

5. Yang, Y., Zhang, Z., Yang, X., Yeo, J.H., Jiang, L., and Jiang, D. (2004) Blood cell counting and classification by non-flowing laser light scattering method. *J. Biomed. Opt.*, **9**, 995–1001.
6. Semyanov, K.A., Zharinov, A.E., Tarasov, P.A., Yurkin, M.A., Skribunov, I.G., and van Bockstaele, D.R. *et al.* (2007) in *Optics of Biological Particles* (eds A.G. Hoekstra, V.P. Maltsev, and G. Videen), Springer, Dordrecht, pp. 269–280.
7. Zharinov, A.E., Tarasov, P.A., Shvalov, A.N., Semyanov, K.A., van Bockstaele, D.R., and Maltsev, V.P. (2006) A study of light scattering of mononuclear blood cells with scanning flow cytometry. *J. Quant. Spectrosc. Radiat. Transfer*, **102**, 121–128.
8. Strokotov, D.I., Yurkin, M.A., Gilev, K.V., van Bockstaele, D.R., Hoekstra, A.G., Rubtsov, N.B., and Maltsev, V.P. (2009) Is there a difference between T- and B- lymphocyte morphology? *J. Biomed. Opt.*, **14**, 064036–064012.
9. Hoekstra, A.G., Grimminck, M.D., and Sloot, P.M.A. (1998) Large scale simulation of elastic light scattering by a fast discrete dipole approximation. *Int. J. Mod. Phys. C*, **9**, 87–102.
10. Loiko, V.A., Ruban, G.I., Gritsai, O.A., Gruzdev, A.D., Kosmacheva, S.M., Goncharova, N.V., and Miskevich, A.A. (2006) Morphometric model of lymphocyte as applied to scanning flow cytometry. *J. Quant. Spectrosc. Radiat. Transfer*, **102**, 73–84.
11. Dunn, A.K. (2007) in *Optics of Biological Particles* (eds A.G. Hoekstra, V.P. Maltsev, and G. Videen.), Springer, Dordrecht, pp. 19–29.
12. Yurkin, M.A., Semyanov, K.A., Maltsev, V.P., and Hoekstra, A.G. (2007) Discrimination of granulocyte subtypes from light scattering: theoretical analysis using a granulated sphere model. *Opt. Express*, **15**, 16561–16580.
13. Orlova, D.Y., Yurkin, M.A., Hoekstra, A.G., and Maltsev, V.P. (2008) Light scattering by neutrophils: model, simulation, and experiment. *J. Biomed. Opt.*, **13**, 054057–054057.
14. Livesey, S.A., Buescher, E.S., Krannig, G.L., Harrison, D.S., Linner, J.G., and Chiovetti, R. (1989) Human neutrophil granule heterogeneity: immunolocalization studies using cryofixed, dried and embedded specimens. *Scanning Microsc. Suppl.*, **3**, 231–239.
15. Bainton, D.F. (1993) Neutrophilic leukocyte granules: from structure to function. *Adv. Exp. Med. Biol.*, **336**, 17–33.
16. Bjerrum, O.W. (1993) Human neutrophil structure and function with special reference to cytochrome b559 and beta 2-microglobulin. *Dan. Med. Bull.*, **40**, 163–189.
17. Ruban, G.I., Kosmacheva, S.M., Goncharova, N.V., Van Bockstaele, D., and Loiko, V.A. (2007) Investigation of morphometric parameters for granulocytes and lymphocytes as applied to a solution of direct and inverse light-scattering problems. *J. Biomed. Opt.*, **12**, 044017.
18. Brederoo, P., van der Meulen, J., and Mommaas-Kienhuis, A.M. (1983) Development of the granule population in neutrophil granulocytes from human bone marrow. *Cell Tiss. Res.*, **234**, 469–496.
19. Brunsting, A. and Mullaney, P.F. (1974) Differential light scattering from spherical mammalian cells. *Biophys. J.*, **14**, 439–453.
20. Metcalf, W.K., Metcalf, N.F., and Gould, R.N. (1978) Lymphocyte cytoplasmic refractive index (LCRI). *Antibiot. Chemother.*, **22**, 149–154.
21. Mourant, J.R., Canpolat, M., Brocker, C., Esponda-Ramos, O., Johnson, T.M., Matanock, A. *et al.* (2000) Light scattering from cells: the contribution of the nucleus and the effects of proliferative status. *J. Biomed. Opt.*, **5**, 131–137.
22. Barer, R. and Joseph, S. (1954) Refractometry of living cells: Part I. Basic principles. *Quart. J. Microsc. Sci.*, **95**, 399–423.
23. Dunn, A. and Richards-Kortum, R. (1996) Three-dimensional computation of light scattering from cells. *IEEE J. Select. Topics Quant. Electr.*, **2**, 898–905.
24. Dunn, A., Smithpeter, C., Welch, A.J., and Richards-Kortum, R. (1997)

Finite-difference time-domain simulation of light scattering from single cells. *J. Biomed. Opt.*, **2**, 262–266.
25. Chaumet, P.C. and Belkebir, K. (2009) Three-dimensional reconstruction from real data using a conjugate gradient-coupled dipole method. *Inv. Probl.*, **25**, 024003.
26. Bohren, C.F. and Huffman, D.R. (1983) *Absorption and Scattering of Light by Small Particles*, John Wiley & Sons, Inc., New York.
27. Yang, W. (2003) Improved recursive algorithm for light scattering by a multilayered sphere. *Appl. Opt.*, **42**, 1710–1720.
28. Mishchenko, M.I., Travis, L.D., and Lacis, A.A. (2002) *Scattering, Absorption, and Emission of Light by Small Particles*, Cambridge University Press, Cambridge.
29. Doicu, A., Wriedt, T., and Eremin, Y.A. (2006) *Light Scattering by Systems of Particles: Null-field Method with Discrete Sources: Theory and Programs*, Springer, Berlin.
30. Yurkin, M.A. and Hoekstra, A.G. (2007) The discrete dipole approximation: an overview and recent developments. *J. Quant. Spectrosc. Radiat. Transfer*, **106**, 558–589.
31. Taflove, A. and Hagness, S.C. (2005) *Advances in Computational Electrodynamics: The Finite-Difference Time-Domain Method*, 3rd edn, Artech House, Boston.
32. Ngo, D., Videen, G., and Chýlek, P. (1996) A FORTRAN code for the scattering of EM waves by a sphere with a nonconcentric spherical inclusion. *Comput. Phys. Commun.*, **99**, 94–112.
33. Ding, H., Lu, J.Q., Brock, R.S., McConnell, T.J., Ojeda, J.F., Jacobs, K.M., and Hu, X.H. (2007) Angle-resolved Mueller matrix study of light scattering by R-cells at three wavelengths of 442, 633, and 850 nm. *J. Biomed. Opt.*, **12**, 034032.
34. Yurkin, M.A., Hoekstra, A.G., Brock, R.S., and Lu, J.Q. (2007) Systematic comparison of the discrete dipole approximation and the finite difference time domain method for large dielectric scatterers. *Opt. Express*, **15**, 17902–17911.
35. de Grooth, B.G., Terstappen, L.W.M.M., Puppels, G.J., and Greve, J. (1987) Light-scattering polarization measurements as a new parameter in flow cytometry. *Cytometry*, **8**, 539–544.
36. Yurkin, M.A., Maltsev, V.P., and Hoekstra, A.G. (2007) The discrete dipole approximation for simulation of light scattering by particles much larger than the wavelength. *J. Quant. Spectrosc. Radiat. Transfer*, **106**, 546–557.
37. Semyanov, K.A., Tarasov, P.A., Zharinov, A.E., Chernyshev, A.V., Hoekstra, A.G., and Maltsev, V.P. (2004) Single-particle sizing from light scattering by spectral decomposition. *Appl. Opt.*, **43**, 5110–5115.
38. Yurkin, M.A. (2007) Discrete dipole simulations of light scattering by blood cells. PhD thesis. University of Amsterdam.
39. Jones, D.R., Perttunen, C.D., and Stuckman, B.E. (1993) Lipschitzian optimization without the Lipschitz constant. *J. Optim. Theor. Appl.*, **79**, 157–181.
40. Seber, G.A.F. and Wild, C.J. (2003) *Nonlinear Regression*, Wiley-Interscience, Hoboken, NJ.
41. Bates, D.M. and Watts, D.G. (1988) *Nonlinear Regression Analysis and its Applications*, John Wiley & Sons, Inc., New York.
42. Fuller, W.A. (1990) in *Statistical Analysis of Measurement Error Models and Applications* (eds P.J. Brown and W.A. Fuller), American Mathematical Society, Providence, Rhode Island, pp. 41–58.
43. Maltsev, V.P., Chernyshev, A.V., Semyanov, K.A., and Soini, E. (1996) Absolute real-time measurement of particle size distribution with the flying light-scattering indicatrix method. *Appl. Opt.*, **35**, 3275–3280.
44. Maltsev, V.P. and Lopatin, V.N. (1997) A parametric solution of the inverse light-scattering problem for individual spherical particles. *Appl. Opt.*, **36**, 6102–6108.
45. Soini, J.T., Chernyshev, A.V., Hanninen, P.E., Soini, E., and Maltsev, V.P. (1998) A new design of the flow cuvette and optical set-up for the scanning flow cytometer. *Cytometry*, **31**, 78–84.

46. Maltsev, V.P. (2000) Scanning flow cytometry for individual particle analysis. *Rev. Sci. Instrum.*, **71**, 243–255.
47. Maltsev, V.P. and Chernyshev, A.V. (1997) US Patent Number 5,650,847. Date of patent: July 22, 1997.
48. Collett, E. (1993) *Polarized Light: Fundamentals and Applications*, Marcel Dekker, New York.
49. Yurkin, M.A., Semyanov, K.A., Tarasov, P.A., Chernyshev, A.V., Hoekstra, A.G., and Maltsev, V.P. (2005) Experimental and theoretical study of light scattering by individual mature red blood cells by use of scanning flow cytometry and a discrete dipole approximation. *Appl. Opt.*, **44**, 5249–5256.
50. Terstappen, L.W.M.M., de Grooth, B.G., Visscher, K., van Kouterik, F.A., and Greve, J. (1988) Four-parameter white blood cell differential counting based on light scattering measurements. *Cytometry*, **9**, 39–43.
51. Terstappen, L.W., De Grooth, B.G., Ten Napel, C.H., Van Berkel, W., and Greve, J. (1986) Discrimination of human cytotoxic lymphocytes from regulatory and B-lymphocytes by orthogonal light scattering. *J. Immunol. Methods*, **95**, 211–216.
52. Doornbos, R.M.P., Schaeffer, M., Hoekstra, A.G., Sloot, P.M.A., de Grooth, B.G., and Greve, J. (1996) Elastic light-scattering measurements of single biological cells in an optical trap. *Appl. Opt.*, **35**, 729–734.
53. Watson, D., Hagen, N., Diver, J., Marchand, P., and Chachisvilis, M. (2004) Elastic light scattering from single cells: orientational dynamics in optical trap. *Biophys. J.*, **87**, 1298–1306.
54. Murray, A. and Hunt, T. (1993) *The Cell Cycle*, Oxford University Press, Oxford.
55. Niwa, M., Kanamori, Y., Kohno, K., Matsuno, H., Kozawa, O., and Kanamura, M. *et al.* (2000) Usefulness of grading of neutrophil aggregate size by laser-light scattering technique for characterizing stimulatory and inhibitory effects of agents on aggregation. *Life Sci.*, **67**, 1525–1534.
56. Ehrengruber, M.U., Deranleau, D.A., and Coates, T.D. (1996) Shape oscillations of human neutrophil leukocytes: characterization and relationship to cell motility. *J. Exp. Biol.*, **199**, 741–747.
57. Sklar, L.A., Oades, Z.G., and Finney, D.A. (1984) Neutrophil degranulation detected by right angle light scattering: spectroscopic methods suitable for simultaneous analyses of degranulation or shape change, elastase release, and cell aggregation. *J. Immunol.*, **133**, 1483–1487.
58. Yuli, I. and Snyderman, R. (1984) Light scattering by polymorphonuclear leukocytes stimulated to aggregate under various pharmacologic conditions. *Blood*, **64**, 649–655.
59. Carulli, G. (1996) Applications of flow cytometry in the study of human neutrophil biology and pathology. *Haemapathol. Mol. Hematol.*, **10**, 39–61.
60. Terstappen, L.W.M.M., Johnson, D., Mickaels, R.A., Chen, J., Olds, G., Hawkins, J.T., Loken, M.R., and Levin, J. (1991) Multidimensional flow cytometric blood cell differentiation without erythrocyte lysis. *Blood Cells*, **17**, 585–602.
61. Lavigne, S., Bosse, M., Boulet, L.P., and Laviolette, M. (1997) Identification and analysis of eosinophils by flow cytometry using the depolarized side scatter-saponin method. *Cytometry*, **29**, 197–203.

5
Optical Properties of Flowing Blood Cells

Martina C. Meinke, Moritz Friebel, and Jürgen Helfmann

5.1
Introduction

Detailed information on the light-scattering and absorption properties of human blood plays an important role in many diagnostic and therapeutic applications in laser medicine, hematology, and routine medical diagnostics. The optical properties of blood are required for a number of optical methods in order to calculate the light distribution in blood perfused tissues, for example, optical tomography, photodynamic therapy, and laser-induced thermotherapy. In clinical medicine, optical measurement methods have the advantage of not requiring sampling, that is, a lower risk of infection, and giving diagnostic results immediately. Applications are pulse oxymetry [1], blood perfusion measurements using Doppler laser technique [2] or online monitoring of relevant blood parameters during extracorporeal circulation or, for example, the components of blood [3, 4].

The optical behavior of human blood is more complex than that of single blood cells. Light is absorbed and scattered mainly by the discoid shaped red blood cells due to their highly concentrated hemoglobin content, which causes a higher real and imaginary part of the refractive index compared to the surrounding plasma [5]. The mean distance between red blood cells in blood of physiological concentration is of the order of their diameter. Therefore, interfering waves from neighboring cells cannot be neglected ("collective scattering" [6]). The distance between the cells varies strongly with the flow conditions. In addition, the absorption, shape, volume, orientation, and alignment of these cells changes with a number of other physiological parameters such as flow-induced shear forces, plasma osmolarity, cell density, or hemoglobin oxygen saturation. Even at a constant hematocrit (Hct), the blood parameters, such as red blood cell and hemoglobin concentration (Hbc), mean corpuscular volume (MCV), and mean corpuscular hemoglobin (MCH) content, show an individual variability that can influence the absorption and scattering properties. Because it is not possible to prepare these parameters for experiments, the biological variability always leads to an uncertainty in the determination of the optical parameters at a defined Hct. Furthermore, the optical

Advanced Optical Flow Cytometry: Methods and Disease Diagnoses, First Edition. Edited by Valery V. Tuchin.

behavior is influenced by the existence of other blood cells such as platelets (PLTs) or the widely variable optical properties of the blood plasma [7].

According to the transport theory, the optical properties of blood can be described by the intrinsic optical parameters such as absorption coefficient μ_a, scattering coefficient μ_s, and anisotropy factor g. The double integrating sphere measurement technique combined with inverse Monte-Carlo simulation (iMCS) has been established [8, 9] as a useful means of determining the optical parameters of undiluted blood. The iMCS calculates the photon trajectories on the basis of a given scattering phase function, which describes the statistical angle distribution of a scattering event. To describe the optical behavior of flowing blood, it is necessary to determine the appropriate effective phase function that allows the description of the real radiation distribution within the investigated medium [10].

Several goniometric measurements of single cells or highly diluted blood samples with a very thin sample thickness have been carried out to evaluate phase functions of the single red blood cells [11, 12]. It is clear that single scattering of individual red blood cells cannot be extrapolated for the scattering of blood [13] of physiological concentration. An effective phase function may depend on the Hct and wavelength especially at high absorption levels.

An optimized iMCS based on Roggan *et al.* [14], which simulates the exact geometry of the experimental set-up, including all kinds of radiation losses, was used to obtain information on the effective scattering phase function and to determine the optical parameters of blood depending on various physiological and biochemical parameters.

5.2 Blood Physiology

5.2.1 Blood Composition

The volume of human whole blood is made up of 60% by plasma and 40% by blood cells of which ~99% are red blood cells. The optical properties of whole blood in the visible and near infrared spectral region are determined by the red blood cells (erythrocytes). This is only, in part, due to the high concentration (4–5.5 million μl^{-1}) of these cells as their optical properties are even more significant than those of the blood contents. The red blood cells in humans have a biconcave shape and no nucleus (Figure 5.1). The diameter is 7–8 μm with a thickness of about 2 μm. The cell has a flexible but not very stretchable membrane (thickness 6–7 nm) and includes a liquid with a very high concentration of the blood dye, hemoglobin. Hemoglobin is a tetramer protein which is responsible for oxygen transport and partly, for that of carbon dioxide. It exhibits strong absorption in the visible range and gives the blood its color.

In addition to the absorption properties, the high concentration of hemoglobin leads to a high refractive index inside the cells compared to that of the surrounding

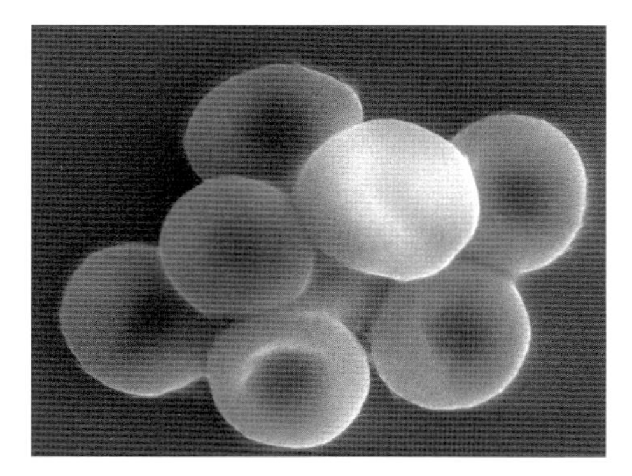

Figure 5.1 Scanning electron microscopy image of a red blood cell.

plasma. The erythrocyte cell membrane also exhibits a slightly higher refractive index than plasma [15]. The scattering effect by the membrane is almost negligible due to the very small thickness. However, photons that are irradiated into the blood medium are not only absorbed by the hemoglobin within the erythrocytes, but are also reflected and refracted by the erythrocyte, that is, considerable scattering occurs. Fresh blood has a glossy bright red appearance, which contrasts with the cherry red translucent appearance of the cell-free solution with the same hemoglobin content. This is because no scattering occurs in the cell-free solution, other than some weak Rayleigh scattering.

Leukocytes or the white (colorless) blood cells are spherical (diameter 6–20 μm), nucleated, hemoglobin-free cells, which serve the body's immune defense system. In a healthy person, the concentration ranges between 4000 and 10 000 μl^{-1}, which is a factor 10^3 lower than the red blood cell concentration. The leukocytes can be divided morphologically and functionally into three groups: granulocytes (60–70%, diameter 10–17 μm), monocytes (2–6%, diameter 12–20 μm), and lymphocytes (25–40%, diameter 7–9 μm). Infections and inflammatory processes can bring about a manifold rise in the leukocyte concentration.

Thrombocytes, also known as **platelets**, are cells without nuclei with a flat, irregular round shape. They are a component part of the blood clotting system. Thrombocytes are 1–4 μm long and 0.5–0.75 μm thick. Healthy adults have between 150 000 and 350 000 μl^{-1} thrombocytes, which corresponds to just 5% of the erythrocyte concentration.

Human blood **plasma** contains 900–910 g of water per liter, 65–80 g protein, and 20 g of substances with low molecular weight. Plasma has a specific density of 1.025–1.029 g cm^{-3} and the pH is neutral at 7.37–7.43. Albumin makes up the largest protein fraction and mainly functions as a transport vehicle. The low molecular weight substances mostly comprise electrolytes such as sodium, potassium, calcium and chlorides or bicarbonates. The viscosity of plasma is approximately double that of water.

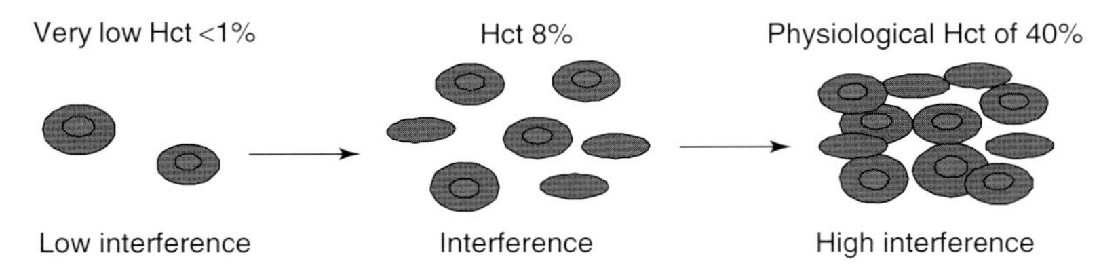

Figure 5.2 Illustration of different Hct.

5.2.2
Clinical Parameters

The **hematocrit** is defined as the proportion of the blood cells by volume in the total blood volume. Physiological Hct values range between 35 and 45%.

When considering the optical influence of the Hct, an increase in Hct would result in higher absorption and scattering, but the interference of the cells should also be kept in mind. At a physiological Hct the interference is high and single cell scattering cannot be ensured (Figure 5.2).

The **red blood cell concentration** (RBCc) indicates the concentration of the erythrocytes in blood. The value for adults is between 4 and $5.5 \times 10^6\,\text{mm}^{-3}$.

The MCV gives a value for the mean erythrocyte volume in cubic micrometers. The typical values are 80–90 μm^3. Changes in the plasma osmolarity, that is, osmotic pressure can effect a change in the MCV.

The MCH is the mean hemoglobin content of an erythrocyte. The normal value is 25–35 pg. This value may be either lower or higher for a given anemic condition.

The **mean corpuscular hemoglobin concentration** (MCHc) is derived from MCH and MCV and describes the mean Hbc within the erythrocytes. The normal value lies between 25 and $35\,\text{g dl}^{-1}$, which represents an exceptionally high protein concentration.

The Hbc describes the hemoglobin content related to the total blood volume and gives a clinical measure of the oxygen transport capacity. Normal values range between 14 and $16\,\text{g dl}^{-1}$.

The **white blood cell concentration** (WBCc) is the concentration of leukocytes in blood. The normal values lie between 4 and $10 \times 10^3\,\text{mm}^{-3}$. However, during infections, the values can rise upto $20 \times 10^3\,\text{mm}^{-3}$.

The **platelets concentration** (PLTc) is the concentration of thrombocytes in blood and a typical value is between 150 and $350 \times 10^3\,\text{mm}^{-3}$.

The relationship of the oxygenated hemoglobin (HbO_2) to the total hemoglobin (HbO_2 + Hb) is known as the **oxygen saturation ($SatO_2$)**

$$\text{SatO}_2 = \frac{[\text{HbO}_2]}{[\text{HbO}_2] + [\text{Hb}]} \times 100\,[\%] \tag{5.1}$$

During the uptake of oxygen, an oxygen molecule is attached to a hemoglobin molecule, whereby no chemical reaction occurs. This reaction is called *oxygenation* and the release of oxygen is called *deoxygenation*.

A number of effects (osmotic, mechanical, immunological, aging, etc.) can lead to the destruction of erythrocytes resulting in **hemolysis** and subsequent leakage of the hemoglobin into the plasma.

5.2.3
Physiological Conditions

In the living body, usually blood is flowing. One important parameter is the flow-induced shear force that can cause characteristic changes in cell morphology and organization of red blood cells.

The *position-dependent shear force* $\tau(x)$ can be described by the shear rate γ which is defined as the velocity gradient in the direction normal to the flow and independent of the blood viscosity η.

$$\tau(x) = \gamma \times \eta = -\frac{dv}{dx} \times \eta \qquad (5.2)$$

where $v(x)$ is the position-dependent flow velocity.

The influence of the flow or **shear rate** is illustrated in Figure 5.3.

The composition of the surrounding medium of the cells can disturb the iso-osmotic balance of 300 mosmol l^{-1}, as is illustrated in Figure 5.4. In a medium of low osmolarity (<300 mosmol), the cells start to swell by diffusion of water into the cell, leading to spherically formed cells, which become full to the point of bursting. In a hyperosmolar medium (>300 mosmol), cells begin to shrink due to water outflow, resulting in characteristically shaped cells called *spinocytes*. These variations in **osmolarity** not only change the shape and the volume of the cells but

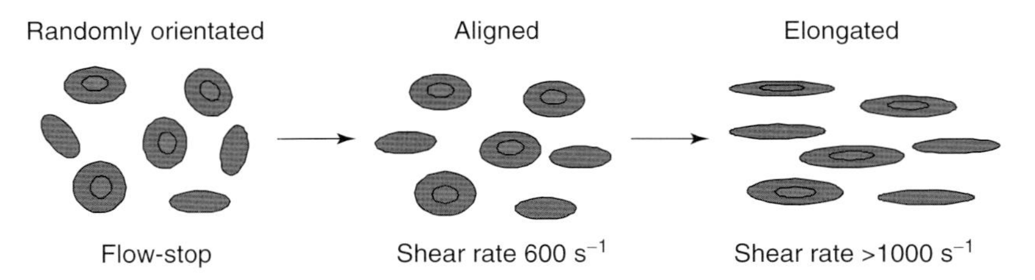

Figure 5.3 Illustration of shear rate influences on red blood cells in physiological saline solution.

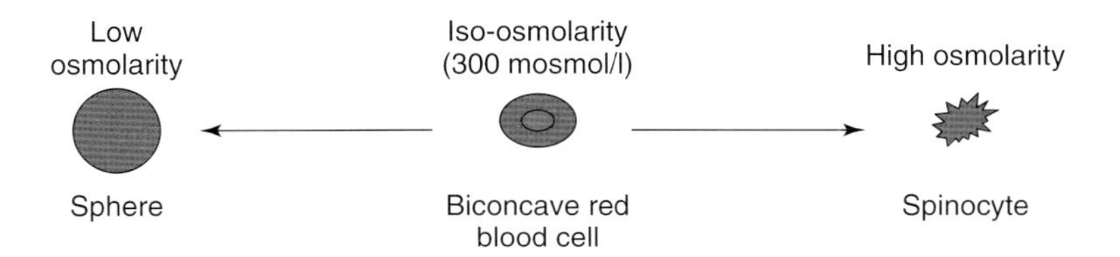

Figure 5.4 Osmolarity influences on red blood cells.

also the inner cell Hbc. Owing to an increase in the Hbc, the complex refractive index increases.

5.3
Complex Refractive Index of Hemoglobin

The optical behavior of blood is mainly determined by the optical properties of the red blood cells. These are determined by the complex refractive index $\tilde{n}$ of the cells which is different from that of the surrounding plasma:

$$\tilde{n}(\lambda) = n(\lambda) - \mathrm{i} \times k(\lambda) \tag{5.3}$$

where n is the real part of the refractive index and k the absorption constant

$$k = \mu_a \lambda / 4\pi \tag{5.4}$$

with absorption coefficient μ_a and the corresponding wavelength λ.

The complex refractive index is essential to obtain information about the scattering and absorption cross section of the single red blood cell [11, 16]. This is possible using methods based on classical electrodynamics such as Mie theory or finite difference time domain (FDTD) method. However, the complex refractive index of the red blood cell is also useful for the interpretation of the results obtained by Monte-Carlo simulations (MCSs) [8]. Therefore, exact knowledge about complex refractive index, dependent on Hb concentration and wavelength, is of particular interest.

The absorption coefficients of a Hb solution at a concentration of 2.67 g dl^{-1} obtained directly from human erythrocytes are shown in Figure 5.5 for oxygenated and deoxygenated hemoglobin. The Soret absorption band appears at 415 nm; other typical absorption maxima are at 274, 344, 542, 577, and 915 nm for oxygenated hemoglobin. At 1130 nm, the values approach those of the water absorption and above 1250 nm the absorption curves are equal. The absorption of deoxygenated hemoglobin shows the expected shift of the Soret peak, the fusion of the two maxima at 542 and 577 nm to one peak at 555 nm and an enhanced absorption in the red light. The absorption of the deoxygenated hemoglobin already approximates that of the water absorption at about 1000 nm and is identical above 1130 nm. The isosbestic point, where absorption is independent of the oxygenation level, is clearly visible at 805 nm.

There are a number of investigations on the absorption behavior of hemoglobin [5, 17, 18] from which the imaginary part of the refractive index can be calculated.

The real part of the refractive index n of Hb solutions of the concentration 0, 10.4, and 28.7 g dl^{-1} calculated from the Fresnel reflectance [5] is shown in Figure 5.6. The value for n increases with Hb concentration and decreases with increasing wavelength (normal dispersion). In the region of 415 nm, where k is maximum, n shows anomalous dispersion, that is, n increases with wavelength. For pure water, the determined value for n is within the error limits equal to the literature values for water [19].

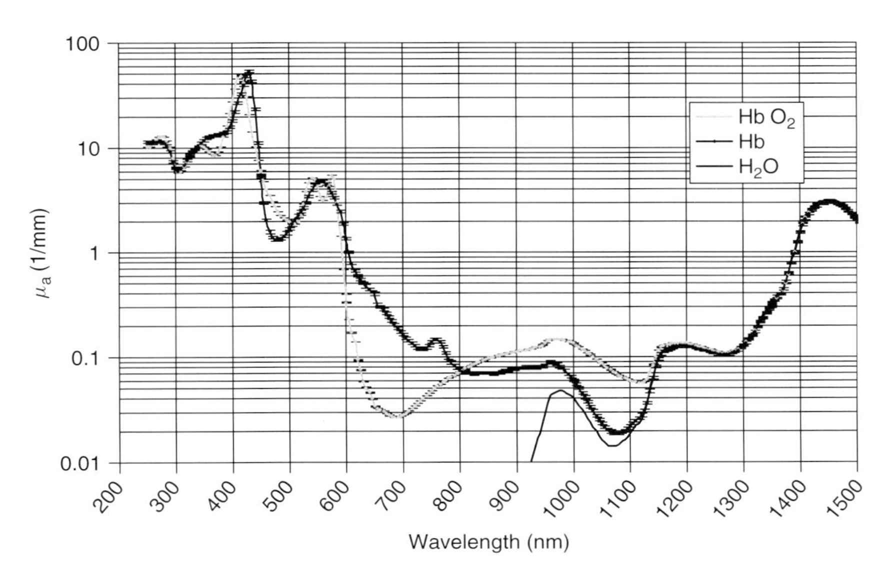

Figure 5.5 Absorption coefficient of oxygenated and deoxygenated hemoglobin solution at a concentration of 2.67 g/dl (measured with different sample thicknesses and calculated using Beer's law).

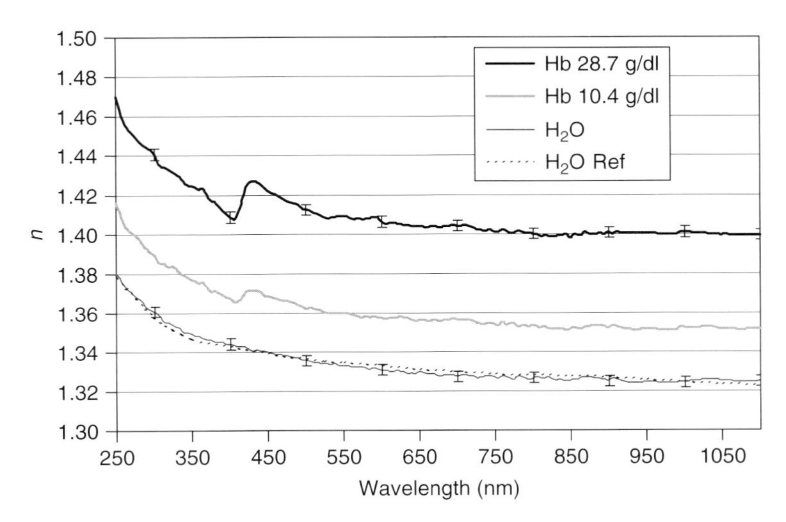

Figure 5.6 Real part of refractive index n for two Hb concentrations and H_2O calculated according to Friebel and Meinke [5] compared to pure water [19].

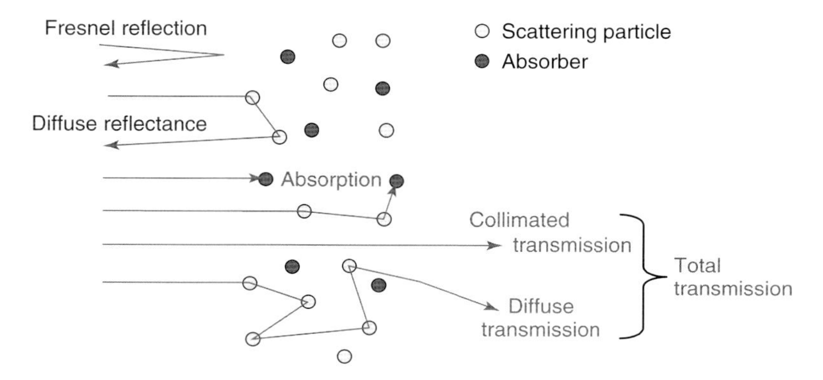

Figure 5.7 Interaction of light with turbid media.

5.4 Light Propagation in Turbid Media

The transmittance in turbid media does not follow the Beer-Lambert law as it does in transparent medium where no scattering appears. The calculation of the light distribution and propagation in turbid media, such as blood or tissue, is more complex.

Figure 5.7 illustrates that the light can be reflected on the interface of the sample (Fresnel reflection) or when entering the sample can be scattered backwards (diffuse reflectance) or forwards (diffuse transmission) or can be absorbed directly or after some scattering events. Ballistic photons are photons which have not been scattered or absorbed and can be measured as collimated transmission if the sample is transparent and/or thin enough.

The sum of collimated and the diffuse transmission results in the total transmission.

The distribution of light in turbid media can be described by the stationary radiation transfer, Equation 5.5. In the past, the equation was developed for the transport of neutrons in solids and liquids [20] and afterward adopted to the photon transport in turbid tissue (atmosphere, biological tissue) [21]

$$\frac{\mathrm{d}L(\vec{\mathbf{r}},\vec{\mathbf{s}})}{\mathrm{d}s} = -(\mu_a + \mu_s)L(\vec{\mathbf{r}},\vec{\mathbf{s}}) + \frac{\mu_s}{4\pi}\int_{4\pi} p(\vec{\mathbf{s}},\vec{\mathbf{s}}')L(\vec{\mathbf{r}},\vec{\mathbf{s}})\,\mathrm{d}\Omega' + S(\vec{\mathbf{r}},\vec{\mathbf{s}}) \tag{5.5}$$

with $L(\vec{\mathbf{r}},\vec{\mathbf{s}})$ (W cm^{-2} sr^{-1}) the radiation density, "radiance," at the position $\vec{\mathbf{r}}$ in direction $\vec{\mathbf{s}}$, $p(\vec{\mathbf{s}},\vec{\mathbf{s}}')$ the scattering phase function, μ_a (mm^{-1}) the absorption coefficient, and μ_s (mm^{-1}) the scattering coefficient. Ω (sr) is the solid angle and $S(\vec{\mathbf{r}},\vec{\mathbf{s}})$ (W cm^{-2} sr^{-1} mm^{-1}) describes the source term (fluorescence, light source).

It should be mentioned that the transport equation describes only the particle behavior of the photons; effects of the wave character, such as interference or polarization, are not considered.

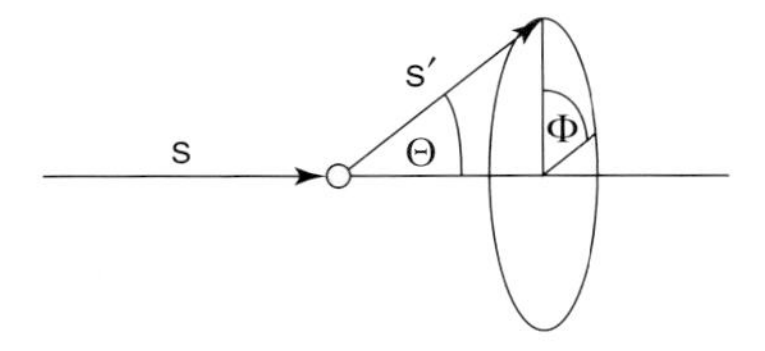

Figure 5.8 Angular definition of a scattering event.

The *absorption and the scattering coefficients* are defined as the product of absorption σ_a or scattering cross section σ_s and the concentration of absorber c_a or scattering particles c_s:

$$\mu_a = \sigma_a \ c_a \tag{5.6}$$

$$\mu_s = \sigma_s \ c_s \tag{5.7}$$

To describe the particular scattering event, the respective scattering phase function $p(\mathbf{s},\mathbf{s}')$ is relevant. It describes the probability that a photon with direction **s** is scattered in direction **s**′ [22, 23] (Figure 5.8):

The scattering phase function can be determined by goniometric measurements on very thin samples or single particles [24], but often it is necessary to adapt it to the real situation.

The Reynolds–McCormick (RM) phase function is often used for blood because it is adaptable by the fit of the variation factor α:

$$p_{GK}(\mathbf{s},\mathbf{s}') = \alpha\hat{g}\frac{1}{\pi}\frac{\left(1-\hat{g}^2\right)^{2\alpha}}{\left((1+\hat{g})^{2\alpha}-(1-\hat{g})^{2\alpha}\right)\left(1+\hat{g}^2-2\hat{g}\cos\Theta\right)^{(\alpha+1)}} \tag{5.8}$$

where $\alpha > -0.5$ and $|\hat{g}| \leq 1$ is a parameter from which the anisotropy factor g can be gained.

If α is 0.5, the RM phase function is equal to the Henyey–Greenstein function [25], which is often used for tissue.

For isotropic media, the scattering can be assumed to be independent of Φ. To describe such scattering processes, the anisotropic factor g is used which is the expectation value of the cosine of the scattering angle Θ:

$$\int_0^{2\pi} \mathrm{d}\Phi \int_0^{\pi} p(\vec{s},\vec{s}\,')\cos\Theta\,\sin\Theta\,\mathrm{d}\Theta = \langle\cos\Theta\rangle = g \tag{5.9}$$

Backward scattering is described by negative values up to -1, forward scattering by positive values up to $+1$, and isotropic scattering by an anisotropy factor of about 0 (Figure 5.9).

The effective scattering coefficient $\mu_s{}'$ can be calculated by

$$\mu_s' = \mu_s\,(1-g) \tag{5.10}$$

For complex problems, such as the light distribution in biological tissue, there is no analytical solution for the radiation transport equation [21]. Different approaches were developed such as the diffusion theory, and the Kubelka–Munk approximation

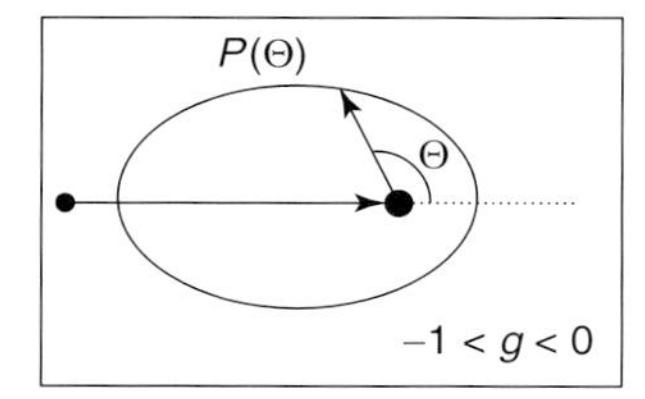

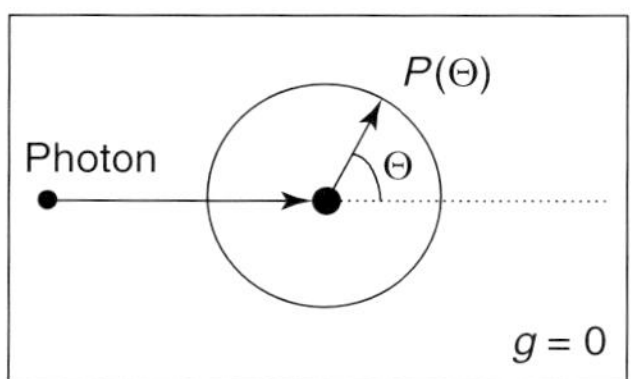

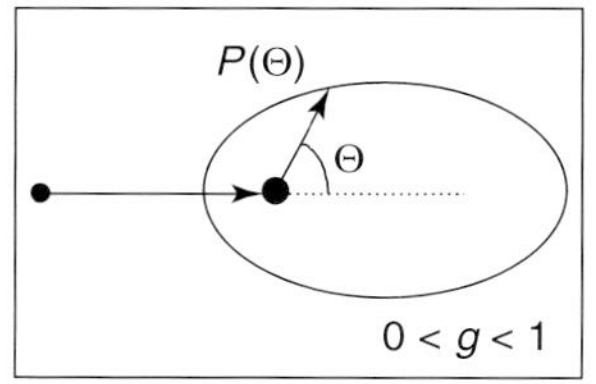

Figure 5.9 Anisotropy factor *g*.

[26]. However, these theories can only be used under special conditions such as isotropic scattering and weak absorption compared to the scattering [27].

The most flexible and exact method to calculate the light distribution in scattering and absorbing media with complex geometric and boundary conditions is the MCS, which can therefore be used for blood and blood compounds.

5.4.1 Monte Carlo Simulation

The MCS is an indirect method which uses the approach of the radiation transfer equation and the optical properties μ_a, μ_s, and the phase function to follow the light trajectories through the tissue. It is a numerical method based on the probability of the interaction of the medium and the light [28–30]. The principle of the MCS is a computer simulation which consecutively injects a number of photons into the medium. The medium can be characterized by probability density functions of the optical properties described by the parameter (μ_a, μ_s, $p(\mathbf{s},\mathbf{s}')$ or g) which are identical to those used in the radiation transport equation (Equation 5.5) [31].

The parameters of the simulation, such as the starting position of the photon, initial direction, free path length between the interactions (scattering, absorption), scattering angle, and the probability of reflection or transmission at an interface, were calculated using random numbers given by the probability density functions. The randomly generated trajectories of the individual photon are calculated until it is absorbed or leaves the sample (Figure 5.10).

5.5 Method for the Determination of Optical Properties of Turbid Media

As direct measurements of the optical parameters are not possible, a combination of integrating sphere measurements followed by iMCS has to be performed.

5.5.1 Integrating Sphere Measurements

Spectra of the diffuse reflectance R_d, the total transmission T_t, and the diffuse transmission T_d are obtained with an integrating sphere spectrometer (Lambda 900, Perkin Elmer, USA). The light source is made up of a deuterium lamp for

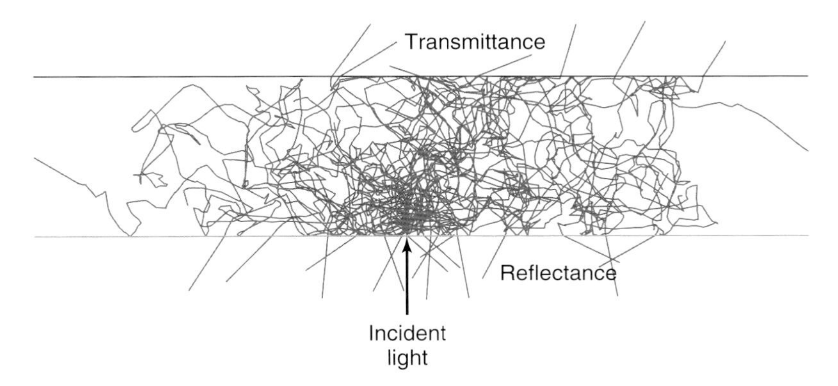

Figure 5.10 Light distribution in turbid medium calculated by Monte Carlo simulation.

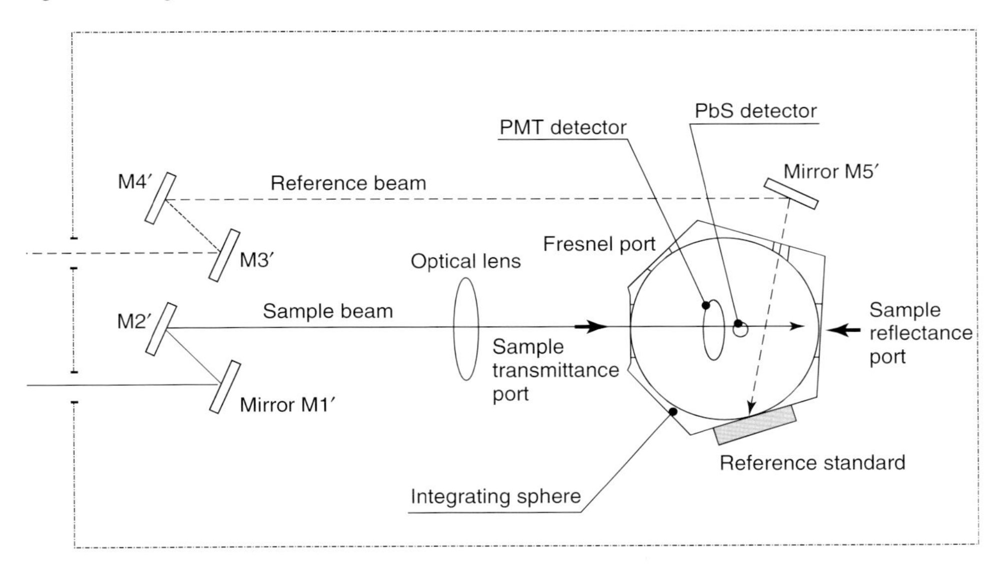

Figure 5.11 Schematic diagram of the integrating sphere spectrometer. The reference beam goes via mirror M3–M5 into the integrating sphere. The sample beam reaches the sample via mirror M1 and M2.

the UV-range and a tungsten halogen lamp for the visible/NIR range. Monochromatic light is generated with a double monochromator system and spectra can be measured sequentially in the wavelength range from 250 to 2500 nm. The monochromatic light is alternately split into a sample beam and a reference beam (Figure 5.11), which enters through a reference port of the sphere and hits a reference sample (Spectralon). This procedure allows one to correct for most of the light source fluctuations and sphere reflectivity inhomogeneities.

T_t is measured by placing the sample at the transmittance port. The reflectance port is closed with a Spectralon standard. T_d is measured by opening the reflectance port. The open reflectance port allows nonscattered light and scattered light within a cone with an angle of 5.3° to leave the sphere. The backscattered light R_d is measured by placing the sample at the reflectance port. The sample is inclined at an angle of 8° to the incoming light giving the specular reflex of the sample the possibility to either leave the sphere if the Fresnel port is open, or to be measured together with the diffuse reflectance.

With this arrangement, measurement accuracies of less than 0.01% can be obtained for the transmittance and reflectance. For comparison with MCSs, it is additionally necessary to know the precise position of the sample, the size of the illumination spot, and the collection geometry of the sample.

5.5.2 Principle of Inverse Monte Carlo Simulation

The iMCS uses forward MCSs iteratively to calculate the optical parameters μ_a, μ_s, and g on the basis of the measured values diffuse reflectance R_d, total transmission

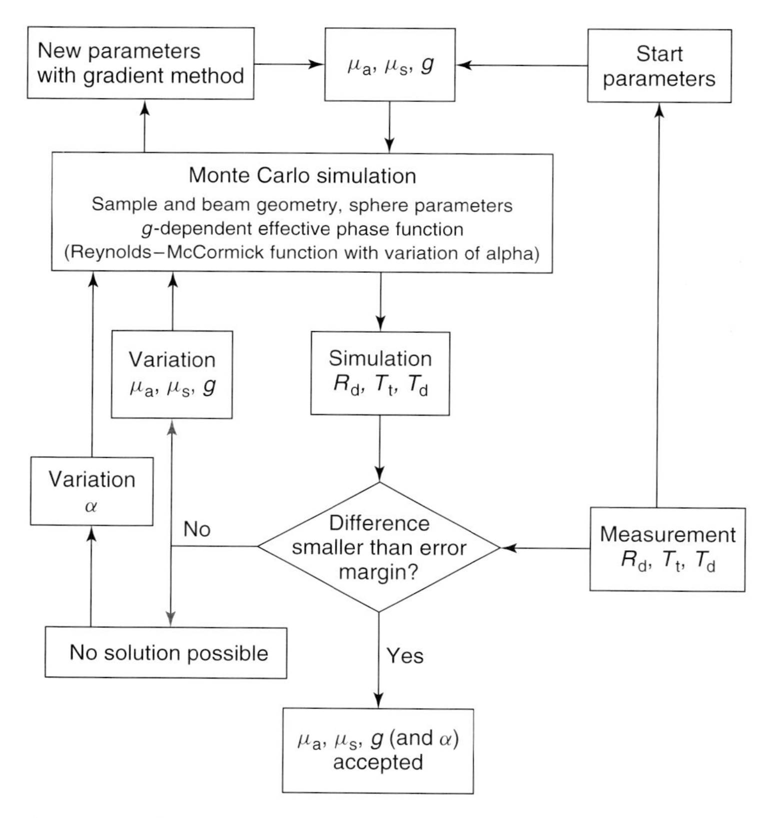

Figure 5.12 The iMCS was carried out using the Reynolds–McCormick (RM) phase function with a variation factor α.

T_t, and diffuse transmission T_d and of a given phase function [8]. The iMCS, shown schematically in Figure 5.12, uses an estimated set of start parameters μ_a, μ_s, and g to calculate the resulting values R_d, T_t, and T_d. These are then compared to the R_d, T_t, and T_d values which have been measured experimentally. In the case of a significant deviation, all three parameters are varied slightly and a new forward simulation is carried out. After this procedure, a gradient matrix is built up, allowing the calculation of a new set of optical parameters that makes a better fit to the measured quantities. This procedure is repeated until the deviation between measured and calculated R_d, T_t, and T_d values is within the given error threshold.

To get valuable results, it is important to use an appropriate effective scattering phase function in contrast to phase functions for the single red blood cell [11, 12], as independent single scattering of individual red blood cells cannot be extrapolated for the collective scattering of blood [13] at physiological concentrations. An effective phase function may depend on the Hct and wavelength especially at high absorption.

5.5.3 Possibility of Determining the Intrinsic Parameters

The intrinsic parameters determined can be seen as the optical properties of the sample with the prerequisite that the transformation function $f(\mu_a, \mu_s, g) \Rightarrow (R_d, T_t, T_d)$ is in "one-to-one" correspondence. This underlying transformation function is a priori unknown and does not only depend on the phase function but also on the geometric position of the detectors, the geometry of the illuminating light beam, the geometry and the optical properties of the sample and as a result, also on the wavelength. Therefore, the "one-to-one" correspondence must be evaluated over the whole spectral range.

The high precision of the intensity measurement combined with the "high resolving" simulation leads to the system registering small deviations in the specified boundary conditions such as the refractive index of the cuvette glass or the space between the cuvette and aperture of the integrating sphere. Under certain conditions, this means that additional information is registered. As a result, it is possible using all three measurement parameters to evaluate the most appropriate effective phase function which leads to the lowest error or gives the most successful simulations at a fixed error threshold. This iMCS makes it possible to compare different phase functions such as Mie [32] and Henyey–Greenstein [25] or to fit an adaptable phase function to the optimal form such as the RM phase function with the variation factor α. From these investigations, it can be deduced that the RM phase function with α between 1.2 and 1.3 is the best effective phase function for describing the real scattering behavior of flowing blood with Hct 0.84%. In the Hct range 5–20%, the appropriate α values are 1.4–1.6. For undiluted blood with Hct 41.2%, the best effective phase function can be obtained with α between 1.6 and 1.8; a significant wavelength dependence was not observed.

5.5.4
Preparation of the Blood Samples

5.5.4.1 Red Blood Cells

For the investigations of red blood cells, fresh human erythrocytes from healthy blood donors were washed with isotonic phosphate buffer (300 mosmol l^{-1}, pH 7.4) to remove the blood plasma and free hemoglobin. To examine the biological variability, erythrocyte concentrates from 10 different blood donors were diluted with buffer to a Hct of 8.6%. To investigate all physiological influences such as Hct, oxygen saturation, shear rate, and osmolarity erythrocytes from one blood donor were used in each case to exclude the interindividual changes.

To keep the blood flowing and to change the oxygen saturation, a miniaturized blood circulation set-up was used with a roller pump and a blood reservoir, which was aerated with a gas mixture of O_2, N_2, and CO_2. Unless otherwise noted, all samples were oxygenated to in excess of 98% and the temperature was kept constant at 20 °C. To avoid sedimentation or cell aggregation, the blood was gently stirred within the reservoir. For the optical measurement, a specially designed, turbulence-free cuvette was used with a laminar flow and a sample thickness of 1020 μm for Hct 0.84% and 116 μm for Hct 4.0–42.1%. Unless otherwise noted, the flow was regulated for each sample thickness to a constant wall shear rate of 600 s^{-1} at the cuvette windows and the osmolarity was 300 mosmol l^{-1}. For RBCs suspended in plasma, the shear rate was reduced relative to the higher viscosity, as compared to water, to give equal shear forces and flow conditions.

The **Hct** was varied by diluting the sample with buffer. The **oxygen saturation** was adjusted by equilibration with a gas mixture of 96% O_2 and 4% CO_2 (100%) or with a gas mixture of 96% N_2 and 4% CO_2, after addition of 0.3% sodium dithionite ($Na_2S_2O_4$) (complete deoxygenation) [33].

To investigate the influence of **flow-induced changes** on the cell shape, native RBCs and fixed RBCs of Hct 42.1% in saline solution were measured at different shear rates of 0, 50, 100, 200, 600, and 1000 s^{-1} at the cuvette windows. The fixed RBCs were prepared by treating with glutaraldehyde (0.15%). Because these steps were performed in saline solution, no aggregation could occur at flow stop [34].

To investigate the influence of **osmolarity**, RBCs were prepared with an Hct of 42.1% in different buffer solutions with osmolarities between 225 and 400 mosmol l^{-1} [35, 36].

The **influence of the medium** on the optical properties of RBC was investigated using one blood sample of Hct 40%, washed with isotonic phosphate buffer or cell-free blood plasma [7].

5.5.4.2 Plasma

A fresh human **plasma** concentrate was centrifuged at 62 000 g for 1 h to remove remaining blood cells [7].

5.5.4.3 Platelets

A fresh leukocyte-depleted **PLT** concentrate from one human donor suspended in plasma ($1075 \times 10^3\,\mu l^{-1}$) was diluted with cell-free plasma to a concentration of $100 \times 10^3\,\mu l^{-1}$ [7].

5.6 Optical Properties of Red Blood Cells

For red blood cells with a Hct of 8.6%, the RM phase function with $\alpha = 1.6$ was evaluated as the effective scattering phase function [37]. The optical properties (μ_a, μ_s, g, and $\mu_s{}'$) were determined for every 5 nm in the wavelength range from 250 to 1100 nm (Figure 5.13). The data are mean data of 10 donor blood samples each of which were measured three times. For comparison, the absorption coefficient of a hemoglobin solution was plotted with the respective concentration. The standard deviation (SD) of the three measurements was small for each individual blood sample. For μ_a, the mean relative SD averaged over all 171 wavelengths is 1 and for g it is 0.85%, leading to a mean SD of 1.6% for $\mu_s{}'$.

Theoretical calculations using Mie theory are applied to the model of erythrocyte spheres with a radius giving the same MCV and having a complex refractive index according to the hemoglobin absorption coefficient and the real part of the

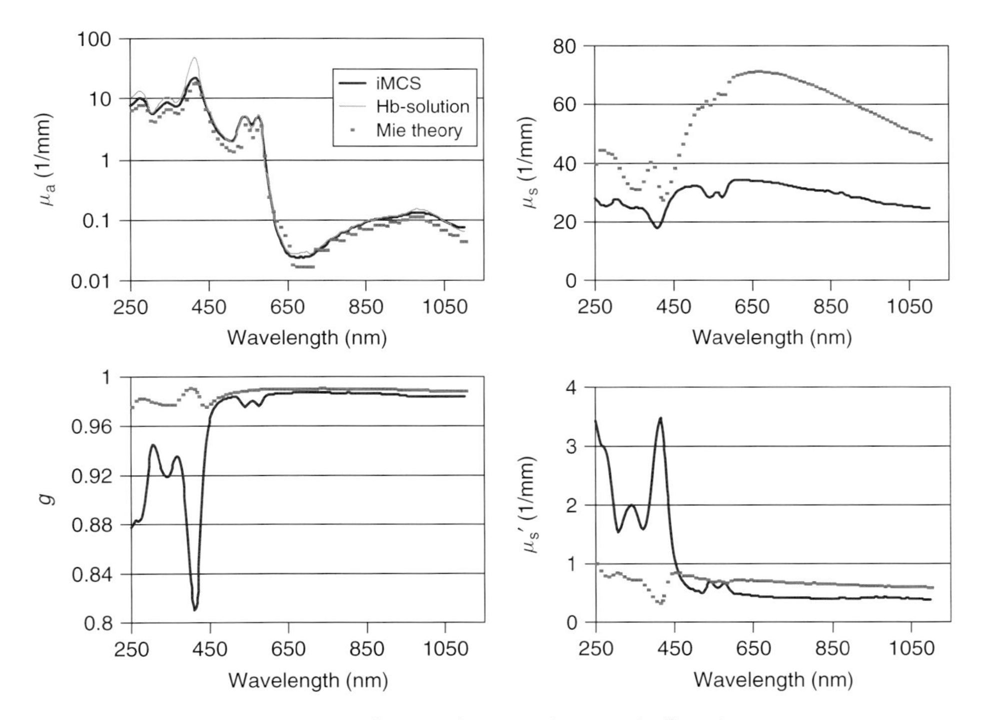

Figure 5.13 The optical properties of RBC with a Hct of 8.6% in buffer solution.

refractive index as determined by measurements [5] (see also Section 5.3). The resulting optical parameters are plotted together with the measured data as shown in Figure 5.13.

The absorption spectrum of RBC shows the typical maxima of oxygenated hemoglobin (Hb) at 415 nm and the double peak at 540 and 575 nm. The Mie curve is comparable to the measured absorption in the range of low absorption above 600 nm. At higher absorptions, Mie values are lower with a maximal difference of about 30% in the range below 600 nm.

The absorption of the Hb solution shows only distinctly higher values at the highest μ_a peaks. If absorption decreases, the difference between the μ_a values of the Hb solution and blood can be neglected. This discrepancy in the blue and ultraviolet absorption maxima can be explained by the Sieve effect. When light passes through a suspension of absorbing particles, such as RBC, photons that do not encounter red blood cells pass unattenuated by absorption. As a result, the transmitted light intensity is higher than it would be if all the hemoglobin was uniformly dispersed in the solution. The magnitude of this effect increases with increasing μ_a and decreasing Hct [38]. This phenomenon is also known as "*absorption flattening*" because the heights of peaks in an absorption spectrum for a suspension are depressed relative to those for a homogeneous solution of the same average concentration as can be seen in Figure 5.13.

The spectrum of the scattering coefficient exhibits, in principle, a similar shape to that calculated by the Mie theory, but is much lower. The calculated values (not shown) of Hammer *et al.* [11] using anomalous diffraction theory do not appear to give better agreement although he approximates the real RBC shape much better.

The anisotropy factor g between 600 and 1100 nm is in the range of 0.99 and is similar to the one calculated with the Mie theory. The g value of 0.9818 determined by Steinke and Shepherd [39] at 633 nm for blood of a similar concentration is a little lower than the measured one. Below 600 nm, the measured values start to decrease compared to those calculated with the Mie theory. The curve of g seems to be inversely related to the absorption spectrum, leading to a distinct minimum of 0.81 at 415 nm.

The μ_s' spectrum looks like the mirror image of the g spectrum. Analogous to μ_s, the effective scattering coefficient μ_s' in the range 460–1100 nm is on average 24% lower than the Mie theoretical values. Below 460 nm, the influence of the strongly decreasing g factor becomes dominant and μ_s' increases similarly, far above the values given by the Mie theory. There is currently no other data available in this high absorption region.

5.6.1
Standard Red Blood Cells

The influence of the biological variability on the statistical distribution of the optical parameters was measured. The RBC parameters investigated were RBCc, Hbc, size

Table 5.1 Statistical distribution of RBC parameters for 10 donors.

	Mean	Minimum	Maximum
RBCc (10^6 cells μl^{-1})	1.008	0.96	1.05
MCV (μm^3)	86.5	82	92.7
MCHC ($g\, l^{-1}$)	32.9	30.4	35.3
Hbc ($g\, dl^{-1}$)	2.87	2.72	3.22

Table 5.2 Statistical distribution of optical parameters for 10 donors.

Relative standard deviation (SD)	Mean of 10 donors	Minimum of 10 donors	Maximum of 10 donors	Measurement variation
SD(μ_a) (%)	4.9	1.9	8.5	1.9
SD(μ_s) (%)	8.4	3.7	13.7	0.85
SD(1−g) (%)	12.8	6.5	25.1	4.0
SD(μ_s') (%)	16.0	5.3	23.3	1.6

(MCV) and shape, and MCH of the RBC. The blood samples from 10 different donors were diluted to have an identical Hct of 8.6%.

Each of the 10 samples with a statistical distribution of blood parameters as shown in Table 5.1 was measured and simulated a total of three times. Figure 5.13 shows the averaged values of μ_a, μ_s, and g in the wavelength range 250–1100 nm. The resulting distribution of the optical parameters, which were averaged over the whole spectrum, is shown in Table 5.2.

The smallest variation is in the absorption of the 10 samples of blood with identical Hct. A variation is observed in μ_s' which is three times greater. Compared to the measurement uncertainty, the variations in the optical parameters of blood from different donors are much greater.

An uncertainty in the optical parameters is induced by the biological variability of the blood parameters RBCc, MCV, MCHC, and Hb concentration which cannot be prepared independently for experiments but are routinely determined by commercial clinical blood analyzers. Therefore, it was attempted to correct the measured optical parameters theoretically on the basis of the determined blood parameters. The absorption coefficient was directly normalized to the mean Hb concentration of all 10 samples of 2.87 g l^{-1}. However, the SDs of the theoretically corrected parameters were not significantly lower than those of the original ones. It should be mentioned that even the mean values of MCHC, MCH, and MCV themselves show broad statistical distributions. The latter, for example, is normally between 80 and 100 μm^3 but the distribution includes cells with 20 and 200 μm^3. Therefore, the optical variability of blood of defined Hct is due to the variability in the mean values of the other blood parameters and their statistical distributions.

To minimize the unavoidable statistical uncertainty when measuring one blood sample, the biologically averaged parameters μ_a, μ_s, g, and $\mu_s{}'$ obtained are used in this chapter as statistical standard parameters to which the Hct-dependent optical parameters determined for that one blood sample will be related.

5.6.2
Optical Parameters of Red Blood Cells Dependent on Hematocrit

The optical parameters of RBCs diluted to different concentrations were determined using the respective Hct-dependent effective phase functions [37] (see Section 5.5.3). Figure 5.14 shows the biologically averaged μ_a, μ_s, g, and $\mu_s{}'$ dependent on wavelength for four different blood concentrations between Hct 0.84 and 42.1%.

The optical parameters μ_a, μ_s, and $\mu_s{}'$ increase continuously with Hct over the whole wavelength range whereas g decreases. The fundamental wavelength dependence of the optical parameters has been discussed before. To quantify the dependence on Hct, the relative optical parameters calculated as the parameter

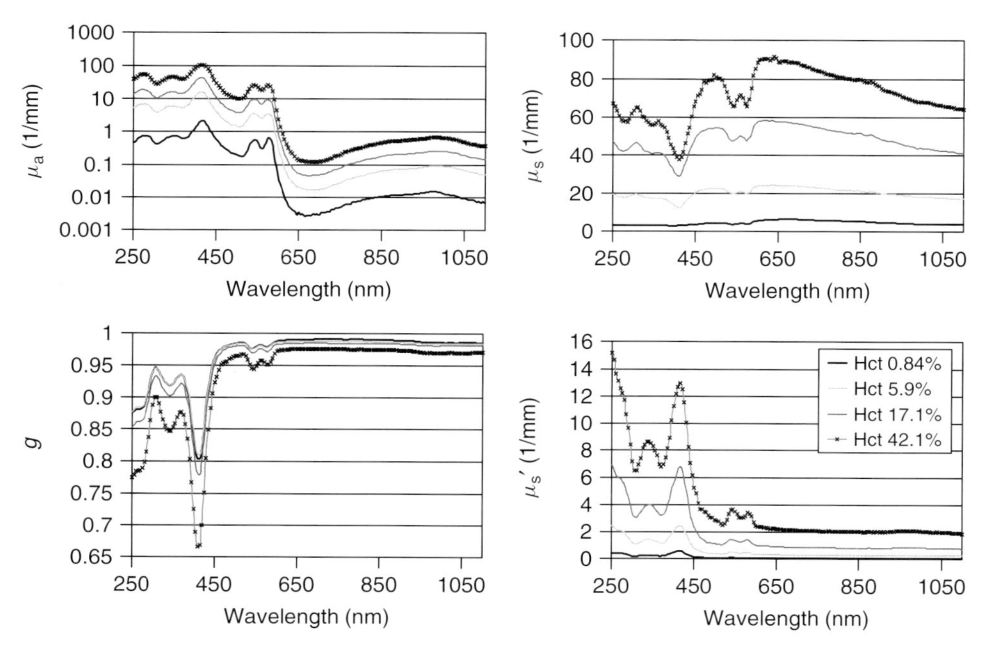

Figure 5.14 Biologically normalized values of μ_a, μ_s, g, and μ_s' of RBC in saline solution of four different blood concentrations between Hct 0.84 and 42.1% in the wavelength range 250–1100 nm.

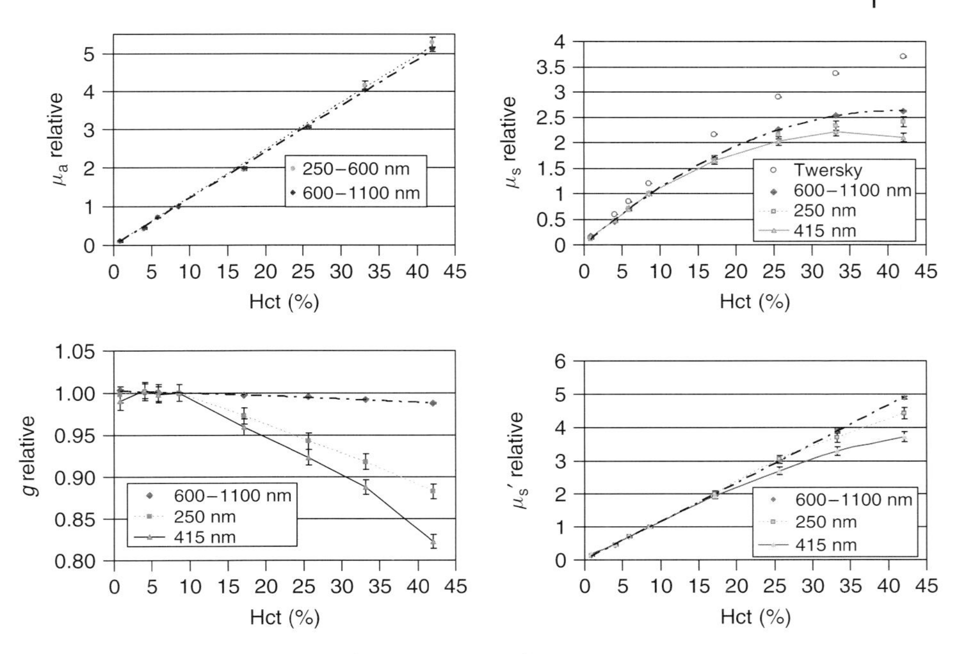

Figure 5.15 Averaged relative optical parameters μ_a_rel, μ_s_rel, g_rel and μ_s'_rel dependent on Hct for selected wavelength ranges.

of a given Hct divided by the parameter of the standard blood at Hct 8.6% were investigated in more detail. The relative optical parameters μ_a_rel, μ_s_rel, g_rel, and $\mu_s{}'$_rel are plotted in Figure 5.15 for certain wavelength regions.

Up to Hct 25.6%, μ_a_rel shows no indication of wavelength dependence in the whole wavelength range. At Hct 33.2%, and more distinctly at Hct 42.1%, μ_a_rel is slightly higher than in the range 600–1100 nm in the region below 600 nm. In other words, the linear increase of absorption with Hct is slightly steeper in the wavelength range below 600 nm than in the range above 600 nm. This is probably due to the Sieve effect. The Sieve effect describes the attenuated absorption when the absorbing particles are not homogeneously distributed. It increases markedly with the absorption within the cell but decreases with increasing Hct. The influence of the Hct increases with the absorption [38]. Therefore, the attenuation of the absorption induced by the Sieve effect occurs mainly in the wavelength range 250–600 nm, where the Hb absorption is high, and decreases with increasing Hct. The linear increase of μ_a_rel with Hct indicates that the absorption cross section σ_a can be seen as being constant over the whole Hct range.

μ_s_rel and $\mu_s{}'$_rel show no wavelength dependence below Hct 17.1% and are only linearly dependent on Hct below 8.6%. Above Hct 17.1%, μ_s_rel and $\mu_s{}'$_rel are only independent of wavelength between 600 and 1100 nm. Above Hct 17.1%, the deviation from the scattering parameters in the range 250–600 nm increases with Hct and with absorption strength as can be seen from the plotted data at 250 and 415 nm.

The relative anisotropy factor g_rel is only wavelength independent above 600 nm for all Hcts. Below 600 nm, g_rel shows a strong dependence on Hb absorption which increases with Hct. Increasing absorption leads to a decrease of g_rel. The g_rel spectrum at higher Hcts looks like the mirror image of the absorption coefficient. A decrease of g with increasing Hct is also a consequence of the decrease of the mean distance between the cells [37]. The increase in the complex refractive index in regions of high absorption results in an increase in reflection and a decrease in transmittance resulting in increased backscattering of the photons, which is identical with a decrease of g.

It should be noted that the Hct dependent saturation effect of μ_s_rel and the Hct dependent decrease of g_rel starts at much lower Hcts, where the effective scattering coefficient $\mu_s{}'$_rel still increases linearly with the Hct. Overall, it should be recognized that the Hct dependence of all optical parameters differs from the linear relationship if the Hct is high and the absorption is strong.

The deviation from linear behavior of μ_s_rel was theoretically described by Twersky [40]. This saturation-like effect should behave, according to his calculation, in the following way:

$$\mu_s = \frac{\text{Hct}(1 - \text{Hct})}{\text{MCV}} \times \sigma_s \tag{5.11}$$

where σ_s is the scattering cross section.

Twersky's formula is adjusted to the measured μ_s at the Hct 0.84% and then plotted for the other values (Figure 5.15). The calculation shows a smaller saturation effect and would yield a maximum scattering coefficient with the Hct of 50%. Contrary to this, the measurement leads to a maximum around a Hct of 33% in the case of maximum absorption at 415 nm. A similar saturation effect above Hct 10% was observed by Roggan *et al.* [8] at 633 nm and Faber *et al.* [41] at 800 nm.

These saturation effects of the scattering parameters correspond, in principle, to the theoretical considerations that the scattering of blood of Hct 100% would be zero due to the disappearance of the refractive index difference between cells and plasma. As a consequence, there must be a maximum between Hct 0 and 100% including a saturation effect in a certain Hct range.

The Hct-dependent nonlinear effects of μ_s, $\mu_s{}'$, and g may be explained by the decrease in independent single scattering and an increase in collective scattering because of the reduced mean distance between the cells. But the spatial distribution is not entirely homogeneous leading to an increasing occurrence of cells which are very close or even in contact with their neighbors. This "double" or "triple" cluster of cells may act as one large scattering cell. This leads to a reduction in the scattering particle concentration combined with an increase in the scattering cross section. The concentration is a three-dimensional parameter whereas the scattering cross section is greatly influenced by the geometrical cross section which is two-dimensional. Therefore, the increase in the scattering cross section does not entirely compensate the reduction of cell concentration, resulting in a reduced μ_s.

The reduction of the scattering efficiency per cell induced by contact areas studied on RBC aggregates was described theoretically by Enejder *et al.* [6] showing, in principle, the observed effects.

5.6.3
Influence of Oxygen Saturation

Investigations on the influence of oxygen saturation have been performed on blood from one donor to exclude interindividual changes and the relative changes were adapted to the standard blood as described in Section 5.6.1.

Figure 5.16 shows μ_a, μ_s, g, and μ_s' of red blood cells of Hct 8.6% with an oxygen saturation of 100 and 0% in the wavelength range 250–2000 nm.

At complete oxygenation, the **absorption spectrum** shows the characteristic peaks at 415 nm and the double peak at 540/575 nm. Above 1100 nm, the μ_a curve increasingly approximates that of water absorption, and above 1360 nm the RBC absorption follows exactly the shape of the water curve. The curve of the deoxygenated sample shows characteristic changes in the absorption behavior with a wavelength shift of the maximum peak to 430 nm and a unique peak at 550 nm. This result also corresponds to the absorption curve of deoxygenated hemoglobin from the literature, taking into account that the Sieve effect induces a lower absorption of the red blood cell suspension compared to the hemoglobin solution.

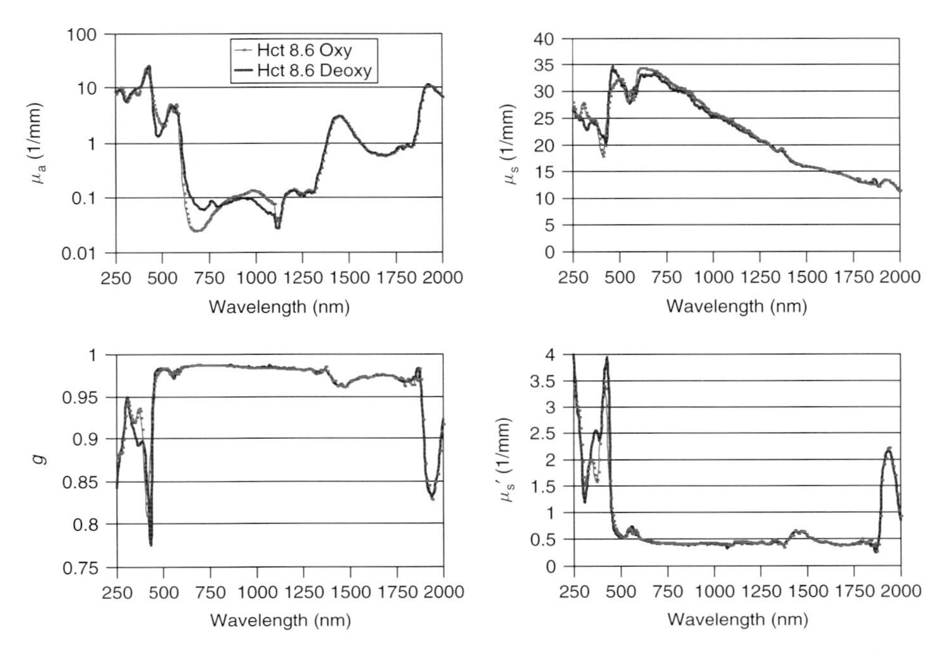

Figure 5.16 μ_a, μ_s, μ_s', and g of RBC in saline solution with an Hct 8.6% dependent on wavelength with an oxygen saturation of 100 and 0%.

In the spectral range above 1150 nm, μ_a is independent of the hemoglobin oxygen saturation.

In the spectral range up to 600 nm, the **scattering coefficient** shows the characteristic decreases correlated with the absorption peaks at 415 and 540/575 nm as described earlier. According to the phenomenon of the absorption-induced decrease in μ_s, the curve of completely deoxygenated RBC shows the corresponding changes. The wavelength of the local minimum shifts from 415 to 430 nm, and the double peak at 540/575 is replaced by the single peak at 550 nm. After reaching the maximum in the region of 600 nm, μ_s decreases continuously with increasing wavelength independent of the oxygen saturation. This decrease is with $\lambda^{-0.93}$ and is near to the theoretically expected value of λ^{-1} according to Mie theory when the sphere size is large compared to the wavelength [22].

The **anisotropy factor** follows the absorption spectrum inversely with the distinct minimum at the maximum absorption, which shifted from 415 nm for 100% oxygenation to 430 nm in the completely deoxygenated state. The local absorption minimum in the oxygenated state at 370 nm results in a relative g maximum, whereas this g maximum disappears, as expected, in the deoxygenated state. The double minimum at 540/575 nm in the oxygenated state turns into a single minimum at 550 nm with complete deoxygenation. Above 600 nm the g spectrum of the oxygenated RBCs and the deoxygenated RBCs are equal. The value of g decreases continuously up to 1800 nm, with the exception of a slight minimum at 1480 nm and is followed by a distinct local minimum at 1900 nm. In contrast to the scattering parameter μ_s, g appears to be influenced not only by the hemoglobin absorption in the wavelength range below 600 nm, that is, absorption within the scattering cells, but also by the absorption maxima of water in the region of 1480 and 1900 nm, that is, absorption within the medium and a reduced absorption within the scatterer.

As compared to the anisotropy factor, **the effective scattering coefficient** shows inverse spectra for both 100 and 0% oxygen saturation, which reflects the absorption spectrum. Above 600 nm, μ_s' decreases continuously and reached the same values independent of oxygen saturation at 1880 nm. At 1950 nm, μ_s' shows the water absorption induced maximum. Corresponding to the minimum value of g at 1480 nm, a local maximum is visible but not significant.

The results show that differences in the oxygen saturation induce changes not only in absorption but also in the optical scattering parameters μ_s, g, and μ_s'. This is mainly due to the characteristic change in the absorption behavior of hemoglobin. The hemoglobin absorption exhibits strong saturation dependence in the range 250–600 nm but is invariant against the oxygen saturation between 1300 and 2000 nm. The saturation induced changes of g and μ_s' occur only between 250 and 600 nm whereas the induced changes of the scattering parameter μ_s occur mainly between 250 and 1100 nm.

The strongest oxygen saturation dependence of g, in relation to the oxygenated state, is about 12% and the maximum change in μ_s is 15%. Apart from absorption, μ_s' shows the strongest oxygen dependence of about 35%, which means that, significant changes have to be considered when the scattering behavior of RBCs is investigated and there are changes in the oxygen saturation.

These results are partly in agreement with the data presented by Faber *et al.* [16] who calculated the optical parameters in the wavelength range 250–1000 nm via Mie theory from the complex refractive indices of oxygenated and deoxygenated hemoglobin solutions taken from porcine blood. There are differences in the spectral range 250–400 nm possibly due to differences in the calculated complex refractive index of hemoglobin by Faber *et al.* The constant higher level of g of the deoxygenation curve compared to the oxygenation curve above 500 nm together with the behavior of g in the wavelength range 250–400 nm could not be observed in the experiments presented here. Also, the absolute g values of Faber *et al.* are higher than the ones presented here, which can also be explained by assuming independent scattering within the calculation.

It is of great interest that the changes in water absorption above 1400 nm induce changes in g and μ_s' in the same way as the strong changes in hemoglobin absorption do. In both cases absorption is increased within the scatterer. In the case of water absorption there is an additional and stronger absorption increase within the surrounding medium. A classical Mie theory does not describe the scattering behavior of spheres in an absorbing medium. However, by using a modified Mie theory it could be shown that under special conditions, similar to the blood situation for spheres and cylinders absorption within the surrounding medium, a decrease in μ_s and g, could be induced [42]. These results, at least partly, correspond to the presented data.

The results are in agreement with the data obtained for a Hct of 33.2%. The differences are due to the changes in Hct, as expected [43].

Therefore, it follows that the oxygen saturation has to be taken into account in order to estimate the scattering properties of blood. Moreover, not only the hemoglobin absorption within the blood cell but also the water absorption within the cell and the medium can influence the scattering properties of RBCs.

5.6.4 Influence of Shear Rate

Figure 5.17 shows the mean relative optical parameters in the wavelength range 250–1100 nm of native and fixed RBCs suspended in saline solution with Hct 42.1% dependent on wall shear rates.

The mean relative absorption of native and fixed RBCs is at a maximum at flow stop. With increasing shear rates, the mean μ_a_rel decreases continuously up to 200 s^{-1}, whereas above 200 s^{-1}, there is no significant change. Compared to μ_a the mean μ_s_rel is at a maximum at flow stop and decreases substantially with increasing shear rate up to 100 s^{-1}. Above 100 s^{-1}, μ_s_rel decreases slightly up to 1000 s^{-1}. For fixed RBCs μ_s_rel is slightly higher at 1000 s^{-1}.

The g_rel shows wavelength dependence only below 600 nm. Nevertheless, a mean g_rel averaged over this spectral range was calculated for each shear rate

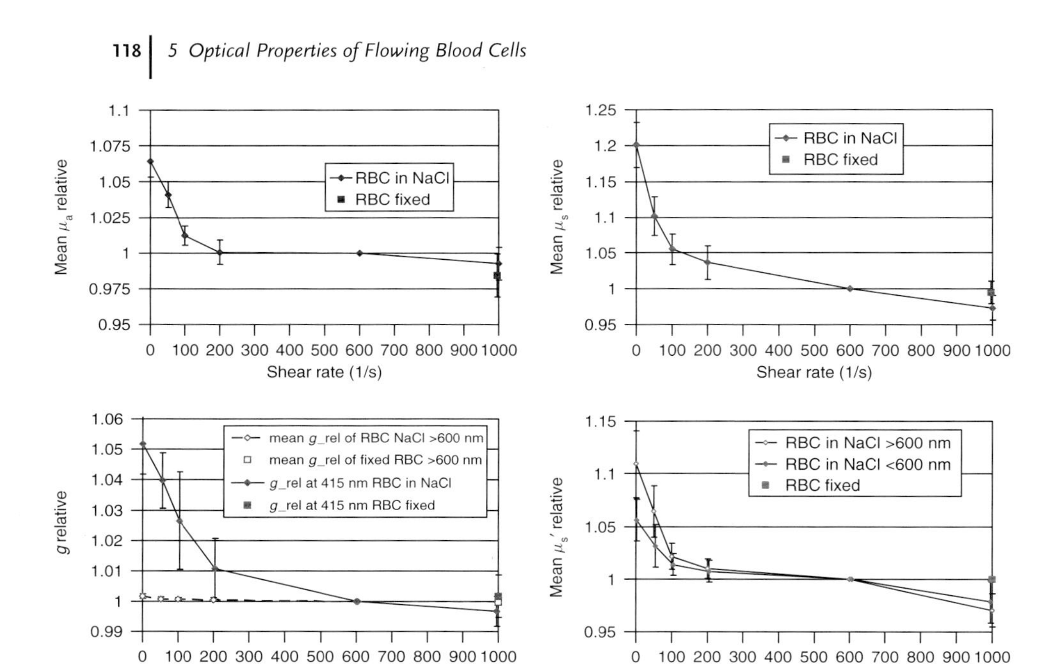

Figure 5.17 Mean and selected values of the optical parameters (μ_a_rel, μ_s_rel, μ_s'_rel, and g_rel) of native and fixed RBC (Hct 42%) in saline solution dependent on wall shear rate.

including g_rel at 415 nm. Above 600 nm, g_rel of native and fixed RBCs is almost shear rate independent.

Corresponding to g, μ_s'_rel shows a wavelength dependence. In the range above 600 nm the mean μ_s'_rel is at a maximum at flow stop and decreases significantly with increasing shear rate until 100 s^{-1}. Above 100 s^{-1}, there is a slight decrease until 1000 s^{-1}. Below 600 nm, μ_s'_rel shows at flow stop and at 50 s^{-1} μ_s'_rel shows significantly lower values than above 600 nm. At a shear rate of 1000 s^{-1} there is no further significant difference. The fixed RBCs show slightly higher μ_s'_rel values at 1000 s^{-1} for both spectral ranges.

5.6.4.1 Shear Rate Range 0–200 s^{-1}

Native RBCs suspended in saline solution are randomly oriented at flow stop and have a maximum absorption and scattering cross section per cell. Cell morphology associated with higher degrees of organization, such as aligned RBCs, exhibited lower scattering coefficients as explained by T-matrix computations [13]. At increasing shear rates, a decrease in the random cell orientation reduces this effect and as a consequence, the cross sections decrease asymptotically. Above a shear rate of 200 s^{-1} only minimal changes can be observed indicating that most of the effect has vanished. The same effect shows μ_s' and g in the wavelength range below 600 nm.

The Fahraeus effect, that is, the dynamic decrease of the Hct, should also cause a decrease in the cell concentration dependent parameters μ_a_rel and μ_s_rel with increasing shear rate [34, 44].

The anisotropy factor is proportional to the degree of absorption. At 415 nm where the absorption is at a maximum, g_rel has the highest value for randomly oriented cells. In spectral regions where μ_a is very low ($<0.1\,\text{mm}^{-1}$) g is almost shear rate invariant. The principle shape of the anisotropy spectra has been discussed in [37].

$\mu_s{}'$ also shows a maximum at flow stop but the maximum decreases with increasing absorption, which is the expected inverse behavior of the anisotropy factor.

5.6.4.2 Shear Rate Range 200–600 s^{-1}

In this shear rate range, the red blood cells start to align with the flow [45]. The absorption cross section does not change within this shear rate range indicating that cell alignment has no significant influence on μ_a and the dynamic Hct decrease is only effective in the shear rate range from 0 to 200 s^{-1}. As the Fahraeus effect has the same influence on μ_a and μ_s, it can be concluded that the small additional decreasing effect on μ_s_rel of 3.7% is only evoked by a decrease in the cross section induced by increasing cell alignment and decreasing random cell orientation. The latter also causes the slight decrease in the relative anisotropy factor in the spectral range below 600 nm, because g is not directly dependent on cell concentration. As a consequence, a decrease of $\mu_s{}'$_rel of about 1% is also induced by cell alignment and decreasing random cell orientation.

5.6.4.3 Shear Rates at 1000 s^{-1}

At these high shear rates, RBCs start to elongate. The decrease in μ_s_rel above 600 s^{-1} does not occur when the cells are fixed, indicating that cell elongation has an additional decreasing effect on the scattering cross section. Cell elongation also reduces g_rel, the effect vanishes after cell fixation, which suggests that elongation causes a further higher degree of organization. At a shear rate of 1000 s^{-1}, the absorption coefficient of untreated and fixed cells shows no significant change compared to 600 s^{-1}, indicating that cell elongation as well as alignment has no influence on the absorption cross section [8, 13].

Measurements on diluted blood with Hct 8.4% showed no significant differences in the shear-rate dependence of the optical parameters. From these results it follows that in order to estimate the optical properties of blood, the flow conditions have to be taken into account.

5.6.5 Influence of Osmolarity

The investigated red blood cell samples had an Hct of 42.1%, an MCV of 89.3 μm^3 and an MCHC of 34.4 g dl^{-1}. An increase in osmolarity resulted, via cell shrinking, in a reduced Hct; a decrease resulted, via cell swelling, in an increased Hct.

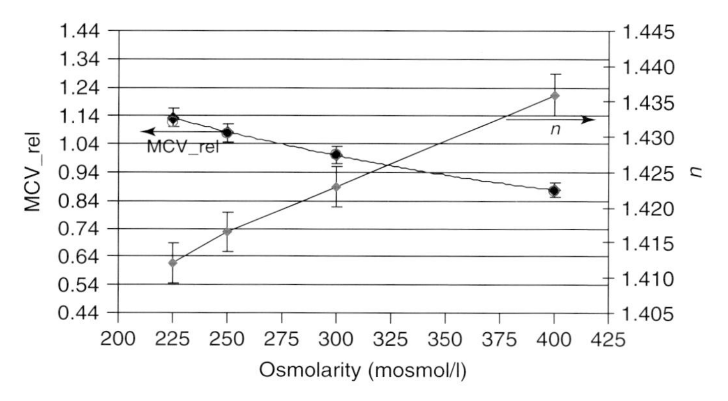

Figure 5.18 Relative mean cell volume (MCV_rel, black diamonds) measured and calculated according to Zhestkov *et al.* [47] (grey circles) and the mean refractive index (*n*, grey diamonds) of the red blood cells dependent on osmolarity, calculated according to Friebel and Meinke [46].

Changes in cell volume lead to a reciprocal change in Hbc within the cell (MCHC) and therefore also in the refractive index [46] (Figure 5.18). The measured MCV changes are in agreement with calculated values based on a formula presented by Tuchin and colleagues [47]. The refractive index was estimated using the formula of Friebel and Meinke [46]. For comparison, effects on absorption and scattering cross sections and the anisotropy factor were estimated from these refractive index changes via Mie theory for spheres of the same volume as a simplified RBC model.

Since the relative optical parameters μ_a_rel, μ_s_rel, and $\mu_s{'}$_rel could be shown as wavelength independent in the observed spectral range within the error tolerance, the mean values over this range were determined and shown to be dependent on the osmolarity (Figure 5.19).

The absorption coefficient increases with increasing osmolarity which means that shrinking of the cells increases μ_a and swelling decreases it. The values calculated with Mie theory show a comparable increase at 400 mosmol l^{-1} whereas the decrease of the osmolarity to 225 mosmol l^{-1} does not lead to a significant change. As found for μ_a_rel, the relative scattering coefficient also increases with increasing osmolarity. In contrast to μ_a, Mie theory shows an inverse change. The g_rel shows distinct wavelength dependence below 600 nm. Therefore, a mean g_rel only averaged over the spectral range 600–1000 nm was calculated for each osmolarity. In addition, the osmolarity dependence at 415 nm was determined to characterize the wavelength dependence in the range of high absorption. The decrease with increasing osmolarity is more pronounced in the high absorption region of hemoglobin and has its maximum at 415 nm.

Above 600 nm the mean relative g value is linearly dependent on osmolarity and within the error tolerance in the range of the values calculated with the Mie theory. The influence of the osmolarity on g is significantly more pronounced at 415 nm.

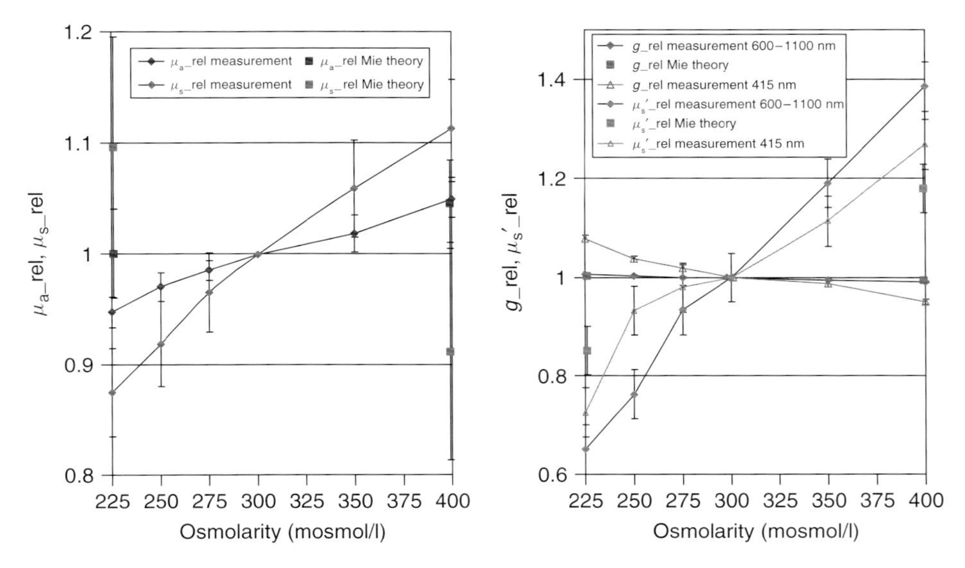

Figure 5.19 Mean values of μ_a_rel, μ_s_rel, g_rel, and μ_s'_rel of the spectral range 250–1100 nm of RBC in saline solution dependent on osmolarity and values for 225 and 400 mosmol/l calculated by Mie theory relative to 300 mosmol/l.

Corresponding to the anisotropy factor, μ_s'_rel shows a wavelength dependent behavior. In the spectral range 600–1100 nm μ_s'_rel increases linearly with increasing osmolarity. At 415 nm μ_s'_rel shows a lower increase. The Mie theory also yields an increase with osmolarity but to only half the extent of the increase in the range 600–1100 nm.

At a constant RBC concentration the absorption coefficient and therefore, the absorption cross section, increases continuously with osmolarity; that is, cell shrinking leads to an increase of μ_a, cell swelling to a decrease. According to this, the absorption cross section is mainly determined by the Hbc, which changes inversely with the cell volume and determines the complex refractive index within the cell, whereas the geometrical cross section of the cells is of minor importance.

This is partly confirmed by the Mie theory predicting a comparable increase of the absorption cross section for the increasing imaginary and real part of the refractive index within the cell for decreasing volume. Conversely, Mie theory does not show the inverse phenomenon.

The increase in the refractive index induced by osmotic cell shrinking leads to an increase of the scattering coefficient, in spite of the decreased geometrical cross section and vice versa.

The results disagree with the expected decrease of the scattering cross section according to Mie theory. Following this theory, the scattering cross section of a sphere is determined in the observed spectral region mainly by the sphere size. As shown in experiments for the Hct dependence of the scattering coefficient

[37, 43] the scattering cross section per cell is not constant at increasing Hct, but continuously decreasing. The cause of this phenomenon is the decreasing distance between the cells leading to an increasing superposition and interference of the single scattering events ("collective scattering" [6]). The osmolarity induced change in Hct influences the distance between the cells accordingly, not by changing the cell concentration but the cell volume.

On the basis of these data, a correction of the relative scattering cross section per cell for the osmotically changed Hct values leads to an almost complete compensation of the μ_s changes predicted by Mie theory at constant cell concentration.

In addition, the change of the cell shape which is not considered by the Mie theory has to be taken into account. The change in cell shape from the sphere via the normal discoid shape to the radiating spinocyte shape possibly leads to an increase in the scattering cross section.

The anisotropy factor is mainly determined by the difference between the real parts of the refractive index within the cell and in the surrounding medium. An osmolarity-induced decrease in volume leads to a decrease in g at wavelengths >600 nm [37] via a refractive index increase. This is in agreement with the theoretical results of Bashkatov *et al.* [48] and with the predictions of the Mie theory. This osmolarity dependence increases strongly with absorption. The very strong increase of g at increasing approximation of the erythrocyte to the spherical shape, compared to the spectral range of low absorption, means a reduction in the decrease of g in the range of high absorption, which could not be explained by Mie theory [37, 43]. The shrinking of the cell induces the opposite. Possibly, this absorption dependent decrease of g compared to the prediction of Mie theory may be connected with the nonspherical shape of the erythrocyte.

The relative changes in the effective scattering coefficient are qualitatively in agreement with the predictions of Mie theory. However, the μ_s' increase at osmolarity increase is underestimated by Mie theory. When correcting the μ_s' values with Hct dependent scattering cross section as is done for the scattering coefficient above, the predicted relative osmolarity induced changes in μ_s' are within the error tolerances of the measured values. This may be also connected with the changes in the geometrical shape.

The optical parameters can be drastically influenced by cell volume and cell shape changes due to osmolarity variations. Therefore, osmolarity-induced changes have to be considered when optical measurements for the determination of blood properties are used [49].

5.7 Optical Properties of Plasma

The refractive indices of plasma were calculated to be between 1.3577 at 400 nm and 1.3438 at 700 nm [46]. For the iMCS, the calculated refractive indices of plasma were used for the refractive indices of the blood samples containing plasma as a

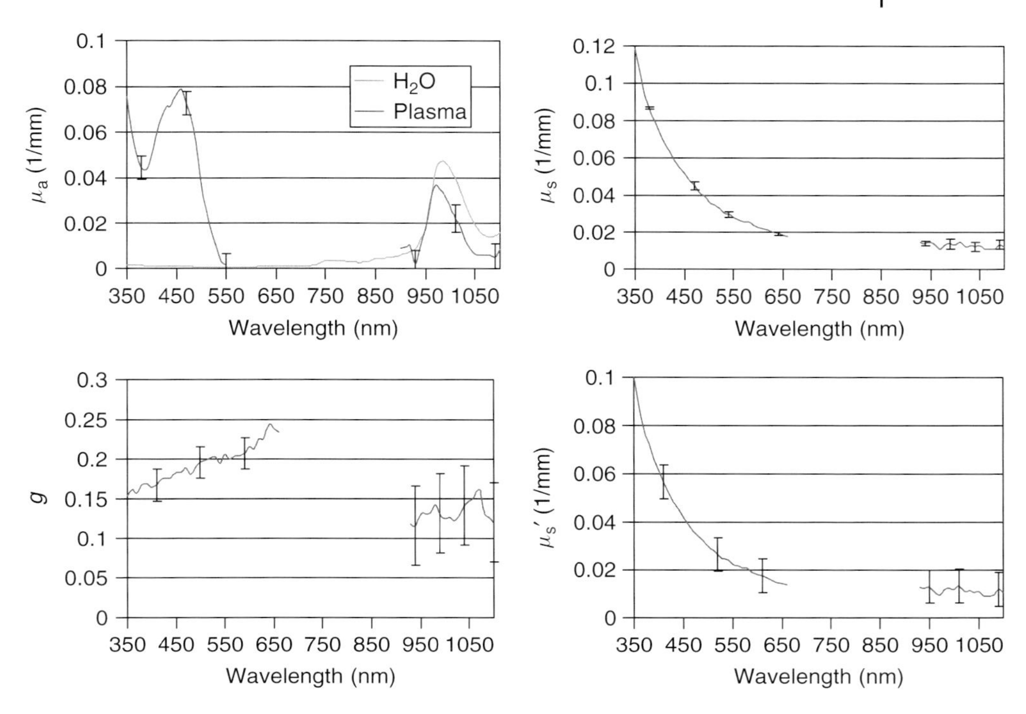

Figure 5.20 Absorption coefficient μ_a, scattering coefficient μ_s, anisotropy g, and effective scattering coefficient μ_s' of a plasma example.

medium, and for the samples containing saline solution, the refractive index of water was used.

For iMCS, the RM phase function with $\alpha = 1.5$, was estimated to be the most appropriate effective scattering phase function for human plasma in the range of 350–1100 nm. Figure 5.20 shows the optical parameters of a plasma sample which was colorless with no detectable cell concentration. In some cases the absorption and scattering properties of the plasma sample were too low for a succesfull simulation and are not present in the figures. The shown error bars are SDs for selected wavelengths resulting from preparing, measurement, and simulation errors.

The absorption coefficient μ_a of plasma shows a peak at about 450 nm, with a maximum at 460 nm. Above 750 nm the absorption of plasma generally follows that of water. The absorption peak of water is smaller and shifted by 15 nm.

The scattering coefficient μ_s decreases continuously from 350 to 1100 nm. The anisotropy factor g in the spectral range of 350–1100 nm is between 0.12 and 0.25. The curve of the effective scattering coefficient $\mu_s{}'$ shows similar behavior to μ_s. The absorption of plasma in the near infrared is dominated by the water absorption. The plasma absorption in the UV–VIS range can exhibit a wide individual variability brought about by various protein concentrations, nutritive compounds or pharmaceuticals, for example, contraceptives. The absorption peak at about 450 nm could be associated with the bilirubin absorption. Possible absorption peaks at 415 and 550 nm are due to the remaining RBCs or free hemoglobin from the plasma production. Lipid contaminations can result in turbid plasma samples that are usually excluded in routine plasma quality control. The plasma shown here is only an example of how plasma can be.

The Rayleigh scattering should describe the scattering properties of pure plasma of protein molecules resulting in a scattering cross section that decreases with increasing wavelength ($\sim\lambda^{-4}$) and isotropic scattering corresponding to a g-factor of zero. For the plasma presented here the g-factor is about 0.2 in the investigated wavelength region, indicating a slightly forward scattering. The plasma μ_s is much higher than expected by the Rayleigh scattering, calculated for a 10% albumin solution. To explain the deviation from Rayleigh scattering, particles with a size of approximately 100 nm and a total concentration of 10% must be present if calculations based on Mie theory are used. These particles might be protein aggregates and lipids or cell components which could not be removed by the given centrifugation procedure.

This could lead to the assumption that the plasma samples are composed of different constituents such as molecules, aggregates of molecules or cells of different sizes. The absorption or scattering of samples is given by the sum of the absorption coefficients μ_a^C or of scattering coefficients μ_s^C of each compound. As described by Friebel and colleagues [7] a composite anisotropy factor g^c can be calculated for a scattering medium consisting of two different scatterers. In this case the single anisotropy factors are averaged and weighted by the scattering coefficients of the respective compound:

$$g^c = \frac{\mu_s^1 \times g_1 + \mu_s^2 \times g_2}{\mu_s^1 + \mu_s^2} \tag{5.12}$$

5.7.1 Influence of the Surrounding Medium on Red Blood Cells

To determine the influence of the surrounding medium, red blood cells were investigated suspended in plasma and in saline solution (Figure 5.21). The cell concentrations were $4.8 \times 10^6\,\mu l^{-1}$ (Hct 40%) for both samples. The absorption of the RBCs in saline solution corresponds to that of RBCs in plasma. The μ_s of RBCs in saline solution is higher than that of RBCs in plasma over the whole wavelength range. Up to 490 nm the g-factor of the cells in saline solution is comparable to the one in plasma whereas above this wavelength g is lower in the cells in saline solution. The effective scattering coefficient of RBCs in saline solution is higher compared to RBCs in plasma. Investigations on RBCs at a Hct 8% [7] show similar results but the differences between the optical properties of RBCs in the two media increase relatively.

The scattering properties of particles suspended in a medium depend on the difference in their refractive indices and therefore, on the refractive index of the medium. Therefore, the change in the scattering properties of RBCs when plasma

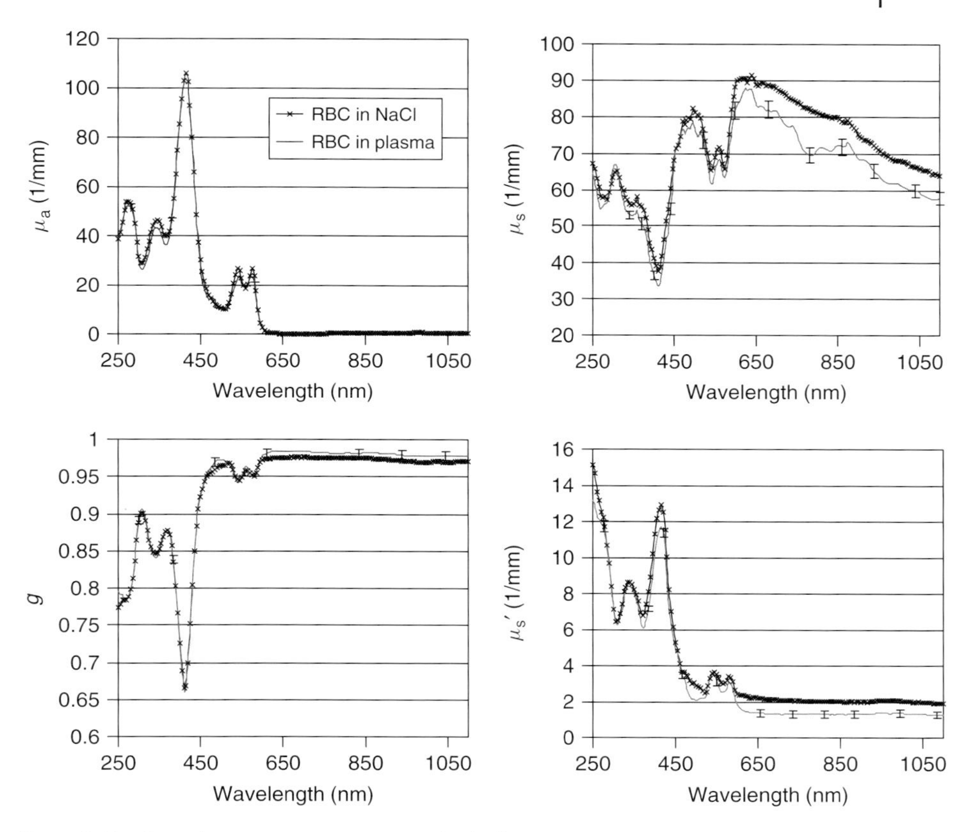

Figure 5.21 Optical parameters μ_a, μ_s, g, and μ_s' of $4.8 \times 10^6\,\mu l^{-1}$ red blood cells in plasma and in 0.9% sodium chlorine solution, corresponding to a Hct of 40%.

is replaced by saline solution is a result of an increase in the difference between the refractive indices of the cells and the medium. A calculation with Mie theory for spheres of the same volume and refractive index as RBCs shows an increase of the scattering cross section and a decrease in the anisotropy factor when replacing plasma by saline solution. If the medium has absorption and scattering properties itself, as is the case for plasma, the composite absorption coefficient, the composite scattering coefficient and the composite g-factor have to be used [7].

Therefore changes in refractive indices, as well as in the composition of the optical parameters, are two different effects. Dependent on the quantities of the optical parameters, one effect can dominate the other.

For the g-factor of RBCs the two effects are in direct competition with one another. Below 600 nm the g-factor of RBCs in saline solution is comparable to that of RBCs in plasma. This effect is due to the low g-factor for plasma and the low g-factor for RBCs in this wavelength region which results in a comparable low composite g-value for RBCs in plasma as RBC in saline solution. For wavelengths over 600 nm the smaller difference in refractive index of RBCs and plasma is the dominant effect, resulting in a higher composite g-factor of RBCs in plasma compared to RBCs in saline solution. For μ_s the changes in refractive index are dominant. These changes in g and μ_s result in an increase of $\mu_s{}'$ of RBCs in saline solution compared to that of RBCs in plasma over the whole wavelength range. This significant change in the scattering behavior must be considered when using RBC concentrates instead of whole blood for optical investigations.

5.8 Optical Properties of Platelets

In whole blood, as well as in PLT concentrates, the PLTs are always suspended in plasma. The RM phase function with α between 1 and 2 was evaluated to be the most effective phase function for the PLTs plasma suspension. The RM phase function was used for all PLT samples and for all wavelengths with $\alpha = 1.5$ [11].

Figure 5.22 shows the optical parameters of two concentrations of PLTs in plasma compared to the corresponding plasma alone. Within the error bars PLTs in plasma have a similar absorption compared to that of pure plasma. This indicates

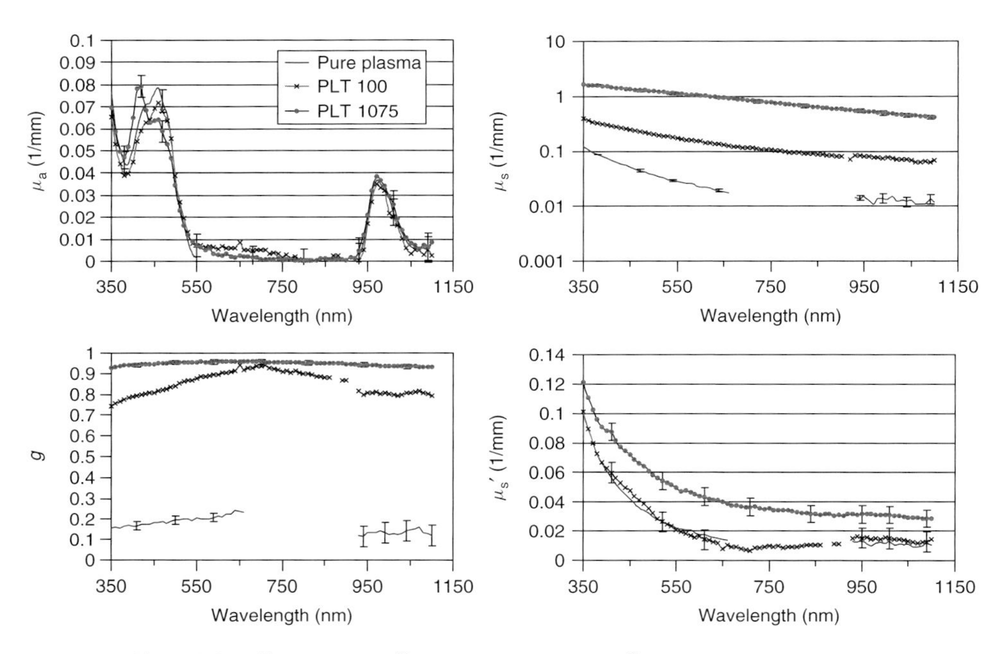

Figure 5.22 Absorption coefficient μ_a, scattering coefficient μ_s, anisotropy g, and effective scattering coefficient μ_s' of two platelet concentrates with concentration $1075 \times 10^3\,\mu l^{-1}$ and $100 \times 10^3\,\mu l^{-1}$ in plasma compared to the corresponding plasma.

that the absorption originates mainly from the plasma. This was demonstrated by substituting the plasma with saline solution. The absorption between 350 and 500 nm was reduced by up to a factor of 10 (data not shown) taking into account that there was still residual plasma present. Only if the PLTs are highly concentrated ($1075 \times 10^3\ \mu l^{-1}$) is the maximum absorption shifted to 420 nm but still with a very low absorption coefficient of 0.079 mm^{-1} compared to erythrocyte concentrates.

The scattering coefficient decreases with wavelength for all PLT concentrations and for pure plasma. The increase of μ_s with PLT concentration is nearly constant for all wavelengths but does not correspond to the concentration increase. The reason for this is the scattering of plasma. The missing data, mainly in the spectra of plasma, have their origin in too high errors in the iMCS.

The g-factor increases with wavelength up to 700 nm and decreases slightly from this point onwards. Furthermore g decreases for all wavelengths with decreasing PLT concentration. The g-factor describes the scattering phase function with a low g value which means that a more isotropic scattering is typical for smaller particles. Therefore the dependence of g on the PLT concentration can be interpreted as being a mean g-factor of a mixture of scattering particles. The g-factor of PLTs (large particles) is large and that of pure plasma (small particles) is small.

The effective scattering coefficient μ_s' is the product of $(1 - g)$ and μ_s, therefore scattering contributions with small g are weighted stronger. This results in an identical μ_s' of pure plasma and the small concentration PLT. μ_s' decreases steeply up to 700 nm and remains constant further on or decreases slightly.

In general, the absorption of PLT concentrates in plasma is dominated by the plasma absorption in the wavelength range from 350 to 1100 nm. The scattering behavior results from the mixture of plasma scattering and PLT scattering. Taking into account that plasma is very different in absorption as well as in scattering (data not shown) this makes it very difficult to extract the PLT concentration from a simple measurement. Only by taking advantage of the strong scattering coefficient of PLTs with strong forward scattering has a small angle scattering measurement any prospect of success.

5.9 Comparison of Optical Influences Induced by Physiological Blood Parameters

The influence of physiological changes is considered in relation to their consequence for the optical parameters [50]. The effect of an equal change of 10% of the physiological blood parameters oxygen saturation 100%, Hct 42%, osmolarity 300 mosmol l^{-1}, and shear rate 600 s^{-1} of a RBC suspension in saline solution is shown in Figure 5.23. The two upper diagrams show the changes in optical parameters at 700 nm which is representative for the low hemoglobin absorption region. The lower panels show the effects of changes in blood optical properties at 415 nm in the high absorption band. The diagrams on the left show the absorption versus the effective scattering coefficient and thus give an overview of the optical behavior.

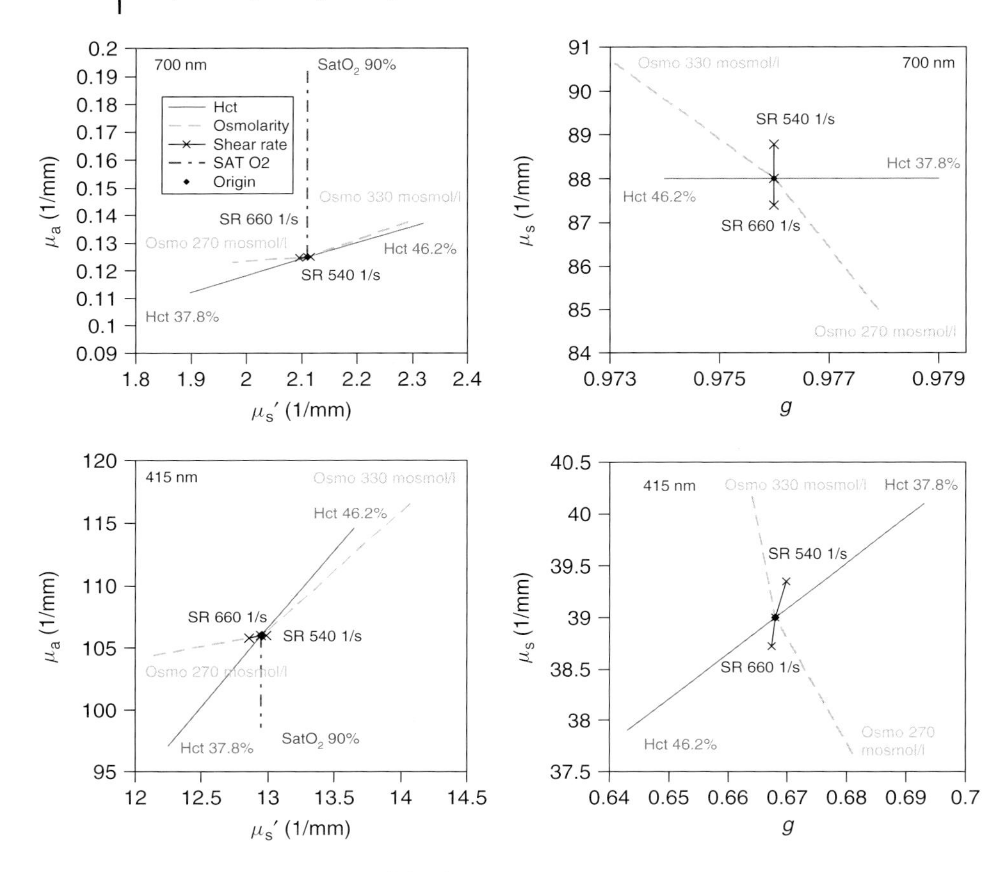

Figure 5.23 Comparison of the changes in optical parameters at 700 and 415 nm induced by 10% changes in the blood parameters oxygen saturation ($SatO_2$), hematocrit (Hct), osmolarity (Osmo), and shear rate (SR).

The diagrams on the right show the scattering coefficient versus the anisotropy factor g and give more details about the scattering properties.

The absorption at 700 nm changes drastically with 10% changes in oxygen saturation, and moderately with 10% changes in Hct and osmolarity. The shear rate effects are negligible. With regard to the effective scattering coefficient, the influence by Hct and osmolarity are strong, whereas the influence of shear rate and oxygen saturation is small.

The influence of osmolarity is relatively great on μ_s and the g-factor, whereas the Hct only changes the g-factor and not μ_s. There is no influence of oxygen saturation and shear rate on the g-factor but an influence of shear rate on μ_s in the lower hemoglobin absorption wavelength range.

At 415 nm osmolarity and Hct both have a strong influence on the absorption and the effective scattering coefficient. Shear rates again effect only small changes and oxygen only affects the absorption. Looking closer at the scattering properties at 415 nm it follows that the Hct equally changes g and μ_s at 415 nm. Osmolarity and shear rate have their main influence on μ_s, and oxygen saturation has a negligible effect on both g and μ_s.

A comparison of the influence of different physiological blood properties on the optical parameters has shown that even when no changes in the prominent parameters RBC concentration or oxygen saturation occur, the optical parameters can still be influenced in a drastic manner by cell shape changes due to osmolarity, and more moderately, by changes in shear rate.

Therefore, flow and osmolarity induced changes have to be considered when optical measurements are used for the determination of blood properties. In whole blood, where plasma proteins are included, aggregation of the red blood cells occurs at flow stop. This behavior has further strong influence on the optical properties [49].

5.10 Summary

The optical properties of blood are dominated by the erythrocytes in the absorption as well as in the scattering. The reason for both is the high hemoglobin content which leads to the strong absorption in the UV, blue, and green spectral range and also is the origin of a high refractive index of erythrocytes leading to the strong scattering. PLTs and plasma both contribute only to a small extent to the optical properties of blood. But plasma has a significant influence on the scattering of erythrocytes. Owing to the smaller difference in the refractive index between erythrocytes and the surrounding liquid, the effective scattering coefficient in whole blood is only about half the value of an erythrocyte concentrate. The influence of the shear rate and osmolarity has shown that they are important parameters that have to be taken into account when the blood is prepared *ex vivo* and the physiological environment cannot be ensured.

Acknowledgments

This work was supported by the Federal Ministry of Education and Research (**Grant No. 13N7522**). The authors thank Lesley Hirst for carefully reading the manuscript. Our thanks also go to the Sorin Group Deutschland GmbH, Munich for placing the equipment at our disposal and to the Department of Transfusion Medicine, Charité-Universitätsmedizin Berlin, Germany for kindly providing the blood bags.

References

1. Rajkumar, A., Karmarkar, A., and Knott, J. (2006) Pulse oximetry: an overview. *J. Perioper. Pract.*, **16**, 502–504.
2. Meinke, M., Schröder, M., Schütz, R., Netz, U., Helfmann, J., Dörschel, K., Pries, A., and Müller, G. (2007)

Frequency weighted laser Doppler perfusion measurements in skin. *Laser Phys. Lett.*, **4**, 66–71.

3. Meinke, M., Müller, G., Gersonde, I., and Friebel, M. (2006) Determination of oxygen saturation and hematocrit of flowing human blood using two different spectrally resolving sensors. *Biomed. Tech. (Berl)*, **51**, 347–354.
4. Meinke, M., Friebel, M., Helfmann, J., and Notter, M. (2005) A novel device for the non-invasive measurement of free hemoglobin in blood bags. *Biomed. Tech. (Berl)*, **50**, 2–7.
5. Friebel, M. and Meinke, M. (2005) Determination of the complex refractive index of highly concentrated hemoglobin solutions using transmittance and reflectance measurements. *J. Biomed. Opt.*, **10**, 064019.
6. Enejder, A.M.K., Swartling, J., Aruna, P., and Andersson-Engels, S. (2003) Influence of cell shape and aggregate formation on the optical properties of flowing whole blood. *Appl. Opt.*, **42**, 1384–1394.
7. Meinke, M., Müller, G., Helfmann, J., and Friebel, M. (2007) Optical properties of platelets and blood plasma and their influence on the optical behavior of whole blood in the visible to near infrared wavelength range. *J. Biomed. Opt.*, **12**, 014024.
8. Roggan, A., Friebel, M., Dörschel, K., Hahn, A., and Müller, G. (1999) Optical properties of circulating human blood in the wavelength range 400–2500 nm. *J. Biomed. Opt.*, **4**, 36–46.
9. Yaroslavsky, A., Yaroslavsky, I., Goldbach, T., and Schwarzmaier, H. (1996) The optical properties of blood in the near infrared spectral range, in optical diagnostics of living cells and biofluids. *Proc. SPIE, Int. Soc. Opt. Eng.*, **2678**, 314–324.
10. Yaroslavsky, A., Yaroslavsky, I., Goldbach, T., and Schwarzmaier, H. (1999) Influence of the scattering phase function approximation on the optical properties of blood determined from the integrating sphere measurements. *J. Biomed. Opt.*, **4**, 47–53.
11. Hammer, M., Schweitzer, D., Michel, B., Thamm, E., and Kolb, A. (1998) Single scattering by red blood cells. *Appl. Opt.*, **37**, 7410–7418.
12. Yaroslavsky, A., Yaroslavsky, I., Goldbach, T., and Schwarzmaier, H. (1997) Different phase function approximations to determine optical properties of blood: a comparison. *Proc. SPIE*, **2982**, 324–330.
13. Nilsson, A., Alsholm, P., Karlson, A., and Andersson-Engels, S. (1998) T-matrix computation of light scattering by red blood sells. *Appl. Opt.*, **37**, 2735–2748.
14. Roggan, A., Minet, O., Schröder, C., and Müller, G. (1993) in *Medical Optical Tomography: Functional Imaging and Monitoring* (ed. G. Müller), SPIE Institute, Bellingham, pp. 149–163.
15. Barer, R. and Joseph, S. (1956) Refractometry and interferometry of living cells. *J. Opt. Soc. Am.*, **47**, 545.
16. Faber, D.J., Aalders, M.C., Mik, E.G., Hooper, B.A., van Gemert, M.J., and van Leeuwen, T.G. (2004) Oxygen saturation-dependent absorption and scattering of blood. *Phys. Rev. Lett.*, **93**, 028102.
17. Assendelft, O. (1970) *Spectrophotometry of Hemoglobin Derivates*, Royal Vangorcum.
18. Prahl, S. (2009) Tabulated data from Various Surces Compiled by S. Prahl, *http://omlc.ogi.edu/spectra/* (accessed 2009).
19. Herman, B., La Rocca, A., and Turner, R. (1978) in *The Infrared Handbook* (eds W. Wolfe and G. Zissis), ERIM, Ann Arbor, MI, pp. 43–45.
20. Chandrasekhar, S. (1950) *Radiative Transfer*, Oxford University Press, London.
21. Ishimaru, A. (1978) Optical scattering and diffusion in turbulence and scatterers. *J. Opt. Soc. Am.*, **68**, 1368–1369.
22. Kerker, M. (1969) *Scattering of Light and other Electromagnetic Radiation*, Academic Press, New York.
23. Van de Hulst, H.C. (1957) *Light Scattering by Small Particles*, John Wiley & Sons, Inc., New York.
24. Forster, F.K., Kienle, A., Michels, R., and Hibst, R. (2006) Phase function

measurements on nonspherical scatterers using a two-axis goniometer. *J. Biomed. Opt.*, **11**, 024018.
25. Henyey, L.G. and Greenstein, J.L. (1941) Diffuse radiation in the galaxy. *Astrophys. J.*, **93**, 70–83.
26. Kubelka, P. (1948) New contributions to the optics of intensely light scattering materials. *J. Opt. Soc. Am. A*, **38**, 448.
27. Farrel, T., Wilson, B.C., and Patterson, M.S. (1992) A diffusion theory model of spatially resolved, steady-state diffuse reflectance for the non invasive determination of tissue optical properties. *Med. Phys.*, **19**, 879–888.
28. Wilson, B.C. and Adam, G. (1983) A Monte Carlo model for the absorption and flux distribution of light in tissue. *Med. Phys.*, **10**, 824–830.
29. Groenhuis, R.A.J., Ferwerda, H.A., and Tenbosch, J.J. (1983) Scattering and absorption of turbid materials determined from reflection measurements. 1. Theory. *Appl. Opt.*, **22**, 2456–2462.
30. Keijzer, M., Jacques, S., Prahl, S., and Welch, A. (2009) Light distribution in artery tissue: Monte Carlo simulation for finite-diameter laser beams. *Lasers Surg. Med.*, **9**, 148–154.
31. Patterson, M.S., Moulton, J.D., Wilson, B.C., Berndt, K.W., and Lakowicz, J.R. (1991) Frequency-domain reflectance for the determination of the scattering and absorption properties of tissue. *Appl. Opt.*, **30**, 4474–4476.
32. Mie, G. (1908) Pioneering mathematical description of scattering by spheres. *Ann. Phys.*, **25**, 337.
33. Friebel, M., Helfmann, J., Netz, U., and Meinke, M. (2009) Influence of oxygen saturation on the optical scattering properties of human red blood cells in the spectral range 250 to 2000 nm. *J. Biomed. Opt.*, **14**, 034001.
34. Friebel, M., Helfmann, J., Müller, G., and Meinke, M. (2007) Influence of shear rate on the optical properties of human blood in the spectral range 250 to 1100 nm. *J. Biomed. Opt.*, **12**, 054005.
35. Meinke, M., Friebel, M., and Müller, G. (2007) Influence of cell shape and orientation on the optical properties of human erythrocytes. *Proc. SPIE*, **6629**, 66290E-1–66290E-10.
36. Friebel, M., Helfmann, J., and Meinke, M.C. (2010) Influence of osmolarity on the optical properties of human erythrocytes. *J. Biomed. Opt.*, **15** (5), 055005.
37. Friebel, M., Roggan, A., Müller, G., and Meinke, M. (2006) Determination of optical properties of human blood in the spectral range 250 to 1100 nm using Monte Carlo simulations with hematocrit-dependent effective scattering phase functions. *J. Biomed. Opt.*, **11**, 34021.
38. Pittman, R. (1986) In vivo photometric analysis of hemoglobin. *Ann. Biomed. Eng.*, **14**, 119–137.
39. Steinke, J.M. and Shepherd, A.P. (1988) Comparison of Mie theory and the light-scattering of red blood-cells. *Appl. Opt.*, **27**, 4027–4033.
40. Twersky, V. (1970) Absorption and multiple scattering by biological suspensions. *J. Opt. Soc. Am.*, **60**, 1084–1093.
41. Faber, D.J., van der Meer, F.J., Aalders, M.C., de Bruin, M., and van Leeuwen, T.G. (2005) Hematocrit-dependence of the scattering coefficient of blood determined by optical coherence tomography. *Proc. SPIE*, **5861**, 58610W.
42. Fu, Q. and Sun, W.B. (2001) Mie theory for light scattering by a spherical particle in an absorbing medium. *Appl. Opt.*, **40**, 1354–1361.
43. Meinke, M., Müller, G., Helfmann, J., and Friebel, M. (2007) Empirical model functions to calculate hematocrit-dependent optical properties of human blood. *Appl. Opt.*, **46**, 1742–1753.
44. Barbee, J. and Cokelet, G. (1971) The Fahraeus effect. *Microvasc. Res.*, **3**, 6–16.
45. Schmid-Schönbein, H., Vogler, E., and Klose, H. (1972) Microrheology and light transmission of blood II. The photometric quantification of red cell aggregate formation and dispersion in flow. *Pflüg. Arch.*, **333**, 140–155.
46. Friebel, M. and Meinke, M. (2006) Model function to calculate the refractive index of native hemoglobin in the wavelength range of 250–1100 nm dependent on concentration. *Appl. Opt.*, **45**, 2838–2842.

47. Zhestkov, D.M., Bashkatov, A.N., Genina, E.A., and Tuchin, V.V. (2004) Influence of clearing solutions osmolarity on the optical properties of RBC. *Proc. SPIE*, **5474**, 321–330.
48. Bashkatov, A.N., Zhestkov, D.M., Genina, E.A., and Tuchin, V.V. (2005) Immersion optical clearing of human blood in the visible and near infrared spectral range. *Opt. Spectrosc.*, **98**, 638–646.
49. Shvartsman, L.D. and Fine, I. (2003) Optical transmission of blood: effect of erythrocyte aggregation. *IEEE Trans. Biomed. Eng.*, **50**, 1026–1033.
50. Friebel, M. (2007) Bestimmung optischer eigenschaften von Vollblut in Abhängigkeit von verschiedenen physiologischen und biochemischen Zustandsparametern. Verlag im Internert: dissertation.de.

6
Laser Diffraction by the Erythrocytes and Deformability Measurements

Sergei Yu. Nikitin, Alexander V. Priezzhev, and Andrei E. Lugovtsov

6.1
Introduction

Among the intrinsic properties of red blood cells (erythrocytes), which enable them to fulfill their major functions in mammalian organisms, is their ability to change their shapes in response to changing environmental conditions including hydrodynamic shear stresses in the blood flow, interactions with other cells and with blood vessel walls, ionic strength and osmolarity of the surrounding medium, and so on.

This property of erythrocytes is usually referred to as *deformability* [1].

Quantitative measurement, assessment of the mechanisms, and monitoring of erythrocyte deformability is an important challenge of cytometry. Progress in its solution provides much valuable information on the cell biophysics and allows for taking important clinical decisions on the microrheologic status of an individual organism, possible alterations of the microcirculation parameters and oxygen supply to the tissue, and other factors determining the physiologic status of the organism. Deformability of erythrocytes can be seen, for example, by means of high-speed and high-resolution vital microscopy of thin vessels in the mesenteries and some other body sites of small animals. However, in humans, such measurements are not practical. Thus, the measurements of human erythrocyte deformability are conducted usually *in vitro* using various techniques either at the level of individual cells or in diluted cell suspensions [2].

In the first case, a classical method of the micropipette aspiration technique, which measures the displacement of an individual cell membrane into a micropipette under a fixed negative pressure, is used [3]. The distance of penetration of the membrane serves as a measure of erythrocyte deformability. Also, a comparatively new method of optical trapping has shown high potential for erythrocyte deformability measurements [4]. Typically, two sharply focused parallel laser beams or two counterpropagating beams from optical fibers are used to stretch the membrane and measure the force needed to obtain a certain deformation, which depends upon both the stiffness of the membrane and the viscosity of the cytozol (solution of hemoglobin) [5–7].

Advanced Optical Flow Cytometry: Methods and Disease Diagnoses, First Edition. Edited by Valery V. Tuchin.

Measurements performed with individual erythrocytes are very important and have a certain superiority; however, they have an evident drawback when compared to similar measurements performed on a large population of cells because they do not allow for large statistical averaging.

This is feasible by means of mechanical filtration methods or optical methods. The latter include two approaches: the so-called ektacytometric methods and the high-speed imaging of the cells in shear flow. The filtration methods are based on the study of erythrocytes penetration through micropore membranes [8]. A good feature of this approach is that it mimics the conditions of microcirculation and allows for accounting for the effect of all factors influencing the erythrocyte deformability. However, these methods are rather slow in comparison with contemporary optical methods.

Ektacytometry is based on the study of the diffraction pattern from a comparatively large population of erythrocytes moving in a shear flow and illuminated with a laser beam [9–12].

Deformation (elongation) of the cells by the shear stress results in the corresponding deformation of the diffraction pattern, which is processed by a computer according to a theoretical model relating the diffraction pattern parameters to the shapes of the illuminated cell [13].

With the advent of fast computer image processing techniques, it has become possible to process real images of cells moving in the shear flow of a diluted suspension and analyze their shapes quantitatively and to select different fractions of cells exhibiting different abilities to deformation without averaging over the whole cell population. This approach is referred to as *selectometry* and is still under development.

In this chapter, we present a theoretical consideration of the formation of the diffraction pattern from a population of deformable particles illuminated by a laser beam. We derive an analytical expression for light intensity distribution on the observation screen of the ektacytometer, accounting for random distributions of the particle sizes and coordinates. The application of the device in experimental and clinical studies is illustrated with two examples showing the potentialities of the method.

6.2 Parameters of the Erythrocytes

In order to make theoretical estimates concerning ektacytometry, one needs to know the optical parameters of the erythrocyte. In our calculations, we have used the following data.

The erythrocyte diameter

$$a = 7.5\ \mu\text{m}$$

the mean thickness of the erythrocyte

$$h = 2\ \mu\text{m} \tag{6.1}$$

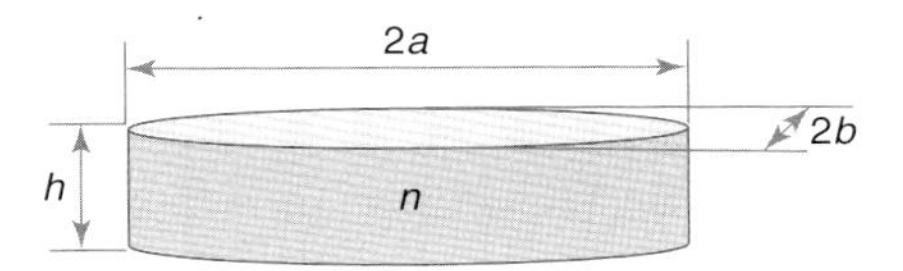

Figure 6.1 Elliptical disc modeling the erythrocyte deformed by a shear flow. Parameters of the model: n – relative refraction index, (a, b, h) – geometrical sizes.

the refractive index of the erythrocyte material relative to that of blood plasma/physiological solution

$$n = 1.05 \tag{6.2}$$

The refractive index of the latter is very close to that of water

$$n_0 = 1.33 \tag{6.3}$$

At the wavelength

$$\lambda = 0.633\ \mu\text{m} \tag{6.4}$$

used in the ektacytometer, the erythrocytes weakly absorb light. Therefore, in our model, we have considered them as transparent particles.

Normal erythrocytes have an intrinsic property to change their shape in shear flows. In these conditions, their shape becomes close to being ellipsoidal. To simplify the calculations in the frames of a simplest model, we consider the erythrocyte as an elliptical disc, that is, by a right cylinder with an elliptical base (Figure 6.1).

6.3 Parameters of the Ektacytometer

In our model, we have used the parameters of our in-house devised ektacytometer. The maximum distance between the erythrocytes in suspension illuminated by the probing laser beam and the beam diameter is

$$A = 1\ \text{mm}$$

The distance from the measurement volume to the observation screen

$$z = 10\ \text{cm}$$

The number of the erythrocytes simultaneously illuminated by the laser beam

$$N \approx 1000$$

These data enable one to calculate such values as wave number $k = 2\pi/\lambda = 0.99 \times 10^5\ \text{cm}^{-1}$, the erythrocyte size parameter $2\pi a/\lambda = 74$, angle of the diffraction divergence for the erythrocyte $\theta = \lambda/a = 0.0844\ \text{rad} = 4.8^\circ$, diffraction length for the erythrocyte $z_a = ka^2/2 = 0.28\ \text{mm}$, size of the diffraction pattern at the

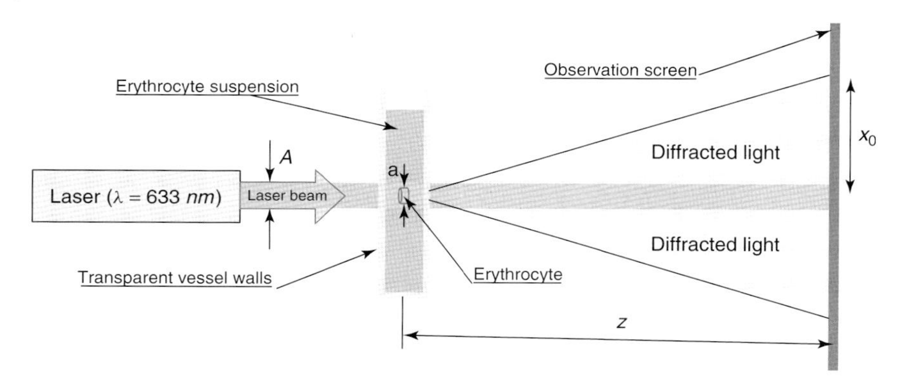

Figure 6.2 Optical scheme of the ektacytometer.

observation screen $x_0 = z \times \theta = 0.84$ cm, and the laser beam diffraction length $z_A = kA^2/2 = 5$ m.

Note that there are strong inequalities between the system parameters

$$z_a \ll z \ll z_A$$

This means that there is an area at the observation screen, which is illuminated by the light diffracted on erythrocytes but is not reached by the direct laser beam (Figure 6.2). The above parameters and relations are sufficient for calculating the light intensity distribution in this area.

Spatial distribution of the erythrocytes in the ektacytometer shear flow enables one to suppose that the bases of the elliptical discs modeling the erythrocytes are located in the same plane, which is perpendicular to the laser beam. Hereinafter, this plane is referred to as the *object plane*, while the observation screen plane is referred to as the *observation plane*.

6.4
Light Scattering by a Large Optically Soft Particle

Data presented above show that the following conditions are fulfilled for an erythrocyte in the ektacytometer

$$a \gg \lambda \tag{6.5}$$

and

$$n - 1 \ll 1 \tag{6.6}$$

This means that an erythrocyte can be considered as a large optically soft particle.

In order to investigate the light scattering by such particles, let a plane optical wave propagating along z-axis illuminate it. We need to calculate light intensity distribution at the observation plane. Condition (6.5) allows one to present the incident light wave as a set of rays. Among them are rays that impinge on the particle as well as rays that pass by the particle. Condition (6.6) allows one to neglect the reflection and refraction of light at the particle boundary and suppose that all rays impinging on the particle traverse it.

Let us characterize a ray by a set of parameters, which is usually associated with a plane wave, namely direction of propagation, amplitude, and phase. Then, the particle causes phase modulation of light, that is, gives rise to an additional phase shift $\Delta\varphi$ for rays that traverse the particle in comparison with the rays that pass by the particle. Modeling the particle by a plate (disc) of constant thickness allows for expressing the phase shift as

$$\Delta\varphi = k \times n_0 \times h \times (n - 1) \tag{6.7}$$

where h is the plate thickness and $k = 2\pi/\lambda$ is the wave number.

According to the Huygens–Fresnel principle, each point of the wave front is a source of an elementary secondary wave, and the diffraction field is a result of interference of the elementary secondary waves. Mathematically, this is expressed by the Huygens–Fresnel diffraction integral (see, e.g., Akhmanov and Nikitin [14]).

$$E(P) = \frac{i}{\lambda} \int_{\Sigma} E_0(M) \frac{\exp(-ik\rho)}{\rho} \mathrm{d}\sigma \tag{6.8}$$

Here, $E(P)$ is the complex amplitude of the light wave at the observation point P, E_0 is the incident wave amplitude, Σ is a surface where the sources of the elementary secondary waves are located, M is a point on the surface Σ, ρ is the distance between the points M and P, and k is the wave number.

Let Σ be a plane $z = 0$, which coincides with the backward surface of the plate, the forward surface of the plate being defined as the plane that faces the light. Let S be a part of the Σ surface, which coincides with the plate surface, and S_1 the remaining part ΣS_1. Assuming that the plate is plane-parallel and the incident field is homogeneous, we present the integral (6.8) in the form

$$E(P) = \frac{i}{\lambda} E_0 \times \exp(-i\Delta\varphi) \times \int_S \frac{\exp(-ik\rho)}{\rho} \mathrm{d}S + \frac{i}{\lambda} E_0 \int_{S_1} \frac{\exp(-ik\rho)}{\rho} \mathrm{d}S_1 \tag{6.9}$$

where $\Delta\varphi$ is defined by Equation 6.7. In particular, if the plate is absent and $\Delta\varphi = 0$, Equation 6.9 should describe the free propagation of the plane wave. In this case, $E(P) = E_0 \times \exp(-ikz)$ and, consequently,

$$E_0 \times \exp(-ikz) = \frac{i}{\lambda} E_0 \int_S \frac{\exp(-ik\rho)}{\rho} \mathrm{d}S + \frac{i}{\lambda} E_0 \int_{S_1} \frac{\exp(-ik\rho)}{\rho} \mathrm{d}S_1 \tag{6.10}$$

It follows from Equations 6.9 and 6.10 that

$$E(P) = E_0 \times \exp(-ikz) + \gamma \times \frac{i}{\lambda} E_0 \times \int_S \frac{\exp(-ik\rho)}{\rho} \mathrm{d}S \tag{6.11}$$

where

$$\gamma = \exp(-i\Delta\varphi) - 1 \tag{6.12}$$

is a parameter depending on the plate material ("material parameter").

Equation 6.11 describes the diffraction of a plane optical wave by a large optically soft plane-parallel plate. Note that this equation expresses the diffraction light field by an integral, which is taken over the plate surface, although the light rays that pass through the particle and the rays that pass by it are all taken into account.

It is not difficult to generalize the above consideration for the case of an arbitrary shape of the particle. Note that in the theory of light scattering by particles, this approach is referred to as the *anomalous diffraction approximation*.

In order to apply the conditions described in Section 6.3 to ektacytometry, the Equation 6.11 should be modified as follows:

$$E(P) = \gamma \times \frac{i}{\lambda} E_0 \times \int_S \frac{\exp(-ik\rho)}{\rho} \mathrm{d}S \tag{6.13}$$

Equation 6.13 describes the diffracted light field at the observation screen in the area, which is not illuminated by the direct laser beam. According to Equations 6.7 and 6.13, $E(P) = 0$, if $n = 1$; that is, the diffracted light field vanishes if there are no erythrocytes in the laser beam. This really takes place in the experimental conditions.

Using Equation 6.13, we obtain the following expression for the light intensity at the observation screen

$$I(P) = |\gamma|^2 \times I_0 \times \left| \frac{i}{\lambda} \int_S \frac{\exp(-ik\rho)}{\rho} \mathrm{d}S \right|^2$$

where I_0 is intensity of the incident light wave and parameter $|\gamma|^2$ is defined as

$$|\gamma|^2 = 4 \times \sin^2(\Delta\varphi/2) \tag{6.14}$$

Equations 6.1–6.4, 6.7, and 6.14 enable one to make a numerical estimate of this parameter. For typical experimental conditions, we obtain $|\gamma|^2 = 1.5$.

Equations 6.13 and 6.8 show that in the area under consideration, the light field scattered by a plane dielectric plate, is similar to the light field diffracted by an opening in a nontransparent screen of the corresponding size and shape.

6.5 Fraunhofer Diffraction

Let a plane monochromatic optical wave of length λ and amplitude E_0 impinge normally on the screen with an opening located in the plane $z = 0$. We need to calculate the optical wave complex amplitude E at point P with coordinates x_0, y_0, and z.

Let the opening size d be much larger than the optical wavelength: $d \gg \lambda$.

Then the amplitude E can be presented in the form of the Huygens–Fresnel diffraction integral [14]:

$$E(P) = \frac{i}{\lambda} \int_S E_0(M) \frac{\exp(-ik\rho)}{\rho} \mathrm{d}S \tag{6.15}$$

Here S is a surface, covering the opening in the screen, M is a point on the S surface, ρ is the distance between the points M and P, and $k = 2\pi/\lambda$ is the wave number.

Using the Cartesian coordinates, we present the integral (6.15) as follows:

$$E(x_0, y_0, z) = \frac{i}{\lambda} \iint_S E_0(x, y) \frac{\exp(-ik\rho)}{\rho} \mathrm{d}x\mathrm{d}y \tag{6.16}$$

where x, y are the Cartesian coordinates of the point M, and ρ is defined as

$$\rho = \sqrt{z^2 + (x_0 - x)^2 + (y_0 - y)^2} \tag{6.17}$$

The distance from the opening center to the observation point is

$$l = \sqrt{z^2 + x_0^2 + y_0^2} \tag{6.18}$$

Consider small angle diffraction when

$$x, y, x_0, y_0 \ll z \tag{6.19}$$

In this case, with good accuracy

$$\rho = l + \frac{x^2 + y^2}{2l} - \frac{xx_0 + yy_0}{l}$$

and the diffraction integral (6.16) takes the form

$$E(x_0, y_0, z) = \frac{i}{\lambda l} \exp(-ikl) \iint_S E_0(x, y) \exp\left[-ik\frac{(x^2 + y^2)}{2l} + ik\frac{(xx_0 + yy_0)}{l}\right] \mathrm{d}x\mathrm{d}y \tag{6.20}$$

Deriving (6.20), we neglected the difference between ρ and l in the denominator of the integrand. This is possible due to conditions (6.19). Expression (6.20) describes diffraction in the Fresnel approximation.

In the far-field zone defined by the condition

$$z \gg kd^2/2$$

the expression for E is reduced to a simpler form

$$E(x_0, y_0, z) = \frac{i}{\lambda l} \exp(-ikl) \iint_S E_0(x, y) \exp\left[\frac{ik}{l}(xx_0 + yy_0)\right] \mathrm{d}x\mathrm{d}y \tag{6.21}$$

or

$$E(x_0, y_0, z) = \frac{i}{\lambda l} \exp(-ikl) \times F(x_0, y_0, z) \tag{6.22}$$

where

$$F(x_0, y_0, z) = \iint_S E_0(x, y) \exp\left[\frac{ik}{l}(xx_0 + yy_0)\right] \mathrm{d}x\mathrm{d}y \tag{6.23}$$

Equation 6.21 describes the diffraction in the Fraunhofer approximation.

From the mathematical point of view, the Equation 6.21 is a two-dimensional spatial Fourier transform. Furthermore, we shall calculate the function $F(x_0, y_0, z)$, which is determined by the amplitude of the field of the incident light wave, as well as by the shape of the opening. This function has a meaning of the Fourier transform of the initial spatial distribution of field.

Assuming the incident field to be homogeneous and using the Equations 6.15 and 6.23, we obtain

$$E(P) = \frac{i}{\lambda} \times E_0 \times \int_S \frac{\exp(-ik\rho)}{\rho} \mathrm{d}S \tag{6.24}$$

and

$$F\left(x_0, y_0, z\right) = E_0 \times \iint_S \exp\left[\frac{ik}{l}\left(xx_0 + yy_0\right)\right] \mathrm{d}x\mathrm{d}y \tag{6.25}$$

It follows from Equations 6.24 and 6.13 that if we need to describe diffraction on a particle instead of diffraction on an aperture, we should multiply the diffraction integral by the parameter γ, defined by Equation 6.12. By this way, we obtain

$$E\left(x_0, y_0, z\right) = \gamma \times \frac{i}{\lambda l} \exp(-ikl) \times F\left(x_0, y_0, z\right) \tag{6.26}$$

Therefore, Equations 6.25 and 6.26 describe the Fraunhofer diffraction by a plane-parallel phase plate.

6.6 Light Scattering by a Transparent Elliptical Disc

In typical conditions of ektacytometry, the erythrocyte shape is close to an ellipsoid or elliptical disc. Here, we consider light wave diffraction by a plane-parallel phase plate of an elliptical shape.

Let us first calculate the diffraction by an elliptical opening. The opening boundary is described by the equation

$$\frac{x^2}{a^2} + \frac{y^2}{b^2} = 1 \tag{6.27}$$

where a and b are the ellipse semiaxes. Let the opening size be much larger than the optical wavelength: $a, b \gg \lambda$, and the observation point be located in the far-field zone, that is, $z \gg ka^2/2, kb^2/2$.

Then the optical field amplitude can be presented in the form of the Fraunhofer diffraction integral (6.21) or in the form (6.22 and 6.23).

Following the paper [13], we introduce the coordinates r, φ in the aperture plane and r_0, φ_0 in the observation plane according to equations

$$x = \sqrt{a/b} \times r \times \cos\varphi, y = \sqrt{b/a} \times r \times \sin\varphi \tag{6.28}$$

$$x_0 = \sqrt{b/a} \times r_0 \times \cos\varphi_0, y_0 = \sqrt{a/b} \times r_0 \times \sin\varphi_0 \tag{6.29}$$

The Jacobian of the transform (Equation 6.28) is

$$\frac{D(x, y)}{D(r, \varphi)} = \begin{vmatrix} \dfrac{\partial x}{\partial r} & \dfrac{\partial x}{\partial \varphi} \\ \dfrac{\partial y}{\partial r} & \dfrac{\partial y}{\partial \varphi} \end{vmatrix} = r$$

According to Equations 6.27 and 6.28, the equation for the aperture boundary becomes $r = \sqrt{ab}$. It follows from Equations 6.28 and 6.29 that $xx_0 + yy_0 = rr_0 \times \cos(\varphi - \varphi_0)$ and, consequently, the integral (6.23) can be written as

$$F(r_0, \varphi_0, z) = \int_0^{\sqrt{ab}} r\mathrm{d}r \int_0^{2\pi} E_0(r, \varphi) \exp\left[\frac{ik}{l} rr_0 \times \cos(\varphi - \varphi_0)\right] \mathrm{d}\varphi$$

where

$$l = \sqrt{z^2 + \frac{b}{a} r_0^2 \cos^2 \varphi_0 + \frac{a}{b} r_0^2 \sin^2 \varphi_0}$$

according to Equations 6.18 and 6.29.

Let the field of the incident optical wave be spatially homogeneous inside the opening, that is, $E_0(r, \varphi) = E_0 = \text{const}$. Then,

$$F(r_0, \varphi_0, z) = E_0 \int_0^{\sqrt{ab}} r\mathrm{d}r \int_0^{2\pi} \exp\left[\frac{ik}{l} rr_0 \times \cos(\varphi - \varphi_0)\right] \mathrm{d}\varphi$$

Using the tabulated integral

$$\int_0^{2\pi} \exp[i\alpha \times \cos(\varphi - \varphi_0)]\mathrm{d}\varphi = 2\pi \times J_0(\alpha)$$

where $J_0(\alpha)$ is the Bessel function of zeroth order, we obtain

$$F(r_0, \varphi_0, z) = E_0 \times 2\pi \int_0^{\sqrt{ab}} r \times J_0\left(\frac{k}{l} rr_0\right) \mathrm{d}r$$

or

$$F(r_0, \varphi_0, z) = E_0 \times 2\pi ab \frac{J_1\left(\frac{kr_0}{l}\sqrt{ab}\right)}{\frac{kr_0}{l}\sqrt{ab}} \tag{6.30}$$

where $J_1(x)$ is the Bessel function of first order. Here, we used the tabulated integral

$$\int_0^x x \times J_0(x)\mathrm{d}x = x \times J_1(x)$$

Now let us come back to the variables x_0, y_0 in the observation plane. The value r_0 in Equation 6.30 is defined as $r_0 = \sqrt{x_0^2 a/b + y_0^2 b/a}$, which follows from Equation 6.29. Consequently, the field Fourier amplitude can be written as

$$F(x_0, y_0, z) = E_0 \times 2\pi ab \frac{J_1\left(\frac{k}{l}\sqrt{x_0^2 a^2 + y_0^2 b^2}\right)}{\frac{k}{l}\sqrt{x_0^2 a^2 + y_0^2 b^2}} \tag{6.31}$$

where l is defined by Equation 6.18. Substituting Equation 6.18 into Equation 6.31, we obtain

$$F(x_0, y_0, z) = E_0 \times 2\pi ab \frac{J_1\left(k\sqrt{\frac{x_0^2 a^2 + y_0^2 b^2}{z^2 + x_0^2 + y_0^2}}\right)}{k\sqrt{\frac{x_0^2 a^2 + y_0^2 b^2}{z^2 + x_0^2 + y_0^2}}}$$

In frames of conditions (6.19), a simpler expression is valid with good accuracy:

$$F\left(x_0, y_0, z\right) = E_0 \times 2\pi ab \frac{J_1\left(\frac{k}{z}\sqrt{x_0^2 a^2 + y_0^2 b^2}\right)}{\frac{k}{z}\sqrt{x_0^2 a^2 + y_0^2 b^2}} \tag{6.32}$$

Inserting Equation 6.32 into Equation 6.22 and taking into account Equation 6.19, we obtain

$$E\left(x_0, y_0, z\right) = iE_0 ab\frac{k}{z} \exp\left[-ik\left(z + \frac{x_0^2 + y_0^2}{2z}\right)\right] \frac{J_1\left(\frac{k}{z}\sqrt{x_0^2 a^2 + y_0^2 b^2}\right)}{\frac{k}{z}\sqrt{x_0^2 a^2 + y_0^2 b^2}}$$

Therefore, the complex amplitude of the light field diffracted by an elliptical aperture is calculated. As the light intensity is the proportional square modulus of the field complex amplitude, the light intensity distribution over the observation plane can be presented in the form

$$I\left(x_0, y_0, z\right) = I_0\left(ab\frac{k}{z}\right)^2 \left[\frac{J_1\left(\frac{k}{z}\sqrt{a^2 x_0^2 + b^2 y_0^2}\right)}{\frac{k}{z}\sqrt{a^2 x_0^2 + b^2 y_0^2}}\right]^2 \tag{6.33}$$

In particular, for round aperture, when $a = b = R$, we have

$$I(\theta, z) = I_0\left(\frac{\pi R^2}{\lambda z}\right)^2 \times \left[\frac{J_1(2\pi R\theta/\lambda)}{\pi R\theta/\lambda}\right]^2 \tag{6.34}$$

where $\theta = r_0/z$ is the diffraction angle, r_0 is the distance from the diffraction pattern center to the observation point, and z is the distance between the screen with the opening and the observation plane. Equation 6.34 expresses the well-known result of the diffraction theory [14].

Consider the points on the observation screen, in which the light intensity has a constant value. We refer to them as isointensity points and to the curves, which are built up by these points, as isointensity curves. The shape of the isointensity curves is determined by the aperture shape. According to Equation 6.33, for an elliptical aperture, the isointensity curves are defined by equation

$$\frac{k}{z}\sqrt{a^2 x_0^2 + b^2 y_0^2} = \text{const} \tag{6.35}$$

or for $z = \text{const}$

$$\frac{x_0^2}{b^2} + \frac{y_0^2}{a^2} = \text{const}$$

that is, they are ellipses of the same shape as the aperture but turned with respect to it at an angle of 90°. The isointensity curves in the far-field zone for the case of plane optical wave diffracted by circular and elliptical apertures are shown in Figure 6.3.

It follows from Equation 6.26 that if we need to describe the diffraction by a particle instead of diffraction by an aperture, we should multiply the complex

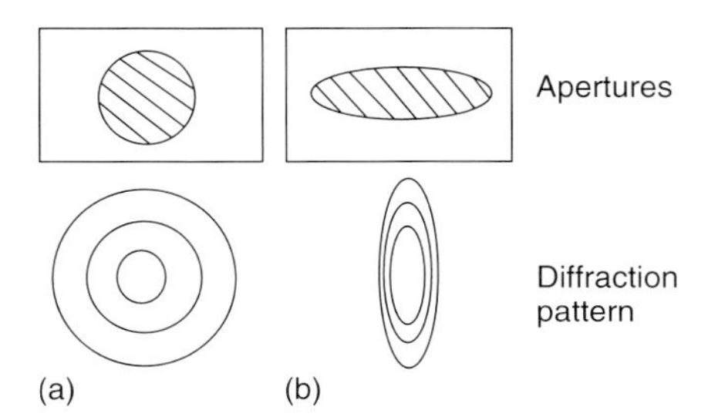

Figure 6.3 The Fraunhofer diffraction by circular and elliptical apertures.

amplitude of the optical field by the parameter, γ, defined by Equation 6.12. Doing so, we obtain the following expression for the light intensity distribution

$$I\left(x_0, y_0, z\right) = |\gamma|^2\, I_0\left(ab\frac{k}{z}\right)^2 \left[\frac{J_1\left(\frac{k}{z}\sqrt{a^2x_0^2 + b^2y_0^2}\right)}{\frac{k}{z}\sqrt{a^2x_0^2 + b^2y_0^2}}\right]^2$$

where parameter $|\gamma|^2$ is determined by Equation 6.14.

6.7 Light Scattering by an Elliptical Disc with Arbitrary Coordinates of the Disc Center

Let us generalize the results obtained for the case when the center of the particle backward surface is located at the object plane in a point with arbitrary coordinates $x = x_1, y = y_1$. In this case, formulae (6.16 and 6.17) remain true but the boundary of the surface is described by equation

$$\frac{(x - x_1)^2}{a^2} + \frac{\left(y - y_1\right)^2}{b^2} = 1 \tag{6.36}$$

Furthermore, we introduce the auxiliary coordinates x', y' into the object plane, defining them by relations

$$x = x_1 + x', y = y_1 + y' \tag{6.37}$$

According to Equations 6.36 and 6.37, the boundary of the disc base in coordinates x', y' is described by equation

$$\frac{x'^2}{a^2} + \frac{y'^2}{b^2} = 1$$

The Jacobian of the transformation (Equation 6.37) is

$$\frac{D(x, y)}{D(x', y')} = \begin{vmatrix} \dfrac{\partial x}{\partial x'} & \dfrac{\partial x}{\partial y'} \\ \dfrac{\partial y}{\partial x'} & \dfrac{\partial y}{\partial y'} \end{vmatrix} = 1$$

Substituting Equation 6.37 into Equations 6.16 and 6.17, we obtain

$$E(x_0, y_0, z) = \frac{i}{\lambda} \iint_S E_0(x', y') \frac{\exp(-ik\rho)}{\rho} dx'dy' \tag{6.38}$$

where

$$\rho = \sqrt{z^2 + (p - x')^2 + (q - y')^2} \tag{6.39}$$

and

$$p = x_0 - x_1, q = y_0 - y_1 \tag{6.40}$$

Equations 6.38 and 6.39 differ from Equations 6.16 and 6.17 only by notation of the parameters and integration variables. This means that the integral (Equation 6.38) can be presented in the form of Equation 6.22 by substituting the parameters x_0, y_0 on the right-hand side of Equation 6.22 by the parameters p, q defined by formula (6.40). Thus, we obtain

$$E\left(x_0, y_0, z\right) = \frac{i}{\lambda l_1} \exp\left(-ikl_1\right) \times F\left(x_0 - x_1, y_0 - y_1, z\right) \tag{6.41}$$

for the diffraction by an aperture and

$$E\left(x_0, y_0, z\right) = \gamma \times \frac{i}{\lambda l_1} \exp\left(-ikl_1\right) \times F\left(x_0 - x_1, y_0 - y_1, z\right) \tag{6.42}$$

for the diffraction by a particle. Here,

$$l_1 = \sqrt{z^2 + (x_0 - x_1)^2 + \left(y_0 - y_1\right)^2} \tag{6.43}$$

the parameter γ being defined by Equation 6.12. The function $F\left(x_0, y_0, z\right)$ in Equations 6.41 and 6.42 is defined by formula (6.23). In the case of an elliptical aperture (disc), it takes the form (6.32).

6.8
Light Diffraction by an Ensemble of Particles

The optical field diffracted by an ensemble of particles is equal to the sum of fields diffracted by each particle individually. This statement is the superposition principle for optical fields. It follows mathematically from Equation 6.15. Using formulae (6.42 and 6.43) for a set of identical particles, we obtain

$$E\left(x_0, y_0, z\right) = \gamma \times \sum_{n=1}^{N} \frac{i}{\lambda l_n} \exp\left(-ikl_n\right) \times F\left(x_0 - x_n, y_0 - y_n, z\right) \tag{6.44}$$

where

$$l_n = \sqrt{z^2 + (x_0 - x_n)^2 + \left(y_0 - y_n\right)^2} \tag{6.45}$$

In Equations 6.44 and 6.45, n denotes the particle number and N is the total number of particles; the values x_n, y_n denote the Cartesian coordinates of the center of the backward surface for the particle with number n.

According to Equation 6.35, the parameters x_0, y_0 characterizing the size of the diffraction pattern in the far-field zone increase proportional to the distance z between the object and observation planes. On the other hand, parameters x_n, y_n characterizing the mutual positions of the particles do not depend on z. Consequently, for a sufficiently large distance z, the following conditions are fulfilled

$$x_n \ll x_0, y_n \ll y_0 \tag{6.46}$$

Relations (6.19 and 6.46) enable one to approximate the expression (6.44) in the following form:

$$E\left(x_0, y_0, z\right) = \gamma \times \frac{i}{\lambda z} F\left(x_0, y_0, z\right) \sum_{n=1}^{N} \exp\left(-ikl_n\right) \tag{6.47}$$

where

$$l_n = L - \frac{x_0 x_n + y_0 y_n}{z} \tag{6.48}$$

In Equation 6.48, the value L does not depend on n and is defined as

$$L = z + \frac{x_0^2 + y_0^2}{2z}$$

Inserting Equation 6.48 into Equation 6.47, we find

$$E\left(x_0, y_0, z\right) = \gamma \times \frac{i}{\lambda z} \exp\left(-ikL\right) \times F\left(x_0, y_0, z\right) \times \sum_{n=1}^{N} \exp\left[i\frac{k}{z}\left(x_0 x_n + y_0 y_n\right)\right]$$

As the light intensity is proportional to the square modulus of the light field complex amplitude, the spatial distribution of the light intensity on the observation screen is described by the formula

$$I\left(x_0, y_0, z\right) = \frac{|\gamma|^2}{(\lambda z)^2} \times \left|F\left(x_0, y_0, z\right)\right|^2 \left|\sum_{n=1}^{N} \exp\left[i\frac{k}{z}\left(x_0 x_n + y_0 y_n\right)\right]\right|^2 \tag{6.49}$$

According to Equation 6.49, the diffraction pattern essentially depends on the number of particles N, as well as on mutual positions of the particles. Let us consider two particular cases.

6.9 Light Diffraction by Particles with Random Coordinates

In the ektacytometer, the erythrocytes continuously move in a shear flow. Consequently, the particles rapidly change their positions with respect to the probing laser beam. In these conditions, it is natural to consider the erythrocytes coordinates as statistically independent random values. The observable variable is the mean intensity of light at the observation screen. The expressions obtained above enable one to calculate this value as a function of the observation point coordinates or as a function of the scattering angle.

Averaging the expression (6.49) over parameters x_n, y_n and assuming the latter as statistically independent random values, we obtain the following equation for the mean light intensity at the observation screen

$$I\left(x_0, y_0, z\right) = \left|\gamma^2\right| \times \frac{N}{(\lambda z)^2} \times \left|F\left(x_0, y_0, z\right)\right|^2 \tag{6.50}$$

Equation 6.50 shows that the mean intensity of light diffracted by an ensemble of chaotically located particles is equal to a sum of light intensities diffracted by each particle individually.

6.10 Light Scattering by Particles with Regular Coordinates

Suppose that the particle centers are located equidistantly with a period d on a direct line parallel to the x-axis. Then

$$x_n = n \times d, y_n = 0 \tag{6.51}$$

Inserting Equation 6.51 into Equation 6.49, we find

$$I\left(x_0, y_0, z\right) = \frac{|\gamma|^2}{(\lambda z)^2} \times \left|F\left(x_0, y_0, z\right)\right|^2 \times |S|^2 \tag{6.52}$$

where

$$S = \sum_{n=1}^{N} q^n \tag{6.53}$$

is the sum of a geometric progression with factor

$$q = \exp\left(i\frac{k}{z}x_0 d\right) \tag{6.54}$$

Let us multiply Equation 6.53 by q, and deduct Equation 6.53 from the equation obtained. Then we have an equation with respect to S, the solution of which is

$$S = q \times \frac{1 - q^N}{1 - q} \tag{6.55}$$

Expression (6.55) is a formula for the sum of a finite geometric progression. It follows from Equations 6.52, 6.54, and 6.55 that

$$I\left(x_0, y_0, z\right) = |\gamma|^2 \times \frac{N^2}{(\lambda z)^2} \times \left|F\left(x_0, y_0, z\right)\right|^2 \times f_N\left(x_0, z\right) \tag{6.56}$$

where

$$f_N\left(x_0, z\right) = \left|\frac{\sin\left(kx_0 Nd/z\right)}{N \times \sin\left(kx_0 d/z\right)}\right|^2 \tag{6.57}$$

is a function describing the multiray interference effect. If $N \gg 1$, the function $f(x_0, z)$ is close to zero everywhere except for the points with coordinates

$$x_{0m} = m \times \frac{\pi z}{kd} = m \times \frac{\lambda z}{2d}$$

in which it is equal to unity.

Comparing Equations 6.56 and 6.50, one can see that the multiray interference effect radically changes the image of the diffraction pattern. It follows from Equations 6.56 and 6.57 that in the case of numerous regularly located particles the diffraction pattern looks like a set of bright spots. The distance between these spots is determined by the period d, while the total number of the spots characterizes the size and shape of an individual particle. Note that the function $F(x_0, y_0, z)$ in Equation 6.56 is defined by formula (6.23). In a particular case of the elliptical discs, it takes the form (6.32).

Similar results can be obtained for the case, when the particle centers are located in knots of rectangular lattice with periods d_1 and d_2. The diffraction pattern on such a set of particles is similar to the one produced by a superposition of two crossed diffraction gratings (see, e.g., Akhmanov and Nikitin [14]). Note that the regular location of erythrocytes in space can be realized with the help of holographic optical traps.

6.11 Description of the Experimental Setup

In order to measure the deformability of the erythrocytes as a function of the cells elongation in shear flow on the shear stress (or shear velocity) in the flow, we used the diffractometer LADE-6 designed and homemade in collaboration with the group of N. N. Firsov at the Russian State Medical University. The schematic layout of the setup is shown in Figure 6.4.

The major parts of the diffractometer are two coaxial thin-walled cylindrical cups made of polished organic glass. The coaxiality is reached by adjusting the inner cylinder with a special precision mechanism. The thickness of the gap between the cylinders walls is 0.8 mm. The dilute suspension of erythrocytes from a freshly drown blood sample is pored into the gap. The height of the cups and their radii are more than 10 times larger than the gap value. This condition allows for assuming the radial distribution of the flow velocities in the gap linear and the shear rate $\overline{\gamma}$ – nondependent on the radial coordinate:

$$\overline{\gamma} = \frac{2\pi RN}{d} \quad \text{where } R = (R_1 + R_2)/2$$

Here R_1 and R_2 are the radii of the cylinder cups, d is the width of the gap between the cups, and N is the rotation rate of the outer cup in rounds per second. At the known viscosity of the suspension η, the shear stress in the gap flow is defined as $\tau = \overline{\gamma} \times \eta$.

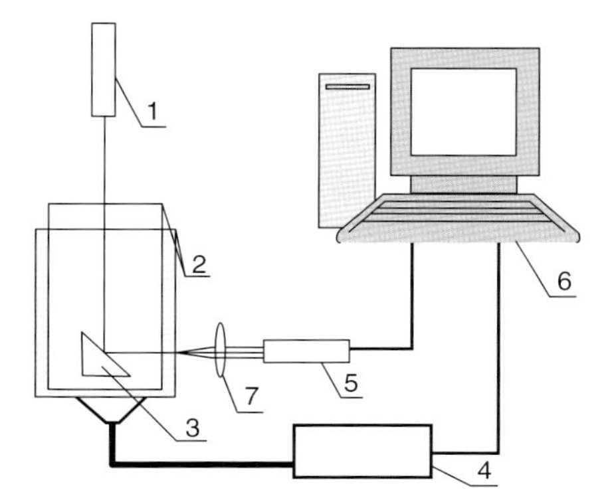

Figure 6.4 Schematic layout of the diffractometer: 1 – He–Ne laser, 2 – couvette chamber, consisting of two coaxial cylindrical cups with submillimeter width rheological gap between the walls; 3 – mirror, 4 – step motor, 5 – video camera, 6 – PC, 7 – lens.

In the course of measurement, the shear rate is changed stepwise from 13.8 up to 1550 s^{-1}. This allows for measuring the coefficients of elongation of the erythrocytes by shear stress in a wide range of shear rates.

The diffraction patterns usually obtained with such a setup clearly show the zero diffraction maximum and the first minimum and maximum of the intensity distribution. It was shown in Stoltz *et al.* [15] that in order to obtain adequate information on the erythrocyte shape, it suffices to measure only the sizes of the zero maximum. The pattern that is obtained as a result of light diffraction by a dilute suspension of erythrocytes looks like a set of concentric isointensity circles at zero or very low shear stress or a set of ellipses with their larger axes oriented $\Pi/2$ relative to the orientation of larger axes of the erythrocytes elongated by the shear stress. Schematically, the diffraction process and pattern obtained from a dilute suspension of erythrocytes with the setup described above is shown in Figure 6.5.

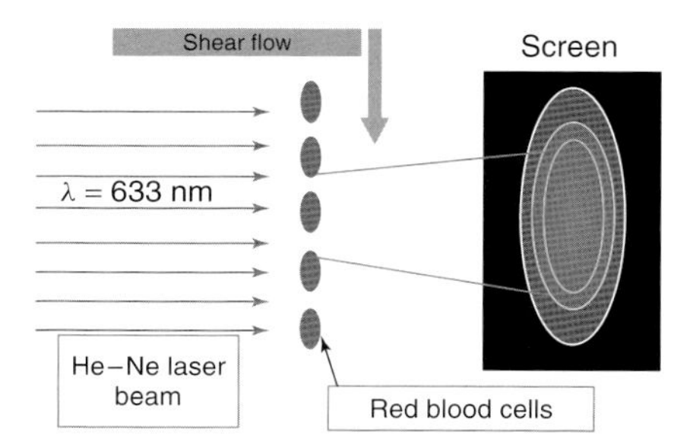

Figure 6.5 Schematic representation of the diffraction process and pattern obtained from a dilute suspension of erythrocytes with the diffractometer LADE-6.

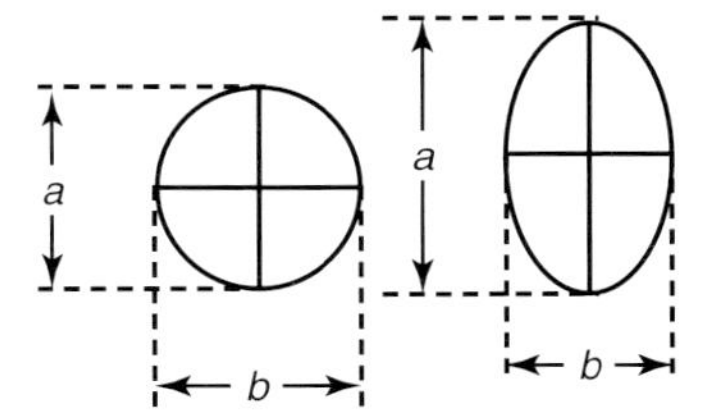

Figure 6.6 The definition of the deformability index.

We have shown above that in the case of highly dilute suspension of noninteracting particles the resulting diffraction pattern corresponds to the diffraction pattern of an individual cell. In the case of a stationary medium, the diffraction pattern looks like a circle, because the resulting field is contributed by scattering from an ensemble of chaotically oriented circularly shaped particle. In this case, the intensity distribution allows for calibrating the device.

Quantitatively, the elongation (deformation) of the erythrocyte at a fixed shear stress can be characterized by a parameter, which is often referred to in literature as the deformability index and is defined as a ratio

$$DI = \frac{a - b}{a + b}$$

where a and b – the long and the short semiaxes of the ellipse – are determined from the level of equal intensities of the diffraction pattern (Figure 6.6).

The actual deformability in shear flow is a function of shear stress and should be characterized by the deformability curve $DI(\eta)$. Such a definition of the deformability is true in the range of the diffraction angles 2–7°. At lower and higher angles, distortion appears due to wider angular dependence of the stray light and nonscattered laser radiation.

Practically, in order to raise the measurement precision, each point of the curve $DI(\eta)$ corresponding to a certain value of shear rate was measured several (minimum three) times. After that, the value of DI was determined for each file with the help of the laser beam analyzer program according to the isointensity level of the zero maximum. Averaging of several measured values yielded the resulting value with a statistical scatter of 5–7%.

Performing such measurements for different values of shear stresses, we obtain the deformability curve $DI(\eta)$ or $DI(\overline{\gamma})$. Typically, when plotted on a linear scale, the latter curve looks like that shown in Figure 6.7.

6.12
Sample Preparation Procedure

The concentration of the erythrocyte suspension is determined from the condition of single scattering at the given gap width. This implies that the whole blood (hematocrit 45%) sample dilution should be around 300-fold.

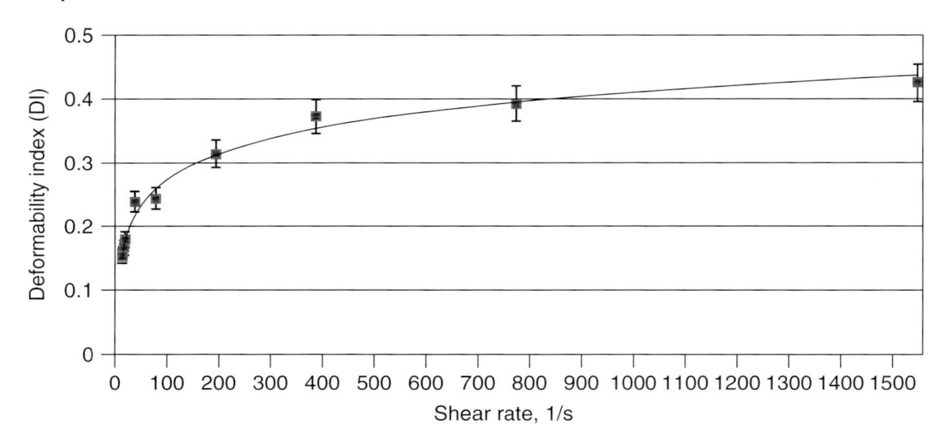

Figure 6.7 Typical dependence of the deformability index on shear rate for the erythrocytes of human blood drawn from a healthy volunteer.

Typically, a 1.4×10^{-2} ml sample of freshly drawn blood is diluted with 6 ml of solution consisting of distilled water, NaCl, and polyox – an isotonic solution of high molecular mass polyethylenoxide.

Indeed, the erythrocyte suspension could be prepared in the authologous plasma. However, in order to achieve the needed shear stress, values of around 600 dyn cm^{-2} and shear rates of 10^4 s^{-1} would be needed, which would not allow the sample to be kept within the rheologic dap. In the experiments, we used the suspending isotonic solution of high molecular weight polyethylenoxide in concentration of 0.5%, yielding the viscosity of 13 cantiPuas. The amount of additional light scattering from the polyox is very low and can be neglected.

All measurements were conducted with blood samples freshly drawn in sterile conditions from healthy volunteers or from human patients suffering from certain diseases along with prescribed blood tests. Also, we experimented with blood samples drawn from small animals, typically rats, with pathologies experimentally induced according to the protocols approved by the ethical committee of the university. All blood samples were stabilized with anticoagulant (7% solution of EDTA) in a concentration of 0.3 ml of EDTA solution per 10 ml of whole blood, which avoided blood clotting during the experiment.

6.13
Examples of Experimental Assessment of Erythrocyte Deformability in Norm and Pathology

Cerebral ischemia is the most spread form of cerebral blood circulation pathology. In our experiments, we aimed at finding the characteristic alterations in the microrheologic properties of blood at the alteration of the cerebral blood circulation during which nonuniform reperfusion takes place, the latter leading in many cases to both brain edema and appearance of vasospastic reactions of the cerebral vessels. Both can lead to the appearance of local sites of tissue blood supply insufficiency (so-called secondary brain ischemia). Thus, the dynamics of alterations of microrheologic parameters may reflect the appearance of the secondary deterioration of the cerebral blood circulations and, thus, serve as a prognostic feature. It cannot be excluded that a decrease in the erythrocyte deformability index in such conditions may not only be a consequence of secondary dysfunction of the cerebral blood circulations but also one of the factors promoting further complications of the stroke.

We used rats of the Krushinsky–Molodkina line [16] to model the alteration of cerebral blood circulation. The model of brain hemorrhage is reproduced in such rats when the high arterial blood pressure induced by a long-lasting acoustic shock (epileptic seizure) leads to a breakage of blood vessels and, consequently, to alteration of blood supply to the brain.

In what follows, we describe a typical experiment, in which the experimental animals ($n = 11$) were put into a chamber supplied with an acoustic generator for sonic stimulation [17]. The standard procedure of inducing brain ischemia included the following steps. The rats were subjected to 1.5 min long loud sonic stimulation (120 dB). Then the sonic action was continued during 15 min with periodic changes in 10 s-long acoustic noise with power levels 120 and 80 dB, respectively, and with 10 s-long intervals without sound. A control group of rats ($n = 7$) was kept during the same time inside a similar chamber without a sound stimulation.

Affecting the hemorrhagic rats, the strong sound induces the development of internal hematomas, which leads to the impairment of the microrheologic properties of the erythrocytes. In order to quantitatively characterize these effects, the measurements of the deformability curves were conducted before, immediately after, and 90 min after the sonic stimulation procedure. Overall, in the described experiment, 18 rats were tested, from which 11 species underwent the sonic stimulus and 7 rats served as a control group (without sonic stimulus). The preparation of rats for the experiment and blood suction from them was performed by certified personnel of the Faculty of Fundamental Medicine of the University who participated in this research project.

Figure 6.8 shows the percent change in the deformability index for the erythrocytes of the rats that were affected by the sonic stimulus and for rats of the control group, measured 90 min after the application of the stimulus. The value of the deformability index measured prior to the sonic action in taken as 100%.

In the experimental group, the deformability index measured at a shear rate $\overline{\gamma} = 789\ \mathrm{s}^{-1}$, 90 min after the exposure to sonic action was reduced by $11 \pm 2\%$ as compared to the control group. In the latter, the deformability index measured at this shear rate remained practically constant during the whole experiment.

The sonic action on hemorrhagic rats induces hematomas in their brain. A hematoma is a source of biologically active substances such as oxyhemoglobin, setorin, and so on. These substances when applied to blood may affect the erythrocyte membranes reducing the deformability index of the cells. Reduced

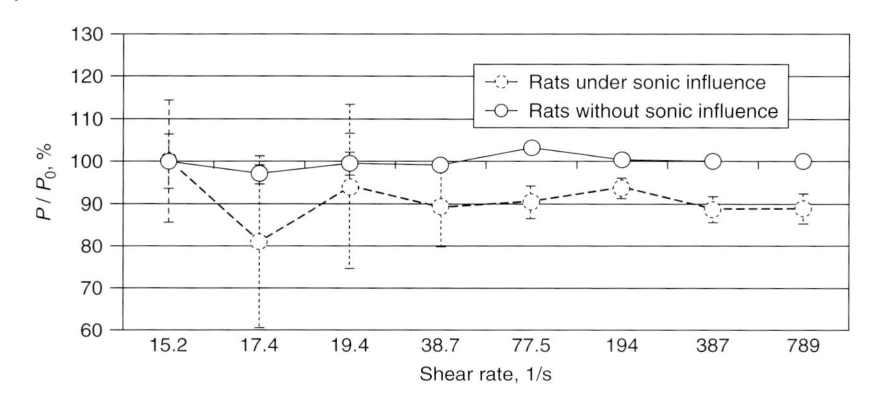

Figure 6.8 Relative alteration in the deformability index for the erythrocytes of hemorrhagic rats under the sonic influence and of the controls measured at different shear rates 90 min after the influence.

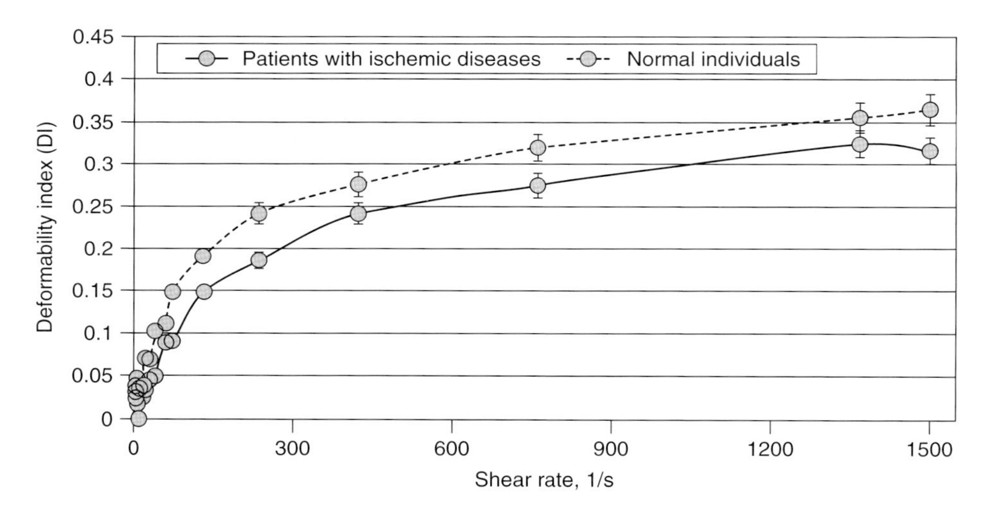

Figure 6.9 Dependences on shear rate of the deformability indices of the erythrocytes for the blood of healthy volunteers (norm) and for the patients suffering from the GDCBC (ischemic patients).

deformability of the erythrocytes of the hemorrhagic rats may lead to further impairment of blood delivery and consumption of oxygen in the brain tissues.

Therefore, we can speculate that in relation to human erythrocytes measuring the deformability index in the blood of stroke patients can also serve as a prognostic feature during medical treatment.

We have also conducted a series of laser diffractometry experiments aiming to study the microrheologic alterations in human patients suffering from general deterioration of the cerebral blood circulation (GDCBC). In one series of experiments, we studied 11 individuals, from which 6 were patients suffering from GDCBC and 5 were a control group of practically healthy volunteers.

Figure 6.9 shows the dependences of the deformability index on shear rate for the erythrocytes of these two groups of people. One can see that the deformability of the erythrocytes of individuals suffering from GDCBC is reduced relative to that of healthy individuals. The decrease in erythrocyte deformability in GDCBC in its turn raises the probability of closure of the capillaries and vessels and may lead to hemorrhages.

6.14 Conclusion

On the basis of the strict theory of diffraction and various approximations of the light scattering, we have provided an analytical description of the distribution of intensities of the diffracted light after illumination of a population of particles mimicking the erythrocytes deformed, while moving in a shear flow in the couvette of the ektacytometer. Doing so, we accounted for random distributions of sizes [18] and coordinates of the moving cells. We illustrated the potentialities of the device and the method with two examples of experimental and clinical studies.

References

1. Pafefrott, C., Nash, G.B., and Meiselman, H.J. (1985) Red blood cell deformability in shear flow. *Biophys. J.*, **47**, 695–704.
2. Principles, C.S. (1977) Principles and techniques for assessing erythrocyte deformability. *Blood Cells*, **3**, 71–99.
3. Reinhart, W.H., Chabanel, A., YaYo, M. *et al.* (1987) Evaluation of a filter aspiration technique to determine membrane deformability. *J. Lab. Clin. Med.*, **110** (4), 483–494.
4. Hénon, S., Lenormand, G., Richert, A., and Gallet, F. (1999) A new determination of the shear modulus of the human erythrocyte membrane using optical tweezers. *Biophys. J.*, **76**, 1145.
5. Guck, J., Ananthakrishnan, R., Mahmood, H., Moon, T.J., Cunningham, C.C., and Kas, J. (2001) The optical stretcher: a novel laser tool to micromanipulate cells. *Biophys. J.*, **81**, 767–784.
6. Grover, S.C., Gauthier, R.C., and Skirtach, A.G. (2005) Analysis of the behaviour of erythrocytes in an optical trapping system. *Opt. Express*, **7**, 533.
7. Gu, M., Kuriakose, S., and Gan, X. (2007) A single beam near-field laser trap for optical stretching, folding and rotation of erythrocytes. *Opt. Express*, **15** (3), 1369.
8. Leblond, P.F. and Coulombe, L. (1979) The measurement of erythrocyte deformability using micropore membranes. A sensitive technique with clinical applications. *J. Lab. Clin. Med.*, **94** (1), 133–143.
9. Bessis, M. and Mohandas, N. (1975) A diffractometric method for the measurement of cellular deformability. *Blood Cells*, **1**, 307–313.
10. Groner, W., Mohandas, N., and Bessis, M. (1980) New optical technique for measuring erythrocyte deformability with the ektacytometer. *Clin. Chem.*, **26** (10), 1435–1442.
11. Besis, M., Mohandes, N., and Feo, C. (1980) Automated ektacytometry: a new

method of measuring red cell deformability and red cell indices. *Blood Cells*, **6**, 315–327.

12. Johnson, R.M. (1994) Ektacytometry of red cells. *Subcell. Biochem.*, **23**, 161–203.
13. Streekstra, G.J., Hoekstra, A.G., Nijhof, E.-J., and Heethaar, R.M. (1993) Light scattering by red blood cells in ektacytometry: Fraunhofer versus anomalous diffraction. *Appl. Opt.*, **32** (13), 2266–2272.
14. Akhmanov, S.A. and Nikitin, S.Yu. (1997) *Physical Optics*, Clarendon Press, Oxford.
15. Stoltz, J.F., Ravey, J.C., Larcan, A., Mazeron, P., Lucius, M., and Guillot, M. (1981) Deformation and orientation of red blood cells in a simple shear flow. Theoretical study and approach at small angle light scattering. *Scand. J. Clin. Lab. Invest Suppl.*, **156**, 67–75.
16. Krushinsky, L.V. (1960) *Formation of the Animal Behavior in Normal and Pathological Condition*, Publishing House of the Moscow University, Moscow.
17. Lugovtsov, A.E., Priezzhev, A.V., and Tyurina, A.Yu. (2003) Rheological behavior of erythrocytes of hemorrhagic rats: examination by the method of laser diffractometry. *Opt. Technol. Biophys. Med. V, Tuchin V.- Ed, SPIE Proc.*, **5474**, 200–203.
18. Nikitin, S.Yu., Lugovtsov, A.E., and Priezzhev, A.V. (2010) On the problem of the diffraction pattern visibility in laser diffractometry of red blood cells. *Quantum Electronics*, **40** (12), 1074–1076.

7
Characterization of Red Blood Cells' Rheological and Physiological State Using Optical Flicker Spectroscopy

Vadim L. Kononenko

7.1
Introduction

The ultimate aim of cytometry as a conceptual approach in cell biology and medical diagnostics is the quantitative characterization of the distinguishing properties of every cell [1–4]. The nomenclature of properties involved in cytometric quantitation includes geometrical, optical, electric, magnetic, mechanical, and other physical–chemical parameters, and it is further expanding. One of the main trends in cytometry is characterization of cellular properties associated with dynamic processes of various natures. Among them, mechanical processes are of primary importance for blood cells, especially for erythrocytes. The point is that adequate functioning of these cells is conditioned in many aspects by their unique mechanical properties, both static and dynamic. Correspondingly, many blood pathologies result from alterations of these properties. Hence, quantitative characterization of erythrocyte mechanical properties, both static and dynamical, provides the basis for medical diagnostic applications.

Some new and promising possibilities in this branch of cytometry are associated with the use of flicker of erythrocytes as a source of quantitative data. The term "*flicker*" was introduced for the phenomenon of spontaneous low-frequency oscillations of the cell membrane of erythrocytes, irregular both in time and space [5, 6]. A detailed review of flicker investigations was given in Ref. [7]. In the following review, the main aspects of cytometric application of optical flicker spectroscopy are presented and discussed. First, theoretical models of erythrocyte flickering and existing registration techniques are described. The relations that connect the shape of both the frequency and the spatial spectra of chaotic membrane oscillations with the geometrical and mechanical parameters of erythrocytes, ambient physical characteristics, and instrumental parameters are given. Next, the existing concepts of excitation mechanisms of membrane flickering are presented. Then the main experimental results of flicker in erythrocytes are reviewed, and flicker interrelations between the physiological state and the membrane dynamics of a cell are considered. The diagnostic applications of flicker in erythrocytes and other cells are also discussed.

Advanced Optical Flow Cytometry: Methods and Disease Diagnoses, First Edition. Edited by Valery V. Tuchin.

For brevity, the term "*erythrocyte*" implies human erythrocytes because the great majority of flicker investigations were carried out on these cells. However, the results and conclusions of general character are valid for any mammal erythrocytes.

7.2
Cell State-Dependent Mechanical Properties of Red Blood Cells

Erythrocytes possesse neither nucleus nor intracellular organelles, and have no three-dimensional cytoskeleton. In a physiologically normal state, they have a characteristic shape of a biconcave disk with a mean diameter of 7.8 μm, a minimal thickness of 0.8 μm at the cell center, and a maximal thickness of 2.6 μm at the cell rim [8]. The cell volume is $V = 94\ \mu m^3$ and the surface area is $S = 135\ \mu m^2$.

From a mechanical standpoint, an erythrocyte can be considered as a thin viscoelastic shell (cell membrane) filled with a viscous fluid (intracellular hemoglobin solution) and placed into another viscous fluid (blood plasma or suspending medium). The observable mechanical behavior of such an object, that is, its deformations under various types of stress applied, depends both on the intrinsic material properties of its constituents (cell membrane and cytoplasm) and on their geometrical characteristics such as, the surface area/volume ratio. Erythrocyte volume is stabilized homeostatically by means of ion pumping [9]. In the normal state, its value is about 0.6 the volume of a sphere with the same surface area. This feature, together with the absence of spatial cytoskeleton, ensures high deformability of erythrocyte. That means mechanical and geometrical freedom for large variations of cell shape under a constant volume and surface area, which is necessary for the physiological functioning of erythrocytes. Those deviations from the normal state, which alter the surface/volume ratio, may change the observable mechanical characteristics of erythrocytes considerably. For example, osmotic swelling or decrease of cell surface area due to the loss of membrane lipids leads to effective "hardening" of erythrocyte even with material properties unchanged.

However, the material characteristics of the cell membrane and the cytoplasm determine the mechanical behavior of erythrocytes most directly. The cell membrane of erythrocytes has a composite multilayered structure. It consists of an outer lipid bilayer with embedded integral proteins, and a membrane skeleton attached to the bilayer on the cytoplasmic side [10, 11]. The membrane skeleton is a two-dimensional quasi-regular network formed by spectrin tetramers, which are linked together by actin oligomers at the network nodes connecting five to seven filament units. The skeleton is attached to the bilayer in a point-like manner by the cytoplasmic domains of transmembrane proteins, mainly by band 3 protein and glycophorin C. There are two types of linkage sites, which are multiprotein complexes [12]. The sites of one type are linked to the spectrin tetramers near dimer–dimer junctions. Here, ankyrin connects spectrin with band 3 protein. The tethering sites of the other type are at the nodes of the spectrin network. Here, protein 4.1R forms a complex with spectrin and actin, thus defining the nodal junction, while its attachment to glycophorin C makes a bond between the bilayer and the skeleton.

The mechanical properties of the cell membrane as a thin shell can be characterized phenomenologically by three elastic moduli and three coefficients of viscosity [13]. They are associated with three basic deformations of a small element of a shell: area dilation (condensation), in-plane extension or shear strain under constant area, and bending. The corresponding elastic moduli are: modulus of dilation K, shear modulus μ, and bending modulus K_c. On a microscopic level, each of these moduli represents specific properties of the relevant structural component of a composite membrane, namely, lipid bilayer or membrane skeleton. This feature leads to several-order difference in their values. The modulus of dilation K, which is determined by the properties of the lipid bilayer as two-dimensional incompressible liquid, possesses a value so high that the cell surface area S remains practically unchanged under all nondestructive deformations of the erythrocyte [13]. In particular, all diversified variations of erythrocyte shape in the circulatory system occur through bending and shear deformations of the cell membrane exclusively. The shear rigidity of the membrane is determined by the entropic elasticity [14] of the spectrin filaments of the membrane skeleton. The shear modulus μ is approximately five orders of magnitude smaller than the dilation modulus: for human erythrocytes, $K = 450$ mN m^{-1}, while $\mu = 6.6\,\mu$N m^{-1} at 25 °C [13]. The membrane bending modulus K_c, which is determined by the bending rigidity of the lipid bilayer, is in turn very much less than the shear modulus: for human erythrocytes, $K_c = 1.6 \div 2 \times 10^{-19}$ J [13], while on a corresponding energy scale $\mu S \approx 0.9 \times 10^{-15}$ J.

Owing to the area conservation of the cell membrane, its dynamic lateral deformations can be characterized by a single viscosity coefficient associated with a flow of lipids under shear deformation of the bilayer. Because of very small membrane thickness, the surface viscosity coefficient η_s is used instead of a bulk one. For human erythrocytes at 25 °C, $\eta_s = (5.1 \div 7.5) \times 10^{-7}$ N s m^{-1}. A description of dynamic bending deformations requires, in principle, a corresponding viscosity coefficient. Its value has been estimated to be 10^{-9} Pa s [5]. However, the bending membrane displacements are always accompanied by the motion of adjacent layers of the intracellular hemoglobin solution and the suspension medium. As a result, the energy dissipation in the course of bending deformations of erythrocytes occurs predominantly in the cytoplasm, with viscosity coefficient $\eta_{Hb} \sim 10^{-2}$ Pa s at 25 °C, and in the ambient medium.

Thus, for description of the static and low-frequency dynamical deformations of erythrocytes, five mechanical moduli are commonly used: K, K_c, μ, η_s, and η_{Hb}. The numerical values as well as the temperature dependence of these moduli for physiologically normal human erythrocytes were measured from quasi-static deformation [13]. However, flicker investigations showed that mechanical interaction between the lipid bilayer and the membrane skeleton should be taken into account for quantitative description of high-frequency bending deformations of the erythrocyte membrane [15, 16]. That adds another few mechanical parameters of the membrane. First, small-wavelength undulations u of the bilayer change its local distance to the skeleton. Their interaction energy varies approximately as $0.5\gamma u^2$, where γ is the binding constant in harmonic approximation. Second, the membrane skeleton of erythrocytes is stretched slightly because of the point-like pinning to the bilayer [14, 17]. That produces the counteracting lateral stress in the membrane. Dynamic bending of bilayer changes the distances between pinning points, thus modulating this membrane stress. Its magnitude $\sigma_{\text{eff}}(q)$ is determined by the entropic elasticity of spectrin filaments of the skeleton, network period, mean time number of filaments attached to the network nodes, and γ value, and depends on the characteristic wavelength $\lambda = 2\pi/q$ of bilayer dynamic deformations, q being the wave number [15, 16]. As a result, dynamic bending deformations of the erythrocyte membrane are characterized by the effective bending modulus $K_c^*(q)$, which can be written in a general form as:

$$K_c^*(q) = K_c + \frac{\sigma_{\text{eff}}(q)}{q^2} + \frac{\gamma}{q^4} \tag{7.1}$$

Particular expressions relating $\sigma_{\text{eff}}(q)$ to the structural and other microscopic parameters of a membrane, as well as estimates of γ are cited in Ref. [7].

There is another fundamental feature of erythrocyte mechanics, which is very important for diagnostic applications also. The magnitudes of the mechanical moduli of a cell depend, in general, on the physiological state of the cell, in particular, on the intracellular concentration of adenosine 5-triphosphate (ATP) [18] and the phosphorylation level of membrane proteins [19, 20]. Such dependence is especially pronounced for the shear elastic modulus of the membrane skeleton μ, which governs major part of erythrocyte deformations in physiological conditions. Thus, $\mu = 2.5 \pm 0.4\ \mu\text{N m}^{-1}$ for erythrocytes with normal ATP level [21] and $5.7 \pm 2.3\ \mu\text{N m}^{-1}$ for isolated membrane skeletons [18]. This may be conditioned by ATP-induced phosphorylation–dephosphorylation of skeletal proteins in erythrocytes [22]. The μ value is proportional to the statistical number of spectrin filaments connected via 4.1R proteins at the nodes of the membrane skeleton [14, 23]. Phosphorylation of these proteins by intracellular ATP may result in dissociation of spectrin filaments from the nodes with delayed restoration of the bonds, that is, in ATP-induced transient defects in the membrane skeleton [22]. Such defects lead to decreased μ value in erythrocytes, compared with the value characteristic of isolated membrane skeletons with no ATP [22, 24].

7.3
Flicker in Erythrocytes

The surface of erythrocyte placed in a physiologically normal medium and observed under a phase contrast (PHACO) high-magnificaton microscope appears in a state of incessant undulations, irregular both in space and time. Owing to such visual impression, the phenomenon was termed the "*flicker*" or "*flickering*" of erythrocyte (see [5] and references therein). The term "*membrane undulations*" is also used. To avoid confusion, note that with all terminological analogy and some common stochastic features, the "flicker phenomenon in erythrocytes" could hardly be included into the family of stochastic phenomena of various nature known as "*flicker noise*" or "$1/f^{\alpha}$-noise," where f is the frequency and $0.5 < \alpha < 2$, α being usually close to 1 [25–27]. The fundamental difference stems from the existence of an upper limit, namely, the cell size for spatial and hence for temporal membrane fluctuations. That results in the saturation of the low-frequency behavior of the frequency spectrum of erythrocyte flickering (Section 7.3.2), contrary to the classical flicker noise, where the $1/f^{\alpha}$ behavior was traced over several frequency decades down to $10^{-6.3}$ Hz [27].

7.3.1
Phenomenology of Cell Membrane Flickering

Membrane flickering appears to be a characteristic feature of erythrocytes. It was observed in various animal species: amphibia (frog), birds (chicken erythrocytes), and mammals [5, 17, 28]. Microscopic observations evidence that flicker is the process of chaotic bending oscillations of erythrocyte surface having a broad distribution of wavelengths in the scale of a cell. Such large-scale bending deformations of the cell membrane are merely due to three main factors characteristic of erythrocytes: low bending rigidity and vanishing lateral stress of a membrane; the geometrical freedom of the shape variations of erythrocytes due to the excess of cell surface area over that of a sphere of equal volume; and lack of three-dimensional cytoskeleton. Remarkably, that membrane flickering is observed also for "giant," about 20 μm in diameter, flaccid unilamellar lipid vesicles [29].

The quantitative characterization of membrane flickering is based on registration of chaotic membrane displacements $u(\vec{r}, t)$ from the time-average (equilibrium) position. Owing to the stochastic nature of flicker, $u(\vec{r}, t)$ is a random variable. Its quantitative parameters, consequently the measured flicker parameters, are statistical characteristics. The most commonly used are the mean square displacement $\overline{u^2}(\vec{r})$, the frequency spectrum (spectral power) $G(\omega, \vec{r}) = \langle |u(\omega, \vec{r})|^2 \rangle$, and the spatial spectrum (more generally, the modal composition) $\langle u(q)^2 \rangle$ of membrane oscillations. Mathematically equivalent correlation functions are also used [28]. Here, $u(\omega, \vec{r})$ is the Fourier transform of $u(\vec{r}, t)$ in time scale, f is the frequency of oscillations, and $\omega = 2\pi f$. Angular brackets mean statistical averaging. The mathematics underlying the notion of $\langle u(q)^2 \rangle$ is more complex. For flat surfaces, those oscillations are a superposition of plane waves, $u(q)$ is the Fourier transform of $u(\vec{r}, t)$ in space, and $q = 2\pi/\lambda$ is the wave vector of a plane wave. For surfaces of complex shape, including that of the erythrocyte membrane, the plane wave decomposition of oscillations is, in general, inadequate. The nature of q is also altered: it becomes a generalized wave number, a kind of multidimensional index of natural modes of thin-shell oscillations (Section 7.3.2). Here, the term "modal composition" is more appropriate for $\langle u(q)^2 \rangle$; its definition refers to the problem of calculation of natural modes (eigenmodes) of shell oscillations.

Three features of $G(\omega, \vec{r})$ and $\langle u(q)^2 \rangle$ spectra are very important for their correct measurement. First, both spectra are statistical characteristics and hence can be obtained only as statistical averages of sufficiently large numbers of individual spectra calculated for different samplings of registered displacement $u(\vec{r}, t)$. Second, the frequency spectrum $G(\omega, \vec{r})$ depends on the position of the registration point $\vec{r}$ on the cell surface because of the complex shape of erythrocytes. Third, the evaluation of spatial spectrum $\langle u(q)^2 \rangle$ by means of decomposition of the measured membrane displacements into plane waves has a rather limited range of validity, and hence should be carefully verified for each particular case.

Experimental studies of flicker in erythrocytes showed that both the frequency and the spatial spectra are very sensitive to the ambient physical conditions (osmotic pressure, viscosity, and temperature), to the influence of various chemical, biochemical, and physiologically active substances, as well as to the physiological state of erythrocytes and intracellular metabolic processes [7].

One of the main and, perhaps, the most intriguing aspect of flicker in erythrocytes comprises the problems of energy sources and excitation mechanisms of membrane oscillations, as well as the possible relations of these oscillations to erythrocyte functioning. Two alternative hypotheses were considered: (i) stochastic membrane oscillations are excited by the thermal motion of molecules in ambient fluid and in the cytoplasm of the cell, that is, thermally and (ii) the flicker arises because of chemical–mechanical conversion of cell energy in some metabolic processes. In a physiological context, these mean passive and active excitation, respectively. The Brownian movement observed for freely suspended erythrocytes, as well as the flicker in flaccid lipid vesicles, indicates that the thermal mechanism should play a major role in flicker excitation in erythrocytes. Indeed, theoretical and experimental studies carried out to date evidence convincingly that thermal agitation is the dominating excitation mechanism [7]. However, some experimental results were interpreted as the manifestation of active mechanism of flicker excitation, and several models of such mechanisms were considered theoretically (Section 7.3.2.3).

7.3.2
Theoretical Models of Flicker

Flicker theory comprises two problems: the oscillatory characteristics of erythrocytes, and the excitation mechanisms of membrane flickering. The oscillations of erythrocytes as thin viscoelastic shells are characterized by the bending eigenmodes $u(\vec{q}, \vec{r})$ and the corresponding eigenfrequencies $\omega(\vec{q})$. Here, $\vec{q}$ is the generalized wave number, which has a continuous value for an infinite surface and assumes a discrete set of allowed values for a finite-sized shell. A particular eigenmode $u(\vec{q}, \vec{r})$ describes a particular stationary distribution of the amplitude of oscillations over the shell (cell) surface determined by the q value as well as by the shell shape and the mechanical properties. The eigenmodes of a flat infinite membrane are plane waves [5], while those of a spherical shell are spherical harmonics [29]. The eigenmodes of shells with a more complex shape have a rather complicated structure, which can be described only numerically. This is the case for erythrocytes [30]. The eigenfrequencies $\omega(\vec{q})$ of a viscoelastic shell filled with some fluid and placed into another fluid depend on the elastic moduli of the shell, viscosity coefficients of both the shell and the fluids, as well as on the shell shape and the characteristic dimensions. They have complex values because of viscous dissipation of the energy of oscillations. Here, the real part of $\omega(\vec{q})$ gives the oscillation frequency, while the imaginary part gives the decay time of the corresponding mode.

The eigenmodes and eigenfrequencies of shell bending oscillations can be calculated using methods of thin-shell mechanics. Here, the set of equations considered includes the elasticity theory equations for bending and shear stresses in the shell arising because of its displacements from the equilibrium shape and the hydrodynamic (Navier–Stokes) equations for the fluid flows induced by these displacements both inside and outside the shell. The system is closed with the boundary conditions equations accounting for the force balance and velocity continuity at the shell–fluid boundaries. Such calculations showed that for erythrocytes (as well as for flaccid lipid vesicles), the eigenfrequencies of membrane oscillations are purely imaginary [5, 29–31]. Physically, that means overdamped oscillations, when the excited mode decays exponentially with decay time equal to the reciprocal modulus of the eigenfrequency. This situation is due to specific mechanical properties of erythrocytes and giant vesicles – low bending rigidity of the membrane together with high viscosity of the enclosed fluid.

After calculating the eigenmodes and the eigenfrequencies of membrane bending oscillations, the flicker amplitude can be written as their superposition: $u(\vec{r},t) = \sum_{\vec{q}} B_{\vec{q}} u(\vec{q},\vec{r}) \exp[j\omega(\vec{q})t]$, where $j = \sqrt{-1}$ and $B_{\vec{q}}$ is the mode excitation amplitude determined by the specific excitation mechanism. Then the frequency $G(\omega,\vec{r})$ and the spatial $\langle u(q)^2 \rangle$ spectra (or other relevant statistical characteristics) of $u(\vec{r},t)$ can be calculated using the methods of statistical mechanics [32, 33] or random noise theory [25]. At this stage, a model of flicker excitation mechanisms is required. The fundamental physical features of thermal excitation are equiprobable excitation of all potentially existing eigenmodes of membrane oscillations and equipartition of thermal energy gained from the ambient medium between all eigenmodes. Correspondingly, the spectra calculations made for thermal excitation either were based on the fluctuation–dissipation theorem [32] of statistical physics [5, 29, 30] or used the energy equipartition law together with the random noise theory [34]. On the contrary, the calculations of spectral contributions from the active mechanisms of excitation [22, 23, 35–38] require the specification of both the driving force and the mechanism of its action on the membrane. Here, the method of Langevin equation with random force [33] was used for flicker calculations in simple model membranes.

Another theoretical approach for description of flicker in erythrocytes is a direct computer simulation of the transient erythrocyte shape, using the methods of molecular dynamics [39–42]. Here, a closed two-dimensional network of triangular geometry, with virtual massive particles as the network nodes, models the cell membrane. These particles do not represent any actual membrane component; they are necessary only for the simulation procedure. This enables incorporation into the model the shape variations of the network through particle translocations, and description of these variations using the motion equations for virtual particles. That also allows introduction of an effective temperature T_{eff} of the network, by equating the mean kinetic energy of a particle to 1.5 times the thermal energy $k_B T_{eff}$ according to the equipartition law. Each particle is assumed to interact with each of its six neighbors via elastic spring potential, thus mimicking the shear elasticity of the membrane skeleton. To account for the membrane bending rigidity, the adjacent triangles of the network are assumed to interact one with another under the potential depending on the dihedral angle between their planes. The simulation procedure starts from some initial global shape of the network, with some initial velocities of particles, and proceeds in time under the restrictions of constant surface area, enclosed volume, and effective temperature [42]. After an equilibration stage, the network system attains a stationary time-average shape approximating the shape of the normal erythrocyte, and further remains in a steady state of fluctuating discocyte (DC). This enables calculation of the fluctuation spectra required as well as other flicker characteristics [42].

7.3.2.1 Models with Various Cell Shape under Thermal Excitation

The theory of thermal flicker of erythrocytes has been developed most completely, because it is based on the general results of statistical physics. According to these results, the power frequency spectrum $G(\omega)$ of deviations of simple overdamped systems from an equilibrium state due to thermodynamic fluctuations has a Lorentzian-type shape $G(\omega) \propto (\omega^2 + \tau^{-2})^{-1}$, where τ is the relaxation time [32]. For a multimodal overdamped system, such as erythrocyte, the summation over all the independent modes gives the general expression for the power frequency spectrum of membrane displacements:

$$G(\omega, \vec{r}) = 2 \sum_{\vec{q}} \frac{g_{\vec{q}} \left\langle u_{\vec{q}}(\vec{r})^2 \right\rangle |\omega(\vec{q})|}{\omega^2 + |\omega(\vec{q})|^2} \tag{7.2}$$

Here $\vec{q}$ is the generalized wave number, $\omega(\vec{q})$ is the imaginary eigenfrequency, $\langle u_{\vec{q}}(\vec{r})^2 \rangle$ is the mean square eigenmode amplitude of membrane oscillations at the surface point $\vec{r}$, and $g_{\vec{q}}$ is the mode multiplicity factor (number of modes with different $\vec{q}$ but the same $\omega(\vec{q})$).

The function $\langle u_{\vec{q}}(\vec{r})^2 \rangle$ describes actually the local modal composition of membrane flickering. It is closely related to the spatial flicker spectrum $\langle u(q)^2 \rangle$ introduced in Section 7.3.1: whenever the deviations of eigenmodes from the plane waves are small enough, $\langle u_{\vec{q}}(\vec{r})^2 \rangle$ coincides with $\langle u(q)^2 \rangle$. Otherwise, $\langle u(q)^2 \rangle$ is the surface-average of $\langle u_{\vec{q}}(\vec{r})^2 \rangle$.

A fundamental property of $\langle u_{\vec{q}}(\vec{r})^2 \rangle$ is that for thermally induced fluctuations, $\langle u_{\vec{q}}(\vec{r})^2 \rangle \propto k_B T$ regardless of the particular shape of the oscillating membrane, k_B being Boltzman constant and T absolute temperature. This follows in a general way from the energy equipartition law and from the proportionality of the elastic energy of membrane bending to the square of its displacement [5, 32]. Thus, for thermal excitation, the intensity of membrane flicker is proportional to absolute temperature.

In practice, the calculations of flicker spectrum (Equation 7.2) for erythrocytes encounter great mathematical and computational difficulties because of the complicated shape of erythrocytes. So, several models with simplified cell geometry

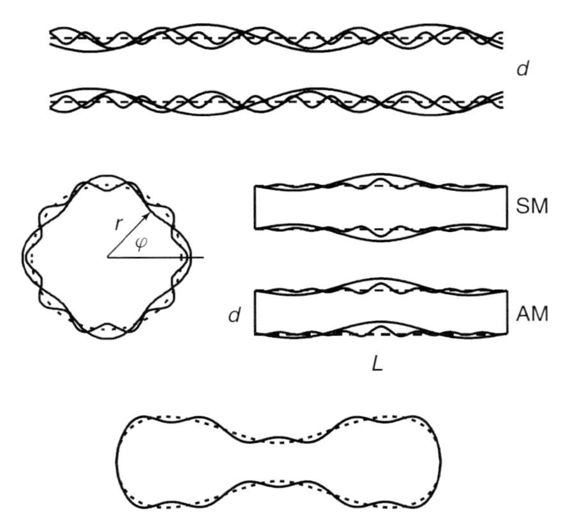

Figure 7.1 Axial cross sections of cell shape used in flicker models (from top to bottom): system of two unbounded plane membranes separated by a layer of viscous fluid [5], spherical flaccid membrane [29, 30, 43], flat disk (upper – symmetric modes, lower – antisymmetric modes) [7, 34], exact biconcave dick [30]. Dashed lines – mean time shape, solid lines – instantaneous shape, d – characteristic cell thickness, L – cell diameter.

were considered (Figure 7.1). They are: the infinite system of two plane membranes separated by a layer of viscous fluid [5]; a flaccid spherical vesicle filled with a fluid and placed into a fluid differing in viscosity [29–31, 43]; and a flat disk containing a fluid and surrounded by a fluid with different values of density and viscosity [7, 34]. Owing to the fixed geometry of the first two models, they did not allow considering the influence of the cell shape on the parameters of flicker spectra. Meanwhile, shape influence is very important both fundamentally and because of alterations of the equilibrium shape of erythrocytes under various influences. Historically, however, just plane and spherical flicker models were employed in the interpretation of major part of experimental data. On the contrary, the flat disk model enables approximate calculation of the influence of cell shape alterations, using the thickness/diameter ratio as the shape parameter.

The main advantage of approximate models of flicker is analytical expressions for eigenfrequencies and eigenmodes, and thereby for the flicker spectrum and modal composition. They give a clear insight into the relation between the spectral parameters and characteristic parameters of erythrocytes and the ambient medium. It is instructive to present here the results [30] for a flaccid spherical membrane with material properties characteristic of the erythrocyte membrane (bending-shear spherical model) and to compare them with the results for a flaccid vesicle with purely bending rigidity [29, 31, 43]. The eigenmodes of spherical membrane oscillations are surface spherical harmonics of degree n and order m, with $-n \leq m \leq n$ [29, 30, 43]. Thus, the mode wavenumber has two components,

$\vec{q} = (n, m)$. According to [30], for erythrocyte-type membrane

$$|\omega_n| = \omega_f \frac{(n-1)(n+2)[n(n+1)-6] + \frac{2\mu a^2}{K_c}\left[1 - \frac{2}{n(n+1)}\right]}{\frac{2\eta_s}{a\eta_i}\left[1 - \frac{2}{n(n+1)}\right] + \frac{(n-1)(2n+3)}{n} + \frac{\eta_e}{\eta_i}\frac{(n+2)(2n-1)}{n+1}}, \quad n \geq 2$$

$$\langle u_n^2 \rangle = \frac{Sk_B T}{4\pi K_c}\left\{(n-1)(n+2)[n(n+1)-6] + \frac{2\mu a^2}{K_c}\left[1 - \frac{2}{n(n+1)}\right]\right\}^{-1} \quad (7.3)$$

$$\omega_f = \frac{K_c}{\eta_i a^3}$$

Here, $S = 4\pi a^2$ and a are the area and mean time radius of the spherical membrane, K_c and μ are its bending and shear moduli, η_s is the surface viscosity, η_i and η_e are the viscosities of the internal and external fluids, and ω_f is the characteristic flicker frequency. Note that eigenfrequencies ω_n depend only on the value of n. They are the same for all eigenmodes with different l values corresponding to the same n. So, the mode multiplicity factor $g_{\vec{q}}$ in Equation 7.2 is equal to $2n + 1$ for a spherical membrane.

The model lipid membrane considered in [29, 31, 43] was assumed to possess neither shear rigidity nor any viscosity. Their results correspond, with minor difference, to expressions (7.3) with $\mu = 0$ and $\eta_s = 0$. The eigenfrequencies $|\omega_n|$ calculated for the mechanical parameters of erythrocytes according to the formulae from [43] with inclusion of medium viscosity [31] (open circles), and from [30] (solid circles) are plotted in Figure 7.2. These plots demonstrate clearly the strong influence of shear rigidity of membrane skeleton on flicker characteristics.

As Equation 7.3 shows, flicker can be characterized by effective frequency ω_f, which is proportional to the bending modulus and inversely proportional both to viscosity of enclosed fluid and to the cubed characteristic size of a cell. For eigenmodes with high $n \gg (2\mu a^2/K_c)^{1/4}$ (short-wavelength oscillations), their eigenfrequencies (inversed relaxation times actually) increase with n as $|\omega_n| \approx K_c n^3/2(\eta_i + \eta_e)a^3$ (Figure 7.2), while the mean square amplitudes decay as $\langle u_n^2 \rangle \approx Sk_B T/4\pi K_c n^4$. The same occurs in the model [29, 43] for $n \gg 1$. On the contrary, the lowest oscillations modes, which actually prevail in the spectrum, depend on n in a complicated manner that is quite different for the two models. The factor $2\mu a^2/K_c$ in bending-shear model (Equation 7.3) accounts for the influence of in-plane shear deformations of a membrane on its bending oscillations. For geometrical reasons, the bending deformations of a membrane (normal displacements u) are accompanied by its shear strains [5, 30, 44]. These strains and the shear stress arising are especially large for lowest bending modes with long wavelengths. The ratio $2\mu a^2/K_c \gg 1$ for erythrocyte parameters. As a result, the shear stress hampers the bending oscillations of the erythrocyte-type membrane at lowest n [30]. The magnitudes of $|\omega_2|$ and $|\omega_3|$ are drastically increased, while the corresponding values of $\langle u_n^2 \rangle$ are decreased compared with the purely lipid membrane with $\mu = 0$ (Equation 7.3 and Figure 7.2). That makes a vesicle effectively "harder." Such an influence of membrane skeleton on observed flicker intensity in erythrocytes is of great importance (Section 7.6).

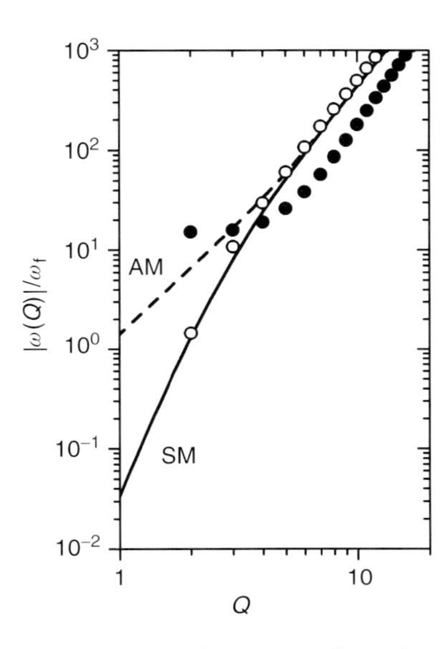

Figure 7.2 Eigenfrequencies of membrane bending oscillations calculated with characteristic parameters of erythrocyte according to various models of flicker. Open circles – purely bending spherical model [29, 31, 43]; solid circles – bending-shear spherical model [30]; solid and dashed lines – dispersion laws $\omega_s(Q)$ and $\omega_s(Q)$ for symmetric (SM) and antisymmetric (AM) modes according to the disk model [31].

Using the energy equipartition law, the spatial flicker spectrum $\langle u(q)^2\rangle$ (as well as functional dependence of $\langle u_{\vec{q}}(\vec{r})^2\rangle$ in Equation 7.3) for thermal excitation can also be obtained [5, 29, 30]. This requires the specifying eigenmodes $u(\vec{q},\vec{r})$ and the wavenumber q for particular cell shapes. The results are well established for a class of flat membranes with composite structure, which model erythrocyte cell membrane [5, 15]. Here, $u(\vec{q},\vec{r})$ are plane waves, and $\langle u(q)^2\rangle$ has the form:

$$\langle u(q)^2\rangle = \frac{k_B T}{S[K_c q^4 + \sigma_{\text{eff}}(q)q^2 + \gamma]} \tag{7.4}$$

In Equation 7.4, $\sigma_{\text{eff}}(q)$ is the lateral tension of a membrane, and γ is the binding constant in harmonic potential of the bilayer–skeleton interaction (Section 7.2). A similar result for a spherical membrane is the expression (7.3) for $\langle u_n^2\rangle$. These results show that the spatial flicker spectrum under thermal excitation can be written in a general form as $\langle u(q)^2\rangle \propto k_B T/K_c^*(q)q^4$. Here, $K_c^*(q)$ is a q-dependent dynamic bending modulus described by an expression similar to Equation 7.1 and obtained from its static prototype K_c by means of some model-specific renormalization.

Along with approximate flicker models, the rigorous theory was developed treating exactly the biconcave shape of erythrocytes [30]. It used the formalism of differential geometry in tensor representation to formulate the governing equations of elasticity theory and hydrodynamics. Of course, all final results can be obtained in this theory only in a numerical form. Both the analytical and computational procedures employed are very complicated, and prevented the theory [30] from being used widely in practical applications it deserved. Unfortunately, the calculations neither of frequency spectrum nor of modal composition of membrane flickering in DC were made in the framework of this theory, as well as of variation of these spectra due to alterations of the mean cell shape.

7.3.2.2 Flat Disk Model

A compromise between exact accounting of erythrocyte mean shape and the possibility of obtaining analytical expressions for the flicker spectrum and its parameter dependences gives an approximate theory based on the disk model for the cell shape [7, 34]. Here a thin flat disk of thickness d and diameter $L \gg d$ approximates the normal biconcave shape of erythrocytes. The disk surface area S remains constant because of conservation of the erythrocyte membrane area during physiological deformations (Section 7.2). Variations of cell mean shape are described in the model by parameter $\kappa = d/L$. For a flat disk the relations $S = (\pi/2)(1 + 2\kappa)L^2$ and $V = (\pi/4)\kappa L^3$ are valid, connecting variations of cell volume V and shape parameter $\kappa(V)$.

Using the discoid model, one may consider approximately the eigenmodes of membrane bending oscillations to be analogous to the natural oscillations of a round membrane fixed over its circumference. The latter are the products of radial-type and angular-type components. The first type is represented by Bessel functions of the first kind of integer order $J_n(2M_{nm}r/L)$, the second, by a periodic function of the polar angle φ. Here (r, φ) is the polar coordinate system at the disk face with the origin at its center and $n = 0, 1, 2 \ldots, M_{nm}$ is the m-th root of $J_n(x)$, $m = 1, 2, 3 \ldots$ The quantity $q_{nm} = 2M_{nm}/L$ serves as the generalized wavenumber of oscillations modes of a flat round membrane. Such choice of eigenmodes supposes a negligibly small contribution to the registered spectrum from the oscillations of the disk rim. So, the approximation is justified under $\kappa = d/L \ll 1$, which holds true for normal erythrocytes, and for registration area far enough from the cell rim, that is, for the central part of cell disk.

With the mode wavenumber set q_{nm} thus defined, the corresponding set of mode eigenfrequencies $\omega(q_{nm})$ can be calculated, provided the relation $\omega(\vec{q})$ (dispersion law) for membrane bending oscillations is known. To obtain this law, a special model system was considered in a general way [31]. The system consisted of two parallel flat membranes separated by a layer of viscous fluid of thickness d and embedded into outer fluid differing in density and viscosity. Such system is a generalization of the model considered in [5]. It was assumed that component membranes possessed the bending elasticity, but they have no shear rigidity and are not prestressed. Such approximation is well justified for the frontal surfaces of normal erythrocytes [5]. The eigenmodes of such system, which are plane waves, form naturally two manifolds due to the symmetry of the system relative to its central plane (Figure 7.1). The first one comprises the symmetric (s) (SM) modes, characterized by mutually opposite displacements of component membranes in the oscillation process. The second one consists of antisymmetric (a) (AM) modes, where both oscillating membranes move in the same direction. In the first case, the intermembrane distance oscillates, while in the second it remains constant.

Using this model, the dispersion law was obtained both for SM and AM modes, for arbitrary values of membrane bending modulus as well as the density and viscosity of inner and outer fluids [31]. In a range of comparatively low bending rigidity of component membranes and sufficiently high fluid viscosities, which is the case for erythrocytes, this equation has purely imaginary solutions $\omega_s(q)$ and $\omega_a(q)$. It was also shown that solutions obtained could be expressed with high precision in analytical form [7, 31]:

$$\omega_s(Q) \approx j\frac{\omega_f Q^6}{2}\left[12 + \left(1 + \frac{\eta_e}{\eta_i}\right)Q^3\right]^{-1}$$

$$\omega_a(Q) \approx j\frac{\omega_f Q^3}{2}\left(\frac{\eta_e}{\eta_i} + \frac{Q}{Q+4}\right)^{-1}, \omega_f = \frac{K_c}{\eta_i d^3} \tag{7.5}$$

Here $Q = qd$ is dimensionless wavenumber of bending mode, d is the distance between component membranes, K_c is membrane bending modulus, η_i and η_e are viscosity coefficients of inner (i) and outer (e) fluids, ω_f is characteristic flicker frequency, $j = \sqrt{-1}$. Relations (Equation 7.5) include as a particular case the results [5] obtained with $\eta_e = 0$ for the limiting cases $qd < 1$ and $qd \gg 1$.

The eigenfrequencies $\omega_{s,a}(q_{nm})$ are obtained from Equation 7.5 by substitution $q_{nm} = 2M_{nm}/L$ or $Q_{nm} = 2\kappa M_{nm}$. Considering that for a disk $d = [2S/\pi(1 + 2\kappa)]^{1/2}\kappa$, relations (Equation 7.5) evidence strong dependence of eigenfrequencies of membrane flickering on the cell shape parameter κ through both ω_f and Q_{nm}. A fundamental feature of dispersion law (Equation 7.5) following from hydrodynamics ([7] for details) is a radical difference between the wavenumber dependences of the SM $\omega_s(Q)$ and AM $\omega_a(Q)$ branches of flicker modes in a long wavelength, low frequency range $qd < 1$: $\omega_s(Q) \propto Q^6$, while $\omega_a(Q) \propto Q^3$. These two branches merge at a high Q (short wavelength) range only. Here, only adjacent layers of fluid influence membrane motion, so the physical difference between the s-modes and the a-modes disappears: $\omega_s(Q) \approx \omega_a(Q) \approx K_c Q^3/2d^3(\eta_i + \eta_e)$. This feature is very important for adequate description of measured spectra, because the lowest modes of erythrocyte flicker fall well within a low-frequency range. The plot of $|\omega_s(Q)|$ and $|\omega_a(Q)|$ calculated for parameter values typical of human erythrocytes at 25 °C is presented in Figure 7.2, together with the eigenfrequencies $|\omega_n|$ for spherical models with solely bending rigidity [43] and bending-shear rigidity [30] of the membrane. For erythrocytes, $K_c = 1.7 \times 10^{-19}$ J, $\mu = 6.6\ \mu\text{N m}^{-1}$, $\eta_i = 10^{-2}$ Pa s, $\eta_e = 10^{-3}$ Pa s, $\eta_s = 7.5 \times 10^{-7}$ N s m^{-1}, and $S = 135\ \mu\text{m}^2$. The average thickness of s normal biconcave erythrocyte $d \approx 2\ \mu$m, giving $L = 7.45\ \mu$m, $\kappa = 0.268$, and $\omega_f = 18$ rad s^{-1} for a model flat disk. Wavenumbers $Q_{nm} = 2\kappa M_{nm}, n \geq 1$, corresponding to several lowest flicker modes, are equal to 2.05, 2.75, 2.96, and 3.42, while the $|\omega_s(Q)|$ and $|\omega_a(Q)|$ branches of the dispersion law are merging only after $Q \sim 4$ (Figure 7.2). It is worth noting also that eigenfrequencies $|\omega_n|$ for purely bending spherical model [43] fall right onto dispersion curve $\omega_s(Q)$ obtained for the disk model (Figure 7.2). Altogether, this means insufficient correctness of the dispersion law $\omega(q) \propto q^3$ for description of erythrocyte flicker at lowest frequencies, which can be found in some papers [6, 17, 45–47].

With thermal excitation, membrane flicker can be considered as a superposition of all possible oscillation eigenmodes, which are excited at some random time moments t_k with random amplitudes B_k. These modes decay exponentially after excitation because of their overdamped character, with decay times $|\omega_{s,a}(Q_{nm})|^{-1}$ defined by Equation 7.5. Mathematically, such a time-dependent displacement $u(r, \varphi, t)$ of a membrane from its equilibrium position can be described by a succession of random pulses. For eigenmodes of discoid model specified above, it can be written as [7, 34]:

$$u(r,\varphi,t) = \frac{1}{\pi}\sum_{s,a}\sum_{\substack{n=0,\\ m=1}}^{\infty}\sum_{k=1}^{N_{nm}} \frac{B_{s,ak}(Q_{nm})}{|J_{n+1}(\mathrm{M}_{nm})|} J_n\left(\frac{2\mathrm{M}_{nm}r}{L}\right)\sin[\mathrm{n}\varphi + \psi_{s,ak}(Q_{nm})]$$
$$\times \exp\left\{-|\omega_{s,a}(Q_{nm})|\left[t - t_{s,ak}(Q_{nm})\right]\right\}\theta[t - t_{s,ak}(Q_{nm})] \quad (7.6)$$

Here, $Q_{nm} = 2\kappa M_{nm}$ is the mode wavenumber, $\psi_{s,ak}(Q_{nm})$ is the phase angle of the eigenmode, and $\theta(t) = 0$ for $t < 0, \theta(t) = 1$ for $t \geq 0$. The summation in Equation 7.6 is taken over all the eigenmodes (indices s, a, n, m), and over all the excitation events of each mode ($k = 1 \div N_{nm}$) preceding the time moment t. The SM mode $J_0(2M_{01}r/L)$ does not enter the sum (Equation 7.6), because it is suppressed mechanically because of changing the disk, that is, the cell, volume.

The power frequency spectrum of chaotic displacements (Equation 7.6) can be obtained using the random noise theory [25] together with the energy equipartition law for calculation of mean square mode amplitudes. It can also be obtained by direct substitution of expressions (7.5) for $\omega_{s,a}(Q_{nm})$ and other relevant quantities into the general spectral equation (Equation 7.2). The result for thermal mechanism of excitation is as follows [34, 48]:

$$G(\omega, r) = \frac{2Sk_BT}{\pi^2(1+2\kappa)}\sum_{s,a}\sum_{\substack{n=0,\\ m=1}} \frac{|\omega_{s,a}(Q_{nm})|}{K_c^*(Q_{nm})(\omega^2 + |\omega_{s,a}(Q_{nm})|^2)[J_{n+1}(\mathrm{M}_{nm})]^2\mathrm{M}_{nm}^4}$$
$$\times \left[J_n\left(\frac{2\mathrm{M}_{nm}r}{L}\right)\right]^2 \quad (7.7)$$

Here $K_c^*(Q_{nm})$ is the dynamic bending modulus of the membrane (Equation 7.1) and $\omega_{s,a}(Q_{nm})$ are defined by Equation 7.5. As in (Equation 7.6), the sum does not include the SM mode $J_0(2M_{01}r/L)$.

For evaluation of spectrum (Equation 7.7), it is convenient to use three integral-type parameters: the area under spectral curve $G_{\text{int}} = \int_0^\infty G(\omega)d\omega$, the "zero-frequency" amplitude $G(0)$, and an effective spectral width $\Delta\omega_{\text{eff}} = G_{\text{int}}/G(0)$. Note that G_{int} is proportional to the square membrane displacement $\overline{u^2}$ averaged over the total frequency range. Taking into account that several lowest modes give the main contribution to the spectrum, one may consider $K_c^*(Q_{nm}) \approx K_c^* = const$ while evaluating the integral parameters. Then,

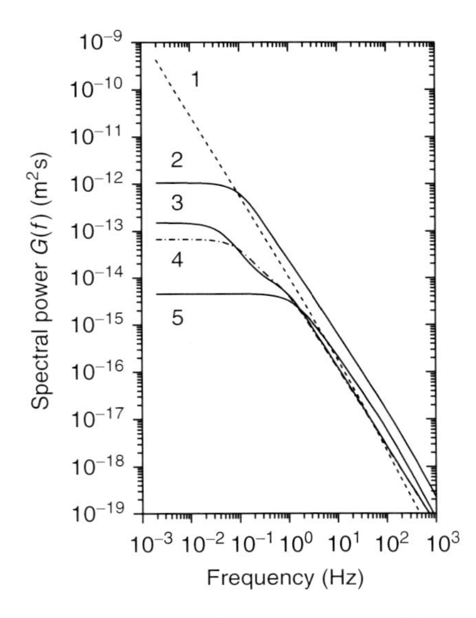

Figure 7.3 Power frequency spectra of thermally excited membrane flickering calculated according to various models of erythrocyte shape (Figure 7.1) and membrane mechanical properties. 1 – unbounded plane model [5]; 2 – spherical flaccid membrane with bending rigidity [29, 43]; 3, 4 – disk model with bending rigidity, for $d = 2$ μm, $L =$ 7.45 μm ($\kappa = 0.268$) and $d = 2.5$ μm, $L =$ 7.07 μm ($\kappa = 0.354$); and 5 – spherical flaccid membrane with bending and shear rigidity [30].

Equations 7.5 and 7.7 lead to following expressions [7, 34]:

$$G_{\text{int}}(r) = \frac{2Sk_BT\text{X}_1(r)}{\pi(1+2\kappa)K_c^*}$$

$$G(0,r) = \frac{3}{8\pi}\left[\frac{2S}{\pi(1+2\kappa)}\right]^{\frac{5}{2}}\frac{k_BT\eta_i\text{X}_2(r)}{K_c^{*2}\kappa^3}\left\{1+\left[\frac{\eta_e}{\eta_i}\text{X}_3(r)+\text{X}_4(r,\kappa)\right]\frac{\kappa^3}{\text{X}_2(r)}\right\} \quad (7.8)$$

$$\Delta\omega_{\text{eff}}(r) = \frac{8\pi}{3}\left[\frac{\pi(1+2\kappa)}{2S}\right]^{\frac{3}{2}}\frac{K_c^*\kappa^3\text{X}_1(r)}{\eta_i\text{X}_2(r)}\left\{1+\left[\frac{\eta_e}{\eta_i}\text{X}_3(r)+\text{X}_4(r,\kappa)\right]\frac{\kappa^3}{\text{X}_2(r)}\right\}^{-1}$$

Here $X_1(r) \div X_4(r,\kappa)$ are expressed analytically through $J_n(r)$ and M_{nm} [34].

Spectral curves calculated according to Equation 7.7 for the central part of erythrocyte disk ($r = 0$) and different values of cell shape parameter κ are presented in Figure 7.3. They are compared with frequency spectra calculated for other approximate models discussed above. The calculations used Equation 7.2 with formulae (Equation 7.3) and analogous formulae from [43], for typical parameter values of human erythrocytes cited after Equation 7.5. These plots, together with expressions (7.3), (7.7), and (7.8) elucidate the main features of frequency spectra of thermal flicker in erythrocyte.

Three general-type features follow from the comparison of curves in Figure 7.3. First, for a closed membrane, the spectral curve saturates at $\omega \to 0$, being nonlinear in a log–log scale at higher frequencies contrary to classical $1/f^\alpha$ systems. Second, both the shape and the magnitude of the low-frequency part of the spectral curve strongly depend on the shape of the erythrocyte and the shear rigidity of the membrane skeleton. Third, in a high-frequency range, the spectrum shape is rather insensitive to the cell geometry, but its magnitude remains very sensitive to the skeleton rigidity.

Analytical calculations based on Equations 7.5 and 7.7 show that $G(\omega) \propto \omega^{-4/3}$ in the intermediate low-frequency range, where the contribution from SM modes prevails, while $G(\omega) \propto \omega^{-5/3}$ in a high-frequency range [7]. The amplitude of modal contribution $G_{nm}(0) \propto |\omega_{s,a}(Q_{nm})|^{-1} M_{nm}^{-4}$ strongly decreases with the increase of the mode number, that is, with wavelength $\lambda \approx \pi L/M_{nm}$ shortening, because of the increase of both $|\omega_{s,a}(Q_{nm})|$ (Equation 7.5) and M_{nm}. Therefore, the main contribution to the spectrum gives a few lowest modes, with predominance of SM modes [31], whose oscillations are characterized by smallest values of $|\omega_s(Q_{nm})|$ (Figure 7.2).

Characteristic parameters G_{int}, $G(0)$, and $\Delta\omega_{\text{eff}}$ (Equation 7.8) of the frequency spectrum are simply related to the mechanical parameters of erythrocytes: G_{int} measures the effective bending modulus K_{c}^{*}, $G(0)$ and $\Delta\omega_{\text{eff}}$, taken together, allow evaluation of the membrane bending modulus and viscosity of the internal hemoglobin solution. However, such rather simple analytical relations between the spectrum parameters and cell mechanical moduli seem to be valid mainly for the central part of the cell disk. Here, the influence of the shear membrane strains accompanying its bending deformations can be well neglected [5, 44]. A more complex situation is seemingly characteristic of erythrocyte rim, where the membrane is always strongly curved. Here the shear stress arising may restrict considerably the flicker amplitudes at low frequencies, as Figure 7.3 shows for the bending-shear model. Hence, characteristic parameters of flicker spectra registered within the rim domain of the erythrocyte membrane seem to be very sensitive to the value of the shear modulus (Section 7.6.3).

A fundamental feature of thermal flicker following from equilibrium thermodynamics is that the area under spectral curve G_{int} (Equation 7.8) does not depend on the viscosity of both cytoplasm and the ambient medium – that means the same independence for the square amplitude of chaotic membrane oscillations $\overline{u^2}$ averaged over total frequency range, in practice, over sufficiently wide range comprising the most part of spectral curve in log–log scale (Figure 7.3). However, partial spectral amplitudes of thermal flicker $G(\omega)$ (7.7), as well as amplitudes averaged over rather short frequency bands, do depend on both viscosities involved. This follows from the proportionality of flicker eigenfrequencies ($\omega(\vec{q})$ in Equation 7.2, ω_n in Equation 7.3, $\omega_{s,a}(Q_{nm})$ in Equations 7.5 and 7.7) to characteristic frequency $\omega_f \propto \eta_{\text{i}}^{-1}$, and from the medium viscosity η_{e} entering the dispersion law (Equation 7.5) of flicker eigenmodes.

Another fundamental feature of the frequency flicker spectrum is the dependence of its shape and intrinsic parameters on the shape of erythrocytes, as well as on the position of the registration point $\vec{r}$ on the cell surface. Such dependence follows in a general way from the shell mechanics; it is quite evident also from the spectrum (Equations 7.7 and 7.8) and from curves 3 and 4 in Figure 7.3. As elucidated by the discoid model, the shape dependence results from two factors: variation of mode wavenumbers $Q_{nm} = 2\kappa M_{nm}$ with shape parameter κ and alteration of the modes' eigenfrequencies $\omega_{s,a}(Q_{nm})$ (Equation 7.5) which are proportional to $\omega_f \propto [2S/\pi(1+2\kappa)]^{-3/2}\kappa^{-3}$. The dependence on registration position has two main origins: nonhomogeneous distribution of eigenmodes amplitudes over the erythrocyte surface and position-dependent influence of the shear rigidity of the membrane skeleton on the bending deformations of the biconcave membrane discussed above.

7.3.2.3 **Active Excitation Mechanisms**

Along with the inevitable thermal excitation, some active driving forces of membrane flickering are expected for a living cell. As far as flicker spectra are concerned, the active mechanism may manifest itself in two ways. First, it may increase the cell temperature by some value ΔT, leading to more intense flicker excitation. In this case, both the frequency $G(\omega)$ and the spatial $\langle u(q)^2\rangle$ spectra will keep their characteristic "thermal" shape (described, e.g., by Equations 7.7 and 7.4) but will be registered for an effective temperature $T_{\text{eff}} = T + \Delta T$, different from the ambient temperature T. Another possibility is that an active mechanism distributes its energy between the flicker eigenmodes according to some specific law, which is essentially different from the equipartition law. That would change the shape of both the flicker modal composition (spatial spectrum) $\langle u(q)^2\rangle$ and the frequency spectrum $G(\omega)$, compared with their "thermal" shape. However, spectral contributions from various excitation mechanisms are summed independently. Therefore, to be detectable against a background of thermal contribution, the spectrum alterations due to active mechanism should possess some pronounced (e.g., nonmonotonic) dependence on the mode wavenumber, while the magnitude of alterations should depend on some nonthermodynamic parameter.

Nowadays, three models of active mechanisms of flicker excitation are considered quantitatively. They are rather schematic and essentially less elaborated compared with the thermal mechanism. The first model considers local membrane displacements due to the fluctuations of the local force applied normally to the membrane in the vicinity of ion pumps, which transport various substances through the membrane [35, 36, 38, 49, 50]. The model is a plane [35, 36, 49] or spherical [50] lipid membrane, which contains a random array of transmembrane protein "pumps" and is surrounded by a fluid with viscosity coefficient η. The proteins transport a substance through the membrane in chaotic [35] or stationary in-time [36] flows and execute diffusive motion in the membrane plane. The reaction force applied to the membrane as a result of the substance momentum transfer leads to the membrane local displacement. The fluctuations of this force due to stochastic functioning of protein "pumps" and (or) variations of their local concentration induce chaotic membrane oscillations. The model considered in [38] accounts both of lateral diffusion of proteins and of the fact that due to specific orientation in the membrane, some proteins may exert forces preferentially in one of two normal directions.

The second active mechanism considered is the ATP-induced stochastic process of temporal dissociation of spectrin filaments of the membrane skeleton from its nodes, which leads to fluctuations of the lateral stress in the nodes-pinned bilayer and its normal displacements [22–24, 37]. The point is that the membrane skeleton of a normal erythrocyte is slightly stretched [14, 17]. The stretched compartments of the spectrin network produce a compressive lateral stress in the bilayer through their attachment points. That builds a humpy bilayer relief governed by the network structure. ATP-dependent phosphorylation of 4.1R protein anchoring the spectrin filament to a network leads to filament dissociation from a node [20] and to a partial release of this node from the network. This results in perpendicular displacement of adjacent bilayer patch. A spontaneous restoration of the broken bond generates an opposite patch displacement. Thus, stochastic membrane oscillations can be excited due to ATP consumption [22].

The third active flicker mechanism proposed considers fluctuations of the bilayer curvature and the lateral mechanical stress near the integral proteins which transport ions through the membrane, in particular, near band 3 proteins [24, 37]. The fluctuations are induced by variations of protein conformation during the transportation cycle [51].

In the framework of the models proposed, various analytical expressions were obtained for both spatial and frequency spectral contributions from the active driving forces considered. A survey of these results was presented elsewhere [7]; here, the main conclusions concerning the spatial spectrum $\langle u(q)^2\rangle_{\text{act}}$, as well as the mean square amplitude of active membrane oscillations are presented. First, various types of q-dependence were obtained for $\langle u(q)^2\rangle_{\text{act}}$. Monotonically decreasing behavior, both with the q^{-4} law characteristic of thermal excitation (Equation 7.4) as well as with other powers of q, was found for the "active pumps" mechanism [35, 36, 49]. Nonmonotonic dependence of $\langle u(q)^2\rangle_{\text{act}}$, with the maximum at $q_m \sim (\gamma/K_c)^{1/4}$, was predicted for ATP-induced oscillations of model plane membrane with effective bending modulus $K_c^*(q)$ described by Equation 7.1 [24]. Second, it was shown that the magnitude of active contribution should depend not only on mechanical and geometrical cell parameters, but also on specific parameters characterizing the functioning of cell components (e.g., membrane proteins) and cell metabolic activity, in particular, on the intracellular ATP level. Third, a fundamental feature of active flicker mechanisms considered is the dependence of their spatial spectrum intensity on the viscosity of the surrounding medium, $\langle u(q)^2\rangle_{\text{act}} \propto 1/\eta$ [24, 35], unlike the spectrum of thermal flicker (Equation 7.4). Physically this is due to the fact that the stationary amplitude of membrane oscillations is determined here not by the equipartition of the thermal and mechanical energy, but from the balance between the active force pushing a membrane piece and the viscous drag from the adjacent fluid. Last, the theoretical arguments together with dynamic simulations evidenced that in certain regimes of active excitation, the probability distribution of fluctuation amplitudes is essentially non-Gaussian, contrary to the case of thermal excitation [38].

7.4 Experimental Techniques for Flicker Measurement in Blood Cells

Measurement of frequency and spatial spectra as well as other relevant flicker characteristics is accomplished using optical registration of the membrane displacements while observing a single cell under a microscope with a high-magnification objective $(40 \div 100)\times$. To minimize contributions from translational and rotational Brownian movements of the whole cell, the measured erythrocyte should be slightly adhered, at one or few points, to the bottom of the measuring chamber. This is ensured usually by poly-L-lysine coating of the glass bottom of the chamber [45]; it can also be achieved using erythrocytes unwashed from blood plasma [52, 53].

Depending on what is to be measured – frequency spectrum or modal composition – two different approaches are typically used. To obtain the frequency spectrum, a small area of the cell membrane is selected usually, and its chaotic displacements are registered using some optical technique. The fluctuating component of the detector signal is further analyzed to calculate its power spectrum $G_{\text{reg}}(\omega)$, time correlation function, or other required statistical characteristics. In the second approach used to obtain the spatial spectrum (modal composition) of flicker $\langle u(q)^2 \rangle_{\text{reg}}$, the instantaneous maps of 3-D relief of erythrocyte surface or the momentary contours of the cell equator are registered using some appropriate optical method. These maps and contours are further analyzed by means of a spatial Fourier decomposition into a series of plane waves or angular harmonics, respectively, to obtain the $\langle u(q)^2 \rangle_{\text{reg}}$ dependence after averaging the amplitudes of the squared modes.

These most used approaches are considered in detail in the next three sections. A few other measuring techniques were also used for flicker registering: dynamic light scattering of coherent (laser) radiation either by a single erythrocyte [28] or by erythrocytes suspension [30], and recently developed technique termed *defocusing microscopy* [54–56]. The latter technique enables direct registration of membrane curvature and its fluctuation spectra.

It should be pointed out that the measured flicker spectra $G_{\text{reg}}(\omega)$ and $\langle u(q)^2 \rangle_{\text{reg}}$ are in general differ in shape, modal composition, and other parameters from the corresponding intrinsic spectra $G(\omega)$ and $\langle u(q)^2 \rangle$ of membrane flickering. This is due to both method-specific and instrumental distortions of the intrinsic spectra in the registration process, discussed in Section 7.4.4.

7.4.1 Measurement of Frequency Spectra of Membrane Flickering

Optical techniques employed for the measurements in the frequency domain include PHACO microscopy [5, 34, 45, 47], reflection laser probing (LASCO) [34], point dark-field microscopy [57], and laser-based edge position detection [58].

7.4.1.1 Phase Contrast and Laser Probing of Central Part of Erythrocyte Disk

The relation between $G_{\text{reg}}(\omega, r)$ and $G(\omega, r)$ are most elaborated for PHACO and LASCO registration [34]. Also, the bulk of the frequency spectra measurements

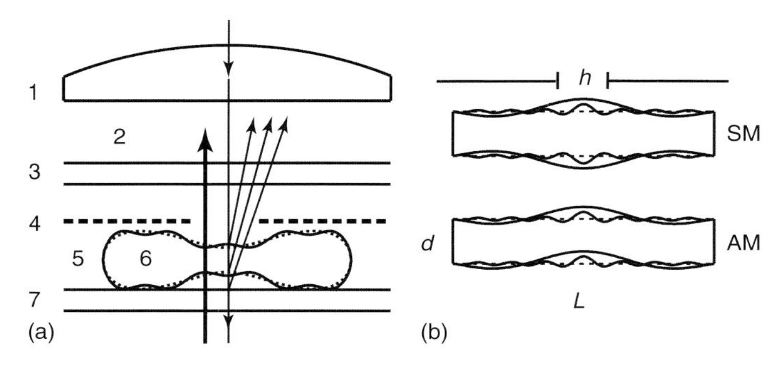

Figure 7.4 (a) schematic drawing of PHACO (thick arrow) and LASCO (thin arrows) registration of frequency flicker spectrum for disk face area of erythrocyte. 1 – objective; 2 – immersion oil; 3, 7 – cover slips; 4 – projection of the measuring diaphragm; 5 – suspending medium; 6 – erythrocyte (dashed line – time averaged, solid line, instantaneous cross-section of a cell). (b) schematic representation of symmetric (SM) and antisymmetric (AM) modes of membrane flickering according to flat disk model. h – projection size of the measuring diaphragm.

was carried out in a PHACO regime. So, it is expedient to consider this regime in detail as a general example. The optical setup for the PHACO and LASCO registration of membrane displacements is shown in Figure 7.4. For measurement, a much-diluted suspension of erythrocytes is placed into a microchamber with transparent windows (e.g., formed by a microscopic slide and a coverslip) at the object plane of the microscopic objective. After most of cells adhere to the chamber bottom, some loosely attached erythrocyte is chosen for measurement. The location, shape, and dimensions of the membrane patch whose oscillations will be registered are controlled by positioning the microscopic stage and by the geometry of the measuring diaphragm of the microscope. To monitor this adjustment, the measuring diaphragm is projected onto the cell surface by special illumination giving a corresponding light spot. The quantity registered in the PHACO regime is the intensity of light passed through the area of erythrocyte defined by the projection of the measuring diaphragm.

The theory of PHACO registration of thickness fluctuations of a transparent microobject was developed [34] using the general theory of PHACO imaging [59]. Let (x, y) be the rectangular coordinates on the object plane of a microscope. If $d(x,y)$ and $u(x,y,t)$ are, respectively, the time averaged and the fluctuating components of the object thickness, then for $|u| \ll d$, the intensity of the registered light $I(t)$ is written as:

$$\begin{aligned} I(t, x_0, y_0) &= I_0 \int_{S_{ap}} \left[a_{PC}^2 + A(x,y) + \frac{4\pi \Delta n}{\lambda_0} B(x,y) u(x,y,t) \right] \mathrm{d}x\, \mathrm{d}y \\ A(x,y) &= 2\left\{(1-\cos\alpha)[1-\cos\varphi_0(x,y)] + a_{PC}\sin\alpha \sin\varphi_0(x,y)\right\} \\ B(x,y) &= 2\left[(1-\cos\alpha)\sin\varphi_0(x,y) + a_{PC}\sin\alpha\cos\varphi_0(x,y)\right] \\ \varphi_0(x,y) &= \frac{2\pi \Delta n}{\lambda_0} d(x,y) \end{aligned} \tag{7.9}$$

Here I_0 is probing light intensity, a_{PC} and α are transmission coefficient and phase shift introduced, which characterize the phase plate of the microscope PHACO objective, Δn is refraction index difference between the object and the ambient medium at the wavelength λ_0 of probing light, and $\varphi_0(x, y)$ is the local phase shift of probing light introduced by an object. The coordinates x_0 and y_0 refer to the center of the projection of the diaphragm onto the object surface, and integration area S_{ap} in Equation 7.9 is the area of this projection. If a phase object is optically "thin" ($\varphi_0 \ll 1$), then the choice $\alpha = \pm\pi/2$ simplifies the expression (7.9) considerably. Here, independently of the value of a_{PC}, the choice $\alpha = \pi/2$ leads to a negative PHACO, while $\alpha = -\pi/2$ gives a positive contrast simulating light absorption regime. However, quite an opposite situation is typical of biological objects: thus, for human erythrocytes, $\Delta n \sim 0.05, d_{min} \sim 0.8\ \mu m, d_{max} \sim 2.5\ \mu m$, which gives $0.5 < \varphi_0(x, y) < 1.6$ at $\lambda_0 \sim 0.5\ \mu m$. For such optically "thick" objects, the positive and the negative contrast are achieved with some specific values of α and a_{PC}, which depend on φ_0 magnitude.

The signal of registering photodetector $i(t)$, which is proportional to the light intensity (Equation 7.9), consists also of the time average and the fluctuating components. Consider measurement with a point-like diaphragm. In practice that means the size h of the measuring diaphragm projection to be small compared with characteristic lateral distance for large enough variations of the object thickness. Then the photodetector signal can be expressed approximately as:

$$i(t, x_0 y_0) = i_0 \left[1 + \frac{A(x_0 y_0)}{a_{PC}^2} + \frac{4\pi\, \Delta n}{a_{PC}^2 \lambda_0} B(x_0, y_0) u(x_0, y_0, t)\right] \tag{7.10}$$

Here i_0 is the detector signal with no object ($\Delta n = 0$).

According to Equation 7.10, the registration of spatial variation of the time-average part of the signal (first two terms in Equation 7.10) enables reconstruction of the distribution of object thickness $d(x,y)$, provided the phase plate parameters entering $A(x,y)$ in Equation 7.9 are known. The third term in Equation 7.10 is proportional to fluctuations of object thickness, for example, due to the flicker. This enables calculation of the frequency spectrum of fluctuations and to follow variation of its intensity over the object surface. For a finite-size diaphragm the relations discussed become integral-type and are hence more complicated.

The expression for the flicker spectrum $G_{reg}(\omega, r)_{PHACO}$ registered by PHACO method is derived by substitution an appropriate series decomposition of membrane displacement $u(x,y,t)$ into eigenmodes of membrane oscillations in the general Equation 7.9 or 7.10 and by subsequent calculation of the power spectrum of the random succession thus obtained [31, 48]. In the framework of the discoid model of flicker, $u(x,y,t)$ is described by the series (Equation 7.6), which is written for an arbitrary position of the registration point $\vec{r}$ at the cell surface. We consider the case $r = 0$, because most of the measurement of spectra in a PHACO regime is made with the axially symmetric setup. Here the axis of erythrocyte disk coincides with the optical axis of the microscope, and the opening of circular measuring diaphragm is projected at the erythrocyte center as a circle of diameter $h < L$

(Figure 7.4). According to the principle of the PHACO method, only those oscillations eigenmodes can be registered that give nonzero variation of cell thickness averaged over the projection of the measuring diaphragm. This feature excludes from the registration all AM modes, which keep erythrocyte thickness constant, as well as all eigenmodes with the membrane displacement dependent on the polar angle, because in the axially symmetric setup, they give zero thickness variation averaged over a circular diaphragm. Mathematically, this means that after substitution of series (Equation 7.6) for $u(x,y,t)$ into the signal expressions (7.9) or (7.10) and subsequent integration, the only components remaining in the series will correspond to SM modes $J_0(2M_{0m}r/L)$ with exclusively radial dependence [31, 48]. The resulting expression for power spectrum $G_{\text{reg}}(\omega, 0)_{\text{PHACO}}$ of fluctuations of photodetector current produced by membrane thermal flickering described by intrinsic spectrum $G(\omega, 0)$ (Equation 7.7) has the form [31, 48]:

$$G_{reg}(\omega, 0)_{\text{PHACO}} = N_{tr} \left\{ \left[1 + \frac{2A(d)}{a_{PC}^2} \right]^2 \delta(\omega) + 128 \left[\frac{B(d) L \Delta n(\kappa)}{\lambda_0 a_{PC}^2 h} \right]^2 \right. \\ \left. \times \frac{S k_B T}{(1 + 2\kappa)} \sum_{m=1} \frac{|\omega_s(Q_{0m})| \left[J_1\left(\frac{M_{0m} h}{L}\right) \right]^2}{K_c^*(Q_{0m})[\omega^2 + |\omega_s(Q_{0m})|^2][J_1(M_{0m})]^2 M_{0m}^6} \right\} \\ + G_{NF}(\omega) \tag{7.11}$$

Here N_{tr} is the gain factor of the signal detection and processing circuit, $\delta(\omega)$ is the delta-function, functions $A(d)$ and $B(d)$ are defined by expressions (7.9) for $d(x, y) \equiv d$, eigenfrequencies $|\omega_s(Q_{0m}, \kappa)|$ are given by Equations 7.5 with $Q_{0m} = 2\kappa M_{0m}$ and $M_{0m} = 2.40, 5.52, 8.65, 11.79, 14.93$, so on, being the roots of Bessel function of zero order $J_0(x)$. The component $G_{NF}(\omega)$ stands for contribution to the registered spectrum from all noise sources other than membrane flickering. Note that the refraction indice difference $\Delta n(\kappa)$ varies with erythrocyte shape because of the accompanying variation of hemoglobin concentration [48]. The integral parameters of the flicker component of the signal spectrum (Equation 7.11) differ from those of the intrinsic spectrum (Equation 7.8) at $r = 0$ mainly by the product in square brackets before the sum in Equation 7.11.

The principle of LASCO registration of membrane flickering is the optical setup with the reference beam (Figure 7.4). Here, a laser beam illuminates the object (erythrocyte) through microscopic objective, and backscattered radiation collected by the same objective is registered by a photodetector [52]. This radiation consists of three components: two comparatively weak light waves reflected from the top and the bottom membrane sheets of the erythrocyte resting on a glass support, and a far more intense wave reflected from that glass (thin arrows in Figure 7.4). The interference of these three waves at the sensitive area of detector results in a signal consisting of a large stationary component defined by light reflection from the glass support, and two weak independent components related to reflection from the upper and lower membrane sheets. Their intensities depend on transient displacements $u_1(x,y,t)$ and $u_2(x,y,t)$ of the corresponding sheets from time-average positions. For $|u_1(x, y, t)|, |u_2(x, y, t)| \ll d$, the expression describing

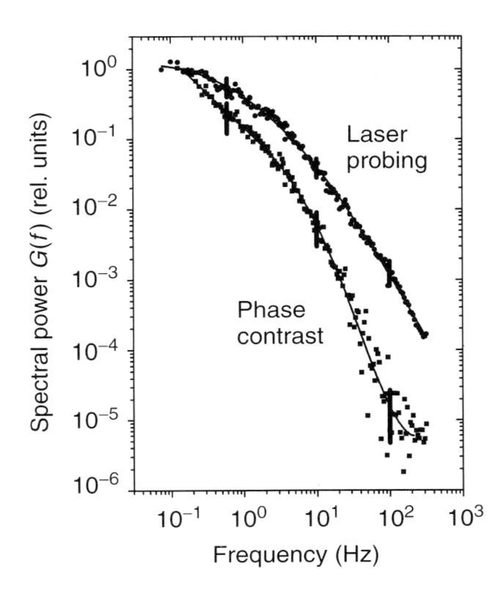

Figure 7.5 Power frequency spectra of flicker measured for the central part of the same erythrocyte using phase contrast regime (squares) and reflection laser probing (circles) with $h = 1.3$ μm [34]. Averaging over 200 samplings. Approximating polynomials (curves): $y = 0.0069x^4 - 0.0032x^3 - 0.2983x^2 - 0.7160x - 0.3898$ (LASCO regime) and $y = 0.0877x^5 - 0.1415x^4 - 0.2795x^3 - 0.0797x^2 - 0.9794x - 0.8299$ (PHACO regime), $x = lg(f), y = lg(G)$. Spectra are normalized to the same level at the lowest frequency. Vertical dashes show power determination errors. High-frequency slopes of spectra are −2.3 (PHACO) and −1.5 (LASCO).

detector signal has a structure similar to Equation 7.9, but it contains two small terms proportional to $u_1(x,y,t)$ and $u_2(x,y,t)$, instead of one proportional to $u(x,y,t)$ in Equation 7.9. Hence, reflection laser probing enables registration of both SM and AM flicker modes, contrary to the PHACO regime. The expression for frequency spectrum $G_{\text{reg}}(\omega, 0)_{\text{LASCO}}$ registered with axially symmetric setup in LASCO regime is obtained similar to Equation 7.11. It has an analogous structure and contains the same instrumental factors $J_1(M_{0m}h/L)$. The main difference is that the summation is carried out over both s- and a-modes. Also, the instrumental factor before the sum is determined by other parameters describing light reflection and transmission at the boundaries involved (Figure 7.4).

Figure 7.5 shows power frequency spectra $G_{\text{reg}}(\omega)_{\text{PHACO}}$ and $G_{\text{reg}}(\omega)_{\text{LASCO}}$ registered for a normal erythrocyte, using PHACO and laser probing with axially symmetric setup [34, 52]. Individual samplings of detector signal necessary to obtain the averaged power spectrum were registered interchangeably in PHACO and LASCO regimes for the same cell. This ensured adequate account of possible time drifts during the registration process in averaged spectra obtained in these regimes. Figure 7.5 clearly illustrates the fundamental feature of flicker registration procedures pointed out at the beginning of Section 7.4: the modal composition of the spectrum registered in a PHACO regime is strongly reduced compared with that measured by reflection laser probing.

7.4.1.2 Point Probing of Cell Edge Fluctuations

For local registration of frequency spectra of erythrocyte edge flickering two techniques were employed: point dark-field microscopy [57, 60] and laser-based detection of cell edge position [58] (Figure 7.5). In point dark-field regime, the edge of measured cell is illuminated by a narrow light beam, which is formed from halogen or mercury lamp radiation using an off-axis rectangular field diaphragm and a high-magnification (100×, NA 1.3) objective as a dark-field condenser [57]. The beam direction is adjusted to miss the observation microscope objective, so the image field is dark in the absence of a cell. If a cell edge is introduced into the beam path, then some light is directed into the objective because of diffraction or reflection from this edge, giving the photodetector signal. Accordingly, edge oscillations generate oscillating signal component, which enables registration of their frequency spectrum, as well as measurement of their amplitude provided a preliminary calibration has been done. In practice, the light beam emerging at the angle $\sim 15-20^{\circ}$ relative to the bottom glass of measuring microchamber is focused at the edge of erythrocyte attached to this bottom, the light spot area being $0.25 \div 1\ \mu m^2$ [57]. A calibration procedure using glutaraldehyde-fixed erythrocytes and special vibrator evidenced a linear dependence between the displacement of the cell edge and the scattered light intensity in the range 10–300 nm, and minimal detected edge displacements of 6–8 nm [60]. It should be noted that point dark-field measurements require very fine initial adjustment of cell distance to the probing beam, because of very strong dependence of light reflection angle on both this distance and cell shape (Figure 7.5). It remains unclear from original description of the method, what are the main steps and criteria of such adjustment, what is the level of its reproducibility for various cells in preparation and among different operators, and what is the validity range of results of a separately done calibration procedure.

No quantitative connection between $G(\omega)$ and $G_{reg}(\omega)$ is formulated for point dark-field regime because of the very complex geometry of light reflection from the curved outer edge of the erythrocyte. Such connection requires precise relating the light intensity collected by observation objective with the instantaneous position and orientation of a small reflecting element of the erythrocyte rim, which is translocated and rotated simultaneously in the process of cell surface oscillations. As is evident from Figure 7.5, such relation is, in general, nonlinear and nonmonotonic because some combinations of a turn and a displacement of reflecting element may hardy alter the intensity of collected light. From geometrical optics, it is also clear that the theory of registration should necessarily consider the actual shape of erythrocyte, as well as the influence of this shape alterations on initial optical adjustment hence on intensity of registered signal in point dark-field regime.

Another technique for local registration of erythrocyte edge flickering [58, 61] uses laser-based edge position detection developed initially in optical trapping of microobjects [62]. Here a laser beam aligned with the optical axis illuminates a probing object (erythrocyte) through a high-magnification (100×, NA 1.3) objective of inverted microscope (Figure 7.5). The position detection scheme employs the interference between the forward-scattered light from the object and the nonscattered light at the back focal plane of the microscope condenser (second oil-immersion objective, 63×, NA 1.32, placed on top of measuring chamber). The interference intensity is monitored with a four-quadrant photodiode positioned in this plane. The magnitude of photodiode signal is proportional to the displacement of the object from the laser beam axis [62]. This enabled calibration of the signal value versus the object (erythrocyte edge) position either by step-wise moving the object through the laser using a piezoelectric stage [58], or by means of step-wise deflection of laser beam using acousto-optical deflectors [61], and the step size was 4 or 21 nm. Typical response curve for erythrocyte edge exhibited a linear region of about 300–400 nm [61]. This allowed registration of time-dependent fluctuations of erythrocyte edge position at a single chosen point with an accuracy of ± 2 nm with microsecond time resolution [58, 61].

Both techniques described above enable fine monitoring of cell edge displacements. However, this high sensitivity is combined with some specific features, which should be considered necessarily while interpreting the measurement results. Both techniques register displacements of a single small spot chosen from the whole cell circumference. For this reason, they cannot distinguish between the local membrane oscillations, which leave the mean cell contour untouched, and the global displacements or deformations of the cell body. During the measurement, the erythrocyte is attached to the bottom of the measuring chamber only at few points or at some annular zone of the lower part of its surface. Hence, the upper part of the cell body is free to execute confined lateral displacements ("swinging") due to Brownian motion (marked by bidirectional arrow in Figure 7.5). Such swinging is accompanied by global changes of the shape of the erythrocyte, with pronounced shear deformations of cell membrane at the rim area. As a result, the registered fluctuation spectrum is, in essence, the sum of two independent spectral contributions with seemingly comparable amplitudes: one from the membrane flickering and the other from the cell "swinging." The shape of these spectrum components and their parameters relation to mechanical parameters of erythrocyte are essentially different.

It should be pointed out that under axially symmetric PHACO and LASCO registration, such "swinging" and displacements do not contribute to the registered spectrum. This is due to a nearly flat geometry the central part of erythrocyte within the measuring diaphragm projection. Its thickness remains practically unchanged in this process (Figure 7.5).

7.4.2
Quantitative Phase Imaging of Entire Cell

The frequency spectra of membrane flicker have an essentially local character due to the complex shape of erythrocytes and hence they can be registered locally as well. On the contrary, the measurement of spatial flicker spectra (or modal composition of flicker) requires full-field registration of cell membrane oscillations. This Section considers the registration of instantaneous thickness "maps" $d(x,y,t)$ for erythrocytes resting horizontally on a glass support. This goal can be achieved by optical measurement of corresponding phase shift $\varphi(x, y, t) = (2\pi \Delta n/\lambda_0)d(x, y, t)$ introduced by a measured object to the probing light, provided the refraction index difference Δn between the object and ambient medium remains constant both in space and time. According to general considerations and experimental data, the latter assumption holds true for flickering erythrocytes.

Modern optics offers a wide variety of techniques for such measurements. It begins with PHACO microscopy, where dynamic phase shift distribution is described by the integrand expression in Equation 7.9. However, this technique has not yet been employed for registration of full-field thickness maps of erythrocytes, maybe partly because of the absence of the a_{PC} and α values for commercial objectives.

Another technique, used for original registrations of spatial flicker spectrum in erythrocytes, is dynamic reflection interference contrast [6, 63, 64]. Here, a measuring chamber with erythrocyte suspension is placed at the stage of inverted microscope equipped with an antiflex oil immersion objective (e.g., Antiflex Neofluar 63×, NA 1.25) to minimize the strong reflection at the air–measuring chamber interface. The measured cell is illuminated through this objective with monochromatic light of $\lambda = 546$ μm filtered from high-pressure mercury lamp radiation. Then the brightness distribution is registered with CCD camera for a cell image that has resulted from the interference of light waves reflected from the two interfaces: the upper boundary of the glass plate supporting the erythrocytes, and the lower part of erythrocyte membrane facing this plate. If an object possesses rotational symmetry and the distance u between its lower boundary and the supporting plate is stationary, $u = u_0(r)$, such image is a system of concentric dark and bright rings (Newton rings). If the object surface undulates, then $u = u(r, \varphi, t)$ and the rings system also undulates, the instant circumferences of the "rings" not being circles any more. The distribution of object image brightness obeys the equation [63]:

$$\begin{aligned} I(r, \varphi, t) &= A_0 + 0.5A_1 \cos 2ku(r, \varphi, t) \\ A_0 &= E_0^2 + E_1^2, \; A_1 = E_0 E_1 \end{aligned} \tag{7.12}$$

Here k is the light wavenumber and E_0 and E_1 are the amplitudes of light waves reflected from the glass support and the object surface, respectively.

Making the expansion $u(r, \varphi, t) = u_0(r) + \delta u(r, \varphi, t)$ with $|\delta u(r, \varphi, t)| \ll u_0(r)$, it is possible to split Equation 7.12 into the stationary and dynamic parts, similar to Equation 7.9 used in the PHACO theory. This enables reconstruction of the stationary shape $u_0(r)$ of the lower sheet of the erythrocyte membrane, which faces the glass support, and the spatial Fourier analysis of the fluctuating component of membrane displacement $\delta u(r, \varphi, t)$ to obtain its spatial spectrum $\langle u(q)^2 \rangle$, using interference images of erythrocytes registered for successive time moments [63]. This method gave a possibility to register membrane oscillation modes in the range 0.25 μm $< \lambda <$ 4 μm with an amplitude resolution better than 0.3 nm and time resolution better than 20 ms [64].

The basic principle of measuring techniques described above is the transformation of phase shift distribution $\varphi(x, y, t)$, which characterizes the object structure, into intensity distribution $I(x,y,t)$ of the registered object image. As a result, the phase distribution is obtained from the measurements indirectly, by solving an inverse mathematical problem. Owing to nonlinear relationship between $I(x,y,t)$ and $\varphi(x, y, t)$, as well as to limited accuracy of $I(x,y,t)$ digital evaluation, such approach is both complicated and approximate.

During the past decade, a whole family of quantitative phase imaging techniques emerged, which enable direct precise measurement of phase shift distribution $\varphi(x, y, t)$ characterizing a transparent object ([65] for review). These techniques use quite different, namely, interferometric principles for phase shift measurement, which are direct and far more precise. It is essential that many of them enable simultaneous acquisition of phase shifts from the whole object area (full-field phase measurements). That gives a unique possibility to measure time-dependent, whole-object thickness distribution for transparent objects of various geometries with nanometer sensitivity and millisecond time resolution [65].

One of such novel approaches for spatial analysis of erythrocyte membrane flickering is digital holographic microscopy [66, 67]. This technique combines traditional holography, which is accomplished here using an inverted microscope, with the digital recording of the hologram obtained and numerical reconstruction of this hologram to obtain instantaneous phase "maps" of measured erythrocyte. According to [66], this enables exploration of time variations of cell shape which occur on a microseconds-to-hours time scale, with lateral resolution $\sim 0.1\ \mu m$ and displacements resolution of few nanometers.

At present, diffraction phase microscopy (DPM) seems to be the most adequate technique for membrane flicker measurement in the disk face area of erythrocyte [65, 68]. Here a standard inverted microscope is combined with a laser as a source of coherent radiation for sample illumination, and a phase diffraction grating is placed at the path of the light transmitted by the measured object. The grating serves for isolation of the zeroth and the first diffraction orders from transmitted light, for subsequent use them as the reference and the sample fields in a standard interferometric optical setup combined with the microscope also. The interference pattern is formed optically at the sensitive area of a detector (CCD camera). The resulting interferogram is a fringe system, which is distorted from an ideal periodicity in accordance with instantaneous phase shift $\varphi(x, y, t)$ introduced by the object at the registration moment $t = const$. The interferogram registered by the CCD camera is further processed to retrieve a momentary $\varphi(x, y, t)$ distribution using the approach termed Hilbert phase microscopy (HPM) [65, 69]. The main advantage of HPM processing procedure is that it enables a quantitative phase image $\varphi(x, y, t)$ to be obtained from only one spatial interferogram recording ("single-shot" registering). Hence, the acquisition speed of $\varphi(x, y, t)$ is determined in the DPM–HPM technique by the frame acquisition rate of the recording device. For the CCD camera used in [69], it was 291 frames/s, but with the recent advances in two-dimensional array detectors it may amount to thousands of frames per second.

Thus, DPM appears to be rather universal technique for investigations of flicker in the disk face area of erythrocyte. It enables registration of both the modal composition $\langle u(q)^2 \rangle_{\text{reg}}$ and the amplitude "maps" $\overline{u^2}(\vec{r})$ of chaotic membrane oscillations, as well as calculation of their frequency spectra $G_{reg}(\omega, \vec{r})$ [70–72]. Considerable limitations to accuracy of such measurements arise within equatorial area of erythrocyte only. First, because of the cell rim geometry, the membrane displacements gradually assume the lateral character as they approach the cell edge, while phase microscopy registers their transverse component only. Second, the accuracy of theoretical description of registered interference pattern is decreased considerably near the cell edge because of diffraction effects, which are difficult to treat exactly [59].

A concluding remark concerns the whole family of quantitative phase imaging techniques. Similar to the PHACO regime, they are sensitive only to those oscillation modes, which alter the object thickness (Figure 7.4). Hence, in theory they do not allow registering the AM modes of erythrocyte flicker (Section 7.4.4 also).

7.4.3 Registration of Fluctuations in Cell Circumference Shape

Flicker investigations for equatorial contour of erythrocyte are accomplished by registration of transient shape of this contour using bright field [73] or PHACO microscopy [42, 47, 74–76] with digital acquisition of time-resolved cell images by means of CCD arrays. The main problems are associated with the choice of adequate (method-dependent) algorithm for estimation of actual lateral position of the cell edge at the registered image, and with precise evaluation of this position from the image. This task is rather straightforward for bright-field registration with illumination at wavelength 415 nm corresponding to hemoglobin absorption maximum [73]. Here the lateral distribution of image contrast is conditioned by light absorption proportional to the local thickness of erythrocyte, so cell edge position can be defined naturally as the maximum position of the radial dependence of the contrast gradient.

The situation is essentially more complex with the PHACO registration [47, 74, 75]. Owing to the edge diffraction effects inherent to phase contrasting of objects with finite area [59], the PHACO image of erythrocyte acquires a specific "halo" around the cell equatorial contour, which is absent at the light absorption image (pictures in [47, 74]). With positive PHACO, this halo appears like a very bright ring, enabling the shape and position of its maximal brightness contour to be determined using an algorithm similar to that described above for bright-field images. Then the position of cell edge was obtained by tracing halo intensity distribution to lowest values near the edge [47, 74]. The lateral resolution of cell edge position thus obtained was estimated as 5 nm [74]. However, it was argued that owing to diffraction nature of the halo its displacements are related to the cell edge displacements in essentially nonlinear manner leading to inaccurate determination of transient cell edge position [75]. An alternative algorithm of erythrocyte PHACO image processing was developed on the basis of theoretical calculation of intensity distribution for a step-like phase object [75]. Here, erythrocyte edge position was detected by fitting the measured intensity distribution for the line-scan along the erythrocyte diameter to the theoretical curve, which inherently includes the halo effect. The resulting accuracy was estimated to be $\leq 10^{-2}$ of the probing light wavelength with samplings interval 4.4 ms [75].

With the edge detection algorithm adopted, each video frame from the registering camera is processed to obtain the instantaneous shape of the equatorial contour of the erythrocyte with subpixel resolution [42, 73, 74]. The succession of these initial data is further processed to calculate the mean cell contour and diameter and the position of cell center, which in general is time-dependent because of erythrocyte "swinging" and possible diffusion, and to obtain the local membrane displacement from the mean cell contour $u(\varphi, t)$ (Figure 7.1). Fourier transforms of $u(\varphi, t)$ in angle and time give the mean-square flicker amplitude as a function of wave number (modal composition) and frequency [42, 73]. It should be stressed that $u(\varphi, t)$ is intrinsically the local membrane displacement due to the flicker, because translational displacements of entire cell contour were automatically subtracted during the calculations. Hence, this registration technique helps to distinguish between the membrane flickering and the "swinging" or translations of entire cell, contrary to the point probing of cell edge fluctuations (Section 7.4.1.2). If the aim is to measure the frequency spectra or the amplitudes of lowest few modes, it can also be achieved using only four-point registration of contour flickering, with registration points set at two mutually perpendicular cell diameters [74].

7.4.4
Fundamental Difference between the Registered Spectra and the Intrinsic Flicker Spectrum

Each particular optical-physical technique of flicker registration exploits some specific type of cell shape variations and uses specific optical mechanism for their transformation into corresponding changes of the amplitude and (or) phase of the probing light. As a result, different registration techniques possess unequal, or even selective sensitivity to different spatial modes of membrane flickering [7]. Thus, the PHACO regime is insensitive to those modes, which do not change cell thickness because of their specific symmetry or induce zero thickness variation averaged over the projection of the measuring diaphragm (Section 7.4.1.1 and Equation 7.11). Reflection laser probing (LASCO) enables registration of both SM and AM flicker modes, but similar to PHACO reduces the amplitudes of certain modes because of the measuring diaphragm effect (Section 7.4.1.1). Single-point probing techniques of cell edge position do not distinguish between the membrane flicker modes and "swinging" modes of entire cell attached to a support (Section 7.4.1.2 and Figure 7.6). For point dark-field probing, this uncertainty in the evaluation of intrinsic flicker modes is aggravated additionally by the very complicated influence of lateral position and orientation of reflecting membrane spot on the registered light intensity. Quantitative phase imaging does not register the AM modes of disk face flicker, as well as the flickering of erythrocyte equator due to the perpendicularity of its displacements to the optical axis of registration

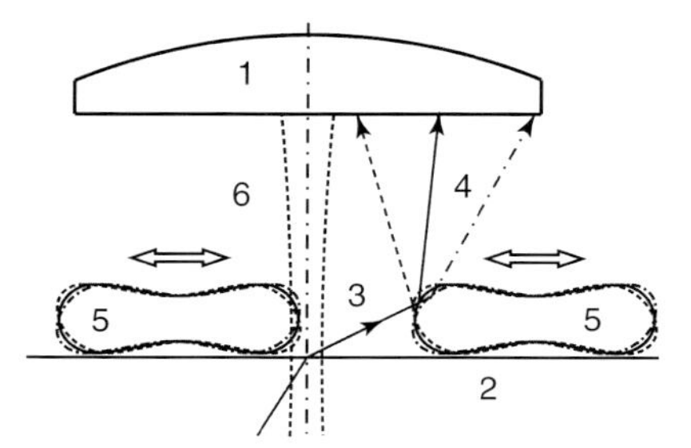

Figure 7.6 Schematic drawing of point probing of erythrocyte edge fluctuations using laser-based detection of edge position (left cell) and dark-field microscopy (right cell). 1 – microscope objective; 2 – glass support; 3 – axial ray of probing light cone from microscope condenser in point dark-field regime; 4 – corresponding rays reflected by a small area of erythrocyte rim at different stages of displacements; 5 – erythrocytes; and 6 – focused probing laser beam in laser-based detection. Bidirectional arrow shows Brownian "swinging" of upper part of adhered cell.

system (Section 7.4.2). Direct contour imaging needs method-specific algorithms for reconstruction of the contour shape and actual displacements from the image structure (Section 7.4.3). Further comparison with intrinsic flicker spectra requires mathematical consideration of the projection of surface eigenmodes of flicker onto the cell equatorial contour [76].

These examples illustrate the fundamental feature of flicker spectroscopy: flicker spectra measured with any particular technique differ from intrinsic flicker spectrum of erythrocytes. In general, this difference may be both in modal compositions of these spectra and in amplitudes distribution of detected modes compared with their distribution in intrinsic spectrum. There are two main origins of such distortion of intrinsic spectrum. First, according to the underlying physical principle as well as to the geometry of registration, any particular technique selects corresponding, method-specific, subset of eigenmodes from the total multitude of excited modes of erythrocyte shell oscillations. This can be termed the *method-specific selectivity*. Such selectivity, together with particular relations between the membrane displacement and induced alterations of probing light parameters, results in method-specific distortions of flicker spectra. Second, different modes from the registered subset are, in general, detected with different instrumental weight because of various instrumental effects. These effects are associated with (i) finite size of planar and angular apertures of the measuring setup, (ii) the dependence of projection of displacement of a membrane patch on some relevant instrumental axis on the surface position of the patch, (iii) nonlinear detection of different flicker modes those amplitudes may differ in 2 orders of magnitude (Figure 7.5), and so on. That can be termed the *instrumental distortion of intrinsic spectrum*.

It is fundamentally important to take into account both types of spectra distortions while comparing the results obtained using different measurement techniques. It is of equal importance to describe these effects quantitatively for adequate comparison of the measured data with the flicker theory. In particular, a direct shape comparison between the measured spectra and theoretically calculated intrinsic spectrum is basically incorrect. The plots from Figure 7.5 clearly demonstrate this fundamental feature of flicker spectroscopy.

At present, a consistent treatment of instrumental effects in flicker spectroscopy was done for registration of the frequency spectra in a PHACO regime [7, 34, 48]. It is instructive to elucidate these effects by comparing expression (7.7) at $r = 0$ for intrinsic flicker spectrum with expression (7.11) for the registered spectrum of corresponding detector signal fluctuations. First, the registered spectrum lacks the contributions from major part of intrinsic flicker modes, namely, from all AM modes and those with angular dependence ($n \neq 0$). Second, the contributions from short-wavelength modes (large m) are reduced effectively by the factor (instrumental function) $IF(h) = [2J_1(M_{0m}\, h/L)L/M_{0m}\, h]^2$ because of the averaging of modal variation of cell thickness over the projection of measuring diaphragm, h being projection diameter, L – cell diameter. Instrumental function $IF(h) \rightarrow 1$ at $h \rightarrow 0$, and decays with the increase of h, the decay rate increasing with M_{0m}, that is, with the mode number m. The instrumental narrowing of registered spectrum due to reduced contribution from the short-wavelength modes was demonstrated by special measurements with increasing size of the measuring diaphragm (Figure 7.7) [52]. It was pointed out also in [6]. Further, along with intrinsic dependence on erythrocyte shape characterized by parameter $\kappa = d/L$ in Equations 7.7 and 7.8, the registered spectrum and its integral parameters have an additional shape dependence of instrumental nature. It is accounted for by the instrumental function $IF(h)$, as well as by the product in square brackets before the spectral sum in Equation 7.11. This product includes the method-specific factor $B(d)$, which directly depends on cell thickness (Equation 7.9), and the refraction indice difference $\Delta n(\kappa)$, which depends on hemoglobin concentration and hence on cell volume and cell shape. At last, in some cases the measured spectrum and its integral parameters may manifest a phantom additional dependence on the influencing external parameter. A characteristic example is the influence of the viscosity of the ambient medium on the flicker spectrum in the experiments where the coefficient of viscosity $\eta_{\rm i}$ was increased by addition of dextran [6]. The point is that the medium refraction index also increases with dextran concentration resulting in a decrease of Δn in Equation 7.11 concurrent with $\eta_{\rm i}$. This distorts the actual dependence of spectrum parameters (Equation 7.8) on ambient viscosity (Section 7.6.1).

The alteration of the refraction indice difference Δn due to both variation of the stationary erythrocyte shape (hence volume V) and molecular composition of the medium is fundamentally an important feature for optical registrations of erythrocyte geometry and flicker spectra evaluation. Refraction index of hemoglobin solution is related to the hemoglobin concentration c_{Hb} as $n_{\rm Hb} = 1.335 + 0.1823 \times c_{\rm Hb} + 8.6526 \times 10^{-4} \times c_{\rm Hb}^2$ [77]. For erythrocyte $c_{\rm Hb}(V) = {\rm Hb}/V$, where ${\rm Hb} \approx 3 \times 10^{-14}$ kg is weight amount of intracellular hemoglobin. This dependence should be considered in the transformation of "phase-shift maps" $\Delta\varphi(x, y)$ obtained by PHACO, LASCO, and quantitative phase imaging techniques into the cell thickness maps $d(x,y)$. In particular, for large shape alterations, some iteration procedure

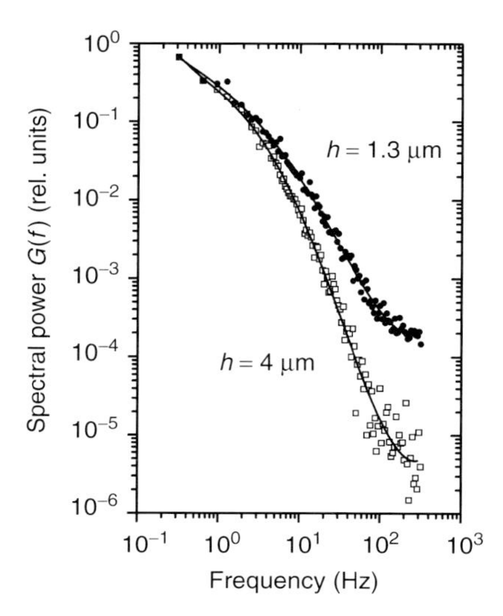

Figure 7.7 The influence of the size h of measuring diaphragm projection on the shape of flicker spectrum registered for the central part of erythrocyte disk using PHACO regime [52]. The curves are brought into coincidence at their first points for comparison.

accounting for $\Delta n(V)$ dependence should be used instead of direct linear scaling of $\Delta\varphi(x, y)$ to $d(x,y)$ commonly employed [65].

There is another kind of instrumental distortions of intrinsic flicker spectrum, which is very important for adequate data processing and interpretation. During the measurements, erythrocyte is slightly adhered to a glass support to prevent cell translational motion. The glass–cell contact area visualized by reflection interference contrast appears usually as an annular ring [63, 73, 74] with the characteristic width varying within $\sim 1.2 \div 2\,\mu\text{m}$ for different cells [74]. Such adhesion alters appreciably the normal shape of the erythrocyte, and hence the flicker eigenmodes and the shape of intrinsic flicker spectra characteristic of the biconcave cell. In particular, owing to the loss of central plane symmetry of erythrocyte, SM and AM flicker modes acquire mixed character in regard of membranes displacements (consider Figure 7.4). The long-wavelength modes of the disk face become the most distorted compared with a normal cell, while modes with shorter wavelengths remain relatively unchanged. Such transformation of eigenmodes results in partial and uncontrollable lift of the above-mentioned restrictions on AM modes registration by phase shift measuring techniques. This feature, together with uncontrollable alteration of intrinsic flicker spectrum complicates theoretical description of the spectra measured for disk face area, as well as cell parameters evaluation from these spectra. Thus, according to the measurements [74], there is a strong correlation between the width of cell-substrate contact ring and the estimated values of membrane bending and shear moduli.

On the contrary, the equatorial contour of erythrocyte, as well as its flicker amplitude [73], is practically insensitive to alterations of cell shape induced by adhesion. Hence, this artifact may have minor influence on the flicker spectra measured for the equatorial contour.

7.5 The Measured Quantities in Flicker Spectroscopy and the Cell Parameters Monitored

Three types of flicker characteristics in erythrocytes described in Section 7.3.1 were measured in various experiments: the shape and integral parameters of frequency spectrum $G(\omega, \vec{r})$, the shape of spatial spectrum $\langle u(q)^2 \rangle$ (modal composition of flicker), mean square membrane displacement $\overline{u^2}(\vec{r})$, and fluctuation amplitude histogram. Owing to strong anisotropy of both erythrocyte shape and the effect of membrane shear rigidity on the flicker amplitude (Section 7.3.2.1), the measurements results as well as cell parameters monitored are, in general, different for the disk face area and for the rim of erythrocyte. The choice of the most adequate registration technique differs accordingly (Section 7.4).

7.5.1 Frequency Spectra

Owing to fast decay of spectral amplitude with frequency, a major part of measurements starting from pioneering work [5] were made in comparatively narrow range $f \sim 0.2$–50 Hz. Here the shape of measured spectral curve $G_{reg}(f)$ can be characterized rather good by a linear slope in the log–log scale, which is very convenient for presentation and evaluation of experimental results. However, the measurements in broad frequency range [52] evidenced essentially nonlinear shape of frequency spectra in a log–log scale, and demonstrated the need of additional parameters, namely, the integral spectrum parameters of type (Equation 7.8) for data evaluation.

7.5.1.1 Disk Face Area of Erythrocyte

Here the measurements of $G(f)$ were made using PHACO [5, 17, 45, 47, 52] and LASCO [52] regimes for the disk center and quantitative phase imaging for the whole disk face [70, 71]. The log–log spectrum slope observed for normal erythrocytes was $-(1.30 \div 1.45)$ in the range 0.2–30 Hz [5], and was close to $-5/3$ in the range ~1–30 Hz with saturation tendency for lower frequencies according to [47]. The measurements [71] evidenced a strictly linear log–log behavior of the spectral curve, the negative slope being 1.54 for DCs, 1.23 for echinocytes (ECs), and 1.15 for spherocytes (SCs). Note that thermal flicker theory predicts the slope $-4/3$ for intermediate frequencies and $-5/3$ for high frequencies (Section 7.3.2.2).

The shape of flicker spectra $G_{reg}(f)$ registered for normal erythrocytes in most broad range $0.05 \div 500$ Hz [34, 52] is illustrated by Figure 7.5. As it was considered in detail in Sections 7.4.1.1 and 7.4.4, the measured shape depends both on particular technique employed and instrumental parameters values used. In the log–log scale, the magnitude of negative slope of $G_{\text{reg}}(f)$ registered in a PHACO regime for different DCs increased from $1.0 \div 1.3$ in the range of lowest measurement frequencies up to $1.7 \div 2.6$ in the range of tens and hundred hertz [34, 52]. The slope magnitude averaged over eleven normal cells was 1.46 ± 0.19 within the referent range 0.2–30 Hz, and 2.16 ± 0.32 in the range $f > \sim 80$ Hz. For spectra measured with the LASCO technique, the slope magnitude increased along the curve from $0.5 \div 0.7$ up to $1.6 \div 2.3$. Accordingly, cells-average slope was 1.22 ± 0.09 within the range 0.2–30 Hz, and 2.04 ± 0.19 in the range $f > \sim 80$ Hz [34, 52]. The scatter of slope magnitudes measured for different erythrocytes appeared as normal DCs under microscopic observation is conditioned seemingly by a strong dependence of flicker spectrum shape on cell thickness d (Equation 7.8 and Figure 7.3), as well as on particular patterns of the erythrocyte–substrate interaction (Section 7.4.4).

The shape of measured spectra $G_{\text{reg}}(f)$, both with the PHACO and LASCO techniques, was well described in the range from ~10 Hz up to highest measured frequency ~500 Hz using flat disk model for erythrocyte shape (Section 7.3.2.2), an appropriate theory for registration procedures (Section 7.4.1.1), and assuming thermal flicker excitation [48, 53]. However, in the range 0.03–0.5 Hz the registered spectral curves run above the curves calculated for the thermal excitation. Special measurements proved also the absence of resonant frequencies in flicker spectra of normal erythrocytes [53].

Assuming thermal mechanism of flicker excitation, the measurements of frequency flicker spectra for disk face area of erythrocyte enable evaluation or monitoring of two diagnostically important parameters: bending elastic modulus of cell membrane K_c, and viscosity of internal hemoglobin solution η_i. They can be obtained from the integral parameters of spectrum (Equation 7.8) with due account of method-specific and instrumental distortions of intrinsic spectrum (Section 7.4.4). Such measurements gave the following K_c values: $(1.3 \div 3) \times 10^{-20}$ J [5], 5×10^{-20} J [6], $(2 \div 3) \times 10^{-19}$ J [17], and $(1.0 \div 1.9) \times 10^{-19}$ J [47].

Methods of quantitative phase imaging opened another diagnostically promising possibility, namely, the registration of flicker "frequency maps" over disk face of erythrocyte [70, 78].

7.5.1.2 **Equatorial Contour and Rim Area**

Flicker spectra for circumference part of normal erythrocyte were measured in a narrow band ~0.2–30 Hz by point dark-field microscopy [60] and by equatorial contour imaging [47, 73]. The measurements in a broad range ~0.2–10^3 Hz were accomplished using laser-based edge position detection [58, 61].

According to [47], flicker spectra for equatorial contour and for the disk face have practically the same shape. The log–log spectrum slope is close to $-5/3$ in the range ~2–30 Hz, and has a smaller absolute value at lower frequencies. The measurements in [73] gave ~0.9 for slope magnitude in the range 0.2–10 Hz. The spectrum shape measured in a broad range [58] could be described surprisingly well using spectral formula derived for thermal flickering of spherical vesicle with purely lipid membrane [43], if three adjustable parameters, K_c, η_i, and surface tension σ, were varied. One may conclude that owing to rotational symmetry of erythrocyte, flicker eigenmodes of its equatorial contour are rather close to those of a spherical vesicle. With such assumption, the use of formulae from [30] derived with due account of membrane shear rigidity, membrane viscosity as well as medium viscosity, seems more adequate for quantitative description of flicker spectra of highly curved erythrocyte rim.

7.5.2 Modal Composition

Quantitative assessment of flicker modal composition from the measured maps of membrane displacements is complicated by the fundamental difference of lowest flicker eigenmodes from both the plane waves and spherical harmonics (Section 7.3.2). The distribution of oscillation amplitude over the whole cell surface, with exact account of erythrocyte shape, is not yet calculated for these modes, and the notion of corresponding wavenumber, as well as its relation to wavenumber q of high-order modes remains rather unclear. An exception presents cell equatorial contour, where all eigenmodes are the circular harmonics due to rotational symmetry of erythrocyte. On the other hand, it is natural to decompose the displacement maps measured for erythrocyte disk faces into the series of plane waves, because of the nearly flat geometry of the central part of erythrocyte and for mathematical convenience. Here each flicker mode is characterized precisely by its wavelength λ and wavenumber $q = 2\pi/\lambda$. For wavelengths small enough, $\lambda \ll L$ (high q values), such approximation is justified by the theory also. Hence, unambiguous interpretation of measured displacements maps in terms of spatial spectrum (modal composition) is possible for short enough flicker wavelength, in practice, less than $\approx 3\ \mu m$ [64].

7.5.2.1 Disk Face Area of Erythrocyte

The spatial spectra of flicker over disk faces of erythrocyte resting on a glass support were measured using reflection interference contrast for the lower disk face [63, 64], and for superposition of both faces oscillations using quantitative phase imaging techniques [71, 78] (Section 7.4.2). Figure 7.8 shows typical spatial spectrum of flicker measured for the lower disk face of normal erythrocyte [64]. A major part of the spectrum demonstrates $\langle u(q)^2 \rangle \propto q^{-4}$ dependence, which is characteristic of thermal flicker (Equation 7.4) in a membrane with negligibly small tension, for example, for flaccid lipid bilayer. On the contrary, the measurements [71] for DCs evidenced $\langle u(q)^2 \rangle \propto q^{-2}$ dependence in the range $\sim 0.3-1.6\ \mu m^{-1}$ and $\langle u(q)^2 \rangle \approx const$ in the range $\sim 0.1-0.24\ \mu m^{-1}$. Such behavior is also in the framework of Equation 7.4; however, it characterizes the composite membrane, where lipid bilayer is confined by a point-like attachment to a cytoskeleton (Section 7.2 and Equation 7.1).

Hence, assuming that the measured spatial spectrum corresponds to thermal flicker, it is possible to evaluate erythrocyte parameters entering Equation 7.4

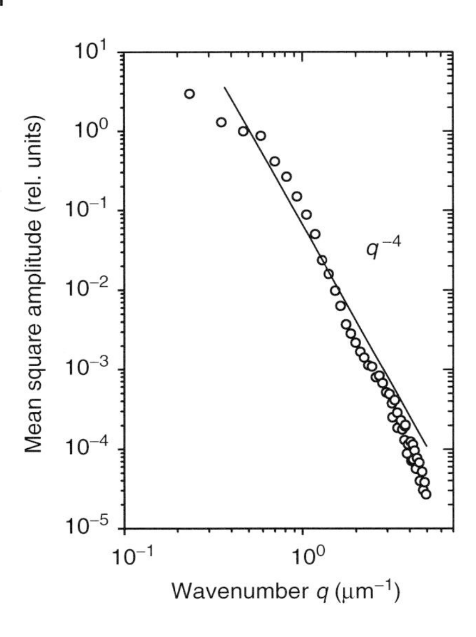

Figure 7.8 Modal composition (spatial spectrum) of disk face flickering of normal erythrocyte measured by reflection interference microscopy (open circles) [64], averaged here for two measured cells; line – plot of $\langle u(q)^2\rangle \propto q^{-4}$ dependence.

for $\langle u(q)^2\rangle$. Such evaluations, made with different assumptions, gave for normal erythrocytes $K_c = (3.4 \pm 0.8) \times 10^{-20}$ J [63] and $(2.3 \pm 0.5) \times 10^{-20}$ J [64]. The lateral tension of a membrane σ_{eff} was estimated as $(1.5 \pm 0.2) \times 10^{-6}$ J m^{-2} for DCs, $(4.05 \pm 1.1) \times 10^{-6}$ J m^{-2} for ECs, and $(8.25 \pm 1.6) \times 10^{-6}$ J m^{-2} for SCs [71]. The estimates of σ_{eff} and γ were obtained from theoretical analysis of results [63] using different models of bilayer-skeleton coupling in erythrocyte membrane: $\sigma_{\text{eff}} \sim 5 \div 12 \times 10^{-7}$ J m^{-2}, $\gamma \sim 1 \div 8 \times 10^{7}$ J m^{-4} [15, 22], and $\sigma_{\text{eff}} \approx 1.6 \times 10^{-6}$ J m^{-2}, $\gamma \approx 5.2 \times 10^{6}$ J m^{-4} [16].

An alteration of erythrocyte mechanical properties in the course of morphological transition discocyte (DC)-echinocyte (EC)-spherocyte (SC) was monitored as well [78]. The results reported are $K_c = (5.6 \pm 1.7)\ k_B T$, $\mu = (7.4 \pm 0.9)$ μN m^{-1} for DC, $K_c = (9.6 \pm 3.2)\ k_B T$, $\mu = (10.4 \pm 2.9)$ μN m^{-1} for EC, and $K_c = (23.9 \pm 6.7)\ k_B T$, $\mu = (12.6 \pm 2.1)$ μN m^{-1} for SC. It should be noted that owing to specificity of theoretical model used to evaluate erythrocyte mechanical moduli from the measured membrane displacement map [78], the obtained values characterize actually the rigidity (or compliance) of erythrocyte as entire body rather then material properties of its membrane. Indeed, the flicker amplitude depends in pronounced manner both on the magnitudes of membrane mechanical moduli and on the cell's "geometrical freedom" characterized by surface area/volume ratio (Sections 7.2 and 7.3). The decrease of this ratio in DC–SC transition may result in cell's effective hardening even under moduli magnitudes unchanged, for mere geometrical reasons (Section 7.3.2.1). The theory should account for the area/volume ratio, which can be readily measured using quantitative phase imaging techniques. In particular, the bending-shear spherical model [30] (Section 7.3.2.1) seems to be adequate for evaluation of the data obtained with SCs.

7.5.2.2 **Equatorial Contour**

Owing to rotational symmetry of equatorial contour, the modal composition of its fluctuations can be determined from the measured data in more rigorous and consistent manner compared with other parts of erythrocyte membrane (Sections 7.4.3 and 7.4.4). Such measurements were done in [42, 47, 73, 74] for oscillations harmonics numbers up to $n \sim 30$ [42]. According to the measurements [42], the log–log dependence $\langle u_n^2 \rangle$ is essentially nonlinear in the whole range of modes numbers: it decreases comparatively slowly in the range of small $2 \le n \le 6$, then shows a steep linear drop in the range $7 \le n \le 18$, then turns rather abruptly to a constant value at $n \sim 20$. The averaging between 121 normal erythrocytes measured gave $\langle u_n^2 \rangle \propto n^{-(3.95 \pm 0.36)}$ for $7 \le n \le 18$, which corresponds formally to thermal excitation law (Section 7.3.2.1). The measured modal distribution $\langle u_n^2 \rangle$ coincided almost completely with the theoretical one calculated for $2 \le n \le 10$ using the direct computer simulation of flicker in a model membrane endowed with both bending and shear rigidity and possessing the shape of erythrocyte [42] (Section 7.3.2). The estimates $K_c \approx 7.5 \times 10^{-19}$ J and $\mu \approx 3.6 \times 10^{-6}$ N m^{-1} were obtained from this comparison.

The measurements [73] evidenced a different law for modal composition, $\langle u(q)^2 \rangle \propto q^{-2.85}$ in the range $6 < q < 20$, q being the number of mode wavelengths over equatorial contour. It is argued however that such dependence is just corresponding to the projection of the thermal oscillations of cell surface onto equatorial contour fluctuations [73, 76]. The evaluation of measured $\langle u(q)^2 \rangle$ spectrum gave $K_c = (9 \pm 0.3) \times 10^{-19}$ J and $\sigma_{\text{eff}} \sim 5 \times 10^{-7}$ J m^{-2} [73].

The theoretical evaluation of measurements results made in [74] gave $K_c = 4 \times 10^{-19}$ J and an estimate of membrane shear modulus, which is about 100 times smaller than inferred from static experiments [13].

7.5.3
Mean Square Amplitude Distribution

Mean square amplitude $\overline{u^2}(\vec{r})$ is the simplest characteristics of membrane flickering; furthermore, it is physically evident and convenient for diagnostic applications. However, serious difficulties in absolute measurements of membrane displacements in nanometer scale prevented widespread use of these characteristics up to quite a recent time. The situation changed radically with the emergence of new measuring techniques in flicker spectroscopy of erythrocytes. The methods of quantitative phase imaging (Section 7.4.2) enable measurement of instantaneous "thickness maps" of a whole cell with nanometer accuracy [65, 72], while laser-based detection of cell edge position (Section 7.4.1.2) ensures the registration of time-dependent displacements of cell equatorial contour with a subnanometer precision [61].

According to thermal flicker theory with exact account of erythrocyte shape [30], the radial profile of cell thickness fluctuations $\overline{u^2}(r)$ has the minimum at the center of DC ($r = 0$) and the maximum at some intermediate distance smaller then the cell radius [44]. An exact maximum position depends on the magnitude of the shear/bending energy ratio $\mu S/K_c$ (Section 7.2), shifting from the central area toward the cell rim with this ratio tending to zero [44]. The measurements of fluctuations profile of DC using PHACO technique (Section 7.4.1.1) confirmed the calculated dependence and gave $(\sqrt{\overline{u^2}})_{\min} \approx 58$ nm at the cell center and $(\sqrt{\overline{u^2}})_{\max} \approx 158$ nm near the radius of maximal cell thickness [44]. Similar results however with smaller displacement values were obtained using Fourier phase microscopy: $\sqrt{\overline{u^2}}$ was measured equal to 48.8 nm at the cell center, 67.0 nm at half of cell radius, and 98.0 nm at the radius of maximal cell thickness [70]. The measurements using digital holographic microscopy gave surface average value 35.9 ± 8.9 nm [66]. The radial distribution $\sqrt{\overline{u^2}(r)}$ was measured also for normal, ATP-depleted and ATP-repleted erythrocytes [72]. For normal cells the measured fluctuations profile corresponded to the calculated in [44] with $\mu S/K_c$ ratio somewhere between 1 and 10. Simultaneous measurements of $\sqrt{\overline{u^2}(r)}$ profile and cell thickness profile compared with exact theoretical calculations enable determination of both K_c and μ moduli [44]. According to [44], $K_c = 1.4 \times 10^{-19}$ or 4.3×10^{-19} J, depending on particular mechanical model of lipid bilayer of the membrane.

The root mean square amplitude $\sqrt{\overline{u^2}}$of erythrocyte equatorial contour displacements was measured using point probing of cell edge position and cell contour imaging (Sections 7.4.1.2 and 7.4.3). The following values of $\sqrt{\overline{u^2}}$ were obtained for normal erythrocytes: 30 ± 5 nm [74], 52.6 ± 10.1 nm [75], $\sim$50 nm [47], $(22.5 \div 24.0) \pm 0.8$ nm (depending on adhesive properties of cell supporting plate) [73], and 33.0 ± 1.3 nm [58]. The measurements of $\sqrt{\overline{u^2}}$ made by point dark-field microscopy gave, approximately, an order of magnitude higher values [60, 79, 80].

Thus, flicker spectroscopy enables the measurement and monitoring of main mechanical parameters of erythrocytes (Section 7.2). At present, considerable scatter of moduli values measured with different techniques exists (Section 7.5). For instance, the measured K_c values vary in order of magnitude, many of them differing from the values obtained using the micropipette technique, 1.8×10^{-19} J [81] and $(1.64 \pm 0.25) \times 10^{-19}$ J [82]. This is due to a strong dependence of evaluation results on two factors: first, on approximate character of different theoretical model used for flicker description, second, on uncertainties in the shape and geometrical parameters of particular measured erythrocytes.

7.6
Flicker Spectrum Influence by Factors of Various Nature

Good application prospects of flicker spectroscopy in cell diagnostics are associated in many respects with high sensitivity of flicker amplitude and spectra to a wide diversity of influencing factors [7]. First of all, they include ambient physical-chemical conditions.

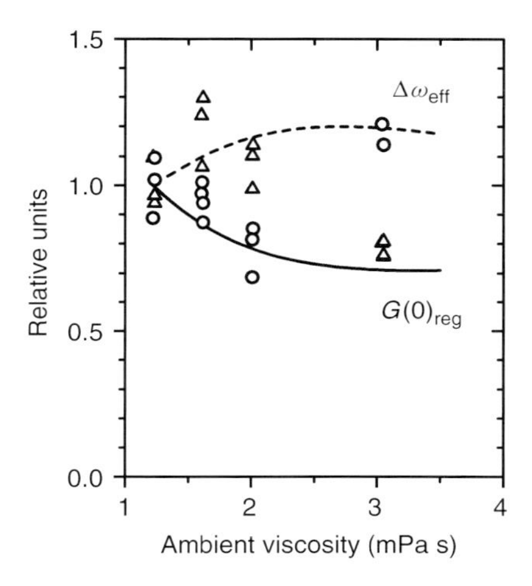

Figure 7.9 Dependences of "zero frequency" amplitude $G(0)_{reg}$ (circles) and effective width $\Delta\omega_{eff}$ (triangles) of flicker frequency spectrum at the central part of erythrocyte on ambient viscosity, recalculated from measurements [6]. Cirves – calculation of $G(0)_{reg}$ and $\Delta\omega_{eff}$ using disk model with $d = 2$ μm at $\eta_e = 1.2$ mPa s [7]. The quantities are normalized to unity at their magnitudes at $\eta_e = 1.2$ mPa s.

7.6.1 Ambient Physical Conditions

The influence of these conditions on flicker characteristics in erythrocytes was investigated in [6, 17, 28, 46, 47, 58, 60, 73]. The paper [6] reported systematic studies of the influence of ambient viscosity, osmotic pressure and temperature upon "zero-frequency" spectral amplitude $G(0)_{reg}$ and effective width $\Delta\omega_{reg}$ of the frequency flicker spectrum registered for central part of disk face using PHACO technique (Figure 7.9–7.11). A detailed analysis of these measurements including the treatment of instrumental effects was published elsewhere [7, 48].

7.6.1.1 Effect of Medium Viscosity

Effect of medium viscosity η_e was investigated for disk face flickering [6] and for random displacements of erythrocyte edge [46, 58]. Figure 7.9 shows the data [6] recalculated to $G(0)_{reg}$ and $\Delta\omega_{reg}$, together with viscosity dependence of these quantities calculated using the spectrum expression (7.11) [7]. The measured dependencies $G(0)_{reg}(\eta_e)$ and $\Delta\omega_{reg}(\eta_e)$ are surprisingly nonmonotonic, while monotone variation of oscillations parameters with the increase of energy dissipation is expected from general considerations. Second, in the range $\eta_e \sim 1.2$–2.5 mPa s they are inverted to those expected both from general results of statistical physics and particular expressions (7.8) [7]. The calculations (Figure 7.9) accounting the influence of both cell shape (Section 7.3.2.2) and instrumental effects (Section 7.4.4) showed that inverted character of registered dependences can be explained by alteration of erythrocyte shape due to dextran adsorption and corresponding increase of cell volume with the rise of dextran concentration in surrounding medium, which is added to enhance its viscosity [7, 48]. However, their nonmonotonic behavior still needs an explanation.

The measurements [46] made by point dark-field microscopy (Section 7.4.1.2) evidenced a linear relation in the range $\eta_e = 1-10$ mPa s between the inversed medium viscosity $1/\eta_e$ and the amplitude of erythrocyte edge displacements measured at 0.3 Hz. The decrease of this referent amplitude with the increase of viscosity was concurrent with a narrowing of the histogram of displacement amplitude. These results were interpreted as the manifestation of some metabolic driving force of flicker excitation; however, an alternative explanation based on specific instrumental features of point dark-field microscopy is possible [7].

The influence of medium viscosity on mean square amplitude, frequency spectrum, and histogram of fluctuation amplitude of erythrocyte edge was investigated also using laser-based edge position detection [58]. According to the measurements, the increase of η_e in the range 1–10 mPa s resulted in systematic variation of the spectrum shape and its integral parameters: the increase of the spectrum slope and "zero frequency" amplitude $G(0)_{reg}$, and the decrease of effective width $\Delta\omega_{reg}$. Contrary to the measurements [46], the width of Gaussian-shaped histogram of fluctuation amplitude did not vary with η_e. It should be noted that just this type of dependences is predicted by the thermal theory of flicker (Equation 7.8 for $G(0)$ and $\Delta\omega_{eff}$). Also, a particular spectral amplitude $u(f)$ may either decrease or increase with viscosity, depending on the referent frequency.

7.6.1.2 Effect of Ambient Temperature

Effect of ambient temperature $t\,^\circ$C on integral parameters of frequency flicker spectrum [6] is illustrated by Figure 7.10. According to general-type relations (e.g., Equation 7.8), the product $G(0)\Delta\omega$ gives the area under the spectral curve G_{int}. From biophysical arguments, G_{int} of intrinsic flicker spectrum is expected to increase with the temperature. No direct measurements of $G_{int}(T)$ seem to be published; however, the plot of $G(0)_{reg}\Delta\omega_{reg}$ values measured in [6] decreases with the temperature [7]. In the framework of thermal model (Equation 7.8) $G_{int}(T) \propto T/K_c^*$. Therefore the decreasing character of observed $G_{int}(t\,^\circ\mathrm{C})_{reg}$ dependence may result from both instrumental effects and the increase of K_c^*. The analysis showed that $G_{int}(t\,^\circ\mathrm{C})_{reg}$ dependence could be well described in the whole range 10–40 °C taking into account small shear strains in the membrane accompanying the bending modes, primarily the lowest flicker modes with the largest amplitudes prevailing in the spectrum [7]. Owing to entropic character of membrane shear elasticity, it is natural to expect an increase of shear modulus μ with the elevation of temperature [14], which is manifested as the increase of effective bending modulus K_c^*. Using such assumption for evaluation of the temperature dependence of the product of $G(0)_{reg}$ and $\Delta\omega_{reg}$ values measured in [6], the dependence $K_c^*(t\,^\circ\mathrm{C}) \approx [1.8 + 2.8 \times 10^{-2} \times (t\,^\circ\mathrm{C} - 25] \times 10^{-19}$ J was obtained [7] with $K_c^*(25\,^\circ\mathrm{C}) = 1.8 \times 10^{-19}$ J [81].

The use of K_{eff} ($t\,^\circ$C) dependence enables calculation of $G(0)(t\,^\circ\mathrm{C})$ and $\Delta\omega(t\,^\circ\mathrm{C})$. That was accomplished using Equation 7.11 with account of temperature dependences of viscosity of intracellular hemoglobin solution η_i, erythrocyte volume,

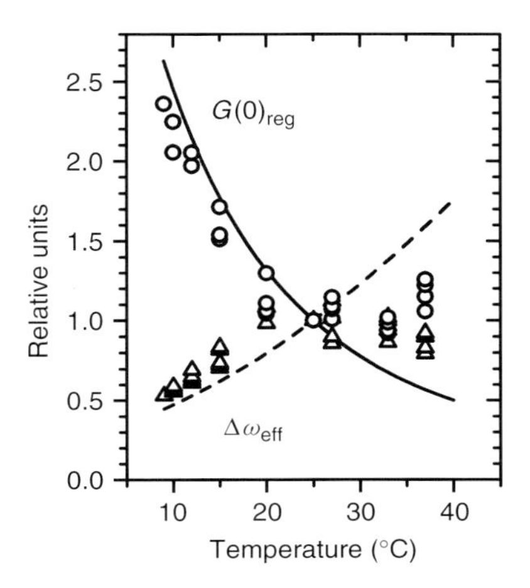

Figure 7.10 Dependences of amplitude $G(0)_{reg}$ (circles) and width $\Delta\omega_{eff}$ (triangles) of flicker frequency spectrum at the central part of erythrocyte on ambient temperature, recalculated from measurements [6]. Cirves – calculation of $G(0)_{reg}$ and $\Delta\omega_{eff}$ using disk model with $d = 2$ μm at 25 °C [7]. The quantities are normalized to unity at their magnitudes at 25 °C.

and membrane area [7]. The calculated curves shown in Figure 7.10 describe well enough the measurements [6] for $t\,^\circ\mathrm{C} < 25$, but strongly disagree with the data registered for higher temperatures. A possible explanation is nonmonotonic variation of parameter $\eta_i\,(t\,^\circ\mathrm{C})$ in spectrum Equation 7.11, if we consider this parameter as a general index of energy loss by membrane oscillations due to all dissipation mechanisms. Such additional dissipation, developing with the elevation of temperature, may result from interaction of bending modes of lipid bilayer with membrane spanning proteins [7]. The model advanced in [83] considered an enhancement of in-plane diffusive motion of transmembrane proteins (e.g., band 3 proteins) by bilayer bending fluctuations. We supposed that such proteins protrude into the cytoplasm far enough to interact with filaments of membrane skeleton, which results in additional damping of bilayer oscillations [7]. An effective thickness of membrane skeleton, being entropic in its nature, should increase with the temperature thus enhancing this damping.

7.6.1.3 **Effect of Medium Tonicity**

Effect of medium tonicity on both frequency and spatial flicker spectra was studied in [6, 17, 28, 45, 47, 73, 80]. Actually, this is the best way to study the influence of cell shape and volume on flicker characteristics in erythrocyte, because the relation between erythrocyte volume V and medium osmolarity U is well explored. A universal empirical equation $V(U)/V_n = AU_n/U + B$ was established for the range $U = 150 \div 750\,\mathrm{mOsm\,kg^{-1}}$, where V_n is erythrocyte volume under normal osmolarity $U_n = 300\,\mathrm{mOsm\,kg^{-1}}$, $A = 0.565$, and $B = 0.435$ are empirically determined constants [84]. It should be remembered, however, that change in erythrocyte volume alters automatically the cytoplasm viscosity η_i and

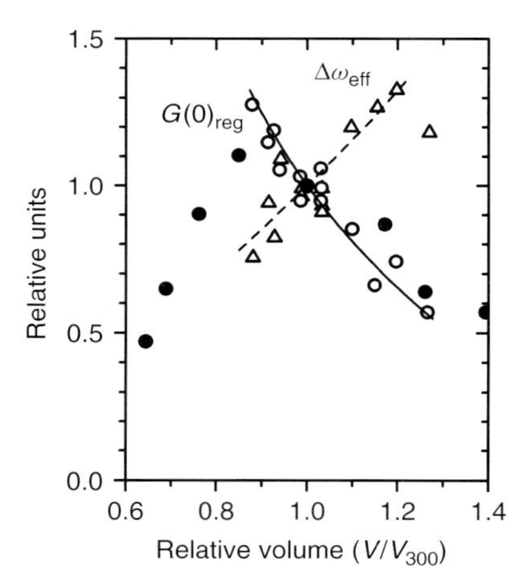

Figure 7.11 Dependences of characteristic parameters of flicker spectra on erythrocyte volume. Open circles – amplitude $G(0)_{reg}$, *triangles* – width $\Delta\omega_{eff}$ of flicker frequency spectrum for the central part of erythrocyte recalculated from measurements [6]. Solid circles – amplitude of the first mode of spatial flicker spectrum of erythrocyte equatorial contour according to [73]. Cirves – calculation of $G(0)_{reg}$ and $\Delta\omega_{eff}$ using disk model with $d = 2$ μm at 300 mOsm kg^{-1} [7]. The quantities are normalized to unity at their magnitudes at 300 mOsm kg^{-1}.

refraction indice difference Δn, leading to additional alterations in registered flicker spectrum (Section 7.4.4).

Two physical mechanisms underlie the effect of osmolarity. The decrease of medium osmotic pressure from physiologically normal value leads to the increase of erythrocyte volume and reduction of S/V ratio. This reduces geometrical freedom to change cell shape without substantial mechanical stresses in the membrane (Section 7.2), thus suppressing flicker amplitude due to the shear and dilatation stresses arising. The elevation of medium osmolarity from the normal value results in diminishing cell volume because of water efflux from the cell, hence in increasing intracellular hemoglobin concentration. This leads to exponential growth of cytoplasm viscosity [48] and to a strong viscous damping of membrane flickering. The measurements of mean flicker amplitude in a narrow band near 0.3 Hz by point dark-field microscopy for medium osmolarities 180–675 mOsm kg^{-1} revealed nonmonotonic dependence of this type [80]. Mean displacements amplitude was ~150 nm at both hypotonic and hypertonic boundaries of osmolarity variation, reaching the maximum ~225 nm in the range 310–470 mOsm kg^{-1}. This dependence correlated positively with the measured erythrocytes filterability in the range ~220–550 mOsm kg^{-1} [80]. The measurements of spatial flicker spectrum of erythrocyte equatorial contour revealed similar nonmonotonic dependence for the amplitudes of first four oscillations modes [73]. The results for the first mode, a most dependent one, are shown in Figure 7.11 in relative units.

The influence of medium tonicity of the shape and parameters of flicker spectra was investigated in [6, 28, 47, 73]. According to [73], both frequency and spatial spectra measured in hypotonic conditions became considerably more flat in the log–log scale. Similar flattening tendency, in succession high-normal-low tonicity, was observed for a low frequency part of flicker spectra [47]. Flattening effect was quite pronounced for the disk face area, while fluctuations spectra of the cell equator diameter were essentially less sensitive to medium osmolarity [47].

The integral parameters of frequency spectrum also depend on medium osmolarity. Using dynamic light scattering from a single cell, the correlation time τ_{corr} of membrane fluctuations was measured for erythrocytes in hypertonic buffer with osmotic strength of 2.5× isotonic [28]. The measured values of τ_{corr} were $\sim 5 \div 11$ of the time value obtained for isotonic conditions. Since $\tau_{corr} \propto 1/\Delta\omega_{eff}$, that means corresponding narrowing of frequency spectrum. The systematic measurements of $G(0)_{reg}(U)$ and $\Delta\omega_{reg}(U)$ [6] are shown in Figure 7.11 together with calculated curves. Data were scaled to variation of erythrocyte volume according to dependence $V(U)/V_n = AU_n/U + B$ (above). The calculations used Equation 7.11 with $S = 134\ \mu m^2$, $\eta_i = 9.7$ mPa s at 25 °C and 300 mOsm kg^{-1}, $K_c = 1.8 \times 10^{-19}$ J. They considered variation of disk shape parameter κ, cytoplasm viscosity η_i, and refraction indice difference Δn with the cell volume [7]. As Figure 7.11 shows, for quantitative description of measured dependences $G(0)_{reg}(U)$ and $\Delta\omega_{reg}(U)$ in the range $V(U)/V_n \approx 0.9 \div 1.2$ or $U/U_n \approx 400 \div 200$ mOsm kg^{-1} it is quite sufficient to take into account the influence of osmotic variation of cell volume and shape, with no hypotheses concerning the change of membrane elastic moduli. It illustrates also a validity range of disk model used in calculations.

7.6.2 Modification of Cell Mechanical Properties

In order to get a deeper insight into relations between the flicker characteristics and membrane microstructure, a whole number of experiments were carried out where erythrocyte mechanical properties were modified specifically by adding appropriate chemical agents to cell suspending medium. It was established that various kinds of cross-linkage created by such agents between both membrane-embedded and cytoplasmic proteins results in severe degradation or even in cessation of membrane flickering. Such substances as diamid, which creates cross-links in proteins via oxidation of SH-groups [6, 17], thiol toxins forming S-S bridges in membrane proteins [60], and glutaraldehyde (beginning with small concentrations ~0.01%) [47, 60, 73, 75] produce an abrupt and irreversible drop of flicker amplitude. The exposure of erythrocytes to hydrogen peroxide, which is known to induce the formation of hemoglobin–spectrin complexes and possibly aggregation of band 3 proteins, resulted in strong amplitudes degradation of lowest ($n \le 6$) fluctuation modes of cell equatorial contour [42].

Influence of more delicate alterations in erythrocyte membrane, associated with the change of its cholesterol content was studied also [6, 28]. The measurements of spatial flicker spectrum of giant monolayer lipid vesicles showed that cholesterol addition to lipid bilayer formed by dimyristoylphosphatidylcholine increased substantially its bending rigidity [85]. A similar measurements of flicker in erythrocytes showed that gradual reduction of cholesterol content in the membrane from the normal value was accompanied by monotonic decrease of "zero-frequency" flicker amplitude $G(0)_{\mathrm{reg}}$, with biconcave shape of erythrocyte preserved till the moment of cell's abrupt transition to EC [6]. In another measurements made using quasi-elastic light scattering by erythrocyte [28], it was found that the rise of cholesterol/phospholipid ratio in the membrane from the normal value $\sim$1.1 up to 3.8 resulted in 30–50% increase of correlation time τ_{corr} of membrane fluctuations, that is, in corresponding spectrum narrowing $\Delta\omega_{\mathrm{eff}} \propto 1/\tau_{\mathrm{corr}}$. The analysis of results [6, 28, 85] showed that variation of cholesterol concentration in erythrocyte membrane led both to change of membrane bending rigidity and to substantial alteration of cell shape [7]. The elevation of this concentration beyond the physiologically normal value causes the thinning of central part of erythrocyte and *vice versa*.

7.6.3
Physiologically Active Substances and Medicinal Drugs

The fundamental aim of such investigations is the influence of cellular physiological processes on flicker parameters, and elucidation of possible involvement of these processes in flicker excitation. For this purpose, the membrane flickering effect of several main regulators of cell functioning and metabolism was investigated. The factors studied include: Ca^{2+} ions, which play a pivotal role in cell physiology [6, 28]; ATP, the main energy storage and transfer molecule in the cell [57, 58, 72, 73, 86]; cyclic adenosine monophosphate (cAMP), a secondary messenger important in numerous biological processes [87]; hormones [88] and oxygen [79]. The results of these studies were analyzed in detail elsewhere [7]; the main results and conclusions are presented below. Flicker characteristics were measured for different parts of cell membrane (disk face, cell rim, and equatorial contour) using different registration techniques. It is well established now that mean flicker amplitude and spectral parameters depend on erythrocyte shape and bilayer-skeleton interaction, such dependences being essentially different for disk faces area and rim area of erythrocyte (Section 7.3.2). It was also shown that different registration techniques are unequal in their sensitivity to membrane displacements of various nature and direction (Section 7.4.4). Hence, for adequate interpretation of observed influence it is fundamentally important to take into account both these features.

Cyclic variation of intracellular concentration of ionophore-injected Ca^{2+} ions (buffer concentration 1 mM) resulted in corresponding reversible alteration of the amplitude and half-width of frequency flicker spectrum measured for disk face of erythrocyte [6]. The injection of Ca^{2+} ions into erythrocyte is known to induce pronounced morphological, structural, and mechanical transformations. The sphero-echinocytosis develops [89] accompanied by cell thickness increase; a polymerization of membrane proteins occurs [90] and the membrane stability alters [91]; an efflux of K^{+} ions from erythrocyte begins, mediated enzymatically by Ca^{2+} ions (Gardos effect [92]), which leads to osmotic dehydration of cytoplasm and increase of its viscosity. Thus, the results of [6, 28] evidence high sensitivity of flicker spectrum parameters to alterations in intracellular Ca^{2+} concentration, which is conditioned by calcium influence on the shape and mechanical properties of erythrocyte.

The influence of ATP on flicker in erythrocytes was studied by three quite different techniques: point probing of cell edge fluctuations [57, 58, 86] (Section 7.4.1.2), registration of fluctuations in cell equatorial contour [73] (Section 7.4.3), and quantitative phase imaging of disk face of erythrocyte [72] (Section 7.4.2). A relation of rim displacements amplitude measured in the range 0.3–35 Hz using point dark-field microscopy, with ATP level in normal erythrocytes, erythrocyte "ghosts," as well as in specially prepared membrane "skeletons" of erythrocyte was studied in [57, 86]. A decrease of rim displacements amplitude with ATP depletion, and subsequent amplitude recovery after restoration of physiologically normal ATP level was observed for all three types of objects. A decrease both in mean square amplitude of erythrocyte edge displacements and in spectral amplitudes over full measured frequency range 0.1–1000 Hz was registered for ATP-depleted erythrocytes by laser-based edge position detection also [58]. On the contrary, the registration of the whole fluctuating contour of erythrocyte revealed no difference in fluctuations amplitude between the normal and ATP-depleted cells [73].

A systematic study of ATP influence on both erythrocyte morphology and disk face flickering was done using quantitative phase imaging [72]. The ATP-depleted cells changed their thickness profile from the concave curve registered for normal cells to nearly flat one. Concurrently, the amplitude of cell thickness fluctuations averaged over entire cell surface dropped from 41.5 ± 5.7 to 32.0 ± 7.8 and 33.4 ± 8.7 nm for irreversibly and metabolically ATP-depleted erythrocytes. The restoration of normal ATP level in depleted erythrocytes resulted in even more pronounced concavity of cell thickness profile and 48.4 ± 10.2 nm fluctuations amplitude [72]. A concurrent reversible change of non-Gaussian parameter (NGP) of fluctuation amplitude histogram was registered also (Section 7.7).

The effect of ligand-receptor interactions and the stimuli influence through signaling pathways on erythrocytes flickering was studied in [87]. The 45% increase in mean amplitude of chaotic displacements of erythrocyte rim in the range $0.3 \div 30$ Hz was observed under the action of β-adrenergic agonists – adrenaline and isoproterenol. The conclusion, based upon cumulative experimental data, was that stimulating action of β-adrenergic agonists was transduced via a cAMP-dependent metabolic pathway, altering eventually viscoelastic properties of the whole membrane complex of erythrocyte [87].

The possibility of direct hormonal influence on flicker amplitude was studied also [88]. The increase of both mean amplitude of chaotic displacements of erythrocyte rim and filtration index of erythrocytes by 50 and 36% was observed under the influence of this hormone at the corresponding concentrations 10^{-9} and 4×10^{-9} M. Interestingly, this effect could be mimicked by membrane-permeant cyclic guanosine monophosphate (GMP) analogs [88].

The oxygen effect on erythrocyte membrane fluctuations was studied in oxygenation-deoxygenation–reoxygenation cycle [79]. Reversible changes of mean amplitude of erythrocyte rim fluctuations from 290 ± 49 to 160 ± 32 nm and back to 268 ± 63 nm was registered. Special experiments showed that the major influencing factor was intracellular concentration of Mg^{2+} ions, which are known as required for association of tropomyosin with actin. Proceeding from this finding, it was suggested that tropomyosin-actin binding is increased in deoxygenated state of erythrocyte when the concentration of Mg^{2+} ions is elevated, thus increasing the rigidity of membrane skeleton and suppressing the amplitude of membrane displacements [79]. Note that in the studies of cAMP effect the isoproterenol-stimulated relative increase of fluctuations amplitude was 2.2 times higher for deoxygenated erythrocytes than for normal ones [87].

There are few studies of medicinal drugs influence on erythrocyte flickering. Irreversible decrease of "zero-frequency" flicker amplitude $G(0)$ by ~39%, which was concurrent with reversible increase of effective spectrum width $\Delta\omega_{\mathrm{eff}}$ by ~78%, was established in the course of ~50 min incubation of erythrocytes with anti-tumor drug actinomycin D in concentration 10^{-6} M [6]. In another study, the effect of actin depolymerizing drug latrunculin A on frequency fluctuation spectrum of erythrocyte edge was registered. The spectrum of latrunculin-treated cells differed from the spectra of both normal and protein kinase C-activated erythrocytes, suggesting a yet unknown effect of latrunculin A on the cytoskeleton-membrane interaction [58].

7.6.4 Erythrocyte Flicker at Human Pathologies

The possibilities of medical diagnostics applications of flicker spectroscopy were studied in several works [6, 17, 45, 60, 93–95]. It was observed that various kinds of hereditary hemolytic anemia – microspherocytosis, elliptocytosis, and stomatocytosis led to strong alteration of mean flicker amplitude and integral parameters of frequency spectrum [17, 60]. The registering of erythrocyte rim displacements in the range 0.3–25 Hz was applied for diagnostics of different stages of diabetic retinopathy (DR) [93, 94]. At the stage of non-prolilferative DR, the mean amplitude of displacements did not differ from the control group for 80% from observed patients, although its scatter within erythrocytes population exceeded the control value. However, with the development of proliferative DR this amplitude decreased by ~25% for 90% patients [94].

Flicker spectroscopy was used also for monitoring intraerythrocytic stages of development of malaria-inducing parasite *Plasmodium falciparum* [95, 96]. In the sequence of stages "ring form-trophozoite-schizont," a systematic decrease of the mean square fluctuation amplitude of cell membrane of invaded erythrocyte was registered; the amplitude became approximately 2.8, 5.3, and 12.9 times smaller than that of healthy erythrocyte [95]. Under the temperature rise from 37 to 41 °C corresponding to the episode of high fever, the fluctuation amplitude of healthy erythrocytes increased reversibly by ~46%, while for infected cells it dropped by ~38% at trophozoite stage and by ~32% at schizont stage, being practically irreversible at schizont stage [95]. Progressive alterations to dynamic light scattering by invaded erythrocyte arising from the development of malaria-inducing parasites were registered also [96]. The spectrum of the scattered light could be described approximately as Lorentzian, having a peak frequency ω_0(~2–8 Hz) and a line width Γ(~1–8 Hz). The extracted values of Γ showed significant differences between healthy erythrocytes and the cells at each progressive stage of development of *Plasmodium falciparum* for the scattering angles $0 \div 25$. At the ω_0, Γ−coordinate plane, the data points measured for each stage including healthy cells formed well separated domains [96].

The results of these pilot investigations, together with other experimental and theoretical studies [7, 80] showed positive correlation between mean flicker amplitude in a range near ~10 Hz with erythrocytes filterability index or, more generally, with erythrocyte deformability [97, 98]. Various deformability parameters of erythrocytes are widely used in clinical applications [97–99]. So, the correlation observed shows tentatively the field of medical diagnostics applications of flicker of erythrocytes, which are possible at contemporary comprehension level of this phenomenon.

7.7 Membrane Flicker and Erythrocyte Functioning

An interrelation between the flicker and cell functioning has two aspects. The first aspect concerns a possible involvement of any intracellular physiological process in flicker excitation. The second one addresses possible physiological functions of membrane flickering.

Cumulative experimental data evidences prevailing character of thermal flicker excitation, as well as a pronounced and diversified influence of physiologically active substances on the amplitude of chaotic membrane displacements (Section 7.6.3). However, the question exists whether cellular physiological and metabolic processes just alter oscillatory characteristics of erythrocyte by influencing its geometric and mechanical properties, or some of them may directly excite membrane fluctuations. The answering this question is complicated by the fact that registered displacements amplitude depends both on geometrical-mechanical parameters of erythrocyte and on the magnitude of acting driving force, which includes thermal agitation and some possible active force. That means a necessity for measurements of additional flicker parameters, first of all, precise measurements of spatial and frequency spectra in order to compare them with those calculated for thermal excitation. Another problem, which is specifically methodical, is the separation in the measured signal the contributions from the local bending oscillations and from the "swinging" of cell body as a whole (Section 7.4).

The most advanced field in answering this question comprises investigations, both experimental and theoretical, of ATP influence on mechanical properties of membrane skeleton and it's binding to lipid bilayer, which are dependent on protein phosphorylation level. Here, the models both for ATP-dependent shear rigidity of spectrin network (Section 7.2) and for flicker excitation due to dynamic remodeling of this network (Section 7.3.2.3) were developed. However, cumulative experimental investigations do not yet allow a final judgment on the presence of the active component in flicker excitation. The reversible decrease of cell edge displacements amplitude in ATP-depleted erythrocytes [57, 58, 86], as well as effects on the edge displacements of other physiologically active substances (Section 7.6.3) can be explained alternatively by stiffening of the whole cell due to the increase in the shear rigidity of membrane skeleton [7]. Precise measurements of flicker spectra in the range 0.1–1000 Hz [58] revealed pronounced difference between the spectra measured for normal, ATP-depleted and protein kinase C-activated erythrocytes. Such spectral difference can be considered as manifestation of non-thermal excitation. However, these spectra were fitted with theoretical dependence for purely bending deformations [43], instead of model [30] precisely considering the influence of shear rigidity of membrane skeleton on flicker spectrum. So, the worst fit was obtained just for the spectrum of ATP-depleted erythrocytes, which should be considered to be a purely thermally driven system due to the lack of ATP. Another precise measurements of ATP effect on the flicker used NGP of membrane displacement histogram [38, 72] for elucidation of non-thermal character of membrane flickering in disk face area [72]. For Gaussian-type distribution of displacement amplitude characteristic of thermal excitation, index NGP = 2, increasing with the deviation of distribution considered from the normal shape [38]. The measured NGP values were 2.8 for healthy erythrocytes, 2.06 and 2.19 for the irreversibly depleted and metabolically ATP-depleted cells, returning to 2.98 value with ATP repletion [72]. These data seem to confirm the action of active driving force associated with dynamic dissociation-association of spectrin filaments from the nodes of membrane skeleton (Section 7.3.2.3). However, the measurement of NGP index is again rather indirect way for identification an active mechanism of flicker excitation. Actually, non-Gaussian distribution of displacements may result from temporal fluctuations in membrane mechanical properties due to ATP-induced remodeling of its structure, not from additional driving force. Additionally, some instrumental distortions may possibly be involved in NGP determination. Indeed, diversity exists in reported shapes of fluctuation amplitude distribution. Thus, the Gaussian distributions were observed for fluctuations in equatorial contour of normal erythrocytes [42, 58, 61] as well as for cells treated with hydrogen peroxide [42] or placed in solutions of increasing viscosity [58]. Perhaps, precise comparative measurements of both amplitude histograms and spatial spectra of flicker in erythrocytes and in giant unilamellar vesicles will be useful in further investigations.

Concerning physiological functions of membrane flickering, three such possibilities were considered using simplified models ([7] for details). The first one is associated with the phenomenon of steric (contact) repulsion of fluctuating membranes [100–104]. The amplitude of fluctuations, hence repulsion potential is governed by the membrane tension [101–103] (Section 7.3.2). For erythrocyte such tension is regulated physiologically via alteration of surface-to-volume ratio. It depends also on instantaneous cell shape in circulation system. So, this mechanism may possibly be employed in regulation of erythrocytes adhesion to the blood vessel walls.

The second possibility is flicker influence on lateral diffusion of transmembrane proteins, for example, band 3 proteins. It employs variation of local distance between lipid bilayer and spectrin filaments of membrane skeleton in the course of flickering. Commonly, free diffusion of protein globule protruding from bilayer into the cytoplasm is hindered by its collisions with spectrin filaments pinned locally to bilayer [105]. Bending fluctuations of bilayer change the local distance between the globule and neighboring filament thus facilitating lateral diffusion of a globule [23, 83].

A third possible physiological function of flicker is associated with in-plane proteins aggregation, as well as with separation of lipid phases in the membrane. An aggregation results from interaction between bilayer-embedded objects via membrane bending oscillations [106]. Flicker amplitude is decreased near an object with bending rigidity higher then host bilayer rigidity. This induces entropic pressure from the fluctuating membrane on the outer parts of boundaries of two neighboring objects. Numerical simulation of dynamics of fluctuating membrane containing more rigid inclusions showed the possibility of membrane separation into inclusion-rich and inclusion-poor phases [107]. These theoretical considerations are adjoined by experimental observations of dynamical existence of two kinds of lipid domains in multicomponent lipid bilayer of a giant vesicle [108], and by observation of dynamic subdomains in erythrocytes [70].

7.8
Flicker in Other Cells

The membrane flickering is highly pronounced in erythrocytes owing to the absence of three-dimensional cytoskeleton and because of large surface-to-volume ratio (Section 7.3.1). Here, the mean amplitude of chaotic membrane displacements amounts to $\sim 0.1\ \mu m$ (Section 7.5.3). White blood cells (as well as other cell types) are essentially less flexible cells. This is due to both more thick and dense scaffold attached to lipid bilayer compared with erythrocyte membrane skeleton, and to formation of three-dimensional fiber structure in activated cells [104]. However, chaotic membrane displacements were registered for white blood cells, as well as for many other cell types also. The nomenclature includes lymphocytes, monocytes, fibroblasts and cardiomyocytes [60], murine lymphoma cells [109, 110], and macrophages [54, 104]. Here, owing to stronger geometrical and mechanical confinements, oscillations wavelengths are shorter and characteristic fluctuations amplitude is nearly order of magnitude smaller compared with erythrocytes. For instance, $\lambda \leq 0.5\ \mu m$ and $\sqrt{\overline{u^2}} \sim 10$ nm for macrophages [104]. Theoretical basis for flicker description in these cells is considerably less developed; however some general type results from the theory of flickering surfaces was used successfully in data evaluation [104]. Flicker phenomenology is essentially more diverse here because of more complex physiological functions of such cells. In particular, there are many interrelations of flicker amplitude with the active cellular processes. Thus, a positive correlation was demonstrated between the relative level of cell membrane fluctuations in BW5147 lymphoma derived cell lines and their metastatic potential [110]. The monitoring of flicker amplitude enabled the dynamics of macrophage cytoskeleton to be followed in the process of phagocytosis [54]. The influence of structural and physiological state of membrane-associated scaffold of macrophages on their adhesion properties, implemented by alteration of flicker amplitude, was observed also [104]. Therefore, the introduction of quantitative phase imaging (Section 7.4.2) in laboratory practice, which enables the measurement of cell thickness maps with nanometer sensitivity and millisecond temporal resolution, opens wide and very promising perspectives for flicker spectroscopy applications in cell cytometry.

7.9 Conclusions

Flicker spectroscopy is a fine instrument, which enables sensitive monitoring, both qualitative and quantitative, of static and dynamic mechanical parameters of composite cell membrane and cytoplasm of erythrocytes. It also enables monitoring of cell physiological state in various conditions and under the influence of external factors of various nature. This is achieved by registering frequency and spatial spectra, mean square amplitude, as well as amplitude histograms of chaotic oscillations of cell membrane. Similar possibilities seem to be realistic for other blood cells, beginning with lymphocytes, and for other cell types. The measured mechanical parameters are associated with detailed molecular organization and composition of cell membrane and cytoplasm, as well as with cell metabolic activity. This opens good prospects for various diagnostic applications of flicker spectroscopy. A family of optical techniques developed for quantitative registration of flicker enables most of the measuring tasks that arise. Various microscope-based methods of frequency spectra registering and quantitative phase imaging combined with fluorescence microscopy (Section 7.4 and [65]) can be naturally integrated into the slide cytometry techniques [4].

For adequate interpretation of flicker spectra measured for disk faces of erythrocyte, it is fundamentally important to register simultaneously the stationary shape of erythrocyte. This is due to pronounced dependence of flicker spectrum parameters in the face area on the cell shape and surface area/volume ratio. The methods of quantitative phase imaging give such possibility. The registration of cell equatorial contour fluctuations is less sensitive to alterations in erythrocyte shape including those induced by adhesion. Here the intrinsic flicker spectra are less distorted by registration procedures also.

For adequate data processing, the flicker theory should be used with due account of cell shape influence, metabolically dependent membrane mechanics, and instrumental distortions of intrinsic flicker spectra. Despite a substantial progress in development of approximate models, the need for more advanced theoretical basis for flicker spectroscopy of erythrocytes still persists.

References

1. Shapiro, H.M. (2003) *Practical Flow Cytometry*, 4th edn, Wiley-Liss, New York.
2. Sack, U., Tarnok, A., and Rothe, G. (eds) (2009) *Cellular Diagnostics: Basic Principles, Methods and Clinical Applications of Flow Cytometry*, S. Karger AG, Basel.
3. Tuchin, V.V., Tarnok, A., and Zharov, V.P. (2009) Towards in vivo flow cytometry. *J. Biophoton.*, **2** (8–9), 457–458.
4. Mittag, A. and Tarnok, A. (2010) Perspectives in cytometry, in *Advanced Optical Cytometry: Methods and Disease Diagnoses*, Chapter 1, 1–24 (ed. V.V. Tuchin), John Wiley & Sons, Inc.
5. Brohard, F. and Lennon, J.F. (1975) Frequency spectrum of the flicker phenomenon in erythrocytes. *J. Phys.*, **36** (11), 1035–1047.
6. Fricke, K., Wirthensohn, K., Laxhuber, R., and Sackmann, E. (1986) Flicker spectroscopy of erythrocytes. A sensitive method to study subtle changes of membrane bending stiffness. *Eur. Biophys. J.*, **14** (2), 67–81.
7. (a) Kononenko, V.L. (2009) Flicker in erythrocytes. I. Theoretical models and registration techniques. *Biochem. (Moscow) Suppl. Ser. A: Membr. Cell Biol.*, **3** (4), 356–371; (b) Kononenko, V.L. (2009) Flicker in erythrocytes. II. Results of experimental studies. *Biochem. (Moscow) Suppl. Ser. A: Membr. Cell Biol.*, **3** (4), 372–387.
8. Evans, E. and Fung, Y.C. (1972) Improved measurements of the erythrocytes geometry. *Microvasc. Res.*, **4** (4), 335–347.
9. Hoffman, E.K. and Simonsen, L.O. (1989) Membrane mechanisms in volume and pH regulation in vertebrate cells. *Physiol. Rev.*, **69** (2), 315–382.
10. Mohandas, N. and Evans, E. (1994) Mechanical properties of the red cell membrane in relation to molecular structure and genetic defects. *Annu. Rev. Biophys. Biomol. Struct.*, **23**, 787–818.
11. Mohandas, N. and Gallagher, P.G. (2008) Red cell membrane: past, present, and future. *Blood*, **112** (10), 3939–3948.
12. Salomao, M., Zhang, X., Yang, Y., Lee, S., Hartwig, J.H., Chasis, J.A., Mohandas, N., and An, X. (2008) Protein 4.1R-dependent multiprotein complex: new insights into the structural organization of the red blood cell membrane. *Proc. Natl. Acad. Sci. U.S.A.*, **105** (23), 8026–8031.
13. Evans, E.A. and Skalak, R. (1980) *Mechanics and Thermodynamics of Biomembranes*, CRC-Press, Boca Raton, FL.
14. Kozlov, M.M. and Markin, V.S. (1987) Model of red blood cell membrane skeleton: electrical and mechanical properties. *J. Theor. Biol.*, **129**, 439–452.
15. Gov, N., Zilman, A.G., and Safran, S. (2003) Cytoskeleton confinement and tension of red blood cell membranes. *Phys. Rev. Lett.*, **90** (22), 228101(4).
16. Fournier, J.-B., Lacoste, D., and Raphael, E. (2004) Fluctuation spectrum of fluid membranes coupled to an elastic meshwork: jump of the effective surface tension at the mesh size. *Phys. Rev. Lett.*, **92** (1), 018102(4).
17. Zeman, K., Engelhard, H., and Sackmann, E. (1990) Bending undulations and elasticity of the erythrocyte membrane: effects of cell shape and membrane organization. *Eur. Biophys. J.*, **18** (4), 203–219.
18. (a) Lenormand, G., Henon, S., Richert, A., Simeon, J., and Gallet, F. (2001) Direct measurement of the area expansion and shear moduli of the human red blood cell membrane skeleton. *Biophys. J.*, **81** (1), 43–56; (b) Lenormand, G., Henon, S., Richert, A., Simeon, J., and Gallet, F. (2003) Elasticity of the human red blood cell skeleton. *Biorheology*, **40** (1–3), 247–251.
19. Manno, S., Takakuwa, Y., Nagao, K., and Mohandas, N. (1995) Modulation of erythrocyte membrane mechanical function by beta-spectrin phosphorylation and dephosphorylation. *J. Biol. Chem.*, **270** (10), 5659–5665.

20. Manno, S., Takakuwa, Y., and Mohandas, N. (2005) Modulation of erythrocyte membrane mechanical function by protein 4.1 phosphorylation. *J. Biol. Chem.*, **280** (9), 7581–7587.
21. Henon, S., Lenormand, G., Richert, A., and Gallet, F. (1999) A new determination of the shear modulus of the human erythrocyte membrane using optical tweezers. *Biophys. J.*, **76** (2), 1145–1151.
22. Gov, N.S. and Safran, S.A. (2005) Red blood cell membrane fluctuations and shape controlled by ATP-induced cytoskeletal defects. *Biophys. J.*, **88** (3), 1859–1874.
23. Zhang, R. and Brown, F.L.H. (2008) Cytoskeleton mediated effective elastic properties of model red blood cell membranes. *J. Chem. Phys.*, **129** (6), 065101(14).
24. Gov, N.S. (2007) Active elastic network: cytoskeleton of the red blood cell. *Phys. Rev. E*, **75** (1), 011921(6).
25. Buckingham, M.J. (1983) *Noise in Electronic Devices and Systems*, Ellis Horwood, Ltd, Chichester.
26. West, B.J. and Shlesinger, M.F. (1990) The noise in natural phenomena. *Am. Sci.*, **78** (1), 40–45.
27. Milotti, E. (2002) A Pedagogical Review of 1/f Noise, Arxiv preprint physics/0204033, 2002 - arxiv.org.
28. Tishler, R.B. and Carlson, F.D. (1993) A study of dynamic properties of the human red blood cell membrane using quasi-elastic light-scattering spectroscopy. *Biophys. J.*, **65** (6), 2586–2600.
29. Schneider, M.B., Jenkins, J.T., and Webb, W.W. (1984) Thermal fluctuations of large quasi-spherical bimolecular phospholipid vesicles. *J. Phys.*, **49** (9), 1457–1472.
30. (a) Peterson, M.A. (1985) Shape fluctuations of red blood cells. *Mol. Cryst. Liq. Cryst.*, **127** (1–4), 159–186; (b) Peterson, M.A. (1985) Shape dynamics of nearly spherical membrane bounded fluid cells. *Mol. Cryst. Liq. Cryst.*, **127** (1–4), 257–272.
31. Kononenko, V.L. (1994) Flicker spectroscopy of erythrocytes: a comparative study of several theoretical models. *Proc. SPIE*, **2082**, 236–247.
32. Landau, L.D. and Lifshitz, E.M. (1980) *Statistical Physics*, Part 1, Course of Theoretical Physics, vol. 5, 3rd edn, Butterworth-Heinemann, Oxford.
33. van Kampen, N.G. (2007) *Stochastic Processes in Physics and Chemistry*, 3rd edn, Elsevier, Amsterdam, Boston, MA, London.
34. Kononenko, V.L. and Shimkus, J.K. (1999) Coherent versus noncoherent optical probing of dynamic shape fluctuations in red blood cells. *Proc. SPIE*, **3732**, 326–335.
35. Prost, J. and Bruinsma, R. (1996) Shape fluctuations of active membranes. *Eur. Lett.*, **33** (4), 321–326.
36. Manneville, J.-B., Bassereau, P., Ramaswamy, S., and Prost, J. (2001) Active membrane fluctuations studied by micropipet aspiration. *Phys. Rev. E*, **64** (2), 021908(10).
37. Gov, N. (2004) Membrane undulations driven by force fluctuations of active proteins. *Phys. Rev. Lett.*, **93** (26), 268104(4).
38. Lin, L.C.-L., Gov, N., and Brown, F.L.H. (2006) Nonequilibrium membrane fluctuations driven by active proteins. *J. Chem. Phys.*, **124** (7), 074903(15).
39. Leibler, S. and Maggs, A.C. (1990) Simulation of shape changes and adhesion phenomena in an elastic model of erythrocytes. *Proc. Natl. Acad. Sci. U.S.A.*, **87** (16), 6433–6435.
40. Döbereiner, H.-G., Gompper, G., Haluska, C.K., Kroll, D.M., Petrov, P.G., and Riske, K.A. (2003) Advanced flicker spectroscopy of fluid membranes. *Phys. Rev. Lett.*, **91** (4), 048301(4).
41. Marcelli, G., Parker, K.H., and Winlove, C.P. (2005) Thermal fluctuations of red blood cell membrane via a constant-area particle-dynamics model. *Biophys. J.*, **89** (10), 2473–2480.
42. Hale, J.P., Marcelli, G., Parker, K.H., Winlove, C.P., and Petrov, P.G. (2009) Red blood cell thermal fluctuations: comparison between experiment and molecular dynamics simulations. *Soft Matter*, **5** (19), 3603–3606.

43. Milner, S.T. and Safran, S.A. (1987) Dynamic fluctuations of droplets microemulsions and vesicles. *Phys. Rev. A*, **36** (9), 4371–4379.
44. Peterson, M.A., Strey, H., and Sackmann, E. (1992) Theoretical and phase contrast microscopic eigenmode analysis of erythrocyte flicker amplitudes. *J. Phys. (Fr.). II*, **2** (5), 1273–1285.
45. Fricke, K. and Sackmann, E. (1984) Variation of frequency spectrum of the erythrocyte flickering caused by aging, osmolarity, temperature and pathological changes. *Biochim. Biophys. Acta*, **803** (3), 145–152.
46. Tuvia, S., Almagor, A., Bitler, A., Levin, S., Korenstein, R., and Yedgar, S. (1997) Cell membrane fluctuations are regulated by medium macroviscosity: evidence for a metabolic driving force. *Proc. Natl. Acad. Sci. U.S.A.*, **94** (10), 5045–5049.
47. Humpert, C. and Baumann, M. (2003) Local membrane curvature affects spontaneous membrane fluctuation characteristics. *Mol. Membr. Biol.*, **20** (2), 155–162.
48. Kononenko, V.L. (2002) Dielectro-deformations and flicker of erythrocytes: fundamental aspects of medical diagnostics applications. *Proc. SPIE*, **4707**, 134–143.
49. Girard, P., Prost, J., and Bassereau, P. (2005) Passive or active fluctuations in membranes containing proteins. *Phys. Rev. Lett.*, **94** (8), 088102(4).
50. Lomholt, M.A. (2006) Fluctuation spectrum of quasispherical membranes with force-dipole activity. *Phys. Rev. E*, **73** (6), 061914(9).
51. Gimsa, J. and Reid, C. (1995) Do band 3 protein conformational changes mediate shape changes of human erythrocytes? *Mol. Membr. Biol.*, **12** (3), 247–254.
52. Beck, A.M. and Kononenko, V.L. (1991) Frequency spectra of erythrocyte membrane flickering measured by laser light scattering. *Proc. SPIE*, **1403**(Part 1), 384–386.
53. Kononenko, V.L. and Shimkus, J.K. (2000) Spontaneous and forced oscillations of cell membrane of normal human erythrocytes: absence of resonant frequencies in a range of 0.03 – 500 Hz. *Membr. Cell Biol.*, **14** (3), 367–382.
54. Agero, U., Monken, C.H., Ropert, C., Gazzinelli, R.T., and Mesquita, O.N. (2003) Cell surface fluctuations studied with defocusing microscopy. *Phys. Rev. E*, **67** (5), 051904(9).
55. Mesquita, L.G., Agero, U., and Mesquita, O.N. (2006) Defocusing microscopy: an approach for red blood cell optics. *Appl. Phys. Lett.*, **88** (13), 133901(3).
56. Glionna, G., Oliveira, C.K., Siman, L.G., Moyses, H.W., Prado, D.M.U., Monken, C.H., and Mesquita, O.N. (2009) Tomography of fluctuating biological interfaces using defocusing microscopy. *Appl. Phys. Lett.*, **94** (19), 193701.
57. Tuvia, S., Levin, S., Bitler, A., and Korenstein, R. (1998) Mechanical fluctuations of the membrane skeleton are dependent on F-actin ATPase in human erythrocytes. *J. Cell Biol.*, **141** (7), 1551–1561.
58. Betz, T., Lenz, M., Joanny, J.-F., and Sykes, C. (2009) ATP-dependent mechanics of red blood cell. *Proc. Natl. Acad. Sci. U.S.A.*, **106** (36), 15312–15317.
59. Born, M. and Wolf, E. (1984) *Principles of Optics*, Pergamon Press, Oxford.
60. Krol, A.Y., Grinfeldt, M.G., Levin, S.V., and Smilgavichus, A.D. (1990) Local mechanical oscillations of the cell surface within the range 0.2–30 Hz. *Eur. Biophys. J.*, **19** (2), 93–99.
61. Gögler, M., Betz, T., and Käs, J.A. (2007) Simultaneous manipulation and detection of living cell membrane dynamics. *Opt. Lett.*, **32** (13), 1893–1895.
62. Neuman, K.C. and Block, S.M. (2004) Optical trapping. *Rev. Sci. Instrum.*, **75** (9), 2787–2809.
63. Zilker, A., Engelhardt, H., and Sackmann, E. (1987) Dynamic reflection interference contrast (RIC-) microscopy: a new method to study excitations of cells and to measure membrane bending elastic moduli. *J. Phys.*, **48** (12), 2139–2151.

64. Zilker, A., Ziegler, M., and Sackmann, E. (1992) Spectral analysis of erythrocyte flickering in the 0.3-4-μm^{-1} regime by microinterferometry combined with fast image processing. *Phys. Rev. A*, **46** (12), 7998–8001.
65. Popescu, G. (2008) in *Methods in Nano Cell Biology*, 1st edn, vol. 90, Chapter 5 (ed. B.P. Jena), Academic Press, Burlington, pp. 87–115.
66. Rappaz, B., Barbul, A., Hoffmann, A., Boss, D., Korenstein, R., Depeursinge, C., Magistretti, P.J., and Marquet, P. (2009) Spatial analysis of erythrocyte membrane fluctuations by digital holographic microscopy. *Blood Cells Mol. Dis.*, **42** (3), 228–232.
67. Kemper, B. and Schnekenburger, J. (2010) Digital holographic microscopy for quantitative live cell imaging and cytometry, in *Advanced Optical Cytometry: Methods and Disease Diagnoses*, Chapter 8, 211–238 (ed. V.V. Tuchin), John Wiley & Sons, Inc.
68. Popescu, G., Ikeda, T., Dasari, R.R., and Feld, M.S. (2006) Diffraction phase microscopy for quantifying cell structure and dynamics. *Opt. Lett.*, **31** (6), 775–777.
69. Ikeda, T., Popescu, G., Dasari, R.R., and Feld, M.S. (2005) Hilbert phase microscopy for investigating fast dynamics in transparent systems. *Opt. Lett.*, **30** (10), 1165–1167.
70. Popescu, G., Badizadegan, K., Dasari, R.R., and Feld, M.S. (2006) Observation of dynamic subdomains in red blood cells. *J. Biomed. Opt.*, **11** (4), 040503(3).
71. Popescu, G., Ikeda, T., Goda, K., Best-Popescu, C.A., Laposata, M., Manley, S., Dasari, R.R., Badizadegan, K., and Feld, M.S. (2006) Optical measurement of cell membrane tension. *Phys. Rev. Lett.*, **97** (21), 218101(4).
72. Park, Y.K., Best, C.A., Auth, T., Gov, N.S., Safran, S.A., Popescu, G., Sures, S., and Feld, M.S. (2010) Metabolic remodeling of the human red blood cell membrane. *Proc. Natl. Acad. Sci. U.S.A.*, **107** (4), 1289–1294.
73. Evans, J., Gratzer, W., Mohandas, N., Parker, K., and Sleep, J. (2008) Fluctuations of the red blood cell membrane: relation to mechanical properties and lack of ATP dependence. *Biophys. J.*, **94** (5), 4134–4144.
74. Strey, H., Peterson, M., and Sackmann, E. (1995) Measurement of erythrocyte membrane elasticity by flicker eigenmode decomposition. *Biophys. J.*, **69** (2), 478–488.
75. Bitler, A., Barbul, A., and Korenstein, R. (1999) Detection of movement at the erythrocyte's edge by scanning phase contrast microscopy. *J. Microsc.*, **193** (2, Pt 2), 171–178.
76. Pecreaux, J., Dobereiner, H.G., Prost, J., Joanny, J.F., and Bassereau, P. (2004) Refined contour analysis of giant unilamellar vesicles. *Eur. Phys. J. E. Soft Matter*, **13** (3), 277–290.
77. Streekstra, G.J., Hoekstra, A.G., Nijhof, E.J., and Heethaar, R.M. (1993) Light scattering by red blood cells in ektacytometry: Fraunhofer versus anomalous diffraction. *Appl. Opt.*, **32** (13), 2266–2272.
78. Park, Y.K., Best, C.A., Badizadegan, K., Dasari, R.R., Feld, M.S., Kuriabova, T., Henle, M.L., Levine, A.J., and Popescu, G. (2010) Measurement of red blood cell mechanics during morphological changes. *Proc. Natl. Acad. Sci. U.S.A.*, **107** (15), 6731–6736.
79. Tuvia, S., Levin, S., and Korenstein, R. (1992) Oxygenation-deoxygenation cycle of erythrocytes modulates submicron cell membrane fluctuations. *Biophys. J.*, **63** (2), 599–602.
80. Tuvia, S., Levin, S., and Korenstein, R. (1992) Correlation between local cell membrane displacements and filterability of human red blood cells. *FEBS Lett.*, **304** (1), 32–36.
81. Evans, E.A. (1983) Bending elastic modulus of red blood cell membrane derived from buckling instability in micropipet aspiration tests. *Biophys. J.*, **43** (7), 27–30.
82. Linderkamp, O., Nash, G.B., Wu, P.Y.K., and Meiselman, H.J. (1986) Deformability and intrinsic material properties of neonatal red blood cells. *Blood*, **67** (5), 1244–1250.
83. Lin, L.C.-L. and Brown, F.L.H. (2004) Dynamics of pinned membranes with application to protein diffusion on the

surface of red blood cells. *Biophys. J.*, **86** (2), 764–780.

84. Lisovskaya, I.L., Ataullakhanov, F.I., Tuzhilova, E.G., and Vitvitsky, V.M. (1994) Analysis of geometrical parameters and mechanical properties of erythrocytes by filtration through membrane nuclear filters. II. Experimental verification of the theoretical model. *Biofizika*, **39** (5), 864–871 (in Russia).
85. Duwe, H.P., Kaes, J., and Sackmann, E. (1990) Bending elastic moduli of lipid bilayers: modulation by solutes. *J. Phys. Fr.*, **51** (10), 945–962.
86. Levin, S. and Korenstein, R. (1991) Membrane fluctuations in erythrocytes are linked to MgATP-dependent dynamic assembly of the membrane skeleton. *Biophys. J.*, **60** (3), 733–737.
87. Tuvia, S., Moses, A., Gulayev, N., Levin, S., and Korenstein, R. (1999) Beta-adrenergic agonists regulate cell membrane fluctuations of human erythrocytes. *J. Physiol.*, **516** (Pt 3), 781–792.
88. Zamir, N., Tuvia, S., Riven-Kreitman, R., Levin, S., and Korenstein, R. (1992) Atrial natri-uretic peptide: direct effects on human red blood cell dynamics. *Biochem. Biophys. Res. Commun.*, **188** (3), 1003–1009.
89. Smith, B.D., La Celle, P.L., Siefring, G.E., Lowe-Krentz, L., and Lorand, L. (1981) Effects of the calcium-mediated enzymatic cross-linking of membrane proteins on cellular deformability. *J. Membr. Biol.*, **61** (2), 75–80.
90. Lorand, L., Siefring, G.E., and Lowe-Krentz, L. (1978) Formation of gamma-glutamyl-epsilon-lysine bridges between membrane proteins by a Ca^{2+}-regulated enzyme in intact erythrocytes. *J. Supramol. Struct.*, **9** (3), 427–440.
91. Takakuwa, Y. and Mohandas, N. (1988) Modulation of erythrocyte membrane material properties by Ca^{2+} and calmodulin. *J. Clin. Invest.*, **82** (8), 394–400.
92. Gardos, G., Szasz, I., and Sarkadi, B. (1977) Effect of intracellular calcium on the cation transport processes in human red cells. *Acta Biol. Med. Ger.*, **36**, 823–829.
93. Alster, Y., Loewenstein, A., Levin, S., Lazar, M., and Korenstein, R. (1998) Low-frequency submicron fluctuations of red blood cells in diabetic retinopathy. *Archiv. Ophthalmol.*, **116** (10), 1321–1325.
94. Goldstein, M., Leibovitch, I., Levin, S., Alster, Y., Loewenstein, A., Malkin, G., and Korenstein, R. (2004) Red blood cell membrane mechanical fluctuations in non-proliferative and proliferate diabetic retinopathy. *Graefe's Arch. Clin. Exp. Ophthalmol.*, **242**, 937–943.
95. Park, Y.K., Diez-Silva, M., Popescu, G., Lykorafitis, G., Choi, W., Feld, M.S., and Suresh, S. (2008) Refractive index maps and membrane dynamics of human red blood cells parasitized by plasmodium falciparum. *Proc. Natl. Acad. Sci. U.S.A.*, **105** (37), 13730–13735.
96. Park, Y.K., Diez-Silva, M., Fu, D., Popescu, G., Choi, W., Barman, I., Suresh, S., and Feld, M.S. (2010) Static and dynamic light scattering of healthy and malaria-parasite invaded red blood cells. *JBO Lett.*, **15** (2), 020506(3).
97. Chien, S. (1987) Red cell deformability and its relevance to blood flow. *Ann. Rev. Physiol.*, **49**, 177–192.
98. Mohandas, N. and Chasis, J.A. (1993) Red blood cell deformability, membrane material properties and shape: regulation by transmembrane, skeletal and cytosolic proteins and lipids. *Semin. Hematol.*, **30** (3), 171–192.
99. Priezzhev, A.V., Lugovtsov, A.G., and Nikitin, S.Y. (2010) Laser diffraction on RBC and deformability measurement, in *Advanced Optical Cytometry: Methods and Disease Diagnoses*, Chapter 6, 133–154 (ed. V.V. Tuchin), John Wiley & Sons, Inc.
100. Helfrich, W. and Servuss, R.-M. (1984) Undulations, steric interaction and cohesion of fluid membranes. *Il Nuovo Cimento D*, **3** (1), 137–151.
101. Evans, E. (1991) Entropy-driven tension in vesicle membranes and unbinding of adherent vesicles. *Langmuir*, **7** (9), 1900–1908.
102. Helfrich, W. (1995) Tension-induced mutual adhesion and a conjectured superstructure of lipid membranes,

in *Handbook of Biological Physics: Structure and Dynamics of Membranes*, vol. 1, Chapter 14 (eds R., Lipowsky and E. Sackmann), Elsevier, Amsterdam, pp. 691–721.

103. Radler, J.O., Feder, T.J., Strey, H.H., and Sackmann, E. (1995) Fluctuation analysis of tension-controlled undulation forces between giant vesicles and solid substrates. *Phys. Rev. E*, **51** (5), 4526–4536.

104. Zidovska, A., and Sackmann, E. (2006) Brownian motion of nucleated cell envelopes impedes adhesion. *Phys. Rev. Lett.*, **96** (4), 048103(4).

105. Tomishige, M., Sako, Y., and Kusumi, A. (1998) Regulation mechanism of the lateral diffusion of band 3 in erythrocyte membranes by the membrane skeleton. *J. Cell Biol.*, **142** (4), 989–1000.

106. Helfrich, W. and Weikl, T.R. (2001) Two direct methods to calculate fluctuation forces between rigid objects embedded in fluid membranes. *Eur. Phys. J. E*, **5** (4), 423–439.

107. Weikl, T.R. (2002) Dynamic phase separation of fluid membranes with rigid inclusions. *Phys. Rev. E*, **66** (6), 061915(6).

108. Marx, S.A., Schilling, J., Sackmann, E., and Bruinsma, R. (2002) Helfrich repulsion and dynamical phase separation of multicomponent lipid bilayers. *Phys. Rev. Lett.*, **88** (13), 138102(4).

109. Mittelman, L., Levin, S., and Korenstein, R. (1991) Fast cell membrane displacements in B lymphocytes: Modulation by dihydrocytochalasin B and colchicine. *FEBS Lett.*, **293** (1–2), 207–210.

110. Mittelman, L., Levin, S., Verschueren, H., De-Baetselier, P., and Korenstein, R. (1994) Direct correlation between cell membrane fluctuations, cell filterability and the metastatic potential of lymphoid cell lines. *Biochem. Biophys. Res. Commun.*, **203** (2), 899–906.

8
Digital Holographic Microscopy for Quantitative Live Cell Imaging and Cytometry

Björn Kemper and Jürgen Schnekenburger

8.1
Introduction, Motivation, and Background

Cytometry techniques are widely used for cell analysis based on specific labeling of cell components. In addition to fluorescence methods, in view of a minimally invasive analysis, technologies that offer a marker-free detection of cellular processes and cell state alterations are of particular interest. Holographic interferometric metrology is a well-established optical tool for industrial nondestructive testing and quality control [1–3]. In combination with microscopy, digital holography provides label-free, quantitative phase contrast imaging for technical inspection, and quantitative live cell analysis [4–10] that is also suitable for modular integration into commercial microscopes [11]. Digital holographic microscopy (DHM) facilitates a combination with established techniques such as laser scanning microscopy, fluorescence imaging, and optical laser micromanipulation [10]. Compared to other phase contrast methods [12, 13], interferometry-based techniques, [14–17] and optical coherence tomography (OCT) or optical coherence microscopy (OCM) [18–23], DHM provides a quantitative phase contrast with subsequent numerical refocusing (multifocus imaging) from a single recorded hologram. This is of particular advantage for measurements on cellular specimens with high-magnification optics for the detection of dynamic processes as well as for long-term live cell analysis, where focus tracking is required due to mechanical instability or thermal effects. Moreover, the evaluation of quantitative DHM phase contrast images permits an effective detection of lateral object movements. In combination with a calibrated imaging system, digital holographic autofocusing also enables the determination of axial sample displacements. Thus, in combination, 3D tracking of objects is provided [24].

Here, the capabilities of DHM in view of quantitative cell imaging are presented in an overview by the results obtained from a pancreatic tumor cell model. First, the principle of DHM is introduced by a modular setup, to be used in combination with commercial research microscopes. The evaluation of digital off-axis holograms is described for the example of spatial phase shifting reconstruction, as these reconstruction methods have been found, in particular, to be suitable for

Advanced Optical Flow Cytometry: Methods and Disease Diagnoses, First Edition. Edited by Valery V. Tuchin.

applications in live cell imaging. Then, the utilization of DHM for quantitative phase contrast imaging and cell thickness determination, as well as DHM-based multifocus imaging by subsequent numerical refocusing and two-dimensional cell tracking are explained. The application of DHM in quantitative cell imaging is illustrated by shape measurements on different adherent tumor cell types, results from the dynamic analysis of cell reactions to toxin and drugs, and data from the use of DHM for the determination of the integral refractive index of suspended cells. Finally, data from investigations on fixed cells in comparison to the fluorescence staining prospect new DHM applications for the label-free identification of subcellular components.

The presented results show that DHM is a suitable method for label-free characterization of dynamic live cell processes involving morphological alterations, migration, and for the analysis of cells in three-dimensional environments. These features are highly relevant to novel insights in tumor cell biology and the development of improved systems for drug and toxicity testing.

8.2
Principle of DHM

In this section, the principle of DHM and the evaluation of digital holograms are illustrated by a modular setup to be used in combination with commercial researches microscopes. For numerical reconstruction, spatial phase shifting is applied which has been found, in particular, to be suitable for the applications in live cell imaging.

8.2.1
DHM Setup and Imaging

Digital holography is based on the classic holographic principle, but here the hologram is recorded by a digital image sensor, for example, a CCD camera sensor [25]. Figure 8.1 depicts the concept of an DHM system, designed for the integration in common commercial inverted research microscopes and investigations on transparent specimens such as living cells [11]. The light of a laser (e.g., a frequency doubled Nd : YAG laser, $\lambda = 532$ nm) is divided into an object illumination wave (object wave) and a reference wave. For a variable light guidance, polarization maintaining single mode fibers is applied. The illumination with coherent laser light is performed by using the microscope condenser. Thus, an optimized (Koehler-like) illumination of the sample is achieved. The reference wave is guided directly to an interferometric unit that is adapted to one of the microscope's camera ports. Holographic off-axis geometry is achieved by a beam splitter that effects a slight tilt of the reference wave front against the object wave front. The interferogram that is formed by the superposition of object wave and reference wave is recorded by a CCD camera and transferred to an image processing system for the reconstruction and evaluation of the digitized holograms (Section 8.2.2). The magnification of the

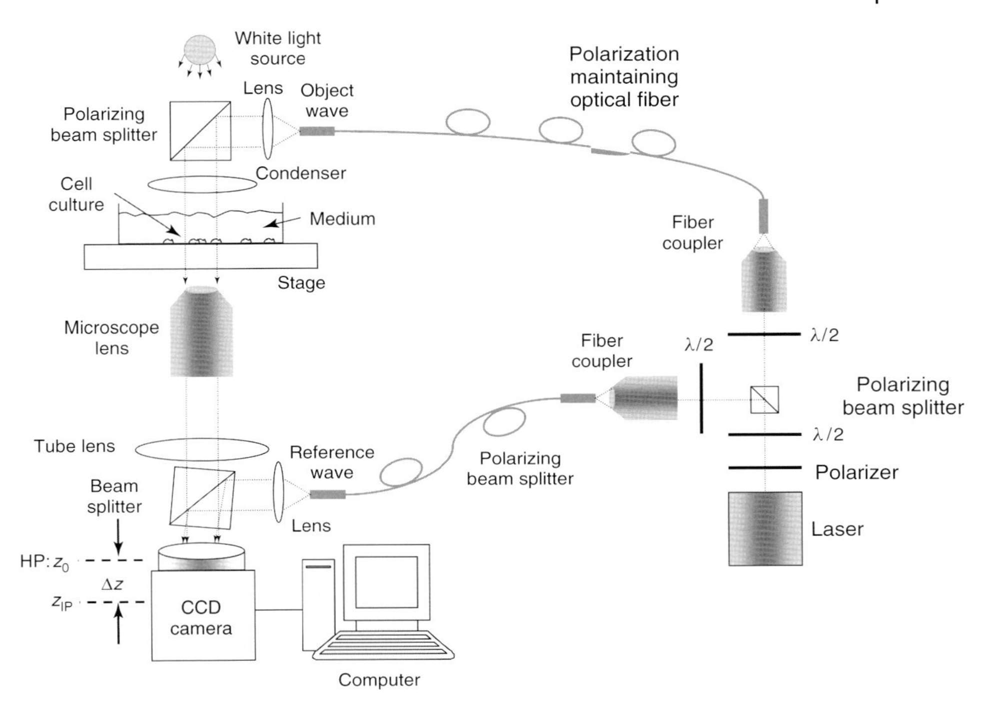

Figure 8.1 Schematic setup of a digital holographic microscopy system for the modular integration in an inverse microscope. HP: hologram (CCD sensor) plane located at $z = z_0$; z_{IP}: location of the image at $z = z_0 + \Delta z$ [11].

microscope lens is chosen in such a way that the smallest imaged structures, given by the restriction of the Abbe criterion, are over sampled by the image recording device. Thus, the maximum diffraction limited resolution of the optical imaging system is not decreased by the numerical reconstruction algorithm described in Section 8.2.2. The typical hologram capturing time depends on the applied imaging device (the CCD camera) and is within the range of 1 ms or below.

Figure 8.2 shows the photographs of a DHM system in combination with an inverse research microscope (Olympus IX81 fluorescence microscope with DHM module in transmission mode). Figure 8.2b shows the DHM unit for coherent illumination that is attached to the condenser of the microscope and a miniaturized climate chamber (ibidi GmbH, Munich, Germany) for temperature-stabilized live cell imaging.

8.2.2
Evaluation of Digital Holograms

Various methods for the numerical reconstruction of digitally captured holograms have been developed (for an overview see, e.g., [27–30]). Spatial phase-shifting (SPS)

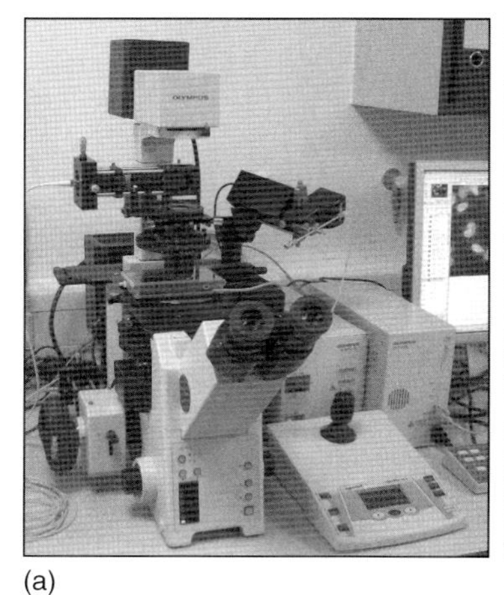

(a)

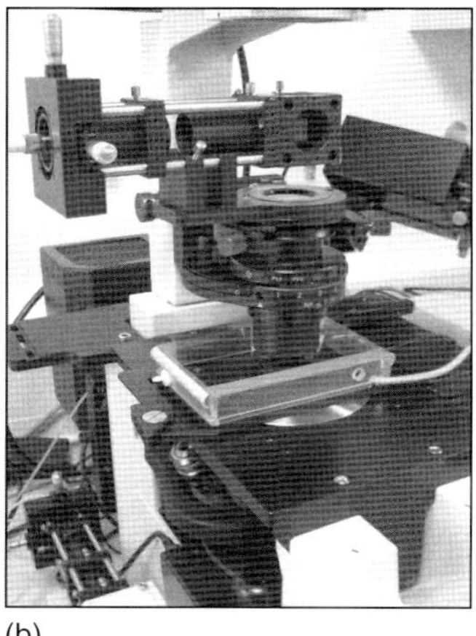

(b)

Figure 8.2 Digital holographic microscopy system in combination with an inverse research microscope. (a) Inverse Olympus IX81 fluorescence microscope with DHM module in transmission mode. (b) The condenser of the microscope with the DHM unit for coherent illumination and a miniaturized climate chamber (ibidi GmbH, Munich, Germany) for temperature-stabilized live cell imaging [26].

holography has been found to be particularly suitable for use in the reconstruction of digital holograms that are recorded in off-axis geometry with DHM as described in Section 8.2.1 and Ref. [6, 31]. An SPS numerical reconstruction offers both the elimination of zero-order intensity and the twin image, as well as the compensation of aberrations of the object wavefront versus the reference wave. Thus, in the following, reconstruction of the digitally captured holograms is illustrated for an SPS technique that has been applied successfully in DHM based live cell imaging [6, 32]. The intensity distribution $I_{\mathrm{HP}}(x, y, z_0)$ in the hologram plane (HP), located at $z = z_0$, is formed by the interference of the object wave $O(x, y, z = z_0)$ and reference wave $R(x, y, z = z_0)$:

$$
\begin{aligned}
I_{\mathrm{HP}}(x, y, z_0) &= O(x, y, z_0)\, O^*(x, y, z_0) + R(x, y, z_0)\, R^*(x, y, z_0) \\
&\quad + O(x, y, z_0)\, R^*(x, y, z_0) + R(x, y, z_0)\, O^*(x, y, z_0) \\
&= I_O(x, y, z_0) + I_R(x, y, z_0) + 2\sqrt{I_O(x, y, z_0)\, I_R(x, y, z_0)} \\
&\quad \cos \Delta\phi_{\mathrm{HP}}(x, y, z_0)
\end{aligned} \tag{8.1}
$$

with $I_O = OO^* = |O|^2$ and $I_O = RR^* = |R|^2$ (*represent the conjugate complex terms). The parameter $\Delta\phi_{\mathrm{HP}}(x, y, z_0) = \phi_R(x, y, z_0) - \phi_O(x, y, z_0)$ is the phase difference between O and R at $z = z_0$. In the presence of a sample in the optical path of O, the phase distribution represents the sum $\phi_O(x, y, z_0) = \phi_{O_0}(x, y, z_0) + \Delta\varphi_S(x, y, z_0)$ with the pure object wave phase $\phi_{O_0}(x, y, z_0)$ and the phase change $\Delta\varphi_s(x, y, z_0)$, which is effected by the sample. For areas without sample, $\Delta\phi_{\mathrm{HP}}(x, y, z_0)$ can be estimated by a mathematical model [6, 31]:

$$\begin{aligned}\Delta\phi_{\mathrm{HP}}(x, y, z_0) &= \phi_R(x, y, z_0) - \phi_{O_0}(x, y, z_0) \\ &= 2\pi\left(K_x x^2 + K_y y^2 + L_x x + L_y y\right)\end{aligned} \tag{8.2}$$

The parameters K_x and K_y in Equation 8.2 describe the divergence of the object wave. The constants L_x and L_y denote the linear phase difference between O and R due to the off-axis geometry of the experimental setup. For quantitative phase measurement from $I_{\mathrm{HP}}(x, y, z_0)$ in a first step, the complex object wave $O(x, y, z = z_0)$ in the HP is determined pixel wise by solving a set of equations that is obtained from the insertion of Equation 8.2 in Equation 8.1. For that purpose, neighboring intensity values, for example, within a square area of 5 × 5 pixels around a given hologram pixel, are considered by the application of an SPS algorithm (for details see Refs [31] and [6]). The utilized algorithm is based on the assumption that only the phase difference $\Delta\phi_{\mathrm{HP}}(x, y, z_0) = \phi_R(x, y, z_0) - \phi_{O_0}(x, y, z_0)$ between the object wave $O(x, y, z_0)$ and the reference wave $R(x, y, z_0)$ varies rapidly spatially in the hologram plane (HP). In addition, the object wave's intensity has to be assumed to be constant within the area around a given point of interest of the hologram in which the numerical evaluation is performed. These requirements are fulfilled for an adequate relation between the magnification of the microscope lens and the image recording device. Therefore, a magnification of the microscope lens is chosen in such a way that the smallest imaged structures of the sample that are restricted by the resolution of the optical imaging system due to the Abbe criterion are over sampled by the CCD sensor. In this way, the lateral resolution of the reconstructed holographic phase contrast images is not decreased by an SPS algorithm [6, 32].

The parameters K_x, K_y, L_x, and L_y in Equation 8.2 cannot be obtained with adequate accuracy directly from the geometry of the experimental setup and for this reason are adapted by an iterative fitting process in an area of the hologram without a sample (for a detailed description see, e.g., Refs [6, 32]).

Digital holographic phase contrast microscopy requires, in correspondence to microscopy with white light illumination, a sharply focused image of the sample. For the case that the object is not imaged sharply in the sensor plane during the hologram recording process, for example, due to mechanical instabilities of the experimental setup or thermal effects, in a second evaluation step, a propagation of the object wave to the image plane can be carried out for subsequent focus correction. The propagation of $O(x, y, z_0)$ to the image plane z_{IP} that is located at $z_{\mathrm{IP}} = z_0 + \Delta z$ in the distance Δz to the hologram plane (Figure 8.1) can be carried out by Fresnel transformation [6, 7] or by a convolution algorithm [32–34]:

$$O(x, y, z_{\mathrm{IP}} = z_0 + \Delta z) = \mathcal{F}^{-1}\left\{\mathcal{F}\left\{O(x, y, z_0)\right\} \exp\left(\mathrm{i}\pi\lambda\Delta z\left(\nu^2 + \mu^2\right)\right)\right\} \tag{8.3}$$

In Equation 8.3, λ is the applied laser light wave length, ν and μ are the coordinates in frequency domain, and $\mathcal{F}$ denotes a Fourier transformation. During the propagation process, the parameter Δz is chosen so that the holographic amplitude image appears sharply, corresponding to a microscopic image under white light illumination. A further criterion for a sharp image of the sample is that diffraction effects due to coherent illumination appear to be minimized in the reconstructed data. As a consequence of the applied algorithms and the parameter model for the phase difference model $\Delta\phi_{HP}$ in Equation 8.2, the resulting reconstructed holographic images do not contain the disturbing terms "*twin image*" and "*zero order*." In addition, the method allows in contrast to the discrete Fresnel transformation, as, for example, in Refs [5, 6], refocusing meat the sensor plane. Furthermore, the propagation effects no change of the image scale during subsequent refocusing. In the special case, the image of the sample is sharply focused in the HP with $\Delta z = 0$ and thus, $z_{IP} = z_0$; the reconstruction process can be accelerated because no propagation of O by Equation 8.3 is required.

8.2.3 Quantitative Phase Contrast Imaging and Cell Thickness Determination

From $O(x, y, z_{IP})$, in addition to the absolute amplitude $|O(x, y, z_{IP})|$ that represents the image of the sample, the phase information $\Delta\varphi_S(x, y, z_{IP})$ of the sample is reconstructed simultaneously:

$$\begin{aligned}\Delta\varphi_S(x, y, z_{IP}) &= \phi_O(x, y, z_{IP}) - \phi_{O_0}(x, y, z_{IP}) \\ &= \arctan\frac{\mathrm{Im}\{O(x, y, z_{IP})\}}{\mathrm{Re}\{O(x, y, z_{IP})\}}\,(\mathrm{mod}2\pi)\end{aligned} \tag{8.4}$$

After the removal of the 2π ambiguity by a phase unwrapping process [1], the data obtained from Equation 8.4 can be applied for quantitative phase contrast microscopy.

In transmission mode as depicted in Figure 8.1, the measured phase information of a semitransparent sample is influenced by the sample thickness the refractive index of the sample, and the refractive index of the surrounding medium. For life cell imaging, information about the cellular refractive index is rare. Thus, several interferometric and holographic methods for the determination of the refractive index of single cells have been proposed. A DHM-based decoupling procedure in which two-cell culture media with slightly different refractive indices was applied to determine the integral refractive index of neurons [35]. In Ref. [32], the integral cellular refractive index is determined with DHM by pressing a coverglass onto the cells to ensure uniform cell thickness. Afterward, the integral cellular refractive index is determined in reference to the phase changes that are effected by air inclusions. Lue *et al.* proposed a similar method [36] based on Hilbert phase microscopy using microfluidic equipment to ensure a constant cell thickness. In Refs [37, 38] (for detailed description see, Section 8.3.5), digital holographic phase contrast images of spherical cells in suspension are recorded. Then, the radius as well as the integral refractive index are determined by fitting the relation between cell thickness and phase distribution to the measured phase data. In Ref. [39], the refractive index and cellular thickness are measured separately by DHM utilizing a two-wavelength method in combination with an enhanced dispersion of the perfusion medium that is achieved by the utilization of an extracellular dye. In Refs [40, 41], a DHM-based refractive index tomography of cellular samples is performed by recording multiple phase contrast images of a rotated sample. Similar results were obtained by Choi *et al.* [42] with a Fourier phase microscopy experimental setup and by Debailleul *et al.* with diffractive microscopy [43] in which an adherent cellular sample is illuminated by transmission from different directions. However, it is not possible to perform an independent determination of the parameter thickness and refractive index for every measurement case.

In the case of cells in cell culture medium with the refractive index n_{medium}, and the assumption of a known homogeneously distributed integral cellular refractive index n_{cell}, the cell thickness $d_{\text{cell}}\left(x, y, z_{\text{IP}}\right)$ can be determined by measuring the optical path length change $\Delta\varphi_{\text{cell}}$ of the cells to the surrounding medium [7, 32]:

$$d_{\text{cell}}\left(x, y, z_{\text{IP}}\right) = \frac{\lambda\Delta\varphi_{\text{cell}}\left(x, y, z_{\text{IP}}\right)}{2\pi} \times \frac{1}{n_{\text{cell}} - n_{\text{medium}}} \tag{8.5}$$

with the wavelength λ of the applied laser light. For adherently grown cells, the parameter d_{cell} is estimated to describe the shape of single cells. Nevertheless, Equation 8.5 has to be handled critically, for example, if toxin- and osmotically induced reactions of cells are analyzed that may effect dynamic changes of the cellular refractive index (see results in Sections 8.3.3 and 8.3.4, and Refs [7, 32, 44]).

Figure 8.3 illustrates the evaluation process of digital recorded holograms. Figure 8.3a,b shows a digital hologram obtained from a living human pancreas carcinoma cell (Patu 8988T [45]) with an inverted microscope arrangement in transmission mode (40× microscope lens, NA = 0.65), and the reconstructed holographic amplitude image that corresponds to a microscopic bright field image at coherent laser light illumination. Figure 8.3c depicts the simultaneously reconstructed quantitative phase contrast image modulo 2π calculated by Equation 8.4. The unwrapped data without 2π ambiguity representing the optical path length changes that are affected by the sample in comparison to the surrounding medium due to the thickness and the integral refractive index, are shown in Figure 8.3d. Figure 8.3e depicts a gray level coded pseudo 3D plot of the data in Figure 8.3d. Figure 8.3f shows the cell thickness along the marked dashed line in Figure 8.3d that is determined by Equation 8.5 for $n_{\text{cell}} = 1.38$, $n_{\text{medium}} = 1.337$, and $\lambda = 532$ nm. Figure 8.3g presents the first derivative of the data in Figure 8.3d in x direction, which is denoted as digital holographic microscopy differential interference phase contrast (DHM DIC). DHM DIC is comparable to the well-known Nomarski DIC with the additional advantages of subsequent numerical focus realignment and a variable sensitivity due to an adjustable digital shear.

8.2.4
DHM Multifocus Imaging and Cell Tracking

Figure 8.4 demonstrates the DHM feature of subsequent numerical focus correction by variation in the reconstruction distance Δz. Figure 8.4a shows the digital

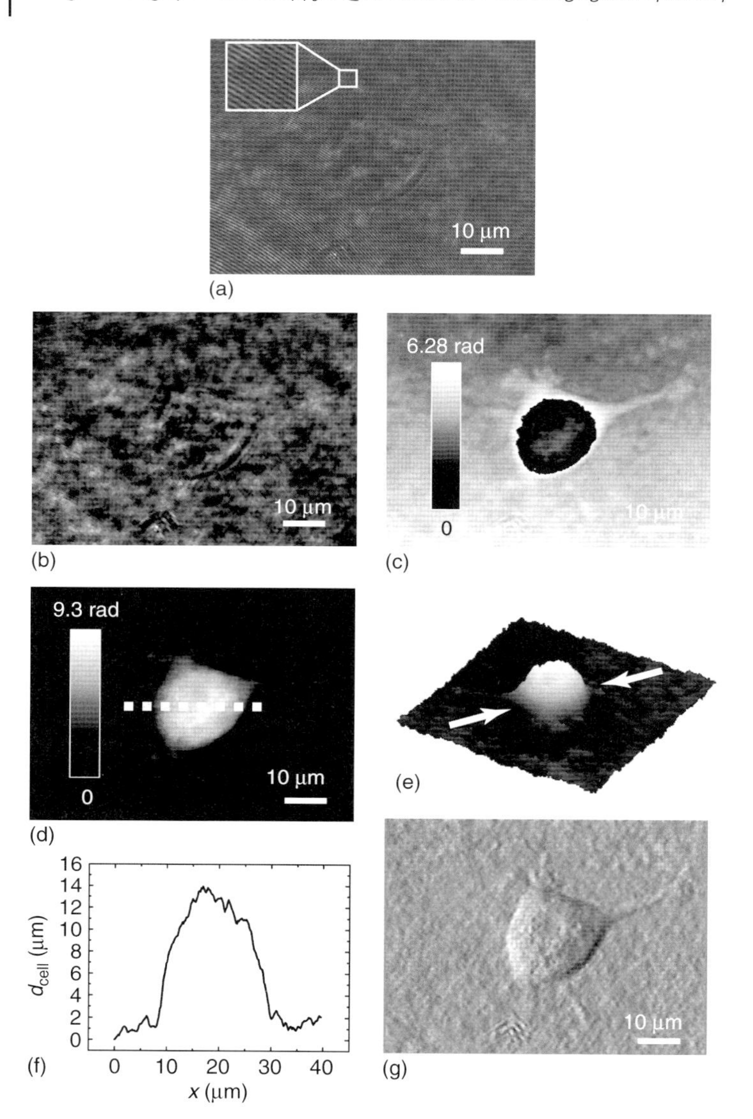

Figure 8.3 Example for evaluation of digital holograms. (a) Digital hologram of a human pancreas carcinoma cell (Patu 8988T), (b) reconstructed holographic amplitude image, (c) quantitative phase contrast image (modulo 2π), (d) unwrapped phase distribution, (e) gray-level coded pseudo 3D plot of the unwrapped phase image, (f) calculated cell thickness along the dashed white line in (d), and (g) DHM DIC image (modified from [11]).

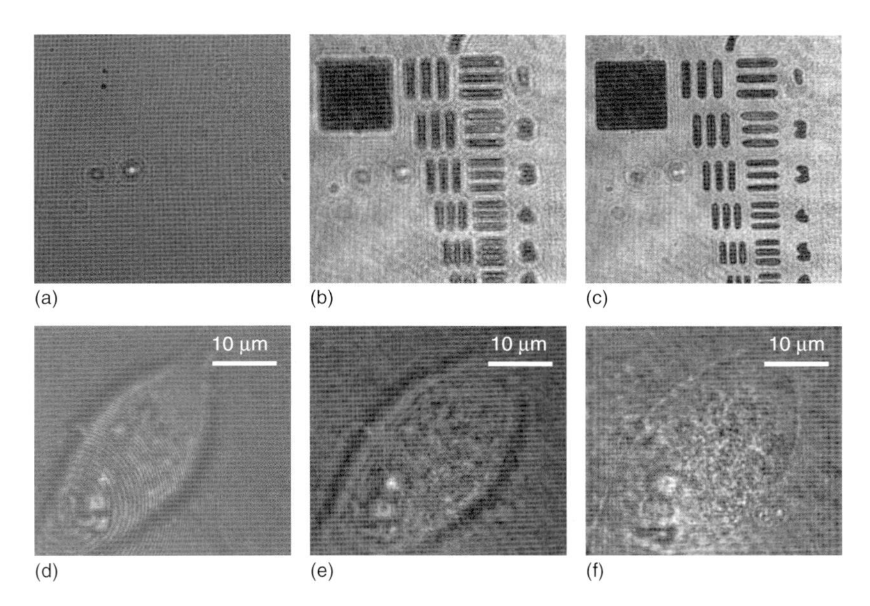

Figure 8.4 Subsequent numerical refocusing. (a) Hologram of a USAF 1951 test chart recorded slightly out of focus by illumination with laser light in transmission, (b) reconstructed object wave intensity (amplitude image) in the hologram plane, (c) numerically refocused holographic amplitude image reconstructed by variation of the reconstruction distance, and (d–f) corresponding results obtained from investigations on a living human pancreas tumor cell in culture medium [11].

hologram of a positive United States Air Force (USAF) 1951 test chart recorded slightly out of focus by illumination with laser light in transmission (40× microscope lens, NA = 0.65). Figure 8.4b depicts the reconstructed holographic image in the hologram plane. The corresponding sharply refocused holographic amplitude is shown in Figure 8.4c. Figure 8.4d–f presents respective results obtained from investigations on a living human pancreas carcinoma cell (Patu 8988S) (transmission mode, 100× oil immersion microscope lens, NA = 1.3). The advantage of subsequent digital holographic focus correction is the avoidance of a mechanical focus adaptation which is particularly convenient, for example, for long-term measurements on living cellular specimen and for automated measurements.

Figure 8.5 illustrates the usage of *subsequent digital holographic focus adjustment* for the evaluation of a digital hologram of pancreas tumor cells (Patu 8988S) in suspension. During the hologram capturing process, the cells were located spatially separated in different focal planes. Figure 8.5a shows the digital hologram. In Figure 8.5b,c, the digital holographic reconstructed amplitude images for each focal plane are depicted. Figure 8.5d,e shows the corresponding quantitative holographic phase contrast images of the cells and illustrates the requirement of subsequent numerical focus correction. The unfocused parts of the phase contrast images contain disturbing phase unwrapping artifacts. The sharply imaged cells appear clearly resolved in the phase distributions and enable a further quantitative

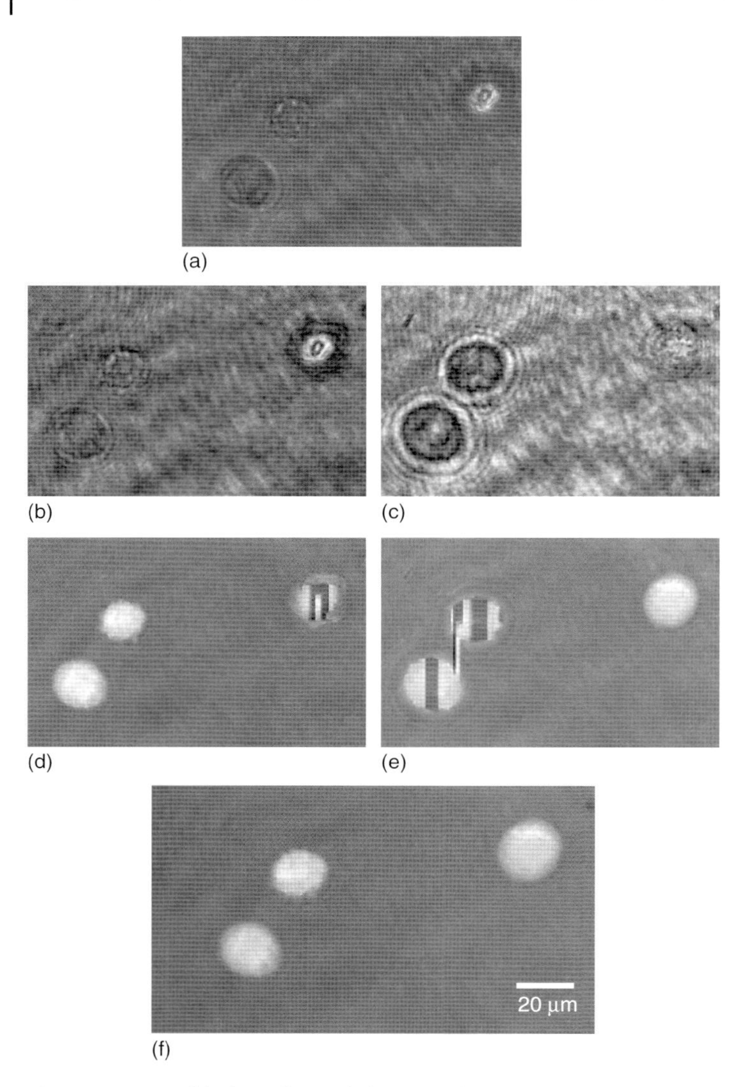

Figure 8.5 Digital holographic multifocus imaging of cells in suspension. (a) Digital hologram of pancreas tumor cells (Patu 8988S) in different focal planes, (b and c) digital holographic amplitude images of the cells in (a) reconstructed in different focal planes, (d and e) unwrapped gray-level coded phase contrast images corresponding to (b) and (c), and (f) phase distribution merged from the sharply focused parts in (d and e) [46].

data evaluation. Figure 8.5f shows the enlarged phase distribution that results from merging the sharply focused images of the cells in Figure 8.5(d,e) in which all cells appear sharply focused.

The numerical digital holographic reconstruction also enables *subsequent autofocusing* from a single recorded hologram. This is performed by determining the image definition of the reconstructed amplitude distributions in dependence of the numerical propagation distance (for a detailed description see Refs [47, 48]). In focus, pure phase objects like cells with negligible absorption appear with least contrast in the amplitude distribution (autofocus criterion) (Figures 8.4 and 8.5). The corresponding quantitative phase contrast images show an optimized lateral resolution and minimized phase noise as already illustrated in Figure 8.5.

Results obtained from experiments on living pancreas tumor cells (PaTu 8988T) show the application of DHM for the visualization and quantitative analysis of dynamic morphological changes and two-dimensional dynamic cell tracking [46]. DHM was therefore used to monitor drug-dependent cell death. Pancreas tumor cells were exposed to the anticancer drug taxol. Digital holograms of selected cells were recorded continuously every 2 min over 14.2 h in a temperature-stabilized environment ($T = 37\,^{\circ}\mathrm{C}$). The upper panel of Figure 8.6 shows representative results for the obtained unwrapped digital holographic phase distributions $\Delta\varphi_{\mathrm{cell}}\left(x, y, z_{\mathrm{IP}}\right)$ of a PaTu 8988T cell at $t = 0, t = 3.5$ h, $t = 5$ h, $t = 8.3$ h, and $t = 14.2$ h after taxol addition. The temporal dependence of the maximum phase contrast $\Delta\varphi_{\mathrm{cell,\,max}}$ that is marked with a white cross in the quantitative phase contrast images in the upper panel of Figure 8.6 as well as the corresponding cell thickness obtained from Equation 8.5 ($n_{\mathrm{medium}} = 1.337$, $n_{\mathrm{cell}} = 1.38$ [32], and $\lambda = 532$ nm) are depicted in the lower left panel of Figure 8.6. The lower right panel of Figure 8.6 shows the two-dimensional cell migration trajectory, that results from the determination of the coordinates with maximum phase contrast. From Figure 8.6, it is clearly visible that taxol rapidly induces morphological changes. First, cell rounding is observed that causes an increase in cell thickness. Afterward, the final cell collapse is detected by a significant decrease of the phase contrast (lower left panel of Figure 8.6).

8.3 DHM in Cell Analysis

The application of DHM in the detection of cellular processes is illustrated by the results obtained from highly characterized isogenetic pancreatic tumor cells with known differences in differentiation and metastatic potential. These cells are an excellent model for closely related epithelial cells with different disease states. Pancreatic cancers are among the leading cancer death cases and remain a major unsolved health problem [49]. All efforts in the past decades had little impact on the course of the disease. Therefore, a detailed understanding of pancreas tumor cell biology is still needed for the development of effective treatments [50].

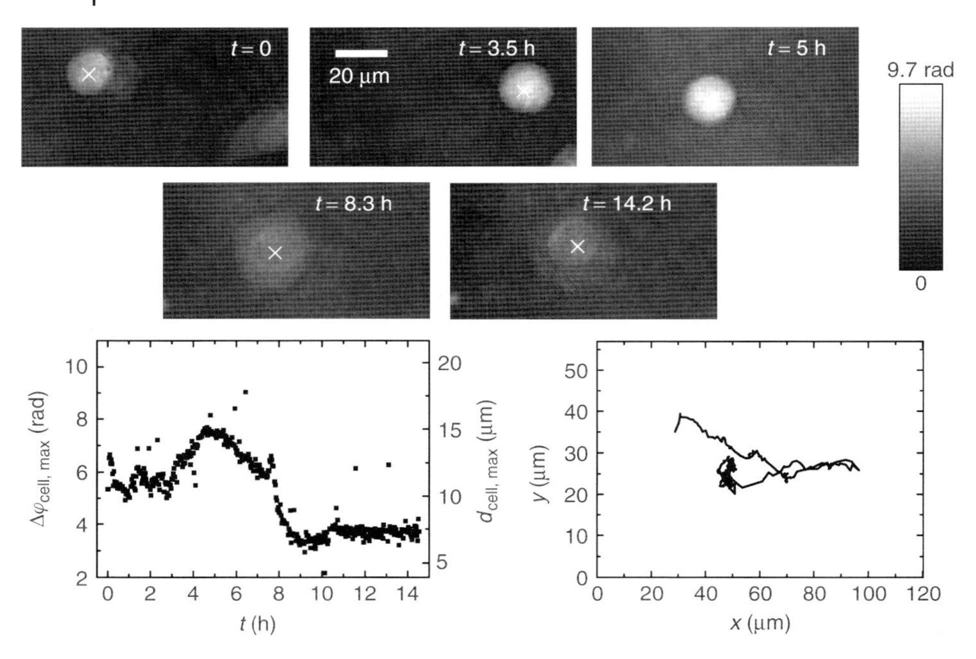

Figure 8.6 Cell thickness monitoring and two-dimensional cell tracking illustrated by results from a living pancreas tumor cell (PaTu 8988T) after addition of taxol to the cell culture medium (40× microscope lens, NA = 0.6). Upper panel: gray-level coded unwrapped phase distributions at $t = 0, t = 3.5$ h, $t = 8.3$ h, and $t = 14.2$ after toxin addition; lower left panel: temporal dependence of the maximum phase contrast and related cell thickness marked with a cross in the upper panel, lower right panel: cell migration trajectory obtained by the determination of the coordinates of the maximum phase contrast in the upper panel [46].

Improved drug screening assays should reflect the complexity of cancers and base on mechanisms and cellular processes rather than single target genes [51].

DHM is mainly a cell thickness sensitive tool and therefore, the primary application in cell analysis is the characterization of outward exposed epithelial cells. Epithelia are tight layers of specialized polar cells composing the surface of skin, lung, or the intestine. Major objectives of DHM development are the characterization and identification of pathologically altered epithelial cells and the measurement of tumor cell reactions to cytotoxic drugs. The potential of DHM to detect altered cells by specific cellular properties is demonstrated by the pancreas tumor cell model.

8.3.1 Cell Cultures

The investigated human pancreatic ductal adenocarcinoma cell lines, PaTu 8988S and PaTu 8988T, were obtained from the German Collection of Microorganisms and Cell Cultures (DSMZ, Braunschweig, Germany). Both cell lines were established in 1985 from a liver metastasis of a primary pancreatic adenocarcinoma in a 64-year-old woman. PaTu 8988S represents a highly differentiated carcinoma cell line, and PaTu 8988T, a poorly differentiated adenocarcinoma with a high-metastatic potential [45]. Both cell lines differ in the expression of integrins and proliferative markers [52, 53] and in the formation of cell–cell contacts. PaTu 8988S cells express the cell–cell contact protein E-cadherin, whereas PaTu 8988T cells lack E-cadherin and display no cell–cell contacts [54]. PaTu 8988T E-cadherin cells were retrovirally transduced with an E-cadherin expression construct containing an E-cadherin cDNA in the expression vector pLXIN (Clontech, Palo Alto, CA, USA). Cells were selected, cloned, and analyzed for E-Cadherin expression [55]. The morphology of the cell lines was examined by scanning electron microscopy (SEM). PaTu 8988T E-cadherin cells showed a flattened fibroblast-like morphology compared to parental rather epithelial PaTu 8988T cells. Pancreas tumor cell lines were cultured in Dulbecco Modified Eagle Medium (DMEM) supplemented with 5% fetal calf serum (FCS), 5% horse serum, and 2 mM L-glutamine at 5% CO_2. For the digital holographic investigations, the cells were trypsinized, seeded subconfluent on glass slides or tissue culture plates, cultured for 24 h, and analyzed in Hepes buffered medium at room temperature or 37 °C at normal atmosphere.

8.3.2 DHM Cell Thickness Measurements

The application of DHM for cell thickness determination is demonstrated by the results obtained from a comparative study on adherently grown living PaTu 8988T cells and PaTu 8988T pLXIN E-Cadherin cells [32]. For the experiments, holograms of PaTu 8988T cells and PaTu 8988T pLXIN E-Cadherin cells are captured in transmission with a DHM setup as depicted in Figure 8.1. From the determined phase data, the thickness d_{cell} of the cells is obtained by the application of Equation 8.5 and estimation of an integral refractive index $n_{\text{cell}} = 1.38$ [32]. Figure 8.7 depicts the results obtained from a PaTu 8988T cell. Figure 8.7a shows the reconstructed holographic amplitude image. Figure 8.7b depicts the unwrapped phase data with included gray-level legend for the phase and the corresponding calculated cell thickness. Figure 8.7d shows a rendered pseudo 3D representation of the cell thickness in comparison to an SEM image of a PaTu 8988T cell Figure 8.7c. The scanning electron micrograph was taken by an LEO 1530 VP (LEO Electron Microscopy Group Ltd., Cambridge, UK) from a sample coated with a thin metal layer followed by a carbon layer. Figure 8.8 represents the corresponding results obtained for two PaTu 8988T pLXIN E-Cadherin cells. For each cell type, the obtained data for the cell thickness is found to be in good agreement with the appearance of the cells in the SEM images.

In Figure 8.9, the phase values as well as cell thickness are plotted for both cell lines corresponding to the cross sections marked by white lines in Figures 8.7b and 8.8b. The different cell types can be differentiated by the phase difference $\Delta\varphi_{\text{cell}}$ as well as by the calculated thickness. For the PaTu 8988T cell, a maximum

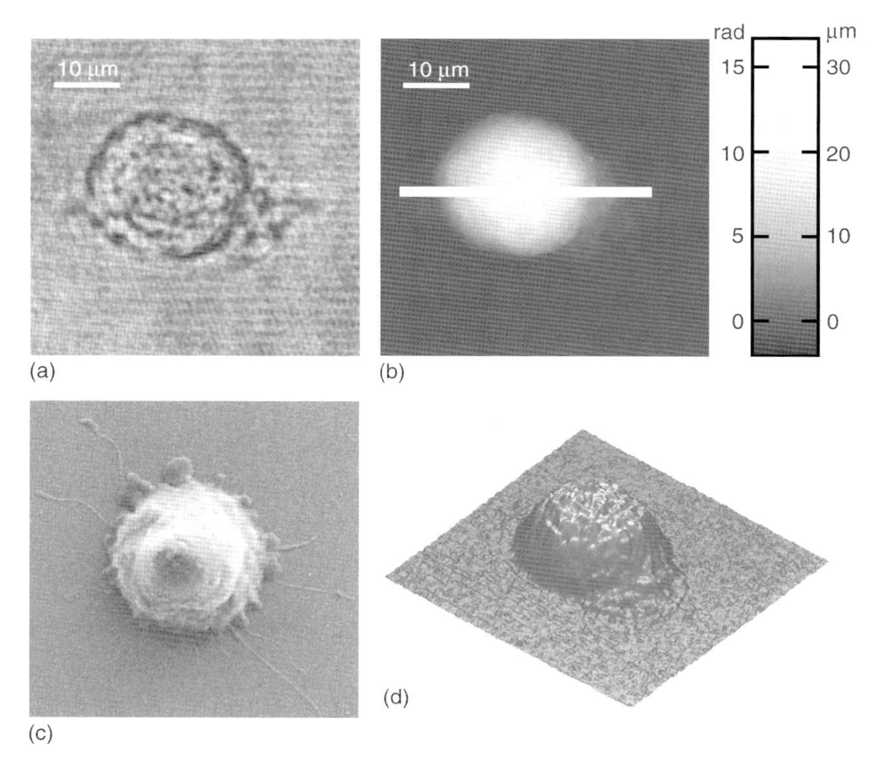

Figure 8.7 Thickness determination of living PaTu 8988T cells by digital holographic microscopy. (a) holographic amplitude image, (b) unwrapped quantitative phase contrast image (for the plot of the cross section along the white line, see Figure 8.9), (c) SEM image of a PaTu 8988T cell, and (d) rendered pseudo 3D plot of the cell thickness obtained from (b) [32].

thickness of $d_1 = (23 \pm 1)$ μm, and for PaTu 8988T pLXIN E-Cadherin a maximum thickness of $d_2 = (7 \pm 1)$ μm, is obtained. The maximum error for d_1 and d_2 is estimated by taking into account the spatial phase noise of the phase distributions that was determined by an average value of 0.26 rad in the area of the sample. In comparison, for areas without cells, the phase noise is determined to be 0.08 rad. Assuming that the investigated cells were grown fully adherently on the carrier glass and taking into account the appearance of the pancreas cells in the SEM images, the results for the cell thickness in Figure 8.9 represent a good estimation about the shape of the cells.

8.3.3
Dynamic Cell Thickness Monitoring in Toxicology

The application of DHM for the detection of dynamic morphological changes within the range of several seconds is illustrated by the results obtained from

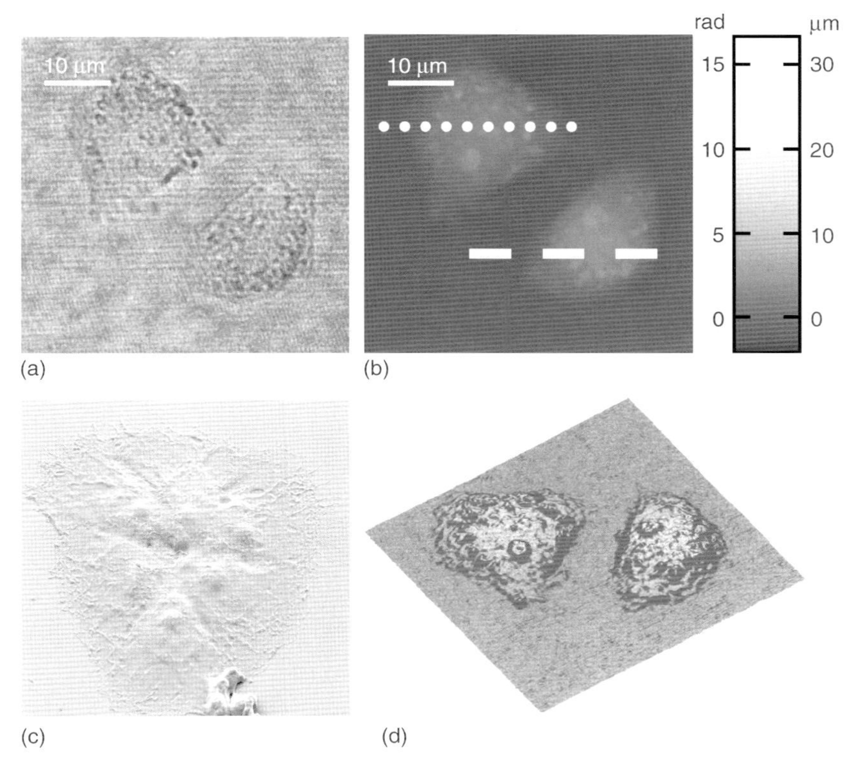

Figure 8.8 Thickness determination of living PaTu 8988T pLXIN E-Cadherin cells by digital holographic microscopy. (a) holographic amplitude image of two cells, (b) unwrapped quantitative phase contrast image (for the plot of the cross section along the white dashed lines, see Figure 8.9), (c) SEM image of a PaTu 8988T pLXIN E-Cadherin cell, and (d) rendered pseudo 3D plot of the cell thickness obtained from (b) [32].

the investigations of living PaTu 8988S cells. Therefore, a hologram series of a single PaTu 8988S cell with a time delay between two recordings of $\Delta t = 10$ s is captured. At the beginning of the experiment at $t = 0$ s, the cells were incubated with the marine toxin Latrunculin B (Calbiochem, San Diego, CA, USA) [56] that was added to a final concentration of 10 μM to the cell culture medium in order to destruct the actin filaments of the cell's cytoskeleton. The upper row of Figure 8.10 represents the resulting wrapped phase distribution modulo 2π after $t = 0$ s (Figure 8.10a), $t = 30$ s (Figure 8.10b), $t = 80$ s (Figure 8.10c), and $t = 180$ s (Figure 8.10d). With increasing time after the addition of Latrunculin B, the phase distributions indicate morphological changes and the cell collapse due to the destruction of the cytoskeleton. The lower row of Figure 8.10 depicts, the corresponding unwrapped phase data along the marked horizontal cross sections of the distribution (see white arrows in the upper row of Figure 8.10) as well as the cell thickness that is calculated for the assumption of negligible refractive index changes of the cell is plotted. During the collapse, $\Delta\varphi_{\mathrm{cell}}$ as well as the cell thickness

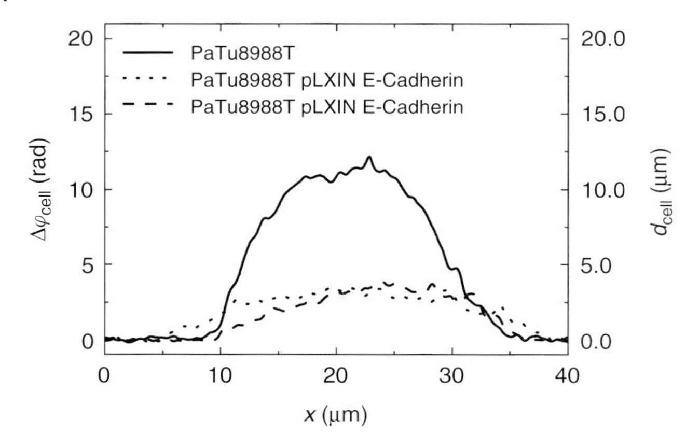

Figure 8.9 Thickness comparison of PaTu 8988T cells and PaTu 8988T pLXIN E-Cadherin cells along the cross section lines in Figures 8.7b and 8.8b [32].

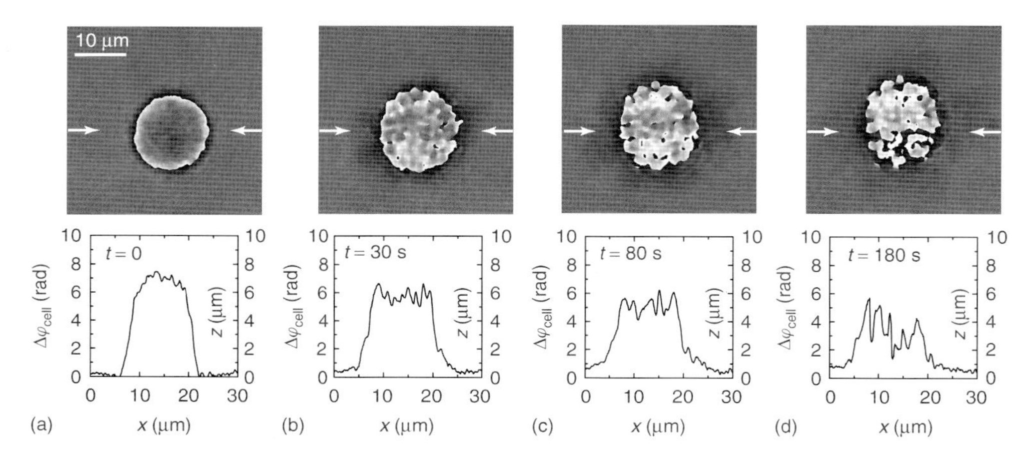

Figure 8.10 Analysis of a living PaTu 8988S cell after addition of a marine toxin (Latrunculin B) to the cell culture medium. Unwrapped phase values and cell thickness along the cross sections (lower row) through the reconstructed quantitative phase contrast images modulo 2π (upper row) after (a) $t = 0$ s, (b) $t = 30$ s, (c) $t = 80$ s, and (d) $t = 180$ s after the addition of Latrunculin B [32].

d_{cell} are locally decreased upto about 50%. Furthermore, the crinkle structure of the collapsed cell is clearly visualized in the phase contrast images as well as in the plotted cross sections. The quantitative values for the cell thickness calculated from these phase measurements have to be handled critically due to the unavailable information about refractive index changes of the cells during the experiment (see detailed discussion in Ref. [32]). However, the obtained information opens up new ways for marker-free dynamic monitoring of morphological changes and the access of new quantitative parameters.

8.3.4
Label-Free Detection of Apoptotic Processes

Established methods for the analysis of cellular reactions upon exposure to drugs or toxic chemicals involve the cost- and time-consuming detection of molecular markers in cell-based assays by fluorescent or luminescent readouts. Thus, investigations on living pancreas tumor cells (Patu 8988T) were carried out to demonstrate the potential of DHM for the label-free temporal visualization of apoptosis processes. PaTu 8988T cells were exposed to cell death inducing concentrations of taxol. Digital holograms of the cells were recorded continuously every 2 min in a temperature-stabilized environment ($T = 37\,^{\circ}\mathrm{C}$) for 12 h. Figure 8.11 shows exemplarily time-dependent results for the obtained unwrapped digital holographic phase distributions of two cells (denoted as A and B) after taxol addition in comparison to the phase contrast images from a control measurement of two cells (denoted as C and D) without the addition of taxol. The phase contrast images in Figure 8.11a show fast morphological changes of the cells that indicate apoptosis while for the results from the control measurements in Figure 8.11b rather necrotic processes at the end of the experiment were observed.

The temporal dependence of the maximum phase contrast $\Delta\varphi_{\mathrm{cell,max}}$ and the corresponding maximum thickness of the cells $d_{\mathrm{cell,max}}$, that is, obtained from Equation 8.5 ($n_{\mathrm{medium}} = 1.337$ and $n_{\mathrm{cell}} = 1.38$ [32]) is depicted for both experiments, with and without taxol, in Figure 8.12. The comparison of the results in Figure 8.12a,b

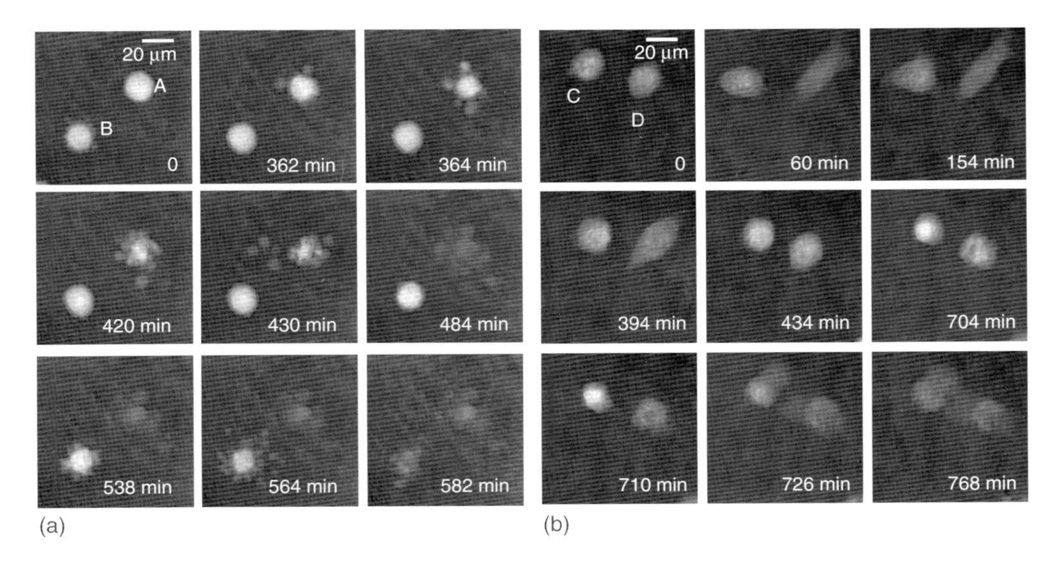

Figure 8.11 Label-free dynamic monitoring of apoptotic processes. Time-dependent observation of living pancreas tumor cells (PaTu 8988T) after addition of taxol. (a) Representative of gray-level coded unwrapped phase distributions after toxin addition and (b) representative phase contrast images from a control measurement without taxol addition [57].

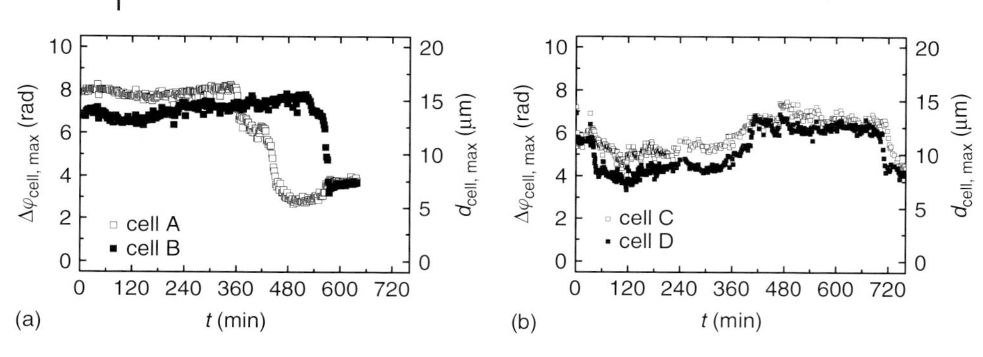

Figure 8.12 Label-free dynamic monitoring of apoptotic processes. Temporal dependence of the maximum phase contrast and related cell thickness living pancreas tumor cells (PaTu 8988T) after addition of taxol; (a) maximum phase contrast $\Delta\varphi_{cell,max}$ and maximum cell thickness $d_{cell,max}$ corresponding to the cells denoted as A and B in Figure 8.11a and (b) maximum phase contrast $\Delta\varphi_{cell,max}$ and maximum cell thickness $d_{cell,max}$ corresponding to the cells denoted as C and D in Figure 8.11b [57].

shows that taxol induces morphological changes like cell rounding that results in an increase of cell thickness. Furthermore, the final cell collapse is detected clearly by a significant decrease of the phase contrast.

8.3.5
Determination of the Integral Refractive Index of Cells in Suspension

As already described in Section 8.2.3, the cellular refractive index is an important parameter in digital holographic phase contrast imaging of living cells. The cellular refractive index influences the visibility of cells and subcellular components in the phase contrast images and represents the main limit of the accuracy for the determination of the thickness of transparent samples [32]. Furthermore, information about cellular refractive indices is useful for the utilization with optical tweezers and related optical manipulation systems as the individual cellular refractive index values influences the resulting optical forces [58, 59].

The refractive index determination of adherent cells as, for example, described in Refs [32, 35, 42, 43] is time consuming or requires complex experimental equipment. Thus, DHM methods for the determination of the integral refractive index of suspended cells have been developed [37, 38], which can be performed without further sample preparation. Another advantage of DHM is that due to the numerical multifocus feature, an increased data acquisition can be achieved by simultaneous recording of suspended cells that are located laterally separated in different focal planes (for illustration, see Figure 8.5).

Here, the determination of the integral refractive index of suspended cells is illustrated by a method in which a two-dimensional model is fitted to the DHM phase contrast data. The measured digital holographic phase distribution $\Delta\varphi_{cell}$ of a cell depends on the cell thickness $d_{cell}(x, y)$ and the difference between the refractive index n_{medium} of the surrounding medium and the integral refractive

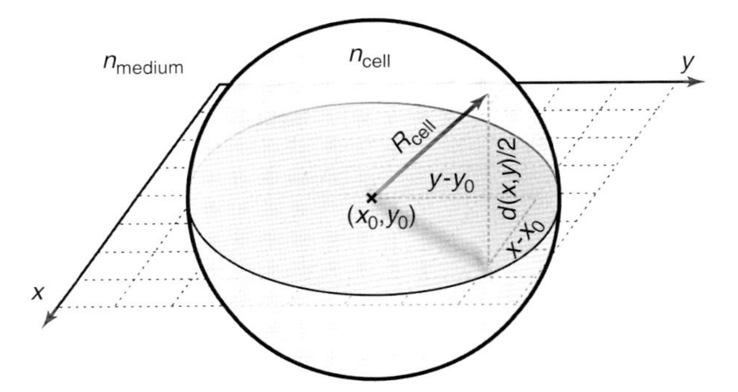

Figure 8.13 Sketch of the relation between radius R_{cell} and thickness $d_{cell}(x,y)$ of a spherical cell with an integral refractive index n_{cell} located at (x_0, y_0) in buffer medium with refractive index n_{medium} [37].

index n_{cell} of the cell:

$$\Delta\varphi_{cell}(x,y) = \frac{2\pi}{\lambda}(n_{cell} - n_{medium})\, d_{cell}(x,y) \tag{8.6}$$

where λ is the wave length of the applied laser light. For spherical cells in suspension located at $x = x_0, y = y_0$ with radius R_{cell} (Figure 8.13) the cell thickness $d_{cell}(x,y)$ is

$$d_{cell}(x,y) = \begin{cases} 2 \times \sqrt{R_{cell}^2 - (x-x_0)^2 - (y-y_0)^2} & \text{for} \quad (x-x_0)^2 + (y-y_0)^2 \leqslant R_{cell}^2 \\ 0 & \text{for} \quad (x-x_0)^2 + (y-y_0)^2 > R_{cell}^2 \end{cases} \tag{8.7}$$

Substitution of Equation 8.7 in Equation 8.6 yields:

$$\Delta\varphi_{cell}(x,y) = \begin{cases} \frac{4\pi}{\lambda} \times \sqrt{R_{cell}^2 - (x-x_0)^2 - (y-y_0)^2} \\ \quad \times (n_{cell} - n_{medium}) & \text{for} \quad (x-x_0)^2 + (y-y_0)^2 \leqslant R_{cell}^2 \\ 0 & \text{for} \quad (x-x_0)^2 + (y-y_0)^2 > R_{cell}^2 \end{cases} \tag{8.8}$$

with the unknown parameters n_{cell}, R_{cell}, x_0, and y_0. To obtain the parameters n_{cell}, R_{cell}, x_0, and y_0 Equation 8.8 is fitted iteratively with the Gauß–Newton Method [60] to the measured phase data of spherical cells in suspension. The required data of the image scale is obtained by calibration of the optical imaging system with a transparent USAF 1951 resolution test chart. The refractive index of the cell culture medium is determined with an Abbe refractometer.

Figure 8.14 illustrates the evaluation process of the phase data for the determination of n_{cell} by a representative result that has been obtained from a trypsinized cell in suspension with spherical shape. Figure 8.14a shows the phase contrast image

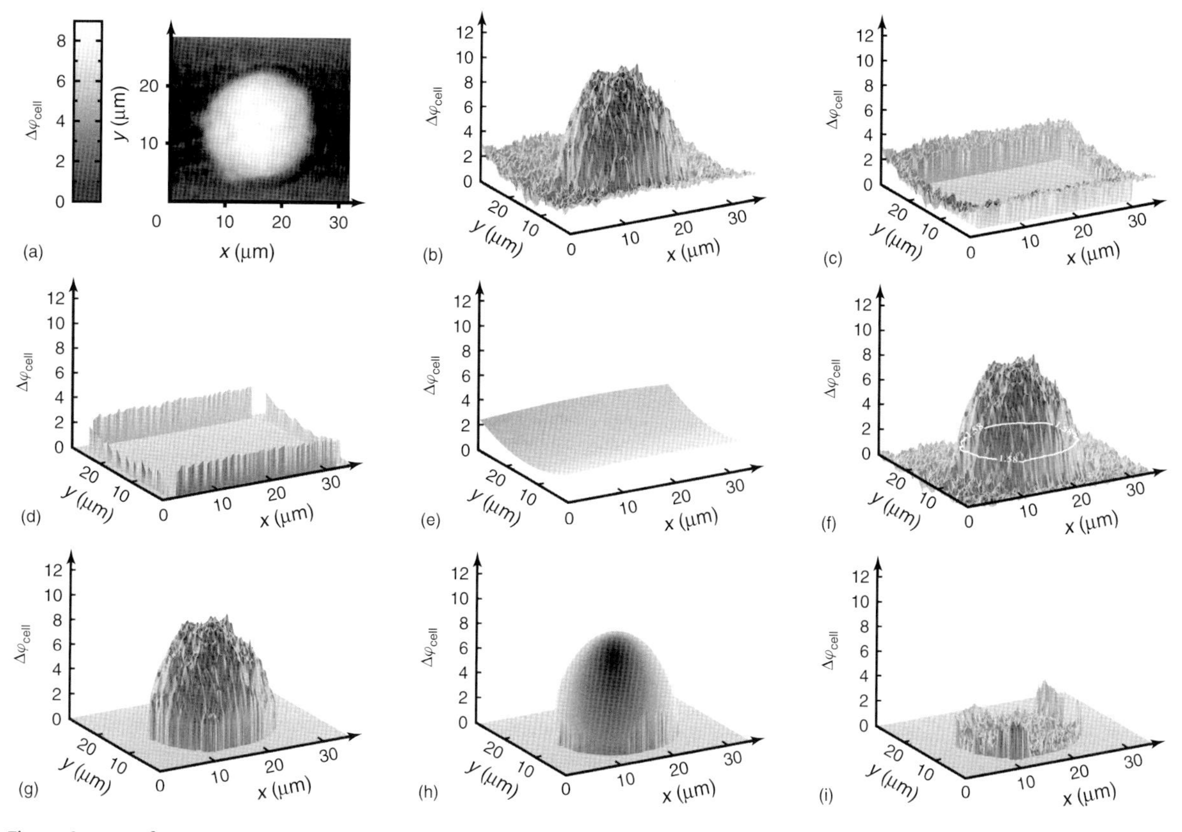

Figure 8.14 Refractive index determination of a spherical cell in suspension. (a) reconstructed quantitative digital holographic phase contrast image ($\Delta\varphi_{cell}$) of a trypsinized cell, (b) pseudo 3D plot of (a), (d) phase background from (c) averaged to a one pixel frame, (e) second degree function fitted background data in (d), (f) difference between (b) and (e), (g) data for the fitting of Equation 8.8, that is, selected from (f) by application of a threshold value, (h) fit of Equation 8.8 to the data in (g), and (i) absolute difference between (g) and (h) [37].

Table 8.1 Average integral refractive indices n_{cell} for different pancreas tumor cell types [37].

Cell type	n_{cell}	N
PaTu 8988T pLIXIN E-Cadherin	1.372 ± 0.003	20
PaTu 8988S	1.376 ± 0.005	15
PaTu 8988T	1.375 ± 0.003	28
Average values (all cells)	1.375 ± 0.004	63

of the cell, coded to 256 gray levels. Figure 8.14b depicts a rendered pseudo 3D plot of the data in Figure 8.14a. Figure 8.14c shows a selected frame of 30 pixels width around the cell that is averaged to a width of 1 pixel (Figure 8.14d). The width of 30 pixels is chosen only for illustration. For the data that is shown in Figure 8.15 and Table 8.1, a width of 3 pixels is applied. In a next step, a second degree function is fitted to the data (Figure 8.14e) and subtracted from the phase distribution as shown in Figure 8.14b. Figure 8.14f represents the resulting phase distribution. Afterward, the data of $\Delta\varphi_{cell}(x, y)$ for the fitting of Equation 8.8 (Figure 8.14g) is selected from the background corrected data by application of a threshold value (white contour line in Figure 8.14f) that considers the phase noise in the area around the cell. Figure 8.14h depicts the resulting fit. The absolute difference between the experimentally obtained data and fit is shown in Figure 8.14i.

In order to demonstrate the determination of the integral cellular refractive index with DHM, three different types of living pancreas tumor cells (PaTu 8988T, PaTu 8988S, and PaTu 8988T pLXIN E-Cadherin) in suspension were investigated. For the experiments the cells were trypsinized and holograms of the cells in culture medium (DMEM containing 5% FCS and 5% horse serum, $n_{medium} = 1.337 \pm 0.001$) were recorded by application of a 40× microscope lens (NA = 0.6). For data evaluation, only cells with spherical shape were selected, which were mainly observed in the recorded holographic phase contrast images for all investigated cell type. After reconstruction, for each cell line, the parameters n_{cell} and R_{cell} were determined as illustrated in Figure 8.14.

Figure 8.15a depicts the resulting values for n_{cell} in dependence of R_{cell}. The error bars represent the uncertainties $\Delta n_{cell} = 0.003$ and $\Delta R_{cell} = 0.1$ for a measurement on a single cell. The values for Δn_{cell} and ΔR_{cell} were obtained by the calculation of the standard deviation from the results of the fitting of Equations 8.8 to 17 independently obtained in the phase contrast images of a single tumor cell [37]. In Figure 8.15b, the histogram of the refractive index for all PaTu 8988 cells is depicted. The maximum number of cells is located near the mean value of the refractive index. The plot in Figure 8.15a shows that n_{cell} decreases with an increasing R_{cell}. The decrease of the refractive index with increasing cell radius, and thus with the cell volume, may be explained by the cellular water content [16].

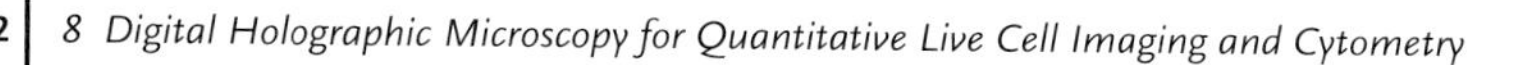

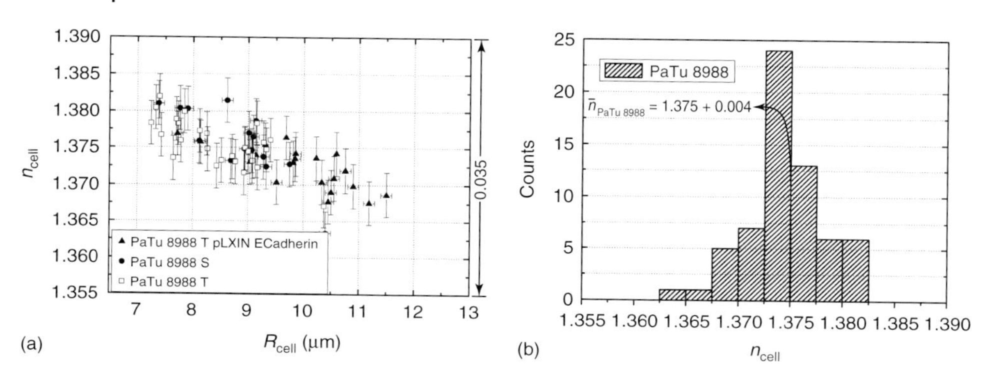

Figure 8.15 (a) Integral cellular refractive index n_{cell} in dependence of cell radius R_{cell} for PaTu 8988 cells. (b) Histogram of the refractive index of PaTu 8988 cells ($N = 63$) [37].

Table 8.1 shows the integral cellular refractive index n_{cell} for all pancreas tumor cell types. Within the range of uncertainty, no significant differences between the refractive index data of the three different types of pancreas tumor cells are observed.

8.3.6
Identification of Subcellular Structures

In Ref. [10], it has already been demonstrated for human erythrocytes that DHM gives complementary information to confocal laser scanning fluorescence microscopy. In this section, it is illustrated that the DHM phase contrast also offers label-free identification of subcellular components. In order to interpret DHM phase contrast images for the analysis of subcellular structures, comparative image acquisitions with fluorescence and bright field microscopy were performed. For the experiment, adherently grown PaTu 8988T cells are fixed with formalin on a glass carrier. Afterward, the cells are stained with DAPI (4′,6-diamidino-2-phenylindol) to identify cell components containing DNA (deoxyribonucleic acid). Triton is used to enhance cell membrane permeability for DAPI staining. Finally, the sample is embedded in glycerol and covered with a cover slip. Bright field images, digital off-axis holograms, and fluorescence images of the focused sample are recorded using an upright fluorescence microscope modified for DHM that is described in detail in Ref. [11] (63× microscope lens, NA = 0.75; frequency doubled Nd : YAG laser: $\lambda = 532$ nm).

Figure 8.16 shows a representative DHM phase contrast image (Figure 8.16b) of the cells in comparison to a bright field image (Figure 8.16a) and a fluorescence image (Figure 8.16c). Cell nucleus components (nucleolus and chromatin) are detected at identical positions in bright field images, fluorescence images, and digital holographic phase contrast images (see enlarged areas in Figure 8.16d and e). Figure 8.16f shows a cross section through the phase contrast image of a

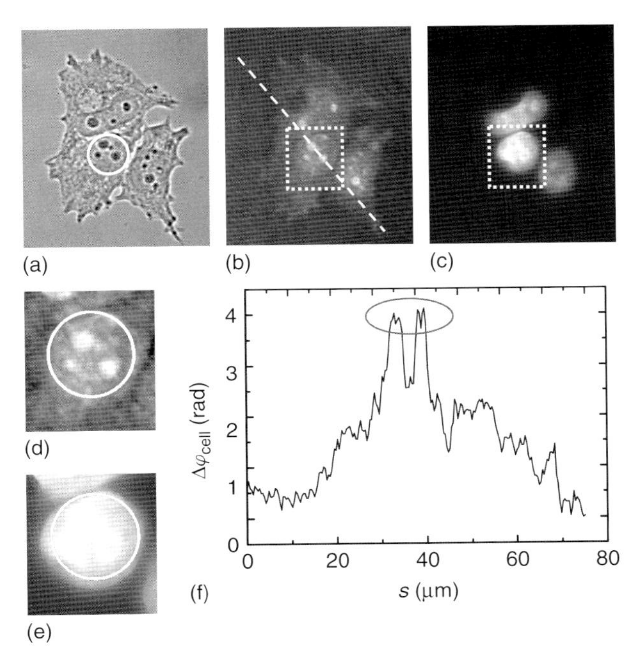

Figure 8.16 Comparison of bright field imaging, fluorescence microscopy and DHM phase contrast. Pancreas tumor cells (PaTu 8988 T) were fixed, stained with the nuclear marker DAPI after Triton treatment and embedded in glycerol. (a) Bright field image, (b) fluorescence image, (c) DHM phase contrast image, (d and e) enlarged DHM and fluorescence images marked with a box in (d) and (c), and (f) cross section through the phase contrast image of the nucleus and two nucleoli along the dashed line in (b) [61].

single cell along the white dotted line in Figure 8.16b that includes the nucleus and two nucleoli (see encircled areas in Figure 8.16d,e). Bright areas in the fluorescence image indicate a high DNA concentration and correspond to bright areas in the digital holographic phase contrast image that indicate cell nucleus structures with high (optical) density. Thus, dark areas in the centers of the nucleoli may be interpreted as a local decrease in the cellular refractive index due to a decreased DNA concentration in comparison to the outer boundaries. This corresponds to results obtained by tomography phase microscopy in Ref. [42]. A statistical evaluation of fluorescence and DHM phase contrast images of DAPI stained PaTu 8988T cells in Ref. [61] shows that for 59% of the components chromatin and nucleoli can be identified by DHM. Furthermore, due to the numerical autofocus feature, for thin adherent cells like PaTu 8988T, a better visibility of cell nucleus components is obtained for the DHM phase contrast than for DAPI staining.

In conclusion, the described results open new applications for label-free identification and analysis of subcellular structures, in particular, in combination with other label-free optical microscopy techniques such as auto fluorescence imaging or Raman spectroscopy. However, further supporting investigations on these topics have to be carried out.

8.4
Conclusion

The presented results show that DHM allows high-resolution multifocus quantitative phase contrast imaging of adherent and suspended cells. Although the presented DHM methods do not achieve the resolution as obtained from atomic force microscopy (AFM), scanning optical near field microscopy (SNOM), or SEM they overcome some particular limitations of these techniques, for example, the scanning process or the requirement of fixed cells in a vacuum. DHM enables a vibration insensitive (hologram capture time in ms range, hologram acquisition rate mainly limited by the digital recording device) and contact less full field measurement of living cells.

The results further demonstrate that the presented DHM concepts can be utilized for investigations of dynamic cellular processes and for the quantitative analysis of cell reactions on drug treatment. The obtained information opens up new ways for label-free dynamic monitoring of morphological changes, and may help access new parameters, for example, for apoptosis recognition or for cell swelling control in microinjection [10, 26]. Further, future prospects include the use of DHM for 3D cell tracking in fluidics and in a three-dimensional environment [24]. In combination with fluorescence microscopy, new possibilities for multifunctional microscopy systems for imaging and analysis of fixed and living cells are opened up. In conclusion, the presented systems have the potential to form a versatile microscopy tool for Life Sciences and Biophotonics.

Acknowledgment

Financial support by the German Ministry for Education and Research (BMBF grants Mikroso, Live Cell Screening, NanoCare, and Cell@Nano) is gratefully acknowledged.

References

1. Kreis, T. (1996) in *Holographic Interferometry: Principles and Methods*, vol. 1 (ed. W. Osten), Akademie-Verlag, Berlin.
2. Shchepinov, V.P. and Pisarev, V.S. (1996) *Strain and Stress Analysis by Holographic and Speckle Interferometry*, Wiley-VCH Verlag GmbH, New York.
3. Beeck, M.-A. and Hentschel, W. (2000) Laser metrology – a diagnostic tool in automotive development processes. *Opt. Lasers Eng.*, **34**, 101–120.
4. Charrière, F. *et al.* (2006) Characterization of microlenses by digital holographic microscopy. *Appl. Opt.*, **45**, 829–835.
5. Cuche, E., Marquet, P., and Depeursinge, C. (1999) Simultaneous amplitude-contrast and quantitative

phase-contrast microscopy by numerical reconstruction of fresnel off-axis holograms. *Appl. Opt.*, **38** (24), 6694–7001.
6. Carl, D. *et al.* (2004) Parameter-optimized digital holographic microscope for high-resolution living-cell analysis. *Appl. Opt.*, **43**, 6536–6544.
7. Marquet, P. *et al.* (2005) Digital holographic microscopy: a noninvasive contrast imaging technique allowing quantitative visualization of living cells with subwavelength axial accuracy. *Opt. Lett.*, **30**, 468–470.
8. Mann, C.J. *et al.* (2005) High-resolution quantitative phase-contrast microscopy by digital holography. *Opt. Express*, **13**, 8693–8698.
9. Kemper, B. and von Bally, G. (2008) Digital holographic microscopy for live cell applications and technical inspection. *Appl. Opt.*, **47**, A52–A61.
10. von Bally, G. *et al.* (2006) in *Biophotonics: Vision for Better Healthcare* (eds J. Popp and M. Strehle), Wiley-VCH Verlag GmbH, pp. 301–360.
11. Kemper, B. *et al.* (2006) Modular digital holographic microscopy system for marker free quantitative phase contrast imaging of living cells, *Proc. SPIE*, **6191**, 61910T.
12. Alexopoulos, L.G., Erickson, G.R., and Guilak, F. (2001) A method for quantifying cell size from differential interference contrast images: validation and application to osmotically stressed chondrocytes. *J. Microsc.*, **205**, 125–135.
13. Barty, A. *et al.* (1998) Quantitative optical phase microscopy. *Opt. Lett.*, **23** (11), 817–819.
14. Popescu, G. *et al.* (2004) Fourier phase microscopy for investigation of biological structures and dynamics. *Opt. Lett.*, **29**, 2503–2505.
15. Ikeda, T. *et al.* (2005) Hilbert phase microscopy for investigating fast dynamics in transparent systems. *Opt. Lett.*, **30** (10), 1165–1167.
16. Farinas, J. and Verkman, A.S. (1996) Cell volume and plasma membrane osmotic water permeability in epithelial cell layers measured by interferometry. *Biophys. J.*, **71**, 3511–3522.
17. Tychinskii, P. (2007) Dynamic phase microscopy: is a dialogue with the cell possible? *Phys.-Uspekhi*, **50**, 513–528.
18. Aguirre, A.D. *et al.* (2003) High-resolution optical coherence microscopy for high-speed, in vivo cellular imaging,. *Opt. Lett.*, **28**, 2064–2066.
19. Swanson, E.A. *et al.* (18) In vivo retinal imaging by optical coherence tomography. *Opt. Lett.*, **1993**, 1864–1866.
20. Zhao, Y. *et al.* (2002) Real-time phase-resolved functional optical coherence tomography by use of optical Hilbert transformation. *Opt. Lett.*, **27**, 98–100.
21. Rylander, C.G. *et al.* (2004) Quantitative phase-contrast imaging of cells with phase-sensitive optical coherence microscopy. *Opt. Lett.*, **29** (13), 1509–1511.
22. Ellerbee, A.K., Creazzo, T.L., and Izatt, J.A. (2007) Investigating nanoscale cellular dynamics with cross-sectional spectral domain phase microscopy. *Opt. Express*, **15**, 8115–8124.
23. Joo, C., Kim, K.H., and de Boer, J.F. (2007) Spectral-domain optical coherence phase and multiphoton microscopy. *Opt. Lett.*, **32**, 623–625.
24. Langehanenberg, P. *et al.* (2009) Automated three-dimensional tracking of living cells by digital holographic microscopy. *J. Biomed. Opt.*, **14**, 014018.
25. Schnars, U. (1994) Direct phase determination in hologram interferometry with use of digitally recorded holograms. *J. Opt. Soc. Am. A.*, **11** (7), 2011–2015.
26. Rommel, C.E., Dierker, C., Schmidt, L., Przibilla, S., von Bally, G., Kemper, B., and Schnekenburger, J. (2010). Contrast enhanced digital holographic imaging of cellular structures by manipulating intracellular refractive index, *J. Biomed. Opt.*, **15**, 041509.
27. Schnars, U. and Jüptner, W.P.O. (2002) Digital recording and numerical reconstruction of holograms. *Meas. Sci. Technol.*, **13**, 85–101.
28. Schnars, U. and Jüptner, W. (2004) *Digital Holography*, Springer-Publishing.
29. Kim, M.K., Yu, L., and Mann, C.J. (2006) Interference techniques in digital holography. *J. Opt. A.*, **8**, 518–523.
30. Poon, T.-C. (ed.) (2006) *Digital Holography and Three-Dimensional*

Display: Principles and Applications, Springer-Publishing.

31. Liebling, M., Blu, T., and Unser, M. (2004) Complex-wave retrieval from a single off-axis hologram. *J. Opt. Soc. Am. A.*, **21** (3), 367–377.
32. Kemper, B. *et al.* (2006) Investigation of living pancreas tumor cells by digital holographic microscopy. *J. Biomed. Opt.*, **11** (3), 034005–034008.
33. Demetrakopoulos, T. and Mittra, R. (1974) Digital and optical reconstruction of images from suboptical refraction patterns. *Appl. Opt.*, **13**, 665–670.
34. Kreis, T.M., Adams, M., and Jüptner, W.P.O. (1997) Methods of digital holography: a comparision. *Proc. SPIE*, **3098**, 224–233.
35. Rappaz, B. *et al.* (2005) Measurement of the integral refractive index and dynamic cell morphotometry of living cells with digital holographic microscopy. *Opt. Express*, **13**, 9361–9373.
36. Lue, N. *et al.* (2006) Live cell refractometry using microfluidic devices. *Opt. Lett.*, **31**, 2759–2761.
37. Kosmeier, S. *et al.* (2008) Determination of the integral refractive index of cells in suspension by digital holographic phase contrast microscopy. *Proc. SPIE*, **6991**, 699110.
38. Kemper, B. *et al.* (2007) Integral refractive index determination of living suspension cells by multifocus digital holographic phase contrast microscopy. *J. Bio. Med. Opt.*, **12**, 054009.
39. Rappaz, B. *et al.* (2008) Simultaneous cell morphometry and refractive index measurement with dual-wavelength digital holographic microscopy and dye-enhanced dispersion of perfusion medium. *Opt. Lett.*, **33**, 744–746.
40. Charrière, F. *et al.* (2006) Cell refractive index tomography by digital holographic microscopy. *Opt. Lett.*, **31**, 2759–2761.
41. Charrière, F. *et al.* (2006) Living specimen tomography by digital holographic microscopy: morphometry of testate amoeba. *Opt. Express*, **14**, 7005–7013.
42. Choi, W. *et al.* (2007) Tomographic phase microscopy. *Nat. Methods*, **4**, 717–719.
43. Debailleul, M. *et al.* (2009) High-resolution threedimensional tomographic diractive microscopy of transparent inorganic and biological samples. *Opt. Lett.*, **34**, 79–81.
44. Klokkers, J. *et al.* (2009) Atrial natriuretic peptide and nitric oxide signaling antagonizes 2 vasopressin-mediated water permeability in inner medullary 3 collecting duct cells. *Am. J. Physiol. – Renal. Physiol.*, **297**, F693–703.
45. Elsässer, H.P. *et al.* (1992) Establishment and characterization of two cell lines with different grade of differentiation derived from one primary human pancreatic adenocarcinoma. *Virchows Arch. B.*, **61**, 295–306.
46. Kemper, B., Langehanenberg, P., and von Bally, G. (2007) Methods and applications for marker-free quantitative digital holographic phase contrast imaging in life cell analysis. *Proc. SPIE*, **6796**, 6796E.
47. Langehanenberg, P., Kemper, B., and von Bally, G. (2007) Autofocus algorithms for digital-holographic microscopy. *Proc. SPIE*, **6633**, 66330E.
48. Langehanenberg, P. *et al.* (2008) Autofocusing in digital holographic phase contrast microscopy on pure phase objects for live cell imaging. *Appl. Opt.*, **47**, D176–D182.
49. Ghaneh, P., Costello, E., and Neoptolemos, J.P. (2007) Biology and management of pancreatic cancer. *Gut*, **56** (8), 1134–1152.
50. Li, D. *et al.* (2004) Pancreatic cancer. *Lancet*, **363** (9414), 1049–1057.
51. Jones, J.O. and Diamond, M.I. (2007) Design and implementation of cell-based assays to model human disease. *ACS Chem. Biol.*, **2** (11), 718–724.
52. Vogelmann, R. *et al.* (1999) Integrin alpha6beta1 role in metastatic behavior of human pancreatic carcinoma cells. *Int. J. Cancer*, **80** (5), 791–795.
53. Erhuma, M. *et al.* (2007) Expression of neutral endopeptidase (NEP/CD10) on pancreatic tumor cell lines, pancreatitis and pancreatic tumor tissues. *Int. J. Cancer*, **120** (11), 2393–2400.
54. Menke, A. *et al.* (2001) Down-regulation of E-cadherin gene expression by collagen type I and type III in pancreatic

cancer cell lines. *Cancer Res.*, **61** (8), 3508–3517.
55. Schnekenburger, J., Keiner, I., and Lerch, M.M. (2003) Pathways of CCK induced reactive oxygen species synthesis in exocrine pancreas cells. *Gastroenterology*, **124** (4), A438–A438.
56. Spector, I. *et al.* (1983) Latrunculins: novel marine toxin that disrupt microfilament organization in cultured cells. *Science*, **219**, 493–495.
57. Kemper, B. *et al.* (2007) Techniques and applications of digital holographic microscopy for life cell imaging. *Proc. SPIE*, **6633**, 66330D.
58. Ashkin, A. (1997) Optical trapping and manipulation of neutral particles using lasers. *Proc. Natl. Acad. Sci.*, **94**, 4853–4860.
59. Guck, J. *et al.* (2000) Optical deformability of soft biological dielectrics. *Phys. Rev. Lett.*, **84** (23), 5451–5454.
60. Björk, Å. (1996) *Numerical Methods for Least Squares Problems*, Society of Industrial & Applied Mathematics, Philadelphia.
61. Kemper, B., Schmidt, L., Przibilla, S., Rommel, C., Ketelhut, S., Vollmer, A., Schnekenburger, J., and von Bally, G. (2010) Influence of sample preparation and identification of subcelluar structures in quantitative holographic phase contrast microscopy, SPIE 7715, 771504.

9
Comparison of Immunophenotyping and Rare Cell Detection by Slide-Based Imaging Cytometry and Flow Cytometry

József Bocsi, Anja Mittag, and Attila Tárnok

9.1
Introduction

High-content analysis of biological specimens requires multiparametric (i.e., multicolor and multifunctional) labeling. In the postgenomic era, multicolor analysis of biological specimens has become increasingly important in various fields of biology in particular, because of the emerging new fields of high-content and high-throughput single-cell analysis for systems biology [1–4] and cytomics [5, 6]. Areas of research and diagnosis with the demand to virtually measure "everything" in the cell include immunophenotyping, rare cell detection and characterization in the case of stem cells and residual tumor cells, tissue analysis, and drug discovery [7, 8]. It is also of particular importance if only small sample volumes are available or if the samples are very precious. In this case, multicolor analysis allows, in the ideal case, to determine all relevant cellular constituents in a single analytical run. Fluorescent assays have the highest versatility for multiparametric qualitative and quantitative characterization of clinical and biological samples, resulting in polychromatic [9] or hyperchromatic cytometry [10, 11].

Cytometry means quantitative single-cell analysis, that is, the quantitation of cellular compounds of the whole cell. The most often used fluorescence-based cytometric techniques are flow cytometry (FCM) and slide-based cytometry, (SBC) also termed *imaging cytometry*. The flow cytometer has a fluidics system transporting the cells in a suspension through the instrument. FCM analyzes cells in high speed with the capability to analyze thousands of cells per second, positioning them at the focus of the illumination point and the focus of the optical systems for an accurate measurement. Flow cytometers use light beams (mostly lasers) to excite fluorescent dyes. Optical filters are used to separate the different spectral ranges of the fluorescent light of the dyes detected by the most sensitive light detection systems – photomultipliers (PMTs) or photodiodes.

On the contrary, in imaging cytometry the sample and the cells are placed on a glass surface (SBC) and they are observed through a light microscope coupled with different essential light sources, optical filters, and photoelectronic detection devices.

Advanced Optical Flow Cytometry: Methods and Disease Diagnoses, First Edition. Edited by Valery V. Tuchin.

9.1.1 Cytometry of Equal Quality?

Do I obtain the same results no matter what cytometer type I use? Are FCM and SBC comparable? These questions have to be asked and will be answered in this chapter. The main applications, pros and cons of both techniques, and a comparison of the systems based on sample analysis are described here.

The following four sections will help to choose the right analysis technique and cytometer for the given biological problem. The following assays will be compared with FCM:

1) Determination of CD4/CD8 T-cell ratio by scanning fluorescence microscope system (SFM)
2) Multiparametric leukocyte phenotyping by laser scanning cytometer (LSC)
3) Rare and frequent tumor cell detection by SFM and LSC
4) Apoptosis analysis on the basis of DNA content measurements by iCyte (LSC instrument).

9.1.2 Fluorescence Analyses

9.1.2.1 Excitation and Emission of Fluorescent Dyes

A plethora of fluorescent dyes with different excitation and emission characteristics and comparable chemical or biological specificity are commercially available. This includes specific DNA-binding dyes (Hoechst 33342, 7AAD, etc.), dyes for analysis of protein content, membrane potential, pH and Ca^{2+} sensitivity, or membrane integrity, as well as additional dyes for functional and structural cell components. Moreover, monoclonal antibodies labeled with different highly fluorescent dyes through the whole spectrum, fluorescein isothiocyanate (FITC), phycoerythrin (PE), Texas Red, cychrome (Cy), allophycocyanine (APC), Cy5.5, or Cy7, and many similar alternatives [12] are commercially available or custom made. Fluorescence detection technologies can be equipped with a broad selection of excitation light sources. Laser light sources have radiation with specific wavelength lines and can excite dyes with matching absorbance characteristic used in multicolor experiments. Generally, available laser lines or other light sources determine the assortment of dyes that can be used for staining of cells and measured by the given technique. Therefore, the number of fluorochromes that are applicable for staining is limited and depends on the given technique. Tandem dyes, relying on fluorescence resonance energy transfer (FRET), increasing the number of measureable dyes were developed in the 1990s. In the case of FRET dyes, one part of the fluorochrome can be excited efficiently by a laser line (donor dye), and fluorescence energy transfer gives the absorbed energy to the other part of the tandem dye (acceptor dye) to emit fluorescence light at a higher wavelength showing a bigger Stokes shift (the difference between the absorption and emission maximum).

Expensive lasers are commonly used in most flow cytometer, whereas mercury or other arc lamps are inexpensive and have a broad light spectrum ranging from the UV through the visible light up to the infrared. The newly developed light-emitting diode (LED) light sources provide discrete stabile excitation light through the whole visible spectrum and also UV and IR, have long life-span and are also relatively cheap. Most fluorescent microscopes use arc lamps because of these advantages. In fluorescence microscopy, fluorescence is excited by high intensity UV and/or visible light. Absorption of the light leads to an excited state of the dye (usually a singlet state) and this leads to fluorescence. However, the excited state of the dye may undergo chemical reactions, leading to its destruction, as evidenced by the fading or bleaching of the fluorescence and subsequent decrease in signal intensity during measurement [13]. This becomes a problem, particularly with SBC, as excitation times are substantially longer (ms) than in FCM (μs).

9.1.2.2 Quantum Dots

Fluorescent semiconductor nanocrystals or quantum dots (QDs) are one example of new developments in fluorescence dyes with high photostability using nanotechnology. QDs were developed in the early 1990s, and in 1998 their use for biological labeling was first reported [14].

The ability to tune broad wavelength ranges makes these materials extremely useful for biolabeling. Colloidal QDs are robust and very stable light emitters and they can be broadly tuned simply through size variation. The inorganic QDs have unique optical properties such as size (5–20 nm diameter, similar size as a typical protein), intensive brightness, and high photostability [15–17], and an extremely high Stokes shift of more than 100 nm. Furthermore, they have a wide absorption range with optimal excitation in the UV range and narrow and symmetric emission spectra. Mean emission of QDs is suitable for most fluorescence systems and filter combinations.

QDs have been covalently linked to biomolecules such as peptides, antibodies, nucleic acids, or small-molecule ligands for applications such as fluorescent probes. QDs are therefore suitable for multiparametric characterization of biological samples and applicable for spectral combination with other "classical" fluorochromes. However, one drawback of QDs is the fact that they "blink," that is, emit light intermittently when excited with high intensity light [18, 19]. This feature can be exploited for a single molecule tracking but may be disadvantageous if quantitative fluorescence signals need to be acquired.

In microscopy and also in cytometric analyses by microscopic systems, selection of dyes is even more important than in FCM. Light exposure in microscopy is usually longer (from 10 up to 10^6-times depending on the instrument) than in FCM, resulting in bleaching of the dye. Light absorption induces an excited state of the dye (usually singlet state), leading to fluorescence. However, the excited state of the dye may undergo chemical reactions such as photo-oxidation, leading to its destruction as evidenced by fading or bleaching of the fluorescence and subsequent decrease in signal intensity during measurement. The commonly used organic fluorochromes are known to bleach relatively fast during illumination.

To overcome these problems, antifading reagents were developed and can be added to mounting media for specimens examined by fluorescent microscopy [20, 21]. Nevertheless, bleaching of fluorochromes is still a problem in fluorescence analysis on microscopes.

Several *in vivo* studies were published with QD staining in the last years: development of tumor vessels [22], tumor distribution in the whole body [23], development of frog oocyte and embryos [24], or lymph node mapping [25]. QDs have been shown to remain fluorescent and retained in live cells and organisms for several weeks with no detectable toxicity [26, 27]. Despite the presence of heavy metals (usually cadmium and selenium) in the core of the QDs, cellular toxicity was not observed in case of systemic injection in mice [23] or pigs [25].

It is always advisable to titrate new antibodies. This is not only important for the determination of optimal concentration (resulting in maximal discrimination of positive and negative cells and, of course, cost-efficient assays) but also for the determination of unspecific staining.

Titration of QD605 was done with CD3-Biotin- and Streptavidin-labeled QD605 in concentrations of 0.2–2 nM. Significant differences in staining were tested by Kolmogorov–Smirnov test (Figure 9.1). Unspecific staining of lymphocytes was not observed, whereas an increased unspecific staining was visible for neutrophils

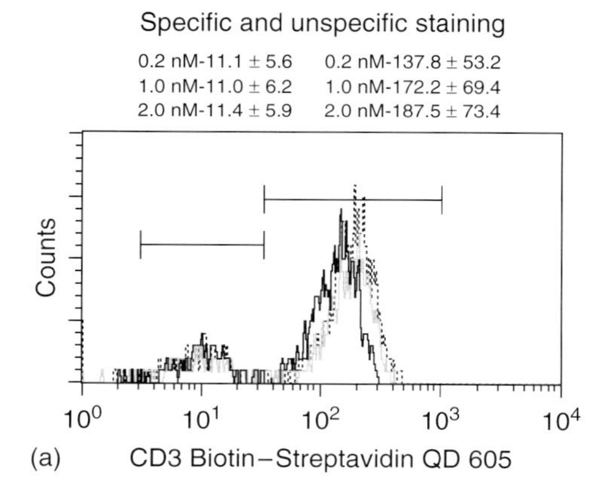

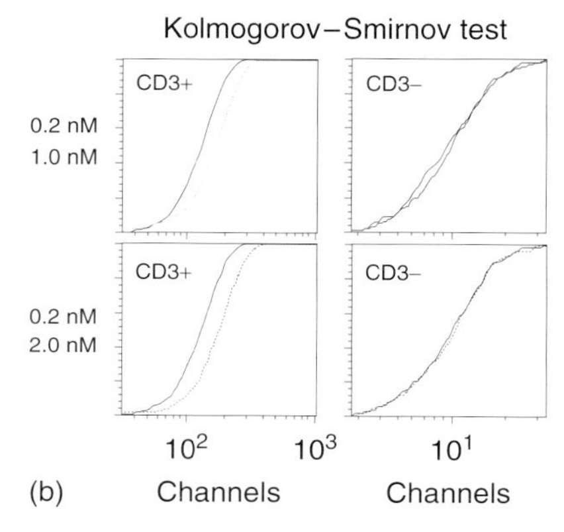

Figure 9.1 Staining characteristics of QDs. (a) Histogram overlay of the measurements with increasing concentrations (0.2, 1.0, and 2.0 nM) of QD605 is displayed. Staining was done by indirect labeling with CD3 biotin antibody and Streptavidin QD605. As can be seen, there is no difference in the fluorescence intensity of negative cells (mean = 11.0–11.4), that is, the background, but intensity of positive cells increases with concentration (mean = from 137.8 to 187.5). (b) Kolmogorov–Smirnov test (CellQuest, Becton-Dickinson, USA) shows the statistical relevance for the comparison of the histograms in (a) obtained after different Qdot concentrations in a fluorescence channel to channel comparison, 0.2 versus 1.0 nM and 0.2 versus 2.0 nM. At CD3+ region, the two curves run far from each other (the difference between the staining intensity is significant) while at CD3− the two curves run together (the difference is not significant, that is, there is no detectable background staining).

at higher dye concentrations (data not shown). Intensity of labeled cells increased with concentration as expected (Figure 9.1).

9.1.2.3 Bleaching Characteristics of Dyes

Bleaching of fluorescence dyes does not play an important role in FCM. In SBC, however, a fluorescence protective medium is needed for analysis. Mounting medium containing antifading reagents is applicable and commonly used for organic dyes.

SBC analysis shows that organic fluorochromes are sensitive for excitation and are bleached at dye-specific rates [13]. In particular, PE was very photosensitive. The decrease of fluorescence intensity of the organic dyes was substantially lower when the specimen was covered with Dako fluorescent mounting medium (DAKO, Denmark) or ProLong mounting medium (Invitrogen, USA) but immense with glycerin-PBS (phosphate buffered saline) (data not shown). QDs, however, showed no substantial bleaching in PBS. This means the use of fluorescence protective media is not necessary but is unavoidable in case of combination with conventional fluorochromes. It turned out that fluorescence intensity of QD decreased rapidly after adding the mounting medium containing free radical scavengers or antioxidants while fluorescence of organic dyes was not affected. Antioxidants, radical scavengers, or other components of the mounting medium probably initiate chemical reactions resulting in fluorescence loss. This fact is an obvious disadvantage in multicolor SBC analyses and makes the combination with other "normal" organic dyes very problematic.

9.1.2.4 Fluorescent Light Detection

The fluorescence emissions generated by the excited fluorochchromes can be analyzed by fluorescent filters characterized by their high light transparency in one particular part of the light spectrum and low light transparency (below 1%) in other spectral regions. Therefore, the fluorescent light spectrum analyzed can be multiplexed (or splitted up) using combinations of complementary band-pass filters.

Fluorescence light can be detected by photodiodes, PMTs, or digital cameras. High quality PMTs are built in cytofluorimeters since the 1970s. From the early 1990s on, as the digital imaging technology increased and higher resolution became available, in combination with parallel developments in image storage and digitization technology, cameras are being used for the analysis of fluorescent samples.

Different filter combinations make it possible to apply multicolor staining for cell analysis. Currently, a multitude of cell-staining techniques are based on such instrument designs. In microscopy, usually the visible 400–700 nm wavelength range is used for emission detection. The development of red light sensitive PMTs, avalanche photodiode detectors (APDs) [28], and scientific digital cameras (up to 1000 nm wavelength, [29]) is opening the IR region [30] for fluorescence technologies.

9.1.2.5 Spillover Characteristics

For implementation of fluorescent dyes in multiparametric panels, spillover (spectral overlap between fluorochrochrome emission spectra) plays an important role. Especially in multicolor experiments, low spillover is desirable for correct interpretation of results. The following example Figure 9.2 demonstrates the spillover characteristic of QD655 in comparison to PE-Cy5 using conventional band-pass (BP) filters.

9.1.3 Two Ways of Cytometric Analysis

9.1.3.1 Flow Cytometry

FCM is routinely used for the analysis of fluorescently labeled cells. FCM proved to be a reliable and powerful technology. However, FCM needs relatively high cell numbers and is mostly incapable of high-resolution morphological analysis.

Therefore, the focus of FCM analyses lies on cell types with high frequency (hematological samples, cell culture, and mechanically dissociated solid tissue cells), cell types that require complex phenotyping for unequivocal authentication, or assays where morphological data are irrelevant or can be obtained from other resources (e.g., pathology).

One key application of FCM is immunophenotyping. This is due to the high-throughput, the capability for multiparametric analysis, the possibility to detect rare events and weak signals, and the standardized preparation and analysis protocols that are accepted worldwide [32]. Particularly, stem cells, dendritic cells, or regulatory T cells with a frequency <0.0001 of all leukocytes are reliably analyzable by FCM. However, there are specific problems that can only be insufficiently solved or not at all by FCM. A major problem in the FCM analysis of clinical and research samples is the fact that the analysis of the morphology of cells is restricted to forward scatter (FSC) and orthogonal side scatter (SSC), in general. In the vast majority of analysis by FCM, morphology can be studied only by cell sorting, which is time consuming but still does not allow correlation with the fluorescence data on a single-cell basis. Morphology, however, is a very important feature, especially under pathological conditions, and therefore morphological information should be available for thorough analysis. This gave rise to the demand for technologies that combine both multiparametric analysis and cell morphology documentation [33, 34].

One step in this direction was recently taken through the implementation of image acquisition into flow cytometers in the image stream instrument [35]. This instrument allows allocation of fluorescence signals and optical verification based on the acquired images of single cells. However, this instrument has another drawback of FCM, namely, that cells cannot be analyzed for a second time on a cell-to-cell basis. Such assays are important for the monitoring of cell physiology or sample restaining for detailed cell characterization. Finally, the use of FCM is very limited, if not impossible, in all cases where only minimal cell numbers can

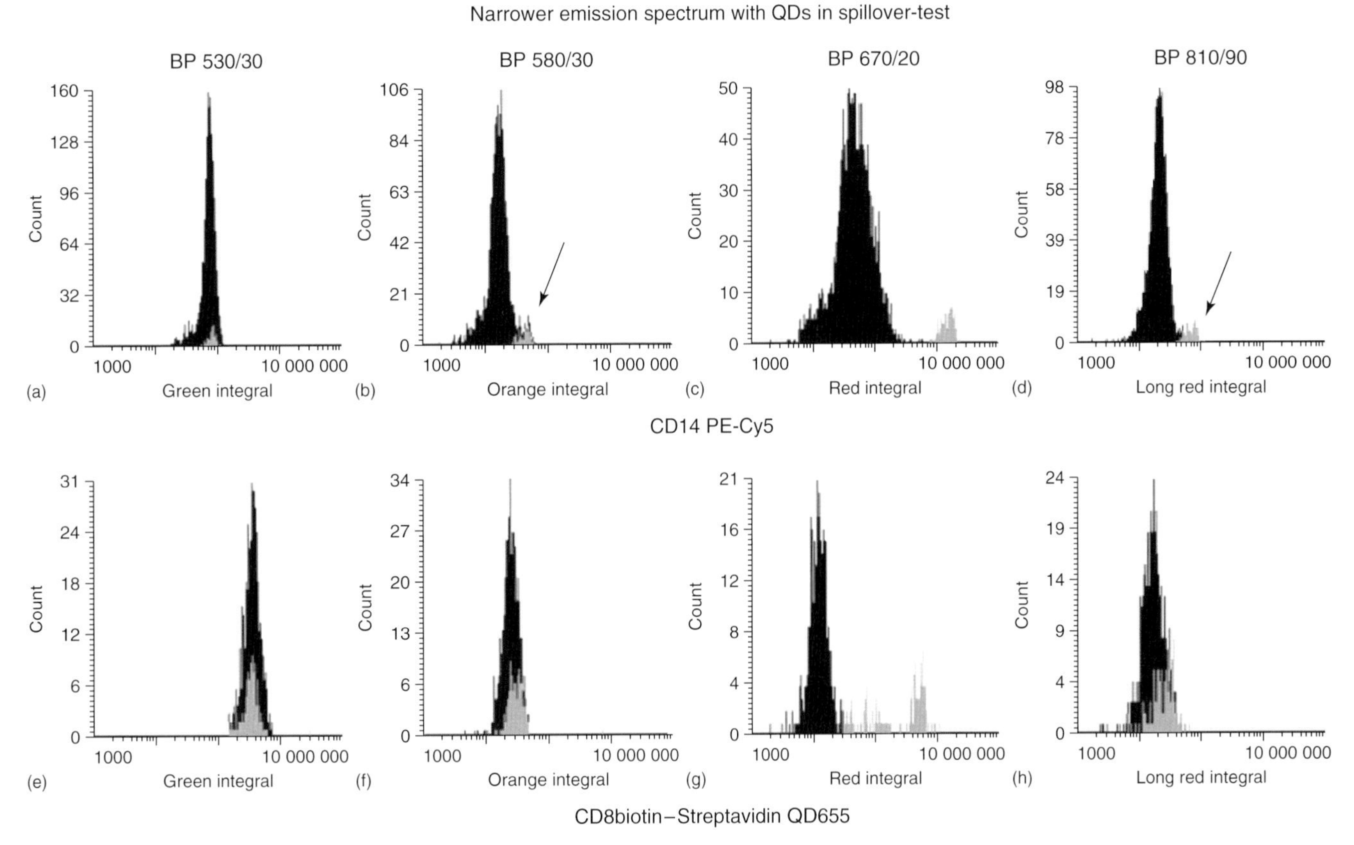

Figure 9.2 Spillover of Qdots. Leukocytes were stained with CD14-PECy5 (a–d) or CD8-QD655 (e–h). Fluorescence intensities for both dyes were comparable. Data were acquired using different band-pass filters (BP). Highest intensity can be expected by spectral characteristics of the dyes in channel 3 (BP 670/20 nm); hence, cells positive for labeling were determined and indicated in gray shading. Fluorescence intensity of these cells was checked in the other fluorescence detection channels. As can be seen, false positive signals are present for PE-Cy5 in channels 2 and 4 (arrow), whereas no substantial spillover could be detected for QD655. Reprint from Ref. [31] with permission from the publisher.

be obtained for analysis, for example, in critically ill neonates or in fine needle aspirate biopsies.

9.1.3.2 Slide-Based Cytometry

SBC is an alternative tool for quantitative cellular analysis, whereas traditional fluorescence microscopy is the instrumentation for qualitative analysis or for documentation ("observe and judge"). Thus, SBC fills the gap between high-throughput multiparametric cytometry and morphological analysis and documentation [36]. SBC applications are highly versatile, ranging from the clinics to drug discovery in the industry. Automated single-parameter scanning was first introduced for histochemistry studies [37, 38] and was soon followed by multiparameter techniques [39, 40]. In the 1990s, image analysis programs were developed for automated fluorescence in situ hybridization (FISH) analysis and rare cell detection [40–42]. Today several SBC systems have hit the market. In these instruments, quantitative imaging is realized in two different ways: (i) laser-based microscopic systems with PMT for fluorescence detection (e.g., iCyte, LSC (both CompuCyte)) or (ii) microscopes with arc lamps and digital cameras for emission detection (e.g., SFM, Zeiss).

LSC, which is applicable on fluorescently labeled specimens, cells, and tissue immobilized on slides, is a state of the art alternative for quantitative fluorescent cell analysis [36, 43–45]. It can be utilized for relatively small specimens and its application for rare cell detection was also demonstrated [46, 47]. The LSC allows analysis of up to 5000 cells/min.

The LSC instrument (CompuCyte Corp., USA) was also the first commercial quantitative microscope system. It is built around a routine epi-fluorescence microscope equipped with a motorized stage and up to three single laser beams (405 nm violet diode laser, 488 nm argon–laser, 630 nm helium–neon–laser and nowadays, mostly diode-laser emitting at similar wavelengths). The cells either grow or are immobilized on a glass slide and scanned. Staining with fluorochromes of interest is done either prior to or after immobilization. The emitted fluorescence is guided to a series of four optical filter sets with PMTs. In addition, the light scattered by the cells is detected by a photodiode underneath the slide, producing a signal comparable to the FSC in FCM analyses. A digital image is created for each PMT on a pixel-to-pixel basis. These images are analyzed by applying a trigger signal for cell recognition defined by the operator, which could be any fluorescence or the FSC signal as well as combinations of different signals. Data on a number of parameters is then acquired per object (cell, cell cluster, subcellular compounds, etc.) such as integral fluorescence intensity and maximum pixel intensity within a cell. These values are equivalent to fluorescence integral and height in FCM, respectively.

Another SBC technique is SFM (Zeiss Imaging Solutions, Germany), which uses an arc lamp for excitation and a charged coupled device (CCD) camera for fluorescence detection. The SFM is a software-controlled new technique for automated fluorescence analysis on motorized microscopes. It can measure multiple fluorescence labels and is suitable for multichannel cytometric analysis [40, 48].

Current developments of digital cameras and computer systems promise as high quality cytometric results as with PMT-based measurements. In addition, quantitative analysis can be done on digital slides. Therefore, the slide is scanned and all acquired images (each field of view) are automatically matched to generate a continuous image in every channel. On the basis of this digital slide, microscopy software simulates the basic functions of the microscope (magnification, stage movement, filter change, etc.). SFM technology proved to be linear and stoichiometric on fluorescent calibration beads using inhomogeneity-compensated Hg arc lamp excitation; also, the feasibility for cytometric measurements could be demonstrated [49]. Image analysis can be done off-line by standard cytometry techniques (gating, data display in scatter plots or frequency histograms, cell galleries, etc.) and even through Internet protocols.

9.2 Comparison of Four-Color CD4/CD8 Leukocyte Analysis by SFM and FCM Using Qdot Staining

The T-helper/T-cytotoxic (CD4+/CD8+) lymphocytes are frequent constituents of the peripheral blood leukocyte population and were selected as a model system for four-color immunophenotyping. T-helper cell count and also CD4/CD8 ratio are important clinical parameters. The determination of these parameters is relevant for testing the immune status or for monitoring disease progress as, for example, in HIV patients. The HIV-1 viral load has been found to correlate with the rate of loss of CD4 T lymphocytes, representing the most commonly used prognostic markers in clinical practice [50]. The rate of loss of CD4 T lymphocytes in a significant proportion of HIV-infected individuals has been documented to be constant; however, considerable individual variation has been observed [51, 52]. Years after the HIV infection, the number of CD4+ lymphocytes decreases, leading to immune decline [51–55]. With the spread of HIV-1, there is an emerging demand for automatic methods to determine the CD4/CD8 cell ratio.

Further, we present a method for the determination of CD4/CD8 ratio by four-color immunophenotyping, simultaneously testing the suitability of QDs for such purposes and for SBC applications.

9.2.1 Analysis of Lymphocytes by SFM and FCM

Peripheral blood was stained with CD4-Alexa488, CD8-Qdot605 and CD3 APC antibodies, and Hoechst 33342 DNA dye in order to determine percentages and ratio of helper and cytotoxic T cells. Aliquots were measured by FCM and SFM by appropriate excitation sources and band-pass filters [31]. The negative and positive cell populations were easily detectable in both systems (Figure 9.3). The obtained CD3+4+ and CD3+8+ cell count ratio values for stained populations were comparable.

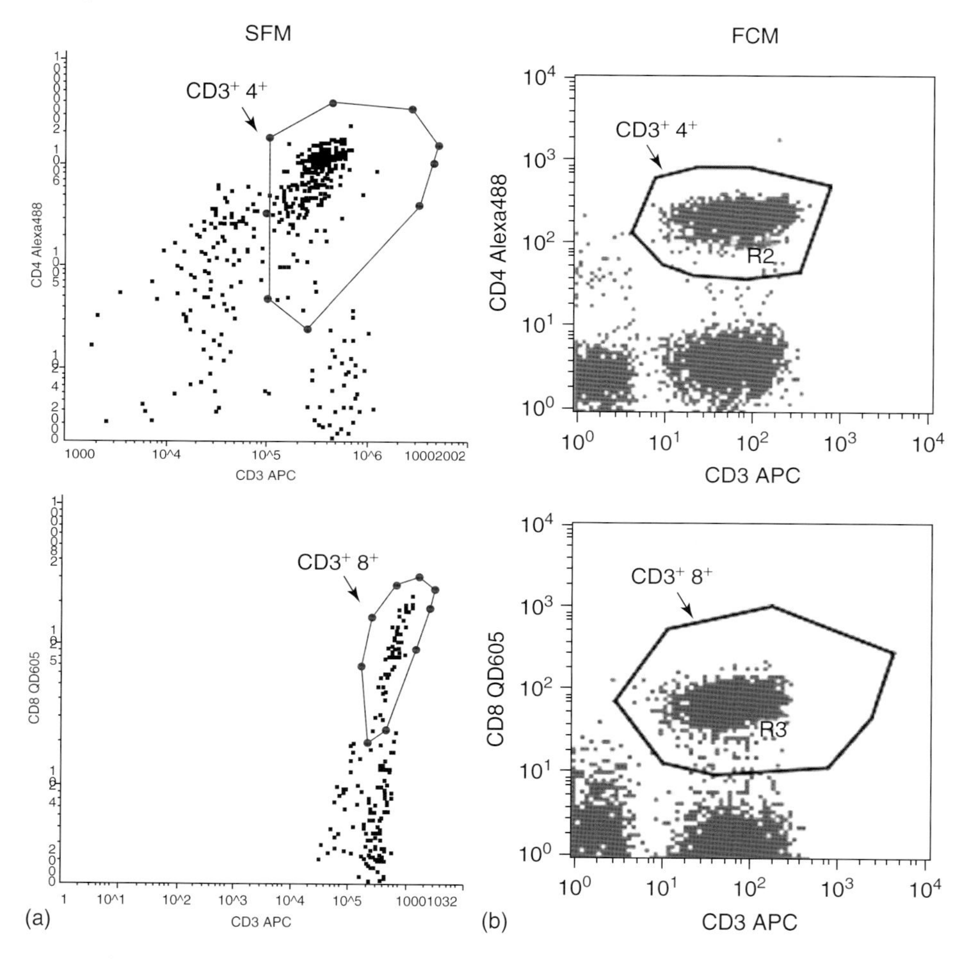

Figure 9.3 Analysis of T-helper and cytotoxic T-cells by SFM and FCM. CD4+ cells were determined in dotplot displaying CD3-APC versus CD4-Alexa488 (a) and CD8+ cells in a CD3-APC versus CD8-QD605 dotplot (b). Although by FCM more cells were analyzed cell populations could be identified unequivocally. Both analyses yielded comparable results. Reprint from Ref. [31] with permission from the publisher.

9.2.2 Comparison of CD4/CD8 Ratio

Both systems yielded comparable results. There was no significant deviation from linearity. Aliquots of 30 different samples were analyzed by both FCM and SFM. CD4/CD8 ratio was determined and compared by linear regression and Passing–Bablok regression (Figure 9.4) (Cusum test for linearity, $P > 0.10$). Passing and Bablok regression of CD4/CD8 ratio ($F(X) = 0.279 + 0.917\times$) showed that all

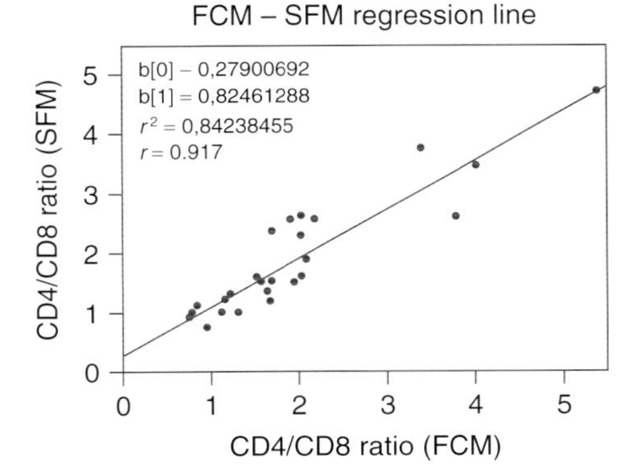

Figure 9.4 Comparison of CD4/CD8 ratio determined by FCM and SFM. Ratio of T-helper (CD4-Alexa488) and cytotoxic T cells (CD8-QD605) were determined in different samples analyzed by SFM and FCM. Results obtained showed a high correlation (Pearson: $R = 0.917, p = 0.01$). Reprint from Ref. [31] with permission from the publisher.

measurement points are within the 95% confidence interval [31]. This comparison indicates that there is no systemic bias between the two different methods. This demonstrates that SBC technology with an instrument such as the SFM is a very useful tool for immunophenotyping, exemplified for the determination of CD4/CD8 ratio. Furthermore, the class of fluorescent nanocrystals can be implemented in multiplexed analyses.

The four-color staining is suitable for identification of designated leukocytes. The trigger used for detection of leukocytes, however, differs from FCM to SBC. While the commonly used FSC and SSC signals are suitable for recognition of cells and inclusion into analysis, DNA staining is preferable in SBC analysis. Nevertheless, T cells (CD3) with their subsets T-helper (CD4) and cytotoxic T cells (CD8) could easily be identified on the basis of their fluorescence signals. Determination of frequencies is standard in cytometric analyses and comparable for both systems. Although the cell numbers analyzed by SFM are limited, results clearly indicate the feasibility of this method for immunophenotyping, and comparability to FCM analyses makes it an alternative to standard methods.

The selected dyes yielded strong signals. They produced neat positive cell populations in FCM and SBC and were bright enough so that exposition times shorter than 1 s per optical filter and image were sufficient in SFM analysis. Fluorescence compensation is a common tool in FCM, whereas it is only rarely used in image analysis. Hence, spillover is much more problematic in SBC than in FCM. If software-based fluorescence compensation is not possible, a careful selection of dyes, that is, marker dye combinations, and gating cascade during analysis are very important.

9.3
Comparison of Leukocyte Subtyping by Multiparametric Analysis with LSC and FCM

The most important feature of LSC is that the exact position of every individual object is recorded together with its fluorescence information and therefore each object can be directly visualized at any time after completing the analysis. Among other capabilities, this allows to verify if objects are single cells, doublets, debris, or artifacts, and to document the cells' morphology. To this end, the slide can be removed from the stage, stained by conventional cytological methods (Giemsa, hematoxylin & eosin), and placed on the stage again [46, 56]. In addition, this "no loss" design is a prerequisite for the analysis of hypocellular specimens. As morphological evaluations are not feasible by FCM, SBC seems to be a very useful tool for the analysis of clinical samples. Concurrently, one of the first clinical applications of LSC was immunophenotyping of peripheral blood leukocytes (PBL) in children [46, 57, 58]. The following example demonstrates the clinical usefulness of LSC analyses. Immunophenotyping of peripheral blood leukocytes was determined by FCM and LSC, and the results were compared. In addition, influence of different parameters on cell analysis such as trigger signals in LSC was tested.

To this end, venous blood samples were stained with directly conjugated anti-CD antibodies, erythrocytes were lysed, and cells were washed. For analysis by FCM at this step, an aliquot of 20 μl was taken and analyzed on a FACScan FCM (BD-Biosciences). Cells for LSC analysis were fixed and stained with 7-AAD (if required) on a slide. After analysis, cover slips were removed, an H&E staining was performed, and slides were covered again; cells from subsets of special interest were relocalized in the LSC and their morphology was investigated and documented, taking transmitted light micrographs.

9.3.1
Different Triggering in LSC Analysis

The trigger in LSC analysis was either set to the 7-AAD signal, CD45, or FSC. Apart from that analysis, settings were identical and comparable to those of FCM (filters, channels, etc.).

Generally, there is no substantial difference in cell analysis results obtained by SBC irrespective of which trigger was used for cell identification. Debris and doublets have to be excluded in every type of trigger signal (Figure 9.5). Further cytometric analysis is not identical. FCM analysis, taken as the golden standard, correlates very well with all three kinds of LSC analysis (Figure 9.6).

The main difference seems to be the deviation in the less frequent cell types, exemplified in Figure 9.7. Here, the correlation of the DNA-based LSC analysis is better than for other triggers. For highly frequent cells (neutrophils), there is no noticeable difference. As FSC signals provided lowest correlation with FCM results, triggering of fluorescence signals (preferably DNA) seems to be superior in SBC analyses [59].

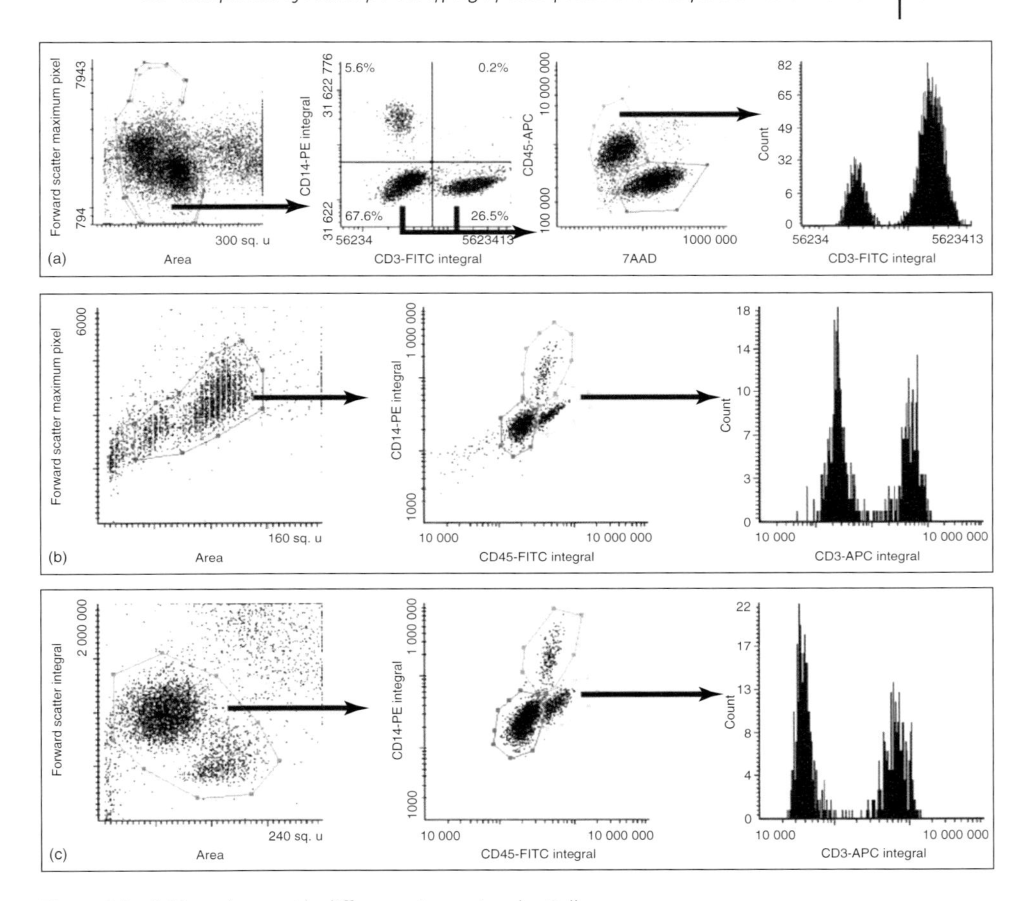

Figure 9.5 LSC analyses with different trigger signals. Cells were stained with anti-CD antibodies as indicated on the axis. Trigger was set either to 7-AAD nuclear staining (a), FSC (b), or CD45 surface antigen staining (c). Analysis, that is, cell identification by gating, was performed as displayed. Reprint from Ref. [59] with permission from the publisher.

Differences between the FCM-light scatter/LSC-7-AAD-triggering concern the quantification of eosinophils, which are slightly underestimated by LSC in a systematic way [59]. This can be explained by the different physical concepts of cell detection: FCM operates with two light scatter characteristics (FSC and SSC), which are applied for triggering of the events and discrimination of the different subsets. In LSC, cells are triggered by their DNA fluorescence after 7-AAD staining and further differentiated by a combination of FSC, 7-AAD fluorescence intensity, and CD45-expression (Figure 9.7). On this basis, eosinophils are detected based on their high FSC maximum pixel values generated by their light-dense granules. Therefore, eosinophils that have degranulated just before fixation would not be detected as

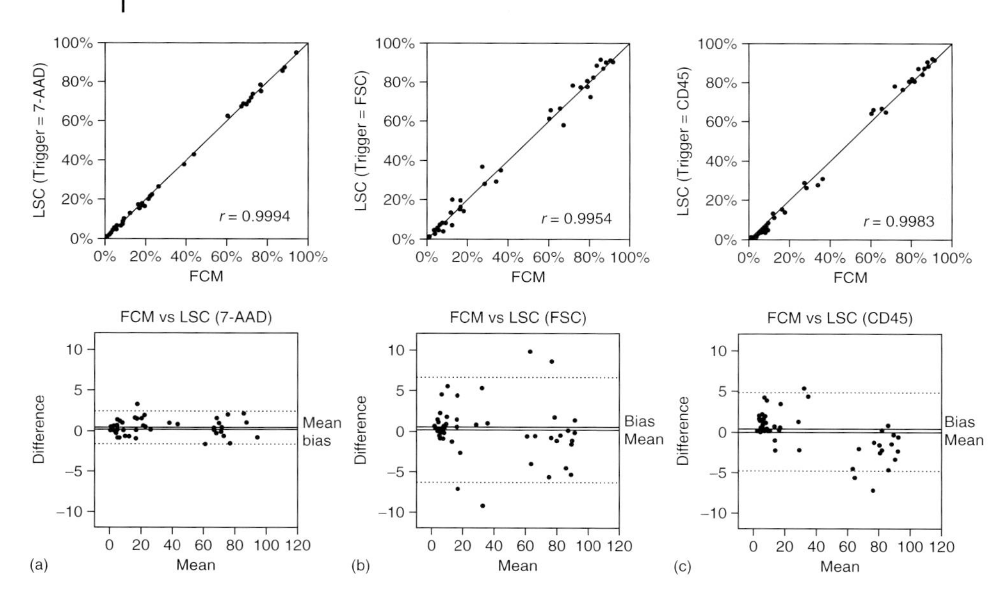

Figure 9.6 Comparison of leukocyte subset percentages obtained by FCM and LSC. Percentages of lymphocytes, monocytes, neutrophils, basophils, and eosinophils were compared for FCM and LSC analysis using different trigger signals (a: 7-AAD, b: FSC, and c: CD45). Bottom row shows corresponding Bland–Altman Plots. Each of the LSC analyses reached a correlation with FCM of $r > 0.99$. If one likes to pick the most suitable trigger, DNA signal would be preferable as dispersion is lowest. Reprint from Ref. [59] with permission from the publisher.

eosinophils. However, since LSC analysis has to be performed on a slide this is hard to circumvent. For lymphocyte subtypes, tight correlation between FCM and LSC analysis with 7-AAD trigger was found (Figure 9.8).

Nevertheless, data presented here show that slide-based immunophenotyping of PBLs yields equivalent data concerning the percentage distribution of cell subsets as compared to standard FCM. The triggering affects the data obtained for the distribution of the PBLs. In the LSC measurements, the trigger on the DNA signal of the cells shows the best correlation with the FCM cell subset percentages with some exceptions.

Triggering on DNA has two advantages: first, neither a nucleated cell is excluded from analysis nor debris with equivalent FSC erroneously included into analysis. Second, cells are partially separated into subsets because of their different 7-AAD intensity (Figure 9.6); unlike for other dyes, such as propidium iodide (PI) or 4′,6-diamidino-2-phenylindole (DAPI), for 7-AAD the integral intensity of the cells depends on the condensation of the nuclear DNA. The higher number of mitochondria leading to a higher amount of mitochondrial DNA in the granulocytes might also contribute to the fact that granulocytes stain brighter than lymphocytes [59].

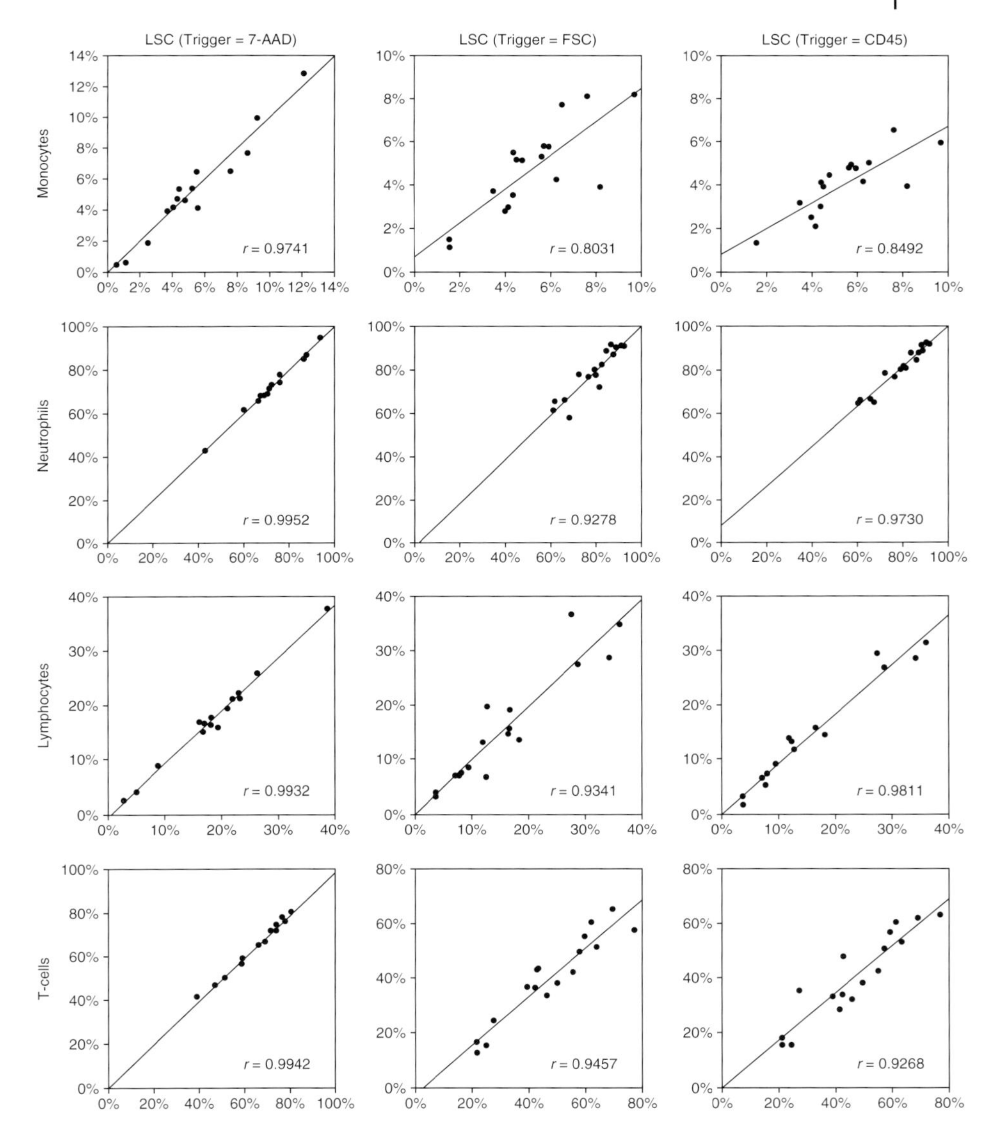

Figure 9.7 Comparison of infrequent and highly frequent cell type quantification by LSC (*y*-axis) and FCM (*x*-axis). Data show correlation of LSC using different trigger signals (columns) with FCM for different cell types (rows). Regression analyses of the percentages on all leukocytes obtained by FCM and by LSC are shown. For LSC, different triggers were applied (7-AAD, FSC, and CD45-FITC). The percent values for T-cells are those with respect to the lymphocyte population. For LSC, monocytes, neutrophils, and lympocytes were gated according to Figure 9.5. Trigger set to 7-AAD (left column) had constantly good results for low (monocytes) and highly frequent cells (neutrophils). CD45 and FSC (central and right column), however, have for monocytes a worse correlation to FCM than for neutrophils. Reprint from Ref. [59] with permission from the publisher.

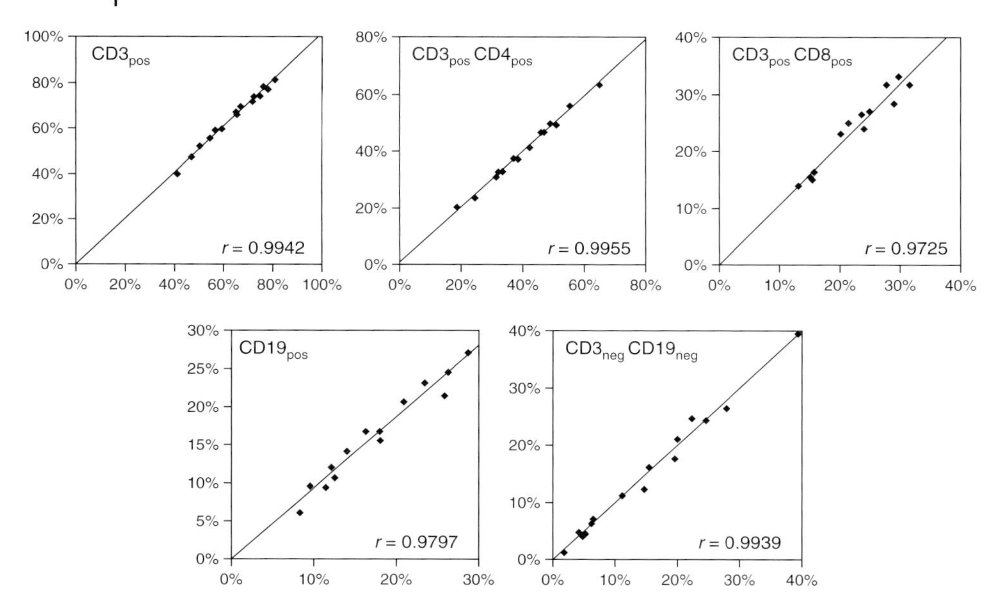

Figure 9.8 Lymphocyte subtypes determined by LSC analysis with 7-AAD trigger versus flow cytometry. Percentages of main lymphocyte subtypes determined in FCM (*x*-axis) and LSC (*y*-axis) are displayed. The percent values for these cells are those with respect to the lymphocyte population. Although the absolute cell numbers for these cell types are comparable to eosinophil cell counts, the correlation with FCM is higher (Figure 9.7).

9.3.2
Immunophenotyping by FCM and LSC

Generally, the same markers and dyes can be applied for both cytometric technologies. There is no substantial difference in the detection of fluorochromes between both instruments. Exceptions are light-sensitive dyes such as PerCP, which are hardly visible on SBC instruments owing to fast bleaching, but are detectable by FCM because of very short excitation time.

The peak ratio, that is, the ratio of the fluorescence intensity values of positive and negative cell populations, indicates how good the separation of the cells is. It could be shown that peak ratios are comparable for both measurements by both instrument types [60, 61]. Hence, LSC can be accepted as equivalent to FCM in respect of fluorescence detection.

An additional advantage of LSC or SBC in general is the ability to visualize what was analyzed. Cells of interest or doubtful events can be relocalized or even restained and further investigated (Figure 9.9). Nowadays, there are techniques available for isolating cells directly out of cell cultures or tissue (laser catapulting [62] or capturing); thereby, selected cells can either be cut and transferred to tubes or all unwanted cells can be destroyed by a laser enabling a further culture with the desired cells only (LEAP technology [63]). Comparable effects can only be

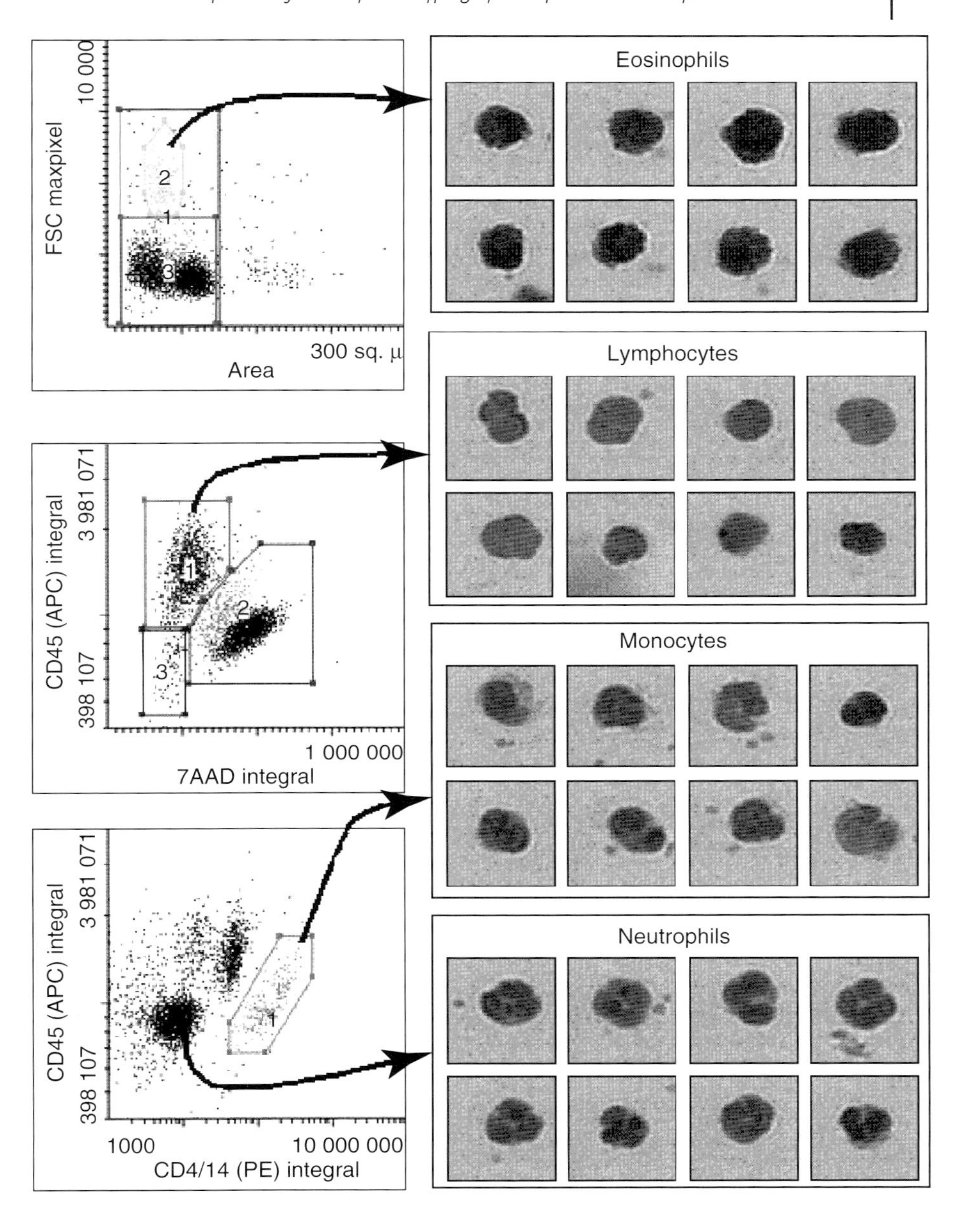

Figure 9.9 Relocalization of cells. Cells of interest can be identified by SBC in analysis and relocalized on the slide. This can be done either by rescanning (LSC) or by recall of images acquired during analysis (SFM). In this example, cells were restained for H&E on the slide and replaced on the microscope stage. Gated cells were relocalized on the slide generating image galleries. Cells can therewith morphologically be examined, which is, for example, important in the analysis of clinical samples. Reprint from Ref. [59] with permission from the publisher.

obtained by cell sorting using fluorescence-activated cell sorting (FACS). However, this method is time consuming, requires high cell numbers, and, above all, needs trypsinization of cultured cells or disintegration of tissue.

9.3.3 Multicolor Analyses

Multiparametric and multicolor analyses can be done on both SBC and FCM instruments. The number of colors in both cases is dependent on the number of detectors and lasers. SBC, however, is capable of extending the number of analyzable dyes (i.e., parameters) by special features. As cells are not lost after analysis as in FCM, in SBC cells can be further investigated and data referred to the previous results of the first analysis on a single-cell basis. Repeated measurement of the same cells is only possible by SBC, and FCM does not provide this option. One of the great advantages of SBC is the opportunity to "ask the cell a second time": a cell on a slide with known $x-y$ coordinates and known fluorescent characteristics, which was measured after a first staining step, can be restained and analyzed again. The resulting "new" fluorescence can then be encoded as extra new color, which, in principle, could yield an n-colored immunofluorescence assay resulting in extended polychromatic cytometry [64].

The measurement by SBC can be made with a slightly different setting. For example, in the usual basic setup for LSC measurements, one filter cube is for the PE-Cy5 and APC fluorescence optimized (670/20 nm) dyes; after the first scan, with a change to a 710/20-nm filter cube the PE-Cy5.5 and APC-Cy5.5 tandem dyes could be detected [9], and by merging these two data files, the presence of four-color antibody staining on the same detector could be detected.

Between the two measurements, the slide can be illuminated for bleaching light-sensitive dyes, and using light-tolerant but sensitive pairs of dyes on the same filter and detector, measurement of two-color staining is still possible [13, 65, 66].

As a general consideration in the LSC, cells are scanned in several stripes by a relatively slow-moving laser beam; this could produce "forward bleaching" in the direction of the stripe as well as "sideward bleaching" to the following neighboring stripes. Together, this could result in a diminished integral fluorescence of the cell. In contrast, in FCM always the entire cell's fluorescence is excited and detected at once in a single very short illumination pulse.

9.4 Absolute and Relative Tumor Cell Frequency Determinations

The cytometric analysis of highly frequent cell types (hematological samples, cell culture, and mechanically dissociated solid tissue cells) is relatively easy to accomplish as in most cases enough material (i.e., cells) is available. More problematic is the analysis of rare cells from, for example, human organs (physiologic and pathologic cells, e.g., stem cells or tumor cells) [67–69]. Detection of circulating tumor cells is of substantial clinical importance and the prognostic value of these results has been demonstrated [48]. A precise and reliable method for analysis, however, is needed. In circulating tumor cell detection without tumor cell enrichment using FCM, sensitivity was found to be unsatisfactory [43].

Highly sensitive assays like real-time polymerase chain reaction (RT-PCR) were found to show false positive results in tumor-free patients [70]. Cytokeratin, a marker of the epithelial cancer cells, is also expressed in gastrointestinal inflammations in lymphocytes [69]. Therefore, RT-PCR or immunohistochemistry alone is not sufficient for reliable micro-metastases detection. FCM is also often used for rare cell detection. In less frequent cell analyses, FCM has some disadvantages in finding cells in the per mille range (below 0.01% of abundance). FCM is a so-called lot measurement. Data from many thousand or millions of cells are acquired and similar signals give a population in a dot plot display. The problem with rare cells is their potential loss in background, that is, artifacts with characteristics (e.g., scatter) similar to those of the cells of interest. Moreover, the minimum number of cells forming a population has to be determined to be able to call it a population and rely on the data as "true positive events." FCM therefore has the disadvantage that measured events are not seen and lost after the acquisition. The individual analyzer therefore has to believe in the dots on the plot. A step in the right direction may be the newly available imaging flow cytometers. Another problem arises while measuring the last drops of the cell suspension (looking for all tumor cells from the patient's sample) because air bubbles could come to the flow channel, causing strong light scattering. Hence, many false signals are acquired, resulting in problematic analysis.

Automated microscopic analysis may be an alternative method of choice although the US Food and Drug Administration (FDA) approved automated microscopy systems only for enzyme-based micro-metastases detection (Chromavision Inc., Applied Imaging Inc.). However, multiparametric fluorescent cell detection is supposed to have higher specificity and sensitivity than immunocytochemistry [71]. With the following experiments, we could demonstrate that the SBC is applicable for rare cell detection.

9.4.1 Comparison of Cell Counts

9.4.1.1 Dilution Series

As a model, HT29 cells were used for rare cell determinations. Cells were stained with CAM5.2-FITC. For microscopic systems, DNA staining (Hoechst) was additionally applied and served as trigger signal. Samples were analyzed by FCM, LSC, and SFM for comparison. The analyses are displayed in Figure 9.10. FITC-labeled HT29 tumor cells provided good signal, resulting in clear discrimination of the cell population in each of the three analysis techniques. Dilution series of HT29 cells was performed, whereby cell suspensions were diluted with peripheral blood leukocytes from 1 : 1 to 1 : 1000 and analyzed by the three instruments (Figure 9.10). For each concentration, at least three samples were prepared and measured.

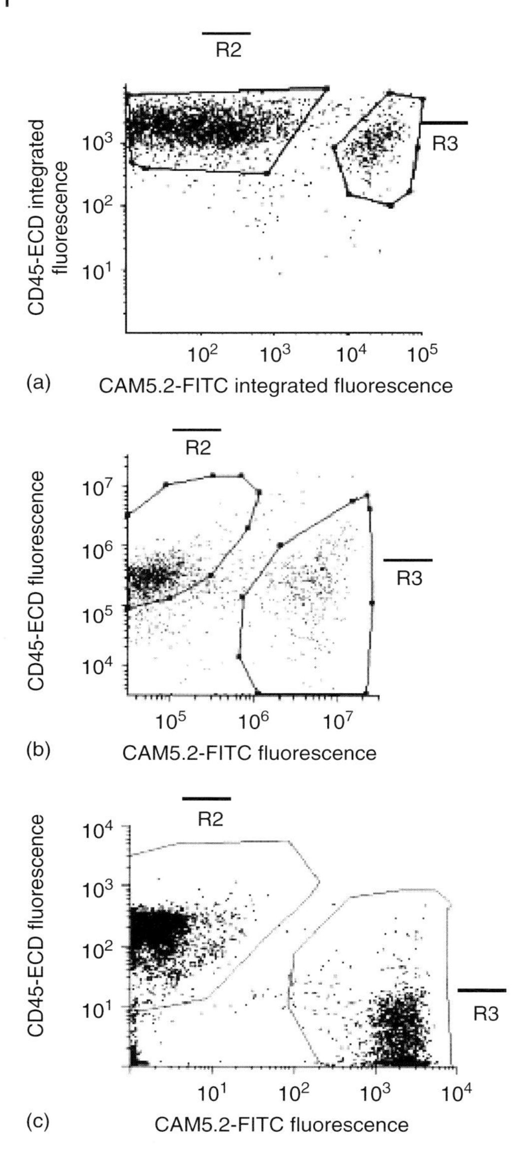

Figure 9.10 HT29 cell line analysis by SFM, LSC, and FCM. Cells were stained with Hoechst, CD45, and CAM 5.2. Fluorescence was compensated in LSC and FCM. CD45 fluorescence intensity was comparable for LSC (b) and FCM (c) whereas in SFM (a) higher values were obtained. This may be the result of the difference in the acquisition method between SFM and the other two systems (illumination time instead of defined laser excitation time). The FITC fluorescence of CAM 5.2., however, is higher in LSC than for the others and may be caused by different PMT settings. Nevertheless, peripheral blood leukocytes (R2) and HT29 cells (R3) were clearly discriminated by SFM, LSC, and FCM analysis. Reprint from Ref. [72] with permission from the publisher.

In this dilution series with rather high cell numbers, FCM showed the most consistent results for detection of fluorescence-labeled HT29 cells (Figure 9.11.). The correlation between the measured and expected cell frequencies was highest ($r^2 = 0.84$) for FCM, followed by the SFM ($r^2 = 0.79$) and LSC ($r^2 = 0.62$).

LSC yielded systematically higher relative cell frequencies than the calculated (expected) cell numbers. SFM showed the highest variation in the measured data. Owing to the high cell numbers analyzed in FCM, there were clearly distinguishable cell populations that could not be obtained by SFM and LSC. In the case of improper gating and/or high amount of artifacts, several events can be classified as false positives. Visual inspection of analyzed cells improves accuracy but needs intervention by the user in the otherwise automatable process. However, the determined cell ratios showed significant correlation with the calculated dilutions ($p < 0.01$) [72].

For clinical analysis, acquisition time is an important factor. Here, FCM analyses are expectedly fastest with 3–5 min for the analysis of 10–20 000 cells. SBC methods require more time for analysis: LSC, 10–15 min, SFM (using autofocus), 40–50 min, but with more reliable results.

9.4.1.2 Rare Cell Analysis

The recovery of a given amount of cells was investigated using the same HT29 cells. A defined amount of cells (5, 10, or 20 FITC-labeled cells) were added to a 100 μl leukocytes cell suspension by micromanipulation. FITC positive cells could be detected either by FCM or by one of the SBC methods. Hence, a cytospin with CD45-PE-TexasRed-labeled (Ficoll separated) peripheral blood leukocytes was performed and the same numbers of cells were placed directly on the slide by a micromanipulator. Slides were analyzed by SFM and LSC.

For analysis of the less frequent cell counts, only SFM and LSC could be used. Recovery in the analysis by SFM and LSC was comparable (Table 9.1). Almost identical cell counts could be measured compared to the inserted cell numbers. Another experiment was done using circulating tumor cells isolated by Carcinoma Cell Enrichment Kit (Miltenyi Biotech, Germany). A cytospin was performed with the enriched fraction and cells were stained for CD45, CAM 5.2, and nuclear DNA. Ten slides were analyzed by SFM, LSC, as well as by manual counting.

SFM showed very high correlation with manual counting ($r^2 = 0.99$), whereas LSC analysis tended to overestimate tumor cell counts. Nevertheless, correlation between LSC and manual counting was also very high ($r^2 = 0.97$). Furthermore, SFM and LSC yielded comparable results ($r^2 = 0.97$). There is a significant correlation between the analysis techniques ($p < 0.01$).

9.4.2 Analysis Documentation

Data storage is important for many analyses. In particular, for pathological investigations documentation of analytical findings is mandatory and in that case images are more preferable than dots in a plot only. SBC analyses provide this

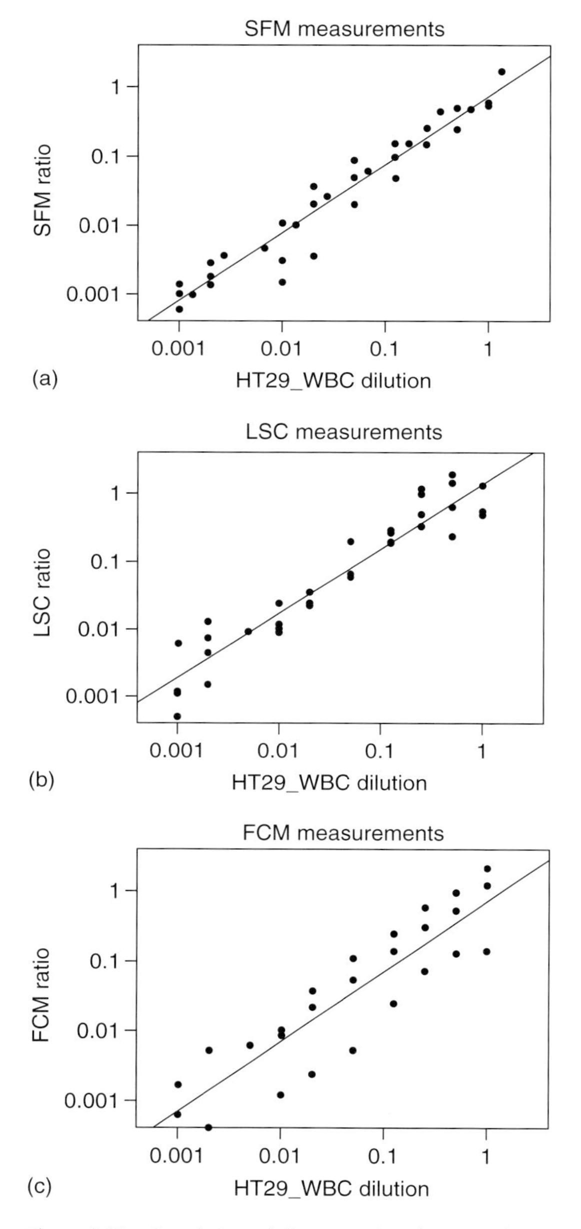

Figure 9.11 Correlation of determined and expected tumor cell ratio among leukocytes. Linear regression shows correlation between the expected (i.e., diluted, *y*-axis) and the determined (*x*-axis) cell frequency of HT29 cells. FCM yielded results with $r^2 = 0.84$ highest correlation. Reprint from Ref. [72] with permission from the publisher.

Table 9.1 Rare cell counting by SFM and LSC. Recovery of FITC positive HT29 cells was comparable good for both methods.

Inserted cells	SFM counting	LSC counting	Number of cases
5	6.5 ± 2.1	5.2 ± 2.6	4
10	9.1 ± 1.5	14.5 ± 4.4	8
20	21.6 ± 3.7	21.3 ± 11.9	3

opportunity. Cytometric data, including cell count and DNA content among others, as well as images or image galleries of analyzed cells and cell types can be archived (Figure 9.12).

9.4.3 Limitations

The examples presented here for rare cell detection showed that fluorescence analyses on cell smears and cell count determinations in dilution series yield similar results for tumor cell counts as well as tumor cell/lymphocyte ratios in high- and low-concentration ranges. Similar studies were performed with comparable results using LSC and FCM [73], or image analysis and FCM [74]. Comparative studies with FCM and LSC have shown that data obtained by both technologies are consistent with each other to study apoptosis or necrosis [75]. LSC is a reliable tool for the intracellular staining of p53 protein [73] and DNA ploidy analysis [76]. Results for SFM analysis showed that SFM is a powerful method for fluorescence-labeled tumor cell analysis and is equivalent to LSC.

In clinical analyses, automation of measurements and data analysis is desirable. In SBC analysis, one of the most problematic issues in automation is autofocusing. Autofocusing in SFM analysis is rather time consuming and should be checked by the observer. Automated slide loading has to be applied for high-throughput analyses as well. However, there are loading robots available for different microscope types. Moreover, algorithms are available for automated scanning and evaluation; new automated and fast scanning hardware systems are expected to be developed very soon. Analysis times of SBC systems have been improved in the past but are still slower than FCM. This needs to be further improved for high-throughput analyses.

Another problem is the total number of cells that can be analyzed on the systems. Rare cells are naturally difficult to find in the sample. SBC analysis is, in that case, restricted to space on the slide. The cell density must therefore be very high to identify the cells of interest. Nevertheless, sample volume is also the limiting factor in FCM analysis although cell numbers in the 10^6 range is very hard to accomplish in SBC.

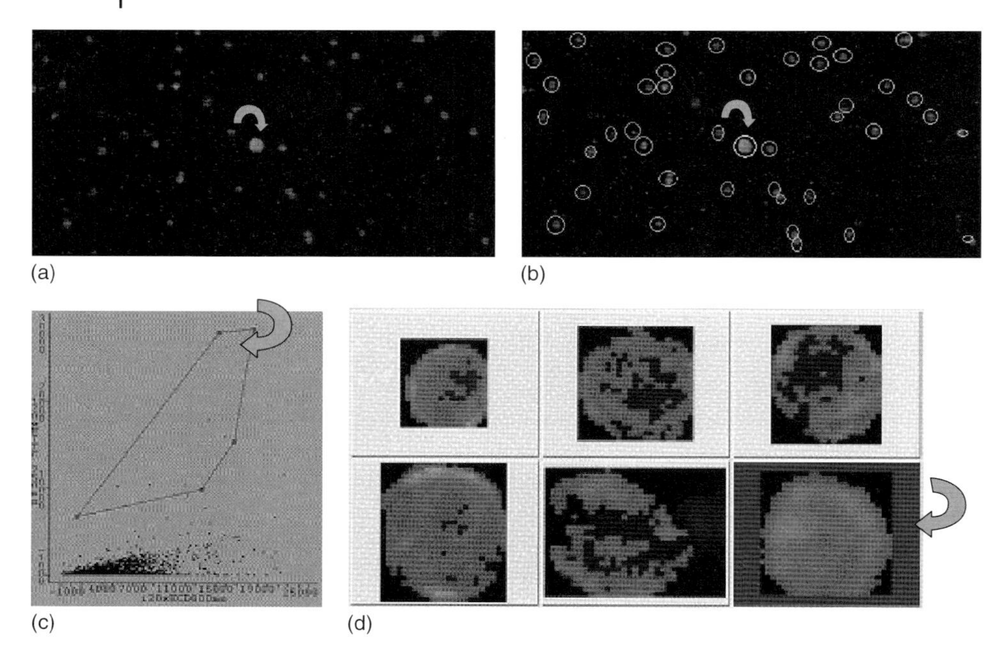

Figure 9.12 SFM analysis of tumor cells. In (a) one field of view of SFM analysis is displayed (100× magnification). Cells are stained for DNA (all cells), CD45 (leukocytes), and CAM 5.2 (carcinoma cells). A single CAM 5.2 positive cell is shown in the middle (arrow) surrounded by CD45 positive cells. (b) Shows the same field of view with trigger contour, that is, cells identified by the software and included into analysis (c). The dotplot displays CD45 versus CAM 5.2 with a gate for tumor cells. Only six cells are inside this gate. Each of those cells can be displayed or image galleries of all gated cells can be generated (d). The nuclei and the CAM5.2-FITC-labeled surface are clearly visible. The cell image (arrow and black underlayed cell image) is thereby linked to the dot in the plot (arrow). This means that each event can be judged by visual inspection and be excluded from analysis when appropriate. (Reprint from Ref. [72] with permission from the publisher).

9.5 Analysis of Drug-Induced Apoptosis in Leukocytes by Propidium Iodide

Cell death analysis is the focus of many fields of life sciences from research on ontogenesis to drug development. In the last decades, cytometric methods and apoptosis analysis developed in parallel and therefore the majority of the apoptotic features can be evaluated. These include cell shrinkage, nuclear condensation, internucleosomal DNA cleavage, and apoptotic signaling cascades that may be detected by multicolor immune fluorescence cytometry. Moreover, cytometry offers the possibility to measure intravitally mitochondrial membrane potential, generation of reactive oxygen species, Ca^{2+} content, caspase activity, and phosphatidylserine externalization.

The gold standard for apoptosis confirmation is the observation of the cell's morphology but this is a time-consuming manual process. Current is for a fast quantitative detection of presence and abundance of apoptotic and living cells after different treatments. The easiest way is the DNA content measurement after removal of the small molecular weight DNA fragments by an alkaline buffer [77]. The ethanol-fixed apoptotic cells lose almost all nuclear DNA, but normal cells do not.

There are different ways of apoptosis induction. According to the study of Tyihak *et al.* [78] formaldehyde (HCHO) can act as a cell proliferation retardation factor through the cellular metabolism generated and mediate the apoptotic process of thymic lymphocytes. Other possibilities, including the blockade of specific binding sites on the surface of lymphocytes, may also be taken into consideration.

The consumption of different polyphenol-containing foods seems to maintain the health of an organism. It is also known that myricetin, one of the polyphenols (a flavonoid compound) inhibits important DNA repair enzymes, the DNA polymerase I [79] and the DNA topoisomerase I and II [80], and the microsomal P450 enzyme activities [81] cause depletion of reduced glutathione [75] and induce the secretion of TNF (tumor necrosis factor)-alpha in the RAW 264.7 macrophages [82]. Supplementation of hepatocyte cultures with myricetin led to the formation of phenoxyl radical intermediates [83]. However, it is unknown how HCHO and myricetin influence apoptosis of lymphocytes.

HCHO can be formed by different endogenous and exogenous compounds in the cells by the well-known demethylation process. The chemical activity of HCHO is nonspecific and there is a wide variety of possible HCHO acceptors. Therefore, it is difficult to ascertain the actual role of HCHO-producing reactions in the cell. As a reactive molecule, HCHO can participate in the degradation of the DNA, which could suggest a possible function for HCHO in the apoptotic process. Myricetin is a polyphenolic compound that can react with HCHO in a Mannich condensation reaction with amines. The three neighboring OH groups on one benzene-ring make myricetin highly reactive with HCHO. The multicyclic polyphenols react with both chains of the DNA helix and create bridges among them. The hypothesis is that myricetin (containing pirogallol group) cross-links the DNA double helix in the presence of HCHO and H_2O_2 and leads to increased apoptosis [84].

9.5.1 Induction of Apoptosis

Peripheral mononuclear cells (PBMCs) were isolated by density gradient centrifugation. Cells numbering 10^5 were seeded in 96-well plates and cultured in RPMI-1640 without any medium. Cells were treated for 24 h with 90 μM myricetin, a mixture of 30 μM H_2O_2, 30 μM HCHO, or a combination of all three compounds. Cells were fixed with 70% ethanol (−20 °C), and low molecular weight DNA was removed by alkaline citrate phosphate solution (pH = 7.8). DNA was stained by

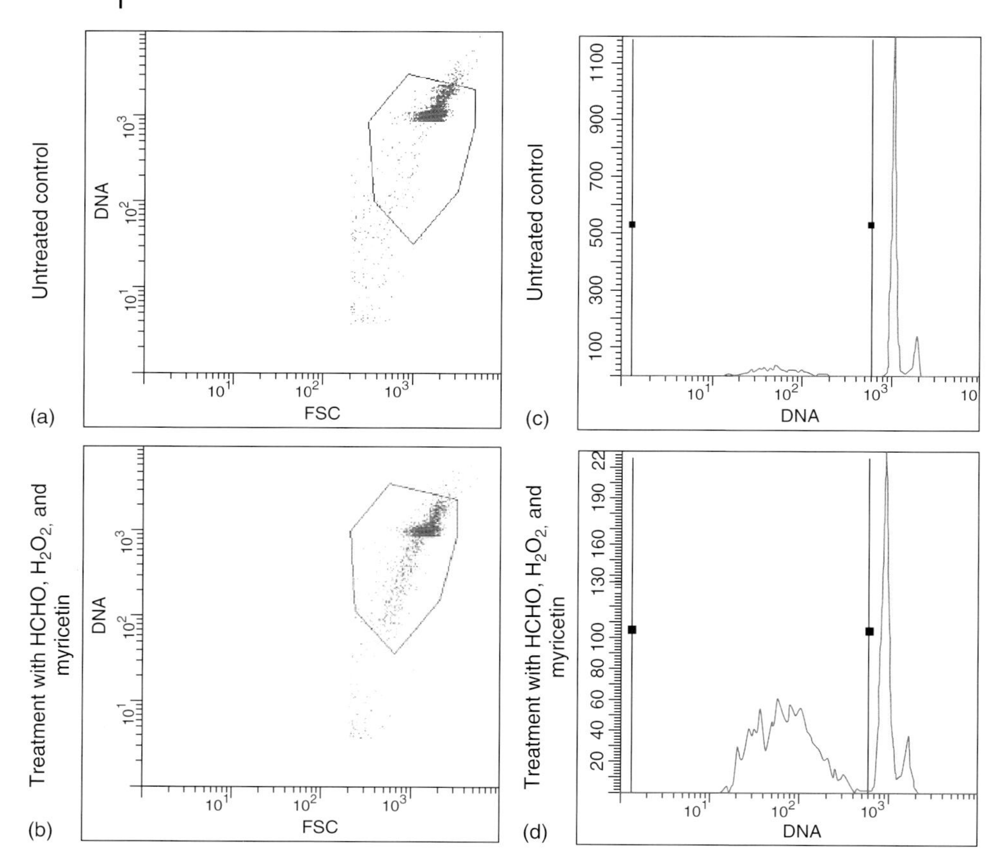

Figure 9.13 FCM analysis of PBMC apoptosis. Debris was excluded by low FSC and DNA signals (a, b). Single cells were than analyzed for DNA content (c, d). Percentage of cells in the sub-G1 region was determined. While only few cells showed apoptotic behavior, that is, exhibited lower DNA content than normal for G1-nuclei, 60% of the treated fraction was apoptotic.

5 μg ml^{-1} PI. The cell suspension was analyzed by FCM (Figure 9.13). For SBC analyses (iCyte Imaging Cytometry, CompuCyte Corp.) cell smears were made on slides and cytometrically analyzed (Figure 9.14).

In both analyses, apoptosis was determined by low (sub-G1) DNA content. Expectedly, untreated control had the lowest amount of apoptotic cells (~5.5% region R2). Treatment of lymphocytes with the combination of HCHO and H_2O_2 surprisingly did not increase apoptosis, contrary to previous findings on tumor cells [78]. The reason for this is not known. Maybe tumor cells have higher sensitivity than mononuclear cells because of metabolic differences. After treatment with myricetin, HCHO, and H_2O_2 combination, however, high amount of cells, 60%, were apoptotic (region R2).

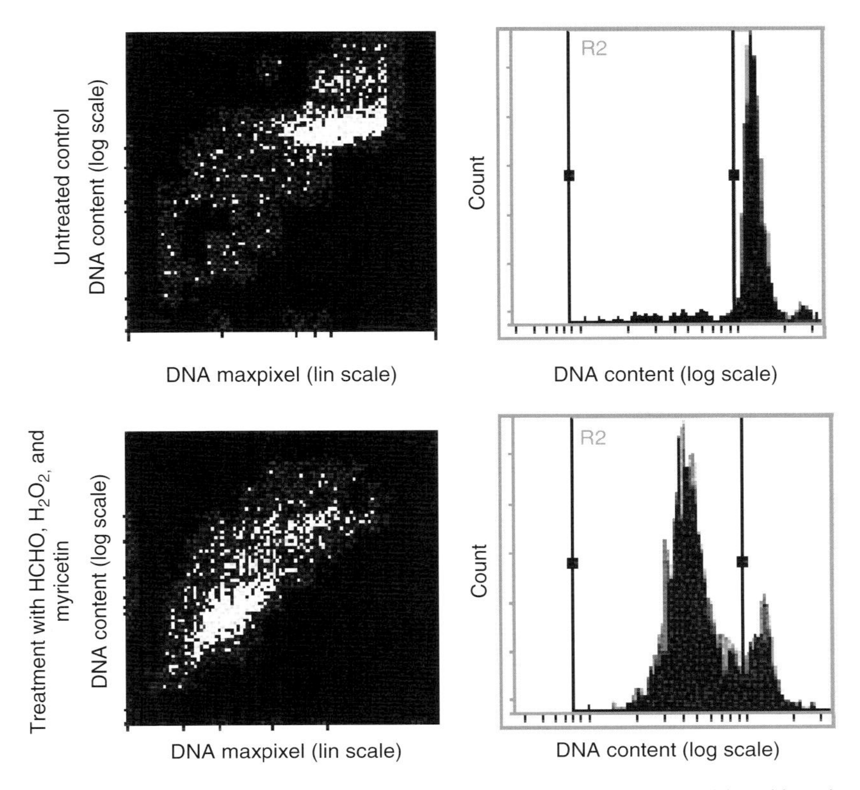

Figure 9.14 iCyte analysis of PBMC apoptosis. Gating is comparable to FCM analysis and was done in the dot plots (Figure 9.13). Debris was excluded by displaying DNA maximum pixel (*x*-axis) versus DNA integral (*y*-axis, a, b) and single cells were analyzed for DNA content in PI intensity histograms (c, d). Untreated cells served as control and contained only few apoptotic cells. Apoptosis was induced by adding the indicated compounds, although the effect of HCHO and H_2O_2 was rather negligible (not shown). Treatment with all three compounds was most effective, as can be seen in the high amount of cells in the sub-G1 region. Reprint from Ref. [84] with permission from the publisher.

9.5.2
Apoptosis Detection by SBC and FCM

Apoptosis measurements are possible with both cytometric systems, yielding very similar apoptosis frequencies. Cell identification by both analyses is different. While in FCM the FSC signal was used, DNA signal served as trigger in SBC analysis. The FSC trigger is useful in FCM measurements. The apoptotic cells have only slightly smaller FSC values than normal lymphocytes and also the discrimination between debris and cells is good. Lymphocytes in the SBC sample lost their FSC signal because of drying and alcohol fixation. Therefore, triggering signal should be the DNA fluorescence [84]. Triggering on DNA in apoptosis measurements could become problematic owing to degradation of DNA and thereby, the loss of fluorescence signal. Therefore, cell identification using a constant threshold for all cells may result in cells being ignored for analysis due to 10–100 times lower fluorescence intensities. (Figure 9.14). However, in the SBC triggering on FSC signals alone produces more artifacts in the analysis. This is similar to the FCM analysis. Although another scatter parameter (SSC) is available in FCM, cell debris, especially in apoptosis measurements, is hardly distinguishable from dirt, bubbles, and so on. Hence, it is difficult to determine the true number of dead cells if the cells are "too dead," that is, if the membrane is already dissolved and the nucleus hyperfragmented.

9.6 Conclusion

The comparison of SBC and FCM analyses has shown that these two techniques provide very similar results in most fields of quantitative cellular analysis. As FCM and SBC are based on the same principles for fluorescence analysis, sample preparation and data acquisition are also comparable. However, the main applications of both techniques are clear. FCM is the golden standard for quantitative multicolor phenotyping of leukocytes or other cells in suspension. SBC is an alternative technology for cytometric analysis mainly used for quantitative analysis of cell cultures and tissue sections. Nevertheless, cell suspensions can also be analyzed by SBC (after fixation on a slide or in cell culture plates where even cell tracking can be performed) and cell cultures and tissue can be analyzed by FCM (after detachment of cells or disintegration of tissues). Both techniques have some pros and cons. The most striking advantage of FCM is its capability of high-throughput and high-content analysis, that is, the fast analysis of high cell numbers. Measuring times in SBC are slower and not as high cell numbers as in FCM analyses can be reached, but SBC is able to analyze even minimal sample volumes with the possibility for morphological evaluation, which is hardly feasible by FCM.

Acknowledgment

This work presented in this paper was made possible by funding from the German Federal Ministry of Education and Research (BMBF, PtJ-Bio, 0313909), Germany and T 034245 OTKA, Hungary.

References

1. Hood, L. and Perlmutter, R.M. (2004) The impact of systems approaches on biological problems in drug discovery. *Nat. Biotechnol.*, **22**, 1215–1217.
2. Hood, L., Heath, J.R., Phelps, M.E., and Lin, B. (2004) Systems biology and new technologies enable predictive and preventative medicine. *Science*, **306**, 640–643.

3. Sachs, K., Perez, O., Pe'er, D., Lauffenburger, D.A., and Nolan, G.P. (2005) Causal protein-signaling networks derived from multiparameter single-cell data. *Science*, **308** (5721), 523–529.
4. Pelkmans, L., Fava, E., Grabner, H., Hannus, M., Habermann, B., Krausz, E., and Zerial, M. (2005) Genome-wide analysis of human kinases in clathrin- and caveolae/raft-mediated endocytosis. *Nature*, **436** (7047), 78–86.
5. Valet, G. (2005) Cytomics: an entry to biomedical cell systems biology. *Cytometry A*, **63A**, 67–68.
6. Kriete, A. (2005) Cytomics in the realm of systems biology. *Cytometry A*, **68A**, 19–20.
7. Tárnok, A., Pierzchalski, A., and Valet, G. (2010) Potential of a cytomics top-down strategy for drug discovery. *Curr. Med. Chem.*, **17** (16), 1719–1729.
8. Pierzchalski, A., Robitzki, A., Mittag, A., Emmrich, F., Sack, U., O'Connor, J.E., Bocsi, J., and Tárnok, A. (2008) Cytomics and nanobioengineering. *Cytometry B Clin. Cytom.*, **74** (6), 416–426.
9. Mittag, A., Lenz, D., Gerstner, A.O., Sack, U., Steinbrecher, M., Koksch, M., Raffael, A., Bocsi, J., and Tarnok, A. (2005) Polychromatic (eight-color) slide-based cytometry for the phenotyping of leukocyte, NK, and NKT subsets. *Cytometry Part A*, **65A**, 103–115.
10. Perfetto, S.P., Chattopadhyay, P.K., and Roederer, M. (2004) Seventeen-colour flow cytometry: unravelling the immune system. *Nat. Rev. Immunol.*, **4**, 648–655.
11. Mittag, A., Lenz, D., Gerstner, A., and Tárnok, A. (2006) Hyperchromatic cytometry principles for cytomics by slide based cytometry. *Cytometry A*, **69** (7), 691–703.
12. Haugland, R.P. (1996) *Molecular Probes' Handbook of Fluorescent Probes and Research Chemicals*, 6th edn, Molecular Probes, Inc., Eugen, OR. ISBN: 0-9652240-1-5.
13. Mittag, A., Lenz, D., Bocsi, J., Sack, U., Gerstner, A.O., and Tárnok, A. (2006) Sequential photobleaching of fluorochromes for polychromatic slide-based cytometry. *Cytometry A*, **69** (3), 139–141.
14. Chan, W.C. and Nie, S. (1998) Quantum dot bioconjugates for ultrasensitive nonisotopic detection. *Science*, **281** (5385), 2016–2018.
15. Niemeyer, C.M., Adler, M., Lenhert, S., Gao, S., Fuchs, H., and Chi, L. (2001) Nucleic acid supercoiling as a means for ionic switching of DNA--nanoparticle networks. *Chembiochem*, **2** (4), 260–264.
16. Alivisatos, A.P., Johnsson, K.P., Peng, X., Wilson, T.E., Loweth, C.J., Bruchez, M.P. Jr., and Schultz, P.G. (1996) Organization of 'nanocrystal molecules' using DNA. *Nature*, **382** (6592), 609–611.
17. Telford, W.G. (2004) Analysis of UV-excited fluorochromes by flow cytometry using near-ultraviolet laser diodes. *Cytometry A*, **61** (1), 9–17.
18. Hohng, S. and Ha, T. (2004) Near-complete suppression of quantum dot blinking in ambient conditions. *J. Am. Chem. Soc.*, **126** (5), 1324–1325.
19. Shimizu, K.T., Woo, W.K., Fisher, B.R., Eisler, H.J., and Bawendi, M.G. (2002) Surface-enhanced emission from single semiconductor nanocrystals. *Phys. Rev. Lett.*, **89** (11), 117401.
20. Florijn, R.J., Slats, J., Tanke, H.J., and Raap, A.K. (1995) Analysis of antifading reagents for fluorescence microscopy. *Cytometry*, **19**, 177–182.
21. Smolewski, P., Bedner, E., Gorczyca, W., and Darzynkiewicz, Z. (2001) "Liquidless" cell staining by dye diffusion from gels and analysis by laser scanning cytometry: potential application at microgravity conditions in space. *Cytometry*, **44**, 355–360.
22. Akerman, M.E., Chan, W.C.W., Laakkonen, P., Bhatia, S.N., and Ruoslahti, E. (2002) Nanocrystal targeting in vivo. *Proc. Natl. Acad. Sci. U.S.A.*, **99**, 12617–12621.
23. Gao, X., Cui, Y., Levenson, R.M., Chung, L.W., and Nie, S. (2004) In vivo cancer targeting and imaging with semiconductor quantum dots. *Nat. Biotechnol.*, **22**, 969–976.

24. Dubertret, B., Skourides, P., Norris, D.J., Noireaux, V., Brivanlou, A.H., and Libchaber, A. (2002) In vivo imaging of quantum dots encapsulated in phospholipid micelles. *Science*, **298** (5599), 1759–1762.
25. Kim, S., Lim, Y.T., Soltesz, E.G., De Grand, A.M., Lee, J., Nakayama, A., Parker, J.A., Mihaljevic, T., Laurence, R.G., Dor, D.M., Cohn, L.H., Bawendi, M.G., and Frangioni, J.V. (2004) Near-infrared fluorescent type II quantum dots for sentinel lymph node mapping. *Nat. Biotechnol.*, **22**, 93–97.
26. Jaiswal, J.K., Mattoussi, H., and Mauro, J.M., and Simon, S.M. (2003) Long-term multiple color imaging of live cells using quantum dot bioconjugates. *Nat. Biotechnol.*, **21** (1), 47–51.
27. Ballou, B., Lagerholm, B.C., Ernst, L.A., Bruchez, M.P., and Waggoner, A.S. (2004) Noninvasive imaging of quantum dots in mice. *Bioconjug. Chem.*, **15**, 79–86.
28. Stewart, C.C., Woodring, M.L., Podniesinski, E., and Gray, B. (2005) Flow cytometer in the infrared: inexpensive modifications to a commercial instrument. *Cytometry A*, **67** (2), 104–111.
29. Leray, A., Spriet, C., Trinel, D., and Héliot, L. (2009) Three-dimensional polar representation for multispectral fluorescence lifetime imaging microscopy. *Cytometry A*, **75** (12), 1007–1014.
30. Harz, M., Rösch, P., and Popp, J. (2009) Vibrational spectroscopy--a powerful tool for the rapid identification of microbial cells at the single-cell level. *Cytometry A*, **75** (2), 104–113.
31. Bocsi, J., Mittag, A., Varga, V.S., Molnar, B., Tulassay, Z., Sack, U., Lenz, D., and Tarnok, A. (2006) Automated four color CD4/CD8 analysis of leukocytes by scanning fluorescence microscopy using quantum dots, in *Colloidal Quantum Dots for Biomedical Applications*, vol. 6096 (eds M. Osinski, K. Yamamoto, and T. Jovon), SPIE, Bellingham. pp. 60960Z01–60960Z11. ISBN: 0-8194-6138-5.
32. Borowitz, M., Bauer, K.D., Duque, R.E., Horton, A.F., Marti, G., Muirhead, K.A., Peiper, S., and Rickman, W. (1998) Clinical Applications of Flow Cytometry: Quality Assurance and Immunophenotyping of Lymphocytes; Approved Guideline, NCCLS document H42-A. NCCLS, Wayne.
33. Kachel, V., Benker, G., Lichtnau, K., Valet, G., and Glossner, E. (1997) Fast imaging in flow: a means of combining flow-cytometry and image analysis. *J. Histochem. Cytochem.*, **27**, 335–341.
34. George, T.C., Basiji, D.A., Hall, B.E., Lynch, D.H., Ortyn, W.E., Perry, D.J., Seo, M.J., Zimmerman, C.A., and Morrissey, P.J. (2004) Distinguishing modes of cell death using the ImageStream® multispectral imaging flow cytometer. *Cytometry A*, **59A**, 237–245.
35. Ortyn, W.E., Hall, B.E., George, T.C., Frost, K., Basiji, D.A., Perry, D.J., Zimmerman, C.A., Coder, D., and Morrissey, P.J. (2006) Sensitivity measurement and compensation in spectral imaging. *Cytometry A*, **69** (8), 852–862.
36. Tarnok, A. and Gerstner, A.O. (2002) Clinical applications of laser scanning cytometry. *Cytometry*, **50** (3), 133–143.
37. Bocker, W., Gantenberg, H.W., Muller, W.U., and Streffer, C. (1996) Automated cell cycle analysis with fluorescence microscopy and image analysis. *Phys. Med. Biol.*, **41**, 523–537.
38. Lockett, S.J., O'Rand, M., Rinehart, C.A., Kaufman, D.G., Herman, B., and Jacobson, K. (1991) Automated fluorescence imaging cytometry. DNA quantification and detection of chlamydial infections. *Anal. Quant. Cytol. Histol.*, **13**, 27–44.
39. Galbraith, W., Ryan, K.W., Gliksman, N., Taylor, D.L., and Waggoner, A.S. (1989) Multiple spectral parameter imaging in quantitative fluorescence microscopy. I: quantitation of bead standards. *Comput. Med. Imaging Graph.*, **13**, 47–60.
40. Galbraith, W., Wagner, M.C.E., Chao, J., Abaza, M., Ernst, L.A., Nederlof, M.A., Hartsock, R.J., Taylor, D.L., and

Waggoner, A.S. (1991) Imaging cytometry by multiparameter fluorescence. *Cytometry*, **12**, 579–596.

41. Jaggi, B., Poos, S.S., Macauly, C., and Palcic, B. (1988) Imaging system for morphometric assessment of absorption of fluorescence in situ stained cells. *Cytometry*, **9**, 566–572.
42. Netten, H., Young, I.T., van Vliet, L.J., Tanke, H.J., Vroljik, H., and Sloos, W.C.R. (1997) FISH and chips: automation of fluorescent dot counting in interphase nuclei. *Cytometry*, **28**, 1–10.
43. Tsavellas, G., Huang, A., Patel, H., McCullogh, T., Kim-Schulze, S., and Mersh, T.G. (2001) Flow cytometric detection of circulating colorectal cells: works in vitro but not in vivo. *Eur. J. Surg. Oncol.*, **27**, 778–779.
44. Kamentsky, L., Burger, G., Gershman, R., Kamentsky, L., and Luther, E. (1997) Slide-based laser scanning cytometry. *Acta Cytol.*, **41**, 123–143.
45. Clatch, R., Walloch, J., Zutter, M., and Kamentsky, L. (1996) Immunophenotypic analysis of hematologic malignancy by laser scanning cytometry. *Am. J. Clin. Pathol.*, **105**, 744–755.
46. Gerstner, A., Laffers, W., Bootz, F., and Tarnok, A. (2000) Immunophenotyping of peripheral blood leukocytes by laser scanning cytometry. *J. Immunol. Methods.*, **246**, 175–185.
47. Darzynkiewicz, Z., Bedner, E., Li, X., Gorzycza, W., and Melamed, M.R. (1999) Laser-scanning cytometry: a new instrumentation with many applications. *Exp. Cell. Res.*, **249**, 1–12.
48. Lindemann, F., Schlimok, G., Dirschedl, P., Witte, J., and Tiethmueller, G. (1992) Prognostic significance of micrometastatic tumor cells in bone marrow of colorectal cancer. *Lancet*, **347**, 649–653.
49. Varga, V.S., Bocsi, J., Sipos, F., Csendes, G., Tulassay, Z., and Molnár, B. (2004) Scanning fluorescent microscopy with digital slide and microscope is an alternative for quantitative fluorescent cell analysis. *Cytometry A.*; **60** (1), 53–62.
50. Masel, J., Arnaout, R.A., O'Brien, T.R., Goedert, J.J., and Lloyd, A.L. (2000) Fluctuations in HIV-1 viral load are correlated to CD4þ T-lymphocyte count during the natural course of infection. *J. Acquired Immune Defic. Syndr.*, **23**, 375–379.
51. Demarest, J.F., Jack, N., Cleghorn, F.R., Greenberg, M.L., Hoffman, T.L., Ottinger, J.S., Fantry, L., Edwards, J., O'Brien, T.R., Cao, K., Mahabir, B., Blattner, W.A., Bartholomew, C., and Weinhold, K.J. (2001) Immunologic and virologic analyses of an acutely HIV type 1-infected patient with extremely rapid disease progression. *AIDS Res. Hum. Retroviruses*, **17** (14), 1333–1344.
52. Alaeus, A., Lidman, K., Bjorkman, A., Giesecke, J., and Albert, J. (1999) Similar rate of disease progression among individuals infected with HIV-1 genetic subtypes A– D. *AIDS*, **13**, 901–907.
53. Urassa, W., Bakari, M., Sandstrom, E., Swai, A., Pallangyo, K., Mbena, E., Mhalu, F., and Biberfeld, G. (2004) Rate of decline of absolute number and percentage of CD4 T lymphocytes among HIV-1-infected adults in Dar es Salaam, Tanzania. *AIDS*, **18** (3), 433–438.
54. Kassu, A., Tsegaye, A., Wolday, D., Petros, B., Aklilu, M., Sanders, E.J., Fontanet, A.L., Van Baarle, D., Hamann, D., and De Wit, T.F. (2003) Role of incidental and/or cured intestinal parasitic infections on profile of CD4+ and CD8+ T cell subsets and activation status in HIV-1 infected and uninfected adult Ethiopians. *Clin. Exp. Immunol.*, **132** (1), 113–119.
55. Sotrel, A. and Dal Canto, M.C. (2000) HIV-1 and its causal relationship to immunosuppression and nervous system disease in AIDS: a review. *Hum. Pathol.*, **31**, 1274–1298.
56. Gerstner, A.O.H., Machlitt, J., Laffers, W., Tárnok, A., and Bootz, F. (2002a) Analysis of minimal sample volumes from head and neck cancer by laser scanning cytometry (LSC). *Onkologie*, **25**, 40.
57. Clatch, R.J. and Walloch, J.L. (1997) Multiparameter immunophenotypic analysis of fine needle aspiration biopsies and other hematologic specimens

by laser scanning cytometry. *Acta Cytol.*, **41**, 109.

58. Clatch, R.J. and Foreman, J.R. (1998) Five-color immunophenotyping plus DNA content analysis by laser scanning cytometry. *Cytometry B*, **34B**, 36.
59. Gerstner, A.O., Mittag, A., Laffers, W., Dähnert, I., Lenz, D., Bootz, F., Bocsi, J., and Tárnok, A. (2006) Comparison of immunophenotyping by slide-based cytometry and by flow cytometry. *J. Immunol. Methods*, **311** (1–2), 130–138.
60. Mittag, A. and Tárnok, A. (2009) Basics of standardization and calibration in cytometry--a review. *J. Biophotonics.*, **2** (8–9), 470–481.
61. Mittag, A., Lenz, D., and Tarnok, A. (2008) Quality control and standardization, in *Cellular Diagnostics; Basic Principles, Methods and Clinical Applications of Flow Cytometry* (eds U. Sack, A. Tarnok, and G. Rothe), Karger, Basel, Freiburg. pp. 159–177. ISBN: 978-3-8055-8555-2.
62. Stich, M., Thalhammer, S., Burgemeister, R., Friedemann, G., Ehnle, S., Lüthy, C., and Schütze, K. (2003) Live cell catapulting and recultivation. *Pathol. Res. Pract.*, **199** (6), 405–409.
63. Szaniszlo, P., Rose, W.A., Wang, N., Reece, L.M., Tsulaia, T.V., Hanania, E.G., Elferink, C.J., and Leary, J.F. (2006) Scanning cytometry with a LEAP: laser-enabled analysis and processing of live cells in situ. *Cytometry A*, **69** (7), 641–651.
64. Laffers, W., Mittag, A., Lenz, D., Tárnok, A., and Gerstner, A.O. (2006) Iterative restaining as a pivotal tool for n-color immunophenotyping by slide-based cytometry. *Cytometry A*, **69** (3), 127–130.
65. Schubert, W. (2007) A three-symbol code for organized proteomes based on cyclical imaging of protein locations. *Cytometry A*, **71** (6), 352–360.
66. Hennig, C., Adams, N., and Hansen, G. (2009) A versatile platform for comprehensive chip-based explorative cytometry. *Cytometry A*, **75** (4), 362–370.
67. Baja, S., Welsh, J.B., Leif, R.C., and Price, J.H. (2000) Ultra-rare-event detection performance of a custom made scanning cytometers on a model preparation of fetal nRBCs. *Cytometry*, **39**, 285–294.
68. Borgen, E., Naume, B., Nesland, J.M., Nowels, K.W., Pavlak, N., Ravkin, I., and Goldbard, S. (2001) Use of automated microscopy for the detection of disseminated tumor cells in bone marrow samples. *Cytometry*, **46**, 215–221.
69. Molnar, B., Ladanyi, A., Tanko, L., Sreter, L., and Tulassay, Z. (2001) Circulating tumor cell clusters in the peripheral blood of colorectal cancer patients. *Clin. Cancer Res.*, **7**, 4080–4085.
70. Ladanyi, A., Soong, R., Tabiti, K., Molnar, B., and Tulassay, Z. (2001) Quantitative reverse transcription-PCR comparison of tumor cell enrichment methods. *Clin. Chem.*, **47**, 1860–1863.
71. Mehes, G., Witt, A., Kubista, E., and Ambros, P.F. (2000) Classification of isolated tumor cells and micrometastases. *Cancer*, **89**, 709–711.
72. Bocsi, J., Varga, V.S., Molnár, B., Sipos, F., Tulassay, Z., and Tárnok, A. (2004) Scanning fluorescent microscopy analysis is applicable for absolute and relative cell frequency determinations. *Cytometry A*, **61** (1), 1–8.
73. Musco, M.L., Cui, S., Small, D., Nodelmann, M., Sugarmann, B., and Grace, M. (1998) Comparison of flow cytometry and laser scanning cytometry for the intracellular evaluation of adenoviral infectivity and p53 protein expression in gene therapy. *Cytometry*, **33**, 290–296.
74. Maciorowski, Z., Klijanienko, J., Padoy, E., Mossrei, V., Fourquet, A., Chevillard, S., El-Nagger, A.K., and Vielh, P. (2001) Comparative image and flow cytometric TUNEL analysis of fine needle samples of breast carcinoma. *Cytometry (Commun. Clin. Cytometry)*, **46**, 150–156.
75. Teodori, L., Grabarek, J., Smolewski, P., Ghibelli, L., Bergamaschi, A., De Nicola, M., and Darzynkiewicz, Z.

(2002) Exposure of cells to static magnetic field accelerates loss of integrity of plasma membrane during apoptosis. *Cytometry*, **49** (3), 113–118.

76. Sasaki, K., Kurose, A., Miura, Y., Sato, T., and Ikeda, E. (1996) DNA ploidy analysis by laser scanning cytometry (LSC) in colorectal cancers and comparison with flow cytometry. *Cytometry*, **23** (2), 106–109.

77. Gong, J., Traganos, F., and Darzynkiewicz, Z. (1994) A selective procedure for DNA extraction from apoptotic cells applicable for gel electrophoresis and flow cytometry. *Anal. Biochem.*, **218** (2), 314–319.

78. Tyihak, E., Bocsi, J., Timar, F., Racz, G., and Szende, B. (2001) Formaldehyde promotes and inhibits the proliferation of cultured tumour and endothelial cells. *Cell Prolif.*, **34** (3), 135–141.

79. Kamendulis, L.M. and Corcoran, G.B. (1994) DNA as a critical target in toxic cell death: enhancement of dimethylnitrosamine cytotoxicity by DNA repair inhibitors. *J. Pharmacol. Exp. Ther.*, **271** (3), 1695–1698.

80. Constantinou, A., Mehta, R., Runyan, C., Rao, K., Vaughan, A., and Moon, R. (1995) Flavonoids as DNA topoisomerase antagonists and poisons: structure-activity relationships. *J. Nat. Prod.*, **58** (2), 217–225.

81. Li, Y., Wang, E., Patten, C.J., Chen, L., and Yang, C.S. (1994) Effects of flavonoids on cytochrome P450-dependent acetaminophen metabolism in rats and human liver microsomes. *Drug Metab. Dispos.*, **22** (4), 566–571.

82. Tsai, S.H., Liang, Y.C., Lin-Shiau, S.Y., and Lin, J.K. (199915) Suppression of TNFalpha-mediated NFkappaB activity by myricetin and other flavonoids through downregulating the activity of IKK in ECV304 cells. *J. Cell Biochem.*, **74** (4), 606–615.

83. Morel, I., Abalea, V., Sergent, O., Cillard, P., and Cillard, J. (1998) Involvement of phenoxyl radical intermediates in lipid antioxidant action of myricetin in iron-treated rat hepatocyte culture. *Biochem. Pharmacol.*, **55** (9), 1399–1404.

84. Bocsi, J., Trezl, L., Mittag, A., Sack, U., Jensen, I., Luther, E., Varga, V.S., Molnár, B., Lenz, D., and Tarnok, A. (2004) in *Purdue Cytometry*, vol. 8, ISAC XXI edn (ed. J.P. Robinson), Purdue University, West Lafayette, pp. 30–40. ISBN: 1-890473-05-7.

10
Microfluidic Flow Cytometry: Advancements toward Compact, Integrated Systems

Shawn O. Meade, Jessica Godin, Chun-Hao Chen, Sung Hwan Cho, Frank S. Tsai, Wen Qiao, and Yu-Hwa Lo

10.1
Introduction

As a workhorse instrument for clinical diagnostics and biomedical research, flow cytometers have been in use and evolving for over half a century. Clinical and research applications include blood cell counting, cell-based fluorescent immunoassays, and DNA stains, among many others. In essence, flow cytometers are designed to measure both scattered and fluorescent light given off by cells or beads as they cross the path of a laser, while flowing through a capillary tube, one by one. Over the years, flow cytometers have been optimized for fast and accurate processing of cell samples – an expense paid for by increasing size, complexity, and cost. To counter this issue, researchers have begun to look to the microelectronics industry for solutions to reduce full-scale systems to compact and integrated ones. A small, inexpensive flow cytometer would allow for more clinical applications as well as better utilization in resource-poor or remote locations such as third world countries where fast and accurate diagnostics are acutely needed. In this chapter, we provide a brief overview of the main components of full-scale flow cytometers, including systems for fluidic control, optical detection and, optionally, cell sorting, each of which is currently being developed into on-chip microfluidic platforms. Additionally, we provide a brief snapshot of microfluidic flow cytometry. Then, in detail, we discuss recent progress in the development of compact and integrated flow cytometers, with an emphasis on on-chip flow confinement, optical detection, and sorting systems.

10.1.1
Main Components of a Full-Scale Flow Cytometer

10.1.1.1 Fluidic Control System

Fluidic control systems of flow cytometers are designed to take up a randomly distributed suspension of cells and focus them into a single-file stream, with uniform velocity from one cell to the next. The stream of cells then flow across the path of a laser, allowing for the collection of scattered and fluorescent light from

Advanced Optical Flow Cytometry: Methods and Disease Diagnoses, First Edition. Edited by Valery V. Tuchin.

each cell. Uniform cell location, and thus speed, is important not only to reduce the chances of multiple cells being detected simultaneously but also to reduce variability in the amount of light measured from each cell. Since the laser beam and the focal region of the detector do not occupy the entire channel, cells that are not centered in the channel may bypass most of the beam, causing less scattered or fluorescence light to be detected. Speed variability also causes variation in the signal because slower cells spend more time in the beam and therefore generate a higher integrated signal (likewise faster particles can yield a lower signal). Speed variability results when cells are not centered in the channel because of the parabolic velocity profile of the fluid flow in a capillary tube. When the stream of cells is centered, this effect is minimized. In commercial flow cytometers, focusing the randomly distributed suspension of cells into a stream with uniform speed is accomplished by coaxial hydrodynamic focusing, and in general is known as *flow confinement* [1]. In this process, the sample of randomly distributed cells is injected into the sample port and pumped through a capillary tube, known as the *core*, which is surrounded by a larger capillary tube, known as the *sheath*, through which buffer is pumped at high pressure. At a point in the fluidics line, before detection, the core wall enclosing the sample terminates, allowing the high pressure sheath fluid to squeeze the sample into the center of the channel. When optimized, this results in a stream of cells, centered in the channel, flowing with uniform speed.

10.1.1.2 **Optical Detection System**

Optical detection systems consist of one, or even multiple lasers of different wavelengths as well as multiple detectors, such as high-sensitivity photomultiplier tubes (PMTs) and very complex and costly optics requiring precise alignment. As mentioned, both scattered and fluorescent light are detected. Most often, scattered light is collected at two angles: forward scatter (FSC), measured between $\sim$0.5 and 5° with respect to the optical axis, and side scatter (SSC), which is collected at 90° with respect to the optical axis [1]. FSC is used as an indicator of cell size and SSC is used to indicate granularity of the cell [1]. Using one or more lasers, in conjunction with dichromic, band-pass, and cut-off filters, as well as multiple detectors, several channels of fluorescent light can be collected simultaneously from the cells, allowing for multicolor assays. The wide range of colors gives the user the ability to implement more selective assays. This, combined with the very high sensitivities achieved by these systems, has contributed to the optical detection system being the most bulky and costly component of a flow cytometer.

10.1.1.3 **Sorting Modules**

Sorting modules lie downstream from where the cells are detected. In most commercial systems, the sorting process is accomplished by droplet sorting [1]. In this process, following detection and analysis, the stream is broken up into droplets, at a frequency on the order of the cell spacing in the flow. After exiting the sample tube, the droplets fall between electromagnetic deflector plates, at which time a voltage is applied across the plates causing the target cell to be collected into a reservoir. All other droplets fall into waste. Sorting speeds of these systems range between 10 000 and 100 000 cells/s.

The main benefits of today's full-scale, commercial flow cytometers are multiparameter detection, high sensitivity, high throughput, and the ability to accurately isolate cells of interest. On the downside, full-scale flow cytometers are expensive, bulky, and complex; they require a skilled operator to ensure proper hardware operation, to minimize cross contamination between samples, and to ensure containment of biohazardous materials.

10.1.2
Microfluidic Flow Cytometry

In order to reduce cost, size, and complexity while increasing operational simplicity of current commercially available flow cytometers, lab-on-a-chip concepts have begun to be developed, resulting in more compact and integrated flow cytometer devices. Of course, the overall goal is to solve these issues while still delivering the user requirements of throughput, sensitivity, and the retention of cell viability. In general, microfluidic flow cytometer devices contain microfluidic channels through which the sample flows. In most cases, fluidic control is achieved by off-chip syringe pumps or peristaltic pumps. In partially integrated microscale devices, lasers are focused from off-chip into the channel. Detectors, such as PMTs and photodiodes, are commonly coupled to the device via a microscope. Later in the chapter, we discuss the progress made in integrating the laser and detector on the chip. Following the practice in the microelectronics industry, the first devices were either etched into silicon or glass, and a glass lid was bonded on top to seal the channels and provide an optically transparent window [2, 3]. However, with the development and widespread adoption of soft lithography [4], polymer replica molding techniques have begun to replace the traditional microelectronic techniques and materials that were originally used for making microfluidic flow cytometer devices. The material of choice for polymer replica molding is polydimethylsiloxane (PDMS), which is able to replicate molds with fine detail. PDMS replica molding allows for rapid, inexpensive, and simpler prototyping cycles. Additionally, after the molding process, PDMS can be bonded to glass, polymers, and other types of materials through processes such as UV/ozone treatment [5], corona activation [6], and partial curing [7]. Using these techniques and materials, researchers have not only built microfluidic channels but have also begun to develop on-chip systems for flow confinement, optical detection, and sorting. However, one concern with PDMS as a material for commercial devices is its permeability to small molecules. Later in this chapter, we discuss techniques to coat the PDMS microfluidic channel with a thin Teflon layer, preventing molecular diffusion into PDMS and producing a microfluidic waveguide structure to guide laser light in the microfluidic channel.

10.2
On-Chip Flow Confinement

As we discussed in the introduction, flow confinement of cells is achieved in full-scale flow cytometers through the use of coaxial sheath flow, driven by hydrodynamic force. Unfortunately, the capillary structures used in that method are very difficult to fabricate in on-chip devices. Therefore, researchers in microfluidic flow cytometry have developed flow confinement schemes that fit within the constraints of planar microfabricated devices. The first set of devices addressed the problem of flow confinement in two dimensions (some researchers say one dimension, citing that the particles are only being focused in one direction). In more recent work, researchers have begun to develop on-chip systems capable of flow confinement in three dimensions (or two dimensions if using the convention mentioned above). In this section, today's most promising and commonly used methods of flow confinement are reviewed. Methods involving both two-dimensional (2D) and three-dimensional (3D) flow confinement using hydrodynamics, electrokinetics, and dielectrophoresis (DEP) are discussed.

10.2.1 A General Discussion of Flow Confinement Forces

Hydrodynamic force used in the coaxial sheath flow method of full-scale flow cytometers can also be used in flow focusing schemes for microfluidic devices. Since hydrodynamic flow focusing was discussed in detail in the introduction, here we will only say that hydrodynamic force merely refers to manipulating fluid flow with pressure gradients. On the other hand, methods such as electrokinetic [8] flow, and DEP [9] use electric field forces to manipulate fluids or the particles (cells or beads) themselves, respectively. Within the context of microfluidics, electrokinetic flow refers to the induced movement of a conductive solution within a microfluidic channel when acted upon by a uniform electric field. One example of an electrokinetic phenomenon is electroosmosis, which describes the net movement of an electrolyte fluid within a negatively charged channel that occurs when a potential is applied across both ends [10]. This causes solvated cations in the diffusible area of an electrical double layer located at the interface of the negatively charged channel walls to be attracted toward the cathode. The solvated ions drag the solvent, in this case water, and everything else, including particles such as cells. One benefit of electroosmotic flow over hydrodynamic flow is that it exhibits a uniform velocity profile across the channel. In contrast, hydrodynamic flow exhibits a parabolic velocity profile, which can contribute to variability in particle speed. DEP refers to the movement of a dielectric particle, such as a latex bead or cell, through a medium when acted upon by a nonuniform electric field [9]. In DEP, both the particles and fluid become polarized; it is the relative difference in polarization between the particles and the fluid that results in movement. DEP can be characterized as either positive or negative. In positive dielectrophoresis (pDEP), the polarizability of the particles is larger than the surrounding fluid, resulting in the particles being attracted toward the location of high electric field strength. In negative dielectrophoresis (nDEP), the opposite is true. nDEP is more widely used in microfluidic flow cytometry than pDEP because there is less chance for damage when the cell is moving away from the region of high field strength [11].

10.2.2
Two-Dimensional Flow Confinement

Jacobson *et al.'s* was one of the first groups to demonstrate on-chip 2D flow confinement [12]. At the heart of the device, which is designed to utilize electroosmotic flow, is the intersection of the sample channel by a cross channel, as shown in Figure 10.1. The left- and right-hand sides of the cross channel are labeled focus 1 and focus 2, respectively. In this device, the sample solution is electroosmotically driven down the main channel, while the sheath liquid is also electroosmotically injected into the main channel from the focus channels, resulting in the sample being squeezed to the center of the main channel and thus focused in two dimensions (*xy*, or the device plane). Electrodes connected to individual power supplies are placed in reservoirs of the sample and focusing channels, while the waste reservoir is grounded. This allows the potentials at the reservoirs of the sample, focus 1, and focus 2 channels to be controlled individually. By setting the potentials of focus 1 and 2 to be higher than that of the sample, the sample flow is compressed and confined to the center of the channel by the resulting pressure difference. As seen in Figure 10.1, three different potential differences were tested, resulting in the sample stream being focused to three different widths. In this early demonstration, only fluids carrying a fluorescent dye were used. With a similar device, Schrum *et al.* demonstrated electrokinetic focusing of a sample stream containing latex particles (<2 μm in diameter) into an 8 μm stream within

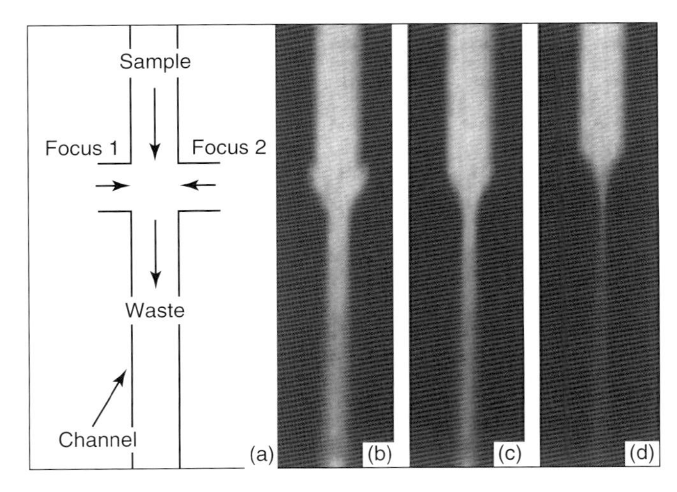

Figure 10.1 (a) Schematic of device designed for electroosmotic flow focusing. The intersection of sample channel and focusing channels is shown. Rodamine 6G dye solution is flown through the sample channel while the potential at the sample channel is decreased from 3.9, 2.9, and 1.9 kV as seen in images (b–d), respectively. The potentials are held constant for the electrodes of focus 1, focus 2, and the waste. The arrows in the channels represent the direction of movement resulting from the electroosmotic flow. Reprinted with permission from Ref. [12]. Copyright 1997 American Chemical Society.

a 50 μm channel with detection throughput of 34 particles/s [13]. Electrokinetic flow confinement methods are accurate and have extremely fast actuation times. However, they require high voltages, on the order of hundreds of volts, which may cause damage to the cells [14, 15].

Using the same cross structure of intersecting channels, Knight *et al.* demonstrated 2D hydrodynamic focusing of a sample stream down to 50 nm inside a 10 μm channel [16]. They developed a mathematical model to relate flow resistance to the inlet, side, and outlet channels. Using both simulation and experiment, Lee *et al.* studied the effects of relative flow rates and channel geometries on the width of the focused stream [17, 18]. Syringe pumps are not the only means to generate hydrodynamic flow confinement. Huh *et al.* and others have shown that vacuum applied to the outlet of the device could also create a sheath fluid flow [19, 20]. They also demonstrated that air could be used as the sheath fluid, which, if controlled, would eliminate the need for sheath fluid reservoir tanks, thus making flow cytometer devices more portable. Unfortunately, the method resulted in an unstable flow; however, one might argue that if this effect were controlled properly, it might be used to duplicate the droplet technique of full-scale flow cytometers, discussed earlier.

10.2.3 Three-Dimensional Flow Confinement

Morgan *et al.* constructed a device using only nDEP to achieve 3D particle confinement [21]. In their device, a suspension of unfocused latex particles is directed down a wide channel, with electrodes both above and below, as shown in Figure 10.2. Both the top and bottom electrodes are composed of two strips that converge toward each other. As the two strips of each electrode begin to approach each other, the particles are forced toward the center of the channel (in both the vertical and horizontal directions) due to the nDEP force. Using this technique, the authors were able to focus latex particles as small as 40 nm in diameter to the center of the channel. Unlike both electroosmotic and hydrodynamic flow focusing, DEP particle confinement moves the particles through the fluid instead of acting on both the particles and the fluid at the same time. DEP also requires much lower voltages than electroosmosis.

Lin *et al.* designed a device capable of 3D flow confinement by utilizing hydrodynamic flow focusing in the device plane and nDEP particle focusing in the vertical plane (normal to device plane). For confinement in the device plane, hydrodynamic flow focusing via a standard cross structure between the main channel and focusing channels was employed. For vertical confinement, the device was fabricated with planar electrodes above and below the channel, allowing for nDEP particle confinement. With this device, they were able to focus and detect both polystyrene beads and human red blood cells. Additionally, they were able to turn the DEP module on and off and observe an increase in signal variability when vertical focusing was turned off. The downside of DEP is that it places constraints on the polarizabilities of the particles (cells or beads) and fluid that can be used.

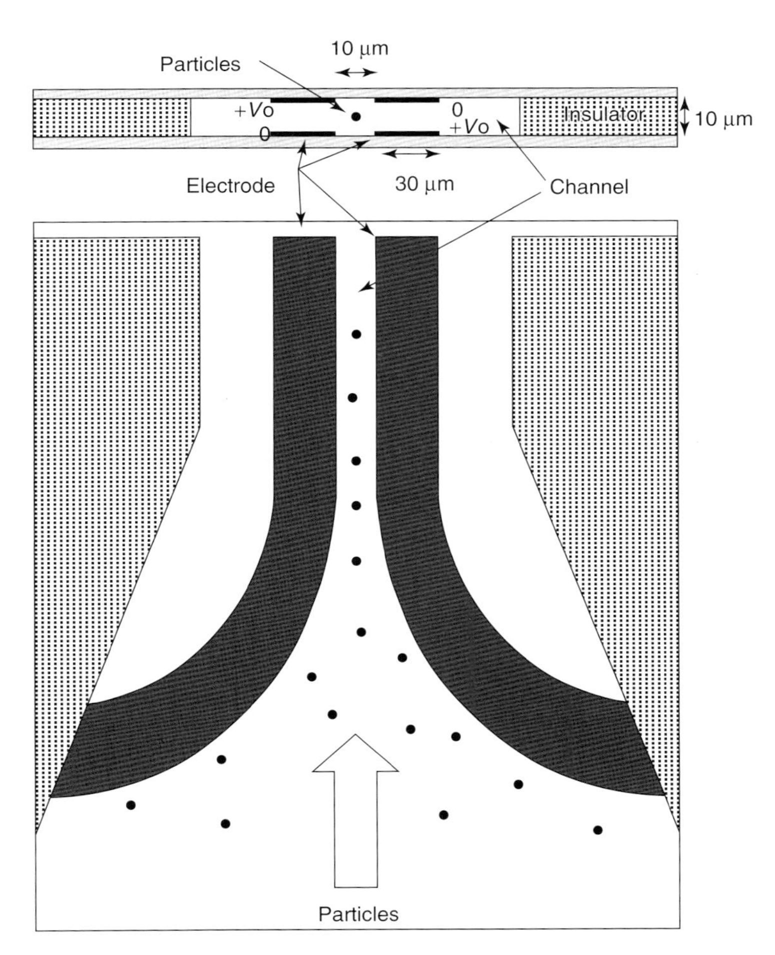

Figure 10.2 Schematic diagram of device built by Morgan *et al.*, utilizing negative dielectrophoresis to achieve 3D particle confinement. The side view, seen above, shows the electrodes as black lines, placed on the top and bottom of the channel and the focused particle stream as a black dot. The top-down view shows the curved electrodes used to focus the random distribution of particles into a single-file stream. Reprinted with permission from Ref. [21]. Copyright 2003 Institution of Engineering and Technology (IET).

Several groups have designed devices able to achieve 3D flow confinement using only hydrodynamic focusing. Klank *et al.* simulated and fabricated a chimney-like device in silicon via reactive ion etching [22]. They were able to achieve an effect similar to coaxial sheathing by injecting the sample through the chimney structure (placed perpendicular to the device plane) and into the main channel. Simonnet and Groisman were able to construct a single-cast soft-lithography device capable of 3D flow confinement [23]. In addition to the standard cross structure for 2D hydrodynamic flow focusing, the device also contained a set of shallow cross

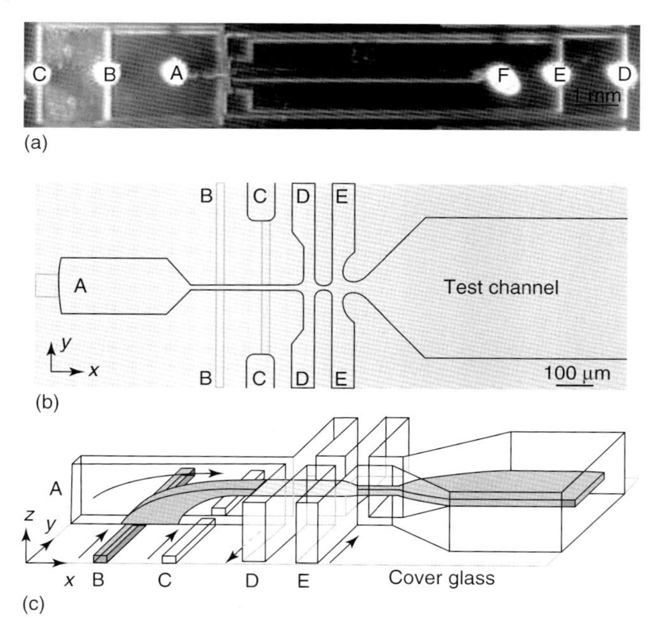

Figure 10.3 Device developed by Simmonet *et al.* capable of 3D hydrodynamic flow focusing. (a) Top-down view of the entire device. (b) Close up of the active area. (c) Schematic illustrating the device working principle. The sample, whose flow pattern is depicted by the shaded region, is injected into the main channel from port B. Reprinted with permission from Ref. [23]. Copyright 2005, American Institute of Physics.

channels for sample injection and vertical focusing. Figure 10.3 shows the sheath fluid flowing through the main channel A from the left of the schematic, and sample being injected into the main channel by the shallow channel B. This results in the sample occupying the lower half of the channel until the intersection with another shallow cross channel C where the sample becomes focused in the vertical direction. Finally, the main channel is intersected by a cross channel D, which is of the same height as the main channel andprovides for 2D focusing, resulting in the sample being focused completely in three dimensions. In this work, the velocity distribution of polymer beads was measured to have a standard deviation of 0.25%. The strengths of this device architecture are that it requires no complex electrical components, as required by the electrokinetic and DEP methods and that it can be cast as a single layer using soft lithography. However, the design requires additional syringe pumps for it to work, which tends to make the system bulky.

Mao *et al.* were able to demonstrate 3D flow confinement using a single-cast single-layer device made by soft lithography [24, 25]. As seen in Figure 10.4, the sample is injected into the main channel B from channel A, and the merging of both streams results in all the particles being pressed to one side of the channel, distributed from top to bottom. The main channel then takes a 90° turn, which, when geometric and flow velocity parameters are optimized, results in the Dean

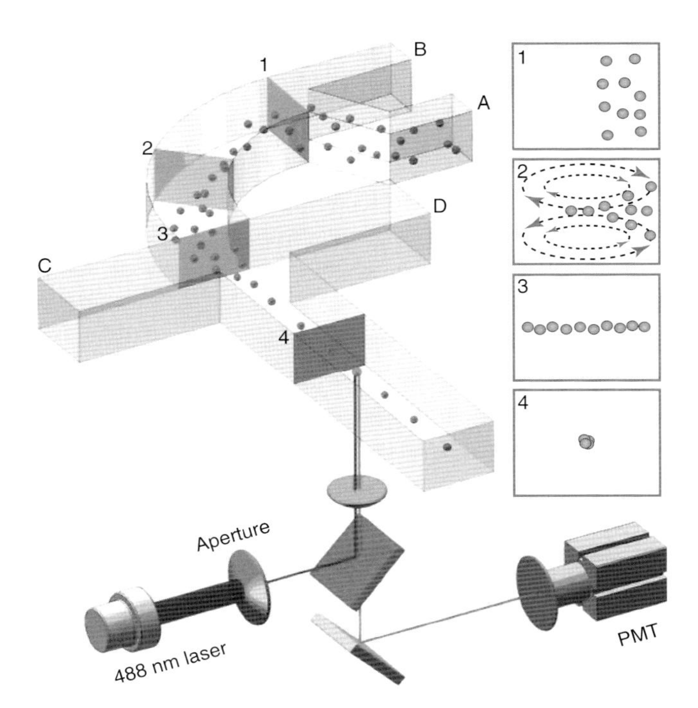

Figure 10.4 Schematic of device built by Mao *et al.*, able to achieve 3D flow confinement through standard hydrodynamic flow focusing in the horizontal plane and the Dean Flow effect for vertical focusing. Cross sections of the main channel are shown to the right, labeled with their corresponding position in the channel [25]. Reproduced by permission of The Royal Society of Chemistry.

effect, shown in insets 1–3 of Figure 10.4. Essentially, the centrifugal force causes vortices to form that accelerate the particles to the center plane of the channel, thus providing vertical focusing. Standard hydrodynamic sheath flow then focuses the particles laterally, as seen in inset 4 of Figure 10.4. The authors were able to focus and detect microparticles, similar in size and density to human lymphocytes, at high flow velocities ($\sim$3.6 m s^{-1}) with a detection throughput of 1700 particles/s. A mixture of two particle types, both 8.32 μm diameter polystyrene beads (one type exhibiting 30% of the fluorescence intensity of the other) was measured to have coefficients of variation (% CVs) of 15.2 and 9.3%, respectively. The % CV is the standard deviation divided by the mean, multiplied by 100. Compared to other microfluidic flow cytometers, this is a good result. The ability to mold the device in one step makes it attractive for mass production, and, unlike the device of Simonnet and Groisman, the master only needs to be one layer and needs only one mask (no precision alignment steps), making the prototyping process much easier. However, the Dean effect needed for vertical focusing requires a very high flow rate (over 3 m s^{-1}) and, to date, the effects of acceleration through the Dean vortexes on cell viability have not been determined.

Finally, the Ligler group has demonstrated 3D flow confinement through a combination of traditional 2D hydrodynamic sheath flow in conjunction with the use of groves placed in the channel, which are designed to wrap the sheath flow around the core sample stream as seen in Figure 10.5a [26, 27]. The design employs a cross structure between the sample channel and the two sheath flow inlets for the generation of hydrodynamic 2D flow confinement in the *xy* device plane. As discussed previously, this allows for adjustment of the width of the core sample stream by tuning the velocity ratio between the sample inlet and the sheath inlets. Flow confinement in the *z* direction is achieved by the placement of chevron-shaped grooves set into the top and bottom of the channel. The chevrons are designed to direct streamlines of sheath flow on each side of the sample stream to move underneath (and above) the sample stream – along each diagonal of the chevron until the apex is reached, where the two diverted streamlines meet and cause a net outward displacement. The displacement force from the chevrons placed above and below the sample channel causes the sample stream to be focused in the *z* direction. Serial placement of multiple chevron structures along the channel can be used to control the height of the core sample stream, or equivalently, the degree

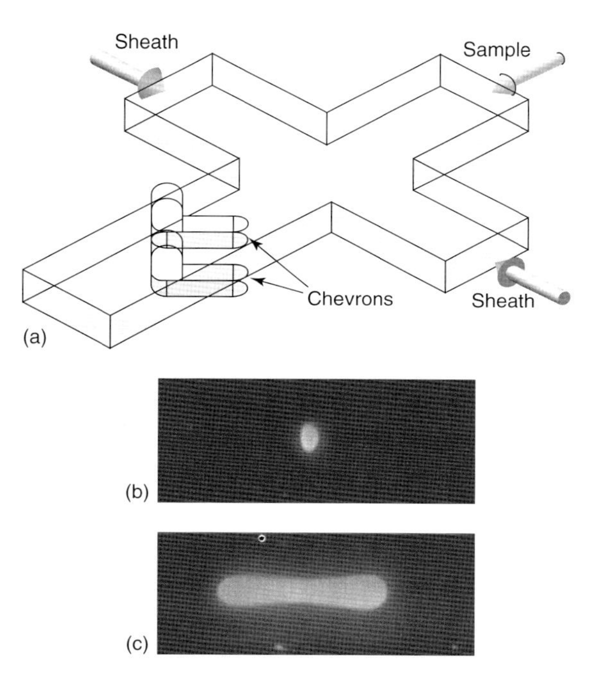

Figure 10.5 (a) Device utilizing chevron design for vertical flow focusing. Three-dimensional flow focusing is achieved by combining the chevron-based flow focusing method with traditional 2D hydrodynamic sheath flow in the *xy* plane. (b) Fluorescence image showing a head-on cross section of the stream where the sheath inlet velocities are much faster than sample stream velocity. (c) Sample and sheath flow velocities are all equal [26]. Reproduced by permission of The Royal Society of Chemistry.

of focusing in the z direction. Images (b) and (c) of Figure 10.5 show the effect of varying the flow ratio between the sample stream and the sheath streams. The upper image (b) was acquired while running the device at a sheath : sample : sheath flow ratio of 29.5 : 1 : 29.5, and the bottom image was acquired while running the device at the ratio of 1 : 1 : 1. The effect of sheath flow velocity on the core stream width can be clearly seen, while the height does not change (for a single device, of course, the number of chevrons is constant). This means that once a device is built, the degree of z focusing cannot be adjusted. In addition to using laser ablation, the group also made devices using soft lithography. The device is relatively simple to make and, as the authors point out, any researcher with the ability to fabricate two-layer microfluidic devices should be able to fabricate this device [26].

In summary, promising results have been demonstrated in on-chip, 3D flow confinement, allowing for lower signal variability during detection. Methods based on electrokinetic, DEP, and hydrodynamic focusing have been implemented, all showing merits and drawbacks. Electrokinetic focusing allows for fast actuation times; however, it requires high voltages which may harm cells. DEP particle confinement can be achieved with lower voltages than electroosmotic methods; however, to achieve the DEP effect, constraints on both the fluid and particle's polarizability must be taken into account. So far, the simplest devices to achieve 3D flow confinement utilize hydrodynamic force.

10.3 Optical Detection System

The optical portion of the flow cytometer is where the information about the sample population is learned and sorting decisions are made. The optical system must provide a sufficient amount of information (multiple parameters for discrimination) of high quality (low noise, highly repeatable, sensitive to low light signals, and small differences between signals) and at an acceptable throughput (i.e., >1000 cells/s, generally) to make the device useful to researchers and clinicians. For a microfluidic flow cytometer, the basic needs are no different. The ways in which we achieve these goals, however, may change dramatically as we shrink the device size and learn to exploit some of the true gains of miniaturization and integration. Miniaturization is poised to revolutionize the field of flow cytometry, especially from an optics standpoint.

10.3.1 The Many Benefits of Integrated Optics

We look into on-chip optics technologies because of the obvious advantages in cost, size, weight, and system maintenance that integration brings. By employing molding techniques, roll-to-roll processing techniques, or similar techniques, the entire optical system can be made quickly, inexpensively, and perfectly integrated with and aligned to the fluidic portions of the device. The shapes, sizes, and locations of each optical component can be designed independently and fabricated together, allowing for custom lens shapes and easy system assembly. The whole device could be made smaller, lighter, and more robust. Instead of its current role as a core-facility device, every clinic and research lab can have its own low-cost, high-performance microfluidic cytometer, expediting diagnosis and treatment, and revolutionizing medical research.

But there are also a host of less obvious reasons for integration that are equally, or perhaps even more, exciting. Integrated refractive optics can help address issues related to the signal-to-noise ratio (SNR) of the system, as well as collimation of the illumination beam and similar technical aspects related to reproducing the quality optical system typically employed in a flow cytometer. The reader is referred to other works for a more detailed discussion on the need to transition from the exclusive use of light-guiding optics to systems including refractive optics [28]. In addition, the compact nature of waveguided and/or miniaturized optics allows designers more flexibility in terms of design footprint: many features can be packed into a very small space, and at little to no added cost (except when they require additional lasers or detectors). The result is an ability to exploit repetition and/or oversampling methods in order to gain more information or improve signal quality through novel or unconventional methods. More complex architectures can be employed for optical systems than would likely ever be found in bulk systems. We can begin to play with algorithms not previously available to flow cytometry signals due to constraints on total space for optical detection lines or the traditional methods of collecting signals. Later in this section we provide examples to show how new architectures (e.g., on-chip sorting verification, individual cell speed measurement, space–time coding, and color–space–time (COST) coding) can enhance device performance and offer functionality not possible with conventional designs. Of course, we can also seamlessly integrate a variety of other functions, such as sample filtration and preparation or downstream cell sorting, alongside our optical components. In this section we review the technological advances enabling the development of sophisticated on-chip optical systems, touching on some of the new possibilities that such systems can provide.

10.3.2 Developing the Tools of the Trade

Waveguides and in-plane lenses are two essential optical elements serving as the analogs of optical fibers and lenses in bulk optics. Other useful components for on-chip optics include beam stops and apertures to suppress stray light, as well as various diffractive elements such as prisms and gratings, optical filters, and power dividers.

10.3.2.1 Light-Guiding Elements

Fiber optics and integrated waveguides (Figure 10.6) are excellent means of transporting light onto and off of the chip, allowing easier integration with the on-chip optical system by exploiting total internal reflection (TIR) to guide light

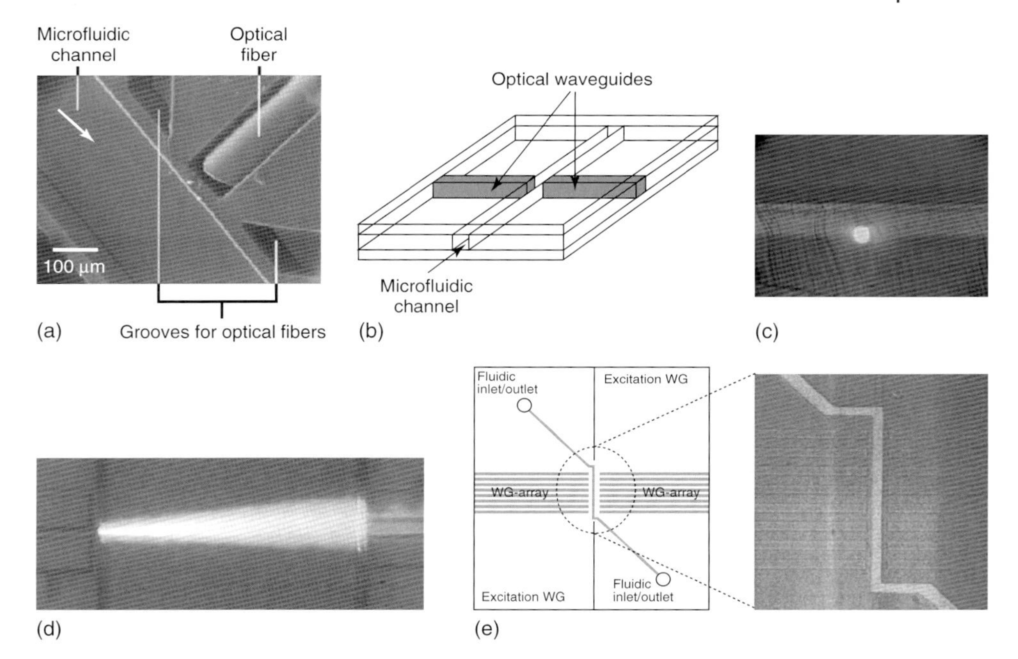

Figure 10.6 Light-guiding elements employed in microfluidic devices. (a) A demonstration of how fibers fit into fiber sleeves created in the device body for precise alignment. Reprinted from Ref. [30], with permission from Elsevier. (b) Schematic showing the easy integration of microfluidic channels with waveguides [34] © 2004 IEEE. (c) A demonstration of the light-guiding performance of a waveguide by looking at light exiting the end facet of a square waveguide [34] © 2004 IEEE. (d) A waveguide made by creating the cladding (rectangles above and below the core, from which light is exiting on the left) [35]. Reproduced by permission of The Royal Society of Chemistry. (e) Small waveguides allow for multiplexing without taking up a lot of space. Reprinted with permission from Ref. [43]. Copyright 2005 American Institute of Physics.

along the desired path in the chip (i.e., to illuminate the sample and to carry away detected light, either scattered or fluorescence, to the detectors). The resulting system can be of fairly low loss (fibers can have losses of $< 0.01\ \mathrm{dB\,cm^{-1}}$; the much shorter integrated waveguides in PDMS can have losses around $1-2\ \mathrm{dB\,cm^{-1}}$ [29] or lower) while maintaining a small footprint (no off-chip lenses should be needed). The fibers may be hand-placed and aligned on the device [5], but often the placement of light-guiding elements is facilitated during the chip fabrication by using a fiber sleeve [5, 30, 31] (Figure 10.6a). Such sleeves are fabricated in perfect alignment to the channel during mold fabrication by effectively molding a channel into which the fiber can be inserted. Some authors choose to fix their waveguides with epoxy after insertion to keep them in place [32]. Index-matching or waveguide-coupling has also been demonstrated [29, 33]. Waveguides can also be integrated into the device in a manner similar to the integration of fluidic channels: a waveguide "channel" is created at the desired position relative to the flow channel [34] (Figure 10.6b–e).

In the molded device, this channel (the waveguide core) is filled with a material of refractive index higher than that of the surrounding body, enabling it to guide light by TIR [31]. The same effect can be obtained, instead, by creating two cladding channels of a low refractive index relative to the device body [35, 36]. Liquid-core waveguides, another interesting means of creating a more tunable optical system, are discussed in a later section of this chapter [37–42] (Figure 10.9f). In either scenario (waveguides or fibers), the light-guiding elements are created in a fixed position, removing the need for alignments necessary when using off-chip optics and theoretically improving repeatability from device to device (due to this lack of critical manual alignment). In both cases, the device footprint has typically shrunk considerably.

10.3.2.2 **Two-Dimensional Refractive Elements**

Light-guiding elements alone will generally not be enough to transform the optical system of the full-scale cytometer into an integrated system on a chip without suffering a significant performance hit. In fact, Shapiro and his colleagues came to the analogous conclusion using guided optics for full-scale flow cytometers some years ago [44]. Again, more detail is given on this subject in other works [28]. The simplest solutions to integrating refractive elements involve lensing a pre-existing surface, using a lensed fiber, a lensed fiber sleeve [45, 46] (Figure 10.7b), a lensed waveguide, or a lensed channel wall [47] (Figure 10.7a). Each of these approaches affords us a focus point, helping to reduce stray light in a collection line or producing an illumination beam with a lower divergence angle (or even a converging beam) and a reduced possibility of cross-talk resulting from the simultaneous illumination of two cells.

At the next level, freestanding 2D lenses offer greater design flexibility (in terms of placement) and allow for the inclusion of a greater number of elements, often with a higher index contrast (i.e., more powerful lenses). Air-filled 2D lenses have been demonstrated by Seo and Lee [49, 50] (Figure 10.7d). In this work, a series of several on-chip lenses were used to focus the illumination beam for fluorescence measurements in a microfluidic channel. These were air-filled lens-shaped chambers created in the PDMS body of the device by the molding process, and thus they too were monolithically integrated with and aligned to the microfluidic channel, just like waveguides or fiber sleeves. Since the lens body has a lower refractive index ($n_{\text{air}} = 1$, $n_{\text{PDMS}} = 1.41$), the lens shapes are negative, that is, the reverse of typical lens designs for focusing. Similar ball lenses have been demonstrated by Llobera *et al.* in 2004 [48] (Figure 10.7c). In 2006, Godin *et al.* demonstrated fluid-filled ($n = 1.67$) 2D lenses [51] integrated with waveguides for standardized light coupling (Figure 10.7f). The use of a fluid enables a refractive index contrast high enough to give a somewhat fast lens but without the high surface reflection losses expected to be incurred by an air lens (~3% per surface with air vs <1% per surface with fluid). These are positive lenses, as the fluid has a higher refractive index than the PDMS. The lenses were encased in a slab waveguide, preventing vertical light loss due to the increase in optical path length needed to accommodate lenses (relative to devices based purely on light-guiding

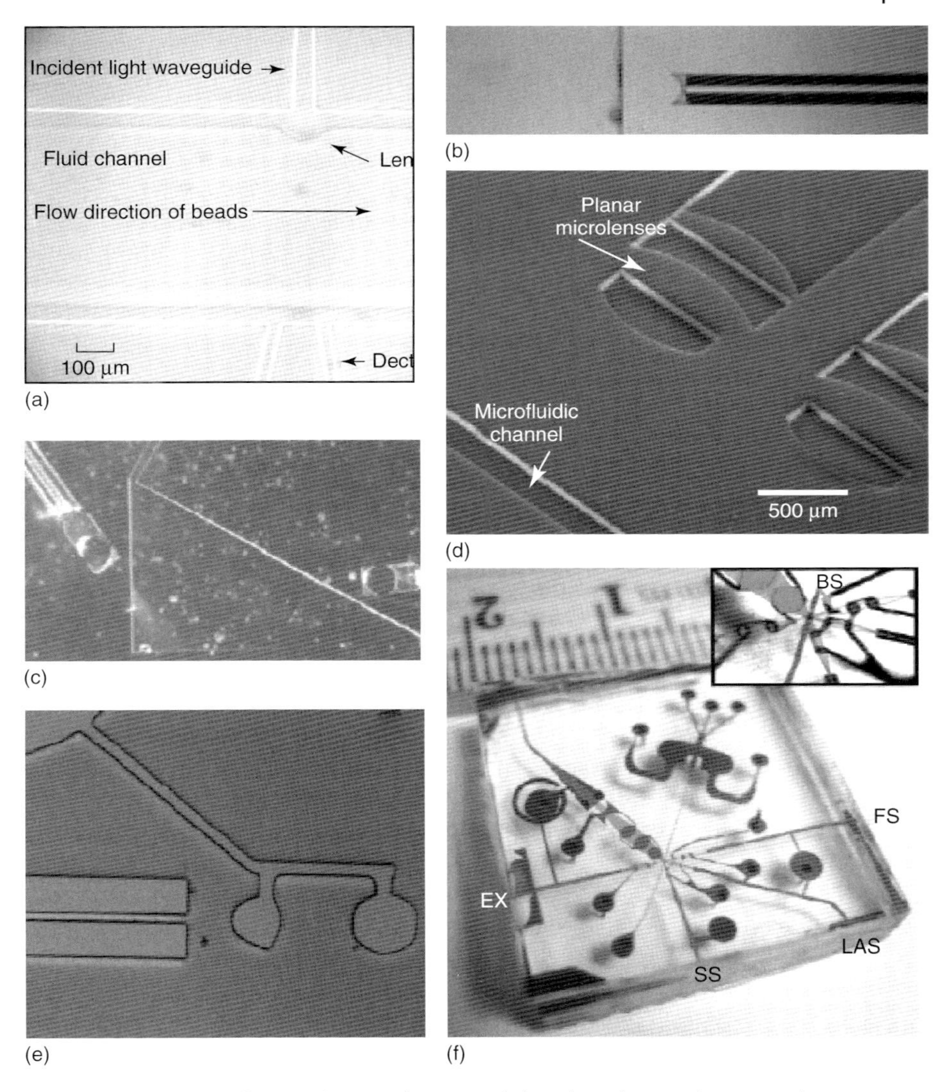

Figure 10.7 Various refractive elements that have been demonstrated in microfluidic devices. (a) A lensed waveguide can be used to collimate or focus light for interrogation in the channel [47]. Reproduced by permission of The Royal Society of Chemistry. (b) A lensed fiber sleeve performs similar functions [45]. Reproduced by permission of The Royal Society of Chemistry. (c) Two lensed air pockets for a ball lens to collimate light exciting a fiber [48]. Reproduced by permission of The Royal Society of Chemistry. (d) Stand-alone lensed air pockets create lenses in the PDMS body of a microfluidic device. Reprinted from [49], with permission from Elsevier. (e) Fluid-filled lens chambers with nonspherical surface curvatures demonstrate the highly customizable surface profiles available for microfabricated 2D lenses [36]. (f) An integrated device demonstrating monolithic integration of lenses, waveguides, and fluidic channels. Reprinted with permission from [28]. Copyright 2008 John Wiley & Sons, Inc.

elements). This slab waveguiding approach has also been demonstrated in air-clad polymer devices [47]. High-index polymers have also been investigated for creating a more stable, all-solid device [52]. Microfabricated 2D lenses, unlike their bulk counterparts, can be easily and inexpensively fabricated in custom shapes, such as aspheric or parabolic profiles (Figure 10.7e), affording the designer a degree of flexibility not typically available in bulk optical systems.

Several devices integrating microfluidic channels, light-guiding elements, and in-plane lenses have been demonstrated (Figure 10.8). Godin *et al.* demonstrated one such device for detecting light scatter [53]. In experiments with polystyrene beads, this device yielded CVs approaching those of a benchtop cytometer (10–15%) for the same sample, and considerably smaller than those from previously demonstrated microfluidic devices for detecting light scatter (25–30%) [30, 47, 54].

10.3.2.3 Improving Quality of On-Chip Optics

With all on-chip optics (especially refractive optics), two potentially important issues are those of (i) surface reflections and (ii) sidewall profile. Surface reflections arise simply because of a refractive index contrast (Fresnel reflections). We can approximate surface reflections through the use of the Fresnel reflection coefficient R from unpolarized light at near-normal incidence as

$$R \approx [(n_2 - n_1)/(n_2 + n_1)]^2$$

where n is the refractive index of the lens material. For a PDMS–air lens, we find reflections ~3%; for fluid-filled lenses ($n = 1.67$) in PDMS, we find surface reflections <1%. These losses will increase as the incidence angle increases (relative to normal). When multiple components are present in the light path, these reflections produce not only significant losses but also background noise, necessitating a careful tradeoff between the lens power and the resulting reflective losses. The second issue, sidewall profile (verticality and roughness), has to do with the fabrication method. A tilt to the sidewall can degrade the lens performance as light is directed away from the plane in which the optical axis lies. Vertical sidewalls may not be easily obtained in processes utilizing photoresist as the mold structure (such as SU8 processes), but are readily achievable via deep-reactive ion etching (DRIE) processes or via cryogenic etching [55]. Sidewall roughness is an important consideration for the same reason as surface reflections: the potential for both light loss and increased background levels. An "optical" surface should have sidewall roughness below $\lambda/10$; for flow cytometry at 488 nm, this means ~35 nm root mean-square (rms) roughness for devices made of PDMS. A few authors have characterized the roughness of their optical features and demonstrated promising means of reducing the roughness. Seo *et al.* reduced the roughness of their DRIE molds from 100 to 20 nm by a chemical etching process [49]. Godin *et al.* demonstrated promise for cryogenically etched molds to achieve similarly low sidewall roughness [36]. In general, when creating integrated optical systems, much more attention must be paid to the mold fabrication techniques than would typically be warranted in a microfluidic device without optics.

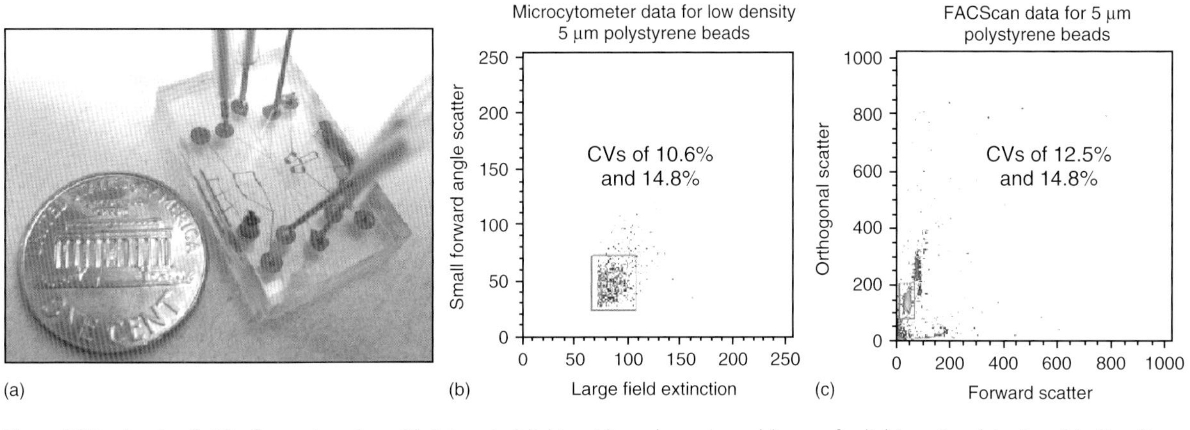

Figure 10.8 A microfluidic flow cytometer with integrated light guiding elements and lenses for light scatter detection (a). Results from the microfluidic cytometer (b) and the benchtop cytometer (c) show comparable CVs for the samples [53]. © 2007 IEEE.

10.3.2.4 Light-Stopping and Reflecting Elements

Beyond refractive and light-guiding elements, there are a few other functions we might like to have for an integrated optical system. The desired components include light-blocking elements (such as apertures, baffles, and beam stops), dispersive elements (such as prisms or gratings), and absorptive elements (for spectral filtering). Apertures have been demonstrated by a few authors. Tang *et al.* created ink-filled channels in PDMS to act as apertures [42]. The resulting aperture was used to restrict light entering the lens to achieve better focusing (more paraxial illumination). Godin *et al.* demonstrated solid apertures by filling aperture chambers with Sylgard 170 (Dow Corning), a black-gray colored PDMS [36] (Figure 10.9a–c). These technologies, or similar approaches, may be capable of creating baffles or absorptive beam blocks. Reflective elements can also be employed on chip. Llobera *et al.* demonstrated TIR-based air mirrors for redirecting light [56], as did Demming *et al.* [57] (Figure 10.9d). Liquid metal mirrors have also been demonstrated using gallium alloys that are liquid at room temperature [58].

10.3.2.5 Spectral Separation

Spectral separation is important for multicolor fluorescence detection. Prisms have been demonstrated for beam steering [59] (Figure 10.9e) and for applications involving refractive index measurements [60] (Figure 10.9d for an example); similar elements with sufficiently strong dispersive properties may find use for spectral separation. Prisms could be useful for spectral separation on chip, especially for applications such as spectral flow cytometers on a chip [61]. Integrated dye-doped PDMS filters have also been demonstrated by at least two groups. In this work, PDMS doped with lysochrome dyes (such as Sudan II or Sudan IV) are used to create long-pass filters with cut-on wavelengths in the 550–580 nm region [62, 63]. One can imagine using other dyes to create filters with different cut-on frequencies, or perhaps even passbands. Tunable fluid-based optical filters were demonstrated by Wolfe *et al.* [64]. Here, liquid dyes are introduced to a microfluidic channel, and input from a tungsten lamp is shown to be modulated by the "filters" created by Congo Red and Napthol Green dyes. The ability to integrate filters can further help to reduce the footprint of the microfluidic cytometer. Long-pass filters can be helpful to reduce background, while integrated passband filters can replace standard glass color filters, keeping the total device size small.

10.3.2.6 Tunable Liquid-Core Waveguides and Lenses

In an approach demonstrated by a few authors, the fluidic path has actually been merged with one of the optical paths (generally the illumination path) to create a truly optofluidic device. This approach can increase illumination efficiency, reduce stray light, and reduce the device footprint by guiding light down the microfluidic channel using the fluid as a waveguide core. Wolfe *et al.* demonstrated a liquid-core and liquid-cladding (L2) waveguide structure [41] (Figure 10.9f). They used a high refractive index solution ($CaCl_2$, $n = 1.445$) to represent the sample flow and to act as the waveguide core, and deionized water as both the sheath flow and optical cladding, due to its lower refractive index ($n = 1.335$). However, since sheath

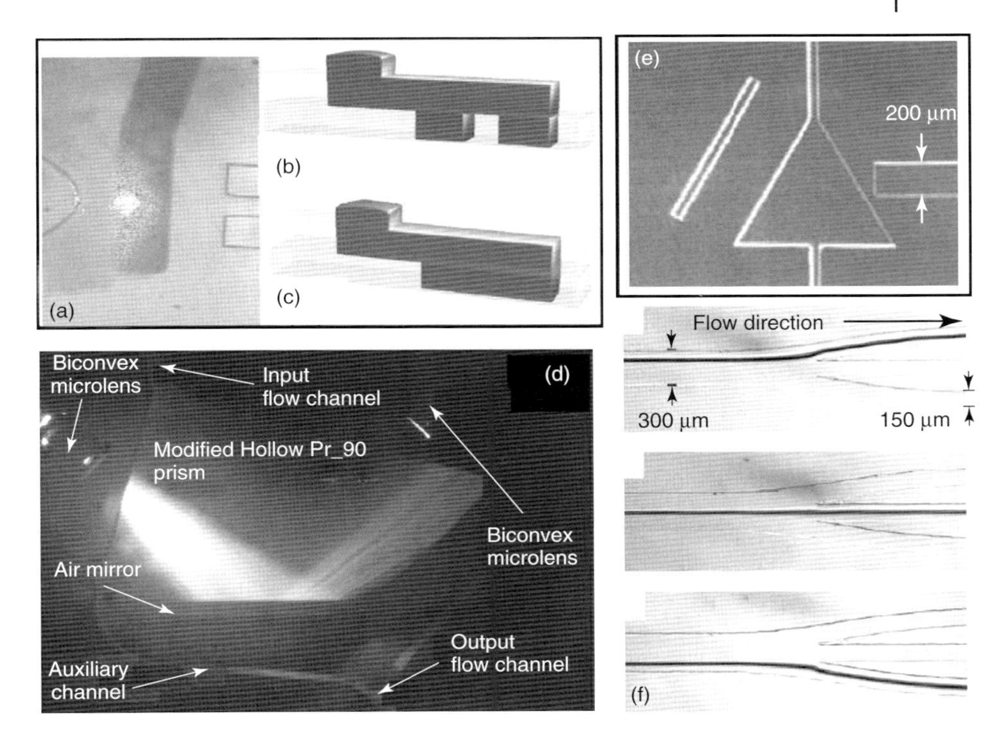

Figure 10.9 Additional optical elements. (a) A beam blocking bar made of black PDMS blocks light exiting the lens (left) from entering a waveguide (right). Side view of schematic for creating (b) apertures and (c) beam blocks show how such light-blocking elements are made [36]. (d) Air mirrors (air pockets) have been demonstrated for reflective light steering (based on TIR). Reprinted from Ref. [57], with permission from Elsevier. (e) Demonstration of beam steering as well as refractive index measurements with a prism. Reprinted from [59], with permission from Elsevier. (f) Tunable fluid-core waveguides have also been demonstrated, allowing changes in light paths. Reprinted with permission from Ref. [41]. Copyright 2009 National Academy of Sciences, U.S.A.

flow was introduced via the standard 2D hydrodynamic flow focusing technique discussed in Section 10.2, waveguiding only occurs in the device plane, allowing for light loss in the vertical direction. The same idea was later used for tunable filters: by changing the fill dye of the channel, one can change the filter passband [64]. Similar light-guiding systems have been demonstrated using air as both the cladding and sheathing layer [65]. However, these devices also suffer from vertical light loss and require an external vacuum pump to maintain air-sheathing flow, which adds to the bulk of the system. The concept has also been modified to create tunable cylindrical lenses [37, 42]. To create the lens, the microfluidic channel opens into an expanded lens chamber with curved walls. The core and cladding flow expand according to the contours of the chamber and their relative flow rates. The resulting curvature of the lenses can then be changed (within the constraints imposed by the chamber shape) by adjusting the relative sheath-to-sample flow rates. In this work, a variety of core and cladding flows were utilized, including a cladding flow that was index-matched to the PDMS sidewalls, effectively removing the effects of sidewall roughness [42].

Tunable components offer flexibility of design as well as the flexibility to accommodate environmental and experimental changes: for example, accommodating temperature drift, adjusting for different particle locations, or adapting to the differing needs of numerous experiments. Each of these qualities is very useful. However, there is often a tradeoff between the level of control and consistency of device tunability and the overall performance reliability in the case of flow cytometers. The speed and accuracy of tuning have important implications for device throughput and stability. Often, the tuning parameter adds unnecessary size, complexity, power draw, and so on, to our microfluidic device. In the case of flow cytometry, one needs to weigh carefully the arguments for and against tunability.

To make a liquid-core waveguide of consistent performance, the most favorable design uses low-index coating materials to line the inside of the channel and act as optical cladding [38, 39, 43, 66]. In these devices, Teflon AF ($n = 1.31$) is the material of choice, acting not only as an optical cladding layer but also as an agent to reduce nonspecific binding. Datta *et al.* demonstrated Teflon AF-coated liquid core waveguides fabricated from silicon substrates [39]. Recently, Cho *et al.* have reported a PDMS-based device incorporating Teflon AF-coated microchannels as liquid waveguides, made not by spin coating the PDMS microfabricated substrate before bonding (as was done by previous groups [66]) but by flowing a dilute solution of Teflon AF through the microchannels after bonding [38]. This method eliminated vertical losses and was shown to guide light even through branched channels [38]. Introducing the Teflon AF solution into the formed channel at a low concentration followed by evaporation also allows for greater control over coating thickness than the spin-coating method. Another benefit of Teflon-coated liquid core waveguide for devices made of PDMS material is the suppression of permeation and adsorption of small molecules due to the high porosity of PDMS.

10.3.3
Opportunities for Significant Improvements over Bulk Optical Systems

By now, researchers have demonstrated a large number of elements to provide a well-stocked toolbox of means for creating various integrated optical elements. The introduction of the optical system on a small, mass-producible platform may offer a number of advantages that designers can exploit to meet and, in some areas, exceed the performance of the benchtop system. In particular, new fabrication methods and novel device architectures offer many possibilities to the enterprising designer.

Microfabricated devices are often created via a molding or etching process using some form of mask, typically created by a computer-aided design program. Often, the fabrication process is developed with mass fabrication in mind. This differs significantly from the processes typically used to create optics, such as diamond turning (for customized optics). On-chip optics offer unique benefits of design-fixed alignment and the ability to possess a highly customizable profile without significant cost increase. Fixed alignment devices, as mentioned before, reduce or eliminate the need for realignment servicing. Furthermore, complex, multielement optical systems can be simultaneously created and perfectly aligned, a time-consuming endeavor in the world of bulk optics. The ability to choose nonstandard lens profiles could also be a major benefit, allowing the designer greater flexibility while keeping cost low. Parabolic, spheric, aspheric, and even highly aspheric lenses are all available to the designer of an integrated lens system [36]. In the same way that highly custom shaped lens can be created, very small, highly custom reflective/TIR optics can also be fabricated. The types of optics used to localize illumination or light collection, for example, can also undergo significant modification. The way we create optical systems on a microfabricated chip, then, can differ greatly from the way we create traditional bulk optical systems to perform the same tasks.

Microfabricated devices are also open to some novel device architecture changes for advancing microfluidic flow cytometry. In particular, the compact chip more readily lends itself to oversampling/repetitive sampling and multiplexing than does the full-scale system (largely due to size considerations). For example, Lien *et al.* demonstrated a multiplexed system of eight waveguides for time-based signal correlation [43]. Eight identically spaced waveguides detected signals from passing fluorescent polystyrene beads. The waveguides were multiplexed into a single waveguide to carry the signals to a detector. The time correlation was then used to de-multiplex and correlate the signals, resulting in both detection and significant signal amplification from a single detector. Chen *et al.* later demonstrated a similar, real-time version of the algorithm for possible use with real-time sorting devices [67]. Tung *et al.* used signals from multiple fiber-based detection lines to enable the use of a lock-in amplifier for detection of fluorescence signals using only a PIN photodiode [30]. Repetitive sampling has also been used for velocity determination as well [68]. In this work, multiple sampling points are employed and fast Fourier transformation (FFT) of the data allows determination of the flow velocity.

Repetitive detection also opens the door to some of the signal processing techniques more common in the realm of electronics, such as spatial and temporal signal coding, to reduce the total number of detectors needed in a system (and thus the cost, size, power draw, etc). For example, Chen *et al.* used a spatial filter to encode a three-lobed detection signal (111) and a downstream sorting-verification signal containing three lobes with a space between the first and second lobe (1011), enabling both to travel to the same detector but still be distinguished from each other [69]. In addition, this scheme enabled the use of a matched finite impulse response (FIR) filter to amplify the signals.

Integrated devices hold exciting possibilities for multicolor fluorescence detection as well. Among the challenges for on-chip fluorescence are light filtering and the need to maximize the number of parameters that can be detected. The basic full-scale cytometer can simultaneously detect at least 12 fluorescence bands with high numerical aperture light collection plus two light scatter angles. Microfluidic devices tend to incorporate one detection line for each parameter, bringing up the issue of available real estate. In addition, the total possible collection angle per line (and thus total photons collected) shrinks as more and more lines are added; as a result, such a device will likely never exhibit performance on par with the full-scale device. To get beyond these problems, microfluidic devices will need to either incorporate spectral separation on chip, spectral separation by the detector, or spectral separation by signal processing to enable a small number of detection lines to yield information about a larger number of emission spectra emitted by various samples. On-chip light dispersion and compact linear detectors may offer one solution. A more elegant solution to this problem lies in animal vision: the use of the relative intensities of a small number of analog signals to infer information about color.

Building upon this principle, our lab has demonstrated a microfluidic flow cytometer device capable of on-chip multicolor fluorescent detection using only one detector. Named COST coding, the method allows the use of only one PMT detector to distinguish at least 11 fluorescent emission wavelengths [70]. Figure 10.10 illustrates how the COST method works. Fluorescent light emitted by a particle is collected in the form of a time-dependent, encoded signal as the particle passes through the sample channel and by an on-chip spatial and color filter waveguide array coupled to only one PMT detector. The COST-coded signal consists of two parts: an intensity reference signal followed by the color-coded portion of the signal. The intensity reference signal is generated as the particle passes by three on-chip apertures that allow all wavelengths of light to pass to the

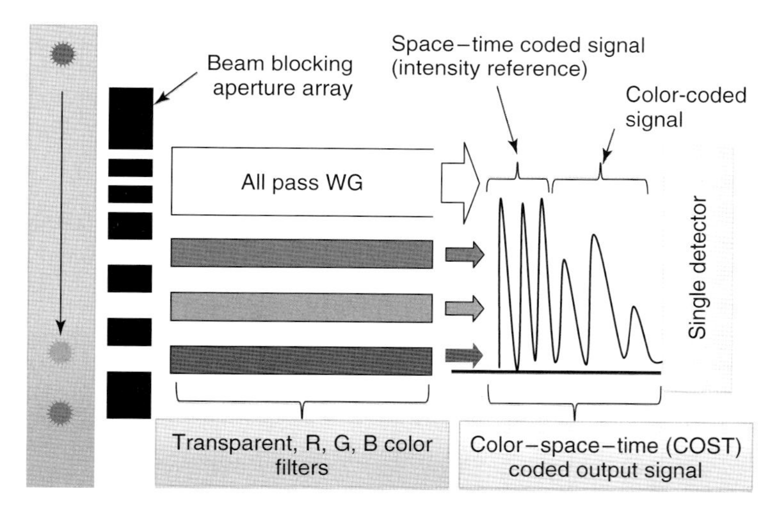

Figure 10.10 Schematic of the color–space–time (COST) encoding device. As fluorescent particles flow through the sample channel (seen to the left), they first pass by three apertures where all emitted wavelengths are transmitted to the detector, forming the reference signal. Next, the particles pass by the three color filters generating the color coded portion of the signal. Reprinted with permission from Ref. [70]. Copyright 2010, American Institute of Physics.

detector, resulting in a three-lobed signal, all roughly equal in intensity. This allows for the normalization to differences in intensity that occur from particle to particle within a given color. The color-coded part of the COST signal is generated as the particle passes by the red, green, and blue filters, where the relative intensity of each signal lobe indicates the amount of red, green, or blue components present within the spectrum emitted by the particle.

The intensity of each color-coded signal lobe is used to form an intensity ratio vector by normalizing green to red and blue to red as seen in Figure 10.11a, in the upper right-hand corner.

Here, the vector [1 : 1.68 : 1.32] indicates particles labeled with Dragon Green fluorescent dye (equivalent to fluorescein isothiocyanate (FITC) with an emission peak at 525 nm). The histograms shown in Figure 10.11b,c shows well-separated bimodal distributions of two particle sets as detected through the green and blue

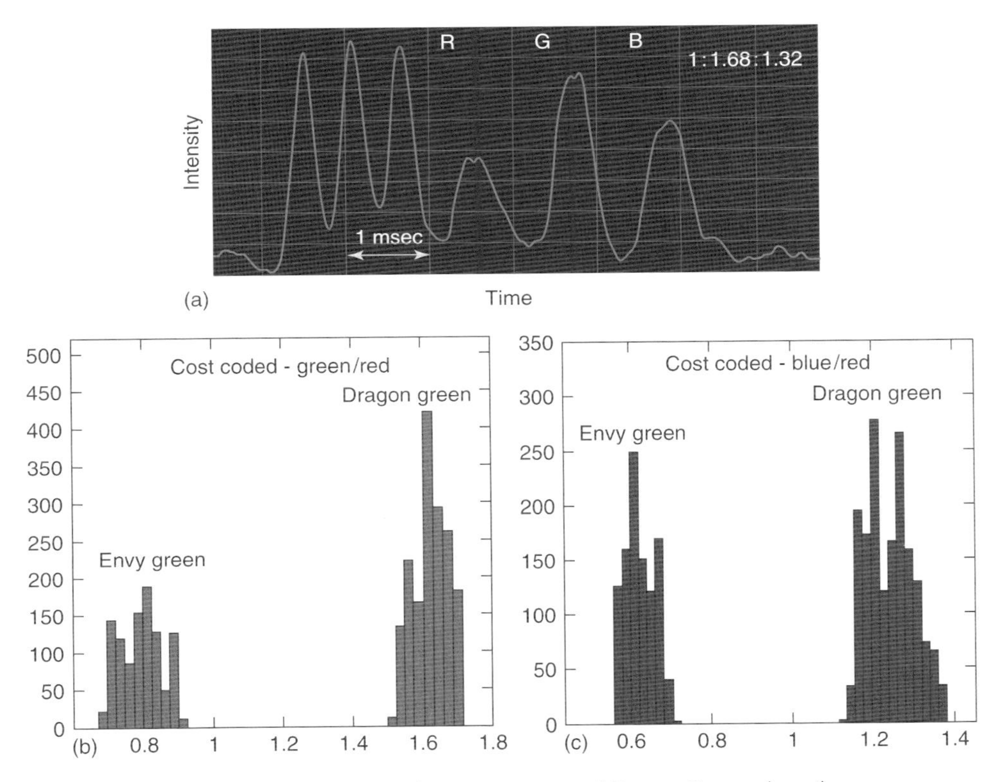

Figure 10.11 (a) Six-lobed signal collected from a particle labeled with Dragon Green fluorescent dye. The intensity ratio vector is generated by normalizing green to red and blue to red. This vector is shown to the upper right. (b) Histogram of a mixture of two particle types, separately labeled with Envy Green and Dragon Green, where the *x*-axis is the green channel normalized to red. (c) Histogram of the same set data set, where the *x*-axis is the blue channel normalized to red. Reprinted with permission from Ref. [70]. Copyright 2010, American Institute of Physics.

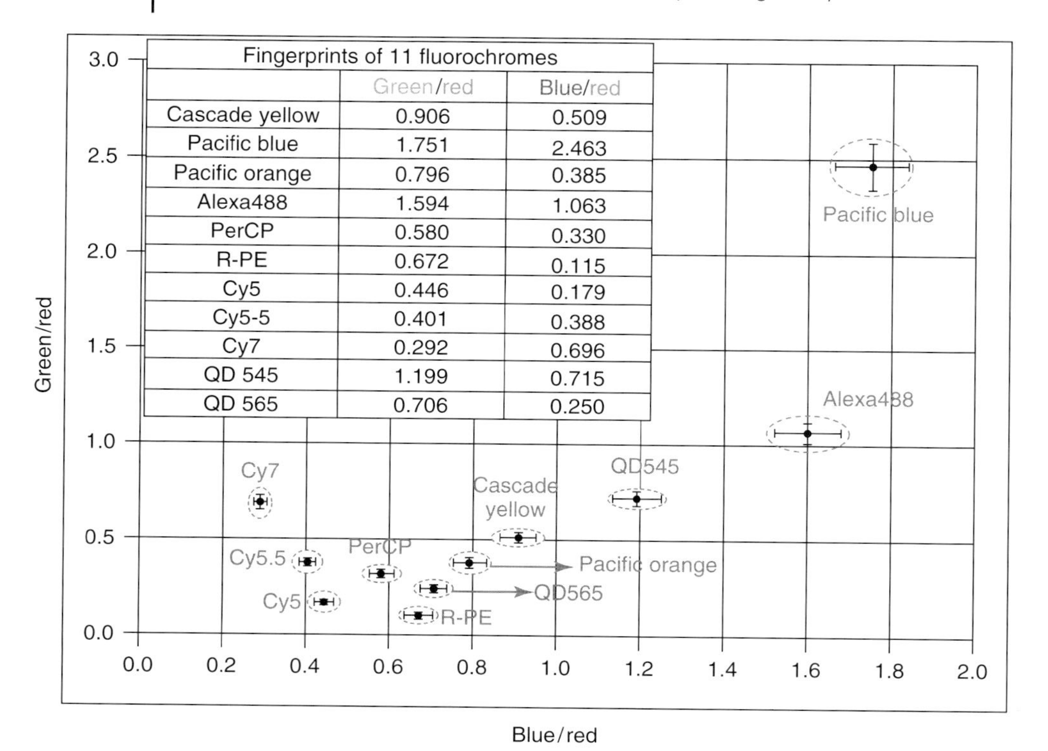

Fingerprints of 11 fluorochromes

	Green/red	Blue/red
Cascade yellow	0.906	0.509
Pacific blue	1.751	2.463
Pacific orange	0.796	0.385
Alexa488	1.594	1.063
PerCP	0.580	0.330
R-PE	0.672	0.115
Cy5	0.446	0.179
Cy5-5	0.401	0.388
Cy7	0.292	0.696
QD 545	1.199	0.715
QD 565	0.706	0.250

Figure 10.12 The particle types, each separately labeled with the commonly used fluorophore listed in the inset table, are plotted with respect to the green/red and blue/red ratio value. The dashed elliptical boundary represents a 5% fluctuation. Discrimination between the 11 different fluorescent colors can be clearly seen. Reprinted with permission from Ref. [70]. Copyright 2010, American Institute of Physics.

channels, respectively. The mode on the left of Figure 10.11b,c contains particles labeled with Envy Green (equivalent to phycoerythrin (PE) with an emission peak at 565 nm) and the mode on the right of Figure 10.11b,c contains particles labeled with Dragon Green. This demonstrates clear discrimination between two particle types labeled with similar colors using only one detector. Figure 10.12 further demonstrates the discrimination of 11 different fluorophores commonly used in biological assays.

The optical toolbox for integrated flow cytometry chips has been steadily growing for some time now. Most of the basic optical elements have been demonstrated in some capacity. Combined into a well-designed system, these components have the potential to create a sophisticated integrated chip. When redesigned into novel architectures to exploit the benefits of a compact, inexpensive device platform, these elements have the potential to revolutionize the field of flow cytometry, bring a lower cost cytometer to every lab with a suite of functions beyond the reach of today's flow cytometer.

10.4
On-Chip Sorting

Along with on-chip integration of the fluidic and optical systems, integration of the sorting module is also needed to create a fully functional microfluidic flow cytometer. In sorting applications, integrated devices are especially valuable because they can easily be designed as closed-loop systems, reducing the risk of contaminating sorted cells. When evaluating new microfluidic cell sorter devices, we must consider several criteria. First, the devices must show the ability to enrich samples with high levels of purity. Second, in most applications, cell viability must be maintained. Third, for compact and integrated devices to be competitive with full-scale instruments, sample throughput must also be maintained. Finally, we must consider the level of difficulty, or complexity, in the fabrication process and device architecture in order to bring down cost. In general, the fewer the processing steps the better. Here, there is an advantage for microfabricated devices over full-scale ones, because devices can be made by polymer injection molding or by silicon elastomer replication, during the prototyping phase. This of course speeds up the R&D cycle. In this section, we evaluate select works of the most popular sorting actuation methods, including work based on electrokinetics, DEP, optical force, and hydrodynamics.

10.4.1
Electrokinetic Sorting

As with flow confinement, several groups have taken advantage of electrokinetic strategies to develop sorting devices [71–73]. Dittrich *et al.* have demonstrated a device, utilizing an electrokinetic flow switching mechanism, made from a microfabricated silicon elastomer (PDMS) layer that is bonded to a glass substrate [73]. In this work, 2D hydrodynamic flow confinement was used to focus the sample into a confocal/fluorescence detection volume as seen in Figure 10.13. After passing through the detection volume, the cells are sent to the sorting module consisting of the sample channel, intersected by a cross channel. Electrodes are placed in reservoirs at each end of the cross-flow channels to generate the electroosmotic flow, which deflects the entire sample stream into one of two directions based on the assigned fluorescence signal of the detected particle. If no voltage is applied, then the sample stream is split at the Y-branch leading to both outlet channels.

In order to demonstrate sorting, a mixture of two bead types, labeled with red and green fluorescence dye, respectively, was processed by the system. In one run, an initial mixture of 0.9 : 0.91 green/red beads was enriched to a ratio of 84.6 : 15.4, yielding an enrichment factor of over 9.3, at a sorting rate of 0.68 events/s. The authors reported typical processing rates, able to yield accurate sorting, to be between 0.3 and 1 event/s. They reported the theoretical sorting rate to be 10 events/s, which is limited by detection and activation time for the electroosmotic flow. In a separate experiment, the cell survival rate was found to be 80–95%. This rate benefited from the relatively low voltages needed to drive the electroosmotic

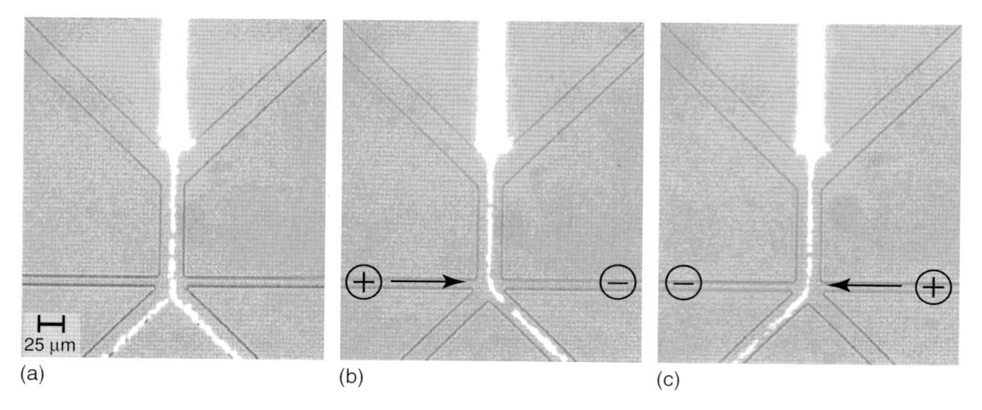

Figure 10.13 Images illustrating the operation of an electrokinetic sorting device utilizing an electroosmotic driven cross flow used to direct the sample stream down one of two channels. When voltage is not applied, the sample stream is split equally across both output channels (a). Changing the polarity of the voltage applied allows the flow to be directed to the right (b) or to the left (c). Reprinted with permission from Ref. [73]. Copyright 2003, American Chemical Society.

flow, which was 40 V peak to peak – lower than previous electroosmotically driven microfluidic sorters. In summary, the fabrication process, including electrode insertion, is relatively easy. The operating voltage is relatively low, as compared to other electroosmotic systems, resulting in a higher cell viability, or survival rate. However, the reported throughput of 1 particle/s, as well as the reported theoretical sorting rate of 10 particles/s, is quite low as compared to commercial systems, where user applications call for rates at least 100–1000 times faster.

10.4.2 Sorting by Dielectrophoresis

Holmes *et al.* developed a sorting device using DEP force for both focusing and sorting of particles in the sample stream [74]. A schematic of the device can be seen in Figure 10.14a, where the sample stream containing a random distribution of the particles (which can be cells or polymer beads) is sent through the focusing module, consisting of a pair of electrodes lining both the top and bottom of the channel. Each set of electrodes is fashioned to taper in toward the center of the channel. When an AC voltage is applied, nDEP force causes the dielectric particles to be focused to the center of the channel (Figure 10.14b). After focusing, the particle stream passes through a detection zone where the particle's fluorescence signal is read out by a fluorescence microscope. After detection, a real-time digitized signal, encoded with information on whether the detected particle was above or below a preset fluorescence intensity threshold, is sent to the sorting module. The sorting module consists of three electrodes, where two are placed on either side of the sample channel and one is placed at the apex of the junction, acting as a ground.

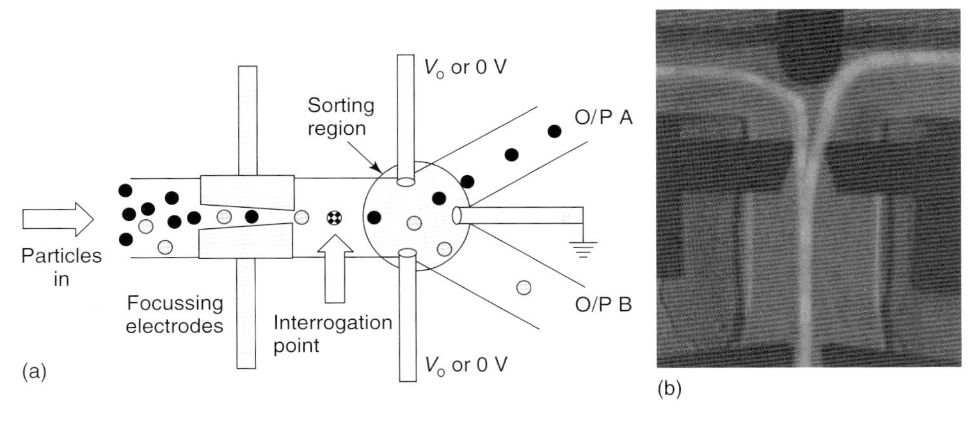

Figure 10.14 (a) Schematic illustrating the working principle of a device that uses negative dielectrophoresis for sorting. (b) Composite image of 200 particle deflections (~100 in each direction) actuated by negative dielectrophoretic force oriented perpendicular to the sample flow. Reprinted with permission from Ref. [74]. Copyright 2005 Institution of Engineering and Technology (IET).

The particle is then deflected to the left or right channel branch by the polarity of the signal sent to the initial pair of electrodes.

The sample flow, generated by an external pump, allowed for flow velocities within the range of $1-10\,\text{mm}\,\text{s}^{-1}$. Using $6\,\mu\text{m}$ diameter Cy5-labeled beads, the device was found to achieve a particle deflection rate of 300 particles/s. The reported automated sorting throughput was 10 particles/s.

The system showed feasibility as a compact and integrated flow cytometer, by displaying a relatively high deflection rate, indicative of the potential for high-throughput automated sorting. However, electronic control issues need to be worked out. Another aspect of the device that indicates its feasibility is that it runs on the tried-and-true method of fluorescence-activated cell sorting (FACS). Additionally, the sorting actuation only requires 20 V peak to peak, which is a relatively low value and unlikely to cause cell damage. However, the two biggest problems for this sorting technique are the low throughput due to the weak dielectrophoretic effect, and critical dependence of the effect on the size, shape, and dielectric properties of the objects. The above setting may work for beads of well-known dielectric properties, but may not work well for cells, proteins, and DNAs.

10.4.3 Sorting by Optical Force

In 1970, researchers demonstrated that micrometer-scale particles suspended in liquids and gases could be trapped or moved by the radiation pressure of a focused optical beam, without the aid of mechanical contact [75]. From this concept, Wang *et al.* developed a micro-FACS (μFACS) device utilizing optical force switching for the sorting actuation mechanism [76]. The system consists of a microfluidic cartridge in which the sample is injected and focused by a 2D hydrodynamic scheme. The single-file particle stream first passes through an optical detection region and then through a sorting region. A 488 nm laser beam is used for fluorescence excitation, and a PMT is used to collect the signals. The optical axis of the off-chip optics is normal to the device plane. Gated fluorescence signals trigger a second near-infrared (NIR) laser to be steered downstream by an acoustooptic modulator to the sorting junction, where the beam is used to deflect the particle into the collection channel. The sorting junction is Y-branched, leading to either a waste channel or a collection channel. The Y-branch is designed asymmetrically so that only cells targeted by the optical switch laser flow into the collection channel. All other particles flow into waste. To demonstrate the device's feasibility for sorting, green fluorescent protein (GFP) expressing HeLa cells were mixed with nonexpressing parental HeLa cells. Three different samples were made up at initial ratios of 50, 10, and 1% of GFP-expressing target cells to nonexpressing cells. Throughputs were determined to be on the order of 20–100 cells/s, depending on both cell density and flow velocity. For the 1% ratio sample, enrichments of 63- to 71-fold were reported. No other enrichment values were reported. In addition to throughput and enrichment, the researchers tested the system's effect on cell viability and cell stress. As an assay for cell viability, trypan blue exclusion was measured for samples that were processed by the system with and without the NIR laser. From that experiment, it was confirmed that no loss in cell viability occurred. The authors also mentioned that NIR lasers have been found to cause cell damage only when used at many times the power used in this experiment. Finally, to measure any effects of transitory stress applied on the cells by the system, the authors monitored the expression of FOS and HSPA6 genes, which are indicators for shear stress and heat shock, respectively. Again, the samples were run through the system with and without the NIR laser. From this data, along with some other experimental controls, it was found that all expression levels were normal, thereby confirming that sorting with optical force switching did not have any negative effects on the cells. On the downside, sorting by optical force switching was found to be quite slow, especially when compared to the work of Wolff [35] using external check valves, which we discuss in the next section.

10.4.4 Hydrodynamic Sorting

An alternative to both electric field and optical sorting techniques is hydrodynamic sorting, in which the sample flow is focused and then deflected into desired branches of the main channel by pressure pulses. The channel architecture in these devices is similar to Dittrich's device that utilized electroosmotic flow, seen in Figure 10.13. Methods using hydrodynamic force eliminate constraints put on buffer composition (i.e., ionic strength) in contrast to the electric-field-based methods. Hydrodynamic sorting also gets around the problems of ion depletion and cell damage due to electric fields. External check valves [35, 77], integrated

PDMS-based valves [7, 78, 79], and syringe pumps [80, 81] have been implemented in hydrodynamic-based sorting devices.

10.4.4.1 Hydrodynamic Sorting with External Check Valves

Using external check valves as the sorting actuator along with fluorescence detection, Wolff *et al.* were able to demonstrate a respectable level of enrichment at an astounding throughput rate for a μFACS device [35]. The device consisted of a sample inlet and two sheath liquid inlets, following the common 2D hydrodynamic flow confinement strategy discussed earlier. After the detection region, which corresponds to the field of view and focal plane of an external fluorescence microscope, the main channel then splits into a waste channel and a collection channel. The microchannels were microfabricated from a Si wafer by reactive ion etch, and a glass plate was anodically bonded on top to seal the channels and act as a viewing window. A high-speed check valve, with an ~2.5 ms response time, was attached to the reservoir of the collection channel. When the fluorescence intensity of a particle of interest is detected, a signal is sent to the check valve which is then activated to pull a plug of fluid containing the particle into the collection channel. Syringe pumps were used to drive the sample and sheath buffer. To demonstrate the system, fluorescent beads were added to a concentrated sample of chicken red blood cells (~10^8 cells/ml) at an initial mixture ratio of 2.4×10^{-5}. After sorting out the fluorescence beads, the final mixture ratio was found to have a respectable 100-fold enrichment in beads, similar to other μFACS systems at the time, and an astounding throughput of 12 000 cells/s, rivaling even commercial systems. However, it should be underscored that in order to achieve these numbers, external check valves and off-chip optics were used.

10.4.4.2 Hydrodynamic Sorting with Piezoelectric Actuators

By integrating a piezoelectric (PZT) actuator on-chip, Chen *et al.*, moved closer to achieving a fully integrated μFACS system utilizing hydrodynamic flow switching [69]. The operating principle of the device can be seen in Figure 10.15a. When the fluorescence intensity of a particle of interest has been detected above a predefined threshold, a voltage signal is sent to the PZT actuator, where, depending on the polarity of the voltage, the PZT actuator is then deflected up or down. The up or down pull translates into a pull or push of the cross stream, respectively, which, in turn, deflects the segment of the sample stream containing the particle of interest into one of two collection channels. If no deflection occurs, the particle stream flows down the center (waste) channel.

Images of the sorting junction show a stream of rhodamine dye being deflected into one of the two channels when acted upon by the PZT (Figure 10.15b,d). The middle image shows the dye stream flowing straight into the center channel when the PZT is not actuated (Figure 10.15c). The authors reported that this type of controlled switching was achieved at up to 1.7 kHz (at 20 Vp−p) and suggested that a deflection rate of >1000 particles/s should be possible. Additionally, the authors suggested that the magnitude of the deflection could be fine-tuned to allow for five sorting branches, or perhaps more.

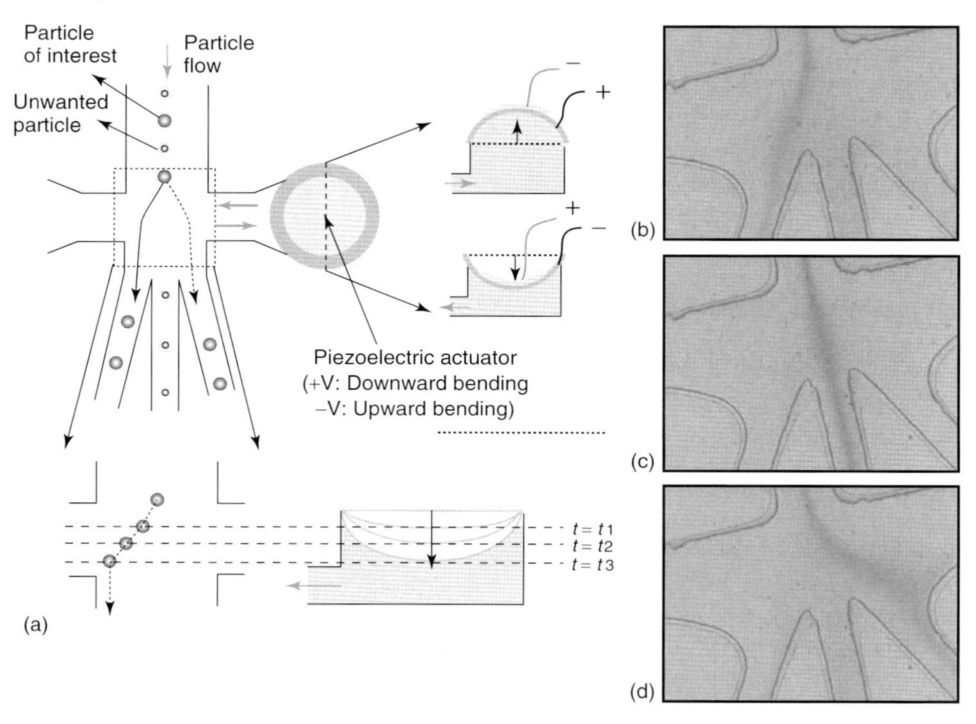

Figure 10.15 (a) Schematic of piezoelectric (PZT) actuated sorting device. Images showing deflection of the sample stream (containing rhodamine dye as an imaging agent) as a result of PZT actuation. Voltage is applied to PZT at opposite polarities in images (b) and (d). (c) No voltage is applied to PZT. Reprinted with permission from Ref. [28]. Copyright 2008 John Wiley & Sons, Inc.

To demonstrate automated sorting, Cho *et al.* designed and implemented a closed-loop control system for the PZT-based device described above, as seen in Figure 10.16 [82]. At the heart of the system is a specially designed spatial filter (photolithography mask, Figure 10.16b) for light collection and a field programmable gate array (FPGA) electronic control system for digital signal processing (DSP) and triggering of PZT actuation. The spatial filter (Figure 10.16b) is placed at the image plane of the objective lens to permit only fluorescence from select regions of the device to reach the PMT detector (Figure 10.16a). The slits of the mask (excluding alignment markers) are the only transparent areas that allow fluorescent signals to reach the detector (Figure 10.16b). When a fluorescent particle (bead or cell) of sufficient intensity passes through the detection region (i.e., detection slits), it registers a three-lobed signal (111), causing the FPGA to actuate the PZT which, in turn, deflects the particle down either the left or right channel. As the particle passes under the other set of slits, another three-lobed confirmation signal (1011 for the right channel and 1101 for the left channel) is detected. Therefore, if the PZT is fired, but the cell is not directed down the targeted collection channel, only the initial three-lobed signal (111) will be detected and a sorting error will be registered.

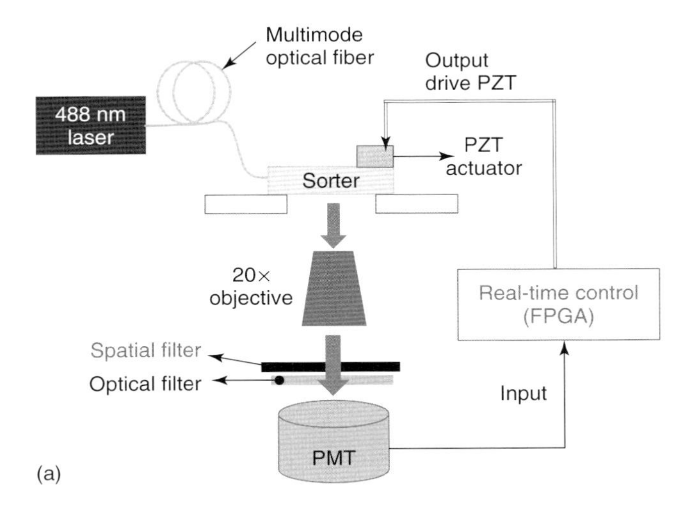

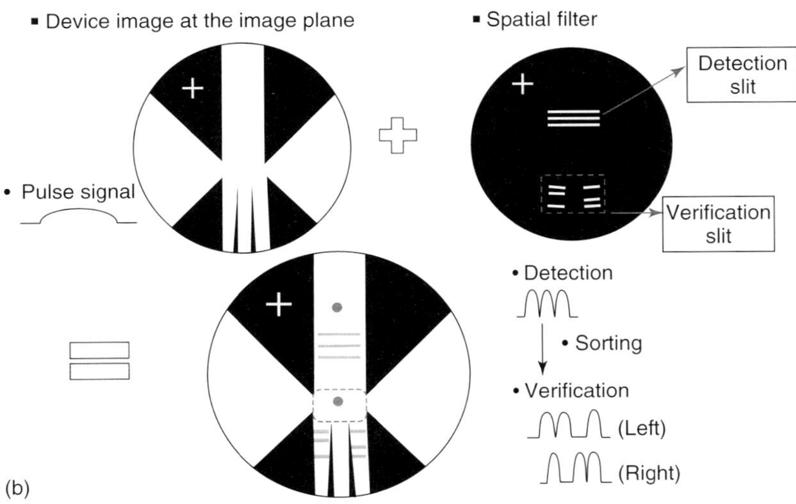

Figure 10.16 (a) Optical setup for PZT actuated sorter device. (b) Schematic and working principle of spatial filter used for verifying sorting events. Note the different triple- and double-lobed signals used for detection and verification, respectively [82]. Reproduced by permission of The Royal Society of Chemistry.

Digital processing of the oversampled (multilobed) signals using the FPGA-based processor offers several advantages. First, the lobed signals can be amplified in real time by an FIR match filter algorithm [83]. In the frequency domain, a match filter is essentially a filter that possesses the identical frequency component as the signal. Thus, signals with similar frequency components (the lobed signals) will be preserved while the noise is suppressed, improving the overall SNR. The authors estimate an improvement in the SNR of ~18 dB using this system. Second, the extra confirmation signal made possible by the coding scheme allows for real-time evaluation of the sorting performance, giving users the ability to monitor the sorting process and make necessary adjustments if needed during an experiment. Third, since the DSP and actuation decisions are carried out by the FPGA chip, the timing jitter for the system is only ~10 μs, enabling high-purity sorting.

The authors demonstrated the automated sorting system's ability to purify a subpopulation of particles within a heterogeneous sample by programming the system to isolate fluorescently labeled 10 μm diameter beads from a binary mixture also containing 5 μm unlabeled beads [82]. Additionally, the authors demonstrated the isolation of labeled human mammalian cells from a binary mixture also containing unlabeled cells. The bead mixture and the cell mixture were characterized using a BD FACScan system (Beckton–Dickinson). Figure 10.17a shows gated scatter plots for unsorted (left) and sorted (right) mixtures, and Figure 10.17b shows bar plots of the same dataset. When analyzed, the results showed a 200-fold enrichment in the labeled 10 μm diameter beads, the highest enrichment factor among all other μFACS systems reported in this section. Figure 10.17c illustrates the enrichment of the initially rare population of labeled cells by 230-fold. Additionally, the device was shown to achieve 70–80% sorting efficiency, using a sample concentration of $\sim 10^5$ particles/ml and traveling at 8 cm/s [82]. In summary, the sorter is not only low-powered (<0.1 mW) and low-cost (PZT material is cost effective and the device requires only simple fabrication) but it also offers a quicker response (0.1–1 ms) than conventional check or membrane valves.

Successful proof-of-principle demonstrations of microfluidic sorting devices based on electroosmotic, DEP, optical force switching as well as hydrodynamic flow switching actuation mechanisms have been demonstrated on microfluidic platforms. Electric-field-based devices, based on electroosmosis and DEP, show potential for a reduced footprint and an increase in portability. However, electroosmotic devices, in particular, require high operation voltages that may be damaging to cells. And both technologies put constraints on the electrochemical properties of the sample (cells) and the buffers used. Optical force switching was found not to harm the cells; however, the throughput rate was quite low. Devices employing high-speed external check valves have achieved sorting rates comparable to full-scale systems with respectable levels of enrichment. However, the system relied on both off-chip sorting actuation (check valves) and optical components. Sorting devices with integrated, on-chip, PZT sorting actuators have shown respectable throughput and enrichment levels with respect to other microfluidic devices. However, these devices still rely on off-chip optical, pumping, electronic control, and detection systems. So, we can safely say that there is still room for research and development in fully integrated, compact microfluidic cell sorting devices.

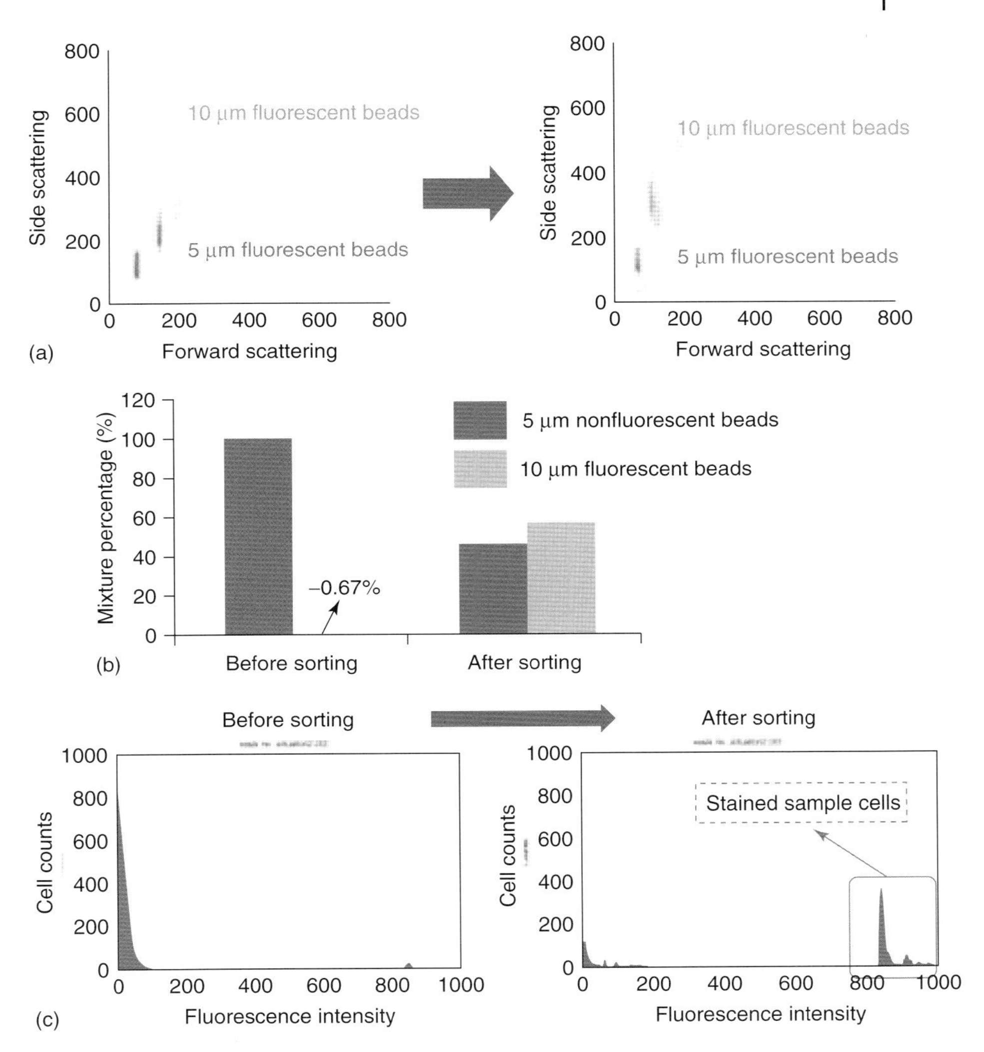

Figure 10.17 (a) Scatter plots showing the sorting results of a bead mixture containing two populations (10 μm diameter fluorescently labeled beads and 5 μm diameter unlabeled beads), before (left) and after (right), using the PZT actuated system. (b) The ratio of labeled beads to nonlabeled before sorting was ~0.0067, while the ratio after sorting was 1.28, reflecting an enrichment of ~200-fold. (c) Histogram (left) showing a mixture of labeled and nonlabeled cells at a ratio of 1 : 150. After sorting, the histogram (right) shows an enrichment factor of 230 [82]. Reproduced by permission of The Royal Society of Chemistry.

10.5 Conclusion

Considerable progress is being made toward more compact and integrated microfluidic flow cytometer devices, which may someday lead to hand-held devices that rival current full-scale flow cytometers in performance, at only a fraction of the cost. This could open the door to a multitude of new applications, while expanding the presence of flow cytometry into areas of the world in great demand for low-cost yet accurate clinical diagnostic devices. A cheap, portable, and accurate flow cytometer would be ideal for monitoring diseases such as HIV in remote areas of third world countries. However, a lot of work is still needed to achieve such lofty visions. Even today, after the advancement of the last several years, the microfluidic flow cytometers we build, and consider to be compact and integrated devices, still rely on bulky, off-chip syringe pumps, electronic control systems, lasers, and detectors. This makes for a system with quite a large benchtop footprint. Current microfluidic flow cytometers are also in need of improvement to meet the needs of the users of full-scale flow cytometers, in terms of sensitivity, linearity, selectivity, and throughput. With respect to detection, this translates into the need for a system with compact optics that allow for multiparameter detection, including multicolor fluorescence and multi-angle scatter measurements with low CVs. However, like most complex devices in the life sciences, it is rare when one subsystem can fully solve a problem, and in this case, optimization of the optical detection system alone will not reduce the CVs. Additionally, for microfluidic flow cytometers to rival full-scale systems, not just in detection performance but also in functionality, on-chip sorting modules must be made to yield highly purified sorted samples at (>90%) at high throughput rates between 1 and 10 kHz: something that has yet to be achieved in a fully integrated on-chip device. Finally, we must carefully consider the performance improvements along with the degree of miniaturization we aim for against the added cost to the end user. In some cases, it may not be worth the added expense and difficulty to make a super-sleek, size-minimized device, so long as the system is fast, accurate, and widely available enough to researchers and clinicians who would otherwise not have access to a full-scale flow cytometer.

Acknowledgments

We would like to acknowledge support from the National Institutes of Health through the following grants: NIH1R01HG004876 and NIH5R21RR024453.

References

1. Shapiro, H.M. (2002) *Practical Flow Cytometry*, 4th edn, Wiley-Liss, Inc., New York.
2. Sobek, D. *et al.* (1993) Proceedings of the Micro Electro Mechanical Systems, p. 219.

3. Altendorf, E. *et al.* (1997) International Conference on Solid State Sensors and Actuators, p. 531.
4. Xia, Y.N. and Whitesides, G.M. (1998) Soft lithography. *Ann. Rev. Mater. Sci.*, **28**, 153–184.
5. Chabinyc, M.L. *et al.* (2001) An integrated fluorescence detection system in poly(dimethylsiloxane) for microfluidic applications. *Anal. Chem.*, **73** (18), 4491–4498.
6. Haubert, K., Drier, T., and Beebe, D. (2006) PDMS bonding by means of a portable, low-cost corona system. *Lab Chip*, **6** (12), 1548–1549.
7. Unger, M.A. *et al.* (2000) Monolithic microfabricated valves and pumps by multilayer soft lithography. *Science*, **288** (5463), 113–116.
8. Lyklema, J. *et al.* (1991) *Fundamentals of Interface and Colloid Science*, Academic Press.
9. Berthier, J. and Silberzan, P. (2006) *Microfluidics for Biotechnology*, Arctech House, Inc., Norwood.
10. Harris, D.C. (1999) *Quantitative Chemical Analysis*, W.H. Freeman and Company.
11. Ateya, D.A. *et al.* (2008) The good, the bad, and the tiny: a review of microflow cytometry. *Anal. Bioanal. Chem.*, **391** (5), 1485–1498.
12. Jacobson, S.C. and Ramsey, J.M. (1997) Electrokinetic focusing in microfabricated channel structures. *Anal. Chem.*, **69** (16), 3212–3217.
13. Schrum, D.P. *et al.* (1999) Microchip flow cytometry using electrokinetic focusing. *Anal. Chem.*, **71** (19), 4173–4177.
14. Yang, R.J. *et al.* (2005) A new focusing model and switching approach for electrokinetic flow inside microchannels. *J. Micromech. Microeng.*, **15** (11), 2141–2148.
15. Huang, K.D. and Yang, R.J. (2006) Numerical modeling of the Joule heating effect on electrokinetic flow focusing. *Electrophoresis*, **27** (10), 1957–1966.
16. Knight, J.B. *et al.* (1998) Hydrodynamic focusing on a silicon chip: mixing nanoliters in microseconds. *Phys. Rev. Lett.*, **80** (17), 3863–3866.
17. Lee, G.B. *et al.* (2006) The hydrodynamic focusing effect inside rectangular microchannels. *J. Micromech. Microeng.*, **16** (5), 1024–1032.
18. Lee, G.B., Hwei, B.H., and Huang, G.R. (2001) Micromachined pre-focused M × N flow switches for continuous multi-sample injection. *J. Micromech. Microeng.*, **11** (6), 654–661.
19. Huh, D. *et al.* (2002) Use of air-liquid two-phase flow in hydrophobic microfluidic channels for disposable flow cytometers. *Biomed. Microdevices*, **4** (2), 141–149.
20. Stiles, T. *et al.* (2005) Hydrodynamic focusing for vacuum-pumped microfluidics. *Microfluid. Nanofluid.*, **1** (3), 280–283.
21. Morgan, H., Holmes, D., and Green, N.G. (2003) 3D focusing of nanoparticles in microfluidic channels. *IEE Proc.-Nanobiotechnol.*, **150** (2), 76–81.
22. Klank, H. *et al.* (2002) PIV measurements in a microfluidic 3D-sheathing structure with three-dimensional flow behaviour. *J. Micromech. Microeng.*, **12** (6), 862–869.
23. Simonnet, C. and Groisman, A. (2005) Two-dimensional hydrodynamic focusing in a simple microfluidic device. *Appl. Phys. Lett.*, **87** (11), 114104.
24. Mao, X.L., Waldeisen, J.R., and Huang, T.J. (2007) "Microfluidic drifting" – implementing three-dimensional hydrodynamic focusing with a single-layer planar microfluidic device. *Lab Chip*, **7** (10), 1260–1262.
25. Mao, X.L. *et al.* (2009) Single-layer planar on-chip flow cytometer using microfluidic drifting based three-dimensional (3D) hydrodynamic focusing. *Lab Chip*, **9** (11), 1583–1589.
26. Howell, P.B. *et al.* (2008) Two simple and rugged designs for creating microfluidic sheath flow. *Lab Chip*, **8** (7), 1097–1103.
27. Golden, J.P. *et al.* (2009) Multi-wavelength microflow cytometer using groove-generated sheath flow. *Lab Chip*, **9** (13), 1942–1950.
28. Godin, J. *et al.* (2008) Microfluidics and photonics for Bio-System-on-a-Chip: a review of advancements in technology

towards a microfluidic flow cytometry chip. *J. Biophoton.*, **1** (5), 355–376.

29. Bliss, C.L., McMullin, J.N., and Backhouse, C.J. (2007) Rapid fabrication of a microfluidic device with integrated optical waveguides for DNA fragment analysis. *Lab Chip*, **7**, 1280–1287.
30. Tung, Y.C. *et al.* (2004) PDMS-based opto-fluidic micro flow cytometer with two-color, multi-angle fluorescence detection capability using PIN photodiodes. *Sens. Actuators, B Chem.*, **98** (2–3), 356–367.
31. Lin, C.H. and Lee, G.B. (2003) Micromachined flow cytometers with embedded etched optic fibers for optical detection. *J. Micromech. Microeng.*, **13** (3), 447–453.
32. Cui, L., Zhang, T., and Morgan, H. (2002) Optical particle detection integrated in a dielectrophoretic lab-on-a-chip. *J. Micromechan. Microeng.*, **12** (1), 7–12.
33. Godin, J., Lien, V., and Lo, Y. (2006) Integrated fluidic photonics for multi-parameter in-plane detection in microfluidic flow cytometry. Lasers and Electro-Optics Society, 2006. LEOS 2006. 19th Annual Meeting of the IEEE, pp. 605–606.
34. Lien, V., Berdichevsky, Y., and Lo, Y. (2004) A prealigned process of integrating optical waveguides with microfluidic devices. *IEEE Photon. Technol. Lett.*, **16** (6), 1525–1527.
35. Wolff, A. *et al.* (2003) Integrating advanced functionality in a microfabricated high-throughput fluorescent-activated cell sorter. *Lab Chip*, **3** (1), 22–27.
36. Cho, S.H. *et al.* (2008) Microfluidic photonic integrated circuits. Proceedings of the SPIE - The International Society for Optical Engineering, p. 71350M (10 pp.).
37. Mao, X.L. *et al.* (2007) Hydrodynamically tunable optofluidic cylindrical microlens. *Lab Chip*, **7** (10), 1303–1308.
38. Cho, S., Godin, J., and Lo, Y.-H. (2009) Optofluidic waveguides in Teflon AF-coated PDMS microfluidic channels. *IEEE Photon. Technol. Lett.*, **21** (15), 1057–1059.
39. Datta, A. *et al.* (2003) Microfabrication and characterization of Teflon AF-coated liquid core waveguide channels in silicon. *IEEE Sens. J.*, **3** (6), 788–795.
40. Lim, J.M. *et al.* (2008) Fluorescent liquid-core/air-cladding waveguides towards integrated optofluidic light sources. *Lab Chip*, **8** (9), 1580–1585.
41. Wolfe, D.B. *et al.* (2004) Dynamic control of liquid-core/liquid-cladding optical waveguides. *Proc. Natl. Acad. Sci. U.S.A.*, **101** (34), 12434–12438.
42. Tang, S.K.Y., Stan, C.A., and Whitesides, G.M. (2008) Dynamically reconfigurable liquid-core liquid-cladding lens in a microfluidic channel. *Lab Chip*, **8** (3), 395–401.
43. Lien, V., Zhao, K., and Lo, Y. (2005) Fluidic photonic integrated circuit for in-line detection. *Appl. Phys. Lett.*, **87**, 194106.
44. Shapiro, H.M. and Hercher, M. (1986) Flow cytometers using optical wave-guides in place of lenses for specimen illumination and light collection. *Cytometry*, **7** (2), 221–223.
45. Camou, S., Fujita, H., and Fujii, T. (2003) PDMS 2D optical lens integrated with microfluidic channels: principle and characterization. *Lab Chip*, **3** (1), 40–45.
46. Ro, K.W. *et al.* (2005) Integrated light collimating system for extended optical-path-length absorbance detection in microchip-based capillary electrophoresis. *Anal. Chem.*, **77** (16), 5160–5166.
47. Wang, Z. *et al.* (2004) Measurements of scattered light on a microchip flow cytometer with integrated polymer based optical elements. *Lab Chip*, **4** (4), 372–377.
48. Llobera, A., Wilke, R., and Buttgenbach, S. (2004) Poly(dimethylsiloxane) hollow Abbe prism with microlenses for detection based on absorption and refractive index shift. *Lab Chip*, **4** (1), 24–27.
49. Seo, J. and Lee, L.P. (2004) Disposable integrated microfluidics with self-aligned planar microlenses. *Sens. Actuators B Chem.*, **99** (2–3), 615–622.
50. Seo, J. and Lee, L.P. (2003) Fluorescence amplification by self-aligned integrated microfluidic optical systems. TRANSDUCERS, Solid-State Sensors, Actuators

and Microsystems, 12th International Conference on, 2003.

51. Godin, J., Lien, V., and Lo, Y.H. (2006) Demonstration of two-dimensional fluidic lens for integration into microfluidic flow cytometers. *Appl. Phys. Lett.*, **89** (6), 061106.
52. Godin, J. and Lo, Y.-H. (2009) Advances in on-chip polymer optics for optofluidics. Lasers and Electro-Optics, 2009 and 2009 Conference on Quantum Electronics and Laser Science Conference. CLEO/QELS 2009, pp. 1–2.
53. Godin, J. and Lo, Y. (2007) Microfluidic flow cytometer with on-chip lens systems for improved signal resolution. IEEE Proceedings Sensors, 466–469.
54. Pamme, N., Koyama, R., and Manz, A. (2003) Counting and sizing of particles and particle agglomerates in a microfluidic device using laser light scattering: application to a particle-enhanced immunoassay. *Lab Chip*, **3** (3), 187–192.
55. Jansen, H. *et al.* (1995) The black silicon method – a universal method for determining the parameter setting of a fluorine-based reactive ion etcher in deep silicon trench etching with profile control. *J. Micromech. Microeng.*, **5** (2), 115–120.
56. Llobera, A., Wilke, R., and Büttgenbach, S. (2008) Enhancement of the response of poly (dimethylsiloxane) hollow prisms through air mirrors for absorbance-based sensing. *Talanta*, **75** (2), 473–479.
57. Demming, S. *et al.* (2009) Single and Multiple Internal Reflection poly(dimethylsiloxane) absorbance-based biosensors. *Sens. Actuators, B Chem.*, **139** (1), 166–173.
58. Li, Z. and Scherer, A. (2009) Optofluidic grating spectrograph on a chip. Lasers and Electro-Optics, 2009 and 2009 Conference on Quantum electronics and Laser Science Conference. CLEO/QELS 2009, pp. 1–2.
59. Kou, Q. *et al.* (2004) On-chip optical components and microfluidic systems. *Microelectron. Eng.*, **73–74**, 876–880.
60. Llobera, A., Wilke, R., and Büttgenbach, S. (2004) Poly (dimethylsiloxane) hollow Abbe prism with microlenses for detection based on absorption and refractive index shift. *Lab Chip*, **4** (1), 24–27.
61. Goddard, G. *et al.* (2006) Single particle high resolution spectral analysis flow cytometry. *Cytometry A*, **69A** (8), 842–851.
62. Hofmann, O. *et al.* (2006) Monolithically integrated dye-doped PDMS long-pass filters for disposable on-chip fluorescence detection. *Lab Chip*, **6** (8), 981–987.
63. Bliss, C.L., McMullin, J.N., and Backhouse, C.J. (2008) Integrated wavelength-selective optical waveguides for microfluidic-based laser-induced fluorescence detection. *Lab Chip*, **8** (1), 143–151.
64. Wolfe, D., Vezenov, D., and Mayers, B. (2005) Diffusion-controlled optical elements for optofluidics. *Appl. Phys. Lett.*, **97**, 181105.
65. Lim, J. *et al.* (2008) Fluorescent liquid-core/air-cladding waveguides towards integrated optofluidic light sources. *Lab Chip*, **8** (9), 1580–1585.
66. Kang, Y. *et al.* (2008) On-chip fluorescence-activated particle counting and sorting system. *Anal. Chim. Acta*, **626** (1), 97–103.
67. Chen, C.H. *et al.* (2007) Scattering-based cytometric detection using integrated arrayed waveguides with microfluidics. *IEEE Photon. Technol. Lett.*, **19** (5–8), 441–443.
68. Mogensen, K.B. *et al.* (2003) A microfluldic device with an integrated waveguide beam splitter for velocity measurements of flowing particles by Fourier transformation. *Anal. Chem.*, **75** (18), 4931–4936.
69. Chen, C.H. *et al.* (2009) Microfluidic cell sorter with integrated piezoelectric actuator. *Biomed. Microdevices*, **11** (6), 1223–1231.
70. Cho, S.H. *et al.* (2010) Lab-on-a-chip flow cytometer employing color-space-time coding. *Appl. Phys. Lett.*, **97**, 093704.
71. Fu, A.Y. *et al.* (1999) A microfabricated fluorescence-activated cell sorter. *Nat. Biotechnol.*, **17** (11), 1109–1111.
72. Fu, L.M. *et al.* (2004) Electrokinetically driven micro flow cytometers

with integrated fiber optics for on-line cell/particle detection. *Anal. Chim. Acta*, **507** (1), 163–169.

73. Dittrich, P.S. and Schwille, P. (2003) An integrated microfluidic system for reaction, high-sensitivity detection, and sorting of fluorescent cells and particles. *Anal. Chem.*, **75** (21), 5767–5774.
74. Holmes, D. *et al.* (2005) On-chip high-speed sorting of micron-sized particles for high-throughput analysis. *IEE Proc. Nanobiotechnol.*, **152** (4), 129–135.
75. Ashkin, A. (1970) Acceleration and trapping of particles by radiation pressure. *Phys. Rev. Lett.*, **24** (4), 156.
76. Wang, M.M. *et al.* (2005) Microfluidic sorting of mammalian cells by optical force switching. *Nat. Biotech.*, **23** (1), 83–87.
77. Bang, H.W. *et al.* (2006) Microfabricated fluorescence-activated cell sorter through hydrodynamic flow manipulation. *Microsyst. Technol. Micro Nanosyst. Inf. Storage Process. Syst.*, **12** (8), 746–753.
78. Fu, A.Y. *et al.* (2002) An integrated microfabricated cell sorter. *Anal. Chem.*, **74** (11), 2451–2457.
79. Studer, V. *et al.* (2004) A microfluidic mammalian cell sorter based on fluorescence detection. *Microelectron. Eng.*, **73–74**, 852–857.
80. Blankenstein, G. and Larsen, U.D. (1998) Modular concept of a laboratory on a chip for chemical and biochemical analysis. *Biosens. Bioelectron.*, **13** (3–4), 427–438.
81. Kruger, J. *et al.* (2002) Development of a microfluidic device for fluorescence activated cell sorting. *J. Micromech. Microeng.*, **12** (4), 486–494.
82. Cho, S.H. *et al.* (2010) Human mammalian cell sorting using a highly integrated micro-fabricated fluorescence-activated cell sorter (mu FACS). *Lab Chip*, **10** (12), 1567–1573.
83. Kay, S.M. (1993) *Fundamentals of Statistical Signal Processing*, Estimation Theory, Vol. 1, Prentice Hall, Inc., Upper Saddle River, NJ.

11
Label-Free Cell Classification with Diffraction Imaging Flow Cytometer

Xin-Hua Hu and Jun Q. Lu

11.1
Introduction

Flow cytometry has become a primary tool for rapid assay of large cell populations after decades of research and development [1]. A flow cytometer typically transports biological cells in a liquid stream under a laminar flow condition for hydrodynamically focusing cells in a single file. One or more light beams interrogate these cells at rates up to 10^4 cells/s. Data are acquired in two different modes: angularly integrated light signals using discrete detectors such as photomultipliers and photodiodes [2, 3], and angle-resolved light signals using one or more imaging sensors such as CCD cameras through microscope objective [4]. The light signals collected from an interrogated cell can be further divided into two types: elastically scattered light signals at the same wavelength and fluorescence light signals at longer wavelengths in comparison with the wavelength of incident light. These signals offer different information contents, and we introduce them separately below to provide a ground for our later discussion on diffraction imaging of scattered light.

Fluorescence signals reveal relevant molecules or molecular markers involved in various cellular processes and pathways, and thus yield molecular signatures of the interrogated cells. To obtain fluorescence signals, one often has to label cells with specific fluorescent dyes or antibodies attached to target organelles and excite these fluorophores in spectral regions from ultraviolet to visible. The requirement of cell labeling not only increases the cost of flow cytometric measurement considerably but could also interfere with molecular trafficking of interest in living cells. For certain cell studies such as investigation of cytoskeletal changes, labeling agents for living cells simply are not available and complex procedures of gene modification with green fluorescence proteins are required [5]. Most of the existing flow cytometers rely on integrated fluorescence signals as the dominant source of information for cell classification. With multiple (typically less than 10) signals of different wavelengths per interrogated cell, cell classification is achieved on the basis of multivariate analysis of these fluorescence signals combined with two scattered light signals (see discussion below). While this approach allows fast signal

Advanced Optical Flow Cytometry: Methods and Disease Diagnoses, First Edition. Edited by Valery V. Tuchin.

acquisition, it carries the obvious disadvantage of the lack of capacity to acquire detailed morphological information.

The development of imaging flow cytometers has attracted increasing attention; over the last three decades, it is being implemented through the combination of conventional microscopy with flow cytometer to acquire fluorescence and bright/dark-field images [4, 6–8]. This type of imaging flow cytometers provides two-dimensional (2D) morphological features and thus significantly increases the morphological information available for cell classification. However, the current imaging approach employs a noncoherent design and results in 2D "shadow-casting" images of 3D cell structures. Without a scanning scheme used in the confocal imaging to acquire multiple 2D slices, it is impossible to classify cells through their 3D morphological features. For rapid cell classification with a flow cytometer, the confocal imaging method takes too much time to complete. Additionally, automated image analysis, which extracts the useful features for cell classification, is required for the imaging flow cytometry to be practical for processing the extremely large amount of image data in real time. This demand is difficult to meet with the existing methods of image analysis and machine learning given the complex structures of biological cells as shown by one or a few 2D noncoherent images. Solving both problems of accurate extraction of 3D morphological features and rapid image analysis is critical for the development of the next-generation flow cytometer as a high-throughput instrument. Apparently, the existing noncoherent approach of fluorescence and bright/dark-field imaging alone cannot satisfy these requirements. Furthermore, this approach of flow cytometric assay relies on fluorescence images and thus still requires cell labeling.

In contrast to fluorescence, elastic light scattering (or simply light scattering) by a cell occurs due to the heterogeneity in its intracellular refractive index and index mismatch from the host medium and thus requires no fluorescence labeling. More importantly, scattered light can be coherent if highly coherent light sources such as lasers are used for excitation. Coherently scattered light distributes in specific fringe patterns in space as a result of index heterogeneity and thus correlates highly with the 3D morphology of the scatterer [9]. This fact was probably first demonstrated experimentally by Dennis Gabor in his classic 1948 paper on holography, in which he observed "records of three-dimensional" of an imaged object through interference of scattered light with a reference beam [10]. The significant insight by Gabor, however, had to wait for the birth of lasers and other technical advances to be fully appreciated as a high-quality 3D imaging modality more than a decade later [11]. As lasers became readily available, researchers further realized that the 3D morphological features of an object can be extracted from diffraction images, or speckle images, formed by scattered light without a reference beam. These efforts led to the acquisition of coherently scattered light in forward directions as diffraction images from a biological cell [12–14], which can be interpreted with simplified models based on the scalar field model of the scattered light for nonspherical cells [9, 15]. Combining scalar field models with backprojection algorithms further gives rise to optical diffraction tomography in which 3D reconstruction of a weak scatterer can be achieved with multiple diffraction images acquired at different incident angles [16].

These techniques, nevertheless, cannot be applied directly to flow cytometry for its requirement of imaging at numerous incident angles and subsequent data processing that are computationally expensive and time consuming.

The conventional design employs two discrete detectors to measure forward and side scatters integrated over angular ranges, often larger than 10°, for sufficient signal strength over noise [17]. Such angular integration diminishes characteristic fringe patterns of the scattered light and therefore takes no advantage of the high coherence of a laser beam as illumination [18]. Early measurement of scattered light signals at multiple polar scattering angles has been pursued with linear detector arrays as a means for rapid analysis of flowing cells [19–21]. The angle-resolved method has been extended with 2D imaging sensors to record the angular distribution of coherently scattered light [22–25]. Among these efforts, an elegant beam-in-flow design was proposed and demonstrated by Valeri Maltsev and coworkers to obtain polar-angle-resolved distribution of scatters using only one detector through a scanning scheme [26, 27]. In this design, the scattered light signal is acquired by a single detector, after reflection by a concave reflector for averaging over the azimuthal angle, as a function of time as the cell moves in a flow channel along which the excitation laser beam is propagated. For a known speed of the flowing cell, the polar angles of the scattered light can be retrieved from the timing positions of the light signal. The beam-in-flow design has also been used to record distribution of the scattered light in both the polar and azimuthal angles with a CCD camera in a microfluidic device [24, 25]. Compared to the conventional designs of flowing cells perpendicular to an incident laser beam in air or a channeled chamber of glass, the beam-in-flow design has the advantage of eliminating index mismatch between the particle carrying fluid and the air or channel medium to the incident light beam. But the index mismatch at the fluid–medium interface still exists to the scattered light and, more importantly, this design only allows one particle to be carried by the longitudinally illuminated flow. This requirement could severely limit the device's throughput to 500 cells/s or less [26].

Over the past eight years, we have conducted research on accurate modeling of light scattering from biological cells with simulated and realistic cell structures [28–33] and development of a flow cytometer for experimental study of a new diffraction imaging approach [34, 35]. In this chapter, we focus our discussion on a finite-difference time-domain (FDTD) method for modeling light scattering, and a new design of imaging flow cytometer for acquisition of high-contrast diffraction image data with coherent signals of scattered light. Both numerical and experimental results will be presented and their implications to future improvement of the flow cytometry discussed.

11.2 Modeling of Scattered Light

In response to an incident light of wavelength λ, the molecules inside an illuminated cell develop induced electric dipoles which radiate electromagnetic wavefields in all directions as scattered light. The majority of the scattered light wavefields have the same frequency or wavelength as the incident light, which is of interest to our concerns here. Several models of light scattering have been applied to problems related to biological cells over the last two decades [36] and the discussions in this section will be focused on the FDTD method. Since Chapter 3 by Stoyan Tanev *et al.* provides a detailed description of the FDTD method [37], we will limit our discussion on FDTD here to those relevant to diffraction imaging.

11.2.1 The Correlation between Scattered Light Distribution and Cellular Structure

For a biological cell excited with a highly coherent incident light beam, the scattered wavefields from different molecules inside the cell can interfere with each other owing to a phase shift related to pathlength difference as schematically presented in Figure 11.1 by two wavelets a and b. If an array of detecting elements such as a CCD camera is used, the phase shifts of the wavelets vary among the imager pixels and thus lead to recorded fringe patterns of scattered light in the far field as a diffraction image. The angular distribution of the scattered light or the fringe pattern in a diffraction image correlates strongly with the type and density variations of the induced molecular dipoles within the cell, which can be described by a complex refractive index $n(\mathbf{r}, \lambda)$ at location $\mathbf{r}$ within the cell. The angular sizes of fringes in a diffraction image depend on the ratio of cell size d to the wavelength λ which can be assembled into a size parameter $\alpha = \pi d/\lambda$. The angular width of a fringe becomes very large for $\alpha < 1$ and very small for $\alpha > 1000$, and in both cases the fringe patterns become very faint and difficult to observe in a diffraction image acquired with a conventional microscopic objective. For most mammalian cells excited with ultraviolet/visible/near-infrared light, α ranges between 10 and 100 for the sizes of the cells and most large intracellular components such as nucleus,

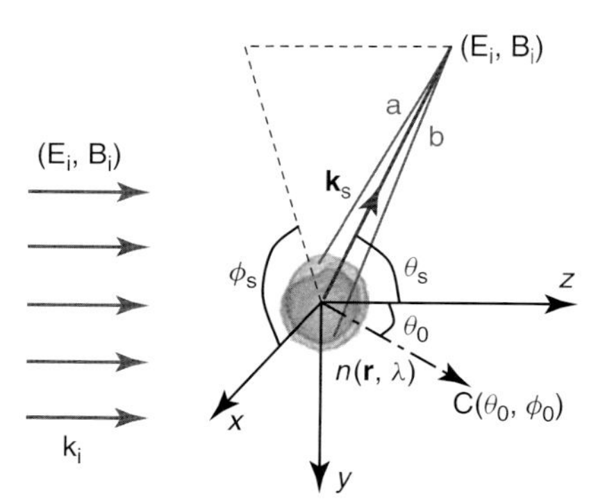

Figure 11.1 Light scattered along the direction of $\mathbf{k}_s$ of (θ_s, ϕ_s) by a cell with orientation direction $\mathbf{C}$ of (θ_0, ϕ_0). The incident light is along the z-axis with a wave vector $\mathbf{k}_0$ and the scatterer cell is described by the intracellular distribution of refractive index $n(\mathbf{r}, \lambda)$. The rays of a and b represent schematically the diffraction due to scattered light originated from different parts of the cell.

mitochondria, and cytoskeleton. As a result, the corresponding fringe patterns are readily observable in diffraction images recorded with an imaging sensor and microscope objective of typical numerical apertures around 0.5 [34, 35].

The spatial distribution of scattered light in the form of diffraction images can be accurately modeled by the classical electrodynamics theory based on the Maxwell equations [38]. In 1908, Gustav Mie published a partial-wave solution of the light scattering problem by a single sphere that had been previously solved by Ludwig Lorenz in 1890 as a boundary-value problem [39]. The Lorenz–Mie results (or simply the Mie theory) can be extended to calculate scattered wavefields by a coated sphere to allow multiple refractive indices to be used to fit scattered light by particles of internal structures [40]. As a fast modeling tool, the Mie theory is useful for the interpretation of certain features in integrated or polar-angle-resolved measurements of scattered light [17, 27]. Their utility reduces drastically, however, if one is concerned with the classification of real biological cells with diffraction images since the intracellular structures possess nearly no symmetry in comparison to the various spherical structures. For cell classification, which is what we are concerned with here, accurate methods must be employed to model angle-resolved light scattering for simulations of diffraction images. Several numerical methods have been applied to these applications, most notable among them being the FDTD method [41–43] and the discrete dipole approximation (DDA) method [44, 45]. In the following, we provide a brief discussion on a parallel FDTD method using reconstructed 3D cell structures for accurate modeling. More details can be found in Chapter 3 of this book or in Ref. [29–31].

Let us consider an incident light of wavefields of $(\mathbf{E}_i, \mathbf{H}_i)$ propagating along the direction of a wave vector $\mathbf{k}_i$. Because of the optical heterogeneity $n(\mathbf{r}, \lambda)$ introduced by a biological cell in a host medium of refractive index n_h, scattered wavefields $(\mathbf{E}_s, \mathbf{H}_s)$ propagate in directions of $\mathbf{k}_s$ described by the scattering angles of (θ_s, ϕ_s). Both indices, $n(\mathbf{r}, \lambda)$ and n_h, are related to their respective dielectric constants of $\varepsilon(\mathbf{r}, \lambda)$ and ε_h through a square-root relation as $n = \sqrt{\varepsilon}$ and for elastic scattering, the wave number $k = |\mathbf{k}_i| = |\mathbf{k}_s| = 2\pi/\lambda$. For a source-free, nonmagnetic, linear dielectric system $(\mu = 1)$, typical for a biological cell embedded within the host medium of carrying fluid, modeling of light scattering starts by solving the Maxwell equations for the total wavefields $(\mathbf{E}, \mathbf{B})$ [46]

$$\nabla \times \mathbf{E} + \frac{\partial \mathbf{B}}{\partial t} = 0, \nabla \times \mathbf{H} = \frac{\partial \mathbf{D}}{\partial t}, \quad \nabla \times \mathbf{D} = 0, \nabla \times \mathbf{B} = 0 \tag{11.1}$$

where $\mathbf{D} = \varepsilon_0 \varepsilon \mathbf{E}, \mathbf{H} = \mathbf{B}/\mu_0, \varepsilon_0$, and μ_0 are the permittivity and permeability of vacuum. Converting these equations into the frequency domain, we obtain the corresponding Helmholtz equations in regions inside and outside of an appropriately selected spatial volume V containing the biological cell

$$\left(\nabla^2 + k^2\right) \mathbf{E}\left(\mathbf{r}, \omega\right) = -4\pi \mathbf{f}\left(\mathbf{r}, \omega\right), \mathbf{r} \in V \tag{11.2}$$

$$\left(\nabla^2 + k^2\right) \mathbf{E}(\mathbf{r}, \omega) = 0, \qquad \text{otherwise} \tag{11.3}$$

where we have $k = (\mu_0 \varepsilon_0 \varepsilon_h \omega)^{1/2} = \omega/v$ with v as the wave speed. The vector function $\mathbf{f}(\mathbf{r}, \omega)$ on the right-hand side of Equation 11.2 is related to the dielectric constants as

$$\mathbf{f}(\mathbf{r},\omega) = \frac{1}{4\pi}\left(k^2 + \nabla\nabla\cdot\right)\left(\frac{\varepsilon(\mathbf{r},\omega)}{\varepsilon_h} - 1\right)\mathbf{E}(\mathbf{r},\omega) \tag{11.4}$$

Consequently, the term $-4\pi\mathbf{f}(\mathbf{r},\omega)$ in Equation 11.2 describes the optical heterogeneity introduced by the cellular structure and is responsible for the scattered wavefields. With Green's second identity in the region [46], the total electric field $\mathbf{E}(\mathbf{r},\omega)$ outside V can be found as

$$\mathbf{E}\left(\mathbf{r},\omega\right) = \mathbf{E}_i\left(\mathbf{r},\omega\right) + \int_V \mathbf{f}\left(\mathbf{r}',\omega\right) G\left(\mathbf{r},\mathbf{r}',\omega\right) d^3r' \tag{11.5}$$

where $G(\mathbf{r},\mathbf{r}',\omega) = \frac{e^{ik|\mathbf{r}-\mathbf{r}'|}}{4\pi|\mathbf{r}-\mathbf{r}'|}$ is the free space Green function. The second term on the right side of Equation 11.5 yields the electric field of scattered light $\mathbf{E}_s$. In the far-field zone satisfying $kr \to \infty$, one can replace $k|\mathbf{r}-\mathbf{r}'|$ by $kr - \mathbf{k}_s \cdot \mathbf{r}'$ in the exponential factor, and the scattered wavefield approaches to an outgoing spherical wave as

$$\mathbf{E}_s(\mathbf{r},\omega) = \mathbf{F}(\mathbf{k}_s,\mathbf{k}_i)\frac{e^{ikr}}{r} \tag{11.6}$$

with a modulated amplitude given by the vector function $\mathbf{F}(\mathbf{k}_s, \mathbf{k}_i)$

$$\mathbf{F}(\mathbf{k}_s,\mathbf{k}_i) = \frac{1}{4\pi}\int_V \left(k^2 + \nabla\nabla\cdot\right)\left(\frac{\varepsilon(\mathbf{r}',\omega)}{\varepsilon_h} - 1\right)\mathbf{E}(\mathbf{r}',\omega)e^{-i\mathbf{k}_s\cdot\mathbf{r}'}d^3r' \tag{11.7}$$

The intensity of scattered light measured by a detector at $\mathbf{r}$ in the far field can be therefore written as

$$I_s(\mathbf{r},\lambda) = \left|\mathbf{E}_s(\mathbf{r},\omega)\right|^2 = \frac{\left|\mathbf{F}(\mathbf{k}_s,\mathbf{k}_i)\right|^2}{r^2} \tag{11.8}$$

The function $\mathbf{F}(\mathbf{k}_s, \mathbf{k}_i)$ is often denoted as the scattering amplitude as a result of its square relation to the scattered light intensity. Equations 11.7 and 11.8 show that the angular distribution of scattered light is determined by $\mathbf{F}(\mathbf{k}_s, \mathbf{k}_i)$ which in turn depends on the cell optical structure represented by $\varepsilon(\mathbf{r}',\omega)$ or $n(\mathbf{r}',\lambda)$ via a volume integral. As can be seen from Equation 11.7 the volume integral contains the unknown total field $\mathbf{E}$, indicating that one must solve a difficult, if not impossible, inverse problem to determine the cellular structure $n(\mathbf{r}',\lambda)$ from the measured light scattering data for cells of complicated morphology. On the other hand, direct modeling of scattered light distribution can proceed accurately if $n(\mathbf{r}',\lambda)$ is known by numerically solving the total field $\mathbf{E}(\mathbf{r}',\omega)$ in the near-field zone from the Maxwell equations within a small volume V containing the biological cell followed by the calculation of the function $\mathbf{F}(\mathbf{k}_s, \mathbf{k}_i)$ through Equation 11.7. One approach is to apply a numerical method such as the FDTD algorithm to solve $\mathbf{E}(\mathbf{r}',\omega)$ from the two curl equations in Equation 11.1 as described in the next section.

11.2.2 The Formulation of Stokes Vector and Muller Matrix

Owing to the vector nature of the electromagnetic wavefields, the scatterer not only redirects the energy flux of the incident light but also modifies its polarization state along the way, as shown in Equation 11.7. Therefore, light scattering problems are in general polarization dependent and most efficiently described with a vector–matrix formalism. Before we introduce the definitions of the Stokes vector and Mueller matrix, let us first cast the Equations 11.6 and 11.7 in an amplitude form on the basis of the linearity of the Maxwell equations. We can resolve the incident and scattering fields into components parallel and perpendicular to the scattering plane defined by $\mathbf{k}_s$ and $\mathbf{k}_i$ as

$$\begin{pmatrix} E_{\|s} \\ E_{\perp s} \end{pmatrix} = \frac{e^{-ikr+ikz}}{-ikr} \begin{pmatrix} S_2 & S_3 \\ S_4 & S_1 \end{pmatrix} \begin{pmatrix} E_{\|0} \\ E_{\perp 0} \end{pmatrix} \tag{11.9}$$

The elements of 2×2 amplitude matrix in Equation 11.9 are complex functions of $\mathbf{k}_s$ or the scattering polar angle θ_s and the azimuthal angle ϕ_s, as shown in Figure 11.1. The amplitude matrix elements can be calculated from the scattering amplitude $\mathbf{F}(\mathbf{k}_s, \mathbf{k}_i)$ using two independent polarization configurations for the incident wave as the input data for the FDTD calculations [47].

Because of the high frequency of the optical wavefields, detected light signals are time averaged and proportional to the squares of the field amplitudes. This prompts the introduction of the real-valued Stokes vectors and Mueller matrix for relating the scattered light to the incident light, up to a multiplicative constant, so that [9, 38]

$$\begin{pmatrix} I_s \\ Q_s \\ U_s \\ V_s \end{pmatrix} = \frac{1}{k^2 r^2} \begin{pmatrix} S_{11} & S_{12} & S_{13} & S_{14} \\ S_{21} & S_{22} & S_{23} & S_{24} \\ S_{31} & S_{32} & S_{33} & S_{34} \\ S_{41} & S_{42} & S_{43} & S_{44} \end{pmatrix} \begin{pmatrix} I_i \\ Q_i \\ U_i \\ V_i \end{pmatrix} \tag{11.10}$$

where I, Q, U, and V are the Stokes parameters for the incident (with subscript i) and scattered (with subscript s) light, and S_{ij} (with $i, j = 1, 2, 3$, and 4) are the elements of the Mueller matrix characterizing the optical properties of the scatterer and are functions of θ_s and ϕ_s. The element S_{11} yields the probability of an unpolarized incident light being scattered into the direction of θ_s and ϕ_s, while other elements provide information on changes of the polarization states of the scattered light. When normalized over the solid angle of 4π steradian, S_{11} is also called the *scattering phase function*.

For a single biological cell as the scatterer, there are at most seven independent elements in the Mueller matrix and each element can be derived from the amplitude matrix of four complex elements [9]. In comparison, the Mueller matrix of a homogeneous sphere contains only four independent elements among the nonzero elements, that is, $S_{11} = S_{22}, S_{12} = S_{21}, S_{33} = S_{44}$, and $S_{34} = S_{43}$. Under the single-scattering condition, it has been shown that a Mueller matrix can be similarly defined for an ensemble of single scatterers at random locations by adding the matrices of individual scatterers as a result of incoherent superposition of the scattered Stokes parameters [48, 49]. In the case of spheres with the same size and refractive index, the Muller matrices are identical for single or sphere ensemble as the scatterer because of the spherical symmetry. For biological cells, however, the shape and optical structure posses no symmetry in general and vary even among the cells of the same phenotype because of their different growth patterns and stages in the cell cycle. Thus, all the 16 Mueller matrix elements of a scatterer consisting of multiple cells are, in general, expected to be independent [50].

11.3 FDTD Simulation with 3D Cellular Structures

We have built parallel FDTD computational codes for numerical simulation of light scattering by single cells on the basis of a serial code developed by Ping Yang and coworkers in Fortran 77 to model light scattering by ice crystals [47, 51]. Our FDTD codes incorporate enhanced structures in Fortran 90/95 for parallel implementation on our computing cluster of 32 CPUs [28, 30, 33, 50] and are continuously improved for performance enhancement, 3D cell structure handling capability, and portability. In this section, we briefly describe the FDTD algorithm and use of realistic 3D cell structures for the calculation of diffraction images.

11.3.1 The FDTD Algorithm

The FDTD algorithms were developed to numerically solve the two curl equations in the Maxwell equations in the time domain for obtaining the wavefields in the near-field region [43, 51]. In this method, the region of interest is divided into rectangular grid units, and the 3D structure of a biological cell is defined by properly assigning $n(\mathbf{r}, \lambda)$ over the grid points denoted by integer indices (i, j, and k) with Δx, Δy, and Δz as the sizes of a basic grid unit along the x, y, and z directions, respectively. Additionally, time is discretized with a step size of Δt. A central-difference scheme is employed to approximate the spatial and temporal derivatives of a field component u as

$$\left.\frac{\partial u}{\partial x}\right|_{x_i,t_n} = \frac{u|_{x_i+\Delta x/2,t_n} - u|_{x_i-\Delta x/2,t_n}}{\Delta x} + O\left[(\Delta x)^2\right] \tag{11.11}$$

$$\left.\frac{\partial u}{\partial t}\right|_{x_i,t_n} = \frac{u|_{x_i,t_n+\Delta t/2} - u|_{x_i,t_n-\Delta t/2}}{\Delta t} + O\left[(\Delta t)^2\right] \tag{11.12}$$

where the assignment of six field components as different u's is illustrated in Figure 11.2 with the components specified at staggered locations. The electric field components, (E_x, E_y, and E_z), are sampled on the edges of the grid unit and computed at multiples of time steps $t = n\Delta t$ while the magnetic field components, (H_x, H_y, and H_z), are sampled at the face centers of the grid unit and calculated at multiples of half time steps $t = (n + 1/2)\Delta t$. After the spatial and temporal derivatives in the curl equations are converted into finite-difference forms, the resultant finite-difference equations are then solved in a time-marching sequence by alternatively computing the electric and magnetic fields on a dual, staggered orthogonal grid within an appropriately selected volume V (white space) containing

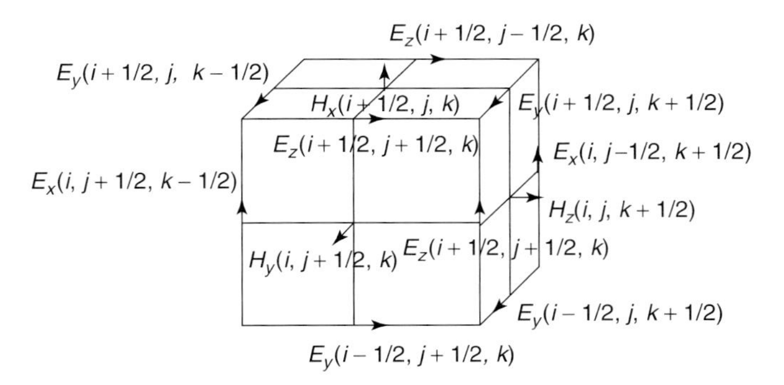

Figure 11.2 The field component assignment on a Yee grid unit centered at (j, j, k).

the scatterer. For example, the field components E_x and H_x can be found from the finite-difference equations as shown below:

$$E_x\Big|_{i,j,k}^{n+1} = E_x\Big|_{i,j,k}^{n} + \frac{\Delta t}{\varepsilon_{i,j,k}}\left(\frac{H_z\Big|_{i,j+\frac{1}{2},k}^{n+\frac{1}{2}} - H_z\Big|_{i,j-\frac{1}{2},k}^{n+\frac{1}{2}}}{\Delta y} - \frac{H_y\Big|_{i,j,k+\frac{1}{2}}^{n+\frac{1}{2}} - H_y\Big|_{i,j,k-\frac{1}{2}}^{n+\frac{1}{2}}}{\Delta z}\right) \quad (11.13)$$

$$H_x\Big|_{i,j,k}^{n+\frac{1}{2}} = H_x\Big|_{i,j,k}^{n-\frac{1}{2}} + \frac{\Delta t}{\mu_0}\left(\frac{E_y\Big|_{i,j,k+\frac{1}{2}}^{n} - E_y\Big|_{i,j,k-\frac{1}{2}}^{n}}{\Delta z} - \frac{E_z\Big|_{i,j+\frac{1}{2},k}^{n} - E_z\Big|_{i,j-\frac{1}{2},k}^{n}}{\Delta y}\right) \quad (11.14)$$

The equations for the remaining electric and magnetic field components are similar. The time and space leapfrog scheme described above yields an efficient, second-order accurate algorithm, and enforces automatically the divergence equations in Equation 11.1 and adopted boundary conditions. Furthermore, the field calculations are local, that is, only the values of the fields in neighboring grid points are needed. This makes the FDTD simulations scalable for efficient parallel execution. To ensure numerical stability of the algorithm, the spatial and temporal increments must satisfy the Courant–Friedrichs–Lewy (CFL) condition given by

$$\Delta t \leqslant \frac{1}{v\sqrt{1/\Delta x^2 + 1/\Delta y^2 + 1/\Delta z^2}} \quad (11.15)$$

For an incident wave of $\lambda = 1$ μm, the time step Δt must be less than approximately 0.1 fs for a grid unit size of $\Delta x = \Delta y = \Delta z = \lambda/20$.

To perform efficient simulations, the computing domain must be limited to just large enough to enclose the scatterer, and a suitable artificial boundary condition on the outer perimeter of the computational domain must be applied to "extend" the finite domain in simulations to infinity without artificial reflections. For this purpose, our codes adopted an absorbing boundary condition of perfectly matched layers (PML-ABC) introduced by Jean-Pierre Berenger [52], which is very efficient in reducing the artificial reflection in comparison to other ABCs by orders of magnitudes [53]. The PML-ABC method is based on splitting the electric and magnetic field components at the absorbing boundary with the possibility of assigning losses to the split components. It creates a fictitious anisotropic absorbing medium adjacent to the boundary of the FDTD domain backed by a perfectly conducting surface that has a wave impedance independent of the angle of incidence and frequency of outgoing scattered wavefields. We further found PML to be suitable suitable for parallel FDTD simulations since no additional interprocessor communication was required. Moreover, they can be generally located very close to the scatterer, which is advantageous over other ABCs for its relatively small memory requirement because of reduced domain size or the "white space."

In the PML region surrounding the computing domain, each field component is split into two subcomponents to achieve a high absorption rate of incident power which is updated according to PML-ABC [52]. For example, H_x is split into H_{xy} and H_{xz}, with H_{xy} updated according to the following relation

$$H_{xy}\Big|_{i+\frac{1}{2},j,k}^{n+\frac{1}{2}} = e^{-\sigma_y^* \Delta t/\mu} H_{xy}\Big|_{i+\frac{1}{2},j,k}^{n+\frac{1}{2}} + \frac{\left(1 - e^{-\sigma_y^* \Delta t/\mu}\right)}{\sigma_y^* \Delta y} \times \left[E_{zx}\Big|_{i+\frac{1}{2},j-\frac{1}{2},k}^{n} - E_{zx}\Big|_{i+\frac{1}{2},j+\frac{1}{2},k}^{n} + E_{zy}\Big|_{i+\frac{1}{2},j-\frac{1}{2},k}^{n} - E_{zy}\Big|_{i+\frac{1}{2},j+\frac{1}{2},k}^{n} \right] \quad (11.16)$$

where σ_y^* is the fictitious magnetic conductivity in the y direction in the PML region. To increase the speed of FDTD simulation, the incident field will only be accounted for in the region very close to the scatterer or V defined in Equation 11.2. The computational domain is separated into two parts: a near-field zone of volume V close to the scatterer where the total (incident and scattered) wavefields are considered and a far-field zone including the PML region where only the scattered wavefield are calculated. This approach is consistent with the linearity of the Maxwell equations which yields

$$\mathbf{E} = \mathbf{E}_i + \mathbf{E}_s \quad \text{and} \quad \mathbf{H} = \mathbf{H}_i + \mathbf{H}_s \quad (11.17)$$

It can be realized by using the equivalence theorem [54] to account for the effect of the incident wavefields in the near-field zone by the equivalent electric and magnetic currents on a virtual surface enclosing this region. This scheme of zoning provides several benefits for optimization of the FDTD algorithm. It allows arbitrary forms of incident wavefields for relatively simple coding of the interaction structures, and a large computational dynamic range. It also minimizes residue reflections from the ABC at the outermost surface of the FDTD domain and allows easy calculation of the far fields.

After establishing the computing grid, one can proceed to calculate the components of total wavefields, $\mathbf{E}(\mathbf{r}, \omega)$ and $\mathbf{H}(\mathbf{r}, \omega)$, in the near-field zone with the discretized finite-difference equation group. This will lead to the solutions of the scattered wavefields in the far-field zone according to Equation 11.7 in terms of the scattering amplitude $\mathbf{F}(\mathbf{k}_s, \mathbf{k}_i)$. Using two incident waves of crossed polarizations, for example, in x and y directions as shown in Figure 11.1, the amplitude matrix defined in Equation 11.9 can be obtained from $\mathbf{F}(\mathbf{k}_s, \mathbf{k}_i)$ in the following fashion [47, 55]:

$$S = \begin{pmatrix} F_{\parallel x} & F_{\parallel y} \\ F_{\perp x} & F_{\perp y} \end{pmatrix} \begin{pmatrix} \cos\theta_s & \sin\theta_s \\ \sin\theta_s & -\cos\theta_s \end{pmatrix} \tag{11.18}$$

where

$$\begin{bmatrix} F_{\parallel,x} \\ F_{\perp,x} \end{bmatrix} = \frac{-ik^3}{4\pi} \int_V \left(\frac{\varepsilon(\mathbf{r}', \omega)}{\varepsilon_h} - 1 \right) \begin{bmatrix} E_{\parallel,x}(\mathbf{r}', \omega) \\ E_{\perp,x}(\mathbf{r}', \omega) \end{bmatrix} e^{-i\mathbf{k}_s \cdot \mathbf{r}'} d^3 r' \tag{11.19}$$

and

$$\begin{bmatrix} F_{\parallel,y} \\ F_{\perp,y} \end{bmatrix} = \frac{-ik^3}{4\pi} \int_V \left(\frac{\varepsilon(\mathbf{r}', \omega)}{\varepsilon_h} - 1 \right) \begin{bmatrix} E_{\parallel,y}(\mathbf{r}', \omega) \\ E_{\perp,y}(\mathbf{r}', \omega) \end{bmatrix} e^{-i\mathbf{k}_s \cdot \mathbf{r}'} d^3 r' \tag{11.20}$$

Finally the amplitude matrix elements along scattering directions of $\mathbf{k}_i$ yield the angle-resolved Mueller matrix elements $S_{ij}(\theta_s, \phi_s)$ based on their relations [38].

11.3.2
Acquisition of 3D Cell Structure through Confocal Imaging

Biological cells possess a complicated and diverse morphology which must be reasonably portrayed to achieve accurate modeling of light scattering by these particles. Nearly all of the previous investigations of light scattering by cells utilize various smooth surface contours for simulation of cell structures. For example, intracellular components can be expressed in contours given by analytical functions such as spheres, coated spheres, ellipsoids, biconcave discs, and their combinations. Despite the fact that these structures allow rapid estimation of the correlations between distributions of scattered light and certain global features of cells such as the cell volume, they are totally inadequate for revealing the complex correlations that are essential for classification of real biological cells. This is especially true of diffraction image-based cell classification in which the fringe patterns are highly sensitive to the scatterer's morphology. For this reason, we have developed a confocal imaging-based method to acquire and reconstruct the 3D structures of cells for accurate modeling of light scattering [30, 35].

Confocal imaging microscopy employs a focused incident beam and applies a spatial filtering technique to the backwardly propagating light from the imaged cell to reduce the focal depth of the acquired 2D image for "optical sectioning." By translating the sample perpendicularly to the image plane a stack of 2D images can be obtained for analysis and reconstruction of 3D structures [56]. Compared to other experimental 3D imaging modalities such as various phase microscopy methods [57, 58], confocal microscopy is easily accessible to most researchers and straightforward to operate. For our study of the effect of nucleus and mitochondria, the challenges are to select specific dyes to obtain high-contrast images of these targeted components and develop an image analysis software for 3D reconstruction from the image stacks. In an earlier study, we developed a confocal imaging method to reconstruct 3D structures with nucleus and cytoplasm for modeling of light scattering by cultured cells derived from human lymphocyte B-cells [30]. This work was recently extended by adding mitochondria as the targeted organelles since they play an important role in cellular functions as the centers of metabolism activities and scattering of light [59].

Several fluorescent dyes were tested to identify targeted organelles with high specificity, emission, and resistance to photobleaching. The Syto-61 and MitoTracker-Orange (Invitrogen) were selected for double labeling of nuclei and mitochondria in various cell lines. After incubation in culture media with the two dyes for about 30 min, the cells were washed twice with phosphate buffered saline PBS solutions and then placed on a depressed glass slide for imaging with a confocal microscope (LSM510, Zeiss). Care was taken to ensure that the cells settled in the media on the glass slide before the images were taken in order to reduce possible cell motion. Since each stack may require up to 80 slices being acquired, image stack acquisition may take minutes to complete. For some cells with relatively low fluorescence intensity the exposure time has to be increased, so the stack acquisition time can be as long as 5 min. In these cases, an imaged cell can move up to 50 pixels between slice images of 512×512 pixels. Without correction, the cell translation can introduce severe distortion in the reconstructed 3D structures. An image alignment algorithm has been developed to correct the above problem by maximizing the image correlation between consecutive slices using a center slice image as Ref. [60].

After alignment, the confocal slice images were imported by an in-house developed software for image pixel sorting and 3D reconstruction. Pixels in each confocal slice image were separated into four bins of background, cytoplasm, nucleus, and mitochondria based on histogram analysis and edge detection in different color channels. Pixels were sorted according to their fluorescence light intensity in either the red (fake-color) channel of Syto-61 emission or the green channel of MitoTracker-Orange emission and to their spatial correlations to reduce the effect of noise. After pixel sorting of all confocal slices, the 3D structure of cells with nucleus and mitochondria was obtained by connecting slice images through interpolation. The 3D structure data files, in a zipped format, can be either converted into the stl format for 3D viewing or imported into the parallel FDTD code (after assignment of refractive indices) for light scattering modeling. Figure 11.3 shows two examples of reconstructed cell structures with different organelles stained.

11.4
Simulation and Measurement of Diffraction Images

Morphological features of biological cells are highly desired in life science research because of the strong correlation between functions and structures. For example, cytoskeleton and its response to various signaling molecules play critical roles in cell adhesion and migration in relation to the early formation of malignancy and metastasis and attract active research efforts. The observation of cytoskeleton

Figure 11.3 The reconstructed cell structures: (a) and (b) a NALM-6 cell derived from human B-cells with only nucleus stained; (c) and (d) a B16F10 cell derived from mouse melanoma cells with both nucleus and mitochondria stained.

in living cells is challenging owing to the difficulty of fluorescence labeling [5] and could be best achieved through diffraction imaging as we discussed above. To illustrate the capability of the diffraction imaging flow cytometry method, we present numerical and experimental results below based on the acquisition of unpolarized scattered light or the element S_{11} in terms of the Mueller matrix (Equation 11.10).

11.4.1 Numerical Results Based on FDTD Simulations

We first investigated numerically the effect of the nucleus on light scattering and the fringe pattern distribution in the calculated diffraction images. These images were obtained by projecting the angular dependence of the S_{11} element in the Mueller matrix (θ_s, ϕ_s) onto a plane representing a CCD camera placed perpendicularly to the y-axis as shown in Figure 11.1. As a result, the images shown in Figure 11.4 show the distributions of side scatters centered along the direction of $\theta_s = 90^\circ$ and $\phi_s = 180^\circ$ with half-cone angles of 20°. With the parallel FDTD code, we carried out simulations of angle-resolved distributions of scattered light with six NALM-6 cells derived from human B-cells to examine the effects of cell structure, nuclear shape and index, and cell orientation on the image patterns. To see the fringe patterns clearly, we set the r^{-2} factor in Equation 11.8 to 1 and the gray scale proportional to $\log(S_{11})$. Among the six cells, the results of three are presented in Figure 11.4 with two in the mitosis or M phase of their cycles, as shown by the cells 9 and 10 of enlarged nuclei in the upper panel. In these simulations, we fixed the refractive indices of hosting medium n_h and cytoplasm n_c at the wavelength $\lambda = 633$ nm and varied the nuclear index n_n and cell orientation angles of (θ_0, ϕ_0). Selected examples of calculated diffraction images are shown in the lower panel of Figure 11.4 to demonstrate the dependence of the image patterns on cell morphology, orientation, and nuclear index of refraction.

Figure 11.4a–d shows the effects of cell orientation and nuclear index on the scattered light distribution by the cell 8. As expected, both affect the fringe patterns in the diffraction images. These changes, however, are very different from those shown in Figure 11.4e–h obtained with the structures of cells 9 and 10 containing the enlarged and dividing nuclei. Comparing the diffraction images of the three cells, one can clearly see that the morphology of the cell nucleus has the most significant effect on the fringe patterns. Also seen from the images for cell 8 at the same orientation angle is that the refractive index of the cell nucleus relative

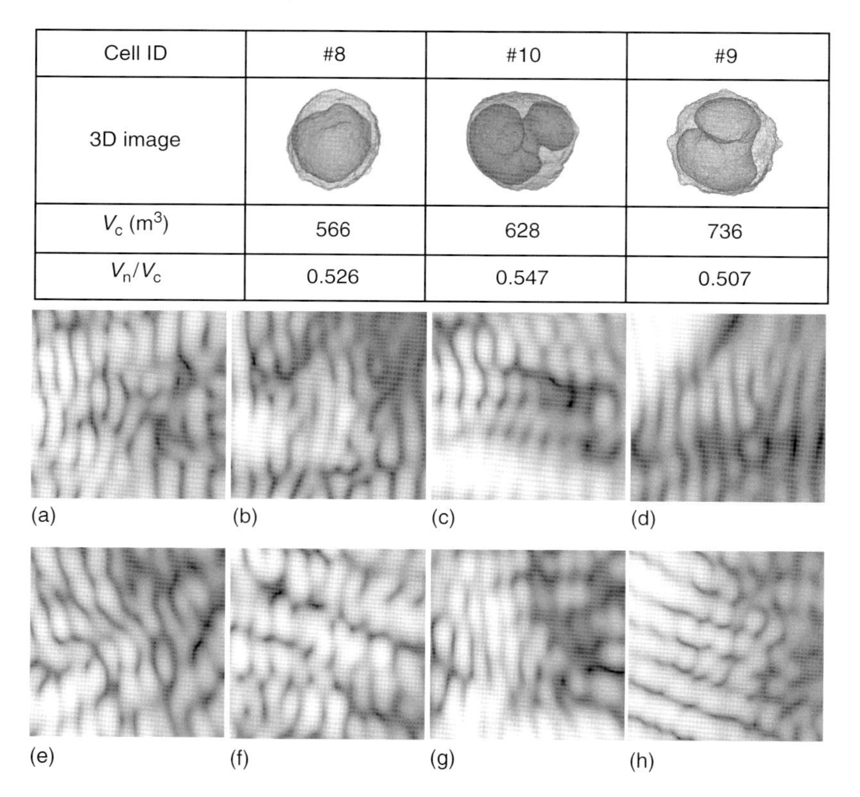

Cell ID	#8	#10	#9
3D image			
V_c (m^3)	566	628	736
V_n/V_c	0.526	0.547	0.507

Figure 11.4 The FDTD simulated diffraction images with three NAML-6 cell structures obtained through confocal imaging. Upper panel: the 3D views and values of morphological parameters with V_c as the volume of cell and V_n as the nuclear volume. Lower panel: diffraction images: (a) cell 8: $n_h = 1.400, \theta_0 = 28^\circ, \phi_0 = 13^\circ$; (b) cell 8 : $n_h = 1.400, \theta_0 = 45^\circ, \phi_0 = 135^\circ$; (c) cell 8 : $n_h = 1.450, \theta_0 = 28^\circ, \phi_0 = 13^\circ$; (d) cell 8 : $n_h = 1.450, \theta_0 = 45^\circ, \phi_0 = 135^\circ$; (e) cell 9 : $n_h = 1.450, \theta_0 = 28^\circ, \phi_0 = 13^\circ$; (f) cell 9 : $n_h = 1.450, \theta_0 = 45^\circ, \phi_0 = 135^\circ$; (g) cell 10 : $n_h = 1.450, \theta_0 = 28^\circ, \phi_0 = 13^\circ$; and (h) cell 10 : $n_h = 1.450, \theta_0 = 45^\circ$, $\phi_0 = 135^\circ$.

to that of the cytoplasm and hosting medium can have a significant effect on the fringe patterns, whereas cell orientation has the least effect. These results confirm that the diffraction images formed by scattered light correlate strongly with the 3D cell morphology and intracellular distribution of the refractive index, which provides a strong motivation for the development of diffraction imaging flow cytometer to rapidly classify cells. We should note here that the cells transported in a flow cytometer are under a constant pressure drop in the core fluid and tend to orient along a preferred direction. The fluid pressure can therefore further reduce the probability of misclassification due to orientation changes. We are currently performing a numerical study of diffraction images with cell structures, including both the nucleus and mitochondria, for the development of pattern recognition algorithms and the results will be reported soon.

11.4.2 Experimental Results Acquired with a Diffraction Imaging Flow Cytometer

We have investigated various configurations of flow chambers to achieve the goal of acquiring high-contrast diffraction images from flowing particles in a flow cytometer, which was accomplished recently with a jet-in-fluid design [34]. Two glass syringe pumps with precisely and independently controlled moving pistons supply two fluids of sheath and core to the flow chamber under an adjustable differential pressure between the two fluids. The core fluid contains the particles to be measured: either polystyrene microspheres of different diameters in deionized water or cultured cells in medium. In contrast to the conventional flow cytometers with laminar flows in air or glass-walled channels, the jet-in-fluid flow chamber allows the two fluids from an in-house built injection nozzle entering into a water-filled cuvette with its glass plane sides 5 mm away from the core fluid. The two fluids form a lamina flow with the core fluid carrying the particles from the injection nozzle to an exit tube, separated by a gap space of 4 mm long at the center of the glass cuvette. This design enables the laminar flow in a water-filled space inside the cuvette and thus eliminates the strong scattering background in a conventional flow chamber due to the fluid–air or fluid–glass interfaces, which are close to the flowing particle, of mismatched refractive index and often of very large curvature. The use of the nozzle and exit tube makes it possible to maintain a laminar flow condition at very small flow speeds ranging from 20 down to 1 mm s^{-1}. In addition, the long gap space allows alignment of multiple light beams for particle interrogation in the field-of-view (FOV) of the imaging unit. With a focused laser beam of diameter the same as that of the core fluid, both at about 30 μm, the positioning of the flowing particles in the laser beam waist can be well controlled. Detailed descriptions of the chamber design and its fluidic performance have been given in a recent publication [34].

The optics of the flow cytometer consists of two light sources and a split-view imaging unit as shown in Figure 11.5. The split-view imaging unit has been constructed for acquisition of diffraction images with an infinity-corrected microscope objective of 50× magnification and 0.55 numerical aperture (M Plan Apo,

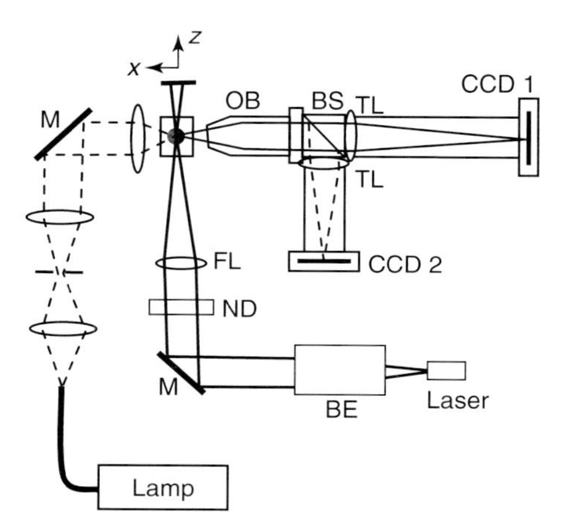

Figure 11.5 The schematic of the diffraction imaging flow cytometer with two light sources and a split-view imaging system: BE, beam expander; M, mirrors; FLs, focusing lenses; ND, neutral density filter; OB, objective; BS, beam splitter; and TLs, tube lens.

Mitutoyo). The use of a microscope objective makes it possible to align the flow chamber with the laser beam without a separate microscope for alignment. A white light Kohler illumination source with a tungsten lamp provides a noncoherent light source, while two lasers provide a coherent beam of wavelength $\lambda = 532$ or 633 nm for diffraction imaging. System alignment is carried out under the noncoherent illumination with a video camera and the coherent scattered light is measured with a cooled 16-bit CCD camera (U2000, Apogee) in the forms of diffraction image data. With an FOV of $144 \times 192\ \mu m^2$ at the focused position, the imaging unit can also be used to measure a flowing particle's speed by recording its blurring shadow under the noncoherent illumination over a long exposure time of 10–50 ms. For alignment, the imaging unit was focused first at the core fluid carrying particles to establish a reference position of $x = 0$ under the noncoherent illumination. This is followed by replacing the noncoherent illumination with a laser beam and translating the imaging unit to a defocused position of $x \neq 0$ for acquisition of diffraction images. Using spheres and cells embedded in gel, we found that fringe patterns in diffraction images acquired at $x < 0$, or away from the flow chamber, are not sensitive to the structure of the imaged object. Rather, only images acquired at $x > 0$ exhibit the fringe patterns depending on the object's 3D structure. It has been found that the diffraction imaging is optimized for $x = 200\ \mu m$ with the imaging unit moved toward the flow chamber. At this position, the fringe patterns of interest remain mostly within the FOV and of good signal-to-noise ratios [35]. Figure 11.6 shows examples of measured diffraction images of spheres (with d as diameter) and cells with an incident beam of $\lambda = 532$ nm and 21 mW in power. The camera's exposure time $\tau = 1$ ms and flow speed $v \sim 2\ mm\,s^{-1}$. These results demonstrate a strong correlation between the diffraction image patterns and the

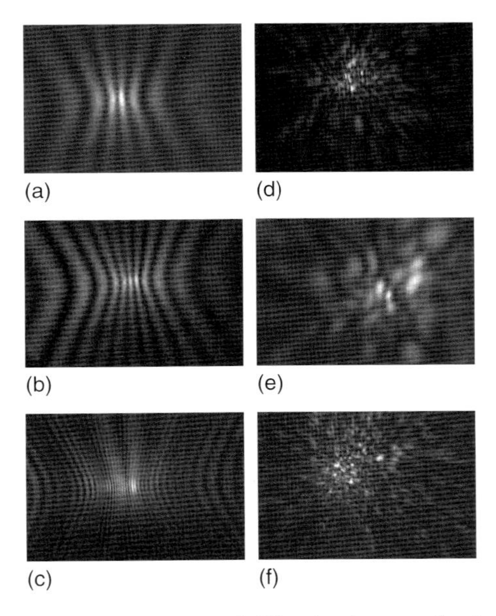

Figure 11.6 Measured diffraction images of polystyrene spheres with diameter d: (a) d = 5.2 μm, (b) d = 9.6 μm, (c) d = 25 μm; and cells (d) B16F10, (e) U-937, and (f) MCF-7.

3D morphology of the imaged particles. In the case of spheres, a Mie theory on light scattering by spheres [38] can be used to obtain numerical diffraction images, which shows an identical correlation between the fringe patterns and diameter d [35]. For cells, the image data exhibit significant differences among the fringe patterns of the three cell lines derived from human leukemic monocytes (U937), human breast adenocarcinoma cells (MCF-7), and mouse melanoma cells (B16F10). Detailed analysis of the fringe patterns and comparison to the cells' 3D morphological features is in progress.

11.5 Summary

A high-throughput method to extract 3D morphological features for cell classification is highly desired and flow cytometer is a tool-of-choice for rapid and quantitative assay of large cell populations. Despite the rich information content of the elastically scattered light signals, existing flow cytometers have very limited capacity to extract 3D morphology features. To incorporate a diffraction imaging method into the flow cytometry, however, two hurdles must be crossed. Firstly, one needs to acquire diffraction images of high contrast or large signal-to-noise ratios by elimination or reduction of the spurious scattering background caused by the index mismatch at optical interfaces near the flowing particles. Secondly, automated image analysis algorithms must be available to handle the extremely large data stream obtained with an imaging flow cytometer. This in turn demands efficient tools to accurately model light scattering by single biological cells and generate data for development and training pattern recognition software. Over the past decade, we have established a research program toward solving the above problems and moving the flow cytometer technology forward. A diffraction imaging flow cytometer has been constructed to demonstrate the feasibility of acquiring high-contrast diffraction image data from biological cells. The experimental results were found to be consistent with the numerical results on modeling light scattering with realistic 3D cell structures. We hope that pursuing these investigations will eventually lead to the realization of a new method of flow cytometry for rapid extraction of 3D morphological features from the scattered light signals and cell classification with no need for fluorescence labeling.

Acknowledgments

The authors thank K.M. Jacobs, L.V. Yang, J. Ding, D.A. Weidner, R.S. Brock, P. Yang, A.E. Ekpenyong, and C.L. Reynolds for their help in numerical simulations, cell imaging, and flow cytometric measurements. The research projects have been supported in part by a National Institutes of Health grant (1R15GM70798-01) and research grants from East Carolina University.

References

1. Melamed, M.R. (2001) in *Methods in Cell Biology* (eds Z. Darzynkiewicz, H.A. Crissman, and J.P. Robinson), Academic Press, San Diego, CA, pp. 3–17.
2. Wyatt, P.J. (1968) Differential light scattering: a physical method for identifying living bacterial cells. *Appl. Opt.*, **7**, 1879–1896.
3. Salzman, G.C., Crowell, J.M., Martin, J.C., Trujillo, T.T., Romero, A., Mullaney, P.F., and LaBauve, P.M. (1975) Cell classification by laser light scattering: identification and separation of unstained leukocytes. *Acta Cytol.*, **19**, 374–377.
4. Lindmo, T. and Steen, H.B. (1979) Characteristics of a simple, high-resolution flow cytometer based on a new flow configuration. *Biophys. J.*, **28**, 33–44.
5. Goldman, R.D. and Spector, D.L. (2005) *Live Cell Imaging: A Laboratory Manual*, Cold Spring Harbor Laboratory Press, Cold Spring Harbor, NY.
6. Steen, H.B. (1983) A microscope-based flow cytophotometer. *Histochem. J.*, **15**, 147–160.
7. Ong, S.H., Horne, D., Yeung, C.K., Nickolls, P., and Cole, T. (1987) Development of an imaging flow cytometer. *Anal. Quant. Cytol. Histol.: Int. Acad. Cytol. Am. Soc. Cytol.*, **9**, 375–382.
8. George, T.C., Basiji, D.A., Hall, B.E., Lynch, D.H., Ortyn, W.E., Perry, D.J., Seo, M.J., Zimmerman, C.A., and Morrissey, P.J. (2004) Distinguishing modes of cell death using the ImageStream multispectral imaging flow cytometer. *Cytometry A*, **59**, 237–245.
9. van de Hulst, H.C. (1957) *Light Scattering by Small Particles*, Structure of Matter Series, John Wiley & Sons, Inc., New York.
10. Gabor, D. (1948) A new microscopic principles. *Nature*, **161**, 777–778.
11. Leith, E.N. and Upatnieks, J. (1964) Wavefront reconstruction with diffused

illumination and three-dimensional objects. *J. Opt. Soc. Am.*, **54**, 1295–1301.
12. Bessis, M. and Mohandas, N. (1975) A diffractometric method for the measurement of cellular deformability. *Blood Cells*, **1**, 307–313.
13. Seger, G., Achatz, M., Heinze, W., and Sinsel, F. (1977) Quantitative extraction of morphologic cell parameters from the diffraction pattern. *J. Histochem. Cytochem.*, **25**, 707–718.
14. Leung, A.F. (1982) Laser diffraction of single intact cardiac muscle cells at rest. *J. Muscle Res. Cell. Motil.*, **3**, 399–418.
15. Streekstra, G.J., Hoekstra, A.G., Nijhof, E., and Heethaar, R.M. (1993) Light scattering by red blood cells in ektacytometry: Fraunhofer versus anomalous diffraction. *Appl. Opt.*, **32**, 2266–2272.
16. Maleki, M.H., Devaney, A.J., and Schatzberg, A. (1992) Tomographic reconstruction from optical scattered intensities. *J. Opt. Soc. Am. A*, **9**, 1356–1363.
17. Salzman, G.C., Singham, S.B., Johnston, R.G., and Bohren, C.F. (1990) in *Flow Cytometry and Sorting*, 2nd edn (eds M.R. Melamed, T. Lindmo, and M.L. Mendelsohn), John Wiley & Sons, Inc., New York, p. 5.
18. Peters, D.C. (1979) A comparison of mercury arc lamp and laser illumination for flow cytometers. *J. Histochem. Cytochem.*, **27**, 241–245.
19. Salzman, G.C., Crowell, J.M., Goad, C.A., Hansen, K.M., Hiebert, R.D., LaBauve, P.M., Martin, J.C., Ingram, M.L., and Mullaney, P.F. (1975) A flow-system multiangle light-scattering instrument for cell characterization. *Clin. Chem.*, **21**, 1297–1304.
20. Jovin, T.M., Morris, S.J., Striker, G., Schultens, H.A., Digweed, M., and Arndt-Jovin, D.J. (1976) Automatic sizing and separation of particles by ratios of light scattering intensities. *J. Histochem. Cytochem.*, **24**, 269–283.
21. Bartholdi, M., Salzman, G.C., Hiebert, R.D., and Seger, G. (1977) Single-particle light-scattering measurements with a photodiode array. *Opt. Lett.*, **1**, 223.
22. Holler, S., Pan, Y., Chang, R.K., Bottiger, J.R., Hill, S.C., and Hillis, D.B. (1998) Two-dimensional angular optical scattering for the characterization of airborne microparticles. *Opt. Lett.*, **23**, 1489–1491.
23. Neukammer, J., Gohlke, C., Hope, A., Wessel, T., and Rinneberg, H. (2003) Angular distribution of light scattered by single biological cells and oriented particle agglomerates. *Appl. Opt.*, **42**, 6388–6397.
24. Singh, K., Liu, C., Capjack, C., Rozmus, W., and Backhouse, C.J. (2004) Analysis of cellular structure by light scattering measurements in a new cytometer design based on a liquid-core waveguide. *IEE Proc.*, **151**, 10–16.
25. Su, X.T., Singh, K., Capjack, C., Petracek, J., Backhouse, C., and Rozmus, W. (2008) Measurements of light scattering in an integrated microfluidic waveguide cytometer. *J. Biomed. Opt.*, **13**, 024024.
26. Maltsev, V.P. (2000) Scanning flow cytometry for individual particle analysis. *Rev. Sci. Instrum.*, **71**, 243–255.
27. Zharinov, A., Tarasov, P., Shvalov, A., Semyanov, K., van Bockstaele, D.R., and Maltsev, V.P. (2006) A study of light scattering of mononuclear blood cells with scanning flow cytometry. *J. Quant. Spectrosc. Radiat. Transfer*, **102**, 121–128.
28. Brock, R.S., Hu, X.H., Yang, P., and Lu, J.Q. (2005) Evaluation of a parallel FDTD code and application to modeling of light scattering by deformed red blood cells. *Opt. Exp.*, **13**, 5279–5292.
29. Lu, J.Q., Yang, P., and Hu, X.H. (2005) Simulations of light scattering from a biconcave red blood cell using the FDTD method. *J. Biomed. Opt.*, **10**, 024022.
30. Brock, R.S., Hu, X.H., Weidner, D.A., Mourant, J.R., and Lu, J.Q. (2006) Effect of detailed cell structure on light scattering distribution: FDTD study of a B-cell with 3D structure constructed from confocal images. *J. Quant. Spectrosc. Radiat. Transfer*, **102**, 25–36.
31. Lu, J.Q., Brock, R.S., Yang, P., and Hu, X.H. (2007) Modeling of light scattering by single red blood cells with a FDTD method, in *Optics of Biological Particles* (eds A. Hoekstra, V.P. Maltsev, and

G. Videen), Springer-Verlag, New York, pp. 213–242.

32. Ding, H., Lu, J.Q., Brock, R.S., Burke, L.P., Weidner, D.A., McConnell, T.J., and Hu, X.H. (2007) Angle-resolved measurement and simulations of mueller matrix of B-cells and HL60 cells. Progress in Electromagnetics Research Symposium, (PIERS).
33. Yurkin, M.A., Hoekstra, A.G., Brock, R.S., and Lu, J.Q. (2007) Systematic comparison of the discrete dipole approximation and the finite difference time domain method for large dielectric scatterers. *Opt. Exp.*, **15**, 17902–17911.
34. Jacobs, K.M., Lu, J.Q., and Hu, X.H. (2009) Development of a diffraction imaging flow cytometer. *Opt. Lett.*, **34**, 2985–2987.
35. Jacobs, K.M., Yang, L.V., Ding, J., Ekpenyong, A.E., Castellone, R., Lu, J.Q., and Hu, X.H. (2009) Diffraction imaging of spheres and melanoma cells with a microscope objective. *J. Biophoton.*, **2**, 521–527.
36. Hoekstra, A., Maltsev, V.P., and Videen, G. (2007) Optics of biological particles, *Mathematics, Physics and Chemistry*, NATO Science Series II, Vol. 238, Springer, Dordrecht.
37. Sun, W., Loeb, N.G., and Fu, Q. (2002) Finite-difference time domain solution of light scattering and absorption by particles in an absorbing medium. *Appl. Opt.*, **41**, 5728–5743.
38. Bohren, C.F. and Huffman, D.R. (1983) *Absorption and Scattering of Light by Small Particles*, John Wiley & Sons, Inc., New York.
39. Logan, N.A. (1965) Survey of some early studies of the scattering of plane waves by a sphere. *Proc. IEEE*, **53**, 773–785.
40. Toon, O.B. and Ackerman, T.P. (1981) Algorithms for the calculation of scattering by stratified spheres. *Appl. Opt.*, **20**, 3657–3660.
41. Yee, S.K. (1966) Numerical solutions of initial boundary problems involving Maxwell's equations in isotropic materials. *IEEE Trans. Antennas Propg.*, **14**, 302–307.
42. Dunn, A. and Richard-Kortum, R. (1996) Three-dimensional computation of light scattering from cells. *IEEE J. Sel. Top. Quantum. Electron.*, **2**, 898–890.
43. Taflove, A. and Hagness, S.C. (2000) *Computational Electrodynamics: The Finite-difference Time-domain Method*, 2nd edn, Artech House, Boston, MA.
44. Draine, B.T. and Flatau, P.J. (1994) Discrete-dipole approximation for scattering calculations. *J. Opt. Soc. Am. A*, **11**, 1491–1499.
45. Yurkin, M.A. and Hoekstra, A.G. (2007) The discrete dipole approximation: an overview and recent developments. *J. Quant. Spectrosc. Radiat. Transfer*, **106**, 558–589.
46. Jackson, J.D. (1999) *Classical Electrodynamics*, 3rd edn, John Wiley & Sons, Inc., New York.
47. Yang, P. and Liou, K.N. (1996) Finite-difference time domain method for light scattering by small ice crystals in three-dimensional space. *J. Opt. Soc. Am. A*, **13**, 2072–2085.
48. Fry, E.S. and Kattawar, G.W. (1981) Relationships between elements of the Stokes matrix. *Appl. Opt.*, **20**, 2811–2814.
49. Mishchenko, M.I., Hovenier, J.W., and Mackowski, D.W. (2004) Single scattering by a small volume element. *J. Opt. Soc. Am. A*, **21**, 71–87.
50. Ding, H., Lu, J.Q., Brock, R.S., McConnell, T.J., Ojeda, J.F., Jacobs, K.M., and Hu, X.H. (2007) Angle-resolved mueller matrix study of light scattering by B-cells at three wavelengths of 442, 633 and 850nm. *J. Biomed. Opt.*, **12**, 034032.
51. Yang, P. and Liou, K.N. (1999) in *Light Scattering by Nonspherical Particles* (eds M.I. Mishchenko, J.W. Hovenier, and L.D. Travis), Academic Press, San Diego, CA, p. 7.
52. Berenger, J.P. (1994) A perfectly matched layer for the absorption of electromagnetic waves. *J. Comput. Phys.*, **114**, 185–200.
53. Katz, D.S., Thiele, E.T., and Taflove, A. (1994) Validation and extension to three dimensions of the Berenger PML absorbing boundary condition for FD-TD meshes. *IEEE Microw. Guid. Wave Lett.*, **4**, 268–270.

54. Schelkunoff, S.A. (1943) *Electromagnetic Waves*, D. Van Nostrand Company Inc., New York.
55. Zhai, P.W., Lee, Y.K., Kattawar, G.W., and Yang, P. (2004) Implemting the near- to far-field transformation in the finite-difference time-domain method. *Appl. Opt.*, **43**, 3738–3746.
56. Pawley, J.B. (1995) *Handbook of Biological Confocal Microscopy*, 2nd edn, Plenum Press, New York.
57. Choi, W., Fang-Yen, C., Badizadegan, K., Oh, S., Lue, N., Dasari, R.R., and Feld, M.S. (2007) Tomographic phase microscopy. *Nat. Methods*, **4**, 717–719.
58. Lue, N., Choi, W., Popescu, G., Yaqoob, Z., Badizadegan, K., Dasari, R.R., and Feld, M.S. (2009) Live cell refractometry using Hilbert phase microscopy and confocal reflectance microscopy. *J. Phys. Chem.*, **113**, 13327–13330.
59. Beauvoit, B., Evans, S.M., Jenkins, T.W., Miller, E.E., and Chance, B. (1995) Correlation between the light scattering and the mitochondrial content of normal tissues and transplantable rodent tumors. *Anal. Biochem.*, **226**, 167–174.
60. Ekpenyong, A.E., Ding, J., Yang, L.V., Leffler, N.R., Lu, J.Q., Brock, R.S., and Hu, X.H. (2009) Study of 3-D cell morphology and effect on light scattering distribution. Advanced Microscopy Techniques/European Conferences on Biomedical Optics, (SPIE Proceedings, 2009), p. 73671J.

12 An Integrative Approach for Immune Monitoring of Human Health and Disease by Advanced Flow Cytometry Methods

Rabindra Tirouvanziam, Daisy Diaz, Yael Gernez, Julie Laval, Monique Crubezy, and Megha Makam

12.1 Introduction

Forty or so years ago, flow cytometry emerged as a critically important and versatile tool, by virtue of its ability to yield data on multiple parameters at the single-cell level and on a large number of cells per second, thereby producing statistically robust results [1]. With the advent of a multitude of monoclonal antibodies against surface and intracellular epitopes [2], as well as the development of fluorescent probes against lipid moieties, small molecules, and enzymes, the possibilities for functional analysis of cells by flow cytometry literally exploded. In keeping with the development of antibodies and probes (by companies such as BD Biosciences, Invitrogen, eBioscience, Abcam, and others), conventional flow cytometry has evolved from a cutting-edge tool, able to measure four simultaneous parameters (two scatters and two fluorescences), to a widely distributed technology (an estimated 50 000 flow cytometers are now scattered around the globe, manufactured by biotechnology giants such as BD Biosciences, Beckman Coulter, and so on), with the ability to measure more than 20 parameters (same 2 scatters and >18 colors).

Besides, newly developed offshoots of flow cytometry have emerged that extend the capabilities for single-cell analysis. Among them, the most promising are perhaps the technology combining high-speed acquisition in flow with full imaging (e.g., the ImageStream platform from Amnis, Inc. [3]) and hyphenated flow cytometry-mass spectrometry enabling highly multiplexed analysis using 50+ mass tag-conjugated antibodies (e.g., the CyTOF platform from DVSSciences, Inc. [4]). At this point in time, flow cytometry-based methods as a whole offer extremely attractive properties for cell analysis that far outweigh those offered by slide-based cytometry methods (Table 12.1). Along with the above-described technological leaps in flow cytometry, computational efforts are also under way to provide better abilities for experimental design and semi- to fully automated data analysis of single samples [5], and complex statistical methods for combined analysis of large sample sets [6].

However, for all the progress that has been made on the technological end, our use of flow cytometry for the monitoring of human health and disease has not reached

Advanced Optical Flow Cytometry: Methods and Disease Diagnoses, First Edition. Edited by Valery V. Tuchin.

Table 12.1 Current methods for flow- and slide-based cytometry.

Technological tool	Measurement method	Rapidity and quantitation	Phenotypic data content	Morphological data content	Subcellular resolution	Potential for automation
Optical microscopy	Slide: RGB, fluorescence	Low	Medium	Medium	High	Low to semi
Manual confocal microscopy	Slide: RGB, fluorescence	Low	Medium	High	**High ($\sim$1000×)**	Low
Laser confocal microscopy	Slide: RGB, fluorescence	Medium	Medium	Medium	High	Semi to full
Conventional flow cytometer	Flow: scatter, fluorescence	**High (up to 10 K events/s)**	High (up to 20 fluorescences)	Low	No image	Semi to full
CyTOF flow/mass spec cytometer	Flow: mass tag quantitation	High but low yield ($\sim$1/50 cells)	**High (>50 mass tags)**	None	No image	Semi to full
Imaging flow cytometer	Flow: scatter, visible colors, fluorescence	High (up to 1 K events/s)	Medium (>10 fluorescences)	**High (100 s of parameters)**	Medium (up to 60×)	Semi to full

For each of the properties listed, the most efficient tool is indicated in bold, with conservative benchmark characteristics (i.e., enabling exploitation of the tool without losing on data quality).

the level of excellence and widespread use as those in mouse developmental biology and immunology, for example. Here, we will delineate key steps to move past the current limitations and truly enable the use of advanced flow cytometry tools for human research. First, we will illustrate how simple precautions can be taken to promote less artifactual methods for sample collection that not only simplify and reduce the cost of human research but also enable better standardization. These lead to the development of true *standard operating procedures* (SOPs) that can be deployed in the most rigorous of research environments, such as studies of samples collected in remote areas of the world (away from the laboratory), and on the other end of the spectrum, multicenter trials.

Second, we will provide a practical overview of reagents available for advanced flow cytometry analysis of human samples and illustrate their use in various contexts, so as to highlight the enormous opportunities for research that they entail. To conclude, we will highlight trends that we believe will constitute the future of advanced flow cytometry technologies, from their application to less obvious corners of cell biology, to the survey of molecular events not usually associated with these technologies. It is our goal to provide the reader with a better handle on this important set of tools, which, if used appropriately, can renew the field of human research by bringing novel insights into the intricate relations between human immunity and complex correlates such as age, gender, ethnicity, environmental exposure, health conditions, and therapies that make each one of us a subject worthwhile of scientific investigation.

12.2 Optimized Protocols for Advanced Flow Cytometric Analysis of Human Samples

12.2.1 Key Limitations of Current Experimental Approaches: Technical and Scientific Biases

Looking back at the initial years of human research using flow cytometry (1970s–1980s), major advances were made by using clever staining combinations of three or four markers to delineate subsets of leukocytes and other large elements (erythrocytes and platelets) in whole blood and bone marrow [7]. In the next two decades (1980s–2000s), however, the impetus given to (i) lymphocyte immunology by human immunodeficiency virus research and (ii) stem cell research by the discovery of specific markers led to the widespread use of separation methods, ever more complex [8], to purify the subsets of peripheral blood mononucleated cells (PBMCs) from the rest of the elements in blood, often deemed unimportant.

Typically, blood is harvested and processed through density gradient centrifugation using Ficoll/Percoll to isolate a PBMC pellet that is then processed through further isolation steps if necessary, until staining, analysis, and sorting for further functional analysis [9]. Studies demonstrating the presence of significant artifacts associated with the use of density gradient centrifugation [10–12] were largely ignored, and till today this method remains widely used. This, despite the fact that PBMCs in the circulation are by all accounts not as informative as those found in tissues (lymph nodes or lymphoid-associated mucosal sites), whereas non-PBMC elements in blood, that is, granulocytes, platelets, and erythrocytes, outnumber them by one to several orders of magnitude and are involved in rapid responses such as anaphylaxis, septic shock, clotting, and so on.

While non-PBMC elements were being largely ignored in mainstream immunological studies, studies of specific diseases and syndromes led to the rediscovery of their important role in human biology. For example, platelets were found to hold major reserves of TGFβ and CCL5, two mediators associated with a number of chronic diseases [13]. Platelets were also found to form temporary aggregates with leukocytes, leading to functional modulation in both metabolic and inflammatory processes [14, 15]. Erythrocytes also form temporary bridges with leukocytes that are physiologically relevant [16, 17].

Most importantly, among non-PBMC elements in blood, granulocytes have been the subject of renewed scrutiny in recent years. Neutrophils, the dominant leukocyte subset in humans (50–70% in blood, produced at 10^9 kg^{-1} day^{-1}), have suffered a reputation of "one-trick ponies," that is, terminally differentiated cells with no flexibility to alter their phagocytic, pro-apoptotic program with which they leave the bone marrow. The dominance of mouse models helped solidify this notion, since neutrophils clearly bear less important functions in mice than in humans [18]. However, work by our group in cystic fibrosis (CF) [19] and by other groups in alternative disease contexts [20] has led to revisiting of conventional concepts about neutrophils and establishing them as prime targets for immune monitoring. Indeed, not only are neutrophils the body's first responders in an immune alert but they also bear considerable importance in shaping the activity of late-coming leukocyte subsets, such as lymphocytes, into various types of second-line responses [21]. Consistently, it is now recognized that neutrophils can undergo profound reprogramming, including an extended lifespan, when finding appropriate conditions in peripheral organs [22].

Other granulocytes, basophils [23] and eosinophils [24], have also been the subject of much attention owing to their ability to act as organizers of immune responses in contexts such as asthma and allergy, two major ailments of modern society. Here again, the one-dimensional concept of basophil and eosinophil function, linked to their ability to degranulate and cause rapid responses, has been significantly augmented by the notion that these cells are apt at presenting antigen and at instructing second-line leukocytes toward the development of sustained responses [25, 26].

12.2.2
Developing Better Protocols for Flow Cytometry Studies of Cells from Human Subjects: Enabling Holistic Studies of Human Health and Disease

On the basis of the above arguments, it is easy to realize why changes in the generic protocols used in flow cytometric analysis of human samples need to undergo drastic changes. At a minimum, such changes should enable researchers to take into account, along with PBMC, the diversity of non-PBMC elements, as well as enable a more holistic survey of functional associations between these diverse elements. Non-PBMC elements (platelets, erythrocytes, and granulocytes) are by nature more reactive than PBMC, which is another reason why most investigators shy away from studying them and opt to separate them from PBMC to start with. Rather than considering this reactivity as a liability, one should in fact consider with scientific appetite the range of opportunities these elements offer toward a more insightful exploration of human biology.

Nonetheless, this reactivity brings a key requirement to protocols aiming to exploit non-PBMC elements. Indeed, degranulation, aggregation, and clotting reactions have evolved such low thresholds for activation that one must take particular care not to trigger them when implementing a whole blood protocol. Thus, whole blood protocols require proper conditions adapted to the task at hand, taking into account (i) time: generally short incubation times, totaling less than 1 h before fixation; (ii) temperature: 4 °C can help alleviate spontaneous cell activation, notably aided by the use of a buffer containing calcium-chelating agents such as ethylene diamine tetra acetic acid; (iii) tubes: some plastic polymers lead to surface activation of some blood elements; and even (iv) shear stress: blood normally circulates and stagnation may lead to stochastic activation. In the last few years, our group has published several studies based on such protocols (Table 12.2), and this chapter offers many illustrations of their power.

Of course, beyond the renewed study of non-PBMC elements, the type of whole blood protocols that we advocate for here bear two important additional advantages: (i) they enable phenotypic and functional profiling of PBMC subsets and (ii) they are applicable not only to blood but also to any other suspension of cells from peripheral organs (the airway fluid is chosen as an example in many figures in this chapter), provided a few relevant adaptations in processing and staining steps are brought about. Another important aspect of the protocols developed for whole blood studies is that although they considerably widen the biological landscape subject to investigation, they are generally greatly simplified (less steps), more standardizable (no variations in yield due to artifacts of cell separation) and less expensive (no cost associated with density gradient preparations) compared with PBMC-based protocols. Old habits die hard, however, and despite this array of advantages, density gradient-based protocols still have a long future in front of them, although they should arguably be limited only to protocols aiming to purify rare subsets of cells for cell culture or other similar downstream assays requiring concentration, rather than for straightforward functional analysis.

Another notion that we have not touched upon yet is that of the considerable yield provided by whole blood when considering most non-PBMC elements. Indeed, human blood contains an average of: (i) $4-6 \times 10^9\,\mu l^{-1}$ erythrocytes; (ii) $1.5-4.5 \times 10^5\,\mu l^{-1}$ platelets; and (iii) $2-6 \times 10^3\,\mu l^{-1}$ neutrophils; eosinophils, T-cells, B-cells, and basophils being much less abundant. Thus, studies of erythrocytes, platelets, and neutrophils can be performed on minute amounts of whole blood, typically one to two orders of magnitude lower than those using blood for PBMC purification (i.e., 50 µl vs 5 ml, respectively). Hence, studies of these non-PBMC elements are particularly easy to implement on populations that are difficult to

Table 12.2 Modular flow cytometry approach for human immune monitoring.

Modular steps	Principles	Timeframe
A – Sample collection/processing	Avoid artifacts: use minimally invasive collection methods, minimal processing	0.5–2 h
B – Viability stain	Favor robust probes (fixable, freeze/thaw proof, e.g., Live/Dead® series)	0.5–1 h (steps B and C may be combined)
C – Surface stain	Mix antibodies/probes extemporaneously, favor directly conjugated reagents	
D – Fixation	Some probes may not withstand fixation. Else, favor Lyse/Fix Phosflow®.	0.5 h to 2 days
E – Permeabilization	Use saponin (regular epitopes) or MeOH (phosphoepitopes, needing denaturing)	0.5 h
F – Intracellular staining	Mix antibodies/probes extemporaneously, favor directly conjugated reagents	0.5 h
G – Freezing/(shipping)/thawing	Lyse/Fix Phosflow® is an adequate medium for long-term freezing/shipping.	Up to several months
H – Sample acquisition	FACS or image cytometry	<0.5 h per sample
Common protocols	**Sequence (parentheses indicates optional steps)**	**Timeframe**
1 – Fresh surface staining	A–B–C–(D, depending on fixability of probes)–H	2 h to 3 days
2 – Batched surface staining	A–B–C–D–G–H	Up to several months
3 – Fresh phosphostaining	A–B–C–D–E, using MeOH–F–H	2 h to 3 days
4 – Fresh intracellular staining	A–B–(C, depending on need)–D–E, using saponin–F–H	2 h to 3 days
5 – Batched intracellular staining	A–B–(C, depending on need)–D–G–E, using saponin–F–H	Up to several months

sample or that are averse to blood draws by venipuncture (e.g., newborns, infants, and toddlers, in particular), since they can rely on blood drops collected by finger pricks. This practical advantage is also critical if one is interested in deploying longitudinal approaches, in which samples are collected at short time intervals, thus making repeated blood draws, but not repeated finger pricks, impractical. Finally, the ability to implement flow cytometry analyses on small volumes means that approximately 20 stains can be run on 1 ml of blood, or, as customary in this era of semi-automated plate-based assays, 5 ml of blood can yield enough material to fill an entire 96-well plate for parallel, mid-throughput analyses of erythrocytes, platelets, or neutrophils. Of course, body fluids with equivalent cell yield, for example, joint fluid from patients with rheumatoid arthritis or airway fluid from patients with chronic or acute lung inflammation can also be used according to similar protocols.

12.2.3
Integrating Flow Cytometry into a Wider Framework for Experimental Research on Human Samples

Beyond the use of cells in flow cytometry, blood and other body fluids can yield important information based on free-floating material including small metabolites, lipid, and protein mediators (as well as resident fungi, bacteria, and viruses) that may be associated with the condition studied. Therefore, it is important to envision the study of cells by flow cytometry in a larger context, whereby cell-based assays become integrated with other platforms with a complementary data output that further informs on the intricate biological mechanisms at play in human health and disease. This approach is conducive to the emergence of scientific concepts that go well beyond cell properties and are more appropriately categorized as "systems biology." While systems biology is generally envisioned in model systems such as yeast [27], *Drosophila melanogaster* [28], and possibly rodents, its application to human health and disease is bound to bring major changes in science, healthcare, and other aspects of society influenced by human biology [29, 30].

An example of integrated research framework placing flow cytometry methods at the heart of the investigation process is shown in Figure 12.1. This example depicts our current framework pioneered in studies of patients with CF, in whom neutrophilic inflammation of the airways is associated with severe proteolysis and oxidation, inducing major changes in the molecular content of body fluids. CF is also characterized by an ecosystemic alteration of the airway mucosa, leading to the selection of resistant, opportunistic pathogens that colonize the airway lumen and cohabit with neutrophils. While idiosyncratic to CF, this complex juxtaposition of immunoinflammatory, infectious, and metabolic processes [31] exemplifies the multidimensional nature of biological processes underlying human health and disease in general. The framework presented in Figure 12.1 also highlights the importance of blood as a control for other body fluids. The latter may indeed feature some of the elements present in blood, but with modifications induced by transmigration and exposure to a different set of environmental cues. Taken together, the various methods called upon for functional analysis of the samples and fractions thereof yield a patchwork of phenotypic information ("phenomics") that hold considerable mechanistic value [27].

The idea of using flow cytometry-based methods (i.e., conventional multiparameter fluorescence-activated cell analysis and sorting – fluorescence-activated cell sorting (FACS) – and to a lesser extent image cytometry) as the basis for systems biology of the human body is novel and should be considered in the light of the large number of FACS machines available worldwide, notably in university hospitals within industrialized nations [32]. With the appropriate attention to details in processing and staining protocols, simple assays surveying the major activation markers expressed by the various elements in blood could revolutionize the use of

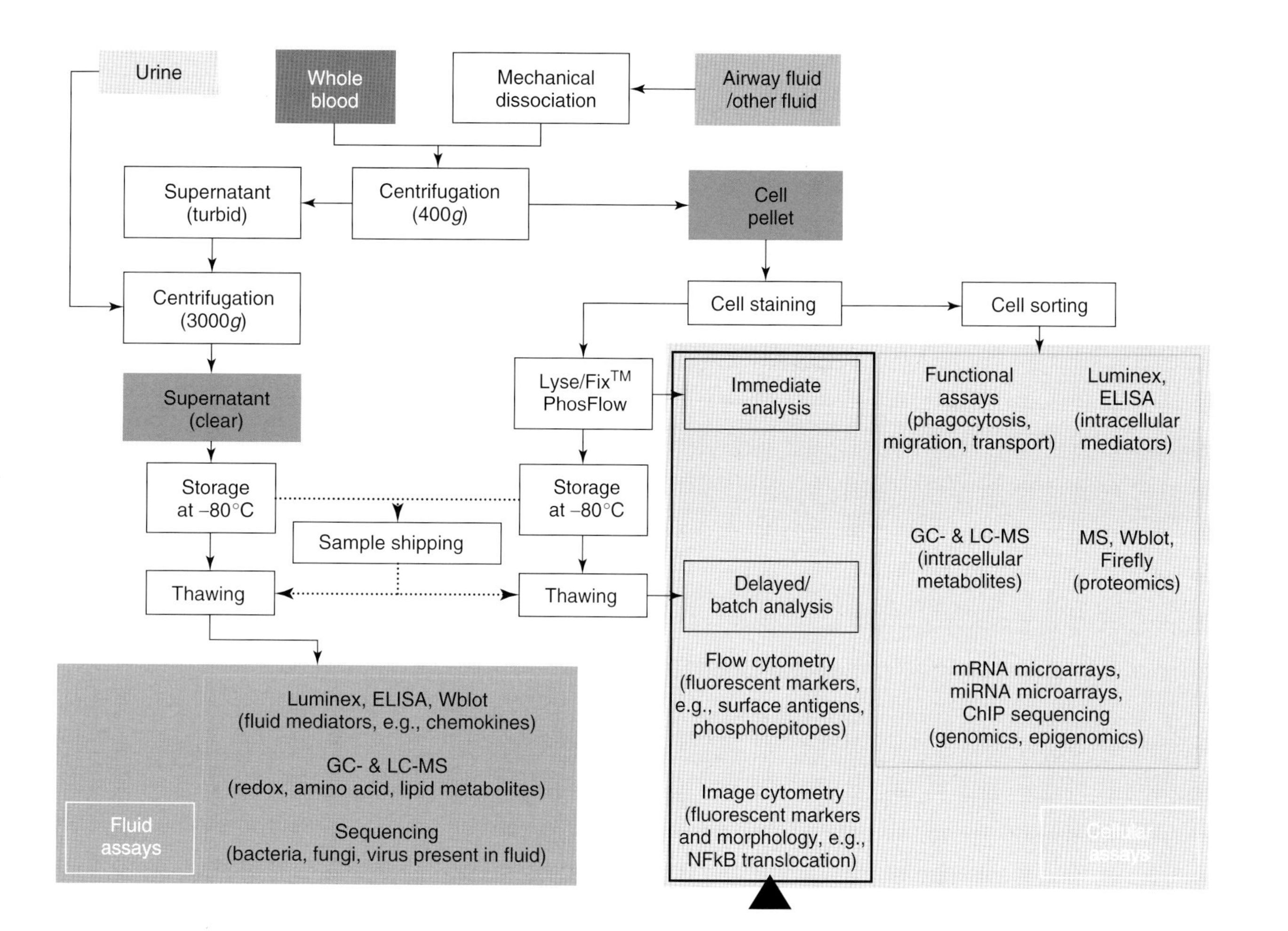

Urine
Whole blood
Mechanical dissociation
Airway fluid /other fluid
Supernatant (turbid)
Centrifugation (400g)
Cell pellet
Centrifugation (3000g)
Cell staining
Cell sorting
Supernatant (clear)
Lyse/Fix™ PhosFlow
Storage at –80°C
Storage at –80°C
Sample shipping
Thawing
Thawing
Immediate analysis
Delayed/ batch analysis
Flow cytometry (fluorescent markers, e.g., surface antigens, phosphoepitopes)
Image cytometry (fluorescent markers and morphology, e.g., NFkB translocation)
Functional assays (phagocytosis, migration, transport)
Luminex, ELISA (intracellular mediators)
GC- & LC-MS (intracellular metabolites)
MS, Wblot, Firefly (proteomics)
mRNA microarrays, miRNA microarrays, ChIP sequencing (genomics, epigenomics)
Cellular assays
Luminex, ELISA, Wblot (fluid mediators, e.g., chemokines)
GC- & LC-MS (redox, amino acid, lipid metabolites)
Sequencing (bacteria, fungi, virus present in fluid)
Fluid assays

Figure 12.1 Overview of the experimental platform for human "phenomics." Our platform relies on minimal *ex vivo* processing of patient samples (see text) to limit artifacts; combined with the use of high-content analytical platforms, including advanced cytometry (both using total fluorescence, that is, conventional flow, and imaging methods, indicated by bold rectangle and arrowhead). Blood is collected by venipuncture (5 ml generally sufficient). Finger pricks (~500 μl) are sufficient to perform limited assays, for example, Luminex, MS and flow/image cytometry (<10 stains). Airway fluid is collected by sputum induction, bronchoalveolar lavage, and tracheal or nasal aspirates. The first two methods generally yield sufficient material for complete fluid and cellular assays (a minimum of two stains can generally be performed even on samples from healthy controls). Other fluids amenable to analysis include joint fluid, peritoneal and pleural effusions, cerebrospinal fluid, and tissue suspensions. Most assays shown here can be run on batches, using samples that have been frozen and, if necessary, shipped to another site (an advantage for multicenter or remote studies).

this body fluid for clinical monitoring, including disease prognosis and diagnosis. For this purpose, assays need to include as few steps as possible and be robust in a clinical context [33].

Currently, flow cytometry assays used in the clinic are limited to the study of particular types of hematopoietic disorders and are usually run by trained technicians. Our group, aided by various collaborators, has developed experimental studies with remote sample collection (e.g., in disease-endemic zones) and multicenter clinical trials, in which flow cytometry and image cytometry assays are being used to collect data on key outcome measures, based on strictly defined SOPs, following the principles highlighted above. The use of these methods is supported by the fact that the ability to capture rich information from body fluids about cellular phenotype (cellular phenomics) is arguably as important as that of collecting information about the molecular phenotype (molecular phenomics). Owing to its ease of collection and richness of embedded information, blood is a critical source of cellular phenomic information as defined here.

12.3 Reagents for Advanced Flow Cytometric Analysis of Human Samples

12.3.1 Antibody Probes

FACS-based cytometry has grown concomitantly with the monoclonal antibody industry [2], leading to a catalog of several thousand antibody probes available worldwide, each capable of detecting a given epitope for cell subsetting and/or functional analysis (Table 12.3). By mixing and matching these antibody probes, one can attain a very high level of precision in identifying even extremely rare cells in complex samples (1–10 in 10^6 cells is typically considered a lower limit). When applied to whole blood, this method can lead to complete subsetting of the major subsets of cells, as well as platelets, using as little as six colors, along with cell

Table 12.3 Nonexhaustive list of reagents for human immune monitoring.

Classes	Instances	Vendors
Surface antibodies for phenotyping assays	B-cells: CD19, CD20, **CD39**, HLA-DR	BD Biosciences, eBioscience, Invitrogen
	Basophils: CD63, CD123, **CD203c**	
	Dendritic cells: CD11c, HLA-DR	
	Eosinophils: **CD16** (lo), CD33 (hi) CD45 (hi), CD66b (hi),	
	Erythrocytes: CD235a	
	Monocytes/macrophages: CD14, **CD33**, HLA-DR	
	Neutrophils: **CBT** (hi), **CD16** (hi), **CD45** (lo), **CD66b** (hi)	
	NK cells: **CD16**, CD56	
	Platelets: **CD41a**, CD42a	
	Stem cells: CD34	
	T-cells: **CD3**, CD4, CD8	
Surface antibodies for functional assays	Degranulation: CD11b, CD14, CD16, CD62P, CD63, CD66b, CD107	BD Biosciences, eBioscience, Invitrogen
	Immune priming/activation/polarization: CD11a, CD11b, CD18, CD24, CD25, CD38, CD45RO, CD47, CD62L, CD71, **CD294**, HLA-DR	
	Sensing: CCRs/CXCRs (chemokine receptors), CD14, CD206, **RAGE**	
Intracellular antibodies for activation/signaling	Phosphoepitopes: Akt, ERK, JNK, p38, PI3K, **Pyk2**, **Src**, Syk (kinases); **4E-BP1**, c-Cbl, **CREB**, **eIF4E**, NFkBp65, **STATs** (effectors)	Cell Signaling Technology, BD Biosciences, Invitrogen
	Cytokines/chemokines/growth factors: several dozens tested	

Intracellular antibodies and probes for organelles	(Auto) phagosomes: LC3 antibody	Invitrogen, Abcam
	Lysosomes: LysoTracker and LysoSensor series	
	Mitochondria: MitoTracker and MitoSensor series, JC-1	
	Polyribosomes: **Vigilin antibody**	
Probes for viability lipid, redox intermediates, intracellular ions, NLR/inflammasome	Viability: 7-AAD, PI, 4′-6-Diamidino-2-phenylindole (DAPI), **Draq5**, H33348, **Live/Dead series**	Invitrogen, Sigma, Axxora
	Lipid: Annexin V, boron-dipyrromethene (BODIPY) series, **CTB**, Merocyanin	
	Redox: **DAF2-DA** (RNS), **DHR** (ROS), **Maleimide series** (surface thiols), MCB (GSH), **ThiolTracker** (small molecular weight thiols	
	Intracellular ions: Fluo-3FF (calcium), FluoZin3 (zinc), SNARF-1 (pH)	
	NLR/inflammasome (caspase-1): **FLICA**	

Abbreviations: 4E-BP1, 4E-binding protein 1; CREB, cAMP response element binding protein; DAF-2DA, diaminofluorescein-2-diacetate; DHR, dihydrorhodamine; eIF4E, eukaryotic initiation factor 4E; ERK, extracellular-regulated kinase; H33348, Hoechst 33348; JNK, cJun N-terminal kinase; NFkB, nuclear factor kappa B; PI, propidium iodide; PI3K, phosphoinositide-3 kinase; SNARF-1, seminaphthorhodafluor-1; STAT, signal transducer and activator of transcription.
Shown are reagents validated for analysis by FACS and/or image cytometry (listed by order of appearance in the text; bold indicates inclusion in one or more of the figure(s) featured in this chapter).

scatter properties (Figure 12.2, panels 1–6). Of course, once the cells of interest are gated, antibody probes can be further used to assess physiological correlates of activation, that is, the modulation of adhesion receptors, receptors for cytokines, chemokines, and growth factors, epitopes from intracellular compartments brought to the surface by virtue of degranulation, and so on (Figure 12.2, panels 7–8).

Antibody-based gating can be implemented not only on fresh cells but also on cells that have been fixed, frozen, and banked until later analysis, as discussed in Table 12.2. To emphasize this notion, Figure 12.3 presents a simple gating strategy for leukocyte subsets in fixed/frozen/thawed whole blood and their subsequent functional analysis by image cytometry [34], including the nuclear translocation of the transcription factor HIF (hypoxia-inducible factor)-1α [35] and the intracytoplasmic formation of Vigilin-positive polyribosomes [36]).

A recent addition to the panel of antibodies used for functional analysis is that of antibodies against phosphoepitopes. These antibodies, developed in large part for use in western blot analysis, have been retooled for use in flow cytometry. While western blot requires a large number of cells and provides a bulk measurement with no ability to differentiate between subsets within the population of interest, flow cytometry measures the amount of phosphoepitopes at the single-cell level, which dramatically increases the depth of the functional analysis thus enabled [37]. As explained in Figure 12.4, phosphoepitopes are integral to phosphorylation cascades, which are ubiquitous in regulatory loops involved in all aspects of cell development, functional modulation, and death. Over the past five years, numerous antibodies against phosphoepitopes in kinases and final effectors in phosphorylation cascades have been validated for use in flow cytometry assays.

In the context of direct analysis protocols, as advocated in this chapter, it is likely that final effectors, rather than kinases, will be the most informative. Indeed, when samples are collected and analyzed without any exogenous culture and stimulation (leading to synchronization of signaling pathways within the population of cells studied), cells are, by and large, asynchronous. Therefore, they offer lesser opportunities to find significant modifications in kinase phosphorylation levels, since kinases are generally characterized by a rapid on/off cycle of their phosphorylated moieties *in vivo*. By contrast, chronically induced phosphorylation cascades generally manifest themselves through a stable increase in the level of phosphorylation of final effectors in these cascades, since these elements are characterized by much slower on/off cycles of their phosphorylated moieties *in vivo*.

Any cell present in blood and other body fluids is amenable to phosphoepitope profiling, provided relevant antibodies are used for subset gating and the subsequent functional analysis. This notion is illustrated in Figure 12.5, where the rare subset of blood basophils is first analytically separated, with the use of a "dump channel" (measurement channel in which bright reagents for unwanted cells, here non-basophils, are combined) and subsequently analyzed for activation markers and phosphoepitopes. It is important to note that whole blood phospho-profiling is particularly well adapted to the study of cells known as *rapid responders*, that is, elements or cells capable of mounting physiologically relevant responses in

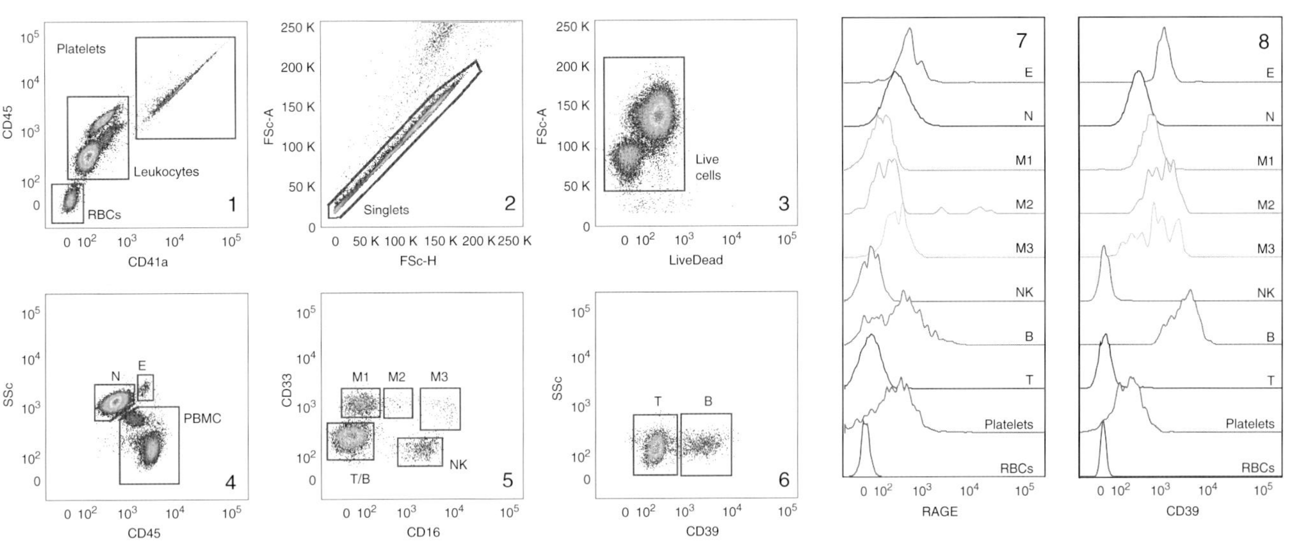

Figure 12.2 Example of whole blood gating strategy leading to simultaneous functional analysis of 10+ subsets from a single stain. Stepwise gating (panels 1–6) yields the populations of interest and these are then compared for their expression of two activation markers of interest. Note, for example, the strong expression of the receptor for advanced glycation end-products (RAGE, panel 7) in eosinophils, neutrophils, B-cells, and platelets, but not T-cells (panel 7); or the difference in expression of the ectonucleotidase CD39 (panel 8) between the three subsets of blood monocytes.

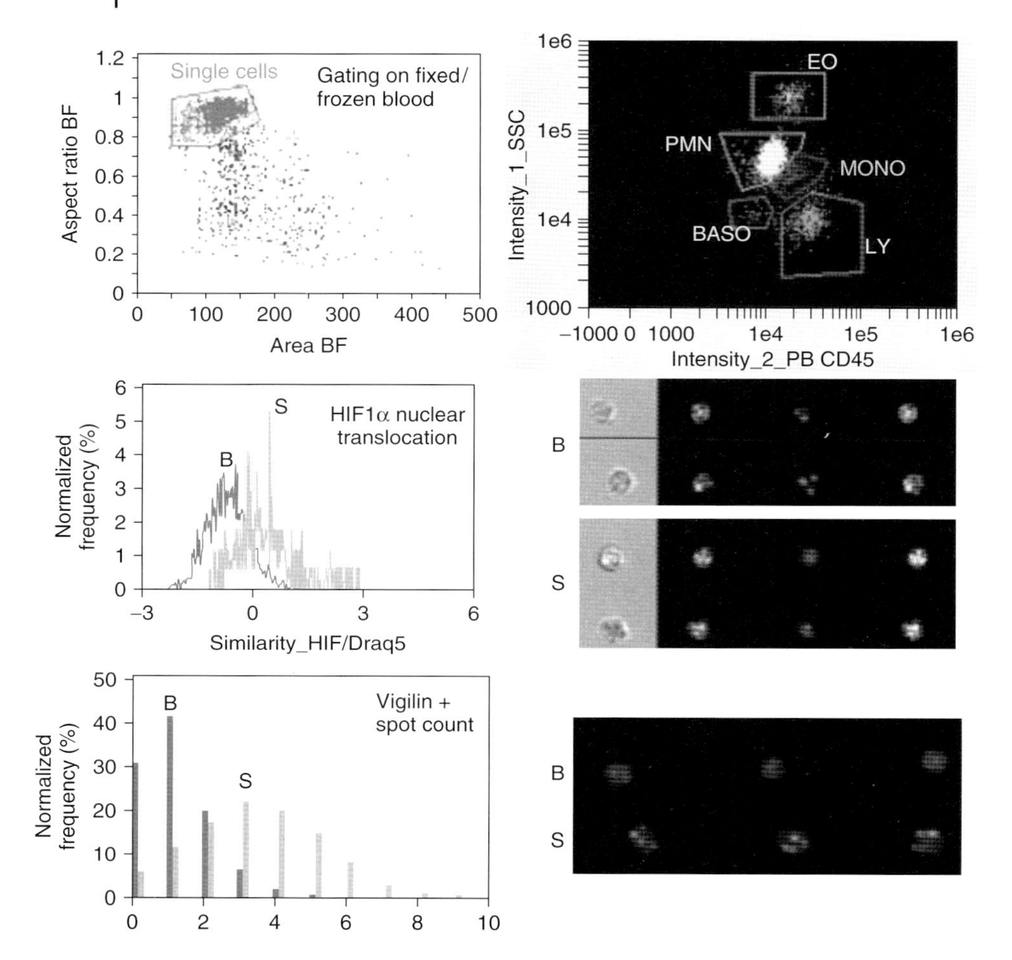

Figure 12.3 Batched assays on banked whole blood samples, as per the ImageStream image cytometry platform. Top panel: two-step gating to obtain major leukocyte subsets from fixed/frozen blood, amenable to various downstream applications. The first step uses morphological parameters, namely, the area and aspect ratio of the brightfield (BF) image. The second step uses side scatter (SSC) and staining intensity for the pan-leukocyte marker CD45. Middle panel: colocalization of HIF1α with the nuclear dye Draq5 (or the absence thereof) is measured by pixel analysis and enables the quantification of the HIF1α nuclear translocation index, increased in sputum (S) compared to blood (B) neutrophils. Bottom panel: morphological analysis combined with fluorescence measurement lead to the definition of the spot count statistics, which show quantitatively increased assembly of Vigilin+ polyribosomes in sputum (S) compared to blood (B) neutrophils.

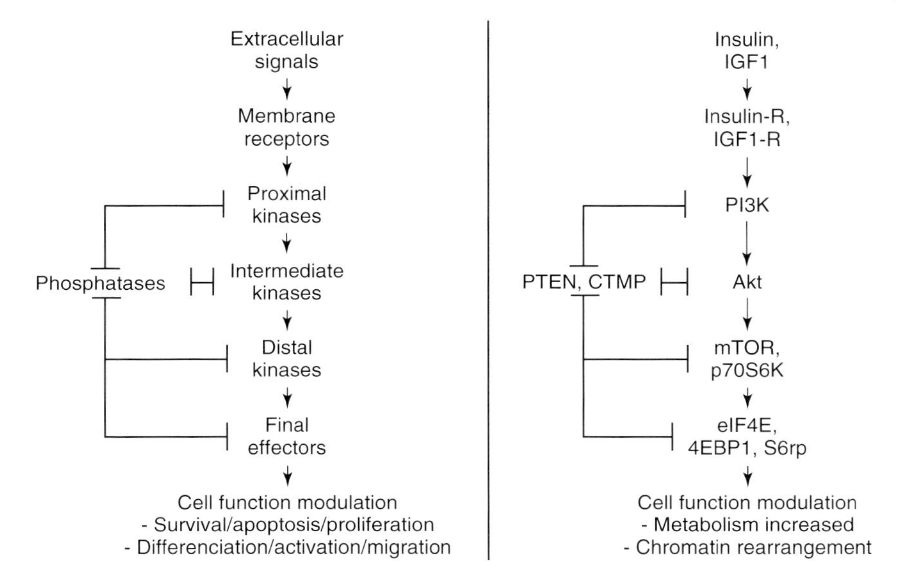

Figure 12.4 Phosphorylation pathways can be interrogated by advanced flow cytometry. Phosphorylation pathways are key to cell function modulation, as shown in the left panel (theoretical model) and right panel (example of the mammalian target of rapamycin (mTOR) pathway). Using antibodies specific to phosphorylated residues on kinases and final effectors (indicated in bold), phosphorylation pathways can be investigated in single cells by FACS-based methods and image cytometry. In the context of protocols using minimal *ex vivo* processing and rapid detection of functional and signaling changes, as advocated in this review, it is likely that final effectors (with stable on/off status), rather than kinases (rapid on/off switch), will yield the most informative measures. By contrast, protocols using cell fractions that have undergone *in vitro* adaptation (cell lines, primary cultures) may also pick up differences in phosphorylation of kinases, since these have been able to undergo synchronization across the culture under sustained exogenous activation.

the context of blood. These include platelets, erythrocytes, and granulocytes, in particular. Our group has exploited this idea in several studies focusing on neutrophils (in CF [19, 38]), T-cells (in asthma [39]), and basophils (in food allergy–Gernez, *et al.*, [40]), as illustrated briefly in Figure 12.6. Taken together, the results shown in Figures 12.2, 12.3, 12.5, and 12.6 illustrate the versatility of our approach for the functional analysis of human whole blood.

12.3.2
Nonantibody Probes

While antibodies form the bulk of the arsenal at the disposal of flow cytometrists, nonantibody probes have also been long used (Table 12.3). Nonantibody probes are based on the use of chemical entities endowed with the ability to recognize and bind to specific molecular motifs in cells. Common examples are (i) viability markers,

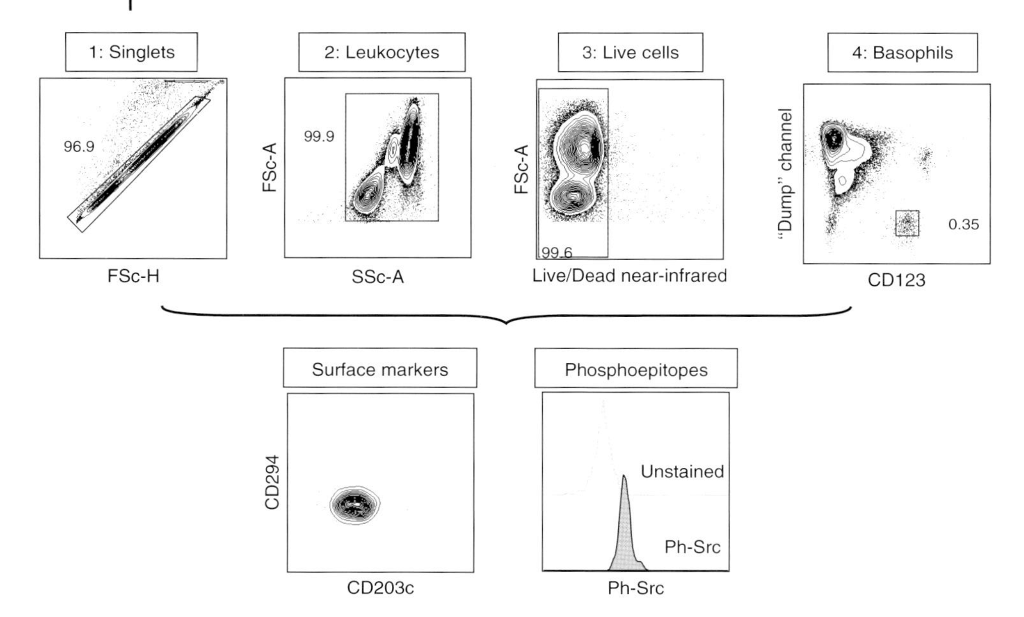

Figure 12.5 Analytical gating and functional analysis of basophils from human blood using antibody probes. This figure shows how rare cells in the circulation, such as basophils, can be readily gated and functionally analyzed from a drop of human whole blood. After the definition of sequential gates for singlets, leukocytes, and live cells, a dump channel is engineered using antibodies against all unwanted cells (here, CD3 for T-cells, CD16 for neutrophils and NK cells, CD20 for B-cells, CD56 for NK cells, CD66b for neutrophils and eosinophils, and HLA-DR for monocytes, dendritic cells, and B-cells) and plotted against a marker expressed highly in basophils, here, CD123. Note that some cells that are positive for the "dump" channel also express CD123 (eosinophils and dendritic cells), which are easily excluded from basophils using this strategy. After this analytical gating, surface markers (CD294 or CD203c) and phosphoepitopes (here, phospho-Src) can be quantified in whole blood basophils. Similar strategies can be used for other rare subsets of interest.

such as propidium iodide (PI), a DNA intercalating agent, or the Live/Dead® series from Invitrogen, which binds to protein moieties at the cell surface (minor staining) or inside the cytosol for compromised cells (major pool of target protein moieties, leading to considerable fluorescence shift); (ii) lipid reagents, such as annexin V, which bind to the exposed phosphatidylserine present on cells undergoing apoptosis and cells undergoing membrane flipping (as in activated phagocytes) or cholera toxin B (CTB), which binds to ganglioside M1, a phospholipid enriched in lipid rafts that mediates cell signaling; (iii) probes for redox intermediates, reacting with small molecular weight thiols, including glutathione, reactive oxygen species/reactive nitrogen species (ROS/RNS), reduced surface thiols; (iv) probes against intracellular ions such as calcium, zinc, or protons; (v) probes against intracellular enzymes, such as collagenase, calpain, or the NOD-like receptor (NLR)/inflammasome-associated caspase-1; and (vi) probes derived from retroviral envelopes, labeling metabolic transporters that are otherwise extremely difficult

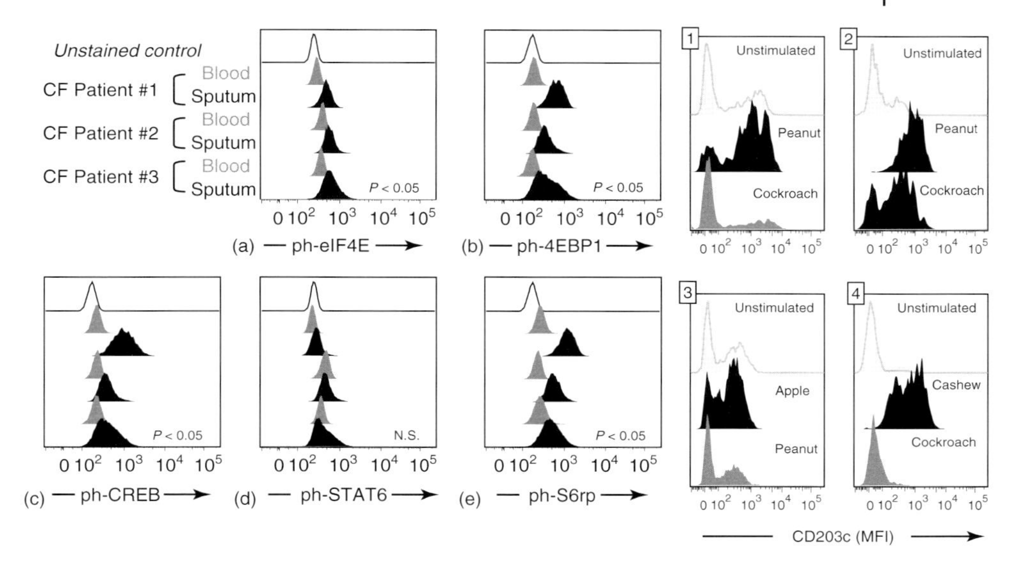

Figure 12.6 Rapid response cells feature key biological signatures measurable directly in blood and other leukocyte-containing body fluids, such as airway fluid. Left panel: Increased phosphorylation of effectors in the mammalian target of rapamycin (mTOR) anabolic cascade (previously illustrated in Figure 12.4) including the eukaryotic initiation actor 4E (eIF4E), 4E-binding protein (4E-BP1), and S6 ribosomal protein (S6rp), as well as of the stress-activated cAMP response element binding protein (CREB) is observed in CF airway neutrophils compared to their blood counterparts, while the signal transducer and activator of transcription (STAT) 6 is unchanged (three representative patients). Right panel: In patients with food allergy, blood basophils undergo dramatic changes in their surface CD203c, when in contact for 2 min with offending allergens (indicated in bold), but not nonoffending allergens (four representative patients).

to raise good antibodies against because of their cross-species conservation and requirement for cell survival [41, 42]. Categories (i)–(v) are well established and widely used while category (vi) has just recently emerged, opening up great opportunities for functional profiling. Indeed, metabolic transporters constitute a large group of molecules with key roles in acute and chronic adaptation of leukocytes, erythrocytes, and platelets to stimuli, and therefore will prove invaluable in the mechanistic study of human health and disease.

In Figures 12.7–12.9, we illustrate the use of three of these nonantibody probes. First, in Figure 12.7, we show how the lipid raft probe, CTB, can be used to label neutrophils in blood and airway fluid samples from human subjects. Neutrophils are ready for activation as they leave the bone marrow to circulate in blood and later, if recruited, to peripheral organs [22]. Two classical correlates of this inherent "priming" state of blood neutrophils are the decreased surface expression and activity of the inhibitory phosphatase CD45 (Figure 12.3), but also, as we demonstrate here in Figure 12.7, the increased basal expression of ganglioside M1 at their surface. The latter, responsible for the strong baseline binding of

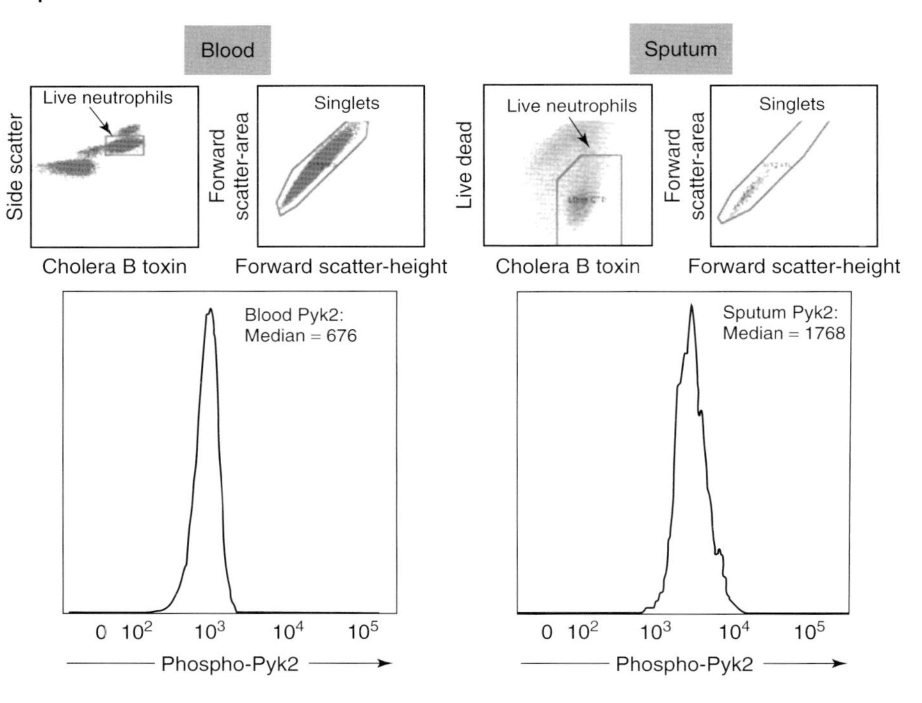

Figure 12.7 Method for gating neutrophils using cholera toxin B (CTB). CTB is a probe that labels lipid rafts, which are enriched in neutrophils at baseline and upon activation (see text for more details). As opposed to antibody probes, CTB staining is not sensitive to fixation, freezing/thawing, and permeabilization treatment with saponin or methanol, and hence can be used as an excellent all-around solution for gating neutrophils.

CTB to blood neutrophils, is an indication of the increased presence of lipid rafts, and therefore of their readiness to undergo immune signaling. A key advantage provided by CTB staining is its resistance to all treatments used routinely in our approach (i.e., fixation, freezing and thawing, and saponin for intracellular staining or methanol (MeOH) treatment for phosphoepitope staining), making it a great all-around probe for use in studies focusing on neutrophils. In addition, when visualized by image cytometry, the discrete appearance of CTB binding (surface vs intracellular; dots vs patches) provides a quantitative measure of lipid raft assembly on and in cells.

As a second example, in Figure 12.8, we illustrate the novel cell-permeable probe against GSH and other small molecular weight thiols, key measures of a cell's redox balance. This probe, named ThiolTracker®, provides the same practical advantages as CTB in that it withstands all treatments used routinely in our approach (i.e., freezing and thawing and saponin for intracellular staining or MeOH treatment for phosphoepitope staining, as laid out in Table 12.2), thereby allowing a robust redox index measurement concomitantly with virtually any other assay. Additionally, ThiolTracker staining is not temperature- or time-sensitive, is

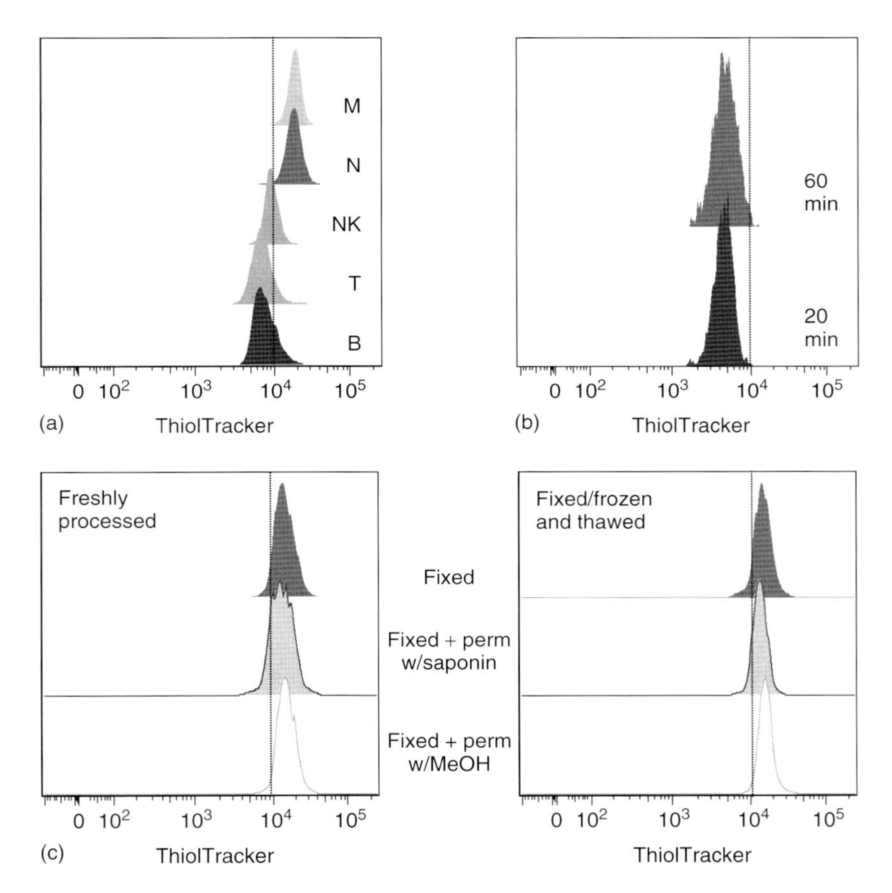

Figure 12.8 Robust measurement of small molecular weight thiols using the ThiolTracker probe. (a) A new cell-permeable probe, ThiolTracker (Invitrogen), reacts with high affinity with small molecular weight thiols, the most abundant of which is glutathione, and shows total fluorescence patterns in various subsets of leukocytes, which are equivalent to that measured with previous thiol probes [43, 44]. (b and c) In addition, ThiolTracker saturates with time and withstands fixation, permeabilization, and even freezing/thawing of samples, making it a very attractive all-around probe for redox measurements.

compatible with fixation, and does not undergo efflux by membrane transporters, four limitations that rendered previous probes for thiols such as monochlorobimane (MCB) impossible to use in nonexpert settings [43, 44].

As a third example, in Figure 12.9, we illustrate the use of the FLICA probe, a cell-permeable chemical that becomes fluorescent upon specific cleavage by the intracellular enzyme caspase-1. Unlike other caspases, whose role is essential to the proteolysis of endogenous targets in the context of apoptosis, caspase-1 is strictly involved in the process of innate immune activation following cell stimulation by ligands of the NLR/inflammasome pathway [45]. The ability to measure the

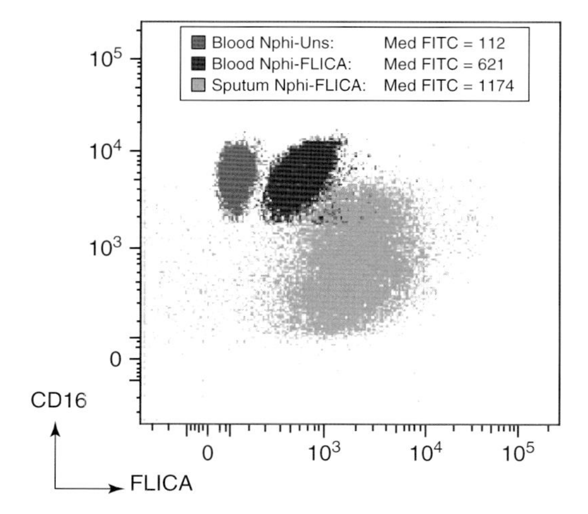

Figure 12.9 Analysis of intracellular caspase-1 activity in blood and airway neutrophils: probing the NLR/inflammasome pathway. A key pathway in innate immunity that has recently emerged depends on the activation of intracellular complexes designated as the inflammasome, upon stimulation by both exogenous and endogenous danger signals by NOD-like receptors (NLR). Activation of the NLR/inflammasome pathway leads to increased caspase-1 activity, here detected by use of a cell-permeable probe (FLICA). Our data demonstrate the increased caspase-1 activity and thus NLR/inflammasome activation in airway neutrophils (light grey), compared to blood neutrophils (black) from a CF patient, even as airway cells undergo cleavage of an important immune adaptor, the surface CD16 molecule (see decreased intensity on the *Y* axis, comparing green to blue subsets). Unstained blood neutrophils are shown as a negative control (dark grey).

activity of this enzyme in cells within the context of whole blood or airway fluid, as illustrated, highlights the considerable opportunities for deep functional analyses of human samples awarded by advanced flow cytometric methods.

12.3.3 Adapting Protocols to the Research Question at Hand: A Few Practical Examples

The rules defined above for the choice of protocols enabling proper sample collection and processing and the selection of appropriate reagents (antibodies and probes) should equip new investigators with sufficient guidelines for the successful setup and implementation of flow cytometry studies of human subjects. In this section, we want to provide additional guidance by illustrating generic rules for staining protocols and by highlighting two common mistakes that can significantly hamper investigations. First, in Table 12.4, we detail three distinct staining combinations used in our group for characterizing human samples.

The first stain, labeled "activation," is a 10-color combination using 9 antibody probes and 1 nonantibody probe for viability (Live/Dead® InfraRed). With this stain, one can detect leukocytes in any body fluid (blood, airway fluid, joint fluid,

Table 12.4 Examples of staining combinations for flow cytometric analysis of whole blood.

	Pacific Blue	Pacific Orange	Quantum Dot 605	FITC or Alexa488	PerCP-Cy5.5	PE or Alexa555	PE-Cy5	PE-Cy7	APC or Alexa647	Alexa700	APC- Cy7
Activation	CD16	CD3	CD14	CD66b	CD45	HLA-DR	Dump	CD39	CD203c	CD107	*LD-IR*
Reactivity	N, NK/ activation	T	M/acti- vation	N, E/ activation	Pan-leuko	B, D, M/ activation	P/R/Ep.	B/acti- vation	Baso/ activation	NK, T/ activation	*Viability*
Redox	*ThiolTracker*			Dump	CD45	*Maleimide*	CML	*Annexin V*	*CTB*	RAGE	*LD-IR*
Reactivity	*Thiols (i)*			P/R/Ep.	Pan-leuko	*Thiols (s)*	DAMP (s)	*PS (s)*	*Rafts (s)*	DAMP-R	*Viability*
Phospho	*CTB*	–	–	Ph-CREB	CD45	Ph-eIF4E	Dump	–	Ph-Stat3	–	*LD-IR*
Reactivity	*Rafts*	–	–	Ph-epitope	Pan-leuko	Ph-epitope	P/R/Ep.	–	Ph-epitope	–	*Viability*

Abbreviations: APC, allophycocyanin; B, B-cells; Baso, basophils; CML, carboxymethyl-lysine; D, dendritic cells; DAMP, damage-associated molecular pattern; DAMP-R, DAMP-receptor; E, eosinophils; EpCAM, epithelial cell adhesion molecule; Ep., epithelial cells; FITC, fluorescein isothiocyanate; LD-IR, Live/Dead® InfraRed; M, monocytes/macrophages; N, neutrophils; NK, NK cells; P, platelets; Pan-leuko, pan-leukocyte; PE, phycoerythrin; R, erythrocytes; and T, T-cells. Nonantibody probes are indicated in italics.

cerebrospinal fluid, pleural or peritoneal effusion, and tissue suspension), using the CD45 pan-leukocyte marker. Further subsetting into all major leukocyte subsets is enabled by a panoply of markers that together delineate B-cell, basophils, dendritic cells, eosinophils, monocytes, neutrophils, natural killer (NK) cells, and T-cells, while using a "dump channel" with markers for contaminating platelets, erythrocytes, and epithelial cells (CD41a, CD235a, and EpCAM (epithelial cell adhesion molecule)). Some of the subset markers (CD14, CD16, CD39, CD66b, CD107, CD203c, and HLA-DR) are significantly modulated upon activation and therefore double as activation markers.

The second stain, labeled "redox," is dedicated to the investigation of surface and intracellular redox properties of leukocytes in any body fluid. As with the first stain, Live/Dead® InfraRed, CD45, and Dump (CD41a, CD235a, and EpCAM) are used to analytically gate on viable leukocytes, while excluding unwanted cells. However, as opposed to the first stain, this second stain makes use of several nonantibody probes to profile redox properties of cells, including (i) intracellular small molecular weight thiols (ThiolTracker, illustrated in Figure 12.8); (ii) surface reduced thiols, measured with maleimide (illustrated in Figure 12.11); (iii) lipids that are modulated by oxidation (phosphatidylserine, measured by annexin V and ganglioside M1, a key component of lipid rafts, measured by CTB, illustrated in Figure 12.7); (iv) oxidation-induced modification of surface protein residues leading to the formation of carboxymethyl-lysine, which is recognized by cells as an endogenous danger signal of the damage-associated molecular pattern (DAMP) category [46]; (v) expression of the DAMP-receptor RAGE (receptor for advanced glycation end-products) (illustrated in Figure 12.3). This stain, combining antibody and nonantibody probes, therefore enables fairly complex investigations of cell-based redox functions, which are often early indicators of stress and disease [47].

The third stain presented, labeled "phospho," is geared at the measurement of several phosphoepitopes of choice on leukocytes. As with the first and second stains, Live/Dead® InfraRed, CD45, and a dump channel (CD41a, CD235a, and EpCAM) are used to analytically gate on viable leukocytes, while excluding unwanted cells. CTB is not only used to provide analytical gating of neutrophils (as illustrated in Figure 12.7) but also to enable visualization of surface lipid rafts, if analyzed by image cytometry. Note that the above probes for cell gating withstand MeOH treatment, a necessary step for the linearization of phosphorylated proteins for recognition by anti-phosphoepitope antibodies. Therefore, any adaptation of this generic protocol for use in phosphoepitope profiling experiments needs thorough testing of probes against MeOH treatment. Combined with the use of appropriate cell gating probes, anti-phosphoepitope antibodies that fluoresce in the brightest channels are used to maximize single-cell detection (generally FITC/Alexa 488 (fluorescein isothiocyanate), phycoerythrin (PE)/Alexa 555, and APC/Alexa 647 (allophycocyanin) are used for this purpose [33]).

While the three examples above highlight successful adaptations of modular protocols (Table 12.2) to develop approaches relevant to specific physiological mechanisms important to human health and disease, one should be aware of potential mistakes that can hamper such efforts. In Figure 12.10, we show how such simple strategies as antibody-based gating of neutrophils can necessitate critical adaptations when transposing them from blood, a well-characterized context, to peripheral organs and other body fluids. Indeed, robust markers such as CD16, used to positively gate on neutrophils, may be unexpectedly downregulated, or, in this case, actively proteolyzed by enzymes present in the airway fluid [38, 48]. The use of a second marker, here CD66b, can mitigate such unexpected changes. Hence, when investigating novel biological contexts, it is advisable to use at least two positive markers for the population of choice.

In Figure 12.11, we provide another telling example of flow cytometric analysis going wrong. This example was specifically designed to illustrate the ability of sample processing methods to obliterate completely physiologically relevant responses. Illustrated here are (i) key changes in redox and activation markers induced by the migration of neutrophils from blood into airways of CF patients and (ii) how these key biological changes can be significantly altered, or even reversed by the use of dithiothreitol (sold as Sputolysin®), a common additive and redox-active agent used in the preparation of airway cell suspensions [38, 49].

Once again, it is important to realize that the results of advanced flow cytometry analyses are as good as the most limiting step in the protocol (Table 12.2). This limiting step is often represented by poor collection/processing methods (Figure 12.11), but it may also be a less than optimal choice of antibody and probes for staining (Figure 12.10). Advanced protocols for FACS and image cytometry analyses of cell function and signaling in human samples should thus be considered in their entirety, as integrative platforms that can truly empower ambitious research endeavors and help them come to fruition. Such efforts can be significantly aided by other investigative techniques (metabolomics, genomics, and proteomics) that complement and enlighten flow cytometric results, as shown in Figure 12.1.

12.4 Conclusion: The Future of Advanced Flow Cytometry in Human Research

Flow cytometric methods have brought about exciting possibilities for basic research but are largely yet to be applied to benefit human translational and clinical research. While the field of human translational research itself is in the process of coalescing and has only now become a trend to be reckoned with, innovative flow cytometric methods and probes continue their march forward and open even more opportunities for exploration. First, the cellular and molecular limits of flow cytometry are being pushed inexorably forward. The CyTOF platform [4], a newcomer in the cytometry field (Table 12.1), is a good example of the technological leaps in front of us. By using reagents conjugated with mass tags, of which several dozens may be multiplexed in a single stain, this technology enables a major leap in the number of measures run on a single cell compared with 20 or so simultaneously allowed by the most advanced FACS-based cytometers available

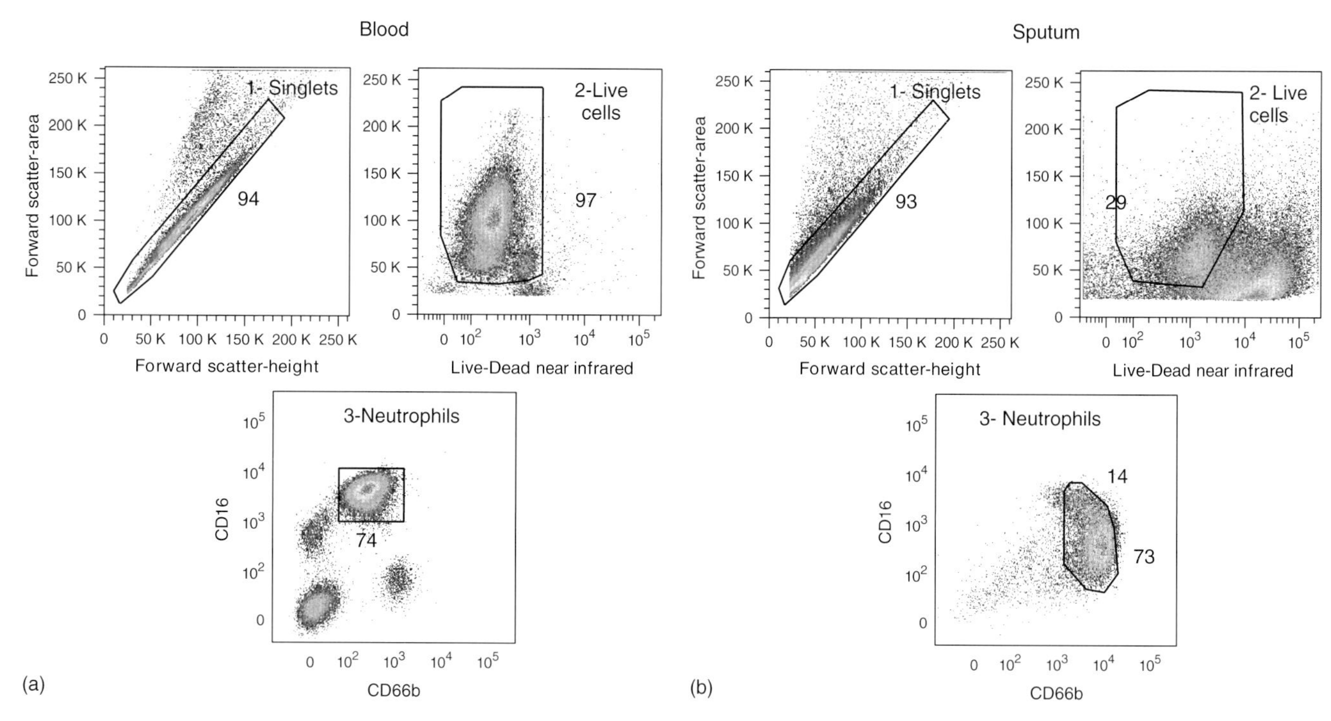

Figure 12.10 Analytical gating of neutrophils from human blood and sputum using antibody probes. Typically, digital flow cytometry enables the simple, stepwise gating/sorting of the population(s) of interest, for example, neutrophils in this example, yielding: singlets (based on scatter properties); then, live cells (based on scatter and LiveDead staining); and neutrophils based on a combination of fluorescent markers, here antibodies against surface CD66b and CD16. Note here that neutrophils dramatically alter the expression of the classical marker CD16 when entering into the airways, here in a patient with the chronic inflammatory airway disease associated with CF. Hence, it is important when surveying various compartments not to assume that expression of even common markers will be maintained.

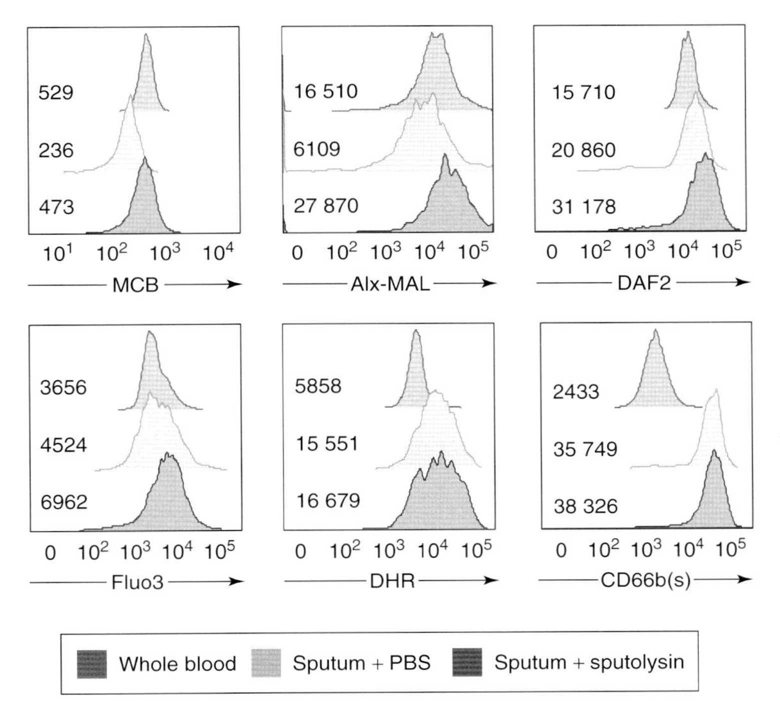

Figure 12.11 Sputolysin treatment alters activation markers on sputum neutrophils. Compared to whole blood (top histogram in each panel), viable sputum neutrophils prepared in the absence of Sputolysin® (middle histogram in each panel, phosphate buffer saline-base method, detailed in [19, 38]) display decreased intracellular glutathione levels (as measured by the monochlorobimane (MCB) probe), as well as decreased thiols (maleimide probe) and increased CD66b at the surface (CD66b(s)), reactive oxygen species (dihydrorhodamine – DHR – probe) and reactive nitrogen species (diaminofluorescein diacetate – DAF2-DA – probe), with moderate intracellular calcium (Ca^{2+}) increase (Fluo3-FF-AM probe). In stark contrast, viable sputum neutrophils prepared by 30 min incubation at 37 °C with the reducing agent dithiothreitol (Sputolysin®, bottom histogram in each panel) show marked signs of reduction (increased glutathione and surface thiols and reactive nitrogen species), as well as increased Ca^{2+} levels, which indicate significant, multifaceted, and totally artifactual alteration of the activation pattern of cells.

today. However, the CyTOF platform is currently very inefficient, since its ability to capture single cells from FACS-type fluidics and vaporize them in the TOF-type analytical module only reaches a mere 1 in 50–100 cells. Thus, CyTOF remains an attractive analytical method only for large samples with low subset diversity, at least for now.

Incremental progress in existing FACS- and image cytometry-based platforms may well bring the most interesting new features for the analysis of complex samples in the near future. For example, the miniaturization and dramatic price drop of lasers have allowed five and now six different light sources to be integrated in state-of-the-art FACS platforms, providing better excitation for a larger set of fluorescent dyes across the UV, visible, and infrared spectra. The same notion of increased laser availability applies to image cytometers, although in this nascent family of analytical platforms, only few have reached a state of usability that put them on par with FACS-based platforms. The ImageStream® platform is such a finished, yet constantly evolving, product, which has been used by a growing number of investigators involved in human research in the last five years [3]. A key add-on to the image stream is that of additional sets of cameras, which might add some confocal-like capabilities for morphological imaging [50] to a platform that currently produces mid-resolution flat images (equivalent to a 60× microscopic objective).

The ability to survey small subcellular structures with more accuracy, be they membrane patches or intracellular organelles, will be greatly enhanced by such technological leaps. Another field that will likely bring great advances to flow cytometry approaches is that of molecular probe development. For example, the discovery of the biological importance of small noncoding RNAs (e.g., microRNAs) has sparked interest in the ability to detect and measure such molecules at the single-cell level [51]. Another important set of molecular events that will likely become available for single-cell detection by flow cytometry in the near future consists of DNA sequence modifications (chromosome rearrangements, single-nucleotide polymorphisms, and copy number variations) and markers of chromatin status and occupancy (methylation and other histone modifications, binding of factors to specific enhancer, silencer, and promoter sequences).

This outlook on the future of flow cytometry approaches for the study of human health and disease would not be complete without a discussion on software capabilities that need to be developed for this type of effort to come to fruition. These software capabilities fall into roughly three categories. First, the democratization of advanced methods requires combined experimental and technological expertise that is difficult to acquire and, therefore, could gain immensely if aided by software solutions making protocol design, including staining combinations and experimental controls, as well as machine interfacing and data tracking more straightforward. One such software is the recently released CytoGenie® (Woodside Logic, Inc.), originating from the Herzenberg group at Stanford, the birthplace of FACS hardware and a pioneering group in integrated experimental and analytical solutions for flow cytometry.

The second group of software consists of semi- to fully automated solutions for multidimensional analysis of cytometry data. Indeed, the inflation of dimensions (20 or more for FACS-based platforms, 50 or more for CyTOF, and several hundreds for image cytometry) prevents the user from exerting simple, manual, histogram, or bi-dimensional plot-based analyses. Many such software use advanced statistics (e.g., clustering methods [5] and Bayesian networks [6]) to provide data visualization and analysis in the *n*-dimensional space. Such analytical software solutions are being developed by a rather large group of commercial and academic entities and users will need to dedicate time and effort to find the most adapted solutions for their use.

The third and final group of software is the one that provides the capability for integration of flow cytometric data into larger human research databases, as well as meta-analysis of large datasets acquired over time to provide insights into complex parameters influencing human health and disease (e.g., age, gender, ethnicity, environmental conditions, and medical history). This last piece in the integrated approach that we wish to delineate here is the most challenging to achieve, since it necessitates a strict definition and enforcement of experimental standards, efficient research coordination, as well as funding toward longitudinal and cross-disciplinary research.

Despite its difficulties, the integrated approach discussed in this chapter is clearly one that researchers are bound to adopt in years to come. By renewing our capacity to observe intricate mechanisms of human health and disease, such an approach should bring about a true revolution in basic understanding of human biology and open opportunities for improved healthcare management and treatments, moving closer to the much-touted era of personalized medicine [29, 30].

Acknowledgments

The authors value greatly the patience and excellent work of Dr. Valery Tuchin, editor of this volume. They are also greatly indebted to Drs. Leonard and Leonore Herzenberg and to colleagues of the Herzenberg Laboratory and Shared FACS Facility at Stanford University, for their guidance and critical advice over the last 10 years of research in the field of flow cytometry. Finally, the authors would like to acknowledge the help of Drs. David Basiji, Thaddeus George, and of Raymond Kong, at Amnis, Inc. for their excellent support and collaboration on experiments run with the ImageStream image cytometry platform.

Abbreviations

AAD	amino actinomycin D
CCR	Cys-Cys chemokine receptor
CF	cystic fibrosis
CTB	cholera toxin B
CXCR	Cys-X-Cys chemokine receptor
DAMP	damage-associated molecular pattern
FACS	fluorescence-activated cell sorting
FLICA	fluorochrome-labeled inhibitor of caspase
GSH	glutathione
HLA-DR	human leukocyte antigen-D related
MAPK	mitogen-activated protein kinase
MeOH	Methanol
MS	mass spectrometry
PBMC	peripheral blood mononucleated cells
RAGE	receptor for advanced glycation end-products
RGB	red-green-blue (color model based on additive color primaries)
RNS	reactive nitrogen species
ROS	reactive oxygen species

SOP	standard operating procedure
TLR	Toll-like receptor
TOF	time-of-flight

References

1. Herzenberg, L.A. (2004) FACS innovation: a view from Stanford. *Clin. Invest. Med.*, **27**, 240–252.
2. Herzenberg, L.A. and De Rosa, S.C. (2000) Monoclonal antibodies and the FACS: complementary tools for immunobiology and medicine. *Immunol. Today*, **21**, 383–390.
3. Basiji, D.A., Ortyn, W.E., Liang, L., Venkatachalam, V., and Morrissey, P. (2007) Cellular image analysis and imaging by flow cytometry. *Clin. Lab. Med.*, **27**, 653–670, viii.
4. Bandura, D.R., Baranov, V.I., Ornatsky, O.I., Antonov, A., Kinach, R., Lou, X., Pavlov, S., Vorobiev, S., Dick, J.E., and Tanner, S.D. (2009) Mass cytometry: technique for real time single cell multitarget immunoassay based on inductively coupled plasma time-of-flight mass spectrometry. *Anal. Chem.*, **81**, 6813–6822.
5. Walther, G., Zimmerman, N., Moore, W., Parks, D., Meehan, S., Belitskaya, I., Pan, J., and Herzenberg, L. (2009) Automatic clustering of flow cytometry data with density-based merging. *Adv. Bioinformatics*, 2009, doi: 10.1155/2009/686759.
6. Sachs, K., Perez, O., Pe'er, D., Lauffenburger, D.A., and Nolan, G.P. (2005) Causal protein-signaling networks derived from multiparameter single-cell data. *Science*, **308**, 523–529.
7. Terstappen, L.W. and Loken, M.R. (1988) Five-dimensional flow cytometry as a new approach for blood and bone marrow differentials. *Cytometry*, **9**, 548–556.
8. Dainiak, M.B., Kumar, A., Galaev, I.Y., and Mattiasson, B. (2007) Methods in cell separations. *Adv. Biochem. Eng. Biotechnol.*, **106**, 1–18.
9. De Rosa, S.C., Herzenberg, L.A., and Roederer, M. (2001) 11-color, 13-parameter flow cytometry: identification of human naive T cells by phenotype, function, and T-cell receptor diversity. *Nat. Med.*, **7**, 245–248.
10. Venaille, T.J., Misso, N.L., Phillips, M.J., Robinson, B.W., and Thompson, P.J. (1994) Effects of different density gradient separation techniques on neutrophil function. *Scand. J. Clin. Lab. Invest.*, **54**, 385–391.
11. Watson, F., Robinson, J.J., and Edwards, S.W. (1992) Neutrophil function in whole blood and after purification: changes in receptor expression, oxidase activity and responsiveness to cytokines. *Biosci. Rep.*, **12**, 123–133.
12. Zahler, S., Kowalski, C., Brosig, A., Kupatt, C., Becker, B.F., and Gerlach, E. (1997) The function of neutrophils isolated by a magnetic antibody cell separation technique is not altered in comparison to a density gradient centrifugation method. *J. Immunol. Methods*, **200**, 173–179.
13. Tanaka, S., Hayashi, T., Tani, Y., and Hirayama, F. (2010) Removal by adsorbent beads of biological response modifiers released from platelets, accumulated during storage, and potentially associated with platelet transfusion reactions. *Transfusion*, **50**, 1096–1105.
14. Elalamy, I., Chakroun, T., Gerotziafas, G.T., Petropoulou, A., Robert, F., Karroum, A., Elgrably, F., Samama, M.M., and Hatmi, M. (2008) Circulating platelet-leukocyte aggregates: a marker of microvascular injury in diabetic patients. *Thromb. Res.*, **121**, 843–848.
15. van Gils, J.M., Zwaginga, J.J., and Hordijk, P.L. (2009) Molecular and functional interactions among monocytes, platelets, and endothelial cells and their relevance for cardiovascular diseases. *J. Leukocyte Biol.*, **85**, 195–204.

16. Goel, M.S. and Diamond, S.L. (2002) Adhesion of normal erythrocytes at depressed venous shear rates to activated neutrophils, activated platelets, and fibrin polymerized from plasma. *Blood*, **100**, 3797–3803.
17. McGee, J.E. and Fitzpatrick, F.A. (1986) Erythrocyte-neutrophil interactions: formation of leukotriene B4 by transcellular biosynthesis. *Proc. Natl. Acad. Sci. U.S.A.*, **83**, 1349–1353.
18. Mestas, J. and Hughes, C.C. (2004) Of mice and not men: differences between mouse and human immunology. *J. Immunol.*, **172**, 2731–2738.
19. Makam, M., Diaz, D., Laval, J., Gernez, Y., Conrad, C.K., Dunn, C.E., Davies, Z.A., Moss, R.B., Herzenberg, L.A., and Tirouvanziam, R. (2009) Activation of critical, host-induced, metabolic and stress pathways marks neutrophil entry into cystic fibrosis lungs. *Proc. Natl. Acad. Sci. U.S.A.*, **106**, 5779–5783.
20. Theilgaard-Monch, K., Knudsen, S., Follin, P., and Borregaard, N. (2004) The transcriptional activation program of human neutrophils in skin lesions supports their important role in wound healing. *J. Immunol.*, **172**, 7684–7693.
21. Whale, T.A. and Griebel, P.J. (2009) A sheep in wolf's clothes: can neutrophils direct the immune response? *Vet. J.*, **180**, 169–177.
22. Tirouvanziam, R. (2005) in *Recent Advances in Immunology* (ed. S. Kundu-Raychaudhuri), Research Signpost, Trivandrum, pp. 1–17.
23. Gibbs, B.F. (2005) Human basophils as effectors and immunomodulators of allergic inflammation and innate immunity. *Clin. Exp. Med.*, **5**, 43–49.
24. Rothenberg, M.E. and Hogan, S.P. (2006) The eosinophil. *Annu. Rev. Immunol.*, **24**, 147–174.
25. Akuthota, P., Wang, H.B., Spencer, L.A., and Weller, P.F. (2008) Immunoregulatory roles of eosinophils: a new look at a familiar cell. *Clin. Exp. Allergy*, **38**, 1254–1263.
26. Sokol, C.L., Chu, N.Q., Yu, S., Nish, S.A., Laufer, T.M., and Medzhitov, R. (2009) Basophils function as antigen-presenting cells for an allergen-induced T helper type 2 response. *Nat. Immunol.*, **10**, 713–720.
27. Vizeacoumar, F.J., Chong, Y., Boone, C., and Andrews, B.J. (2009) A picture is worth a thousand words: genomics to phenomics in the yeast Saccharomyces cerevisiae. *FEBS Lett.*, **583**, 1656–1661.
28. Tirouvanziam, R., Davidson, C.J., Lipsick, J.S., and Herzenberg, L.A. (2004) Fluorescence-activated cell sorting (FACS) of Drosophila hemocytes reveals important functional similarities to mammalian leukocytes. *Proc. Natl. Acad. Sci. U.S.A.*, **101**, 2912–2917.
29. Auffray, C., Chen, Z., and Hood, L. (2009) Systems medicine: the future of medical genomics and healthcare. *Genome Med.*, **1**, 2.
30. Hanauer, D.A., Rhodes, D.R., and Chinnaiyan, A.M. (2009) Exploring clinical associations using '-omics' based enrichment analyses. *PLoS One*, **4**, e5203.
31. Tirouvanziam, R. (2006) Neutrophilic inflammation as a major determinant in the progression of cystic fibrosis. *Drug News Perspect.*, **19**, 609–614.
32. Herzenberg, L.A., Parks, D., Sahaf, B., Perez, O., and Roederer, M. (2002) The history and future of the fluorescence activated cell sorter and flow cytometry: a view from Stanford. *Clin. Chem.*, **48**, 1819–1827.
33. Tung, J.W., Heydari, K., Tirouvanziam, R., Sahaf, B., Parks, D.R., and Herzenberg, L.A. (2007) Modern flow cytometry: a practical approach. *Clin. Lab. Med.*, **27**, 453–468, v.
34. McGrath, K.E., Bushnell, T.P., and Palis, J. (2008) Multispectral imaging of hematopoietic cells: where flow meets morphology. *J. Immunol. Methods*, **336**, 91–97.
35. Zarember, K.A. and Malech, H.L. (2005) HIF-1alpha: a master regulator of innate host defenses? *J. Clin. Invest.*, **115**, 1702–1704.
36. Vollbrandt, T., Willkomm, D., Stossberg, H., and Kruse, C. (2004) Vigilin is co-localized with 80S ribosomes and binds to the ribosomal complex through its C-terminal domain. *Int. J. Biochem. Cell Biol.*, **36**, 1306–1318.

37. Perez, O.D., Krutzik, P.O., and Nolan, G.P. (2004) Flow cytometric analysis of kinase signaling cascades. *Methods Mol. Biol.*, **263**, 67–94.
38. Tirouvanziam, R., Gernez, Y., Conrad, C.K., Moss, R.B., Schrijver, I., Dunn, C.E., Davies, Z.A., and Herzenberg, L.A. (2008) Profound functional and signaling changes in viable inflammatory neutrophils homing to cystic fibrosis airways. *Proc. Natl. Acad. Sci. U.S.A.*, **105**, 4335–4339.
39. Gernez, Y., Tirouvanziam, R., Nguyen, K.D., Herzenberg, L.A., Krensky, A.M., and Nadeau, K.C. (2007) Altered phosphorylated signal transducer and activator of transcription profile of CD4+CD161+ T cells in asthma: modulation by allergic status and oral corticosteroids. *J. Allergy Clin. Immunol.*, **120**, 1441–1448.
40. Gernez, Y., Tirouvanziam, R., Yu, G., Ghosn, E.E.B., Reshamwala, N., Nguyen, T., Tsai, M., Galli, S.J., Herzenberg, L.A., Herzenberg, L.A., and Nadeau, K.C. (2011) Basophil CD203c levels are increased at baseline and can be used to monitor omalizumab treatment in subjects with nut allergy. *Int. Arch. Allergy Immunol.*, **154**, 318–327.
41. Kim, F.J., Battini, J.L., Manel, N., and Sitbon, M. (2004) Emergence of vertebrate retroviruses and envelope capture. *Virology*, **318**, 183–191.
42. Montel-Hagen, A., Kinet, S., Manel, N., Mongellaz, C., Prohaska, R., Battini, J.L., Delaunay, J., Sitbon, M., and Taylor, N. (2008) Erythrocyte Glut1 triggers dehydroascorbic acid uptake in mammals unable to synthesize vitamin C. *Cell*, **132**, 1039–1048.
43. Anderson, M.T., Roederer, M., Tjioe, I., Herzenberg, L.A., and Herzenberg, L.A. (1996) in *Handbook of Experimental Immunology* (eds L.A. Herzenberg, D.M. Weir, L.A. Herzenberg, and C. Blackwell), Blackwell Scientific, Boston, MA, pp. 54.1–54.9.
44. Tirouvanziam, R., Conrad, C.K., Bottiglieri, T., Herzenberg, L.A., and Moss, R.B. (2006) High-dose oral N-acetylcysteine, a glutathione prodrug, modulates inflammation in cystic fibrosis. *Proc. Natl. Acad. Sci. U.S.A.*, **103**, 4628–4633.
45. Kabelitz, D. and Medzhitov, R. (2007) Innate immunity–cross-talk with adaptive immunity through pattern recognition receptors and cytokines. *Curr. Opin. Immunol.*, **19**, 1–3.
46. Lotze, M.T., Zeh, H.J., Rubartelli, A., Sparvero, L.J., Amoscato, A.A., Washburn, N.R., Devera, M.E., Liang, X., Tor, M., and Billiar, T. (2007) The grateful dead: damage-associated molecular pattern molecules and reduction/oxidation regulate immunity. *Immunol. Rev.*, **220**, 60–81.
47. Rubartelli, A. and Lotze, M.T. (2007) Inside, outside, upside down: damage-associated molecular-pattern molecules (DAMPs) and redox. *Trends Immunol.*, **28**, 429–436.
48. Birrer, P. (1995) Proteases and antiproteases in cystic fibrosis: pathogenetic considerations and therapeutic strategies. *Respiration*, **62** (Suppl. 1), 25–28.
49. Woolhouse, I.S., Bayley, D.L., and Stockley, R.A. (2002) Effect of sputum processing with dithiothreitol on the detection of inflammatory mediators in chronic bronchitis and bronchiectasis. *Thorax*, **57**, 667–671.
50. George, T.C., Fanning, S.L., Fitzgeral-Bocarsly, P., Medeiros, R.B., Highfill, S., Shimizu, Y., Hall, B.E., Frost, K., Basiji, D., Ortyn, W.E., Morrissey, P.J., and Lynch, D.H. (2006) Quantitative measurement of nuclear translocation events using similarity analysis of multispectral cellular images obtained in flow. *J. Immunol. Methods*, **311**, 117–129.
51. Guil, S. and Esteller, M. (2009) DNA methylomes, histone codes and miRNAs: tying it all together. *Int. J. Biochem. Cell Biol.*, **41**, 87–95.

13
Optical Tweezers and Cytometry

Raktim Dasgupta and Pradeep Kumar Gupta

13.1
Introduction

Cytometry, the quantification of the physical or biochemical features of cells, is a very active area of research because of its applications in understanding basic biological processes and in clinical diagnosis. The field of cytometry has evolved considerably over the last few decades and the present day flow cytometers can perform simultaneous multiparameter analysis of several thousands of cells per second and are being routinely used in clinical diagnostics [1, 2]. Although extremely useful because of their high throughput and specificity, flow cytometry methods are not amenable for studying temporal changes or spatially resolved studies on cells. Several developments like faster laser scanning techniques, confocal microscopy, 4Pi microscopy, and so on, help address these issues. Another important development is the use of optical forces to manipulate individual cells or microbes. Single beam laser trap, more commonly known as "*optical tweezers*" can work as a precise pressure transducer [3–6]. These have been used to apply mechanical forces at single cell level and thus measure viscoelastic parameters of the cells and understand how these are altered under some disease conditions [7]. Optical tweezers have also been used to orient or rotate whole cells or intracellular objects for three-dimensional visualization of the object [8, 9] or for applications such as measurement of intracellular microviscosity [10, 11]. More recently, the ability of optical tweezers to immobilize cells in their natural environment without any physical contact has been exploited to record good quality Raman spectra from single trapped cell [12]. The ability to have spatially resolved Raman spectra from a single cell can provide a wealth of information about the functioning of an individual cell, its dependence on environmental parameters, and also its cellular interactions. Many interesting results have already been reported on the dynamics of thermal deactivation of cells [13], the effect of mechanical forces on the oxygen-carrying capacity of red blood cells (RBCs) [14], and so on. Raman spectroscopy at cellular level also has great promise for rapid diagnosis of microorganisms or diseased cells [15]. Coupled with microfluidic techniques, optical forces are also being utilized for high-throughput sorting of different cell types [16–19].

Advanced Optical Flow Cytometry: Methods and Disease Diagnoses, First Edition. Edited by Valery V. Tuchin.

In this chapter, we shall provide an overview of the use of optical tweezers for cytometric applications. First, we provide a very brief introduction to optical tweezers and the use of these as force transducer. Thereafter, the use of optical tweezers for the measurement of viscoelastic parameters of cells, in particular RBCs, is discussed. Next we describe the use of optical tweezers for Raman spectroscopic studies at single cell level and discuss a few representative examples to illustrate the potential of this approach for cytometric applications.

13.2 Optical Tweezers: Manipulating Cells with Light

13.2.1 Basics Principles

Optical tweezers owes its origin to the seminal work by Arthur Ashkin who showed that a focused laser beam could be used to trap and manipulate individual microscopic objects [20]. Since light carries momentum, owing to its absorption, scattering, or refraction by an object there is a momentum transfer and a resulting force on the object. While for a collimated beam this force is in the direction of light propagation, for a tightly focused beam there also exists a gradient force in the direction of the spatial gradient of the light intensity. For stable trapping in all three dimensions, the axial gradient component of the force pulling the particle toward the focal region must exceed the scattering component of the force pushing it away from that region. To achieve this, the trap beam needs to be focused on a diffraction-limited spot using a high numerical aperture (NA) objective lens. Although a detailed understanding of optical forces requires rigorous electromagnetic theory treatment [21–23], for objects with size much greater than the wavelength of the trap beam (as is the case for biological cells), a simple ray optics description [24] can be used to explain the basic idea. Referring to Figure 13.1, consider two light rays ("a" and "b") situated at equal radial distances from the beam axis. Owing to the refraction of rays a and b from the sphere, assumed to have a refractive index higher than the surroundings, there will be forces F_a and F_b, respectively, on it. The net force, denoted as F, will try to pull the sphere to the focal point. When at the focal point, there is no refraction and hence no force on the sphere. It can be verified from Figure 13.1 that in all the cases where the sphere is positioned away from the focal point the resultant force acts to pull the sphere onto the beam focus (the equilibrium position). In this ray-optics description, the sphere is assumed to be weakly reflective or absorptive at the trapping wavelength so that the forces arising due to the absorption or reflection of light by the sphere can be neglected.

It is important to note here that for a TEM_{00} laser beam the gradient force is linear with small displacements of the trapped particle [4, 25], and can be expressed as

$$F = -kx \tag{13.1}$$

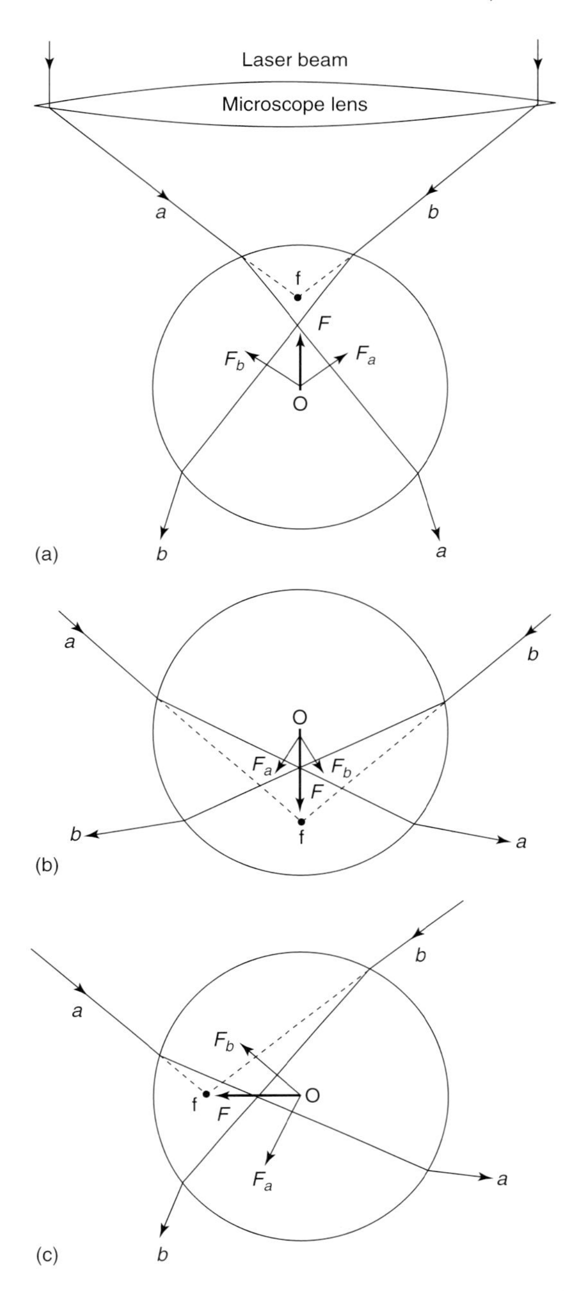

Figure 13.1 A ray diagram explanation of the trapping of a dielectric spherical particle in a focused laser beam. F is the net gradient force. Reprinted with permission from Elsevier B. V.: Ashkin, A. (1992). *Biophys J.*, **61**, 569–582. Ref. [24], copyright (1992).

where x is the displacement of a trapped bead from the trap center and k is known as *trap stiffness*. Several experimental techniques involving viscous drag on the trapped bead, or position spectrum of the Brownian fluctuation of the trapped bead are employed to calibrate the stiffness of an optical trap [5, 26]. Knowledge of trap stiffness allows precise measurement of the restoring force at specific displacement of the trapped bead.

13.2.2
Experimental Considerations

For developing an optical tweezers set-up, the essential elements are a laser, beam expansion and steering optics, a high NA objective lens, a sample chamber, and some means of observing the trapped specimen, that is, an illumination source and a CCD (charge coupled device) camera. Optical traps are most often built around an inverted microscope by introducing the laser beam into the optical path before the objective. Figure 13.2 shows a schematic of a typical optical tweezers set-up. Usually the same objective lens is used for viewing the objects as well as for trapping. A dichroic mirror is used to couple the laser beam onto the objective. The trapped object viewed by the objective lens is imaged onto a CCD camera by the tube lens. An additional lens must be inserted into the trapping beam path before the tube lens to ensure collimation of the coupled laser beam into the objective lens. The choice of the objective lens to be used in optical tweezers is critical. An objective with high NA (typically >1.2) is required to produce an intensity gradient sufficient to overcome the scattering force and produce a stable optical trap. The objective

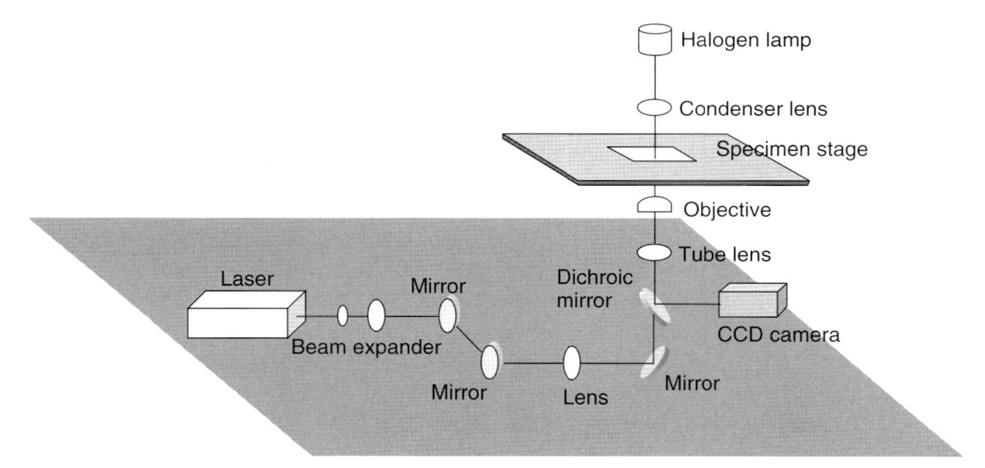

Figure 13.2 Schematic of a typical optical tweezers set-up. The optical beam from a laser source is first expanded using a beam expander assembly. The beam is then coupled to the bottom port of an inverted microscope using mirrors. The objective lens focuses the beam to create the optical trap at the specimen plane. For viewing the trapped cells the in-built halogen lamp of the microscope is used along with a CCD camera.

should be corrected for optical aberrations to ensure a diffraction-limited focal spot and should also have good transmission at the wavelength of the trapping light.

The typical light intensity at the trapping plane can be of the order of several megawatts per centimeter square and may be potentially harmful for living cells and microorganisms depending upon the absorption characteristic of the object at the trapping wavelength. The cellular components have relatively higher transparency in the near infrared (700–1300 nm) portion of the spectrum, because at shorter wavelength, absorption by proteins and nucleic acids becomes important and at longer wavelengths water absorption becomes large. Therefore, lasers operating in the near infrared spectral range like neodymium-doped yttrium aluminum garnet (Nd:YAG) laser ($\lambda = 1064$ nm) or semiconductor diode lasers are popular choices for use as trap beam in optical tweezers. While diode lasers offer the advantage of lower cost and availability at several wavelengths in the near infrared, the power output of these devices in the lowest order transverse mode, the mode required for efficient trapping, is typically limited to less than 250 mW. By far, the most versatile laser option is a tunable cw Titanium:Sapphire laser, a system that delivers high power $\sim$1 W, over a large portion of the near infrared spectrum ($\sim$750–950 nm). This is particularly useful for building Raman optical tweezers since the large tuning range helps excite Raman signals at different wavelengths.

Several applications (e.g., experiments where one needs to immobilize the cell by holding it by optical traps at different points and pull the cell from some other point to probe its viscoelastic characteristics) require multiple traps. Using separate trap beams to generate multiple traps is cumbersome and a more convenient approach to generate multiple traps is time-sharing the single laser beam over the trap points. This, of course, requires that the laser is scanned such that the "off" time of the beam at any trap point is less than the escape time of the objects from the trap. For achieving such time-sharing of the laser beam, galvanometric scanning mirrors [27] or acousto-optic deflector systems [28] are often used. While galvanometric scanning mirrors offer the advantages of large deflection angle and low insertion loss they have the drawback of a comparatively slow temporal response (greater than tens of milliseconds). As a result, the beam can be scanned over a limited number of trap points. In contrast, acousto-optic deflectors have fast response time (of the order of few microseconds), but their insertion loss is large (first-order diffraction efficiency of $\sim$60%).

13.3
Use of Optical Tweezers for the Measurement of Viscoelastic Parameters of Cells

Optical tweezers offer a precisely controllable force transducer in a few to several hundred piconewtons range and have been widely used for investigating viscoelastic parameters of single cells. In particular, there has been considerable interest in the use of optical tweezers for the measurement of the viscoelastic parameters of RBCs. This is motivated by two factors. First, the RBCs are one of the simplest biological cells and therefore easier to model, and second, the viscoelastic parameters of RBCs can provide important medical diagnostic information.

13.3.1 Red Blood Cells

RBCs are biconcave disks (flattened and depressed in the center to result in a dumbbell-shaped cross section) with a diameter of ~8 μm and a thickness of ~2 μm. These have a simple aneucleated structure and are filled with hemoglobin, which plays the role of the carrier of oxygen from lungs to tissues. The RBCs undergo severe elastic deformation as these pass through narrow capillaries whose inner diameter is as small as 3 μm. It is also known that as RBCs age, and also under some disease conditions, their membrane elasticity reduces.

Therefore, there has been substantial work on the measurement of elastic parameters of the RBC with a view to understanding its functional aspects as well as from the point of view of disease diagnosis. Although earlier measurements made use of micropipette aspiration methods involving measurement of the length of RBC aspirated in a micropipette under calibrated pressure difference [29–31], more recently the use of optical tweezers has received much attention for this purpose. While a cell trapped in a buffer suspension can be stretched by moving the specimen stage to exert a viscous drag on it, the use of small beads attached to the cell membrane as handles to pull the membrane has been more widely used as it provides a wider range of the forces that can be applied on the cell. In the bead-based techniques, the range of the trapping force is determined by the trapping laser power and the attached bead size and can be varied from a few to hundreds of piconewtons. The latter is sufficient to induce large deformations as encountered during passage through small capillaries [32, 33].

For small deformations, the cell membrane can be characterized by the area expansion modulus K, the shear modulus μ, and the bending stiffness B [34]. For RBCs, the bending stiffness is too small (less than a few $k_B T$, where k_B is the Boltzmann constant and T is temperature in degree Kelvin), to have any significant influence on the overall membrane deformation [35, 36]. Shear deformation, which is often the most important factor related to the functionality of RBCs, is constrained by the fact that shape change for an intact cell occurs under the condition of minimum change in cell volume [37] and the phospholipid bilayer of the cell membrane has high resistance to change in its surface area. These constant volume and surface conditions impose restrictions to overall deformability of the cell when acted upon by any shear stress and therefore the mechanical properties of the protein cytoskeleton (network of proteins underneath the phospholipid bilayer) cannot be decoupled from those of the phospholipid bilayer in studies performed with intact cells. Therefore, for the estimation of pure cytoskeletal characteristics RBC ghosts (cells prepared by controlled lysis method) have been used instead of intact cells [34, 38].

The values of the viscoelastic parameters of RBCs obtained from the micropipette aspiration methods are $K \sim 300–500\ \text{mN m}^{-1}$ [30, 39, 40] and $\mu \sim 4–10\ \mu\text{N m}^{-1}$

[29–31, 41]. The values estimated with optical forces are often in disagreement with the micropipette aspiration data. One plausible reason for this is that when the cells are aspirated, the cytoskeleton being bound to the phospholipid layer may slip with respect to it [42, 43], resulting in local dilation or compression of the cytoskeleton. The resulting concentration gradient of molecules in cytoskeleton implies that the shear modulus measured in the micropipette experiments is actually a combination of μ and the area expansion modulus K. Moreover, the micropipette technique assumes that the cells do not adhere to the pipette wall, which is often not a very realistic assumption [37].

13.3.1.1 Use of Multiple Optical Traps

Bronkhorst *et al.* [44] were the first to use optical tweezers to monitor the changes in the elastic parameters of RBC membrane as a function of its age. They used three optical tweezers to hold the cell at three equidistant points on its membrane and used these to stretch the membrane (Figure 13.3). When the traps were turned off the cell relaxed back to its original shape. Significant differences in the relaxation time were observed for the density-separated subpopulations of RBCs. While the mean relaxation time was found to be 162 ms for the least dense (young) cells, it was measured to be 353 ms for the most dense population (old) of cells. These results were consistent with previous reports [45, 46]. However, since the stretching forces were not calibrated, no values for the elastic parameters of RBC membrane were reported in this study.

Hénon *et al.* [37] used the force transducer property of optical traps by attaching two small silica beads (radius ~1.05 μm) to the RBC membrane and using these as handles to stretch the RBC (Figure 13.4). In this configuration, by monitoring the displacement of the bead, the applied stretching forces could be precisely calibrated, facilitating measurement of the elastic shear modulus of the RBC membrane. In

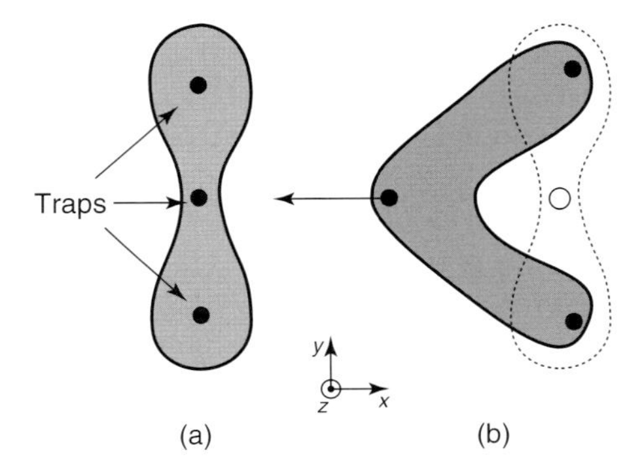

Figure 13.3 The red blood cell is viewed along the trap beam axis (z axis). (a) The red blood cell orients vertically in the traps. (b) The cell is deformed into a parachute shape by moving the middle spot to the left at a standardized low speed. The laser beam was scanned at 50 Hz through the three trap positions. Reprinted with permission from Elsevier B. V.: Bronkhorst *et al.* (1995). *Biophys J.*, **69**, 1666–1673. Ref. [44], copyright (1995).

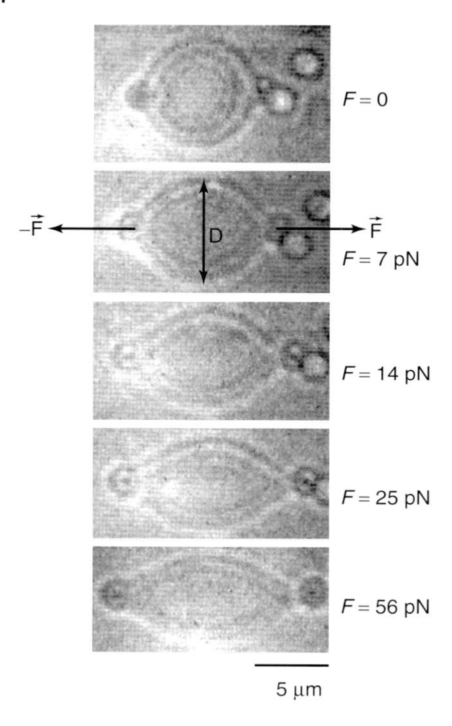

Figure 13.4 Discotic RBC with two silica beads attached in diametrical positions. The deformation increases with increasing force **F**. The variations in diameter **D** are measured in the direction perpendicular to **F**. Reprinted with permission from Elsevier B. V.: Hénon *et al.* (1999). *Biophys J.*, **76**, 1145–1151. Ref. [47], copyright (1999).

the stretching force range used (<15 pN), the value of μ *was* estimated to be $2.5 \pm 0.4\,\mu\text{N m}^{-1}$, which is about half that measured by the micropipette techniques. This was attributed to the fact that while these measurements were carried out in a small deformation regime, the micropipette-based methods operate in the large deformation regime. Indeed, more recent experiments performed in large stretching force regime support this conjecture [32, 33].

Sleep *et al.* [38] used RBC ghosts prepared by saponin lysis for measuring elastic constants of pure protein cytoskeleton. The estimate for the shear modulus obtained in these experiments was $\sim 200\,\mu\text{N m}^{-1}$, which is about an order of magnitude higher than the value reported for intact cells using micropipette aspiration techniques. This discrepancy was attributed to the fact that in the micropipette aspiration method protein cytoskeleton may slip over the phospholipid bilayer in the aspirated part of membrane and therefore lead to a lower estimate of membrane shear modulus.

Lenormand *et al.* [34] used three small silica beads attached at three equiseparated points on the membrane to apply stretching force onto the membrane skeleton

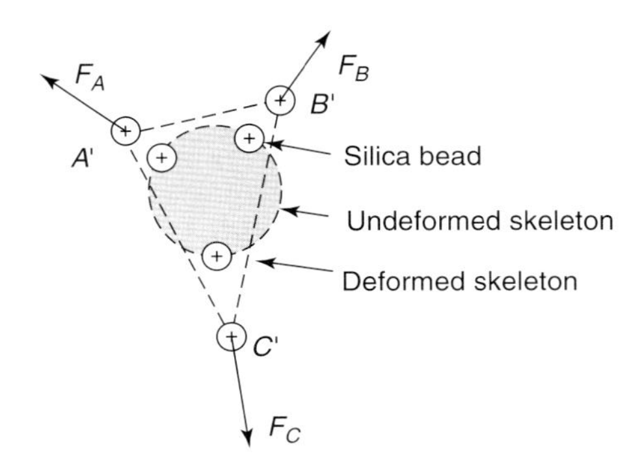

Figure 13.5 A schematic of the experimental procedure used by Lenormand *et al.* to measure μ and K simultaneously. An RBC is seized by trapping each of the three silica beads bound to its periphery with optical tweezers. Its deformation, interpolated from the beads' positions, is a superposition of shear and area expansion. Reprinted with permission from Elsevier B. V.: Lenormand *et al.* (2001). *Biophys J.*, **81**, 43–56. Ref. [34], copyright (2001).

that was prepared by bilayer dissolution with detergent solution. This configuration (Figure 13.5) allows estimation of both K and μ simultaneously. An increase in the distances between the traps simultaneously in two orthogonal directions leads to pure area expansion. By increasing the distance in only one direction, both area expansion and shear will result. Further, by increasing the distance in one direction and decreasing it in the other direction, nearly pure shear could be generated. From these three measurements the values obtained for the elastic parameters in low hypotonic buffer (25 mOsM kg^{-1}) and for small stretching force of up to 8 pN were $K = 4.8 \pm 2.7\ \mu N\ m^{-1}$, $\mu = 2.4 \pm 0.7\ \mu N\ m^{-1}$, and K/μ ratio $= 1.9 \pm 1.0$.

All the measurements discussed so far employed force values of upto ~15 pN, which is insufficient to induce large deformations often encountered *in vivo*. This aspect was addressed by Dao *et al.* [32] and Lim *et al.* [33] by attaching large-sized silica beads to cell surface, which allows the use of higher trap beam power and thus larger stretching forces. One of the microbeads (4.12 μm in diameter) was adhered to the glass surface while the other was trapped using the laser beam. By moving the bead with the trap beam the cell could be stretched up to a stretching force of ~400 pN. At a stretch force of ~340 pN, an increase in the axial diameter of cell by ~50% and the reduction of transverse diameter by more than 40% was observed. The in-plane shear modulus for large deformation was estimated to be $\mu = \sim 20\ \mu N\ m^{-1}$. This value is higher than the range of values obtained from micropipette aspiration experiments. This was attributed to the highly nonlinear stress on the membrane near the bead attachment points.

Suresh *et al.* [48] used optical tweezers to monitor the changes in elastic parameters of the human RBC at different stages of parasite (*Plasmodium falciparum*) infection (Figure 13.6). Apart from the control [the healthy red blood cells (H-RBCs) and RBCs exposed to *P. falciparum* but uninfected (Pf-U-RBC)], the different developmental stages of the *P. falciparum* parasite, such as ring stage (Pf-R-pRBC), *P. falciparum* trophozoite stage (Pf-T-pRBC), and *P. falciparum* schizont stage

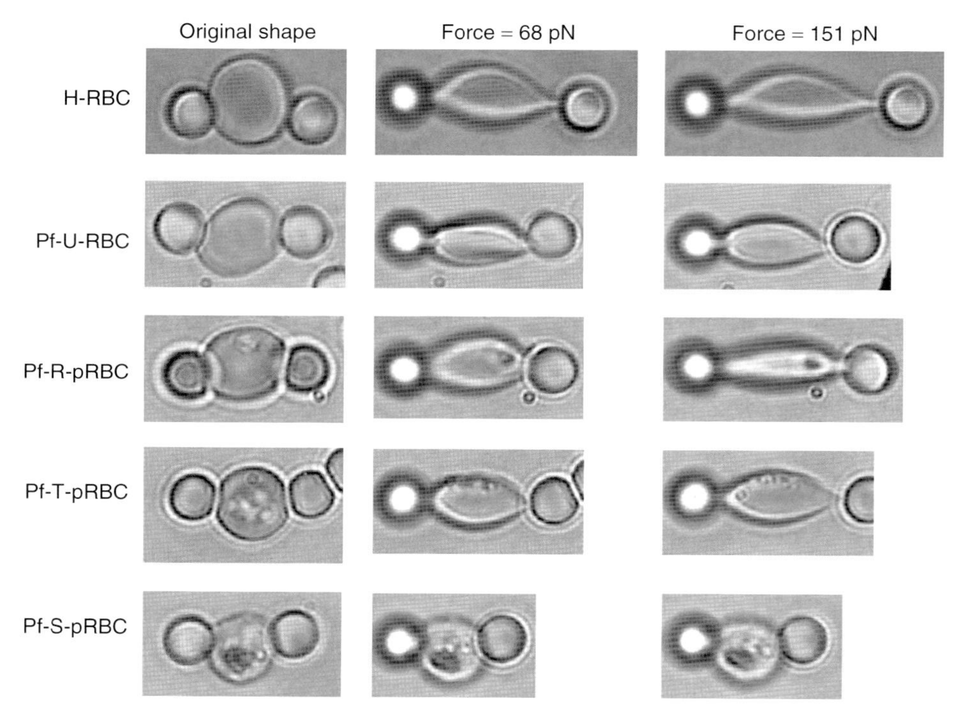

Figure 13.6 Images of H-RBC, Pf-U-RBC, Pf-R-pRBC, Pf-T-pRBC, and Pf-S-pRBC in PBS (phosphate buffer saline) solution at 25 °C: prior to tensile stretching by optical tweezers (left column), stretching at a constant force of 68 ± 12 pN (middle column) and at a constant force of 151 ± 20 pN (right column). Note the presence of the *P. falciparum* parasite inside the infected RBCs. Reprinted with permission from Elsevier B. V.: Suresh *et al.* (2005). *Acta Biomater.*, **1**, 15–30. Ref. [48], copyright (2005).

(Pf-S-pRBC), were also studied. At a fixed stretching force, the deformability of the parasitized cell was significantly reduced compared to that of the control population. Shear modulus for Pf-R-pRBC, Pf-T-pRBC, and Pf-S-pRBC were measured to be 16, 21.3, and 53.3 μN/m, respectively, for large stretching force regime of ~60–150 pN. Notably, the value for Pf-S-pRBC was about an order of magnitude higher than the value measured for H-RBC (5.3 $\mu N\ m^{-1}$).

13.3.1.2 **Use of Counterpropagating Light Beams**

Guck *et al.* [49, 50] showed the advantage of using two weakly diverging counterpropagating beams to stretch cells. In contrast to optical tweezers where gradient force balances the scattering force to ensure three-dimensional trapping of objects, here the scattering forces by the two counterpropagating beams balance each other (Figure 13.7). The stretching force arises because of a change in the momentum of the light inside the object as compared to that outside the object. As light enters a cell, having refractive index higher than the surrounding medium, the light gains momentum so that the cell surface gains momentum in the opposite or backward

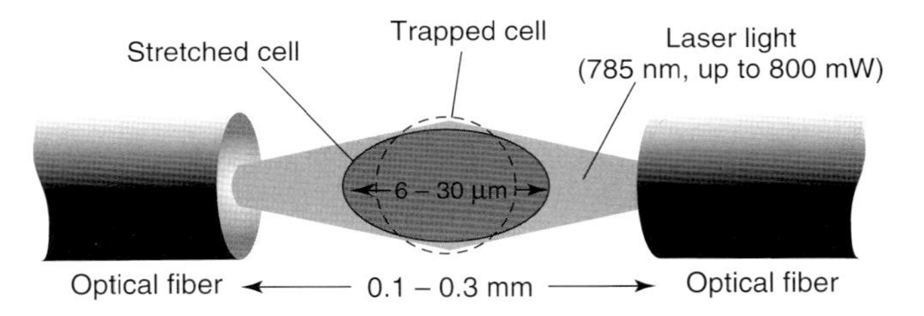

Figure 13.7 Schematic of the stretching of a cell by two counterpropagating beams. The cell is stably trapped in the middle as the net optical scattering forces from the two counterpropagating beams of identical power cancel each other. But depending on the elastic strength of the cell, at a certain light power the cell is stretched out along the laser beam axis owing to the light pressure at its surface. Reprinted with permission from Elsevier B. V.: Guck *et al.* (2001). *Biophys. J.*, **81**, 767–784. Ref. [50], copyright (2001).

direction. Similarly, the light loses momentum upon leaving the cell so that the opposite surface gains momentum in the direction of the light propagation. The resulting surface forces on the front and backside of the cellular membrane are in opposite directions and result in the stretching of the cell along the common axis of the light beams.

Using this approach, shear modulus of osmotically swollen RBCs was measured to be $13.0 \pm 5.0\ \mu N\ m^{-1}$ [50]. Nonlinear distortion was observed above a stress of $2\ N\ m^{-2}$, and a relative increase in radius up to 82% at a stress of $2.55\ N\ m^{-2}$ was noted. A significant advantage of an "optical stretcher" is that stretching forces are distributed over the membrane unlike in bead-based methods that may result in highly nonlinear deformation near the points of attachment of the beads.

13.3.1.3 Use of Evanescent Wave of Light

The use of optical evanescent wave for manipulation of micro-objects has also been investigated. Interestingly, by the use of evanescent wave, a deformation ratio ~53% larger than that achieved with the bead-based method was achieved at a factor of 10 smaller trap power levels. This was attributed to the highly confined nature of the evanescent light field near the interface, which leads to stronger trapping forces at lower light exposure on the sample [51].

13.3.1.4 Use of Viscous Drag on Optically Trapped Cell

The RBC, optically trapped in aqueous buffer suspension, can also be stretched by moving the stage and thus the fluid around the cell (Figure 13.8) [52]. Significant differences in shear modulus and viscoelastic time constant were obtained for normal and aged RBCs and cells infected with *P. falciparum*. An interesting consequence of the difference in membrane rigidity of normal and infected RBCs was that in hypertonic buffer medium (osmolarity $>800\ mOsM\ kg^{-1}$), while a normal RBC rotates by itself when placed in laser optical trap, at the same trap beam power, malaria-infected RBCs, due to their larger membrane rigidity, do not rotate [53]. It was shown [54, 55] that the rotation of the normal RBC in hypertonic buffer arises

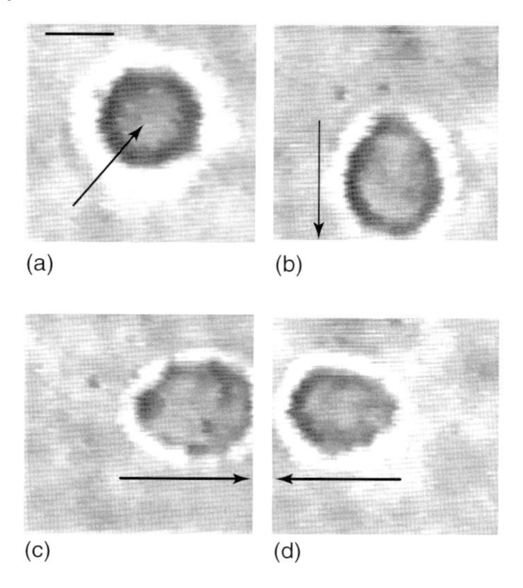

Figure 13.8 Stretching of an optically trapped RBC by the use of viscous drag. RBC (marked by arrow) on a stationary stage (a); stretching of RBC in different directions (b)–(d) by application of viscous force. The arrows indicate the direction of stage movement in panels (b)–(d). All images are of the same magnification. Scale bar: 5 μm. Reprinted with permission from Wiley-VCH: Mohanty *et al.* (2008). *J. Biophoton.* **1**, 522–525. Ref. [52], copyright (2008).

because of the torque generated on the cell by transfer of linear momentum from the trapping beam, similar to the previous reports of rotation of specifically structured nonbiological objects [56, 57]. For a given trap power, the speed of rotation of RBC could be controlled by a change in the osmolarity of the hypertonic buffer that alters the shape of the cell. The larger asymmetry of the shape of RBC at higher osmolarity of the suspension leads to higher torque, resulting in an increase in rotational speed of RBC. At a trapping power level of 80 mW, RBC was observed to rotate at a speed of ~25 rpm when suspended in buffers with osmolarity ~1000 mOsM kg^{-1}, and at the same trap power the rotational speed increased beyond ~200 rpm at an osmolarity of 1250 mOsM kg^{-1}. However, owing to their larger membrane rigidity, the infected cells, when suspended in hypertonic solution, did not get distorted into an asymmetric shape and as a result experienced no torque.

13.3.2 Cancer Cells

It is known that the cell cytoskeleton, apart from providing mechanical rigidity to the cell, also fulfills many important cellular functions including facilitating cell motility, ribosomal and vesicle transport, mechanotransduction, and so on. Consequently, changes to cellular functionalities during diseased condition are mirrored in the cytoskeleton. During disease progression in malignancy the cytoskeleton evolves from an ordered and rigid state to an irregular and compliant state [58, 59]. These changes are also expected to get reflected in the overall mechanical properties of the cell as well. Guck *et al.* [60] used an optical stretcher to monitor the changes in fibroblasts and human breast epithelial cells as these progressed from normal to a cancerous state. To study the cells in a high throughput manner required for statistical validation of the data, Guck *et al.* coupled their optical stretcher with a microfluidic system to capture cells from a flow and center them automatically in the two beam trap. Measurements were carried out on a mouse embryonic fibroblast cell line (BALB/3T3) and SV-T2 cells that were derived from BALB/3T3 cells by transformation with the oncogenic DNA virus SV40. The measured data are shown in Figure 13.9. Since the refractive indices for fibroblasts

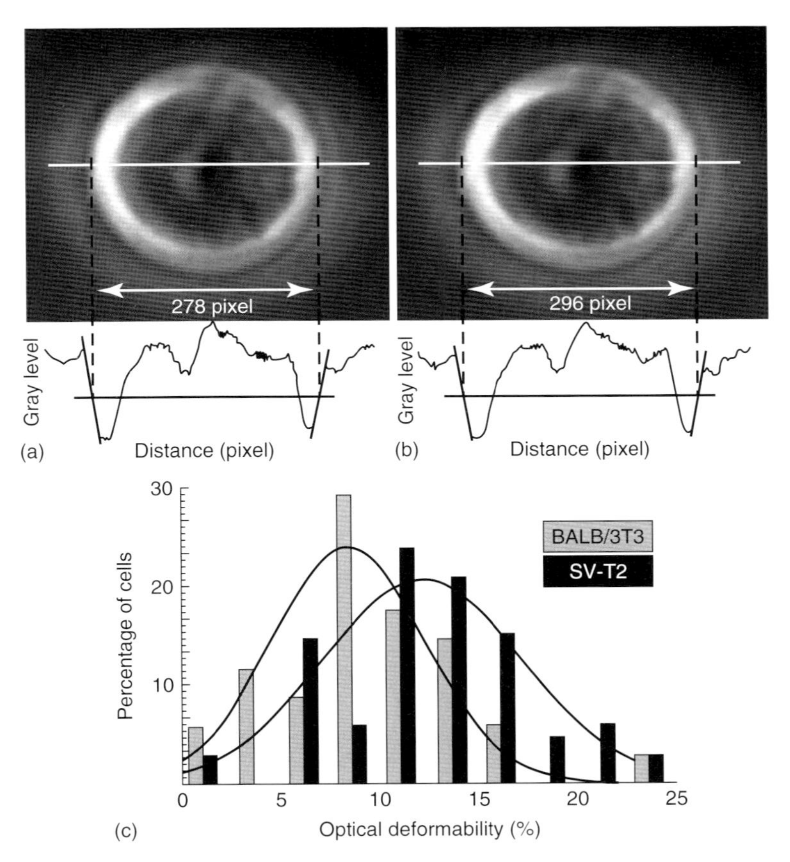

Figure 13.9 Deformation of fibroblasts in an optical stretcher. A BALB/3T3 fibroblast deforms by 6.48 ± 0.36% along the laser axis, when the light power in both the beams is increased from 0.1 W (a) to 1.7 W (b). Measurements carried out on a large numbers of cells reveal that the optical deformability of malignantly transformed SV-T2 fibroblasts is significantly shifted to higher values compared to normal BALB/3T3 fibroblasts (c). $OD_{BALB/3T3}$ (optical deformability) = 8.4 ± 1.0 μm; $OD_{SV\text{-}T2}$ = 11.7 ± 1.1 μm. Reprinted with permission from Elsevier B. V.: Guck *et al.* (2005). *Biophys. J.*, **88**, 3689–3698. Ref. [60], copyright (2005).

BALB/3T3 [$n_{BALB} = 1.3722 \pm 0.0036$ (mean and SD)] and SV-T2 cells ($n_{SV\text{-}T2} = 1.3711 \pm 0.0039$) are almost identical, the difference in optical deformability shown in Figure 13.9 can be solely attributed to the mechanical properties of the cells.

13.3.3 Stem Cells

Optical stretching has found its use in the study of stem cells. Mesenchymal stem cells (MSCs) when treated with appropriate growth factors *in vitro* can be induced to differentiate into osteoblasts, adipocytes, and chondrocytes. Thus it has got vast therapeutic potential in tissue engineering and regenerative medicine. Interestingly, mechanical properties of stem cells such as cytoskeleton organization and elasticity, membrane tension, cell shape, and adhesion strength is believed to play an important role in cell fate and differentiation [61, 62]. Titushkin and Cho [63] used optical tweezers to study the membrane mechanics of MSCs and terminally differentiated fibroblasts by extracting tethers from the outer cell membrane. It was found that tether extraction force did not increase with increasing tether length, suggesting the existence of a membrane reservoir for both types of cells. Further investigation on the role of cytoskeleton integrity over membrane mechanics was performed by disrupting the cytoskeleton. This resulted in a fourfold tether length increase in fibroblasts but had no effect in MSCs. This indicated a weak association between the membrane and cell cytoskeleton for MSCs and suggested that whereas for fibroblasts membrane–cytoskeleton association plays a crucial role in regulating their membrane reservoir, MSC cytoskeleton has only a minor impact on membrane mechanics.

13.4 Cytometry with Raman Optical Tweezers

13.4.1 Raman Optical Tweezers: Basics

Raman scattering is an inelastic scattering of light where the difference between the energy of the scattered photon and the incident photon corresponds to a change in vibrational, rotational, or electronic energy of the scattering medium [64–66]. Since Raman shifts serve as fingerprints of molecules, Raman spectroscopy of cells can provide valuable cytometric information without the need for any exogenous stain.

However, the inherent weak nature of Raman signal demands that cell should be immobilized so that spectra can be acquired over long time (often tens of seconds to few minutes) to achieve good signal-to-noise ratio. The physical or chemical methods used for immobilization of cells in micro-Raman techniques often lead to undesirable surface-induced effects on the cells or lead to strong background spectra originating from substrate/immersion medium [67]. The use of optical

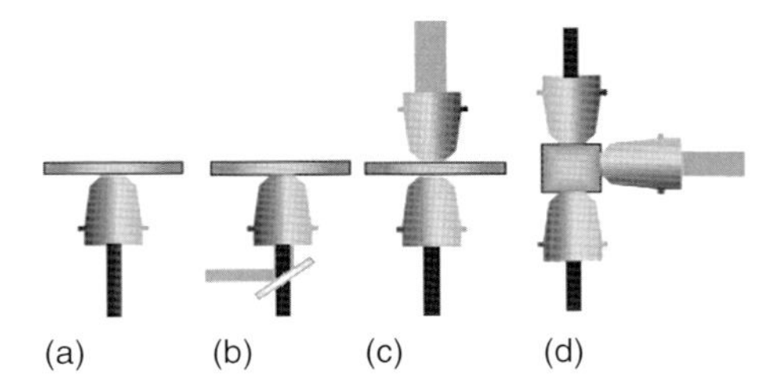

Figure 13.10 Possible configurations of Raman tweezers. The trapping beam(s) is shown in black and Raman excitation beam(s) in gray. Reprinted with permission from IOP Publishing Ltd.: Petrov, D. V. (2007). *J. Opt. A: Pure Appl. Opt.*, **9**, S139–S156. Ref. [12], copyright (2007).

tweezers to immobilize a cell without direct contact helps avoid these problems and therefore Raman optical tweezers, or the set up facilitating acquisition of Raman spectra from an optically trapped cell, have received much attention [12, 68]. In particular, the use of near infrared radiation for Raman studies is gaining rapid interest due to the much reduced fluorescence background and the longer penetration depth of the near infrared radiation.

Several configurations have been used (Figure 13.10) for Raman optical tweezers [12]. The simplest technique uses one microscope objective and one optical beam that is used for both trapping the microparticle and exciting the Raman spectra (Figure 13.10a) [13, 69]. The alignment of the system is comparatively easy but the flexibility of the set-up is limited since the same beam is used to trap the cell and excite its Raman signal.

Using two separate beams for trapping and Raman excitation with a single objective has the advantage that the choice of wavelength and the power of the trapping beam and the Raman excitation beam can be independently chosen (Figure 13.10b) [70, 71]. For example, a high power near infrared laser can be employed for nondestructive trapping of the cells whereas a low power UV/visible laser can be chosen to selectively excite resonance Raman spectra from biomolecules of interest.

The Raman tweezers can be made even more versatile by having the trapping and Raman excitation beams focused on the particle by separate objectives (Figure 13.10c). Although alignment of the system becomes difficult, this configuration allows collection of spatially resolved spectra from the trapped biological system by scanning the Raman beam with respect to the trapping beam.

Two counterpropagating optical beams can also be used to stably trap objects even with low NA objectives, because in this case the scattering force of one beam is balanced by the scattering force exerted by the second beam [72]. Raman spectra are excited by a third beam focused on the trapped particle through a separate objective, which also collects the spectra (Figure 13.10d). This configuration offers more flexibility for studying large cells.

The Raman spectra, being weak, are often overwhelmed by noises and background resulting from luminescence of the cell or suspension. Noise may arise from a variety of sources such as cosmic rays, variations in the intensity of the light source, and dark current in the detector system. Different noise filtering algorithms are employed to improve the signal-to-noise ratio of the acquired spectra [73]. Dominant source for the background in Raman spectra is fluorescence from the cell or from the substrate/medium. Since this background usually has a much broader structure in contrast to the useful Raman bands, it can be separated with techniques such as smoothening filters, baseline fitting, and so on [74].

13.4.2 Cytometry Applications

Since the first report of acquisition of Raman spectra from a single optically trapped biological object (single synaptosomes, terminal particles in neuron cells) by Ajito and Torimitsu [75], several reports have appeared on the use of Raman tweezers for various cytometric applications. We discuss a few representative studies to highlight the potential of the approach.

13.4.2.1 Real-Time Study of Dynamic Processes in Single Cell

Raman tweezers have proved very useful for monitoring the response of an individual living cell to external stresses such as change in temperature, pH, saline concentration, and so on, *in situ*.

Xie *et al.* [13] reported acquisition of Raman spectra from optically trapped living cell during a heat-denaturation process. Large changes in the intensity of the phenylalanine band (1004 cm^{-1}), assigned to the breathing vibration mode of the aromatic ring of Phe molecules were observed, between healthy and heat-affected cells (yeast, *Escherichia coli*, and *Enterobacter aerogenes*). The results are shown in Figure 13.11. In the functional form, cellular proteins fold into an ordered structure and the intensity of the Raman peak at 1004 cm^{-1} is low. Due to hydrogen bonds disruption during heat denaturation the proteins unfold and a significant enhancement in the intensity of this peak was observed. The authors showed that intensity of the Raman band at 1004 cm^{-1} could characterize the thermal death process of individual cells. Furthermore, the results show the potential of Raman optical tweezers for studying conformational changes of proteins inside a single cell in real time.

Rao *et al.* [14] used Raman optical tweezers to affect mechanical stress on the trapped RBC and monitor the changes in the oxygenation state *in situ* by observing the Raman spectra. The cells were stretched using a dual beam trapping configuration. The Raman excitation beam was positioned at the center of the stretched RBC. By comparing the Raman spectra of equilibrium and stretched RBCs, it was concluded that cells with a significant oxygen concentration were pushed to a deoxygenated state when stretched with optical tweezers. The mechanically induced change in oxygenation state was attributed to the enhanced interaction between neighboring hemoglobins or between hemoglobin and the membrane, arising as a result of stretching the cell.

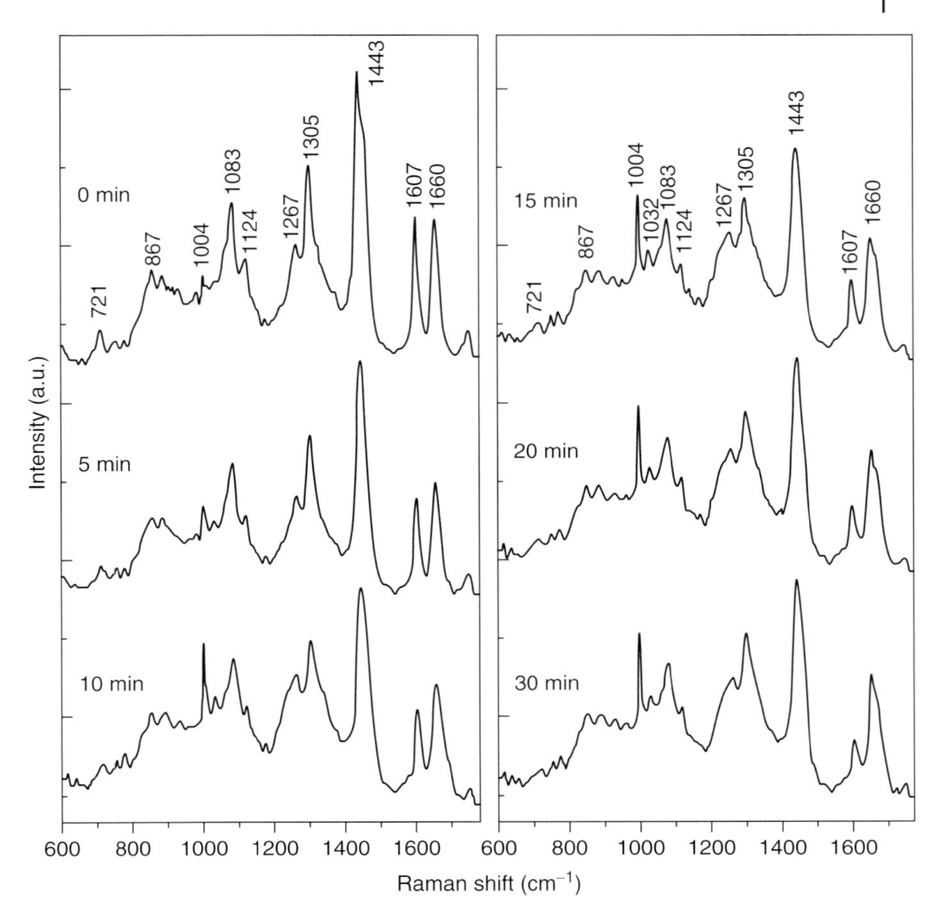

Figure 13.11 Raman spectra of a single trapped yeast cell at different time intervals after the temperature of the culture medium was raised and maintained at 80 °C. Measurements were repeated with 30 cells and averaged. Reprinted with permission from American Institute of Physics: Xie *et al.* (2003). *J. Appl. Phys.*, **94**, 6138–6142. Ref. [13], copyright (2003).

13.4.2.2 Identification and Sorting of Microorganism

The characterization of microorganisms is one of the most frequent tasks in clinical microbiology laboratories. These analyses are performed using different techniques such as microscopy, serology, biochemical tests, and so on. However, most of these methods are time consuming and sometimes it may take weeks to identify the microorganism, delaying the choice of the most appropriate treatment. This may not only result in deterioration of the patient's condition but may also contribute to the growing problem of drug resistance in microorganisms.

The use of Raman tweezers for rapid identification of single trapped bacterium inside a complex mixture of living and nonliving objects looks very promising. In Figure 13.12 we show the Raman spectra for three types of bacterial cells, *Bacillus*

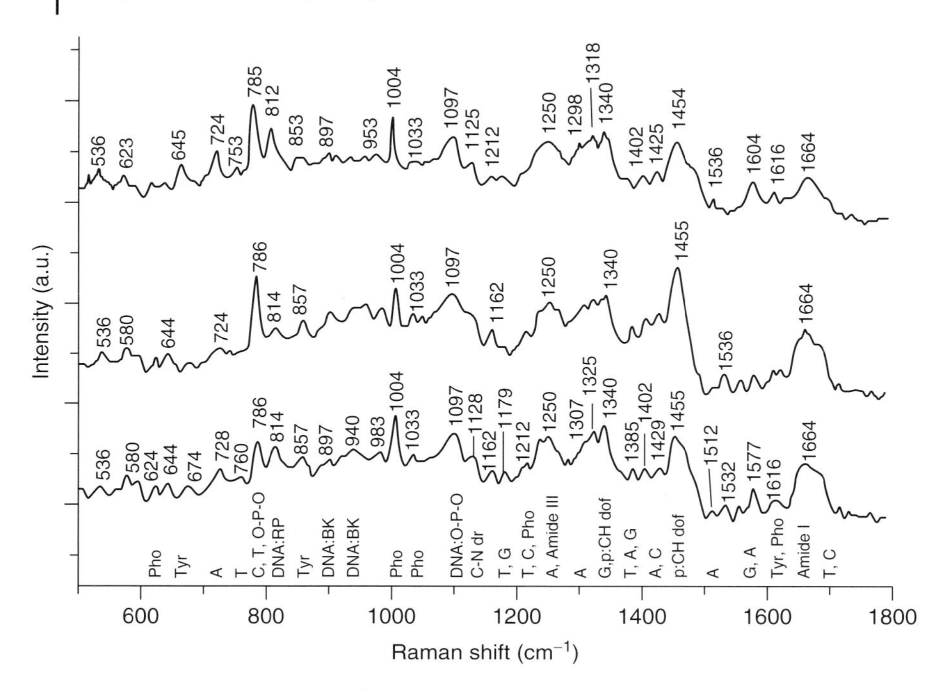

Figure 13.12 Raman spectra of *Bacillus sporeformer*, (top) *Streptococcus salivarius* (middle), and *E. coli* bacterium (bottom). Reprinted with permission from Optical Society of America: Xie *et al.* (2004). *Opt. Exp.*, **12**, 6208–6214. Ref. [15], copyright (2004).

sporeformer, Streptococcus salivarius, and *E. coli* [15]. The ratio of the intensity of the 785 cm^{-1} band (assigned to nucleic acids) and 1004 cm^{-1} band (assigned to phenylalanine) is significantly different for the three (~1.35 : 1.0 for *B. sporeformer*, 1.45 : 0.72 for *Streptococcus salivarius,* and 0.72 : 1.09 for *E. coli*) and can be used to discriminate these. It is pertinent to note here that the rod-like bacteria *E. coli* and *B. sporeformer* are about 2 μm in length and 0.5 μm in diameter, and are therefore most unlikely to be visually separable under microscopic observation.

Xie *et al.* [76] used Raman tweezers not only for identification of single biological particles but also to sort them (Figure 13.13). A sample mixture that contains both live and dead cells was loaded in the sample chamber. Cells were thereafter randomly captured by the focused laser beam, and their Raman spectrum was taken. By comparing the spectrum with previously loaded database, the cell was identified as live or dead and, moved to a collection chamber through a microchannel (10 μm wide and 20 μm deep) by the same laser tweezers. In this manner, the sorted dead cells could be placed in a specific row and live cells in the other row. The possibility of a cell migrating from one chamber to another one by Brownian motion was made very low due to the narrow channel width used for connecting the chambers.

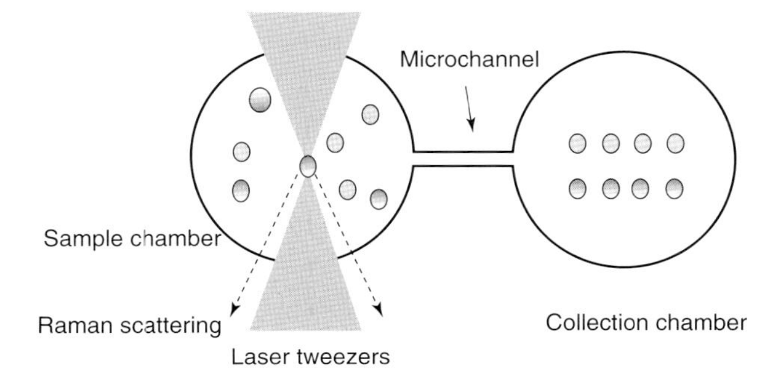

Figure 13.13 Schematic of a setup used to sort live and dead cells based on their Raman signature. Reprinted with permission from Optical Society of America: Xie *et al.* (2005). *Opt. Lett.*, **30**, 1800–1802. Ref. [76], copyright (2005).

13.4.2.3 **Studies on Disease Diagnosis**

Raman tweezers can provide a fingerprint for the biochemical status of the biological cells and thus can be used to differentiate between normal and abnormal cells.

Recently, De Luca *et al.* [47] reported the use of Raman tweezers for monitoring signatures of thalassemia (a disease of RBC, caused by genetic defect) at single cell level. Thalassemia results in reduced rate of synthesis of one of the globin chains present in hemoglobin and is classified according to the chain of the globin molecule affected: in α-thalassemia, the production of α globin is deficient, while in β-thalassemia the production of β globin is deficient. De Luca *et al.* used Raman optical tweezers to characterize RBCs from β-thalassemic patients. A significant reduction in the intensity of Raman bands corresponding to oxygenated Hb was observed in these cells suggesting a lower oxygen transport efficiency for RBCs from β-thalassemia patients.

Raman spectroscopic studies carried out by us on optically trapped RBCs collected from healthy volunteers and patients suffering from malaria (*Plasmodium vivax* infection) show significant alteration in the spectra averaged over ~50 cells per sample [77, 78]. As compared to RBCs from healthy donors, in cells collected from patients with malaria a significant decrease in the intensity of the low spin (oxygenated-Hb) marker Raman band at 1223 cm^{-1} (ν_{13} or ν_{42}) along with a concomitant increase in the high spin (deoxygenated-Hb) marker bands at 1210 cm^{-1} ($\nu_5 + \nu_{18}$) and 1546 cm^{-1} (ν_{11}) was observed (Figure 13.14). The results look promising because several methods (like acridine orange or giemsa staining) used for diagnosis of malaria are based on the presence of the *Plasmodium* parasite in the RBCs, the frequency of which is typically few in 1000 cells in the infected patients [79].

13.5
Cell Sorting

Though flow cytometery-based schemes are well established for high-throughput, multiparameter cell separation there is a strong interest in the use of microfluidic

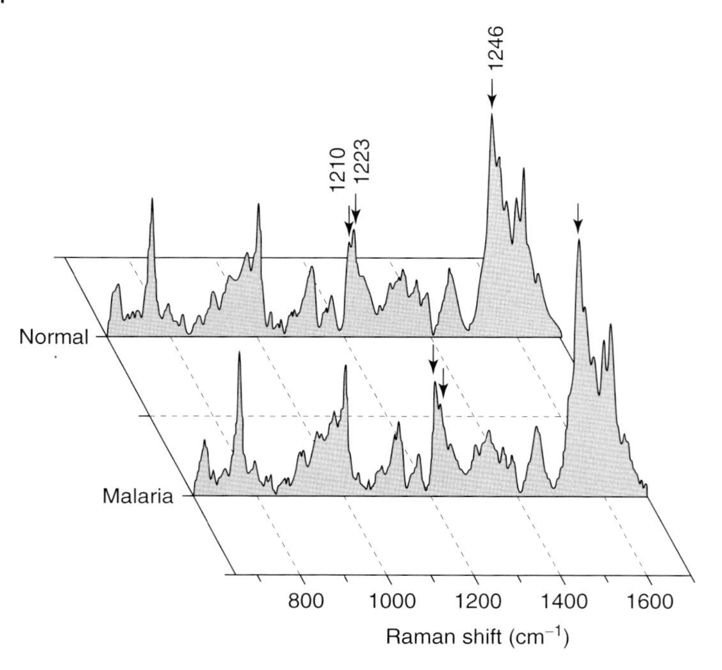

Figure 13.14 Mean Raman spectra of RBCs collected from healthy donors and malaria patients. The spectra were acquired with ~2 mW of excitation/trapping laser power at 785 nm. Acquisition time for each contributing spectrum from single trapped RBC was 30 s. The arrows indicate Raman bands where significant changes are present.

chip-based systems nowadays. The miniaturized systems are expected to offer higher speed and reliability, and should also prove to be cheaper because of less consumption of costly samples or reagents.

Optical cell sorting schemes may be divided into two major categories: active and passive sorting. In the active sorting scheme, a cell is first identified through spectroscopic or image analysis techniques and thereafter it is sorted by applying an optical force [16, 17, 80]. In passive sorting [18, 19, 81–83], the difference in the magnitude of the optical force experienced by different biological cells is exploited for both the selection and separation. Compared to passive schemes, the discrimination of active sorting methods is higher but the sorting speed is limited by the speed of the detection unit and response of the control electronics. Passive methods for cell sorting offer the advantage that no pretreatment step is required. In passive sorting schemes, the cell properties that can be used for discrimination are their size, shape, and/or refractive index. Therefore, such passive sorting is limited to sorting tasks that involve cell populations consisting of cells that do not show large variation in size/shape or composition, such as a blood sample, and would not be applicable to cells that are subject to considerable changes during their life cycle (e.g., cancer cells).

13.6 Summary

Optical-tweezers-based approaches have shown promise for the measurement of viscoelastic parameters of trapped cells over a force range of a few piconewton to several hundred piconewtons. Apart from providing a basic understanding of the elastic behavior of the cells, these measurements can also provide valuable diagnostic information since several disease conditions lead to alteration in the elastic parameters of the cells. Further, spectroscopic measurements on cells trapped in their normal environment are proving very useful for monitoring real-time response of individual cells to a specific perturbation and thus help in understanding the different aspects of the functioning of cells.

References

1. Givan, A.L. (2001) *Flow Cytometry: First Principles*, Wiley-Liss.
2. Darzynkiewicz, D., Roederer, M., and Tanke, H.J. (2004) *Cytometry: New Developments*, Academic Press, San Diego, CA.
3. Ashkin, A. (2006) *Optical Trapping and Manipulation of Neutral Particles Using Lasers*, World Scientific.
4. Ashkin, A. (1997) Optical trapping and manipulation of neutral particles using Lasers. *Proc. Natl. Acad. Sci. U.S.A.*, **94**, 4853–4860.
5. Greulich, K.O. (1999) *Micromanipulation by Light in Biology and Medicine: The Laser Microbeam and Optical Tweezers*, Birkhäuser Verlag, Basel.
6. Sheetz, M.P. (ed.) (1998) *Laser Tweezers in Cell Biology*, Academic Press, San Diego, CA.
7. Zhang, H. and Liu, K. (2008) Optical tweezers for single cells. *J. R. Soc. Interface*, **5**, 671–690.
8. Mohanty, S.K. and Gupta, P.K. (2004) Laser-assisted three-dimensional rotation of microscopic objects. *Rev. Sci. Instrum.*, **75**, 2320–2322.
9. Dasgupta, R., Mohanty, S.K., and Gupta, P.K. (2003) Controlled rotation of biological microscopic objects using optical line tweezers. *Biotechnol. Lett.*, **25**, 1625–1628.
10. Yanai, M., Butler, J.P., Suzuki, T., Kanda, A., Kurachi, M., Tashiro, H., and Sasaki, H. (1999) Intracellular elasticity and viscosity in the body, leading, and trailing regions of locomoting neutrophils. *Am. J. Physiol. Cell Physiol.*, **277**, C432–C440.
11. Zarnitsyn, V.G. and Fedorov, A.G. (2007) Mechanosensing using drag force for imaging soft biological membranes. *Langmuir*, **23**, 6245–6251.
12. Petrov, D.V. (2007) Raman spectroscopy of optically trapped particles. *J. Opt. A: Pure Appl. Opt.*, **9**, S139–S156.
13. Xie, C., Li, Y., Tang, W., and Newton, R.J. (2003) Study of dynamical process of heat denaturation in optically trapped single microorganisms by near-infrared Raman spectroscopy. *J. Appl. Phys.*, **94**, 6138–6142.
14. Rao, S., Bálint, S., Cossins, B., Guallar, V., and Petrov, D. (2009) Raman study of mechanically induced oxygenation state transition of red blood cells using optical tweezers. *Biophys. J.*, **96**, 209–216.
15. Xie, C., Goodman, C., Dinno, M.A., and Li, Y. (2004) Real-time Raman spectroscopy of optically trapped living cells and organelles. *Opt. Exp.*, **12**, 6208–6214.
16. Buican, T.N., Smyth, M.J., Crissman, H.A., Salzman, G.C., Stewart, C.C., and Martin, J.C. (1987) Automated single-cell manipulation and sorting by light trapping. *Appl. Opt.*, **26**, 5311–5316.

17. Wang, M.M., Tu, E., Raymond, D.E., Yang, J.M., Zhang, H.C., Hagen, N., Dees, B., Mercer, E.M., Forster, A.H., Kariv, I., Marchand, P.J., and Butler, W.F. (2005) Microfluidic sorting of mammalian cells by optical force switching. *Nat. Biotechnol.*, **23**, 83–87.
18. MacDonald, M.P., Spalding, G.C., and Dholakia, K. (2003) Microfluidic sorting in an optical lattice. *Nature*, **426**, 421–424.
19. Imasaka, T., Kawabata, Y., Kaneta, T., and Ishidzu, Y. (1995) Optical chromatography. *Anal. Chem.*, **67**, 1763–1765.
20. Ashkin, A., Dziedzic, J.M., Bjorkholm, J.E., and Chu, S. (1986) Observation of a single-beam gradient force optical trap for dielectric particles. *Opt. Lett.*, **11**, 288–290.
21. Harada, Y. and Asakura, T. (1996) Radiation forces on a dielectric sphere in the Rayleigh scattering regime. *Opt. Commun.*, **124**, 529–541.
22. Lock, J.A. (2004) Calculation of the radiation trapping force for laser tweezers by use of generalized Lorenz–Mie theory. I. Localized model description of an on-axis tightly focused laser beam with spherical aberration. *Appl. Opt.*, **43**, 2532–2544.
23. Lock, J.A. (2004) Calculation of the radiation trapping force for laser tweezers by use of generalized Lorenz–Mie theory. I. On axis trapping force. *Appl. Opt.*, **43**, 2545–2554.
24. Ashkin, A. (1992) Forces of a single-beam gradient laser trap on a dielectric sphere in the ray optics regime. *Biophys. J.*, **61**, 569–582.
25. Neuman, K.C. and Block, S.M. (2004) Optical trapping. *Rev. Sci. Instrum.*, **75**, 2787–2809.
26. Capitanio, M., Romano, G., Ballerini, R., Giuntini, M., Pavone, F.S., Dunlap, D., and Finzi, L. (2002) Calibration of optical tweezers with differential interference contrast signals. *Rev. Sci. Instrum.*, **73**, 1687–1696.
27. Sasaki, K., Koshioka, M., Misawa, H., Kitamura, N., and Masuhara, H. (1991) Pattern formation and flow control of fine particles by laser-scanning micromanipulation. *Opt. Lett.*, **16**, 1463–1465.
28. Visscher, K., Gross, S.P., and Block, S.M. (1996) Construction of multiple-beam optical traps with nanometerresolution position sensing. *IEEE J. Sel. Top. Quantum Electron.*, **2**, 1066–1076.
29. Evans, E.A. (1973) New membrane concept applied to the analysis of fluid shear- and micropipette-deformed red blood cells. *Biophys. J.*, **13**, 941–954.
30. Hochmuth, R.M. and Waugh, R.E. (1987) Erythrocyte membrane elasticity and viscosity. *Annu. Rev. Physiol.*, **49**, 209–219.
31. Lelièvre, J.C., Bucherer, C., Geiger, S., Lacombe, C., and Vereycken, V. (1995) Blood cell biomechanics evaluated by the single-cell micromanipulation. *J. Phys. III Fr.*, **5**, 1689–1706.
32. Dao, M., Lim, C.T., and Suresh, S. (2003) Mechanics of the human red blood cell deformed by optical tweezers. *J. Mech. Phys. Solids*, **51**, 2259–2280.
33. Lim, C.T., Dao, M., Suresh, S., Sow, C.H., and Chew, K.T. (2004) Large deformation of living cells using laser traps. *Acta Mater.*, **52**, 1837–1845.
34. Lenormand, G., Hénon, S., Richert, A., Siméon, J., and Gallet, F. (2001) Direct measurement of the area expansion and shear moduli of the human red blood cell membrane skeleton. *Biophys. J.*, **81**, 43–56.
35. Evans, E.A. (1983) Bending elastic modulus of red blood cell membrane derived from buckling instability in micropipette aspiration tests. *Biophys. J.*, **43**, 27–30.
36. Peterson, M.A., Strey, H., and Sackmann, E. (1992) Theoretical and phase contrast microscopic eigenmode analysis of erythrocyte flicker: amplitudes. *J. Phys. II Fr.*, **2**, 1273–1285.
37. Hénon, S., Lenormand, G., Richert, A., and Gallet, F. (1999) A new determination of the shear modulus of the human erythrocyte membrane using optical tweezers. *Biophys J.*, **76**, 1145–1151.
38. Sleep, J., Wilson, D., Simmons, R., and Gratzer, W. (1999) Elasticity of the red cell membrane and its relation to hemolytic disorders: an optical tweezers study. *Biophys. J.*, **77**, 3085–3095.

39. Evans, E.A., Waugh, R., and Melnik, L. (1976) Elastic area compressibility modulus of red cell membrane. *Biophys. J.*, **16**, 585–595.
40. Waugh, R.E. and Evans, E.A. (1979) Thermoelasticity of red blood cell membrane. *Biophys. J.*, **26**, 115–131.
41. Engelhardt, H. and Sackmann, E. (1988) On the measurement of shear elastic moduli and viscosities of erythrocyte plasma membranes by transient deformation in high frequency electric fields. *Biophys. J.*, **54**, 495–508.
42. Mohandas, N. and Evans, E.A. (1994) Mechanical properties of the red cell membrane in relation to molecular structure and genetic defects. *Annu. Rev. Biophys. Biomol. Struct.*, **23**, 787–818.
43. Discher, D.E., Mohandas, N., and Evans, E.A. (1994) Molecular maps of red cell deformation: hidden elasticity and in situ connectivity. *Science*, **266**, 1032–1035.
44. Bronkhorst, P.J.H., Streekstra, G.J., Grimbergen, J., Nijhof, E.J., Sixma, J.J., and Brakenhoff, G.J. (1995) A new method to study shape recovery of red blood cells using multiple optical trapping. *Biophys. J.*, **69**, 1666–1673.
45. Linderkamp, O. and Meiselman, H.J. (1982) Geometric, osmotic, and membrane mechanical properties of density-separated human red cells. *Blood*, **59**, 1121–1127.
46. Sutera, S.P., Gardner, R.A., Boylan, C.W., Carroll, G.L., Chang, K.C., Marvel, J.S., Kilo, C., Gonen, B., and Williamson, J.R. (1985) Age related changes in deformability of human erythrocytes. *Blood*, **65**, 275–282.
47. De Luca, A.C., Rusciano, G., Ciancia, R., Martinelli, V., Pesce, G., Rotoli, B., Selvaggi, L., and Sasso, A. (2008) Spectroscopical and mechanical characterization of normal and thalassemic red blood cells by Raman Tweezers. *Opt. Exp.*, **16**, 7943–7957.
48. Suresh, S., Spatz, J., Mills, J.P., Micoulet, A., Dao, M., Lim, C.T., Beil, M., and Seufferlein, T. (2005) Connections between single-cell biomechanics and human disease states: gastrointestinal cancer and malaria. *Acta Biomater.*, **1**, 15–30.
49. Guck, J., Ananthakrishnan, R., Moon, T.J., Cunningham, C.C., and Käs, J. (2000) Optical deformability of soft biological dielectrics. *Phys. Rev. Lett.*, **84**, 5451–5454.
50. Guck, J., Ananthakrishnan, R., Mahmood, H., Moon, T.J., Cunningham, C.C., and Käs, J. (2001) The optical stretcher: a novel laser tool to micromanipulate cells. *Biophys J.*, **81**, 767–784.
51. Gu, M., Kuriakose, S., and Gan, X. (2007) A single beam near-field laser trap for optical stretching, folding and rotation of erythrocytes. *Opt. Exp.*, **15**, 1369–1375.
52. Mohanty, S.K., Uppal, A., and Gupta, P.K. (2008) Optofluidic stretching of RBCs using single optical tweezers. *J. Biophoton.*, **1**, 522–525.
53. Mohanty, S.K., Uppal, A., and Gupta, P.K. (2004) Self-rotation of red blood cells in optical tweezers: prospects for high throughput malaria diagnosis. *Biotechnol. Lett.*, **26**, 971–974.
54. Mohanty, S.K., Mohanty, K.S., and Gupta, P.K. (2005) Dynamics of interaction of RBC with optical tweezers. *Opt. Exp.*, **13**, 4745–4751.
55. Mohanty, S.K. and Gupta, P.K. (2007) *Laser Manipulation of Cells and Tissues*, Academic Press, San Diego, CA, pp. 563–599.
56. Galajda, P. and Ormos, P. (2001) Complex micromachines produced and driven by light. *Appl. Phys. Lett.*, **78**, 249–251.
57. Galajda, P. and Ormos, P. (2002) Rotors produced and driven in laser tweezers with reversed direction of rotation. *Appl. Phys. Lett.*, **80**, 4653–4655.
58. Ben-Zéev, A. (1985) The cytoskeleton in cancer cells. *Biochim. Biophys. Acta*, **780**, 197–212.
59. Rao, K.M. and Cohen, H.J. (1991) Actin cytoskeletal network in aging and cancer. *Mutat. Res.*, **256**, 139–148.
60. Guck, J., Schinkinger, S., Lincoln, B., Wottawah, F., Ebert, S., Romeyke, M., Lenz, D., Erickson, H.M., Ananthakrishnan, R., Mitchell, D., Käs, J., Ulvick, S., and Bilby, C. (2005) Optical deformability as an inherent cell

marker for testing malignant transformation and metastatic competence. *Biophys. J.*, **88**, 3689–3698.
61. Settleman, J. (2004) Tension precedes commitment-even for a stem cell. *Mol. Cell.*, **14**, 148–150.
62. Thomas, C.H., Collier, J.H., Sfeir, C.S., and Healy, K.E. (2002) Engineering gene expression and protein synthesis by modulation of nuclear shape. *Proc. Natl. Acad. Sci. U.S.A.*, **99**, 1972–1977.
63. Titushkin, I. and Cho, M. (2006) Distinct membrane mechanical properties of human mesenchymal stem cells determined using laser optical tweezers. *Biophys. J.*, **90**, 2582–2591.
64. Long, D.A. (2002) *The Raman Effect*, John Wiley & Sons, Inc.
65. Ferraro, J.R., Nakamoto, K., and Brown, C.W. (2003) *Introductory Raman spectroscopy*, Elsevier.
66. Smith, E. and Dent, G. (2005) *Modern Raman spectroscopy-A Practical Approach*, John Wiley & Sons, Inc.
67. Ramser, K., Fant, C., and Kall, M. (2003) Importance of substrate and photo-induced effects in Raman spectroscopy of single functional erythrocytes. *J. Biomed. Opt.*, **8**, 173–178.
68. Min, A., Jun-Xian, L., Shu-Shi, H., Gui-Wen, W., Xiu-Li, C., Zhi-Cheng, C., and Hui-Lu, Y. (2009) Application and progress of raman tweezers in single cells. *Chin. J. Anal. Chem.*, **37**, 758–763.
69. Xie, C., Dinno, M.A., and Li, Y.Q. (2002) Near-infrared Raman spectroscopy of single optically trapped biological cells. *Opt. Lett.*, **27**, 249–251.
70. Ramser, K., Logg, K., Goksor, M., Enger, J., Kall, J., and Hanstorp, D. (2004) Resonance Raman spectroscopy of optically trapped functional erythrocytes. *J. Biomed. Opt.*, **9**, 593–600.
71. Creely, C.M., Singh, G.P., and Petrov, D. (2005) Dual wavelength optical tweezers for confocal Raman spectroscopy. *Opt. Commun.*, **245**, 465–470.
72. Jess, P.R.T., Garces-Chavez, V., Smith, D., Mazilu, M., Paterson, L., Riches, A., Herrington, C.S., Sibbett, W., and Dholakia, K. (2006) Dual beam fibre trap for Raman microspectroscopy of single cells. *Opt. Exp.*, **14**, 5759–5791.
73. Lewis, I.R. and Edwards, H.G.M. (2001) *Handbook of Raman Spectroscopy: From the Research Laboratory to the Process Line*, Marcel Dekker, New York.
74. McCreeley, R.L. (2000) *Raman Spectroscopy for Chemical Analysis*, John Wiley & Sons, Inc.
75. Ajito, K. and Torimitsu, K. (2002) Laser trapping and Raman spectroscopy of single celular organelles in the nanometric range. *Lab Chip*, **2**, 11–14.
76. Xie, C., Chen, D., and Li, Y. (2005) Raman sorting and identification of single living micro-organisms with optical tweezers. *Opt. Lett.*, **30**, 1800–1802.
77. Gupta, P.K. (2008) Use of light to study and manipulate blood cells. International Conference on Fiber Optics and Photonics, New Delhi, India, December 13–17, 2008.
78. Dasgupta, R., Verma, R.S., Ahlawat, S., Uppal, A., and Gupta, P.K. (2010) Raman spectroscopic studies on optically trapped red blood cells in malaria infected blood sample, XII International Conference on Laser Applications in Life Sciences, University of Oulu, Oulu, Finland, June 9–11, 2010.
79. Determination of Parasitemia, *http://www.med-chem.com/procedures/DetofPara.pdf* (accessed on 11 January, 2010 from Medical Chemical Corporation website).
80. Grover, S.C., Skirtach, A.G., Gauthier, R.C., and Grover, C.P. (2001) Automated single-cell sorting system based on optical trapping. *J. Biomed. Opt.*, **6**, 14–22.
81. Ladavac, K., Kasza, K., and Grier, D.G. (2004). Sorting mesoscopic objects with periodic potential landscapes: optical fractionation. *Phys. Rev. E* **70**, 010901-1–010901-4.
82. Hart, S.J. and Terray, A.V. (2003) Refractive-index-driven separation of colloidal polymer particles using optical chromatography. *Appl. Phys. Lett.*, **83**, 5316–5318.
83. Hart, S.J., Terray, A., Kuhn, K.L., Arnold, J., and Leski, T.A. (2004) Optical chromatography of biological particles. *Am. Lab.*, **36**, 13–17.

14
In vivo Image Flow Cytometry

Valery V. Tuchin, Ekaterina I. Galanzha, and Vladimir P. Zharov

14.1
Introduction

Blood and lymph flow are pivotal features of the complex lives of multicellular organisms. At every moment, billions of cells in the organism move from one location to another. One of the greatest challenges is to observe how cells flow and function within their natural environment. We ask the question whether it is possible to observe, monitor, and understand a single cell's destiny inside the whole organism, and to realize at least the major achievements of the well-developed conventional *in vitro* flow cytometry (FC) [1–4] under *in vivo* conditions.

Since the invention of the optical microscope, it has become possible to visualize cells within organisms. In the last century, scientists have invested significant efforts to visualize cells. The methods have evolved from visualizing cells as they appear in histological preparations to monitoring cell–cell interactions *in vivo*.

This chapter covers practical and relevant technologies in the field of vital microscopy for monitoring cell flow dynamics *in vivo*. We focus on a single aspect of this field, covering microcirculation within the blood and lymphatic systems. We discuss some routine and useful techniques to study cell behavior *in vivo*, such as vital/intravital microscopy (IM). At the same time, we introduce some modern and state-of-the-art techniques that can resolve single-cell behavior inside living organisms at high spatial and temporal resolution. Recent advances in cell biology, animal models, and optical techniques (e.g., lasers, charge coupled device, CCD cameras, software, etc.) provide the basis for significant improvements of *in vivo* imaging.

In vivo circulating cell imaging is very important for providing anatomic, physiologic, metabolic, and pathologic information at single-cell and molecular levels [5–7]. Imaging of individual cells in flow *in vivo* is potentially important for studying cell–cell and cell–vessel wall dynamic interactions, blood transport (e.g., oxygen delivery), the response of cells to different interventions (e.g., drugs, smoking, radiation), and disease diagnosis and prevention (e.g., metastases, heart attack or stroke alert, diabetic shock, sickle-cell crisis, etc.). In particular, the degree and dynamics of changes in cells' mechanical properties may be viewed as a new biological

Advanced Optical Flow Cytometry: Methods and Disease Diagnoses, First Edition. Edited by Valery V. Tuchin.

marker sensitive to early disease development because even small disturbances at the molecular level (e.g., modification of the protein-spectrin structure of red blood cells (RBCs)) may be accompanied by significant changes in cytoskeletal structure responsible for cell deformability and elasticity (e.g., increased rigidity of cancer cells) [8].

A number of well-developed, highly sensitive methods provide high-resolution imaging of RBCs *in vitro*, including monitoring of RBC deformability dynamics with advanced flexible microchannel [9] (see also Chapters 6, 7, and 10). To date, these and others similar approaches have been tested only in *in vitro* studies [10]. However, there is increasing recognition that *in vitro* data cannot correspond entirely to the physiological situation *in vivo* including complex environment within vessel, multilevel regulation of cell function, and cell–cell and cell–wall interactions [11, 12]. Furthermore, invasive isolation of cells from their native environment and their processing may not only introduce artifacts but also make it impossible to examine the same cell population over long time periods. Some other chapters of this book describe different approaches to *in vivo* and *in situ* FC and intravital imaging microscopy (see Chapters 15–19).

14.2
State of the Art of Intravital Microscopy

14.2.1
General Requirements

In principle, there is very little difference between conventional microscopy and IM. In practice, however, there are tremendous differences [13–35]. Firstly, the setup for IM is optimized for *in vivo* work and in the handling of live animals or even humans rather than fixed tissue or cell cultures. The next major difference is the way image acquisition occurs. In most cases, IM systems monitor physiological functions in real time as opposed to the frizzed tissue sample in conventional microscopy. This requires the use of fast, sensitive detectors such as CCD cameras, intensified charge coupled device (ICCD) cameras, complementary metal oxide semiconductor (CMOS) cameras, electronic multiplying charge coupled devices (EMCCDs), or photomultipliers (PMTs). The optical design usually, but not necessarily, needs to be optimized for the use of long working distance lenses since the target samples are thick.

The available options in the use of the IM technique include methods that vary according to their technical complexity from conventional microscopy that utilizes transillumination, reflected light, or fluorescence [14–27], up to modern multiphoton microscopes (MPMs), allowing acquisition of high-quality images up to a depth of 0.7 mm of tissue [28–33] (see Chapter 19), and photothermal/photoacoustic (PT/PA) FC [34, 35] with potential nanoscale resolution at depths down to a few centimeters (see Chapter 16).

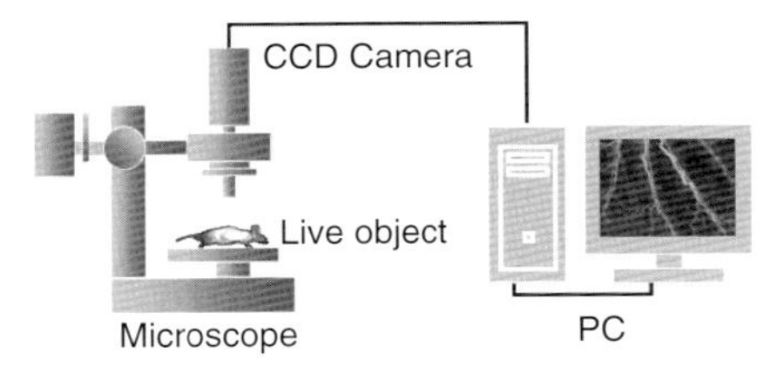

Figure 14.1 Typical setup for intravital microscopy [13].

14.2.2 Intravital Video Microscopy (IVM)

One of the most widely used techniques is intravital video microscopy (IVM) using absorption and scattering phenomena with forward (transillumination) or backward (reflected light) schemes. An example of a conventional setup for IVM is shown in Figure 14.1 [13].

IVM based on an upright or invert microscope provides the following functions [14–21]:

1) real-time monitoring of relatively large structures such as lymph and blood microvessels with relatively low magnification ($\times 4$ to $\times 10$);
2) quantitative dynamic evaluation of blood and lymph microvessel diameter, parameters of lymphatic phasic contractions and valve activity, and cell concentration in flow;
3) determination of cell velocity in flow by video recording cell movement [so-called particle image velocimetry (PIV)];
4) single-cell identification at high magnification ($\times 40$, $\times 60$, and $\times 100$, water immersion);
5) time-resolved monitoring of the cell deformation dynamics in blood flow.

14.2.3 Fluorescent Intravital Video Microscopy (FIVM)

Another modification of IM is fluorescent intravital video microscopy (FIVM) [22–27]. When properly used, FIVM is suitable for the following applications:

1) monitoring gene expression in living animals, using fluorescent protein tags such as green fluorescent protein (GFP), red fluorescent protein (RFP), and others;
2) time-lapse survey and measurement of cell–cell interactions *in vivo* by labeling cells of interest with different fluorescent markers. Marked cells can be of animal or human origin, or infectious pathogens including bacteria or viruses;
3) investigation of the anatomical structure and function of blood or lymphatic vessels, (e.g., evaluation of vascular development by injecting high molecular weight fluorescent substances such as dextran–fluorescein isothiocyanate

(FITC), which is large enough to remain in the circulation for a long time and selectively extravasates from permeable vessels only);

4) monitoring physiological processes at the cellular level in living animals, using special fluorescent indicators sensitive to changes in the microenvironment, such as pH, ion fluxes, membrane electrical potential, redox levels, protease activity, and more;
5) using the above techniques to monitor the response of cells and tissues to pharmacological treatments in research laboratories and pharmaceutical industry.

14.2.4 Experimental Preparations for IVM: Animal Models

To fully exploit the capabilities of IVM, it is extremely important to choose the appropriate experimental model. The acute preparation is widely used and usually requires a surgical procedure on the anesthetized animal while placing it on a specially heated platform for IVM. Such preparations are usually required for studying the microcirculation of internal organs, tumor development, angiogenesis, inflammation, thrombosis, and the progression of infection. IVM utilizes acute preparations in rodent models to expose the mesenteric vessels, cremaster muscle, skin flap, brain, liver, lungs, and so on [19].

Among different animal structures, the mesentery of small animals – rats and mice – is unique for studying cells in blood and lymph flow *in vivo* at single-cell and subcellular levels, as the mesentery consists of thin (10–15 μm), transparent, duplex connective tissue with a single layer of blood, as well as lymph microvessels (Figure 14.2).

There are also some disadvantages associated with the acute preparation, such as surgical trauma, which may introduce artifacts associated with the injury and limit the time available for observation to several hours. However, according to experimental data gathered during at least 2–3 h of acute observation after microsurgery, these procedures do not produce marked changes in microvessel function (diameter, phasic activity, valve function, lymph flow) or in metabolic or respiratory parameters. Coating the intestinal loop with oxygen-impermeable plastic foil helps to maintain stable physiological parameters for up to 5 h [36]. Our data also demonstrate that it is possible to repeat the surgical procedure to periodically monitor the same microcirculation area during the development of chronic pathology, even over a two-month time frame [37].

In order to overcome some drawbacks associated with the acute preparation, new flexible optical mini-probes have recently become commercially available, enabling microscopic fluorescent imaging inside the animal with minimal surgical interventions [38]. Another way to bypass acute surgical trauma is to use the chronic window preparation (Figure 14.3i–l). This technique involves placement of a transparent chamber over the area of interest in order to allow long-term monitoring of processes in the living animal. Models based on a permanent transparent window implantation into the tissue facilitate intravital microscopy

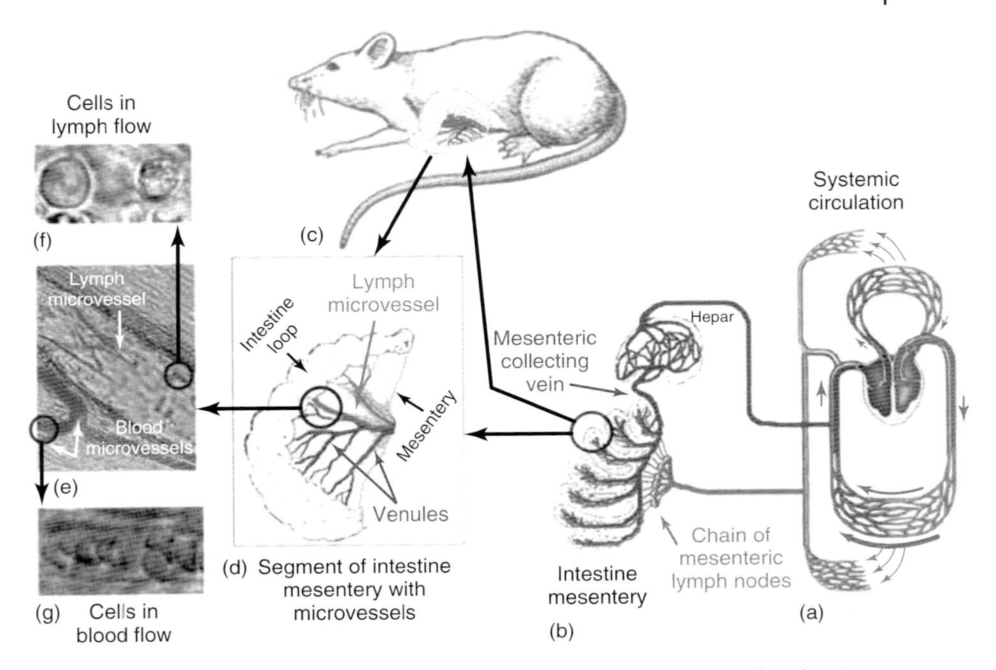

Figure 14.2 Right and middle: schematic of rat vascular systems (a–d) structural organization of systemic blood circulation and lymphatic system; (b) local venous and lymphatic vascular nets of entire intestinal mesentery; (c) and particular mesenteric venules and lymph microvessels of investigated loop. Left: high resolution TDM images of mesentery *in vivo*; (e) lymph microvessel with opening valve and surrounding venules (×10); (f) high resolution images of white blood cells (WBCs) moving in lymph flow (×100, water immersion); and (g) high-speed images of RBCs moving through small venule (×40, water immersion) [13, 19].

capability for the monitoring of biological processes repeatedly, in nonanaesthetized animals, over a period of two to three weeks [39]. Variants of the chronic window are the dorsal skin fold chamber and the cranial, pancreas, and lung windows [40].

Despite its multiple advantages, the chronic window model is not ideally suited for intravital studies since some biological mechanisms can be affected by the surgical implantation of the chamber and the subsequent healing processes. Alternatively, there are several biological processes that can be addressed *in vivo*, without surgical intervention, using the natural transparency of some tissues. When this approach is employed, IVM is conducted on an intact animal.

A widely used model for studying microcirculation is the intact eye [41]. This approach includes visualization of corneal and iridal microcirculation [42]. Another popular model exploits the mouse ear (Figure 14.3a–h) [19, 43]. Image quality in these particular models can be improved in two ways: (i) by using thin hairless skin (e.g., ear of nude mouse ~270 μm thick) or (ii) by decreasing light scattering using a recently developed optical clearing method [44] combined with spectral selection (e.g., use of a green filter to increase blood vessel contrast) [13].

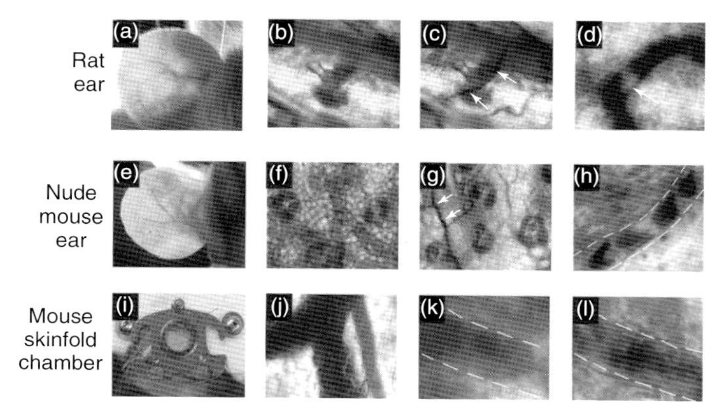

Figure 14.3 Label-free imaging of blood vessels in different animal models. Rat ear (with hair): (a) external view of large vessel; (b, c) transmission images of microvessel at low magnification (×4) before (b) and after (c) topical administration of an optical clearing agent such as glycerol (arrows show microvessel); (d) high-resolution image of rolling WBC (arrow) in a venula (magnification ×40). Ear of nude mouse: (e) external view of large vessels; (f, g) transmission images of microvessels at low magnification (×10) before (f) and after (g) topical administration of glycerol (arrows show microvessel); (h) high-resolution image of individual RBCs in a capillary (magnification ×100); (i) large vessels in skinfold chamber of a mouse (with hair); (j) transmission image of blood microvessel at low magnification (×10); and (k, l) high-resolution image of a venula before (k) and after (l) topical administration of glycerol (×40) [13, 19].

14.2.5 Microcirculation and Cell Flow Examination

14.2.5.1 Microcirculation and Cell Flow

Microcirculation is an important component of live organism functionality and, consequently, understanding and investigation of microvascular phenomena are critical. In the last decades, serious efforts have focused on understanding the finer mechanisms of microcirculation, roles of different factors, behavior of cells, and the molecules that participate in the overall process. Here we briefly introduce some conventional ways to examine microcirculation in general and single-cell behavior inside blood and lymphatic vessel streams in particular.

14.2.5.2 Light Microscopy

Conventional light microscopy is historically the first technique utilized in intravital observation of microcirculation, where the light source (usually a halogen lamp) directs light onto the sample and the objective is focused on blood vessels. Samples can be observed by transmitted or reflected light. Light microscopy also includes dark-field illumination and illumination with polarized light [45, 46] (see Chapters 15 and 18). This technique is based on the optical contrast of RBCs versus the surrounding tissue or the contrast of white blood cells (WBCs) versus RBCs or versus plasma.

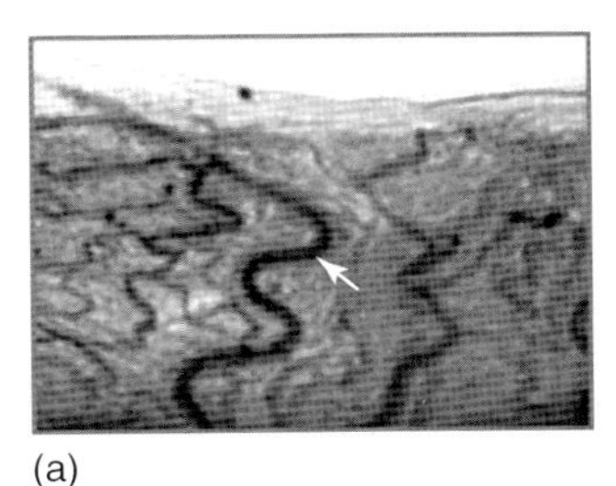
(a)

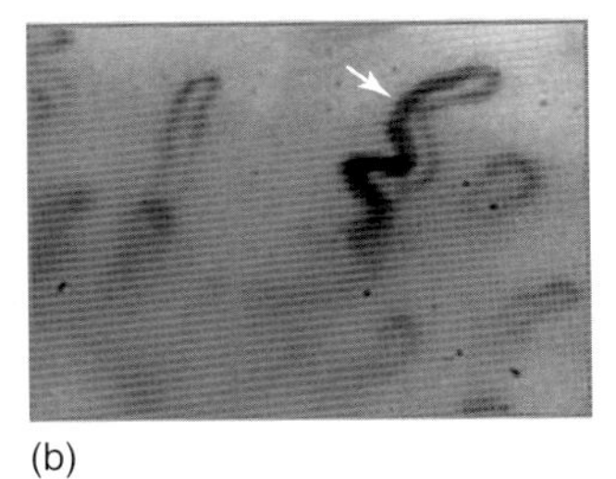
(b)

Figure 14.4 Visualization of blood vessels (white arrow) assisted by conventional light microscopy and reflected light illumination: (a) mouse iridal blood vessels; (b) human nailfold capillaries [13].

Based on a simple contrast technique, several parameters can be acquired, such as microvessel diameter, blood flow (RBC velocity), leukocyte–endothelium interactions, and so on. Applicability of this method has been proven not only in animal preparations, but also in humans. Examples of human applications are microscopy of retinal or iridal blood vessels on the eye, nailfold video capillaroscopy as shown in Figure 14.4a,b, which is a simple noninvasive and proven technique that is currently used in clinical practice [47, 48] (see Chapter 15).

An interesting modification of light microscopy is the use of spectral blood vessel imaging which reveals hemoglobin oxygen saturation *in vivo* at high spatial and temporal resolution [21] (see also Chapter 18). Another novel noninvasive technique is orthogonal polarization spectral (OPS) imaging which utilizes a virtual light source that is created at a depth of approximately 1 mm within the tissue by using polarized light. The light shining back through the tissue is absorbed by hemoglobin, yielding an image of the illuminated vessel in negative contrast with high spatial resolution. Moreover, the method is applicable to humans because no exogenous dyes are needed [22] (see also Chapters 15 and 18).

14.2.5.3 High-Resolution High-Speed Transmission Digital Microscopy (TDM)

The integration of high-resolution and high-speed monitoring enhances spatial resolution and measurement accuracy. In particular, our technique has enabled us to measure the velocity of individual cells up to 10 mm s^{-1} without marked optical distortion of cell images in packed multifile blood flow [18].

In vivo, label-free, high-speed (up to 10 000 with the potential for 40 000 frames per second), high-resolution (up to 300 nm), real-time, continuous imaging with successive framing of circulating individual erythrocytes, leukocytes, and platelets in fast blood flow has been demonstrated. This technique, used in an animal model (rat mesentery) reveals the high deformability of parachute-like RBCs as they are squeezed at 0.6 mm s^{-1} through a narrow gap between the vessel wall and rigid adherent cells on the opposite wall (Figure 14.5a) [18]. It also shows how quickly the relatively fast-flowing RBCs ($\sim$1 mm s^{-1}) change shape as they interact with the much more slowly moving (so-called rolling) leukocytes ($\sim$ 0.1 mm s^{-1}; Figure 14.5b). We also observed significant dynamic deformation of two RBCs

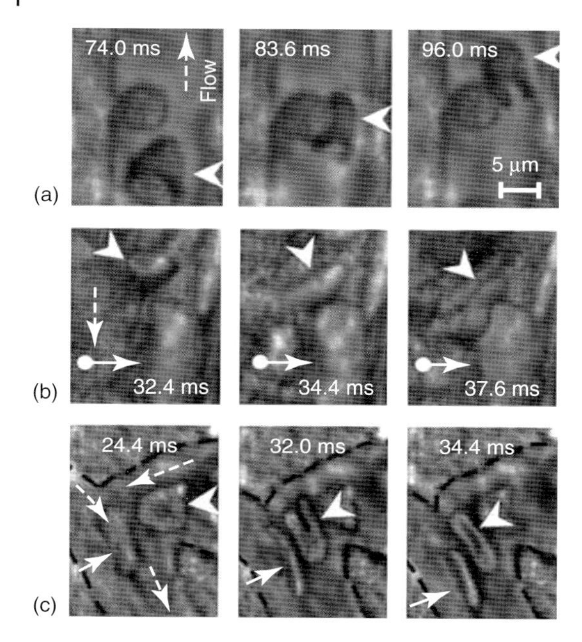

Figure 14.5 (a) Three subsequent frames of parachute-shaped RBC traveling through capillary at 0.4 $mm\,s^{-1}$ (deformation index, defined as the ratio of cell length to width: DI = 1.4, 3.7, and 1.5, respectively; 1250 fps; 40×). (b) Interaction of fast-moving RBC (arrow head) with slow-rolling WBC (arrow with circle) (2500 fps; 40×). (c) Two RBCs in an area of merging flow streams with a velocity of 0.3 $mm\,s^{-1}$; left RBC DI = 3.2, 6.7, and 7.3; right RBC DI = 1.1, 2.2, and 3.4, subsequently (2500 fps; 40×) [18].

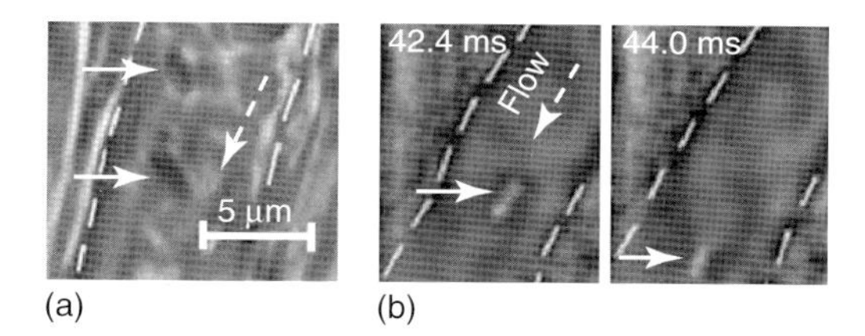

Figure 14.6 (a) Image of platelets in slow flow with velocity of 20 $\mu m\,s^{-1}$ (12 fps; 100×). Two subsequent images (b) of platelets (arrows) in relatively fast flow with a velocity of 2.5 $mm\,s^{-1}$ (2500 fps; 40×); dashed lines show internal margins of microvessel [18].

in merging flow streams in a bifurcation zone (Figure 14.5c), and extremely fast stretching (0.4 ms at 2500 fps) of initially discoid RBCs to ~0.7–0.9 µm. Our technique has provided relatively good contrast images of both slow- (Figure 14.6a) and fast-flowing (2.5 $mm\,s^{-1}$) single platelets in blood microvessels (Figure 14.6b). The best quality images of platelets were obtained in the RBC-free space of the microvessel lumen.

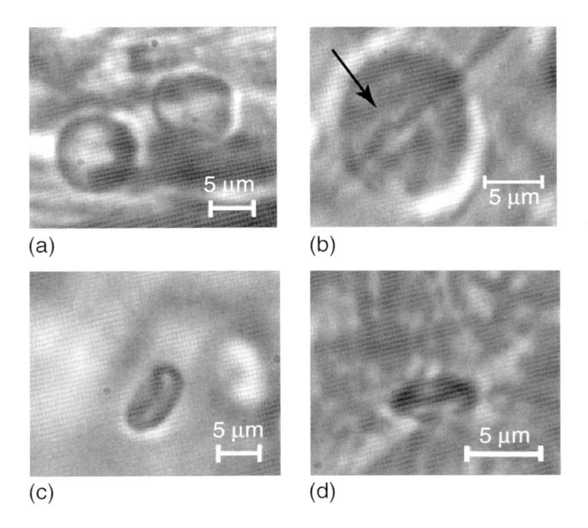

Figure 14.7 High-resolution transmission images of moving single leukocytes (a, b) and RBCs (c, d) in lymph flow at the different position of cells. Magnification ×60 (a, c) and ×100 (b, d). The leukocyte (b) in lymph flow looks transparent: the structure of deep part of lymphatic wall clearly visualized through cell (as shown by arrow) [17].

14.2.5.4 **Monitoring of Cells in Lymph Flow**

In contrast to blood, lymph is colorless with lower concentrations of cells. Because of these features, most existing optical techniques are not quite suitable for imaging mesenteric lymphatics with high resolution, sensitivity, and image contrast.

Several studies have demonstrated that TDM of rat mesentery is a promising model for studying some microlymphatic functions in norm as well as its response to drugs, and for experimental modeling of pathologies [13–17, 19, 20, 37, 49]. To image single cells and their interaction in lymph flow of mesenteric microvessels *in vivo*, we used high magnifications: ×40, ×60, and ×100 with water immersion. With advanced micro-objective (high magnification ×100, high NA 1.25, and water immersion), the obtained *in vivo* resolution was around 350–400 nm.

The moving leukocytes and erythrocytes with their specific shapes (typically, a round shape for leukocytes including lymphocytes, and discoid for normal RBCs) were easily identified (Figure 14.7). However, contrary to strongly absorbing RBCs in the visible range, relatively low absorbing lymphocytes in lymph flow were imaged as smooth, relatively transparent spots, so it was possible to observe the bottom lymph wall elongated structure through the cell (arrow in Figure 14.7).

High-resolution imaging allowed visualizing intralymphatic cell aggregation in intact microlymphatics. As can be seen in Figure 14.8, we observed the aggregates of different sizes.

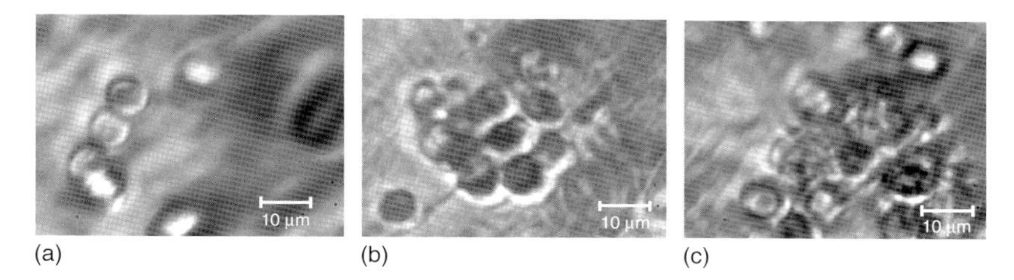

Figure 14.8 Moving aggregates with different sizes in lymph flow of intact lymph microvessel [17].

14.2.5.5 **Fluorescent Image Microscopy**

Fluorescence microscopy is based on the phenomenon that certain substances emit energy detectable as visible light when irradiated with the light of a specific wavelength. The sources of fluorescence in live organisms can be endogenous to an organism such as molecules like collagen or nicotinamide adenine dinucleotide (NADH) [50, 51] or exogenous molecules such as fluorescent dyes like fluorescein [52].

This technique uses an upright or inverted microscope and a light source, which is usually a mercury or Xenogen short-arc lamp. Optical filters are then used in order to select the wavelength of the excitation light (referred to as the *excitation filters*). Excitation light is directed to the sample via a dichroic mirror (i.e., a mirror that reflects some wavelengths but is transparent to others), and fluorescent light is detected usually by a CCD camera.

The FIVM is suitable for a few applications focusing on microcirculation, such as investigation of the anatomical structure and function of blood or lymphatic vessels (e.g., evaluation of vascular development by injecting into the animal high molecular weight fluorescent substances, such as dextran–FITC [53], that remain in the blood circulation or lymphatic system for a long time and selectively extravasate from permeable vessels only). An example of labeling blood and lymphatic vessels with fluorescent molecules that remain in the circulation for extended periods is presented in Figure 14.9.

Another example is shown in Figure 14.10, where rhodamine 6G dye was injected intravenously. The injection of this substance leads to labeling only of nucleated cells or cells containing DNA or RNA. This method allows one to label only WBCs within the circulatory system while RBCs remain unlabeled [54]. Thus, it is possible to observe the kinetics of WBC interactions with the blood vessel walls. Monitoring of WBC kinetics can help to understand some inflammation processes.

A very useful approach is to label or use *ex vivo* labeled cells by fluorescent markers or the use of genetically modified cells (e.g., GFP or RFP [55]). In Figure 14.11a, an example of RBC labeled by the membrane-staining dye DiI that specifically binds to cell membrane lipids and makes them highly fluorescent is presented. In this case, the RBCs can be taken from donor mice or the same animal and later on reinjected into a recipient animal. Thus, labeled RBCs are

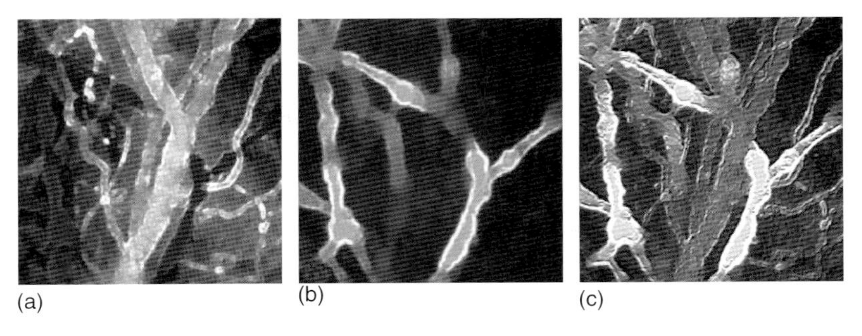

Figure 14.9 Simultaneous visualization of blood and lymphatic vessels of a mouse ear assisted by FIVM. (a) An image of blood vessels labeled by FITC–dextran 500 000. (b) An image of blood vessels labeled by Tetramethylrhodamine B thiocarbamoyl (TRITC)–dextran 150 000. (c) A superimposed image of (a) and (b) [13].

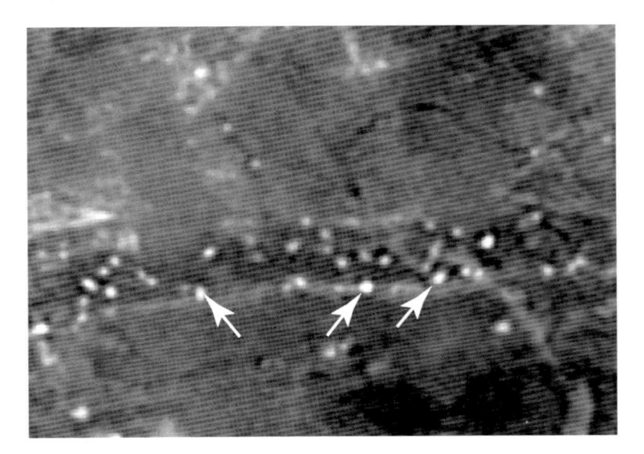

Figure 14.10 Imaging of WBC interaction with endothelium of a mouse ear. WBC labeled by intravenous injection of rhodamine 6G. Arrows show WBC adhered to endothelium [13].

used as fluorescent beads to assess blood flow [56]. In Figure 14.11b, a snap shot of fluorescent tumor cells labeled with the same dye in the microcirculation is presented.

Unfortunately, the use of FIVM in humans is limited. There are only a few fluorescent substances that have been used in clinical practice for many years. Examples of these are fluorescein and indocyanine green (ICG), which are used in ophthalmology for imaging retinal and choroidal microcirculation [57].

14.2.5.6 **Laser Scanning Microscopy**

There are two major forms of laser scanning microscopy, namely confocal laser scanning microscopy (CLSM) and multiphoton laser scanning microscopy (MPLSM). The two forms are very similar on the illumination side. A beam of

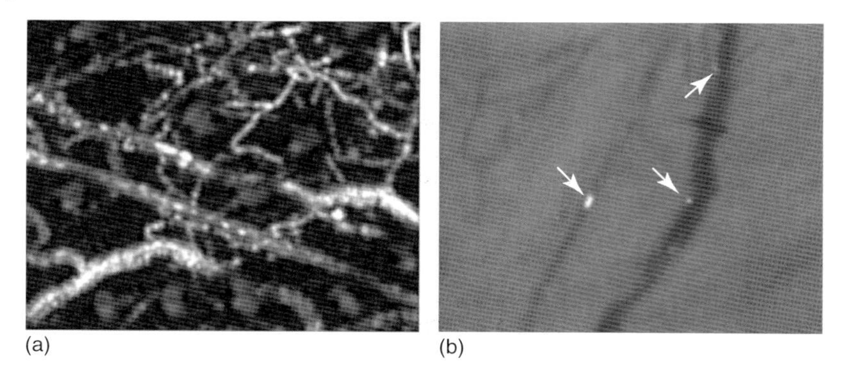

Figure 14.11 FIVM of fluorescently labeled cells in the mouse ear. (a) A superimposed image of DiI labeled RBC (maximum intensity projection, 300 frames). (b) A snapshot of carboxyfluorescein succinimidyl ester (CFSE) labeled cancer cells in the blood vessels (white dots, highlighted by arrows) [13].

laser light is focused into a small point at the focal plane of the specimen, for example, inside a cell loaded with a probe. A computer-controlled scanning mirror can move or scan this beam in the X–Y direction at the focal plane; hence the name scanning microscopy. The fluorescent emission created by the point as it scans in the focal plane is detected by a photomultiplier tube. This detection input is reconstructed by computer software into an image. Both methods have recently grown in demand, especially over the last decade, for intravital imaging of cell flow *in vivo* (see a comprehensive overview in Chapter 16).

The major advantage of CLSM, and especially MPLSM, over conventional FIVM is the penetration depth. For example, MPLSM can resolve single-cell behavior at the depth of 500 μm in tissue that scatters light to a high degree [28, 58]. There is also a modification of laser confocal microscopy that is extremely useful during intravital observation of blood flow. The use of spinning disk laser confocal microscopy for real-time imaging of thrombus formation in microvessels has been demonstrated [59]. This technique allows one to simultaneously multiply fluorescent labeling of different components that play an important role during blood stasis after injury, such as tissue factor, platelets, fibrin, and so on. One example of blood microvessel images obtained with the assistance of CLSM is shown in Figure 14.12. There are also some interesting systems that are based on CLSM technology, dedicated for intravital imaging, already commercially available and have proved their applicability for monitoring flow dynamics *in vivo*.

An example of a commercially available system is the Cell-Vizio from Maunakea, which is a combined CLSM with fiber optics. The technology they use is referred to as fibered confocal fluorescence microscopy (FCFM), in which the objective of a microscope is replaced with a micro-mini optical probe having a diameter as small as 650 μm. This technique allows one to observe cell flow not only inside an animal but in places hardly reachable by conventional objectives without complicated

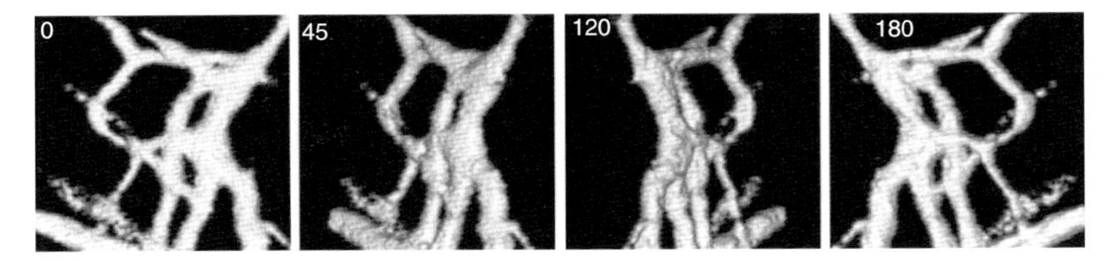

Figure 14.12 A serial 3D rendered image of mouse ear blood vessels. Carbocyanine lipophilic dye (DID) from Invitrogen was used as a fluorescent contrast material to label blood vessels. ImageJ and VolumeJ software were used for image processing [13].

surgical procedures [38]. Another example of a system using CLSM was recently introduced by Olympus. This system utilizes a truly multichannel confocal imaging capability and is called the *IV100*. It is dedicated for use with small animals and is equipped with specially designed ultraslim stick objectives and flexible observation angles in order to reach the areas of interest.

14.2.5.7 Laser Doppler Perfusion Imaging and Laser Speckle Contrast Imaging

Laser Doppler perfusion imaging (LDPI) is a laser-based interferometric technique successfully used for visualization of two-dimensional (2D) microvascular flow maps in a number of biomedical investigations, including peripheral vascular disease, skin irritants, diabetic conditions, burns, and organ transplants. The technical principle is based on the Doppler effect, where light is scattered by the moving particles (e.g., flowing blood cells produce a slight frequency shift, which can be measured by heterodyne detection). A 2D flow map image is obtained by means of successive measurements from a plurality of predetermined points. Classical LDPI systems use mechanical scanning of the area of interest with a narrow collimated or focused laser beam [60].

However, this scanning approach is time consuming and suffers from artifacts caused by the mechanical steering of the irradiating laser beam. In the current commercial LDPI systems, these artifacts are circumvented at the expense of imaging time. An alternative full-field flow imaging technique that uses laser speckle contrast analysis (LASCA) has been proposed [61, 62]. The advantage of the classical LASCA approach is fast image acquisition, which is achieved at the expense of spatial resolution. There is another modification of the speckle techniques [63, 64] (see also Chapter 19) that utilizes laser temporal contrast imaging (LTCI), which we have successfully used in our studies and is shown in Figure 14.13.

Another example of LTCI application is its use for noninvasive investigation of blood clotting *in vivo* during mechanical occlusion [65] (Figure 14.14). A potential advantage of this method is the avoidance of the use of any fluorescent or artificial chemicals for *in vivo* coagulation monitoring in humans. The approach described allows the detection of fine changes in RBC motion during occlusion in the vascular system through the skin, as shown in our experiments down to a depth of 0.3 mm.

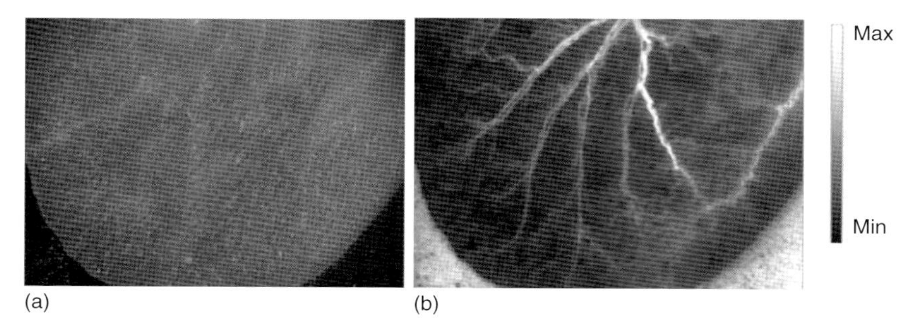

Figure 14.13 (a) An image of a mouse ear illuminated by diffused light from diode laser, 15 mW, 670 nm. (b) An integrated laser temporal contrast image of perfused blood vessels of same ear. Dark areas represent regions of low velocity flow, while white regions represent areas of high velocity flow [13].

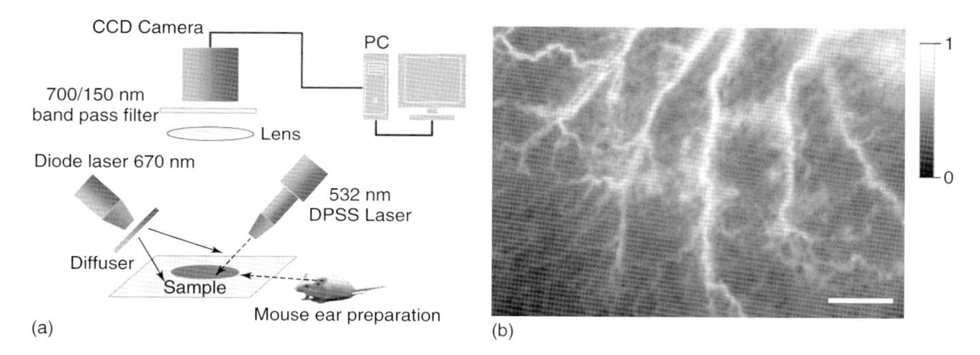

Figure 14.14 LTCI imaging of a mouse ear after complete occlusion of major ear blood vessels. (a) Design of a system for LTCI imaging of laser-induced blood clot formation. (b) LTCI image, intensity scale on the right (0–1) is the value of laser speckle temporal contrast. Scale bar is 1 mm. Large blood vessels appears as white wires. White and fuzzy background corresponds with the signal produced by RBC movements in capillaries [65].

It is quite possible, however, that further development of the system will allow one to achieve better resolution and greater depth of a detectable signal.

14.2.5.8 **Other Intravital Techniques**

There are several additional imaging methods. Among them are the following approaches:

- Optical coherence tomography (OCT) is an interferometric, noninvasive optical tomographic imaging technique offering millimeter penetration (approximately 2–3 mm in tissue) with submicrometer axial and lateral resolution [66, 67].
- Optical Doppler tomography (ODT) is a technique that can be coupled with OCT to measure the velocity of blood flow within biological tissues [68–70].

- Optical microangiography (OMAG) has its origins in Fourier domain OCT and is similar to Doppler optical coherence tomography (DOCT). It is a recently developed high-resolution label-free 3D method of imaging blood perfusion and direction at capillary level resolution within microcirculatory beds [71, 72].
- PA imaging is a noninvasive optical imaging technique used to assess blood vessel anatomy and physiology. This technique is based on acoustic waves that are created by the absorption of light (laser radiation) and detected by a special microphone (transducer) [35, 73–75] (see Chapter 17).
- PT imaging is based on irradiation of microvessels with a short laser pulse and detection of temperature-dependent variations of the refractive index with a second probe laser beam [34, 76]. The PT thermal lens method has demonstrated advantages over the PA method in the sensitivity of label-free detection of individual cells. On the other hand, an advantage of the PA technique is its backward mode (i.e., its laser and transducer are on one side), which is crucial for use in humans. PT and PA methods may beneficially supplement each other, and in combination provide a very powerful tool, especially when integrated with TDM, FIVM, and speckle velocity measurements [14, 19] (see Chapter 17).

14.2.5.9 Conclusion

In the present section, we have presented an overview of the prevalent methods of IM addressing single-cell behavior *in vivo*. These approaches are widely used in laboratories around the world and are being extensively developed. These methods are of serious importance since their final goal is to develop novel noninvasive diagnostic modalities allowing one to evaluate rapidly and accurately the present state of blood and lymph flow in humans.

Recent progress in optical techniques has enabled the development of a multi-module, multifunctional experimental system that integrates transmission digital microscopy (TDM), a PT technique, a PA technique, a highly sensitive fluorescent module, and speckle techniques with advanced CCD or CMOS cameras. This new approach is crucial to realizing free-labeling *in vivo* PT-PA FC, and provides new opportunities to monitor and quantify individual cells in natural biological environments (see Chapter 19). However, TDM for many applications could be considered as a robust and self-consistent approach that allows cell velocity, deformability, and cell–cell interaction measurements in blood and lymph streams *in vivo*.

14.3 *In vivo* Lymph Flow Cytometry

14.3.1 Basic Idea: Natural Cell-Focusing Phenomenon

Adaptation of the FC principles to *in vivo* studies requires precautions related to light scattering by surrounding tissues, and to fluctuations in the position and velocity of cells in a vessel, especially in the "colorless" lymphatics with unstable, turbulent, relatively slow (compared to blood), and oscillating flow [17, 19, 77]. The schematic (Figure 14.15) is derived from our long-term observation of lymphatic function in many (130) rats [77]. Lymph motion results from phasic contractions of the lymphangion, which is the segment of lymph vessels situated between two valves that undergoes cycles of compression (systole) and dilation (diastole) [19, 77–79]. During systole, phasic contractions create a positive pressure gradient ($\Delta P = P_1 - P_2 \geq 0$) (Figure 14.15) that moves lymph forward for 0.5–2 s; during diastole, the pressure gradient decreases to 0, and lymph flow stops for 0.05–0.1 s; and with a negative pressure gradient, lymph motion is retrograde for 0.2 ± 0.02 s (i.e., approximately 12 times shorter in duration than forward motion). Lymph flow then stops again for a longer time, 1–3 s, and the lymphangion fills with lymph from its distal portion. In this stage, the spatial distribution of cells (mainly WBCs) is random, and cells fluctuate in the radial directions up to 70–100 μm. At the end of diastole, the pressure gradient ΔP become positive, the valve begins to open, and lymph moves forward again. This complex cell motion makes it difficult to detect individual cells compared to FC *in vitro* with the well-organized single-file flow by hydrodynamic cell focusing using sheath fluids created by two coaxial tubes acting as an artificial nozzle [1, 2]. Nevertheless, we discovered that nature has also created single-file cell flow in a localized zone near the valve.

In particular, in vessels with diameters of 136 ± 10 μm, the valve leaflets form natural nozzles, measuring approximately 44 ± 7 μm in diameter, that provide an approximately threefold constriction of flow. This constriction significantly increased the flow velocity, on average from 300 to 500 μm s^{-1} up to a few millimeters per second. As a result, cell acceleration, together with sheering forces, leads to hydrodynamic focusing of cells into single-file cell flow, with radial cell fluctuations of just 5–10 μm (Figure 14.15b). The phasic contraction in the central part of the lymphangion between valves provides also secondary cell focusing (Figure 14.15c); however, because of the relatively low degree of constriction (20–30%), single-file cell flow was only observed in small-diameter vessels (<80 μm). We also discovered laser-induced localized squeezing of a lymph vessel (Figure 14.15d). At the laser parameters indicated in figure caption, this effect was reversible, with a vessel recovery time of 30–60 min.

14.3.2
Animal Model and Experimental Arrangement

Among different animal structures (e.g., ears, mouth cavity, stretched skin, etc.), the mesentery of small animals – rats and mice – is unique for studying cells in blood and lymph flow *in vivo* at single-cell and subcellular levels, as the mesentery consists of thin (~10–15 μm), transparent duplex connective tissue with a single layer of blood and lymph microvessels [19]. Additionally, established preparation procedures do not markedly influence the animal's physiological parameters, at least during the short time period of observation (~2–3 h) [36].

Many biological and medical problems are associated with quantitative characterization of microlymphatic functioning (e.g., its role in immune response

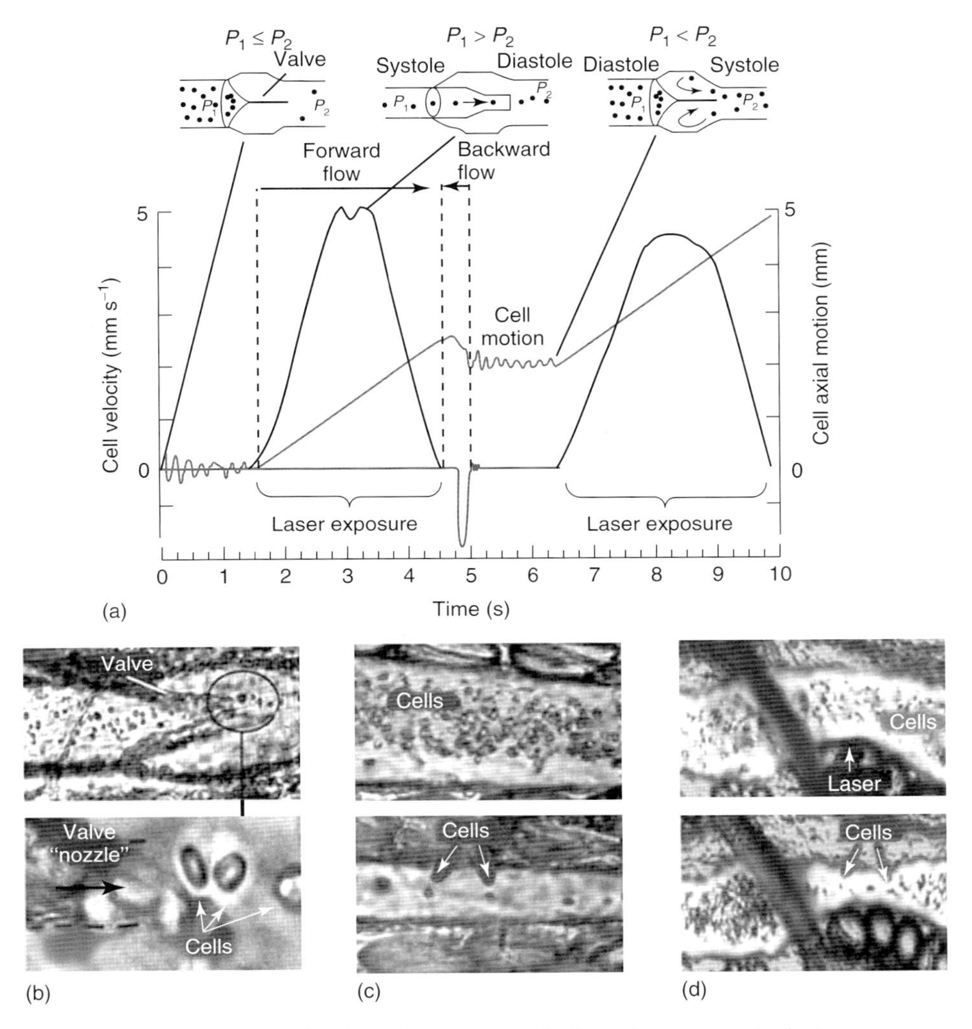

Figure 14.15 (a) Principle of lymph *in vivo* cytometry and dynamics of cell motion in the lymphatics; P_1 and P_2 are the pressure in the distal and in the proximal lymphangion, respectively. (b) High-speed TDM imaging of cells after a valve at ×10 (top) and ×100 (bottom) magnifications. (c) TDM images of cells in the central part of a lymphangion in diastole (top) and systole (bottom). (d) Lymph vessel before (top) and after (bottom) one laser shot (wavelength, 585 nm; pulse width, 8 ns; beam diameter 100 μm, fluence 0.5 J cm^{-2}) [79].

or interstitial fluid balance, and its disturbance in different diseases, such as lymphedema, venous insufficiency, infections, inflammation, lymphatic malformations, metastasing, etc.); however, the lymphatic system has received much less attention than blood microcirculation in the mesentery, particularly within the rat model, which has been studied in appropriate detail (e.g., morphology, rheology, microflow dynamics, velocity profiling, leukocyte rolling, etc.) (see [16–19, 37] and references therein).

In contrast to blood, lymph is colorless with lower concentrations of cells (mainly lymphocytes with a small portion of RBCs) and has a relatively slow flow velocity. Lymph also has periodic oscillations (retrograde/back flow) and the absence of a reverse correlation between vessel diameter and liner flow velocity (as is typical for blood vessels) [80]. Because of these features, most existing optical techniques are not quite suitable for imaging mesenteric lymphatics with high resolution, sensitivity, and image contrast.

For example, phase and confocal diffuse reflectance microscopy is, in general, a very promising tool [81, 82]. However, this technique is not well suited for imaging low-scattering, low-refractive microlymphatics at the single-cell level, especially with scattering and refractive background noise from surrounding connective tissue. Powerful fluorescent imaging can only be applied to fluorescent samples, and most structures in rat mesentery in its native state are non- or weakly fluorescent. The fluorescent labeling may change tissue properties; in particular, obtaining lymph flow velocity by fluorescent imaging using injected fluorescent particles (fluorescein isothiocyanate–dextran) in the interstitial space leads to elevation of interstitial pressure and alteration of lymph viscosity [83]. Potentially powerful combined DOCT can simultaneously monitor blood flow velocity in microvessels while imaging living tissue microstructures [84], especially the novel OMAG technique [71, 72]. However, DOCT has not been used to study low-scattering lymphatics and, in general, the achieved resolution of $\sim 5-10\,\mu m$ is insufficient for analyzing single cells and, especially, intracellular structures [85, 86].

Several studies have demonstrated that conventional TDM of rat mesentery is a promising model for studying some microlymphatic functions in norm as well as its response to drugs, and for experimental modeling of pathologies [49, 87–89]. In particular, it was demonstrated that in parallel with routine imaging (whole lymphangion, neighboring blood microvessels, lymphatic walls, leaflets of valve, and flowing single cells) and obtaining quantitative data (lymph and blood microvessel diameters, amplitude, and rate of lymphatic phasic contractions), TDM allowed us to obtain more detailed parameters of phasic contractions (velocity of wall movement, duration of contraction, and its periodicity), valve function (duration of valve cycle and its periodicity), and lymph flow (mean cell velocity, relation of forward and backward cell velocity, cell concentration) [14, 80]. Additionally, using TDM, the correlations between investigated parameters, including a close correlation between valve function and phasic contractions and increased cell numbers with decreased amplitude of phase contractions and lymph flow velocity at moderate cell concentrations, were determined [14, 17–19, 80, 90, 91]. With TDM, it was also possible to monitor microlymphatic responses to chemicals (sodium nitroprusside, *N*-nitro-L-arginine, dimethylsulfoxide) [14, 90], nicotine [16, 92], lasers [92], and microsurgical interventions (experimental lymphedema) [37, 90–92]. The capability of the combination of TDM with speckle technique to measure lymph flow velocity [14, 15, 19, 93] and highly sensitive PT and PA imaging to visualize intracellular structures of moving RBCs and WBCs, microorganisms,

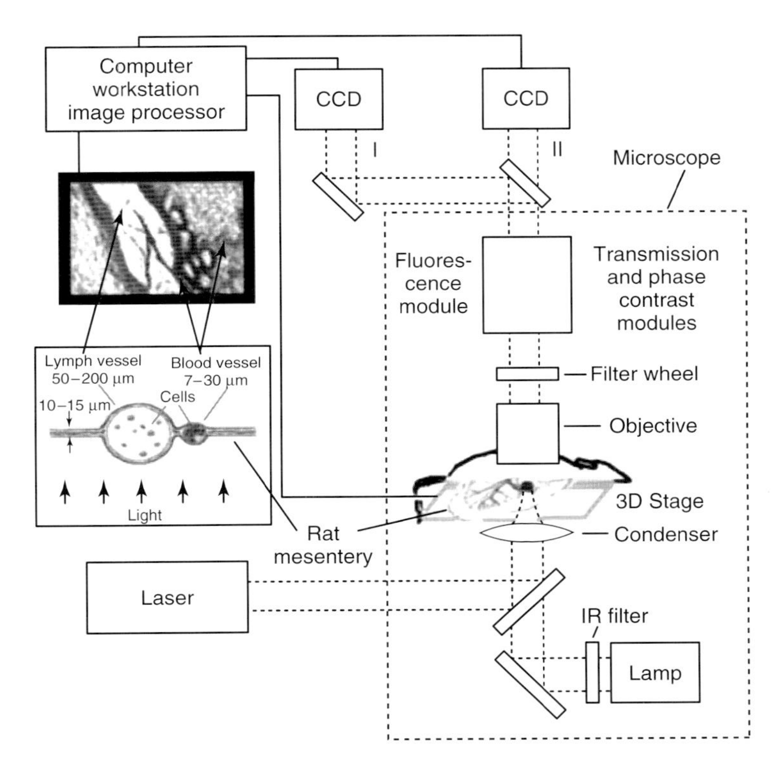

Figure 14.16 Optical scheme of flow transmission image cytometer based on TDM for *in vivo* study of lymph and blood microcirculation in rat mesentery [17].

and cancer cells *in vivo* has been demonstrated [34, 35, 75, 94–97]. Some of these data were obtained with low-resolution and low-speed (thus low-cost) TDM; however, they have demonstrated the simplicity and usefulness for many animal studies. The capabilities of the advanced TDM technique for high-resolution and high-speed imaging of moving cells in rat mesentery with focus on cells in lymph flow and the features of the optical technique are described in [17–19, 77, 96, 97].

The optical scheme of the TDM module as a part of the integrated PT flow cytometer [94, 95] (see Chapter 17) is shown in Figure 14.16. In an integrated cytometer, TDM based on an upright Olympus BX-51 microscope provided the following functions: (i) determination of cell velocity in lymph flow by video recording the cell movement; (ii) simultaneous measurement of vessel diameter, parameters of phasic contractions and valve work, and cell concentration in flow; (iii) navigation of the pulse laser beam in the central part (or other section, if necessary) of a vessel cross section to realize PT or PA imaging; and (iv) single-cell identification (to distinguish leukocytes from RBCs, to estimate their shape, size, and aggregation types, and possibly, their interaction in flow). For the studies described in this section, we focused on the capabilities of just the TDM mode with a conventional illuminating lamp. Depending on the structures' size, different magnifications were used to image relatively large whole lymphangion (×4, ×8, ×10, ×20), as well as single cells (×40, ×60, ×100) in lymph and blood flow.

The images were recorded with several digital cameras: a color Nicon DXM1200 and a black-and-white CCD Cohu 2122 with a speed up to 25 frames s^{-1} and minimal exposure time 0.04 s. In selected experiments, we a used black-and-white high-resolution CCD camera with speed up to 500 frames s^{-1} (Cascade 650, Photometrics), which was sufficient for studying a relatively slow blood flow in capillary. However, the nonoverlapped mode of this camera allowed us to decrease exposure to 0.1 ms and even less (i.e., quite sufficient to study relatively fast blood flows). A high-speed CMOS camera (model MV-D1024-160-CL8; Photonfocus, Switzerland) with speeds of 10 000 fps for an area of 128 × 128 pixels and 39 000 fps for an area of 128 × 16 pixels provided high-resolution imaging of fast-moving RBCs in blood flow (2–10 mm s^{-1} in arterioles and large venules) and WBCs in valvular lymphatics (ranging from 0 to 7 mm s^{-1}). Scion Image (Scion Corp.), Nicon software (ACT-1), and WinView/32 V2.5.18.2 (Roper Scientific) were used for processing, capturing, measuring, and editing images of the moving cells. In addition to TDM, we used the phase-contrast module with phase-contrast objectives (×10 and ×100) to compare different images in real time.

White Fisher rats (F-344) weighing 150–200 g were used in all experiments, in accordance with UAMS protocol #2318 approved by Institutional Animal Care and Use Committee. Rats were narcotized by Nembutal (50 mg kg^{-1} body weight, i.m.), and then a laparotomy was performed. After this procedure, the mesentery of the intestine was exposed. The animal was placed on a heated 37.7 °C microscope stage with an optical window. The mesentery and intestine were preserved by diffusion with Ringer's solution (37 °C, pH 7.4).

14.3.3
Lymph Flow Velocity Measurements

Lymph flow velocity is usually estimated by measuring cell velocity. Measurements of cell velocity with the optical tracking of each cell within the field of view of a video camera allows one to evaluate a velocity of a cell as its path image from frame to frame divided by the time interval between corresponding frames. Evidently, the accuracy of the measurements depends on spatial and temporal resolution of cameras used. The main advantage of such direct measurements of cell velocity is the easy determination of direction of flow of the cells and spatial distribution of cell velocities. The main drawback is that, for fast and highly concentrated cell flows to provide real time measurements, very fast and sensitive cameras should be used and intensive computing should be provided for.

However, for many cases conventional cameras are good; for example, measurements with a conventional video camera (25–30 fps) revealed that mean cell velocity in an axial flow in the nonvalvular central part of mesenteric lymphangions (i.e., approximately equal distance between input and output valves) is 211 − 262 μm s^{-1} on average, with a maximum of 1–2 mm s^{-1} [14, 17, 19, 37, 98]. When lymph flow

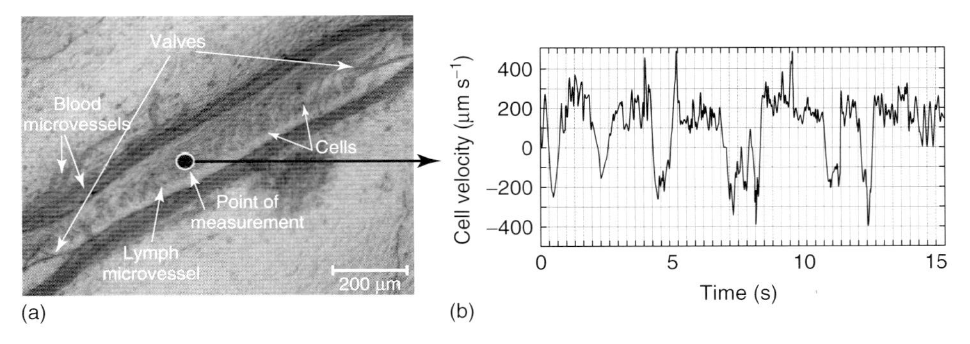

Figure 14.17 Real-time dynamics of cell velocity (a) in axial lymph flow (b) within the nonvalvular segment of a lymphangion (mean diameter of 170 ± 5 μm) without phasic contractions and valve activities, measured by processing the video recording (μm s^{-1}) for 15 s [19].

goes through the valve, the phase contraction leads to acceleration of flow because of the narrow valve nozzle. In turn, this leads to significantly increased cell velocity.

Lymph usually moves in the forward direction for a short period; then, the motion is interrupted and the lymph stops for up to 1–1.5 s. After that, the lymph flow starts in the reverse direction. Usually, cells oscillate at a rate of 50–70 oscillations per minute [91]. Figure 14.17a shows cells in a lymph flow; at small magnification, cells resemble small, but distinct, points. Imaging of moving cells in lymph flow allows one to measure absolute cell velocities and the direction of their motion by video recording (frame by frame). Cell velocity oscillations measured during 15 s in a nonvalvular part of an individual lymphangion (mean diameter in the investigated site is 170 ± 5 μm; mean cell velocity during the investigated time is 168 ± 6 μm s^{-1}) is shown in Figure 14.17b [91]. In some rare microvessels, the time-averaged velocity of lymph flow equaled zero. In this case, lymphocytes only oscillated relative to a position without the leak-back of the lymph.

To increase the ability to measure the absolute value and direction of flow, we also used the speckle correlation method or laser speckle method (LSM), which uses two photodetectors to record the intensity in fluctuations of the speckle field at two points [15, 93]. The described algorithm was verified *in vivo* by measuring lymph flow in a lymphatic vessel concurrently with LSM and TDM. There was relatively good correlation (coefficient of linear regression [$\sim$0.72]) between the two methods; however, the LSM mode had advantages, such as rapid calculation of lymph-flow velocity (compared to TDM). On the other hand, combining LSM with TDM allowed us to (i) monitor the quantitative dynamics of cell velocity in lymph or blood flow, (ii) verify data from the speckle method, (iii) obtain a profile of lymph velocity through the changing shapes of the speckle signal from the center region to the near-wall region, and (iv) determine the dynamic relationships among changes in lymph flow velocity and other functional activities of lymphatics. This integrated schematic also has the potential for speckle imaging of mesenteric structures at cellular and subcellular levels, including detection of individual cell rotation and functional state (e.g., live, apoptotic, and necrotic) in flow [99].

Using a high-speed videotape recorder (2000 fps), Sekizuka *et al.* demonstrated that maximum cell velocity in rat mesenteric lymphatics was 2–3 $mm\,s^{-1}$ [89]. Later, using a relatively high-speed camera (500 fps), Dixon *et al.* determined that the maximum cell velocity in some mesenteric lymphatics may reach 7 $mm\,s^{-1}$ and that their oscillations correlate with phasic contractions [20]. Such big variability of cell velocity may be due to the different experimental conditions.

It is important that the analysis of lymph flow requires that both cellular and plasma velocities (analogous to blood flow) be estimated. In fact, Starr *et al.* found that the blood flow in cat mesenteric microvessels *in vivo* is characterized by a discrepancy between RBC and plasma velocities [100]. This discrepancy is dependent on the elastic properties of RBCs and flow dynamics, including plasma shear and pressure stress field. We expect a similar phenomenon in lymph flow, especially due to its oscillating character; however, these effects require further quantitative verification.

Depending on lymph flow velocity and vessel structure, cell distribution in the cross section of the lymphangion varies [14, 17, 19, 98]. Most often, at relatively low velocities and/or in the nonactive valves (80% of cases), cell distribution was relatively uniform. However, at high velocities and/or in functioning valves (20% of cases), most cells were concentrated near the vessel axis. It is well known that lymph has markedly lower concentrations of cells than blood. However, there is relatively little data about cell concentration in prenodal lymph. In particular, cell concentrations range from 200 to 1000 cells μl^{-1} in prenodal sheep's lymph and 100 cells μl^{-1} in prenodal rabbit's lymph [101, 102]. It was found that approximately 1×10^6 cells h^{-1}(1.7×10^4cells min^{-1}) go through a single prenodal (afferent) sheep's lymph [103, 104].

According to our data, the cell concentration in rat mesenteric prenodal lymphatics is larger. Specifically, we measured the number of cells in a $50 \times 50\ \mu m^2$ square in the central part of a lymphangion from a rat mesenteric microvessel using 2D imaging [14]. Because the depth of field was approximately 28 μm, the number of cells was determined in a $50 \times 50 \times 28\ \mu m^3$ volume (i.e., 7×10^{-5} μl of lymph). Taking into account distribution of cells in a cross section of a lymphangion, the mean concentration of cells in the lymph flow of intact mesenteric microvessels was estimated as $\sim(0.5–1) \times 10^5$ cells μl^{-1} [19]. From these data, the average cell fraction in lymphatics was estimated as 5.5%. This parameter is analogous to hematocrit in the blood system, and may be called *lymphocrit*. It is interesting that a 5–6% level of hematocrit can be found only in blood capillaries, as compared to that in arterioles with diameters from 60 to 70 μm, in which the hematocrit achieved 20–30%. Cell flux in the prenodal lymph was estimated to be 10–50 cells s^{-1}. The cell concentration in flow is usually higher in vessels with higher lymph flow velocities, and is somewhat correlated with the amplitude of phase contractions and rate of valve activity [14].

Data obtained by Dixon *et al.* [105] with a high-speed video system and by capturing multiple contraction cycles in rat mesenteric lymphatic preparations with smaller diameters ($91 \pm 9\ \mu m$) revealed that lymphocyte densities were (1.20 ± 0.52) $\times 10^4$ cells μl^{-1} and the cell flux was 990 ± 260 cells min^{-1} (16.5 ± 4.3 cells s^{-1}).

The relationship between lymphocrit and pathologies is still unknown. There are indications that the leg massage of healthy rabbits stimulates lymph flow and increases cell concentration, while leg edema due to venous pressure rise is not accompanied by an elevation in cell concentration [102].

Data about cell composition in afferent lymph are very limited and have been obtained in *in vitro* tests. In particular, 80–90% of cells in normal prenodal lymph are related to mature lymphocytes, 5–20% to macrophages and dendritic cells, and 0–10% to other cells (e.g., basophilic blast cells, RBCs) [101]. The composition of WBC types changes depending on the pathology. For example, it is estimated that there is a significant appearance of neutrophils in experimental peritonitis in sheep. Other important problems are the mechanisms by which different cells (e.g., metastatic tumor cells) enter into peripheral lymphatics. Studies of tumor-associated absorbing lymphatics in the peritumor area of fixed tissue specimens (adenocarcinoma, melanoma, colorectal cancer) using light microscopy, transmission electron microscopy, and histochemistry with 3D image reconstruction led to the conclusion that cancer cells can enter into lymphatics through intraendothelial channels (1.8–2.1 μm in diameter and 6.8–7.2 μm in length) [106]. These results are very promising, but such a mechanism remains to be proven in living cells *in vivo*.

14.3.4 Imaging of Cells and Lymphatic Structures

Typical transmission images of rat mesentery are shown in Figure 14.18, with two relatively small magnifications [(a) $\times 4$, field of view $250 \times 350\ \mu m^2$ and (b) $\times 8$, field of view $150 \times 200\ \mu m^2$]. Such magnifications allow one to visualize the entire lymphangion (the fragment of lymph vessel between closely located input and output valves), its wall, lymphatic valves, and cells in lymph flow, as well as neighboring blood microvessels. Mean cell velocity was $262 \pm 6\ \mu m\ s^{-1}$, ranging from 0 up to $1–2\ mm\ s^{-1}$.

A comparison of images obtained with phase-contrast and TDM techniques revealed that, in general, the phase-contrast technique yielded lower contrast of the desired lymphatic structures (vessels, valves, cells, etc.), and is deteriorated by small refractive heterogeneities of the connective tissue (Figure 14.18c,d). Thus, the refractive background makes it difficult to distinguish single cells in a lymph flow, and especially their cellular structure. However, the phase-contrast images of selected single cells with high magnification ($\times 100$, with water immersion), without overlapping refractive heterogeneities of connective tissue, sometimes demonstrated better contrast of cell margins and subcellular structures (Figure 14.18e,f).

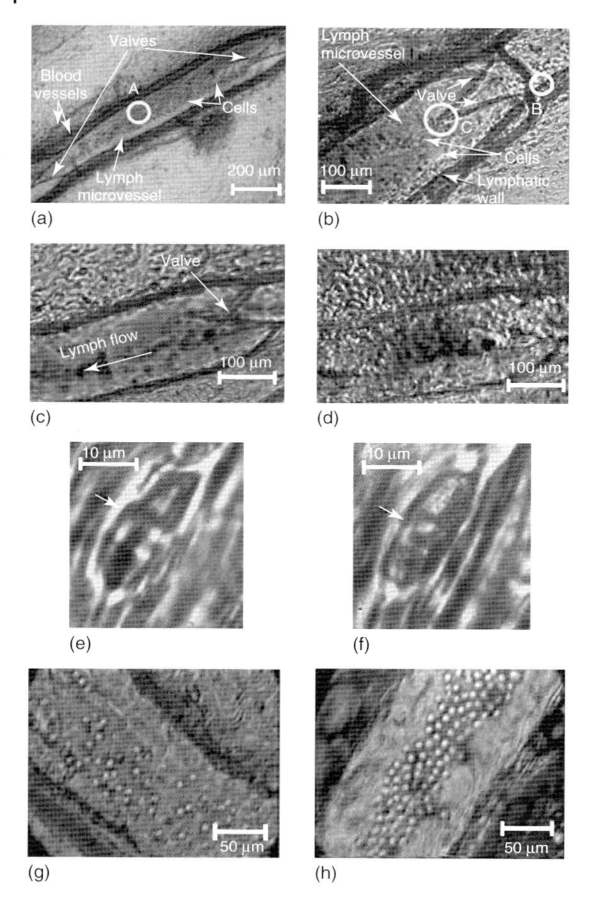

Figure 14.18 Typical transmission images of rat mesentery fragments: (a, b) with one lymph microvessel with closed leaflets of valve and blood microvessels along lymphatic wall (magnification: (a) ×4 and (b) ×8); transmission light (c) and phase-contrast (d) images of the same structure of rat mesentery – lymph microvessel with the open valve at the magnification ×8; transmission light (e) and phase-contrast (f) images of the same cell in interstitial space (shown by arrow) at the magnification ×100; (g, h) the cell distribution in lymph flow in central part of a lymphangion in norm: (g) relatively uniform distribution, and (h) cell concentration along centerline near vessel axial [17].

Depending on lymph flow velocity and vessel structure, distribution of cells in the cross section of microvessel in the central part of the lymphangion was different. Most often, at relatively low velocity and/or at the nonworked valves (typically 80% of cases), cell distribution was relatively uniform (Figure 14.18g). However, at high velocity and/or functioning valves (typically 20% of cases), most cells were concentrated in the central region of the vessel (axial flow) (Figure 14.18h). In the valvular region, the "funnel" shape of valve provided the constriction of lymph flow (Figure 14.19).

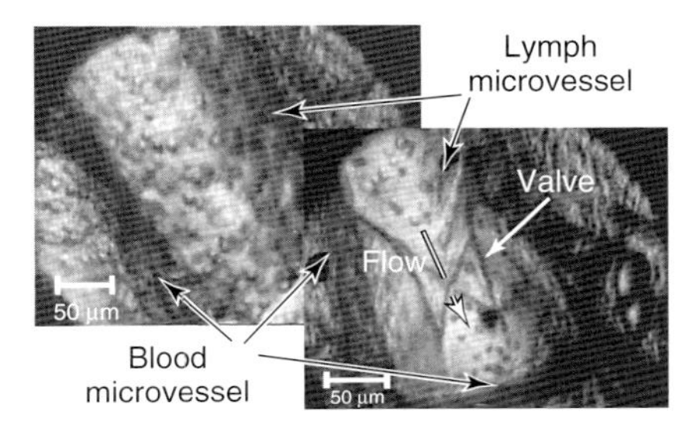

Figure 14.19 The distribution of flowing cells in different parts of a lymphangion: the uniform distribution in central part (upper); concentration of cells near valve area (lower) [17].

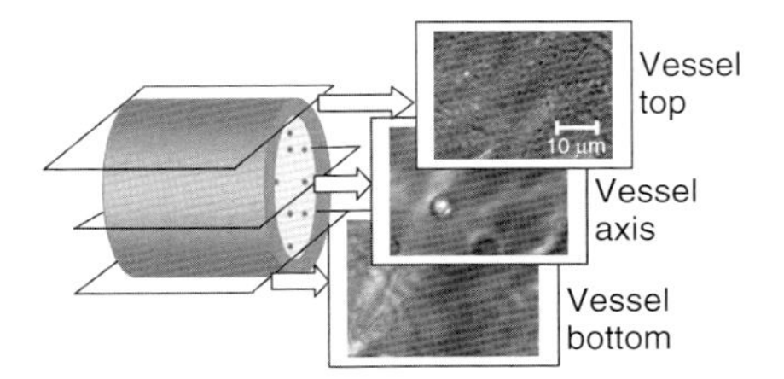

Figure 14.20 Images of different layers of lymph microvessel (diameter 110 μm), magnification of objective ×100, water immersion [17].

The reasonable temporal resolution (~ 40 ms) and size of the field of view ($150 \times 200\,\mu m^2$) permit one to determine directly the linear velocity of lymph flow in small lymphatics by video recording the cell movement (as a ratio of the distance moved to the time of movement). The imaging system allows one to obtain basic information on lymph microcirculation and quantitative data on intravital lymphatic function to determine many indices of phasic contraction and valve function, number of microvessels with lymph flow, number of cells in lymph, and the most objective criteria of lymph drainage function. The most significant data on measured dynamic parameters of microlymphatics in norm are summarized in Table 14.1.

To image single cells and their interaction in flow, high magnifications should be used: ×40, ×60, and ×100 with water immersion. By changing the position of the focal plane, one could monitor different cross sections of lymph microvessels, including the "up" and "down" parts of the lymphatic wall and central cross sections of lymphatic lumen with several cells (Figure 14.20). The approximate distance between the top and bottom walls, as well as cell velocity in different locations of cross section, could also be measured.

With advanced microobjective (high magnification ×100, high NA 1.25, and water immersion), the obtained *in vivo* resolution was around 300 nm. It was verified by analyzing the specific diffraction pattern around particles with known

Table 14.1 Data on measured dynamic parameters of microlymphatics in norm [17].

Parameter	Number of investigated lymphangions	Number of measurements	Mean ± SE
Diameter (D) (μm)	70	205	147 ± 3
Phasic contractions (50% of lymphangions have phasic contractions)			
Amplitude (%) $((D_{max} - D_{min})/D_{max}) \times 100\%$	35	92	29 ± 9
Rate (min^{-1})	35	92	12 ± 1
Duration of contraction cycle (time in systole) (s)	14	51	2.7 ± 0.1
Duration of pause (time in diastole) (s)	14	51	3.8 ± 0.7
Valve function (36% of lymphangions have worked valve)			
Rate of valve work (min^{-1})	25	68	9 ± 1
Duration of valve cycle (s)	9	26	1.5 ± 0.2
Duration of pause (s)	9	26	7.7 ± 2.3
Lymph flow (89% of lymphangions have lymph flow)			
Mean cell velocity ($\mu m\ s^{-1}$)	62	389	262 ± 6
Mean cell velocity of lymphangions with contractions ($\mu m\ s^{-1}$)	32	204	282 ± 9
Mean cell velocity of lymphangions without contractions ($\mu m\ s^{-1}$)	30	185	239 ± 9
Rate of lymph flow oscillations (min^{-1})	2	8	64 ± 8

sizes such as gold nanoparticles with sizes 100 and 250 nm as well as polystyrene beads. Such high resolution allowed us to estimate the size and shape of moving single cells (Figure 14.7). In particular, we recorded the marked amount of RBCs in lymph flow of intact microvessels. The moving leukocytes and erythrocytes with their specific shapes (typically, a round shape for leukocytes including lymphocytes, and discoid for normal RBCs) were easily identified. However, in contrast to the strongly absorbing RBCs in visible range, relatively low absorbing lymphocytes in lymph flow were imaged as smooth, relatively transparent spots, so it was possible to observe the elongated structure of the bottom lymph wall through the cell (Figure 14.7b).

This technique also allowed us to monitor the rotation of a single RBC by capturing its successive images. For example, Figure 14.7c,d shows positions from two different angles of the same RBC.

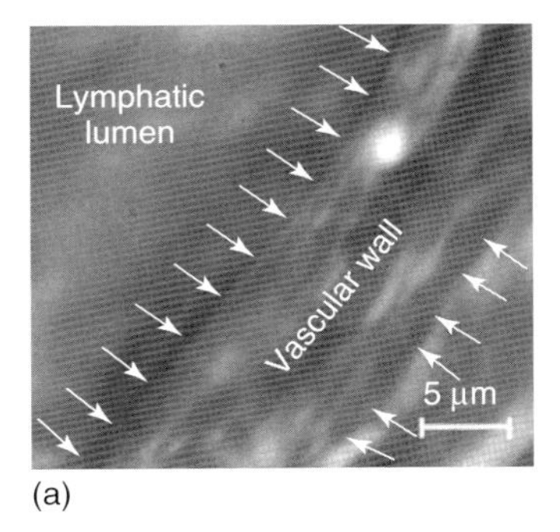

(a)

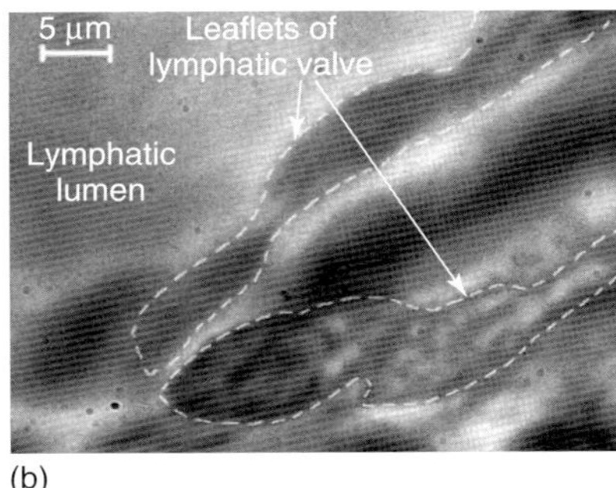

(b)

Figure 14.21 The high-resolution transmission images of vascular structures: (a) lymphatic walls (area B in Figure 14.18b), arrows show margins of wall and (b) lymphatic valve (area C in Figure 14.18b), arrows show valve leaflets, white puncture line – shape of leaflets [17].

The edge, contour, and shape of the lymphatic wall and valve leaflets are visualized with a high contrast, but intracellular structures and holes in the vascular walls cannot be well imaged (Figure 14.21).

High-resolution imaging at the high magnifications of the objectives (×40, ×60, and ×100) allowed us to visualize intralymphatic cell aggregation in intact microlymphatics. As can be seen in Figure 14.8, we observed the aggregates of different sizes. The nature of these moving aggregates is likely related to groups of leukocytes and RBCs, but the cause of their aggregation and their dependence on different factors require additional detailed study [107].

The data presented provide evidence that the combination of TDM with the rat mesentery as an animal model is a unique technical and biological platform for the study of blood cells in blood and lymph microflows with high optical resolution. Besides, this model allows one to study simultaneously microlymphatic function in norm and in pathological states as was first demonstrated with a lower resolution [14, 80, 91]. In particular, the main mesenteric microstructures (lymph vessel diameter in different sections, valve geometry, single moving cells, etc.) and their dynamics (wall motion, valve function/parameters, phasic contractions, cell velocity, etc.) can be imaged and monitored in real time. The data obtained using high magnification (×40 to ×100) allowed us to determine the size, shape, and aggregation of various cell types, as well as cell concentration in lymph flow. Additionally, we will probably be able to study cell–cell and cell–wall interactions, as well as the number of active lymphatics (with lymph flow), rate of flow oscillations, mean lymph flow velocity, and relation of forward and backward flow using TDM. The imaging of rotating cells opens the way to realize projection tomography *in vivo* for reconstruction of 3D cell images when necessary capturing of images at different angles is realized simply during natural cell rotations in a flow.

According to obtained data, there is no doubt that imaging some intracellular structural detail in cells containing relatively strongly absorbing chromophores, such as RBCs with hemoglobin, can be accomplished with TDM. However, studies of much lower absorbing individual cells (e.g., WBCs or various cancer cells) in the visible and near-IR spectral range with TDM are limited because of their relatively low absorption sensitivity due to the short optical path of light in cellular microstructures and the limited accuracy in measuring the small variations in light energy transmitted through these structures [108]. For example, to measure absorption in a 300 nm long optical path (e.g., average diameter elongated mitochondria) with a typical 5% accuracy in light variation, a minimum detectable coefficient of absorption would be approximately 1.6×10^3 cm^{-1}, which is a few orders higher than absorbing cellular structures in WBCs with a typical coefficient of absorption in the visible spectral range of $10^{-2} - 10^1$ cm^{-1} [108]. Evidently, this level of cell absorption is insufficient to get contrast images. However, because of scattering phenomena, the TDM technique allows one to image clearly the boundaries of such cells and even distinguish their shapes. We discovered that the lymph of intact microvessels of narcotized rats contained markedly greater amounts of RBCs. To our knowledge, there was no discussion in the available literature of the appearance of RBCs in microlymphatics, although these data are important for understanding *in vivo* lymph rheology at the microvascular level, or dependence of the vessel's permeability on different factors. However, at present, many questions remain. In particular, what are the reasons for the aggregation and appearance of RBCs in lymph (is this a physiological state or the consequence of anesthesia and preparation procedures?); how does the number of RBCs and their ability to aggregate affect the numbers, sizes, and quality of aggregates; and finally, how does aggregate formation and intravascular changes of viscosity, as a whole, change lymph flow in microcirculation and the local balance of water in tissue? These are still open questions.

The measurement of flow velocity with TDM by video recording of cells movements may be limited by several factors: (i) sometimes high cell concentrations in flow, which prevents tracking of a single cell position with a high accuracy; (ii) change of the transverse position of a moving cell in vessel's cross sections with different velocity profiles; and (iii) mismatch of plasma flow velocity and a cell itself. In particular, in microlymphatics with nonregular flow and change of its direction, the velocity of cells should be less than plasma flow velocity when flow starts to move because of a cell's inertia. On the contrary, on the flow stopping, the cell velocity is higher than that of plasma.

Thus, a high-resolution TDM technique provides important information of cell margins, sizes, and shapes in flows, although it has some limitations with imaging of low-absorbing structures and measurement of flow velocities. To overcome these limitations of TDM, we recently suggested combining TDM with the PT technique [75, 94, 95] (see Chapter 17), which supplement each other beneficially and provide simultaneous study both low and strongly absorbing cells in flow including their imaging without conventional staining and labeling (which might be limited in *in vivo* study) and laser scanning. In addition, this combination provides the potential for characterizing flow without use moving cells as markers, and allows one to identify normal and abnormal cells in flow based on their differences in absorption, sizes, and shapes. For the future study of lymphatics, it is important that well-developed lymphatic valves with a "funnel" shape may act as a natural nozzle for the cells (Figure 14.19), and thus play the role of natural "hydrodynamic" focusing (in analogy to conventional flow cytometer *in vitro*) of a moving cell near a vessel axis. It will allow one to decrease the influence of cell spatial fluctuations on the quality of *in vivo* optical images.

14.3.5 Summary

The use of the high-resolution TDM was demonstrated to be appropriate for *in vivo* imaging of moving and static cells in the lymph system of rat mesentery as an animal model. TDM's capability to measure the main microlymphatic parameters without labeling, with simultaneous imaging of mesenteric structures such as blood and lymph microvessels, their walls and valves, and some cells in interstitial space (mast and fat cells), was explored. Potential applications of this method and animal model may include studies of microlymphatic responses to various therapeutic and microsurgical interventions (drug, laser radiation, etc.) at the single-cell level [19]. The use of this method may considerably increase our knowledge about microlymphatic functions in norm and pathology, including primary or secondary lymphedema, lymphatic malformation, venous insufficiency, as well as the influence of smoking, alcohol, and drugs on organism functioning. Additionally, the capabilities of this technique using the animal model could extend to the following experimental studies related to *in vivo* cytometry: diagnosis in lymph and blood flow of (i) abnormal cells circulating in blood and lymph systems, including cancer cells (e.g., leukemia); (ii) apoptotic cells; (iii) changes in RBC properties; (iv) modified cells after drug, toxic, or X-ray radiation; and (v) pathogenic bacteria and viruses.

14.4 High-Resolution Single-Cell Imaging in Lymphatics

14.4.1 The High-Speed TDM

A summary of the results on high-speed TDM studies of cells in lymphatics with the aim of *in vivo* lymph FC is presented in Figure 14.22 [77]. To better understand cell interactions and pathways of their migration, the relationships between initial lymphatics, afferent lymphatics, lymph nodes, and efferent lymphatics are also shown. The high-speed TDM provides *in vivo* label-free visualization of natural single-cell transport from tissue to regional (e.g., sentinel) lymph nodes, including monitoring of their passage through the endothelial wall to an initial lymphatic vessel (Figure 14.22b,c). On the basis of notable differences in size and shape, TDM provides the identification of individual leukocytes (Figure 14.22d), RBCs (Figure 14.22e), chylomicrons (Figure 14.22f), and

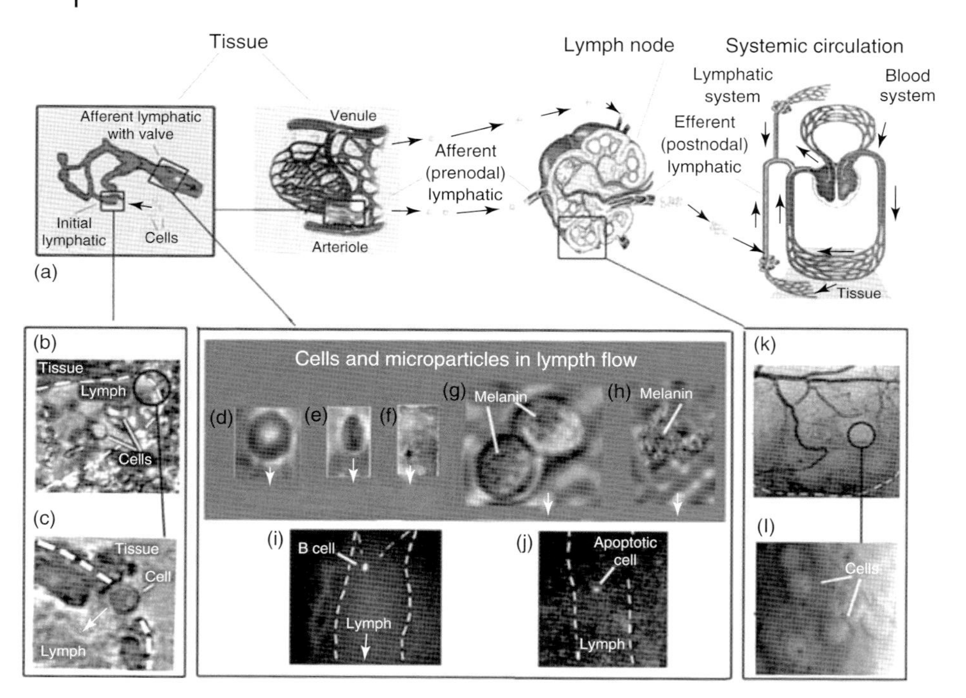

Figure 14.22 Image lymph flow cytometry: (a) the relationships between initial lymphatics, afferent lymphatics, lymph nodes, and efferent lymphatics; (b, c) cells passing through the endothelial wall of an initial lymphatic (magnifications, ×40 and ×100, respectively); (d–h) high-resolution high-speed TDM images of a WBC, a RBC, a microparticle (likely chylomicron), a melanoma cell aggregate containing two cells and a partly lysed single melanoma cell in lymph flow (magnification, ×40; 500–2500 fps), respectively; (i) B lymphocyte exiting a valve aperture; (j) fluorescent image of an apoptotic cell in the lymphatic lumen (dash lines); and (k, l) images of a mesenteric lymph node fragment *in vivo* at different magnifications (×4 and ×100, respectively) [77].

melanoma cells (Figure 14.22g,h) in lymph flow. In particular, in relatively transparent mesenteric lymphatics, the metastatic cells were identified by their larger size (approximately two times more than leukocytes) and black localized pigmentation (Figure 14.22g). Melanoma cells had round shapes, which were likely in a live stage, or an irregular form that was probably in the necrotic stage (Figure 14.22h). They were transported with a velocity similar to that of normal cells. In contrast to lymphocytes, which moved typically as single cells, melanoma cells traveled frequently as small aggregates (Figure 14.22g).

To count lymphocyte's subsets (e.g., B lymphocytes), the fluorescence-labeled antibodies specific to CD45 markers of B lymphocytes were injected into rat peritoneum *in vivo* followed by time-resolved fluorescence monitoring of mesenteric lymphatics (Figure 14.22i). The mean velocity of B lymphocytes in lymph flow was determined to be $1720 \pm 18\,\mu m\,s^{-1}$, and the proportion of B lymphocytes in prenodal lymph was 4.5%. To capture an image of apoptotic cells, apoptosis was induced by intraperitoneal injection of dexamethasone followed 6 h later by the injection of annexin V–FITC, and then by fluorescence monitoring of apoptotic cells. The majority of apoptotic lymphocytes were found mainly in mesenteric tissues. We observed also single lymphocytes adhered to a vessel wall. Only a few apoptotic cells were detected in lymph flow during the 1 h observation (Figure 14.22j).

Compared to the blood test, lymphatics has been paid much less attention, although recent discoveries in the biology of lymphatics (e.g., new receptors, LYVE-1, VEGFR-3, and PROX1) emphasized the key role of lymphatics in the development of many specific diseases such as metastasis, infections, or immunity disorders [78, 79, 109–113]. To date, many questions of lymphatic function are still not satisfactorily answered. For example, despite dramatically growing evidence of the crucial role of the lymphatic route in the spread of metastatic cells such as melanoma or breast cancer to other organs, researchers know almost nothing about transport of these cells through lymph vessels before their invasion into lymph nodes [109, 110].

Another very important issue is transportation of immune-related cells in lymphatics. According to a few *in vitro* animal studies, afferent lymph is mainly composed of immune-related cells: 85% T lymphocytes, 5% B lymphocytes, and 10% other cell types (dendritic, and macrophages, among others) [111, 112]). Besides differences in composition, the total concentration of cells in lymphatics is lower and varies significantly ($10^3 - 10^5$ cells μl^{-1}) than in the blood flow ($\sim 5 \times 10^6$ cells μl^{-1}). Since the sampling of lymph is not well established and the procedure is of low accuracy, the exact composition of cells in lymphatics (e.g., amount RBCs), as well as cell rate, especially, at pathology, is not still clear. Despite significant progresses in diagnostic techniques (e.g., magnetic resonance imaging, lymphoscintography, or fluorescence lymphography), to date, none has demonstrated the ability to real-time label-free detection of individual cells in the lymph flow.

14.4.2
Optical Clearing for *In vivo* Label-Free Image Cytometry

The ability to image individual cells and structures in lymph vessels and nodes through thicker tissue layers and blood by TDM and other optical modalities can be significantly enhanced by the application of immersion OCAs [44, 67, 114, 115]. To decrease light scattering in a lymph node, which prevents getting clear optical images in this rather thick tissue, we tested several OCAs with well-established clearing effect in tissues and blood. *In vivo* assessment of sentinel lymph nodes (SLNs) and lymphatic vessels was performed noninvasively through intact skin and intraoperatively through a small skin incision overlying the SLN and direct attachment of the folder with a fiber to skin and the lymph node, respectively. For *ex vivo* study, lymph nodes were excised without fat and surrounding tissues from the mouse and placed in a well (8 mm in diameter and 2 mm in depth; Molecular Probes) of a microscope slide to obtain optical images at different magnifications (×4, ×10, ×20, ×40, ×100 with oil immersion). To decrease beam blurring due

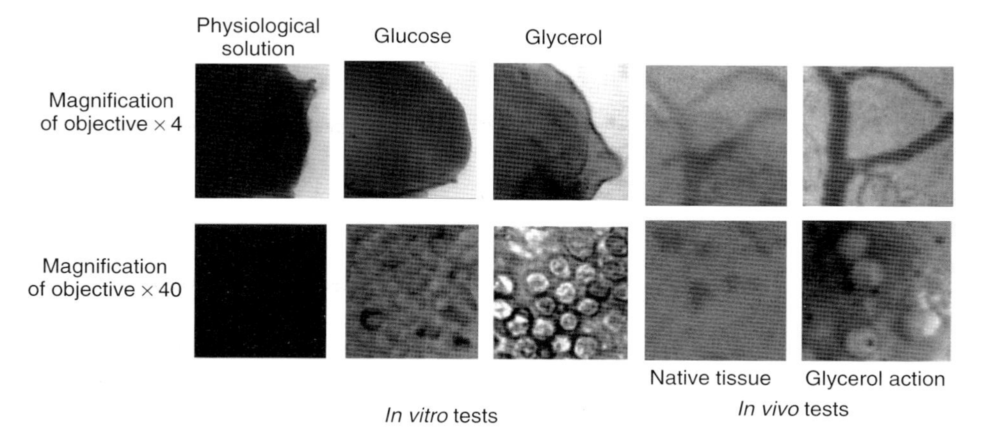

Figure 14.23 The effect of 40% glucose, 100% DMSO, and 80% glycerol on the optical imaging of a mesenteric lymph node at the different microscopic magnifications. The images of the lymph node embedded in PBS served as a control [97].

to light scattering in lymph nodes, the nodes were embedded for 5 min in 40% glucose, 100% DMSO, and 80% glycerol as hyperosmotic OCAs [97, 114].

Comparisons of optical images with different agents and phosphate buffered saline (PBS, control) revealed that glycerol had the maximal optical clearing (Figure 14.23). The treatment of SLNs with 80% glycerol allows the label-free imaging of a fresh node at the cellular level including localization of immunerelated, metastatic, and other cells (e.g, lymphocytes, macrophages, dendritic cells, and melanoma cells) as well as the surrounding microstructures (e.g., afferent lymph vessel, subcapsular and transverse sinuses, the medulla, a reticular meshwork, follicles, and the venous vessels) (Figure 14.24). The combination of PA mapping of SLN metastases with high-resolution TDM mode (up to 300 nm) allowed us to assess the exact distribution and number of melanoma metastasis and laser-induced damage area [97].

14.5 *In vivo* Blood Flow Cytometry

14.5.1 The Specificity of Blood Flow Cytometry

Imaging of individual cells in blood flow *in vivo* is important to study cell interactions, RBC oxygen transport function, blood cell response to different interventions and stresses, and disease diagnoses and monitoring of medical treatments. Imaging of the flowing blood cells was realized at moderate CCD camera speeds, in

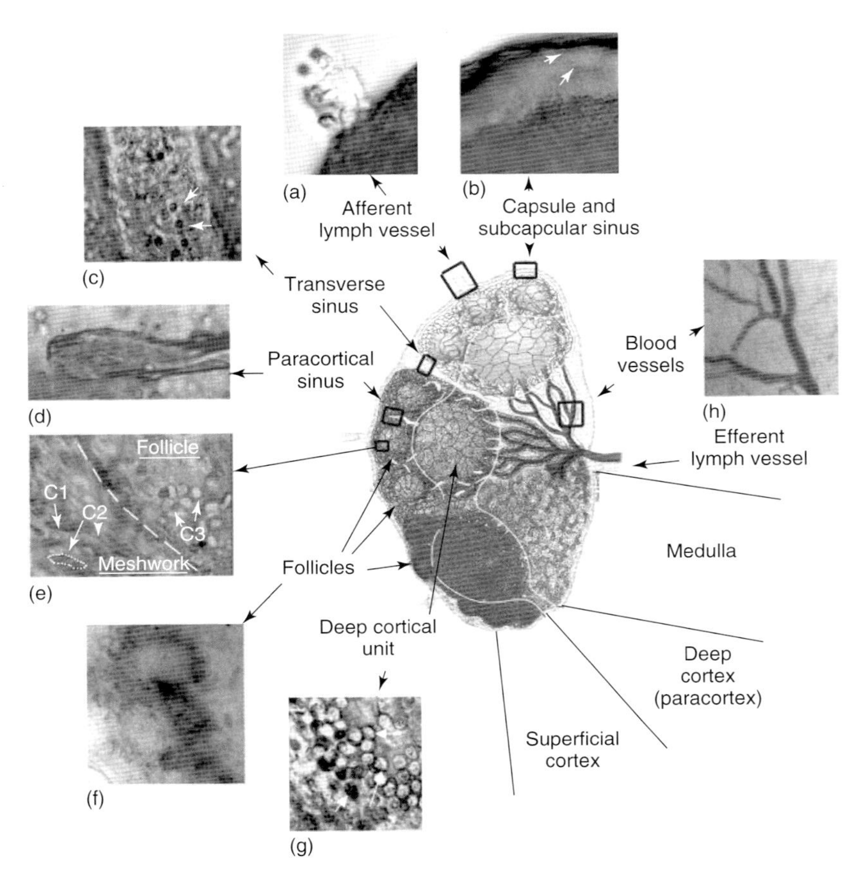

Figure 14.24 Optical imaging of microanatomy of the fresh lymph node *ex vivo* obtained with optical clearing using 80% glycerol. The central schematic designed by David Sabio [116] shows a midsagital section of a lymph node containing three lymphoid lobules with the basic anatomical and functional unit of the lymph node. Top and middle lobules: microanatomical schematics of lobular compartments (superficial cortex, deep cortex, and medulla) without (top) and with (middle) reticular meshwork. Bottom lobule: a lobule of a mouse lymph node as it appears in conventional histological section. C1 shows likely basophilic lymphocytes; C2 shows elongated fibroblastic reticular cells; and C3 shows B lymphocytes and follicular dendritic cells [97].

particular, at 750–2500 fps in blood flow [117, 118]. These speeds are not quite enough to image fast-moving RBCs in the most of blood microvessels. In addition, high optical resolution should be provided to have in mind cytometry aims. The relatively high-resolution vital optical microscopy has been used in studies of different animal models (e.g., rat and mouse mesentery, muscle, and ear) [17, 34, 75, 95, 119, 120] mainly in two modes: (i) a relatively slow, successive-framing mode (<30 fps) by which only slow-moving cells can be imaged (e.g., the so-called rolling leukocytes with an average velocity of $\sim 50\,\mu m\,s^{-1}$) or (ii) a short time-exposure

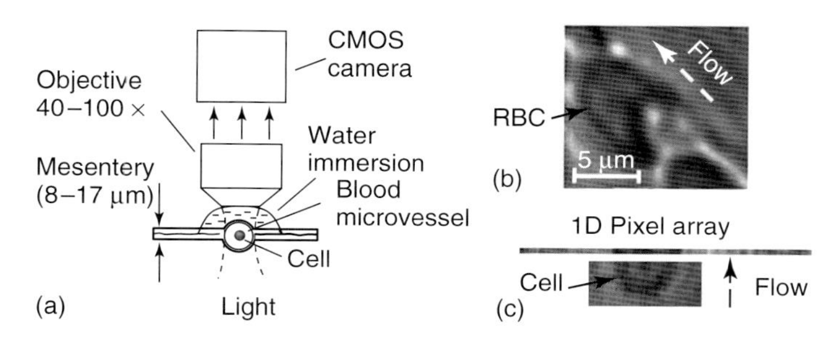

Figure 14.25 (a) Optical scheme of high-speed imaging of cells in blood vessels of rat mesentery with different areas of interest: (b) 128 × 128 pixels; and (c) 512 × 1 pixels with magnification ×100 and ×40, respectively [18].

mode (~0.1–1 ms) by which just single images of fast-moving cells can be captured. Recently, simultaneous high optical resolution and high-speed imaging of fast-moving cells (e.g., RBCs in arterioles with a velocity up to 5–10 mm s^{-1}) in natural conditions with successive high-speed framing, which is crucial to time-resolved monitoring of cells' transient (i.e., dynamic) deformability, has been realized *in vivo* [18].

In vivo flow image cytometery schematics (see Figure 14.16) on the technical platform of an Olympus BX51 microscope was adopted to realize this approach with an advanced CMOS camera (model MV-D1024-160-CL-8, Photonfocus, Lachen, Switzerland) (Figure 14.25a). Rat mesentery was chosen to test high-speed optical imaging of flowing cells.

The minimum frame rate $f_{\min}$ for avoiding optical resolution (δ_{opt}) distortion due to motion was chosen on the basis of the condition $f_{\min} \geq (V_F)_{\max}/\delta_{\text{opt}}$, where $(V_F)_{\max}$ is the maximum flow velocity. For the typical parameters $(V_F)_{\max} \approx 5$ mm s^{-1} and $\delta_{\text{opt}} \approx 0.5$ μm, $f_{\min} \geq 10\,000$ fps. High-speed imaging without loss of spatial resolution was achieved by adjusting the area of interest. For example, four typical areas of interests, 512 × 512, 128 × 256, 128 × 64, and 512 × 1 pixels, corresponded to frame rates of 500, 2500, 10 000, and 40 000 fps, respectively. At ×40 magnification, the sizes of these areas of interest in the sample plane were 136 × 136, 34 × 68, 34 × 17, and 136 × 0.3 μm, respectively. The largest areas of interest were useful for providing an overview of the mesenteric vessel network, while small areas were better suited for high-resolution imaging of single cells (Figure 14.25b). The highest frame speed could be achieved with a linear pixel array, although further reconstruction of 2D cell images was required (Figure 14.25c). Optical resolution, estimated with nanoscale colored polystyrene beads, was approximately 700, 350, and 250 at ×10, ×40, and ×100 magnifications, respectively.

The combination of light absorption and scattering effects on cells made it possible to visualize and identify most blood cells without conventional labeling and vital staining with the use just of bright-field microscopy. In particular, due to relatively strong light absorption by RBCs in the visible spectral range, these single cells in flow appeared mostly as dark objects in the transillumination mode, while weakly absorbing WBCs and platelets appeared either as light objects (e.g., in the presence of many more strongly absorbing RBCs in blood flow or with dominant scattering effects) or, in contrast, as slightly dark objects (e.g., in the transparent plasma without RBCs). In the dense flow, RBCs sometimes exhibited bright margins, or were seen even as light objects due to multiple scattering effects. This is because the light, during propagation through many other RBCs and before reaching the plane of focus, was significantly scattered and attenuated through absorption, resulting in dominant scattering light around imaged cells. The CMOS camera's sensitivity is sufficient to visualize most of these cells with conventional continuous mode illumination in the frame rate range used.

14.5.2
The High-Speed, High-Resolution Imaging Blood Flow Cytometry

First of all, application of high-speed, high-resolution imaging technique confirmed the well-known fact that normal RBCs (basically a 5–7 μm diameter biconcave disk) flowing in single-file through capillaries or postcapillary venules with a velocity of 0.3–0.7 mm s^{-1} are usually deformed into parachute-like shapes [18, 120, 121] (Figure 14.5a,b). High-speed, high-resolution imaging, however, revealed for the first time [18] that RBCs in fast blood flow undergo more significant (than previously believed) deformations in geometrically irregular regions (e.g., bifurcations, curves, narrowed areas, etc.), as well as in their interactions with other cells. In particular, Figure 14.5a shows the high deformability of parachute-like RBCs as they squeeze through a narrow gap between the vessel wall and another cell (rigid RBC or small WBC) adherent to the opposite wall. High-speed imaging also demonstrated how quickly the relatively fast-flowing RBCs (velocity ≥ 1 mm s^{-1}) changed shape as they interacted with the much more slowly moving (~ 20 μm s^{-1}) rolling WBCs (Figure 14.5b). Figure 14.5c shows very considerable stretching up to at least 0.4–0.7 μm (in 2D projection) of two RBCs in the diverging flow streams in a bifurcation zone. The typical value of the deformation index (DI), which is defined as the ratio of a cell's length to width, has been found to be approximately of 7.3 in curved-vessel flow (Figure 14.5c, middle) compared to a DI of 1.4–1.5 in straight-microvessel flow (Figure 14.5a, left) and a DI of 1 for almost static or adherent RBCs (Figure 14.5a).

Slow-moving (30–80 μm s^{-1}), rolling WBCs [18, 122] were reliably visualized at a relatively slow frame rate of 20–30 fps, at which fast-flowing RBCs (velocity $\sim$1–5 mm s^{-1}) looked like the nonstructural dark background due to motion distortion. Only a fast frame rate (>2000 fps) provided time-resolved monitoring of the deformation of fast-moving RBCs as they interacted with rolling WBCs (Figure 14.5b).

Label-free identification of single platelets is difficult to achieve because these weakly absorbing cells are also small (2–3 μm). Nevertheless, our technique has

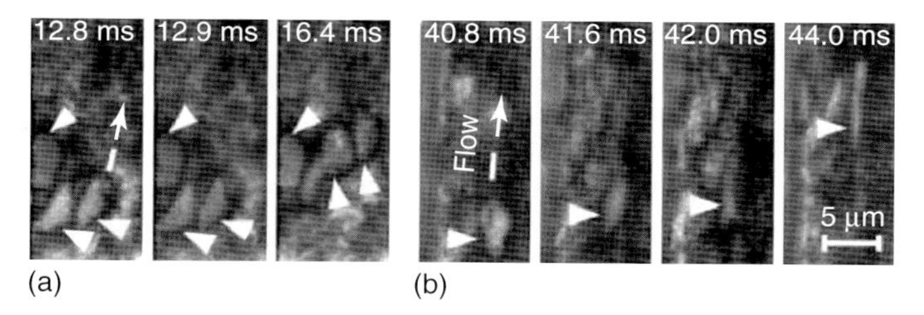

Figure 14.26 (a) Three subsequent images of a few RBCs in different files of fast arteriolar flow with velocities 1 mm s^{-1} for left RBC (nearest to vessel wall), 2 mm s^{-1} for middle RBC, and 2.5 mm s^{-1} for right RBC (nearest to vessel axis) (10 000 fps; ×40). (b) Fast transient deformation of single RBC in flow with a velocity of 5 mm s^{-1}; DI = 2.1, 3.0, 5.9, and 10.3, respectively (2500 fps; ×40).

provided relatively good contrast images of both slow-flowing (Figure 14.6a) and fast-flowing (2.5 mm s^{-1}, Figure 14.6b) single platelets in blood microvessels. The best quality images of platelets were obtained in the RBC-free space of the microvessel lumen (Figure 14.5b).

The most difficult is to obtain images of RBCs in multifile cell flow in 30–40 μm venules and especially in arterioles, which have the highest axial velocity (up to 10 mm s^{-1}). Nevertheless, the monitoring of image sequences at a high frame rate of 10 000 fps by adjustment of the focus locations made it possible to record the size and shape dynamics of selected single RBCs in fast flow (Figure 14.26a). In addition, the continuous video framing has the potential to measure linear velocity of several cells traveling along a cross section of a microvessel simultaneously (Figure 14.26a). High-speed imaging demonstrated for the first time the dynamic changes of the 2D shape of the RBCs, and, correspondingly, transient deformation, in relatively fast arteriolar straight flow (5 mm s^{-1} and higher) [18]. In particular, extremely high cell dynamic deformation (DI up to 10) was observed in a very short time ($\leq$0.4 ms) (Figure 14.26b). In contrast, for blood straight flow in a tube *in vitro* the maximum DI has been reported to be approximately 1.5 in healthy individuals, with a significant decrease to 1.05 in patients with diabetes mellitus [123].

14.5.3
The Limitations and Future Perspectives

As expected, some limitations of the optical imaging used have been revealed in fast multifile flow. In particular, even at a low hematocrit (~20%) [122], several cells simultaneously appeared in a small focal area of detection, and up to 6–8 cells appeared in a whole microvessel cross section, which led to the overlapping of cell images in 2D projection making their identification quite difficult. It seems that this problem could be resolved partly by hemodilution and reduction of hematocrit in analogy to study of RBC aggregation [117, 118]. However, hematocrit may affect significantly on RBC deformability [121]. Overcoming this limitation may require developing of 3D imaging *in vivo*.

In some cases, fast radial displacement of cells in multifile flow prevented continuing high-resolution imaging of the same cell with a strongly focused micro-objective at high magnification. Small arterioles (diameter 20–30 μm) were found to be the most suitable for such measurements with a fixed focal plane because cells in these vessels mainly moved along the flow axis. Furthermore, a high enough flow velocity in such vessels still decreased the probability of transient RBC aggregation (in "rouleaux" or clumps) [124], thus facilitating the study of individual cells.

In a future study, several digital cameras operating at different frame rates and hence sensitivity levels (which usually decrease with increases in speed) might be useful for simultaneous imaging of cells moving at different velocities. This approach might include (i) imaging of static (i.e., adhered) or slow-moving cells (e.g., rolling WBCs) at 10–30 fps; (ii) monitoring of the deformability of RBCs traveling in single-file flow in capillaries and postcapillary venules (velocity, 500–1000 $\mu m\,s^{-1}$) at 500–1000 fps; and (iii) high-speed imaging of individual cells moving in multifile fast flow (velocity, 5–10 $mm\,s^{-1}$) at $\geq$ 10 000 fps. Despite the distinctive optical contrast of RBCs, WBCs, and platelets, the low absorption sensitivity of transmission microscopy makes it difficult to differentiate cells with slightly different absorption properties (e.g., subtypes of WBCs). For this purpose, recently developed label-free, PT FC *in vivo* might be useful because its high absorption sensitivity makes it possible to image weakly absorbing cellular structures with specific "fingerprints" [34, 75, 95] (see Chapter 17).

The approach that we have developed also has the potential to be used *in vivo* in humans. Noninvasive mode can be realized with the use of thin, translucent structures, such as ear, eyelid, or interdigital membrane conjunctiva [75], with optical clearing [44]; an invasive approach might involve incorporation of a fiber-chip-based catheter in vessels.

14.5.4 Summary

In summary, a high-speed (up to 10 000 with the potential for 40 000 fps), high-resolution (up to 300 nm) continuous *in vivo* optical imaging technique allows one to monitor and identify RBCs, WBCs, and platelets in the blood flow of rat mesenteric microvessels without conventional labeling. It is demonstrated that frame rates up to 10 000 fps at the high optical resolution would be sufficient for full estimation of individual cell behavior and cellular biomechanical properties *in vivo* in microvascular net.

Potential applications of this technique may include [18] (i) fundamental study of cell–cell interactions in native flow with a focus on transient RBC deformability during RBC–RBC or RBC–WBC interactions; (ii) estimation of the proportion of fast-moving WBCs traveling with RBCs in axial (core) flow to slow-moving (rolling) WBCs, which may have diagnostic value for the study of some pathologic conditions (e.g., inflammation, when the number of rolling WBCs dramatically increases) [122]; (iii) identification of rare abnormal cells (e.g., cancerous or sickled cells) on the basis of their different deformability in flow; (iv) imaging of platelets during thrombus formation and interaction with metastatic cancer cells; (v) study of the influence of environmental factors (e.g., drugs or radiation) on individual blood cells; (vi) estimation of blood viscosity in high-velocity flow through the principal determinants of blood rheology (hematocrit, cell deformability, and aggregation); (vii) study of the dynamics of intravascular cell aggregation and adhesion to endothelial cells; (viii) estimation of velocity profiles in fast flow; and (ix) imaging and detection of individual cell dynamics in afferent lymph flow with cell velocity up to 7 mm s^{-1} [20].

14.6 Conclusion

The high-resolution (up to 300 nm) and high-speed (up to 10 000 with the potential for 40 000 fps) video imaging cytometry was demonstrated to provide *in vivo* imaging of moving and static cells in lymph and blood systems of typical animal models. Imaging cytometry allows one to identify and monitor RBCs, WBCs, and platelets in the lymph and blood microvessel flows without conventional labeling, including their velocity, size, and shape within complex nonstationary lymph and blood flows caused by their functioning and vessel pathologies (i.e., vessel bifurcations, anevrism). The major attraction of this approach is that, in the framework of a single technique, many dynamic and static or quasistatic parameters of lymph and blood microvascular systems can be concurrently evaluated, such as microvessel diameters, vessel walls and valves behavior and structure, cell migration in interstitial space, and so on.

However, to use the full potential of the optical video imaging method, its integration with other well-established and novel technologies is desirable. Description of many of these technologies can be found in this book (see, e.g., Chapters 16 and 17). Integrated *in vivo* FC approaches give new facilities in real-time diagnosis and immediate therapy by detecting abnormal cells circulating in blood and lymph systems, including cancer and apoptotic cells, pathogenic bacteria and viruses, and changes in WBC and RBC properties, in particular modified by drugs, toxins, or X-ray radiation.

Acknowledgments

This work was supported by: The National Institute of Health grants **R01 EB000873, R01 CA131164, R01 EB009230, R21 CA139373;** The National Science Foundation grant **DBI 085 2737; Grant 224014** PHOTONICS4LIFE of FP7-ICT-2007-2; Projects 2.1.1/4989 and 2.2.1.1/2950 of RF Program on the Development of High School Potential; The Special Program of RF "Scientific and Pedagogical Personnel of

Innovative Russia," Governmental contracts 02.740.11.0484, 02.740.11.0770, and 02.740.11.0879.

The authors are thankful to Vycheslav Kalchenko for reading the manuscript and valuable comments.

References

1. Shapiro, H.M. (2003) *Practical Flow Cytometry*, 4th edn, Wiley-Liss, New York.
2. Givan, A.L. (2001) Principles of flow cytometry: an overview. *Methods Cell Biol.*, **63**, 19–50.
3. De Rosa, S.C., Brenchley, J.M., and Roederer, M. (2003) Beyond six colors: a new era in flow cytometry. *Nat. Med.*, **9**, 112–117.
4. Hogan, H. (2006) Flow cytometry moves ahead. *Biophotonics Int.*, **13**, 38–42.
5. Schmidt, R.F. and Thews, G. (eds) (1989) *Human Physiology*, Springer-Verlag, Berlin, Heidelberg.
6. Cherry, S.R. (2004) *In vivo* molecular and genomic imaging: new challenges for imaging physics. *Phys. Med. Biol.*, **49**, R13–R48.
7. Witte, C.L., Witte, M.H., Unger, E.C. *et al.* (2000) Advance in imaging of lymph flow disorder. *Radio Graph.*, **20**, 1697–1719.
8. Lincoln, B., Erickson, H.M., Schinkinger, S., Wottawah, F., Mitchell, D., Ulvick, S., Bilby, C., and Guck, J. (2004) Deformability-based flow cytometry. *Cytometry*, **A59**, 203–209.
9. Abkarian, M., Faivre, M., and Stone, H.A. (2006) High-speed microfluidic differential manometer for cellular-scale hydrodynamics. *Proc. Natl. Acad. Sci. U.S.A.*, **103**, 538–542.
10. Popescu, G., Ikeda, T., Best, C.A., Badizadegan, K., Dasari, R.R., and Feld, M.S. (2005) Erythrocyte structure and dynamics quantified by Hilbert phase microscopy. *J. Biomed. Opt.*, **10**, 060503-1–060503-3.
11. Chung, A., Karlan, S., Lindsley, E., Wachsmann-Hogiu, S., and Farkas, D.L. (2006) *In vivo* cytometry: a spectrum of possibilities. *Cytometry*, **A69**, 142–146.
12. Scheinecker, C. (2005) Application of *in vivo* microscopy: evaluating the immune response in living animals. *Arthritis Res. Ther.*, **7**, 246–252.
13. Kalchenko, V., Harmelin, A., Fine, I., Zharov, V., Galanzha, E., and Tuchin, V. (2007) Advances in intravital microscopy for monitoring cell flow dynamics *in vivo*. *Proc. SPIE*, **6436**, 64360D-1–64360D-15.
14. Galanzha, E.I., Brill, G.E., Aisu, Y., Ulyanov, S.S., and Tuchin, V.V. (2002) in *Handbook of Optical Biomedical Diagnostics*, vol. PM107, Chapter 16 (ed. V.V. Tuchin), SPIE Press, Bellingham, WA, pp. 881–937.
15. Fedosov, I.V., Ulyanov, S.S., Galanzha, E.I., Galanzha, V.A., and Tuchin, V.V. (2004) in *Coherent-domain Optical Methods: Biomedical Diagnostics, Environmental and Material Science*, vol. 1, Chapter 10 (ed. V.V. Tuchin), Kluwer Academic Publishers, Boston, MA, pp. 397–435.
16. Galanzha, E.I., Chowdhury, P., Tuchin, V.V., and Zharov, V.P. (2005) Monitoring of nicotine impact on microlymphatics of rat mesentery with time-resolved microscopy. *Lymphology*, **38**, 181–192.
17. Galanzha, E.I., Tuchin, V.V., and Zharov, V.P. (2005) *In vivo* integrated flow image cytometry and lymph/blood vessels dynamic microscopy. *J. Biomed. Opt.*, **10**, 54018-1–54018-8.
18. Zharov, V.P., Galanzha, E.I., Menyaev, Y., and Tuchin, V.V. (2006) *In vivo* high-speed imaging of individual cells in fast blood flow. *J. Biomed. Opt.*, **11** (5), 054034-1–054034-4.
19. Galanzha, E.I., Tuchin, V.V., and Zharov, V.P. (2007) Advances in small

animal mesentery models for *in vivo* flow cytometry, dynamic microscopy, and drug screening (invited review). *World J. Gastroenterol.*, **13** (2), 198–224.
20. Dixon, J.B., Zawieja, D.C., Gashev, A.A., and Coté, G.L. (2005) Measuring microlymphatic flow using fast video microscopy. *J. Biomed. Opt.*, **10**, 064016.
21. Japee, S.A., Pittman, R.N., and Ellis, C.G. (2005) A new video image analysis system to study red blood cell dynamics and oxygenation in capillary networks. *Microcirculation*, **12**, 489–506.
22. Langer, S., Born, F., Hatz, R., Biberthaler, P., and Messmer, K. (2002) Orthogonal polarization spectral imaging versus intravital fluorescent microscopy for microvascular studies in wounds. *Ann. Plast. Surg.*, **48**, 646–653.
23. Novak, J., Georgakoudi, I., Wei, X., Prossin, A., and Lin, C.P. (2004) *In vivo* flow cytometer for real-time detection and quantification of circulating cells. *Opt. Lett.*, **29**, 77–79.
24. Georgakoudi, I., Solban, N., Novak, J., Rice, W.L., Wei, X., Hasan, T., and Lin, C.P. (2004) *In vivo* flow cytometry: a new method for enumerating circulating cancer cells. *Cancer Res.*, **64**, 5044–5047.
25. Wei, X., Sipkins, D.A., Pitsillides, C.M., Novak, J., Georgakoudi, I., and Lin, C.P. (2005) Real-time detection of circulating apoptotic cells by *in vivo* flow cytometry. *Mol. Imaging*, **4**, 415–416.
26. Lee, H., Alt, C., Pitsillides, C. *et al.* (2006) *In vivo* imaging flow cytometer. *Opt. Express*, **14**, 7789–7797.
27. Butros, S., Greiner, C., Hwu, D. *et al.* (2007) Portable two-color *in vivo* flow cytometers for real-time detection of fluorescently-labeled circulating cells. *J. Biomed. Opt.*, **12**, 020507-1–020507-3.
28. Padera, T.P., Stoll, B.R., So, P.T., and Jain, R.K. (2002) Conventional and high-speed intravital multiphoton laser scanning microscopy of microvasculature, lymphatics, and leukocyte-endothelial interactions. *Mol. Imaging*, **1**, 9–15.
29. Rubart, M. (2004) Two-photon microscopy of cells and tissue. *Circ. Res.*, **95**, 1154–1166.
30. He, W., Wang, H., Hartmann, L.C., Cheng, J.X., and Low, P.S. (2007) *In vivo* quantitation of rare circulating tumor cells by multiphoton intravital flow cytometry. *Proc. Natl. Acad. Sci. U.S.A.*, **104**, 11760–11765.
31. Tkaczyk, E.R., Zhong, C.F., Ye, J.Y. *et al.* (2008) *In-vivo* monitoring of multiple circulating cell populations using two-photon flow cytometry. *Opt. Commun.*, **281**, 888–894.
32. Zhong, C.F., Tkaczyk, E.R., Thomas, T. *et al.* (2008) Quantitative two-photon flow cytometry – *in-vitro* and *in-vivo*. *J. Biomed. Opt.*, **13**, 034008-1–034008-19.
33. Tkaczyk, E.R., Tkaczyk, A.H., Katnik, S. *et al.* (2009) Extended cavity laser enhanced two-photon flow cytometry. *J. Biomed. Opt.*, **14**, 041319-1–041319-12.
34. Zharov, V.P., Galanzha, E.I., and Tuchin, V.V. (2006) *In vivo* photothermal flow cytometry: imaging and detection of individual cells in blood and lymph flow. *J. Cell Biochem.*, **97** (5), 916–930.
35. Zharov, V.P., Galanzha, E.I., Shashkov, E.V., Khlebtsov, N.G., and Tuchin, V.V. (2006) *In vivo* photoacoustic flow cytometry for monitoring of circulating single cancer cells and contrast agents. *Opt. Lett.*, **31**, 3623–3625.
36. Horstick, G., Kempf, T., Lauterbach, M., Ossendorf, M., Kopacz, L., Heimann, A., Lehr, H.A., Bhakdi, S., Meyer, J., and Kempski, O. (2000) Plastic foil technique attenuates inflammation in mesenteric intravital microscopy. *J. Surg. Res.*, **94**, 28–34.
37. Galanzha, E.I., Tuchin, V.V., and Zharov, V.P. (2007) Optical monitoring of microlympatic disturbances at experimental lymphedema. *Lymphat. Res. Biol.*, **5**, 11–28.
38. Laemmel, E., Genet, M., Le Goualher, G., Perchant, A., Le Gargasson, J.F., and Vicaut, E. (2004) Fibered confocal fluorescence microscopy (Cell-viZio) facilitates extended imaging in the field of microcirculation. A comparison with

intravital microscopy. *J. Vasc. Res.*, **41**, 400–411.

39. Menger, M.D., Laschke, M.W., and Vollmar, B. (2002) Viewing the microcirculation through the window: some twenty years experience with the hamster dorsal skinfold chamber. *Eur. Surg. Res.*, **34**, 83–91.
40. Jain, R.K. (2002) Angiogenesis and lymphangiogenesis in tumors: insights from intravital microscopy. *Cold Spring Harb. Symp. Quant. Biol.*, **67**, 239–248.
41. Paques, M., Tadayoni, R., Sercombe, R., Laurent, P., Genevois, O., Gaudric, A., and Vicaut, E. (2003) Structural and hemodynamic analysis of the mouse retinal microcirculation. *Invest. Ophthalmol. Vis. Sci.*, **44**, 4960–4967.
42. Becker, M.D., Garman, K., Whitcup, S.M., Planck, S.R., and Rosenbaum, J.T. (2001) Inhibition of leukocyte sticking and infiltration, but not rolling, by antibodies to ICAM-1 and LFA-1 in murine endotoxin-induced uveitis. *Invest. Ophthalmol. Vis. Sci.*, **42**, 2563–2566.
43. Barker, J.H., Kjolseth, D., Kim, M., Frank, J., Bondar, I., Uhl, E., Kamler, M., Messmer, K., Tobin, G.R., and Weiner, L.J. (1994) The hairless mouse ear: an *in vivo* model for studying wound neovascularization. *Wound Repair Regen.*, **2**, 138–143.
44. Tuchin, V.V. (2006) *Optical Clearing of Tissues and Blood*, vol. PM154, SPIE Press, Bellingham, WA.
45. McNamara, P.M., O'Doherty, J., O'Connell, M.-L., Fitzgerald, B.W., Anderson, C.D., Nilsson, G.E., Toll, R., and Leahy, M.J. (2010) Tissue viability (TiVi) imaging: temporal effects of local occlusion studies in the volar forearm. *J. Biophoton.*, **3** (1–2), 66–74.
46. Zhu, Q., Stockford, I.M., Crowe, J.A., and Morgan, S.P. (2009) An experimental and theoretical evaluation of rotating orthogonal polarization imaging. *J. Biomed. Opt.*, **14**, 034006.
47. Hanazawa, S., Prewitt, R.L., and Terzis, J.K. (1994) The effect of pentoxifylline on ischemia and reperfusion injury in the rat cremaster muscle. *J. Reconstr. Microsurg.*, **10**, 21–26.
48. Lipowsky, H.H., Sheikh, N.U., and Katz, D.M. (1987) Intravital microscopy of capillary hemodynamics in sickle cell disease. *J. Clin. Invest.*, **80**, 117–127.
49. Shirasawa, Y., Ikomi, F., and Ohhashi, T. (2000) Physiological roles of endogenous nitric oxide in lymphatic pump activity of rat mesentery *in vivo*. *Am. J. Physiol. Gastrointest. Liver Physiol.*, **278**, G551–G556.
50. Schneckenburger, H., Stock, K., Steiner, R., Strauss, W., and Sailer, R. (2002) in *Handbook of Optical Biomedical Diagnostics*, vol. PM107, Chapter 15 (ed. V.V. Tuchin), SPIE Press, Bellingham, WA, pp. 825–8764.
51. Duebener, L.F., Hagino, I., Schmitt, K., Sakamoto, T., Stamm, C., Zurakowski, D., Schafers, H.J., and Jonas, R.A. (2003) Direct visualization of minimal cerebral capillary flow during retrograde cerebral perfusion: an intravital fluorescence microscopy study in pigs. *Ann. Thorac. Surg.*, **75**, 1288–1293.
52. Becker, M.D., Kruse, F.E., Joussen, A.M., Rohrschneider, K., Nobiling, R., Gebhard, M.M., and Volcker, H.E. (1998) *In vivo* fluorescence microscopy of corneal neovascularization. *Graefes Arch. Clin. Exp. Ophthalmol.*, **236**, 390–398.
53. Kalchenko, V. and Plaks, V. (2005) Intravital video microscopy – from simple solutions to a multiuser core facility. *Proc. RMS*, **40** (4), 221–226.
54. Baatz, H., Steinbauer, M., Harris, A.G., and Krombach, F. (1995) Kinetics of white blood cell staining by intravascular administration of rhodamine 6G. *Int. J. Microcirc. Clin. Exp.*, **15**, 85–91.
55. Dadiani, M., Kalchenko, V., Yosepovich, A., Margalit, R., Hassid, Y., Degani, H., and Seger, D. (2006) Real-time imaging of lymphogenic metastasis in orthotopic human breast cancer. *Cancer Res.*, **66**, 8037–8041.
56. Braun, R.D., Abbas, A., Bukhari, S.O., and Wilson, W. III (2002) Hemodynamic parameters in blood vessels in choroidal melanoma xenografts and rat choroid. *Invest. Ophthalmol. Vis. Sci.*, **43**, 3045–3052.

57. Talks, J., Koshy, Z., and Chatzinikolas, K. (2007) Use of optical coherence tomography, fluorescein angiography and indocyanine green angiography in a screening clinic for wet age-related macular degeneration. *Br. J. Ophthalmol.*, **91**, 600–601.
58. Oheim, M., Beaurepaire, E., Chaigneau, E., Mertz, J., and Charpak, S. (2001) Two-photon microscopy in brain tissue: parameters influencing the imaging depth. *J. Neurosci. Methods*, **111**, 29–37.
59. Celi, A., Merrill-Skoloff, G., Gross, P., Falati, S., Sim, D.S., Flaumenhaft, R., Furie, B.C., and Furie, B. (2003) Thrombus formation: direct real-time observation and digital analysis of thrombus assembly in a living mouse by confocal and widefield intravital microscopy. *J. Thromb. Haemost.*, **1**, 60–68.
60. Lauritzen, M. and Fabricius, M. (1995) Real time laser-Doppler perfusion imaging of cortical spreading depression in rat neocortex. *Neuroreport*, **6**, 1271–1273.
61. Briers, J.D. (2001) Laser Doppler, speckle and related techniques for blood perfusion mapping and imaging. *Physiol. Meas.*, **22**, R35–R66.
62. Zimnyakov, D.A., Briers, J.D., and Tuchin, V.V. (2002) in *Handbook of Optical Biomedical Diagnostics*, vol. PM107, Chapter 18 (ed. V.V. Tuchin), SPIE Press, Bellingham, WA, pp. 987–1036.
63. Cheng, H., Luo, Q., Zeng, S., Chen, S., Cen, J., and Gong, H. (2003) Modified laser speckle imaging method with improved spatial resolution. *J. Biomed. Opt.*, **8**, 559–564.
64. Luo, Q., Cheng, H., Wang, Z., and Tuchin, V.V. (2004) in *Coherent-domain Optical Methods: Biomedical Diagnostics, Environmental and Material Science*, vol. 1, Chapter 5 (ed. V.V. Tuchin), Kluwer Academic Publishers, Boston, MA, pp. 165–195.
65. Kalchenko, V., Brill, A., Bayewitch, M., Fine, I., Zharov, V., Galanzha, E., Tuchin, V., and Harmelin, A. (2007) *In vivo* dynamic light scattering imaging of blood coagulation. *J. Biomed. Opt.*, **12** (5), 052002-1–050002-4.
66. Fujimoto, J.G., Boppart, S.A., Tearney, G.J., Bouma, B.E., Pitris, C., and Brezinski, M.E. (1999) High resolution *in vivo* intra-arterial imaging with optical coherence tomography. *Heart*, **82**, 128–133.
67. Tuchin, V.V., Xu, X., and Wang, R.K. (2002) Dynamic optical coherence tomography in studies of optical clearing, sedimentation, and aggregation of immersed blood. *Appl. Opt.*, **41**, 258–271.
68. Seki, J., Satomura, Y., Ooi, Y., Yanagida, T., and Seiyama, A. (2006) Velocity profiles in the rat cerebral microvessels measured by optical coherence tomography. *Clin. Hemorheol. Microcirc.*, **34**, 233–239.
69. Yang, V.X., Tang, S.J., Gordon, M.L., Qi, B., Gardiner, G., Cirocco, M., Kortan, P., Haber, G.B., Kandel, G., Vitkin, I.A., Wilson, B.C., and Marcon, N.E. (2005) Endoscopic Doppler optical coherence tomography in the human GI tract: initial experience. *Gastrointest. Endosc.*, **61**, 879–890.
70. Leahy, M.J. and Nilsson, G.E. (2010) in *Handbook of Photonics for Medical Science*, Chapter 11 (ed. V.V. Tuchin), CRC Press, Taylor & Francis Group, London, pp. 323–342.
71. Wang, R.K. (2007) Three-dimensional optical micro-angiography maps directional blood perfusion deep within microcirculation tissue beds *in vivo*. *Phys. Med. Biol.*, **52**, N531–N537.
72. An, L., Jia, Y., and Wang, R.K. (2010) in *Handbook of Photonics for Medical Science*, Chapter 15 (ed. V.V. Tuchin), CRC Press, Taylor & Francis Group, London, pp. 401–421.
73. Ku, G., Wang, X., Xie, X., Stoica, G., and Wang, L.V. (2005) Imaging of tumor angiogenesis in rat brains *in vivo* by photoacoustic tomography. *Appl. Opt.*, **44**, 770–775.
74. Li, P.C., Huang, S.W., Wei, C.W., Chiou, Y.C., Chen, C.D., and Wang, C.R. (2005) Photoacoustic flow measurements by use of laser-induced shape transitions of gold nanorods. *Opt. Lett.*, **30**, 3341–3343.

75. Zharov, V.P., Galanzha, E.I., and Tuchin, V.V. (2005) Integrated photothermal flow cytometry *in vivo*. *J. Biomed. Opt.*, **10**, 51502-1–515502-13.
76. Zharov, V.P. (2003) Far-field photothermal microscopy beyond the diffraction limit. *Opt. Lett.*, **28**, 1314–1316.
77. Galanzha, E.I., Shashkov, E.V., Tuchin, V.V., and Zharov, V.P. (2008) *In vivo* multispectral photoacoustic lymph flow cytometry with natural cell focusing and multicolor nanoparticle probes. *Cytometry*, **73**, 884–894.
78. Schmid-Schonbein, G.W. (1990) Microlymphatics and lymph flow. *Physiol. Rev.*, **70**, 987–1028.
79. Aukland, K. and Reed, R.K. (1993) Interstitial-lymphatic mechanisms in the control of extracellular fluid volume. *Physiol. Rev.*, **73**, 1–78.
80. Galanzha, E.I., Ulyanov, S.S., Tuchin, V.V., Brill, G.E., Solov'eva, A.V., and Sedykh, A.V. (2000) Comparison of lymph and blood flow in microvessels: coherent optical measurements. *Proc. SPIE*, **4163**, 94–98.
81. Barty, A., Nugent, K.A., Paganin, D., and Roberts, A. (1998) Quantitative optical phase microscopy. *Opt. Lett.*, **23**, 817–819.
82. Pawley, J.P. (1995) *Handbook of Biological Confocal Microscopy*, 2nd edn, Plenum, New York.
83. Berk, D.A., Swartz, M.A., Leu, A.J., and Jain, R.K. (1996) Transport in lymphatic capillaries. II. Microscopic velocity measurement with fluorescence photobleaching. *Am. J. Physiol.*, **39**, H330–H337.
84. Nelson, J.S., Kelly, K.M., Zhao, Y., and Chen, Z. (2001) Imaging blood flow in human port-wine stain *in situ* and in real time using optical Doppler tomography. *Arch. Dermatol.*, **137**, 741–744.
85. Castenholz, A. (1998) Functional microanatomy of initial lymphatics with consideration of the extracellular matrix. *Lymphology*, **31**, 101–118.
86. Kato, S. (2000) Organ specificity of the structural organization and fine distribution of lymphatic capillary networks: histochemical study. *Histol. Histopathol.*, **15**, 185–197.
87. Benoit, J.N., Zawieja, D.C., Goodman, A.H., and Granger, H.J. (1989) Characterization of intact mesenteric lymphatic pump and its responsiveness to acute edemagenic stress. *Am. J. Physiol.*, **257**, H2059–H2069.
88. Benoit, J.N. (1991) Relationship between lymphatic pump flow and total lymph flow in the small intestine. *Am. J. Physiol.*, **261**, H1970–H1978.
89. Sekizuka, E., Ohshio, C., and Minamitani, H. (1995) Automatic analysis of moving images for the lymphocyte velocity measurement. *Microcirc. Ann.*, **11**, 107–108.
90. Galanzha, E.I., Tuchin, V.V., Zharov, V.P., Solovieva, A.V., Stepanova, T.V., and Brill, G.E. (2003) The diagnosis of lymph microcirculation on rat mesentery *in vivo*. *Proc. SPIE*, **4965**, 325–333.
91. Galanzha, E.I., Tuchin, V.V., Solov'eva, A.V., and Zharov, V.P. (2002) Development of optical diagnostics of microlymphatics at the experimental lymphedema: comparative analysis. *J. X-ray Sci. Technol.*, **10**, 215–223.
92. Galanzha, E.I., Tuchin, V.V., Chowdhury, P., and Zharov, V.P. (2004) Monitoring of small lymphatics function under different impact on animal model by integrated optical imaging. *Proc. SPIE*, **5474**, 204–214.
93. Fedosov, I.V., Tuchin, V.V., Galanzha, E.I., Solov'eva, A.V., and Stepanova, T.V. (2002) Recording of lymph flow dynamics in microvessels using correlation properties of scattered coherent radiation. *Quantum Electron.*, **32** (11), 849–867.
94. Zharov, V., Galanzha, E., and Tuchin, V. (2004) Photothermal imaging of moving cells in lymph and blood flow *in vivo* animal model. *Proc. SPIE*, **5320**, 185–195.
95. Zharov, V., Galanzha, E., and Tuchin, V. (2005) Photothermal image flow cytometry *in vivo*. *Opt. Lett.*, **30**, 628–630.
96. Zharov, V.P., Galanzha, E.I., Shashkov, E.V., Kim, J.-W., Khlebtsov, N.G., and Tuchin, V.V. (2007) Photoacoustic flow cytometry: principle and application for real-time detection of circulating single

nanoparticles, pathogens, and contrast dyes *in vivo*. *J. Biomed. Opt.*, **12** (5), 051503-1–051503-14.

97. Galanzha, E.I., Kokoska, M.S., Shashkov, E.V., Kim, J.-W., Tuchin, V.V., and Zharov, V.P. (2009) *In vivo* fiber photoacoustic detection and photothermal purging of metastasis targeted by nanoparticles in sentinel lymph nodes at single cell level. *J. Biophoton.*, **2**, 528–539.
98. Brill, G.E., Galanzha, E.I., Ul'ianov, S.S., Tuchin, V.V., Stepanova, T.V., and Solov'eva, A.V. (2001) Functional organization of lymphatic microvessels of the rat mesentery. *Ross. Fiziol. Zh. Im. I.M. Sechenova*, **87**, 600–607.
99. Zharov, V.P., Menyaev, Y., Shashkov, E.V., Galanzha, E.I., Khlebtsov, B.N., Scheludko, A., Zimnyakov, D.A., and Tuchin, V.V. (2006) Fluctuations of probe beam in thermolens schematics as potential indicator of cell metabolism, apoptosis, necrosis and laser impact. *Proc. SPIE*, **6085**, 10–21.
100. Starr, M.C. and Frasher, W.G. Jr. (1975) *In vivo* cellular and plasma velocities in microvessels of the cat mesentery. *Microvasc. Res.*, **10**, 102–106.
101. Smith, J.B., McIntosh, G.H., and Morris, B. (1970) The traffic of cells through tissues: a study of peripheral lymph in sheep. *J. Anat.*, **107**, 87–100.
102. Ikomi, F., Hunt, J., Hanna, G., and Schmid-Schonbein, G.W. (1996) Interstitial fluid, plasma protein, colloid, and leukocyte uptake into initial lymphatics. *J. Appl. Physiol.*, **81**, 2060–2067.
103. Young, A.J. (1999) The physiology of lymphocyte migration through the single lymph node *in vivo*. *Semin. Immunol.*, **11**, 73–83.
104. Hall, J.G. and Morris, B. (1965) The origin of the cells in the efferent lymph from a single lymph node. *J. Exp. Med.*, **121**, 901–100.
105. Dixon, B.J., Greiner, S.T., Gashev, A.A., Coté, G.L., Moore, J.E., and Zawieja, D.C. (2006) Lymph flow, shear stress, and lymphocyte velocity in rat mesenteric prenodal lymphatics. *Microcirculation*, **13**, 597–610.
106. Azzali, G. (2006) On the transendothelial passage of tumor cell from extravasal matrix into the lumen of absorbing lymphatic vessel. *Microvasc. Res.*, **72**, 74–85.
107. Priezzhev, A.V., Ryaboshapka, O.M., Firsov, N.N., and Sirko, I.V. (1999) Aggregation and disaggregation of erythrocytes in whole blood: study by backscattering technique. *J. Biomed. Opt.*, **4**, 76–84.
108. Tuchin, V.V. (2007) *Tissue Optics: Light Scattering Methods and Instruments for Medical Diagnosis*, 2nd edn, vol. PM166, SPIE Press, Bellingham, WA.
109. Brown, P. (2005) Lymphatic system: unlocking the drains. *Nature*, **436**, 456–458.
110. Alitalo, K., Tammela, T., and Petrova, T.V. (2005) Lymphangiogenesis in development and human disease. *Nature*, **438**, 946–953.
111. Haig, D.M., Hopkins, J., and Miller, H.R. (1999) Local immune responses in afferent and efferent lymph. *Immunology*, **96**, 155–163.
112. Olszewski, W.L. (2003) The lymphatic system in body homeostasis: physiological conditions. *Lymphat. Res. Biol.*, **1**, 11–21.
113. Browse, N., Burnand, K.G., and Mortimer, P.S. (2003) *Diseases of the Lymphatics*, Arnold, London.
114. Galanzha, E., Tuchin, V., and Zharov, V. (2007) Optical clearing in small animals: an application for *in vivo* lymph flow cytometry. *Lasers Surg. Med.*, **19**, 10.
115. Galanzha, E.I., Tuchin, V.V., Solovieva, A.V., Stepanova, T.V., Luo, Q., and Cheng, H. (2003) Skin backreflectance and microvascular system functioning at the action of osmotic agents. *J. Phys. D: Appl. Phys.*, **36**, 1739–1746.
116. Willard-Mack, C.L. (2006) Normal structure, function, and histology of lymph nodes. *Toxicol. Pathol.*, **34**, 409–424.
117. Kim, S., Popel, A.S., Intaglietta, M., and Johnson, P.C. (2005) Aggregate formation of erythrocytes in postcapillary venules. *Am. J. Physiol. Heart. Circ. Physiol.*, **288**, H584–H590.

118. Kim, S., Popel, A.S., Intaglietta, M., and Johnson, P.C. (2006) Effect of erythrocyte aggregation at normal human levels on functional capillary density in rat spinotrapezius muscle. *Am. J. Physiol. Heart. Circ. Physiol.*, **290**, H941–H947.

119. Sipkins, D.A., Wei, X., Wu, J.W., Runnels, J.M., Cote, D., Means, T.K., Luster, A.D., Scadden, D.T., and Lin, C.P. (2005) *In vivo* imaging of specialized bone marrow endothelial microdomains for tumour engraftment. *Nature*, **435**, 969–973.

120. Skalak, R. and Branemark, P.I. (1969) Deformation of red blood cells in capillaries. *Science*, **164**, 717–719.

121. Lipowsky, H.H. (2005) Microvascular rheology and hemodynamics. *Microcirculation*, **12**, 5–15.

122. Firrell, J.C. and Lipowsky, H.H. (1989) Leukocyte margination and deformation in mesenteric venules of rat. *Am. J. Physiol.*, **256**, H1667–H1674.

123. Minamitani, H., Tsukada, K., Sekizuka, E., and Oshio, C. (2003) Optical bioimaging: from living tissue to a single molecule: imaging and functional analysis of blood flow in organic microcirculation. *J. Pharmacol. Sci.*, **93**, 227–233.

124. Schmid-Schonbein, H. (1981) Blood rheology and physiology of microcirculation. *Ric. Clin. Lab.*, **11**, 13–33.

15
Instrumentation for *In vivo* Flow Cytometry – a Sickle Cell Anemia Case Study

Stephen P. Morgan and Ian M. Stockford

15.1
Introduction

In vitro flow cytometry is a tool that has found widespread use in the characterization of cell populations. However, this involves taking blood samples, which is invasive. Removal of cells from the body can also change their properties, for example, red blood cells become oxygenated. Additionally, other characteristic features of cell types, for example, different flow rates between sickled and non-sickled red blood cells [1–3], are difficult to measure in an *in vitro* environment.

Recently, methods for *in vivo* measurements of circulating cells [4–7] have been developed. These are described in more detail in other chapters of this book and so are only mentioned briefly here. Georgakoudi and coworkers [4–6] have developed a system in which a line of laser light illuminates a vessel. The scattered light from fluorescently labeled cells is used as a method for discriminating between different cell types. This method has been used to study circulating prostate cancer cells in rats and mice [4]. It has also been used [6] to monitor the depletion kinetics of circulating green fluorescent protein-labeled breast cancer cells in the vasculature of mice, and the detection of circulating hematopoietic stem cells labeled with two antibodies. Interesting research has been carried out by Tuchin and coworkers, and an excellent review can be found in [7]. One example [7] is where normal rat lymphocytes, rat basophilic leukemia (RBL-1) cells, K562 human leukemia cells, and rat RBCs have been fluorescently labeled and monitored in the mesentery and lymphatic microvessels.

Fluorescently labeling cells is extremely useful, but there are several other methods that can be used to discriminate between cells. For example, label-free properties such as the autofluorescence of white blood cells [8] and the flow rates of different cell types [1] also have the potential for *in vivo* flow cytometry. Other interesting imaging techniques such as photothermal [9], photoacoustic [10], and multi-photon [11] techniques have also been applied.

In this chapter, we discuss the label-free monitoring of the properties of circulating blood cells and the associated instrumentation challenges. Specifically, we have been involved in the *in vivo* monitoring of circulating red blood cells in

Advanced Optical Flow Cytometry: Methods and Disease Diagnoses, First Edition. Edited by Valery V. Tuchin.

sickle cell anemia sufferers. Imaging has been performed either on the lower lip or under the tongue where the superficial tissue above the microcirculation is thinner than on other sites on the body, and blood cells can be visualized. Sickled cells have reduced oxygen-carrying capability, become polymerized, and move more slowly through the microcirculation. There is therefore potential for discriminating between sickled and non-sickled red blood cells using absorption measurements to obtain oxygen saturation, polarization to measure polymerization, or flow to indicate cell adhesion to the vascular walls.

Measurements of the properties of single blood cells in human subjects represent a considerable instrumentation challenge. Reflections from the surface and superficial tissue layers can obscure the blood cells, and illumination methods have been developed to overcome these. In human subjects, the subject is not usually anesthetized and so good design and patient compliance are necessary. Typically, imaging is performed at lower resolutions than for animal studies owing to motion artifacts.

The next section discusses the clinical need for such measurements. The illumination and detection requirements and challenges are discussed in Section 15.3. This is followed by a description of image processing methods for correction of image distortions and a Monte Carlo model of the image formation process in Sections 15.4 and 15.5, respectively. The design of the clinical device is discussed in Section 15.6 and a selection of clinical results is shown in Section 15.7. Discussions and suggestions for future research follow in the final section.

15.2 Clinical Need

Our main clinical interest has been in discriminating between populations of sickled and non-sickled red blood cells. Sickle cell anemia is caused by a point mutation in the beta globin gene [12] giving rise to the production of an abnormal hemoglobin, HbS. This structural abnormality leads to the formation of long chains of polymerized hemoglobin that produces cells of sickle shape plus other abnormal shapes of both permanently and temporarily sickled cells [13].

Rigid polymerized cells have their oxygen-carrying capability impeded, and their ability to deform when traveling through the smallest capillaries is impaired [14]. The reduction in deformability caused by polymerization can lead to vaso-occlusion in the microcirculation and consequently ischemia, which is responsible for the clinical phenotype. Sickle cell anemia is the most common genetic condition worldwide, and it is estimated that 300 000 babies are born with the condition every year [15] and that 5% of the global population ($\sim$340 million) carry the genetic trait [15].

It would be advantageous if a method could be developed to distinguish between sickled and healthy red blood cells. Clinicians would be able to gain a better understanding of the condition and the effects of treatment could be routinely monitored. For characterizing the cell population in routine clinical use, there should ideally be an intrinsic marker of a sickled cell. Three possibilities could be considered: (i) Flow rate – sickled cells are stiffer and flow through the microcirculation is slower, as demonstrated by Lipowsky *et al.* [1]. (ii) Polarization – the polymerization of HbS when deoxygenated leads to cells becoming linearly dichroic [13]. Previous polarization-sensitive measurements of isolated polymerized sickled cells *in vitro* have demonstrated a difference of ~3% in measurements taken in two orthogonal polarization states [13]. (iii) Oxygenation – the reduced oxygen-carrying capability of sickled red blood cells could be monitored by differences in the absorption spectra between sickled and normal cells. In our research, we investigated monitoring the linear dichroism of sickled red blood cells by developing a polarization-sensitive capillaroscope that can be used to determine the presence of hemoglobin polymerization in isolated red blood cells. The flow of cells through capillaries smaller than, or comparable to, the diameter of the observed erythrocytes allows observation of individual cells. Performing measurements on isolated single cells, as opposed to within wider capillaries, means that multiple scattering by red blood cells can be largely neglected and each cell can be treated as a "cuvette" for optical analysis.

Measurement of the oxygenation properties of single cells and vessels would also be of interest in research in other clinical areas. Examples include sepsis, where measuring the ability of red blood cells to supply oxygen to various tissues could help understand why poorly functioning microcirculation leads to hypoxic tissue, a cascade of pathogenic mechanisms, and, in turn, to major organ failure. In diabetes, mitochondrial dysfunction leads to inefficient transfer of nutrients to tissue [16]. Observation of this dysfunction via the inability to extract oxygen from cells could help assess potential treatments. In the development of anti-inflammatory drugs, characterization of the performance of vasodilators that attempt to improve the supply of oxygen could be studied using oxygenation-based approaches.

15.3 Instrumentation

The main instrumentation challenges are associated with appropriate illumination and detection methods. Section 15.3.1 describes three different illumination techniques used for overcoming surface reflections and light scattering from superficial tissues that obscure images. The challenges of detecting small intensity changes on large intensity backgrounds are discussed in Section 15.3.2.

15.3.1 Illumination Methods

For label-free monitoring of cells, imaging is often performed at the same wavelength as the illumination (e.g., for absorbance, flow, or polarization measurements). Reflections from the tissue surface (due to the refractive index mismatch

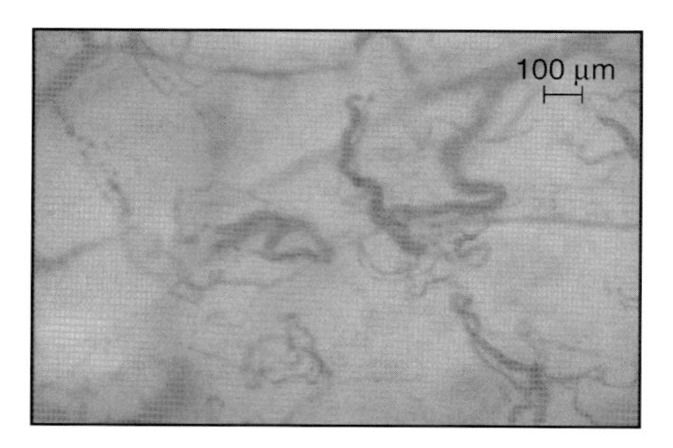

Figure 15.1 A typical image of moving blood cells under the tongue using a Microvision SSDF system.

between air and tissue) and backscattering from cells close to the surface that overlay the blood vessels can obscure the cells and prevent accurate characterization of the cell population.

The three most prominent solutions to overcoming surface reflections are orthogonal polarization spectral (OPS) imaging [17], dark field epi-illumination (DFEI) [18], and sidestream dark field (SSDF) imaging [19]. Commercial examples include the CytoScan (based on OPS) and the MicroScan (based on SSDF), which have enabled monitoring the microvasculature on the floor of the mouth (e.g., Figure 15.1), lip, eye, brain, and muscle in a clinical setting. To date, analysis of recorded images includes assessment of morphological parameters such as capillary density [20], capillary shape/tortuosity [21], various vessel dimensions [20, 22], and red blood cell flow rate [23]. The illumination techniques discussed here are helpful in overcoming relatively weak scattering and are therefore usually applied to sites on the body such as on the nail fold or under the tongue where the scattering between the blood cells and the tissue surface is relatively weak. For other sites on the body, such as the skin, techniques such as optical clearing [24] can be used, where chemical agents are applied to match refractive indices within the tissue to help reduce the effect of scattering.

It should be noted that surface reflections are not particularly a problem for fluorescence measurements, as imaging is performed at a wavelength different to that of the illumination. In this case, surface reflections can be removed by a color filter aligned to the spectral band of the fluorophore, which does not overlap that of the illumination. However, there may be applications where autofluorescence obscures the fluorophore of interest and spatial-filtering-based approaches (DFEI, SSDF) could be applied.

15.3.1.1 Orthogonal Polarization Spectral Imaging (OPS)

OPS was originally described by Groner [17] and eventually led to the development of a commercial device – the Cytoscan. Figure 15.2a demonstrates the basic principle of the technique. Linearly polarized light illuminates the tissue under investigation, and detection is performed in the orthogonal polarization state. In

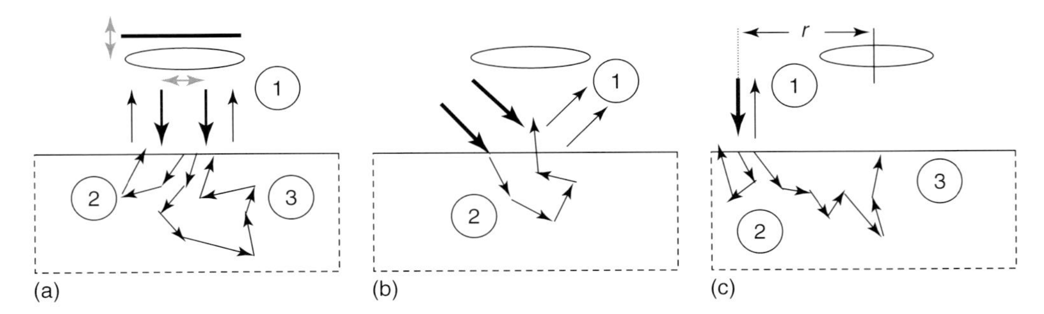

Figure 15.2 Illumination methods for overcoming surface reflections: (a) orthogonal polarization spectral imaging, (b) dark field epi-illumination, and (c) sidestream dark field illumination.

conventional imaging, blood vessels are obscured by reflections at, or just beneath, the surface. By detecting in the orthogonal polarization state, these contributions are significantly reduced as these single backscattered components maintain their original polarization state. Orthogonal polarization detection also significantly reduces reflections from optical components within the system, thus offering an additional benefit to image quality.

A variation on OPS called *rotating orthogonal polarization imaging* has recently been described [25]. In this method, the polarization axes of the illumination and the detection are maintained orthogonal to one another and are synchronously rotated (Figure 15.3). This provides sensitivity to the polarization properties of the underlying tissue free from surface reflections.

15.3.1.2 Dark Field Epi-illumination (DFEI)

Conventional dark field imaging has been extensively applied in microscopy to provide sensitivity to scattered light and remove unscattered zero-order contributions [18]. In microscopy, this generally provides sensitivity to light that has undergone a small number of scattering events from thin samples with unscattered or directly reflected light falling outside the detection aperture. During *in vivo* imaging of tissue, light can be scattered many times within the volume of the tissue and so multiply scattered light can still be detected when DFEI is applied. This is a useful approach for *in vivo* imaging since, as described previously, it is this heavily scattered light that is used in image formation. Figure 15.2b shows the incident light (bold arrows) illuminating the sample at an angle from outside of the detection aperture. Surface reflections are directed away from the detection aperture and are not detected. Light that enters the tissue is scattered and that which falls within the numerical aperture (NA) of the imaging optics will form the observed image. The main rejection mechanism utilized here is angular filtering of surface contributions through off-axis detection. An example of an imaging system applying this approach can be found in Ref. [18].

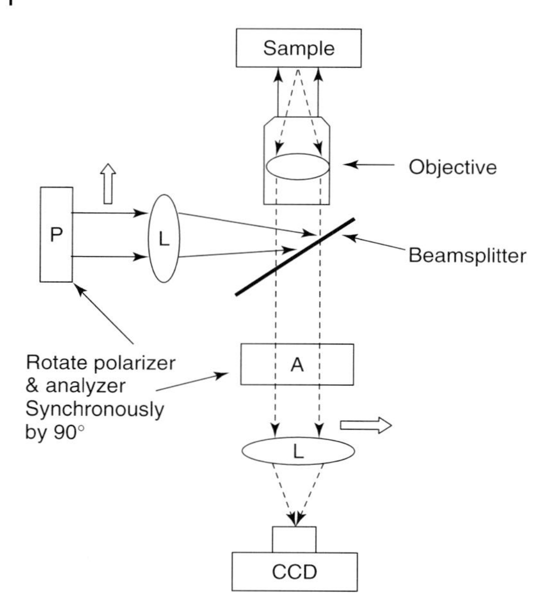

Figure 15.3 Illumination and detection for rotating orthogonal polarization imaging (P – polarizer, L – lens, and A – analyzer).

15.3.1.3 Sidestream Dark Field (SSDF) Illumination

An alternative approach to overcoming surface reflections has been described by Ince and coworkers [19]. In this configuration, light illuminates the tissue in an annulus (Figure 15.2c) and the light backscattered from the surface or superficial tissue does not reach the detector. A proportion of the light propagates beneath the red blood cells and "back-illuminates" them, effectively turning the system into a transmission microscope. A small proportion of the light propagates directly between the source and the detector, but this is significantly reduced compared to that in a conventional system. This technique has led to the development of a commercial device by Microvision Medical.

In both OPS and SSDF, illumination is usually performed using green light as this provides a compromise between achieving sufficient detectable signals and maximizing image contrast. Blue light is absorbed heavily by the blood but is also significantly attenuated by the surrounding tissue. Red or near infrared light would provide more detected light, but the contrast from red blood cells is reduced. For example, the molar extinction coefficients of oxy- and deoxy-hemoglobin are 43 876 and 40 584 cm^{-1} M^{-1}, respectively, at a wavelength of 532 nm and 561 cm and 4931 cm^{-1} M^{-1} at a wavelength of 632 nm [26]. Line illumination [4–6] has been demonstrated to be an effective method of counting and discriminating between different cell types that have been fluorescently labeled, and this approach also has potential in a label-free system.

SSDF illumination was used in our sickle cell anemia imaging system as it allowed two polarization states to be imaged simultaneously while reducing the effects of surface reflections. Rotating orthogonal polarization imaging would have involved taking two or more images sequentially, which results in problems registering consecutive images for comparing the polarization states of individual cells. SSDF illumination also has advantages in terms of image resolution and contrast, as discussed in Section 15.5.

15.3.2 Detection

The illumination approaches described in the previous section are usually applied at fairly low magnification with 5× or 10× microscope objectives. The wider field of view (typically 1.5 mm) and depth of focus (typically 12 μm) allow motion artifacts associated with imaging human subjects *in vivo* to be reduced to an acceptable level. Higher magnifications can potentially be obtained using high-magnification optics coupled with high frame rate commercial complementary metal oxide semiconductor (CMOS) cameras discussed in more detail elsewhere [7]. However, motion artifacts will remain a challenging problem in human subjects.

To analyze the cell population, the detected light needs to be split into different components to examine properties that can be used to discriminate between them. For example, in the line illumination system described by Novak [5], fluorescent light emitted by cells is analyzed by different color filters. Photomultiplier tubes are used to achieve the sensitivity required to detect low light levels.

In the system that our group has developed, different polarization states of the scattered light are analyzed by a polarization-sensitive beam-splitter with the aim being to characterize populations of blood cells in sickle cell anemia sufferers. A conventional SSDF illumination system is shown in Figure 15.4a and the polarization-sensitive system is shown in Figure 15.4b. As discussed previously, SSDF is used to overcome the effects of surface reflections and back-illuminate red blood cells. Light is then imaged through a polarizing beam-splitter, which splits the light into two orthogonal polarization images that can be formed on the same camera. An example is shown in Figure 15.5 where two images (horizontally and vertically polarized) of the lower lip of a sickle cell anemia sufferer are formed on the same camera. Some small distortions are introduced by the different paths that the light follows through the beam-splitter and the imaging lens. The images need to be aligned before difference images are calculated. An image processing method for correction of distortion and subsequent image alignment is described in the next section.

For detection of polarization changes in an SSDF system, it is necessary to detect a small change in light intensity on a large intensity background. For *in vitro* studies of sickled cells, a difference of 3% between the two polarization images has been observed [13]. One of the most important factors governing detection of small signals on large backgrounds is the well capacity of the camera. When the background intensity level of the light is high (i.e., close to the full well capacity of the pixel),

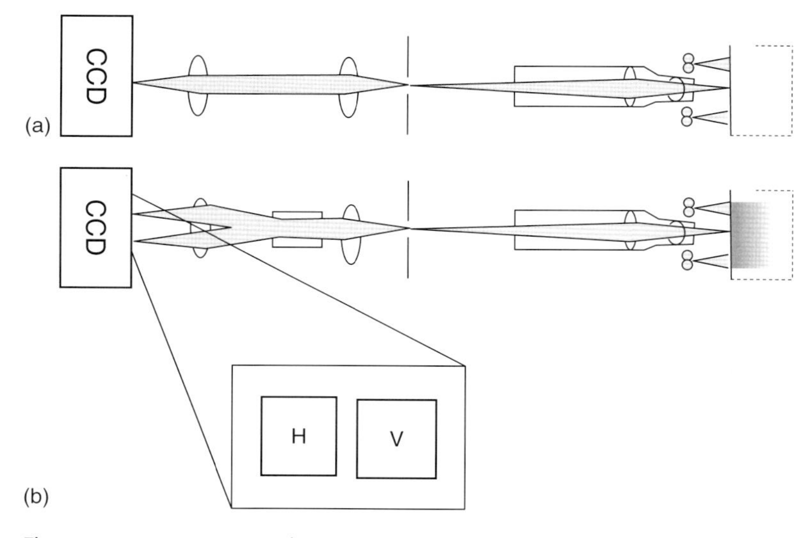

Figure 15.4 (a) Conventional SSDF system and (b) polarization-sensitive SSDF system.

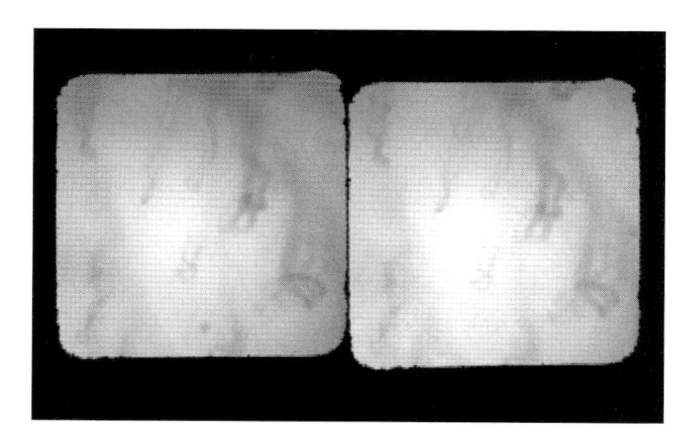

Figure 15.5 Image formed at the square aperture discriminated into orthogonal polarization states for detection.

the dominant source of noise is shot noise, which is calculated by taking the square root of the number of electrons within a well. As a practical example, consider a commercial camera such as the Hamamatsu Orca ERII that has a full well capacity of 18 000 electrons, a dark current of 0.1 electrons/pixel/s, and a readout noise of 6 electrons. The analogue to digital converter on the camera has 12 bit resolution, which provides 4096 gray levels. At full well capacity, the shot noise is 134 electrons, which dominates both dark current and readout noise. One might expect the intensity resolution of the camera to depend on the number of bits in the camera; however, close to full well capacity, this is not the case. For a 12 bit analogue to digital converter with a full well capacity, the number of electrons per gray level is $18\,000/4096 = 4.4$. However, this cannot be achieved in the presence of a large background owing to the presence of shot noise. In the case of measuring small difference signals, the resolution of the analogue to digital converter suggests that the smallest observable difference signal is 1 gray level (4.4/18 000) = 0.024% of the background. In the shot noise limit, this is, in reality, 134/18 000 = 0.74% of the background. However, if the 3% differences in polarization states for sickled red blood cells that were observed *in vitro* [13] were present *in vivo*, then they should be detectable within the sensitivity limits of the camera.

The system could be modified relatively easily to become a two wavelength system. This would open up possibilities to measure the absorbance properties of individual blood cells at both wavelengths, which could be related to the oxygen saturation of cells. As the oxygen saturation (SO_2) of hemoglobin changes, so does its absorption spectrum. Therefore, in the simplest case, taking two measurements of intensity transmitted through hemoglobin at different wavelengths can be used to provide a measure of the level of oxygenation saturation. Application of this approach to determine the oxygenation state of individual *in vivo* red blood cells traveling in the microcirculation is highly challenging for a number of reasons. The primary barrier is the small variation in intensity caused by a change in oxygenation state.

Oxygen saturation imaging of cells is another potentially important marker when discriminating between populations of sickled and non-sickled red blood cells because of the reduced oxygen-carrying capability of sickled cells. Simply replacing the polarizing beam-splitter shown in Figure 15.4b with a dichroic beam-splitter and illuminating with two wavelengths will allow differences in the absorption of cells to be measured. Figure 15.6 provides an impression of what could be observed, provided the necessary detection sensitivity is achieved.

Similar calculations to those provided for polarization difference measurements can be performed for two wavelength measurements of oxygen saturation. To a first approximation, the reduction in intensity caused by absorption within a blood cell can be estimated by applying the Lambert–Beer law to a typical

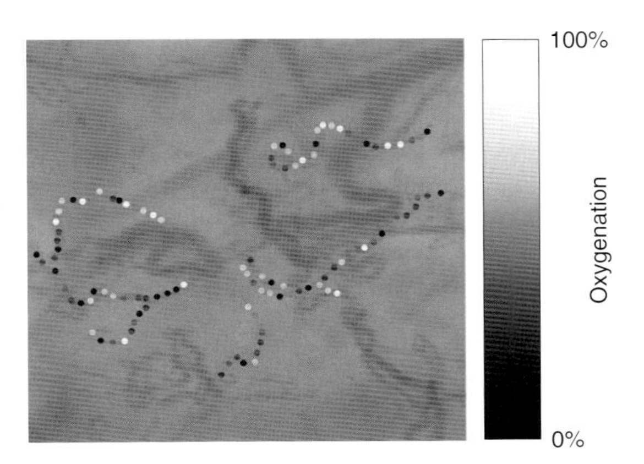

Figure 15.6 Mock-up image of measurement of oxygen saturation in isolated single cells.

optical path length through a red blood cell of 10 μm. Operating in the green region of the spectrum, required for successful formation of images in reflection capillaroscopy, the reduction in intensity due to the presence of a fully deoxygenated single red blood cell is ~5% of the transmitted intensity. For a fully oxygen-saturated red blood cell, this value is ~2.5%. As described in the calculations for polarization difference imaging, the smallest observable difference in the shot noise limit is ~0.7%. Measurement of the oxygen saturation of single cells in the presence of a large DC light background is therefore a considerable challenge. A custom-made CMOS sensor array provides a potential solution to this problem.

15.3.2.1 **Custom-Made CMOS Sensors**

Custom-made CMOS sensors [27] are arrays of photodetectors in which *on-chip* processing is incorporated. They have advantages over conventional cameras in that they can be tailored to the signals of interest. This property means that they offer an interesting new detection method for *in vivo* flow cytometry.

Our group has applied this approach to laser Doppler blood flow imaging of tissue [28, 29]. In this technique, laser light illuminates tissue and a Doppler shift is imparted to the light by moving red blood cells. This Doppler shifted light interferes with light that has not been Doppler shifted (i.e., it has been scattered only by static tissue) to produce a small AC signal on a large DC light background with frequency components up to ~20 KHz. Blood perfusion is calculated through the integral of the frequency weighted power spectrum of the detected light [30]. The high-frequency components associated with laser Doppler blood flowmetry are difficult to image using conventional cameras and so usually an image is built up by scanning [31]. Knowing the signal of interest means that the pixel size, current to voltage conversion gain, and number of digitization bits can be designed to best match those of typical laser Doppler signals. Appropriate anti-aliasing filters can also be added at the pixel level. An important advantage of custom-made sensors is that on-chip processing allows the data bottleneck that exists between the photodetector array and processing electronics to be overcome. This is because the *processed* data can be read out from the image sensor to a PC or display at a low data rate since the on-chip processing has reduced the bandwidth. This provides the potential for a large number of parallel processing channels to be implemented, each being sensitive to high-frequency fluctuations but with a low data readout rate from the sensor. An example of a 64 × 64 pixel CMOS sensor with the on-chip processing required to extract blood flow is shown in Figure 15.7. The raw Doppler signal at the pixel is sampled at 40 KHz. As it has a low modulation depth (typically 1–10%), it is passed through an amplifier that has a gain of 40 for its AC component and 1 for its DC component. This means that a lower number of bits are required for the on-chip analogue to digital converter. The high-frequency flow information is processed by on-chip signal processing to provide a lower bandwidth flow image. Figure 15.8 shows the images obtained by illuminating a finger with laser beam expanded to 2.5 cm by 2.5 cm. The light scattered from the tissue is then imaged onto the custom-made CMOS sensor. Differences in blood

Figure 15.7 A 64 × 64 pixel CMOS sensor for laser Doppler blood flow imaging.

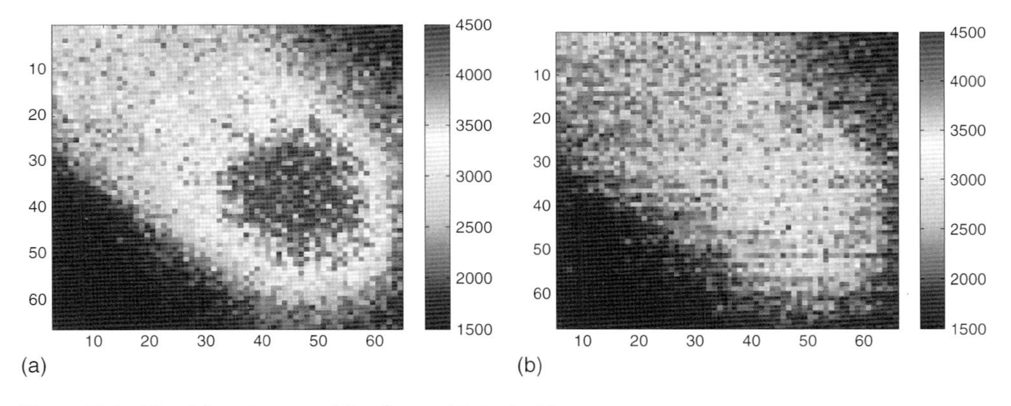

Figure 15.8 Blood flow images of the finger obtained with a custom-made CMOS sensor (a) under normal conditions and (b) under occlusion.

flow can be clearly observed between normal blood flow and the case when the blood flow is occluded.

Similar laser Doppler measurements could be applied to flowing cells in *in vivo* flow cytometry when it is of interest to discriminate between cells based on their flow rate. Alternatively, in the case of the differential measurements required to extract properties such as linear dichroism or oxygen saturation of blood cells, as discussed, the difference signal is relatively small compared to the overall background light. If either the polarization state or color of the illumination was rapidly modulated, this would produce a small AC signal (i.e., the difference signal) on a large DC light background at the detector. The amplitude of this AC signal could be imaged using an array of pixels incorporating lock-in detection [27, 32].

15.4
Image Processing

To achieve accurate measurement of dichroism (or a differential color measurement in spectroscopy), the two polarization images must be precisely overlaid. Ideally, the optical system would provide images that have identical dimensions that are free from distortions, allowing perfect registration of all pixels. Relative distortions between the images are inevitable, as the image beams travel along different optical paths and experience different aberrations. To align these images, a post-capture image correction is applied. The technique is based on the approach applied in Ref. [33], although other approaches such as those used in image de-warping [34] may also be appropriate.

The first stage of the algorithm coarsely locates and overlays the two images. An intensity threshold is applied to the image to extract the boundaries formed by the square aperture (Figure 15.5) and the two halves of the image are cross-correlated. Each of the two images (Figure 15.9a,b) is then divided into subimages (Figure 15.9c,d), which are then cross-correlated with the corresponding subarray in the other image, for example, a 25×25 pixel subarray in a vertically polarized image is cross-correlated with the corresponding subarray in a horizontally polarized subarray. The location of the peak in the resulting cross-correlation function allows any relative shift between the subimages to be determined. For more accurate alignment, it is useful to be able to obtain non-integer (subpixel) pixel shifts. This is achieved by fitting a seventh order polynomial to the cross-correlation function and locating the peak in the polynomial function. There is a trade-off in performance associated with the size of the selected subarray. A smaller subimage provides better resolution on the image shift matrix; however, the sharpness of the cross-correlation function (and therefore the potential accuracy of the peak location) is reduced because of a reduction in common pixels between the subimages. Acceptable performance for this application was found to be provided by a 25×25 pixel subimage. The result of this peak location is to indicate the shift of that whole subarray relative to its equivalent in the orthogonal image. Instead of applying the same shift to each pixel in the subarray, the shift amount is interpolated between subarrays. Once the required x and y shift has been determined for each pixel, this value can be re-binned, based on its individual shift values. This provides a pair of images that can be accurately overlaid. Subsequent processing such as segmentation and application of morphological operators [35–37] can allow identification of individual blood cells (Figure 15.9e) and taking the normalized difference image can help obtain linear dichroism (Figure 15.9f).

The success of this technique relies on the accuracy of the cross-correlation and therefore is most effective on spatially irregular images that contain high

Figure 15.9 (a) Original copolar image, (b) original cross-polar image; (c) and (d) show (a) and (b) split into subimages for alignment, (e) cells isolated through segmentation, and (f) linear dichroism measurements of isolated cells.

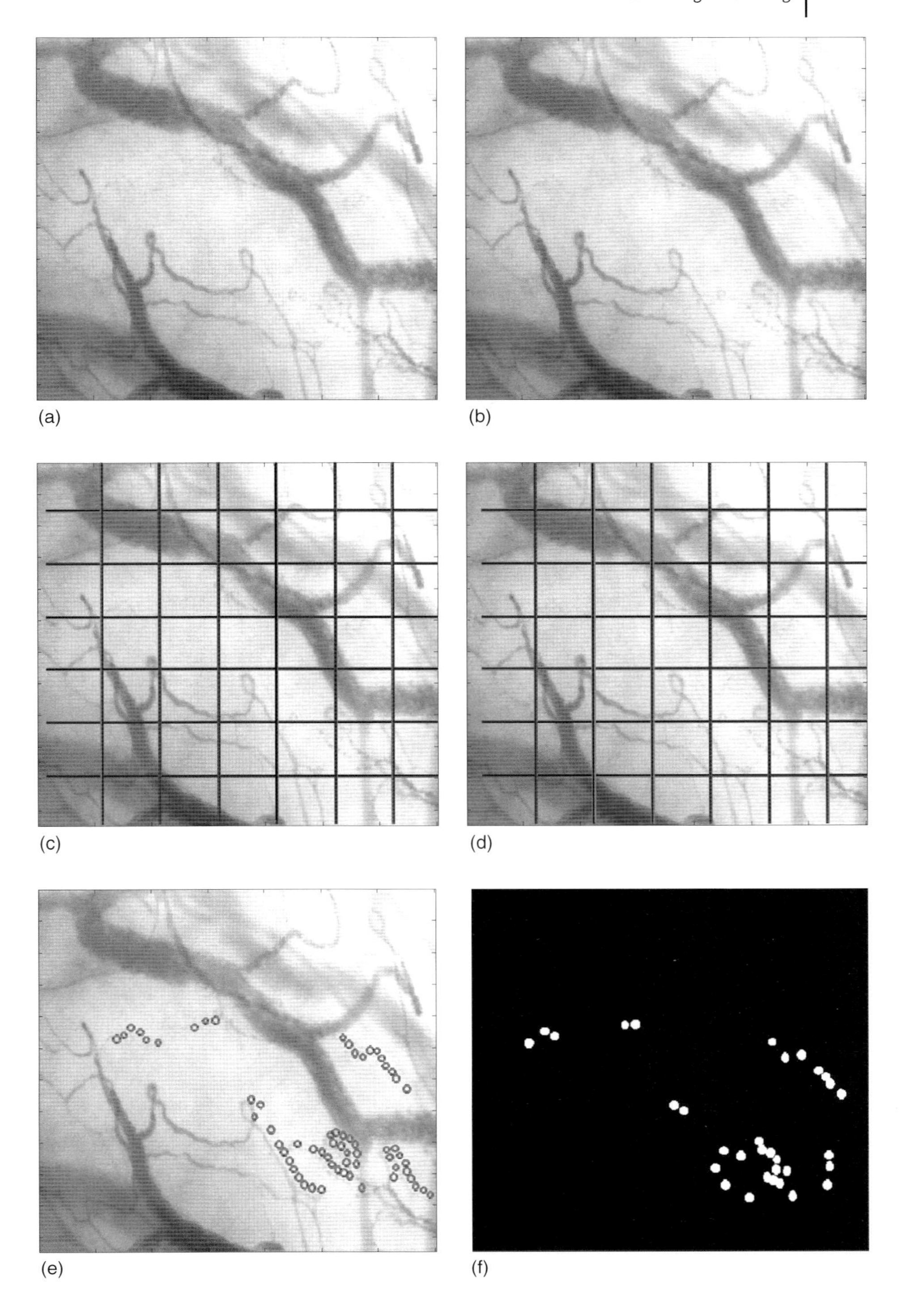
(a)
(b)
(c)
(d)
(e)
(f)

spatial frequencies. For this reason, in practice, the image distortion correction is performed using a high spatial frequency calibration image (in this case, a speckle pattern generated by a separate laser source illuminating a ground glass diffuser). An example of the calibration speckle pattern is shown in Figure 15.10. These calibration images are recorded before the examination of each clinical subject and stored along with the data for post-processing.

Maps of the pixel shifts required to achieve alignment of the images shown in Figure 15.10a are shown in Figure 15.10b,c. Superimposed on both of these plots is also the combined x–y arrow plot showing the required pixel shift at selected intervals.

(a)

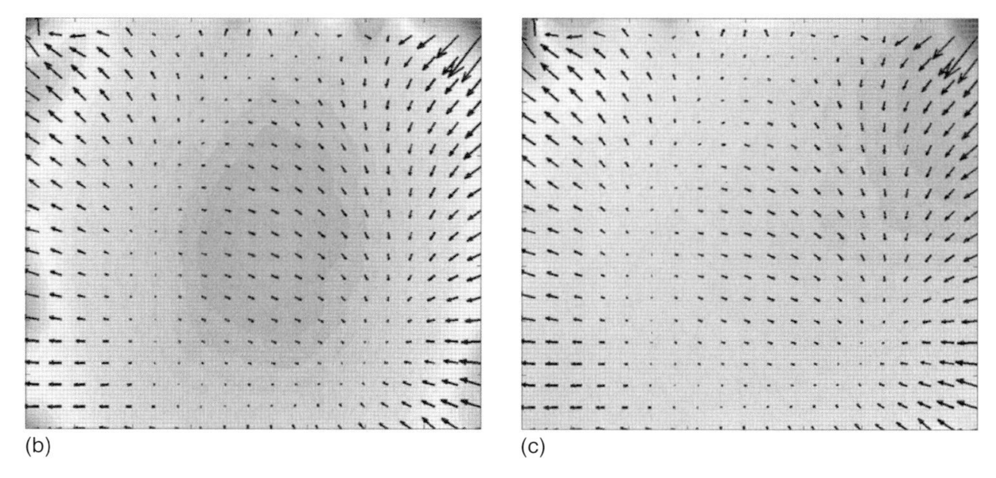

(b) (c)

Figure 15.10 (a) A typical speckle pattern generated for subpixel realignment and the resulting shifts required in the x and y directions represented in (b) and (c) respectively.

15.5 Modeling

A Monte Carlo simulation of polarized light propagation through tissue has been developed, which incorporates polarization information at each scattering event [38]. This type of simulation is frequently used to simulate light propagation, as it is flexible when specifying the geometry of the tissue. The diffusion approximation [38] provides faster analytical solutions but is less flexible and does not allow polarization information to be included. The Monte Carlo simulation applied in this study was developed in-house but has been validated against an independent code from other research groups [39]. Polarized light Monte Carlo simulations are now freely available (e.g., *http://omlc.ogi.edu/software/mc/*). The model is useful in this context as it allows investigation of the most appropriate illumination technique for *in vivo* flow cytometry.

15.5.1 Model Description

The model simulates a beam of light illuminating a scattering medium composed of Mie scattering particles. Photons (or light packets) are individually tracked through the medium and at each photon–particle collision, the direction and polarization state are modified via adjustment of the directional cosines and Stokes' parameters, respectively. For the simulations described, the sample is a single layer semi-infinite medium with a scatterer diameter, $d = 0.96$ μm, scatterer–medium refractive index mismatch, $\Delta n = 1.19$, and wavelength $\lambda = 550$ nm. These properties give rise to a scattering anisotropy, g, of 0.9, which is close to that of tissue. Absorption is added post-simulation, using the Lambert–Beer law based on the recorded path length of each individual photon. The absorption coefficient $\mu_a = 0.033\ \text{mm}^{-1}$ is approximately that of the dermal tissue [40]. This absorption is relevant, as the techniques applied here are dependent on photons that have traveled over a long path length which, by their nature, will be more heavily affected by absorption.

The model has been used to compare the imaging performance of the SSDF, DFEI, and OPS illumination methods described in Section 15.3. To evaluate the imaging performance, the model has been extended from the conventional polarized light Monte Carlo simulations to include (i) full field illumination, (ii) an absorbing object that is infinitesimally thin in the z direction embedded within the medium, and (iii) an imaging lens that allows different imaging planes within the scattering medium to be observed.

Full field illumination could be achieved by performing a number of single-point Monte Carlo simulations across the surface of the sample that contains an object at a particular location. However, this approach would require simulation of an extremely large number of photons to cover wide-field illumination with closely spaced discrete points of illumination. In addition, the simulation would need to be repeated for each geometry, for example, different positions of objects embedded in the medium, which would be prohibitively time consuming.

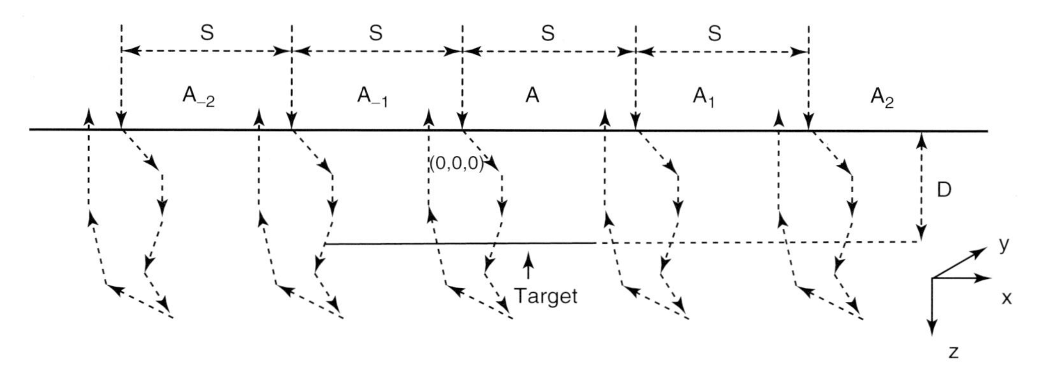

Figure 15.11 The simulated image is formed by indexing the photon paths from a single point "base" simulation to the required illumination points and applying post-processing to determine interactions with a submerged object.

This inefficient approach can be avoided through repeated use of a single simulation applied at a single point that records information about *possible* interactions with an embedded target as it propagates (Figure 15.11). If, within the propagation path of a photon, the photon passes through the depth ($z = D$) where the target is located, then the entrance and exit positions of the photon through the target are recorded. This information can then be decoded post-simulation to take into account a range of illumination and object positions.

For example, if a photon interacts with a totally absorbing target, then it is discarded and does not contribute to the final image. If a photon does not interact with the target, then it has the potential to contribute to the final image. Full field illumination (or any type of structured illumination) can be simulated by simply changing the position of the illumination point relative to the target and summing the overall contribution of all illumination points to the image plane. Different object positions and depths can also be simulated using a similar method. Partially absorbing objects can also be straightforwardly modeled by attenuating the light by a fixed amount each time the photon interacts with the target. We have also presented [25] a model that includes simulation of a polarizing target that can be used as a simple representation of materials such as a collagen.

An image of the *surface* is formed by assuming that a detector bin at the surface of the medium has a one-to-one mapping with the image plane (Figure 15.12a). The action of the imaging lens is approximated as simply filtering the angle of the emerging light, with photons that fall outside of the NA of the lens not being detected. For imaging of planes *within the sample*, light is projected back onto the plane being imaged, based on the angle at which it emerges from the surface (Figure 15.12b). The projection ignores any scattering that may have occurred between the sample's surface and the plane being imaged, as this more accurately

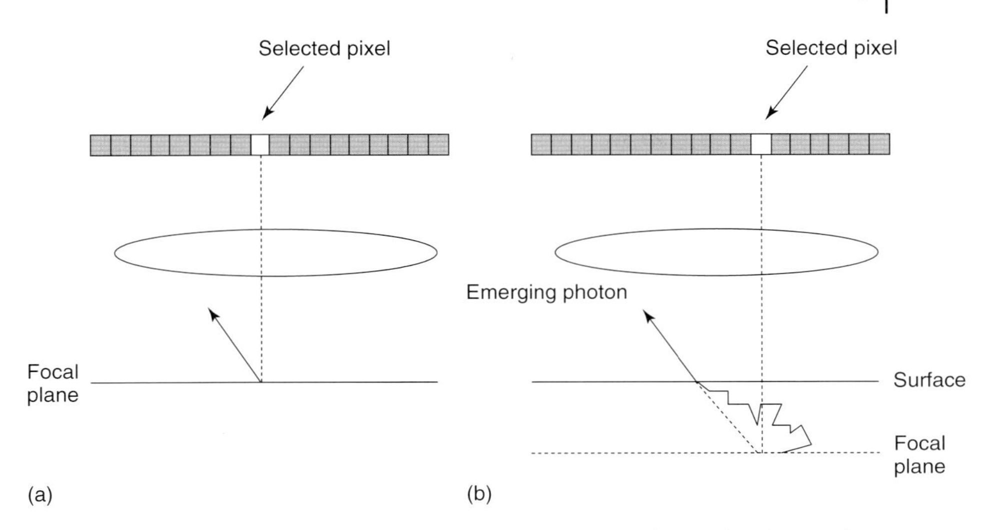

Figure 15.12 Image formation (a) at the top surface and (b) at a plane within the sample.

models the true imaging process. The back-projected image is now mapped onto the image plane in the same way as that described for imaging the surface.

Another feature incorporated in the model is the ability to illuminate the target with a range of angles rather than normal illumination to represent more realistic illumination schemes. The distribution of angles used for illumination is varied, depending on the illumination method.

To simplify the simulation, the position of the target varies in only two dimensions (z and x) and the illumination varies in x only. For the OPS and DFEI arrangements, the area being imaged is directly illuminated. Therefore, in both of these arrangements, the illumination is applied across a 2 mm line centered on the optical axis. In the case of OPS, the illumination is delivered on-axis via a microscope objective and therefore the angles obtained within the illumination are defined by this component's NA. Typically used in capillaroscopy applications are 10× microscope objectives that have NA = 0.25. Therefore, the standard deviation of the Gaussian distribution and maximum angle allowable in the distribution from which the illumination angle is drawn is 14.5°. Figure 15.13a shows a typical distribution of angles defined by selecting 5×10^6 photons from the Gaussian probability density function. DFEI typically illuminates using light-emitting diodes (LEDs). Therefore, using the illumination described in Ref. [18], the mean and standard deviation of the Gaussian distribution from which the illumination angles are chosen are 45 and 10°, respectively (Figure 15.13b). For SSDF illumination, the light is also delivered using LEDs. However, these are typically small surface mount devices that exhibit lower directionality than standard LEDs and, therefore, the distribution from which the angles are selected has a standard deviation of 60° centered on the optical axis as shown in Figure 15.13c. In these simulations, the SSDF illumination is provided at a center position of 4 mm from the optical axis spread over a 2 mm line.

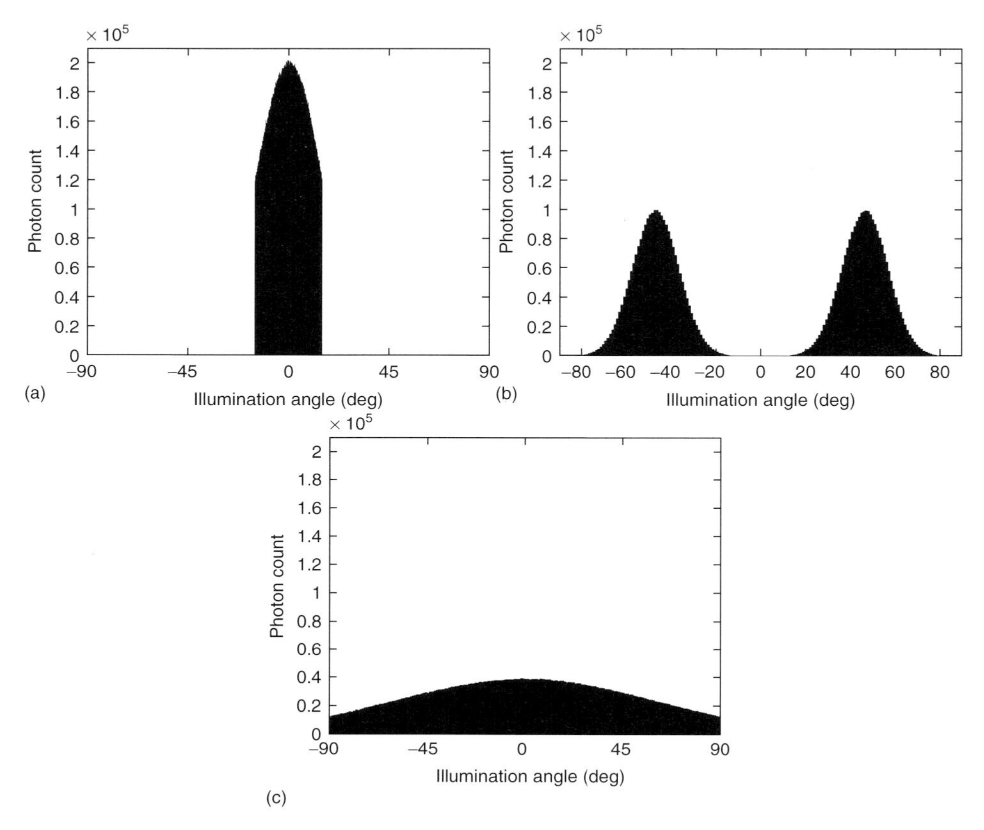

Figure 15.13 Illumination angle distributions for (a) OPS, (b) DFEI, and (c) SSDF. All figures are obtained by drawing 5×10^6 photons from the illumination angle probability density function.

15.5.2
Comparison of Illumination Techniques (DFEI, OPS, and SSDF)

The motivation behind the different illumination techniques is to configure the illumination geometry so that reflections at, or close to, the surface are rejected. If the majority of the light "back-illuminates" the target, then the image quality will be better. This section uses two parameters to briefly compare image quality using the three different illumination approaches: (i) the proportion of detected photons that back-illuminate the target and (ii) resolution of a line scan of an edge embedded with the medium.

15.5.2.1 Proportion of "Useful" Photons

As discussed in Section 15.3, the mechanism of image formation is the back-illumination of the microvasculature by heavily scattered photons. The illumination

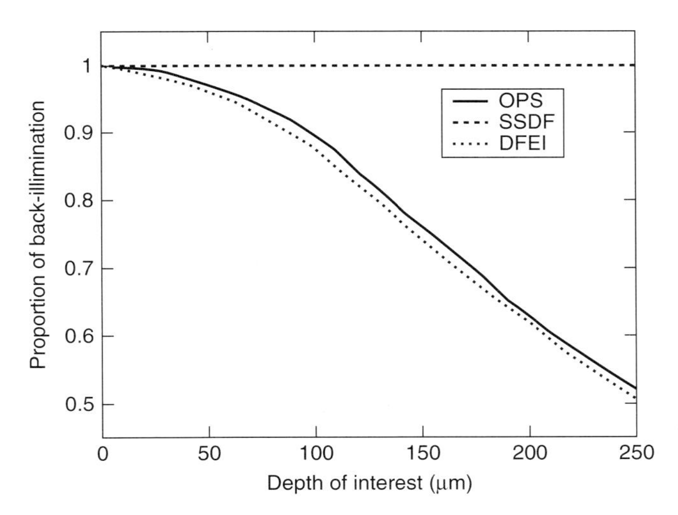

Figure 15.14 Proportion of light back-illuminating the depth of interest using the three different illumination techniques.

methods described previously (Figure 15.2) aim to reject light that is either reflected from the surface or is superficially scattered. None of the techniques are perfect in their exclusion of unwanted contributions. A figure of merit for comparing these approaches is to record the proportion of detected photons that back-illuminate the target plane. Reflections directly from the surface are currently not included in the model and therefore the unwanted component of light detected will be solely due to superficially penetrating scattered photons. Figure 15.14 shows the proportion of back-illuminating light for different depths of interest, D, using the three different illumination techniques. This figure is obtained from Monte Carlo simulations with no target present.

As the depth of interest increases the proportion of light that back-illuminates the target plane is reduced. For all the illumination techniques, at $D = 0$, all detected photons have back-illuminated the target plane. It can be seen that as D increases, the performance of the OPS and DFEI arrangements deteriorates faster than the SSDF geometry. The performance of OPS is slightly better than that of DFEI as weakly scattered light that has scattered to depths less than D will have maintained its polarization state to a certain degree. This will cause it to be rejected on detection in an OPS arrangement and will therefore not reduce the proportion of back-illuminating photons.

This figure of merit is helpful but provides no indication of the amount of light that is detected using the different techniques. In general, extracting "useful" light inevitably results in a reduction of the amount of light detected. For example, for the same number of photons simulated using the Monte Carlo model, OPS detects 49.5% and SSDF detects 1% of the amount of light detected by DFEI. The simulations in this section indicate that, provided sufficient backscattered photons

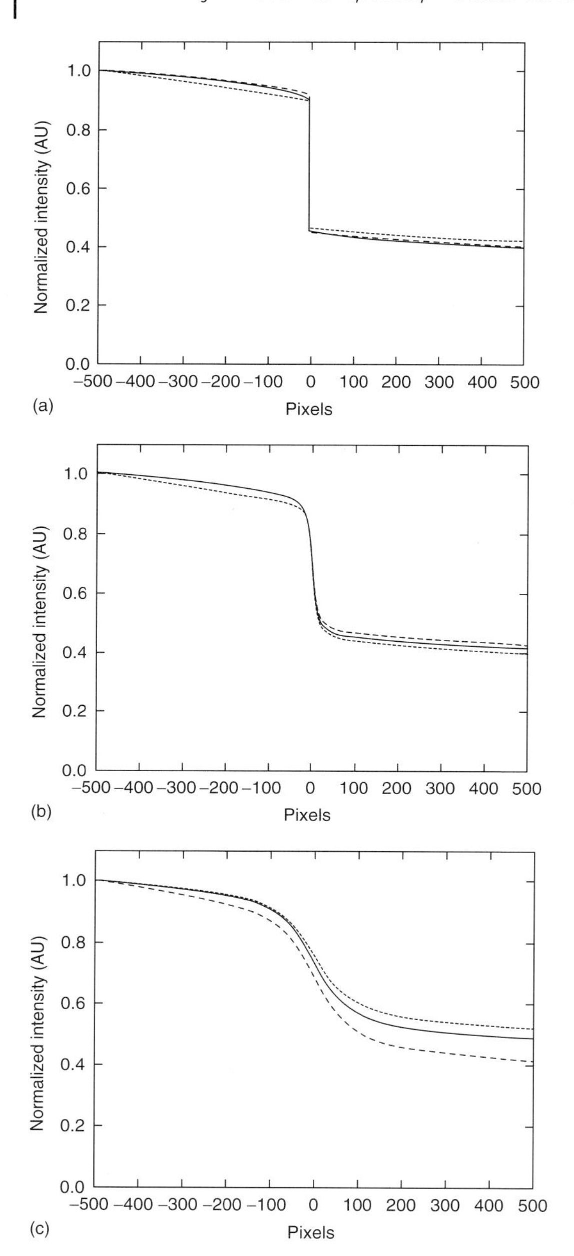

Figure 15.15 Edge response function for three different illumination techniques: (i) DFEI (dotted), (ii) OPS (solid), and (iii) SSDF (dashed) when an edge that has a transmittance of 0.5 is positioned beneath the surface of a scattering medium at depths of (a) 0.01 mfps, (b) 1 mfps, and (c) 3 mfps.

can be detected to form an image, SSDF should provide the best image quality as superficially scattered light that obscures blood cells is most effectively rejected.

15.5.2.2 Imaging Performance

In the previous section, the rejection of superficially penetrating light that does not contribute to the image of the microcirculation was considered. This rejection will clearly impact on the properties of the image formed and the ability to perform reliable quantitative measurements on the desired structures. For the simulations in this section, a partially absorbing edge (transmittance = 0.5) is embedded at different depths (D) within the medium and the response investigated. Figure 15.15 shows line scans detected using the three different illumination techniques for target depths of $D = 0.01$, 1, and 3 mfps (scattering mean free paths, where 1 mfp $= 1/\mu_s$). For ease of comparison, the scans are normalized to the intensity at the first position. As the depth of the absorbing target increases, the gradient of the edge response is reduced because of the presence of scattering between the target and the sample's surface, demonstrating that worse imaging resolution will be achieved at these depths.

It can be seen that, for the most superficially submerged absorber (Figure 15.15a), the edge response is slightly sharper when using SSDF imaging than for OPS. Both SSDF and OPS provide a much sharper response than DFEI. As the edge is increasingly submerged (Figure 15.15b,c), it can be seen that the edge response using the three different techniques becomes more similar in gradient and the gradient of the transition is much lower because of the increased scattering encountered between the absorber and the surface. The reason for the improved resolution in the SSDF arrangement is that the vast majority of the detected photons back-illuminate the absorber as described previously (Figure 15.14).

15.6 Device Design – Sickle Cell Anemia Imaging System

For our sickle cell anemia system, based on the simulation results described in the previous section, SSDF was selected as the illumination method with polarization discrimination incorporated as shown in Figure 15.4b. The illumination consists of 18 surface mount LEDs (organized into six groups of three) around the field of view and mounted at the distal tip of the imaging assembly as shown in Figure 15.16. The imaging assembly is a modified microscope objective (10×, NA = 0.25, Edmund Optics NT36-132) repackaged into a custom-made low-profile carbon-fiber lens tube that can fit easily within the mouth.

As discussed in Section 15.3.1, green light is chosen to provide the balance between maximizing the contrast and achieving an acceptable signal to noise ratio (SNR). The largest absorption in the hemoglobin spectra is in the Soret band (400–440 nm) and using light within this band would provide high contrast to hemoglobin. Also for sickle cell anemia, there is a high linear dichroism signal in this band, which would provide high dichroism contrast. However, overall tissue

Figure 15.16 Modified microscope objective to enable greater comfort for subjects. SSDF imaging is implemented through annular illumination.

absorption (e.g., from melanin) is high in this range and there is insufficient light detected to form a high-quality image. At longer wavelengths (~550 nm), the levels of absorption provided by melanin and hemoglobin are reduced by a factor of ~3 compared to 420 nm. There is also often an increase in detector quantum efficiency, leading to a significant increase in the detected signal levels.

To enable detection of the two orthogonal polarization states, two images are simultaneously imaged onto the charged coupled device (CCD) camera having traveled through different polarization optics. These images can then be overlaid using the method described in Section 15.4 allowing calculation of the diattenuation on a pixel-by-pixel basis.

The objective lens forms an image at a 4.9 mm^2 aperture (which can be seen in the images in Figure 15.5) corresponding to a 490 μm field of view. This aperture is included to ensure that the formed images are well-bounded, eliminating overlap of the polarization images on detection. This intermediate image and aperture are reimaged onto a CCD sensor via a unity magnification relay lens combination ($f = 80$ mm). The polarization-sensitive optics comprises a nonpolarizing beam-splitter, two orthogonally orientated Glan-Thompson prisms (rejection ratio of 1×10^6), and mirrors to provide lateral displacement of the two images on the detector. An example of two recorded polarization images using this setup is shown in Figure 15.5. This approach was selected as opposed to angular separation of images using a Wollaston prism, as there was less image distortion. The numerical methods applied to remove image distortions introduced by transmission through two different optical paths are discussed in Section 15.4.

The camera selected is the PCO2000 from PCO Imaging. This is a 14 bit camera with a 2048 × 2048 array of 7.4 μm wide pixels with a well capacity of 4×10^4 electrons. This camera is limited to a dynamic range of 16 384 owing to the digitization to 14 bits; however, the well capacity of a single pixel limits the available SNR to 200 (<8 bits of information); therefore interpixel or frame averaging is required to make greater use of the available dynamic range. Averaging over a 4.4 μm × 4.4 μm area of a cell (6 × 6 pixels) results in an available SNR of 1200 in a single channel. As the dichroism calculation is a combination of two measurements, the SNR on the combined measurement is 848,[1] thereby meeting the required specification and providing a polarization difference sensitivity of ~1.2%.

As discussed in the introduction, a number of sites on the body can be selected for examination. An initial assessment using a commercially available capillaroscope (Microvision Medical) was performed on sickle cell anemia patients to establish the most clinically suitable site. The clinical protocol was approved by the UK National Ethics Committee and informed consent was obtained from the participants. The aim was to assess patient acceptance, image quality, and ease of applicability. The examined sites included the nail fold, the inside of the cheek, and the inside of the lower lip. Examination of the nail fold provided the greatest patient comfort and ease of examination. However, this suffered because of the relatively poor image quality due to the increase in thickness of the overlying epithelium and high levels of pigmentation in the participating group, leading to a significantly reduced signal. Within the mouth, the absorption due to melanin is low and the lower lip was found to be the preferred site, as the images recorded were of reasonable quality and access to the site was more acceptable to the patients. Example images from different sites are shown in Figure 15.17.

For infection control, the tip of the imaging system (Figure 15.16) was covered with a custom-made replaceable and disposable cover to provide infection immunity and also to fix the required working distance of the objective lens. This is covered with a 1 mm thick silica glass cover slide to ensure that there are no birefringence effects and to aid location of the appropriate working distance for the objective lens. The clinical patient station (Figure 15.18) is based on an adapted Fundus eye camera stand that provided an effective means of positioning of the system.

15.7 Imaging Results – Sickle Cell Anemia Imaging System

To test the ability of the developed imaging system to sense the presence of *in vivo* HbS polymerization, patients with sickle cell anemia were examined using the patient station shown in Figure 15.18. A total of nine participants were monitored while in a clinically stable state with no recent episodes of sickle cell pain. These examinations produced 28 usable video streams. A selection of images is shown in Figure 15.19.

1) Determined by $\left(\frac{1}{(1/S)^2+(1/S)^2}\right)$ where S is the SNR of a single measurement.

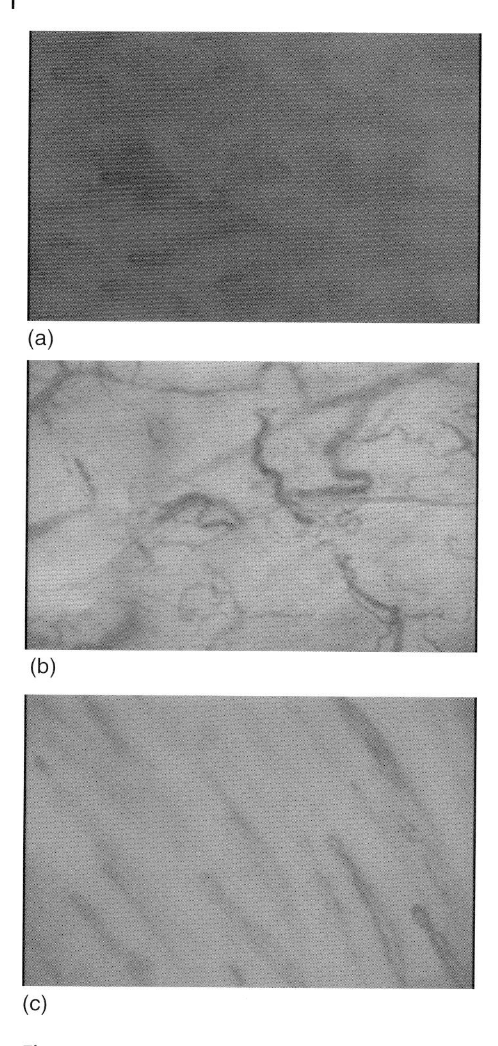

Figure 15.17 Images taken at different sites: (a) nail fold, (b) on the lower lip, and (c) inside the cheek. The field of view is ~1.6 mm × 1.25 mm.

Our data demonstrated that the structure of the microvasculature varies greatly both inter- and intrapatient for different sites on the lower lip. This contrasts with the regular loop structure that occurs on the nail fold (Figure 15.17a). However, the main objective of this research was not in the shape of the vessels but in the investigation of individual red blood cells. For this system an individual cell occupied a region of approximately 6–9 pixels within the image, and for each pixel within an isolated cell, the level of linear dichroism was calculated.

Within the subject group, no significant levels of dichroism outside of those expected from statistical variation of the system noise were observed. Given the

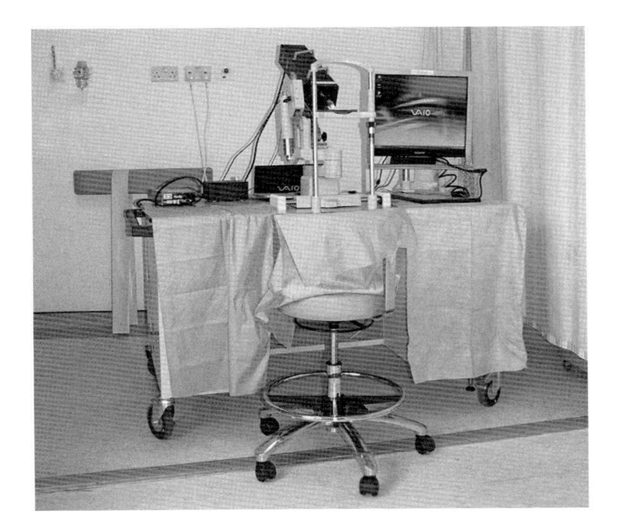

Figure 15.18 Patient station.

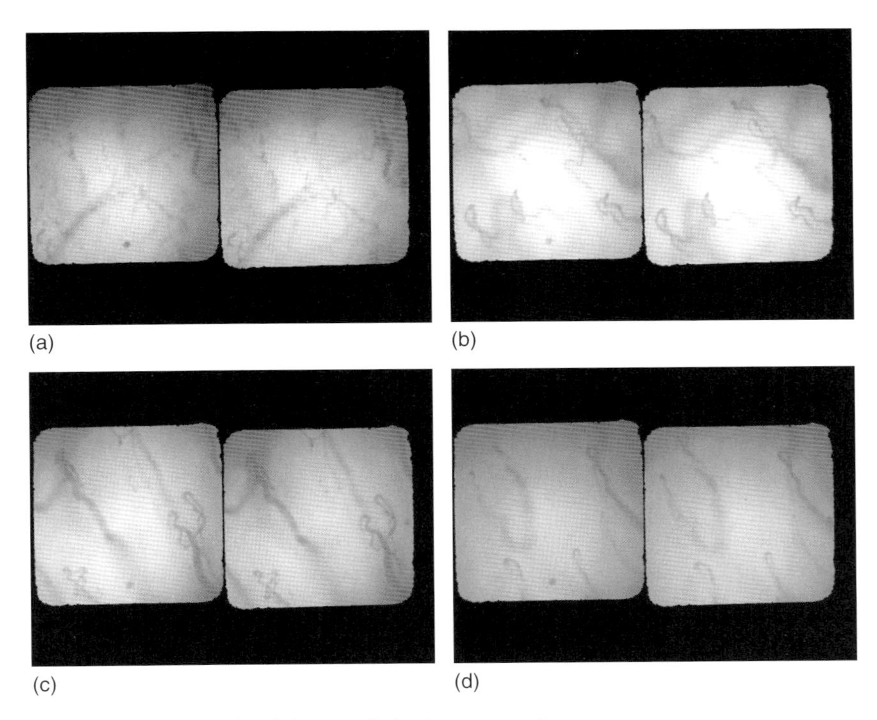

Figure 15.19 A sample of the recorded video streams/images.

sensitivity of the device (1.2%), it can be concluded that in this subject group the linear dichroism values obtained previously *in vitro* [13] have not been observed *in vivo*. There may be several clinical and technical reasons for this.

One of the effects of polymerization on the behavior of sickled erythrocytes is that their deformability is impeded and therefore their ability to enter the narrowest capillaries (where separation into a chain of individual cells occurs) may also be affected, reducing the frequency of their occurrence.

Effects of scale may also contribute for the absence of the anticipated signal. The previous *in vitro* study [13] has been performed at a much higher spatial resolution. The reduced resolution in this study occurs for two reasons: (i) a wide field of view is required for practical *in vivo* application, plus the need for interpixel averaging restricts the number of pixels representing a single cell and (ii) the overlying epithelium, although thin enough to enable detection of the cells, introduces some scattering into the images. The occurrence of dichroic material with different polymer orientations within an individual pixel will have the effect of reducing the strength of the available polarization signal, thus averaging out the signal of interest.

A further barrier to detection of polymerization may also be due to cells behaving differently when they are removed from the body. In addition, we are using a particular site (lower lip) where cells can be observed, but this may not be representative of the behavior of cells in other organs. For example, a recent study [3] showed no significant differences in flow rate under the tongue of sickle cell anemia sufferers.

15.8 Discussion and Future Directions

Many *in vivo* flow cytometry techniques use fluorescent labels for discriminating between cell populations. However, there is potential for label-free flow cytometry in which cell populations are determined on the basis of properties such as flow rate, oxygen saturation, or linear dichroism. This chapter has described some of the instrumentation challenges that researchers in this field face. A particular example, polarization-sensitive measurements of sickle cell anemia sufferers, has been used to demonstrate the fundamental instrumentation principles.

Illumination is important, particularly when the detection wavelength is the same as the illumination wavelength, when surface reflections and superficially scattered light obscure the blood cells. The choice of wavelength needs to balance achieving high contrast against a sufficient amount of detected light. For example, light in the blue wavelength range will provide the greatest contrast to red blood cells but insufficient light will be detected to form a high-quality image. Conversely, in the red wavelength range, sufficient light is backscattered from the tissue, but the contrast is low. Green light provides the optimum performance. In terms of configuring the light sources, SSDF is found to provide the most effective rejection of surface-reflected and superficially scattered light. In this case, annular illumination around the region of interest is applied and the scattered light "back-illuminates" the blood cells, effectively turning it into a transmission system.

Detection of small changes on a large light background requires cameras that have a large well capacity. Alternatively, custom-made CMOS sensors offer potential for making sensitive differential measurements of, for example, linear dichroism or oxygen saturation and other measurements such as the Doppler shifted light from moving cells.

A basic model of the imaging process based on Monte Carlo simulations has been described to compare the performance of three different illumination methods. The main drawback of this model is that the target to be imaged is an infinitely thin, partially absorbing object. Further improvements will be needed to realistically represent the scattering from individual cells [41].

To date, application of this model in sickle cell anemia sufferers has shown no significant differences between sickled and healthy cells. There may be clinical reasons for this, but further instrumentation improvements could also be made. For example, imaging at a higher magnification should be feasible, albeit with the use of faster cameras and an increase in motion artifacts. Custom-made CMOS cameras may allow smaller changes to be detected than is currently possible with CCD cameras.

References

1. Lipowsky, H.H., Sheikh, N.U., and Katz, D.M. (1987) Intravital microscopy of capillary hemodynamics in sickle cell disease. *J. Clin. Invest.*, **80**, 117–127.
2. Cheung, A.T.W., Chen, P.C.Y., Larkin, E.C., Duong, P.L., Ramanujam, S., Tablin, F., and Wun, T. (2002) Microvascular abnormalities in sickle cell disease: a computer-assisted intravital microscopy study. *Blood*, **99**, 3999–4005.
3. van Beers, E.J., Goedhart, P.T., Unger, M., Biemond, B.J., and Inc, C. (2008) Normal sublingual microcirculation during painful crisis in sickle cell disease. *Microvasc. Res.*, **76** (1), 57–56.
4. Georgakoudi, I., Solban, N., Novak, J., Rice, W.L., Wei, X., Hasan, T., and Lin, C.P. (2004) *In vivo* flow cytometry: a new method for enumerating circulating cancer cells. *Cancer Res.*, **64**, 5044–5047.
5. Novak, J., Georgakoudi, I., Wei, X., Prossin, A., and Lin, C.P. (2004) *In vivo* flow cytometer for real-time detection and quantification of circulating cells. *Opt. Lett.*, **29** (1), 77–79.
6. Boutrus, S., Greiner, C., Hwu, D., Chan, M., Kuperwasser, C., Lin, C.P., and Georgakoudi, I. (2007) Portable two-color *in vivo* flow cytometer for real-time detection of fluorescently-labeled circulating cells. *JBO Lett.*, **12**, 020507–020501.
7. Galanzha, E.I., Tuchin, V.V., and Zharov, V.P. (2007) Advances in small animal mesentery models for *in vivo* flow cytometry, dynamic microscopy, and drug screening. *World J. Gastroenterol.*, **13** (2), 192–218.
8. Monici, M., Pratesi, R., Bernabei, P.A., Caporale, R., Ferrini, P.R., Croce, A.C., Balzarini, P., and Bottiroli, G. (1995) Natural fluorescence of white blood cells: spectroscopic and imaging study. *J. Photochem. Photobiol. B: Biol.*, **30** (1), 29–37.
9. Zharov, V.P., Galanzha, E.I., and Tuchin, V.V. (2006) *In vivo* photothermal flow cytometry: imaging and detection of individual cells in blood

and lymph flow. *J. Cell Biochem.*, **97**, 916–932.

10. Zharov, V.P., Galanzha, E.I., Shashkov, E.V., Khlebtsov, N.G., and Tuchin, V.V. (2006) *In vivo* photoacoustic flow cytometry for monitoring of circulating single cancer cells and contrast agents. *Opt. Lett.*, **31**, 3623–3625.
11. He, W., Wang, H., Hartmann, L.C., Cheng, J., and Low, P.S. (2007) *In vivo* quantitation of rare circulating tumor cells by multiphoton intravital flow cytometry. *Proc. Natl. Aacd. Sci.*, **104** (28), 11760–11765.
12. Christoph, G.W., Hofrichter, J., and Eaton, W.A. (2005) Understanding the shape of sickled red cells. *Biophys. J.*, **88**, 1371–1376.
13. Beach, D.A., Bustamante, C., Wells, K.S., Foucar, K.M. and Hemoglobin, S. (1988) Differential polarization imaging III. Theory confirmation. patterns of polymerization of in red blood sickle cells. *Biophys. J.*, **53**, 449–456.
14. Mohandas, N. and Hebbel, R.P. (1994) Erythrocyte deformability, fragility and rheology, in *Sickle Cell Disease: Basic Principles and Clinical Practice* (eds S.H. Embury, R.P. Hebbel, N. Mohandas, and M.H. Steinberg), Raven Press, New York. pp. 205–216.
15. World Health Organisation (2006) Sickle Cell Disease and Other Hemoglobin Disorders, Fact sheet number 308.
16. Lowell, B.B. and Shulman, G.I. (2005) Mitochondrial dysfunction and type 2 diabetes. *Science*, **307**, 384–387.
17. Groner, W., Winkelman, J.W., Harris, A.G., Ince, C., Bouma, G.J., Messmer, K., and Nadeau, R.G. (1999) Orthogonal polarization spectral imaging: a new method for study of the microcirculation. *Nat. Med.*, **10** (5), 1209–1213.
18. Monari, J.J. (1999) Human capillaroscopy by light emitting diode epi-illumination. *Microvasc. Res.*, **59**, 172–175.
19. Ince, C. (2005) The microcirculation is the motor of sepsis. *Crit. Care*, **9** (4), 13–19.
20. Verdant, C. and De Backer, D. (2005) How monitoring of the microcirculation may help us at the bedside. *Crit. Care*, **11** (3), 240–244.
21. Anders, H.J., Sigl, T., and Schattenkirchner, M. (2001) Differentiation between primary and secondary Raynaud's phenomenon: a prospective study comparing nailfold capillaroscopy using an ophthalmoscope or stereomicroscope. *Ann. Rheum. Dis.*, **60**, 407–409.
22. Nagy, Z. and Czirják, L. (2004) Nailfold digital capillaroscopy in 447 patients with connective tissue disease and Raynaud's disease. *J. Eur. Acad. Dermatol. Venereol.*, **18**, 62–68.
23. Cutolo, M., Grassi, W., and Cerinic, M.M. (2003) Raynaud's phenomenon and the role of capillaroscopy. *Arthritis Rheum.*, **48** (11), 3023–3030.
24. Tuchin, V.V. (2006) *Optical Clearing of Tissues and Blood*, SPIE Press, Bellingham, WA.
25. Zhu, Q., Stockford, I.M., Crowe, J.A., and Morgan, S.P. (2009) An experimental and theoretical evaluation of rotating orthogonal polarization imaging. *J. Biomed. Opt.*, **14**, 034006.
26. Oregon Medical Laser Centre (2008) *http://omlc.ogi.edu/spectra/hemoglobin/summary.html* (accessed November 24, 2009).
27. Bourquin, S., Seitz, P., and Salathé, R.P. (2001) Optical coherence topography based on a two-dimensional smart detector array. *Opt. Lett.*, **26**, 512–514.
28. Gu, Q., Hayes-Gill, B.R., and Morgan, S.P. (2008) Laser Doppler blood flow CMOS imaging sensor with analog on-chip processing. *Appl. Opt.*, **47** (12), 2061–2069.
29. Kongsavatsak, C., He, D., Hayes-Gill, B.R., Crowe, J.A., and Morgan, S.P. (2008) CMOS imaging array with laser doppler blood flow processing. *Opt. Eng.*, **47**, 104401.
30. Belcaro, G., Hoffman, U., Bollinger, A., and Nicolaides, A. (1994) *Laser Doppler*, Prentice Hall.
31. Essex, T.J.H. and Byrne, P.O. (1991) A laser Doppler scanner for imaging blood flow in skin. *J. Biomed. Eng.*, **13**, 189–194.

32. Johnston, N.S., Stewart, C., Light, R.A., Hayes-Gill, B.R., Somekh, M.G., Morgan, S.P., Sambles, J.R., and Pitter, M.C. (2009) Quad-phase synchronous light detection with 64x64 CMOS modulated light camera. *Electron. Lett.*, **45**, 1090–1091.
33. Sawyer, N.B.E., Morgan, S.P., Somekh, M.G., See, C.W., Cao, X.F., Shekunov, B.Y., and Astrakharchik, E. (2001) Wide field amplitude and phase confocal microscope with parallel phase stepping. *Rev. Sci. Instrum.*, **72** (10), 3793–3801.
34. Scarano, F., David, L., Bsibsi, M., and Calluaud, D. (2005) S-PIV comparative assessment: image dewarping and misalignment correction and pinhole and geometric back projection. *Exp. Fluids*, **39**, 257–266.
35. Dobbe, J.G.G., Streekstra, G.J., Atasever, B., van Zijderveld, R., and Ince, C. (2008) Measurement of functional microcirculatory geometry and velocity distributions using automated image analysis. *Med. Biol. Eng. Comput.*, **46**, 659–670.
36. Ross, N.E., Pritchard, C.J., Rubin, D.M., and Duse, A.G. (2006) Automated image processing method for the diagnosis and classification of malaria on thin blood smears. *Med. Biol. Eng. Comput.*, **44**, 427–436.
37. Cseke, I. (1992) A fast segmentation scheme for white blood cells. Pattern Recognition, Vol. III. Conference C: Image, Speech and Signal Analysis, Proceedings, 11th IAPR International Conference, pp. 530–533.
38. Tuchin, V.V. (2007) *Tissue Optics: Light Scattering Methods and Instruments for Medical Diagnosis*, 2nd edn, PM 166, SPIE Press, Bellingham, WA.
39. Ramella-Roman, J.C., Prahl, S.A., and Jacques, S.L. (2005) Three Monte Carlo programs of polarized light transport into scattering media: part I. *Opt. Exp.*, **13**, 4420–4438.
40. Meglinsky, I.V. and Matcher, S.J. (2002) Quantitative assessment of skin layers absorption and skin reflectance spectra simulation in the visible and near-infrared spectral regions. *Physiol. Meas.*, **23**, 741–753.
41. Scott Brock, R., Xin-Hua, H., Weidner, D.A., Mourant, J.R., and Lu, J.Q. (2006) Effect of detailed cell structure on light scattering distribution: FDTD study of a B-cell with 3D structure constructed from confocal images. *J. Quant. Spectrosc. Radiat. Transf.*, **102**, 25–36.

16
Advances in Fluorescence-Based *In vivo* Flow Cytometry for Cancer Applications

Cherry Greiner and Irene Georgakoudi

16.1
Introduction

Epithelial tumors are known to be the most common form of cancer and the leading cause of cancer morbidity primarily due to failures in detecting and treating the metastatic disease [1, 2]. The migration of cancer to secondary organs can complicate treatment since different environments can modify tumor cell response to therapy altering therapeutic efficacy [3]. Given the clinical implications of metastasis, there is much interest in the early detection and monitoring of its occurrence. A growing number of studies have shown that the detection of circulating tumor cells (CTCs) is clinically relevant to individual patient prognosis and therapeutic response. Although tumor cells in peripheral blood have been observed over a hundred years ago, only through recent technological advances have researchers been able to identify, characterize, and enumerate CTCs [4]. However, given the rarity of these cells, as few as 1 in 10^6–10^7 nucleated blood cells, current techniques are still challenged by reliable and sensitive quantification of CTCs [5]. Moreover, the detection of such rare events is hindered by the limited volume of the sample available for analysis.

Flow cytometry (FC) is an established technique developed for the rapid enumeration and sorting of cells such as CTCs. In conventional FC, cells in suspension are flowed in a single stream and illuminated by a laser. Fluorescence or scattered light is collected and used to count and characterize each cell. Although many researchers are continually improving FC designs, few are focusing on addressing the major limitation of standard flow cytometers and other *in vitro* CTC detection techniques, which is that they can only analyze cells from drawn blood samples. Not only can blood sampling be painful and difficult in some patients but the sample processing required for *in vitro* CTC detection can also introduce artifacts that can affect sensitivity and specificity. To address this limitation, *in vivo*-based detection techniques, such as fluorescence-based *in vivo* flow cytometry (IVFC), have been developed to directly measure tumor cells in circulation. IVFC eliminates labor-intensive cell enrichment and labeling of blood samples and also provides physicians and researchers with a noninvasive tool for real-time monitoring of

Advanced Optical Flow Cytometry: Methods and Disease Diagnoses, First Edition. Edited by Valery V. Tuchin.

CTCs in animals for research and ultimately in clinical patients. Consequently, IVFC can potentially enhance the clinical detection of CTCs and improve decision making, patient care, and quality of life.

In this chapter we discuss the clinical potential of the enumeration and detection of CTCs in terms of disease progression, patient risk assessment, and treatment response. We review current techniques in CTC detection, as well as the principle of operation of different modes of optical-based IVFC and how they can address limitations of current techniques. Finally, we discuss studies demonstrating the potential application of IVFC for CTC detection *in vivo* and some of the improvements to enable clinical utility of this approach. For details on IVFC techniques based on imaging and absorption sensitivity, refer to Chapters 14 and 17.

16.2
Background: Cancer Metastasis

Metastasis can be described as a series of sequential steps that cancer cells from the primary tumor must successfully complete for tumors to form in secondary organs. Metastasis typically involves angiogenesis, loss of cell adhesion, increased motility and invasiveness, entry and survival in the circulation, invasion into new tissue, and, finally, initiation and growth of a metastatic tumor in the distant tissue. An overview of the metastatic process is shown in Figure 16.1 [6]. Several reviews on details of the metastatic process are available [1–3, 7].

The proliferation of cancer cells is challenged by mechanisms that suppress tumor formation. Factors in the microenvironment that can limit tumor progression include the basement membrane, extracellular matrix (ECM) components, limited availability of nutrients and oxygen, attack from the immune system, and the endothelial barrier into the circulation [2, 7]. However, the tumor cells' response to these external cues, such as hypoxia, can enhance their metastatic potential by promoting angiogenesis, anaerobic metabolism, and genomic instability that promotes cell survival and propensity for invasion [7, 8]. Acquired genomic and phenotypic instabilities of tumor cells can result in a broad spectrum of morphological changes which further increase tumor aggressiveness. For example, the loss of epithelial cell adhesion during epithelial to mesenchymal transition (EMT) results in highly motile cells that are more likely to invade the basement membrane [6, 9]. After traversing the basement membrane, cancer cells must degrade the ECM of the stroma to reach available blood vessels [10]. In the stroma, the tumor cells produce angiogenic factors that bind to the vascular cells, increasing vascular permeability and enhancing cell migration into surrounding vessels [7, 11]. Furthermore, the tumor cells interact with a variety of cells, resulting in the activation of the host stroma [7, 11]. For example, inflammatory cells that are attracted to regions of necrosis and hypoxia induce angiogenesis and release growth factors that facilitate tumor proliferation, survival, and invasion, while activated fibroblasts stimulate tumor cell motility and promote angiogenesis by recruiting endothelial progenitor cells from the blood vessels, giving rise to highly vascularized tumors [7, 11].

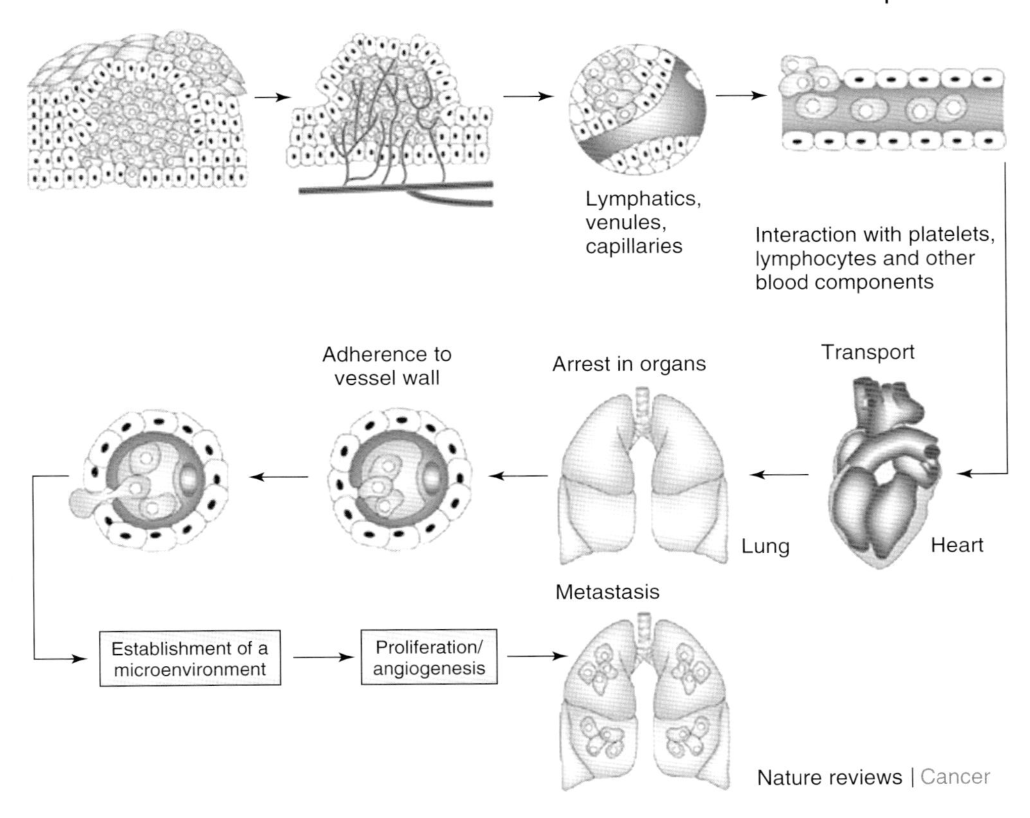

Figure 16.1 Overview of the metastatic process [6]. (a) Primary tumor, (b) proliferation/angiogenesis, (c) detachment/invasion, (d) embolism/circulation, (e) extravasation, and (f) proliferation/angiogenesis in metastatic tumor.

Once in the bloodstream, cancer cells must survive physical damage from hemodynamic shear forces and from the body's defenses before trying to re-attach themselves in a new tissue by extravasation [8]. Only very few ($\sim$1 in 10 000) cells that survive in circulation are able to form metastasis in tissue organs away from the primary site [12]. The circulation of the blood plays a significant role in determining where cancer cells travel. Mechanical lodging in capillaries is the most probable form of tumor entrapment in secondary sites, although homing of cancer cells to secondary sites is also possible through adhesive interactions between surface receptors expressed on malignant cells and cognate ligands expressed on target organs [1, 7]. For example, the chemokine receptor CXCR4, which is highly expressed in breast cancer cells, is important for metastasis in CRCL-12-rich tissues such as the lungs [13]. Cells that disseminate from the primary tumor may also enter the lymphatic system, where they can form a tumor in the lymph node that may eventually disseminate to distant organs [1]. To colonize the new site, the tumor cells must interact with the new organ microenvironment for growth and survival advantages [2, 7].

16.3 Clinical Relevance: Role of CTCs in Cancer Development and Response to Treatment

Various reports on CTC detection have been published. However, with variations in the techniques and tumor markers, as well as patient and control sample sizes, results have been variable and difficult to compare [8]. To date, the American Society of Clinical Oncology has not yet supported the use of CTC assays for clinical management decisions, mainly due to the need for more clinical validation studies [14]. Nonetheless, CTC detection and enumeration has been demonstrated in various cancers including ovarian, lung, and bladder [15–19]. Moreover, numerous studies provide evidence supporting the use of CTCs for clinical decision making, particularly in the treatment of patients with breast, prostate, or colorectal cancer [20–24].

Before resection of the primary tumor, neoadjuvant therapy is provided to reduce tumor size and eradicate any micrometastases. However, without a reliable method of predicting treatment outcome, some patients will be overtreated while others will be undertreated. In one study, a strong correlation was found between the decrease in CTCs during the first few cycles of therapy and primary tumor response, based on the final reduction in tumor size confirmed after surgery on patients with breast cancer receiving neoadjuvant therapy [25]. In patients undergoing resection for colorectal cancer, positivity for CTCs 24 h post surgery was a strong predictor of recurrence [26, 27]. Therefore, CTC enumeration can indicate the success of neoadjuvant therapy, while CTC clearance after resection can indicate the success of surgery and improve management decisions regarding follow-up adjuvant therapy.

However, despite years of clinical research and improvements in early cancer detection and treatment, most patients, particularly those with metastatic cancer, will experience a relapse and die of the disease. To prevent this from occurring, adjuvant therapy is typically provided following removal of the primary tumor. A means of predicting a patient's response to this type of therapy is of significant clinical value. In a prospective multicenter study, newly diagnosed patients with metastatic breast cancer with ≥ 5 CTCs prior to treatment and at first followup were correlated with shorter progression-free survival and overall patient survival [22, 28]. Similarly, elevated CTCs of ≥ 5 in 7.5 ml of blood in patients with metastatic prostate cancer and ≥ 3 CTCs in the same blood volume in patients with colorectal cancer before the initiation of therapy and in subsequent weeks was shown to correlate with poor overall survival [29–32]. Moreover, compared with traditional imaging techniques, such as computerized tomography scans of the chest and abdomen, CTC detection provided a more reliable estimation and earlier indication of disease progression and patient survival [22, 33, 34]. Therefore, these studies indicate that, in general, the number of CTCs above a certain threshold within a specific volume of blood can be used as a guide for assessing the potential success of a given therapeutic regimen or the need for more aggressive treatment.

16.3.1 Detection and Enumeration of Circulating Non-Epithelial Cancer Cells

Leukemia is a cancer of the bone marrow or blood characterized by abnormal proliferation of blood cells. Leukemia is expected to affect more than 44 000 adults and about 3500 children under the age of 15 in the United States [35]. Although leukemia affects more adults than children, it is the most common pediatric cancer, with acute lymphoblastic leukemia (ALL) being the most common form of leukemia in children affecting approximately three out of four children with leukemia [35]. It accounts for approximately 25% of pediatric cancer cases and is the leading cause of disease-related deaths in children younger than 20 years of age [36, 37]. Bone marrow or blood cell analysis is of clinical importance in assessing patient treatment response and risk for relapse in patients with leukemia. For example, the persistence or slow decrease in circulating leukemic blast cells during early treatment is correlated with slow therapeutic response and poor prognosis [38, 39].

Most patients with leukemia enter remission after induction therapy, but some, which can be more than 25% of patients with ALL, will experience a relapse due to the incomplete eradication of leukemic cells [40]. However, the traditional approach for identifying residual leukemia cells or minimal residual disease (MRD) by morphologic examination of peripheral blood or bone marrow cells is subjective and is limited in sensitivity [41]. Because the morphology of leukemic cells closely resembles that of normal hematopoietic progenitors, it can be difficult to identify leukemia cells with confidence. To determine MRD with certainty, the leukemic blast cells must constitute at least 5% of the total nucleated cell population [41]. Therefore a patient with $<5\%$ of leukemia cells is considered to be in complete remission, but may still harbor a considerable number of leukemic cells, in fact as many as 10^{10} [41]. A means of better assessing MRD could improve clinical management of patients and potentially advance cure rates. One approach is to exploit the general observation that most leukemia cells express abnormal phenotypes not seen in normal blood cells. Aberrant antigen profiles generally mean that the cancer cells express one or more antigens that do not fit the normal blood cell lineage commitment or express antigens typically not expressed by normal bone marrow or peripheral blood cells. Studies have found that the number of cells with aberrant phenotypes can be used to estimate the total leukemic cell burden after therapy [42, 43]. The presence of cancer cells in blood samples of patients with leukemia identified to be in complete remission by traditional techniques was associated with poor prognosis and higher risk of relapse [42, 43]. Moreover, since multiple, let alone a single, bone marrow aspirations can be difficult, CTC enumeration from blood samples is preferable and allows for more continuous monitoring of patients. Consequently, CTC detection can be a more sensitive means of diagnosing patients with MRD.

16.4 Current Methods

16.4.1 Enrichment Techniques

Because of the rarity of CTCs in peripheral blood, current detection techniques typically require an enrichment step to increase sensitivity. The most commonly used enrichment techniques are isolation by cell density or specific antibodies. In cell density isolation, mononuclear cells are separated on the basis of their lower density in comparison to other blood components such as polymorphonuclear leukocytes and red blood cells (RBCs). A density gradient of 1.077 $\mathrm{g\,ml^{-1}}$, such as in Ficoll density gradient centrifugation, is typically used. A major drawback of density gradient isolation is the low purity of the enriched sample, since it will contain mononuclear leukocytes (i.e., lymphocytes, monocytes, and macrophages) and tumor cells. Detection methods that rely on cell density isolation must be able to discern contaminating leukocytes from rare CTCs. Aside from low purity of the enriched sample, cell loss can also be an issue. After centrifugation, it is expected that all the CTCs would settle in the interphase between the density gradient and the plasma. The use of Ficoll density gradient centrifugation of normal blood samples spiked with a known number of breast cancer cells resulted in varying recovery rates. Results showed that 13–60% of the cancer cells settled at the interphase, while 30–80% sedimented with heavier density polymorphonuclear leukocytes and RBCs below the density gradient layer [44].

Density-based isolation can be combined with antibody tagging to enhance the isolation of CTCs in peripheral blood. For example, CD45 (pan-leukocyte marker) positive cells can be cross-linked to RBCs forming rosettes, increasing the density of leukocytes that would normally be collected with the tumor cells by standard density gradient isolation. To target other unwanted cells, a more extensive negative isolation can be performed, by targeting against a host of antibodies in conjunction with CD45, including CD2 (T and NK cells), CD16 (NK, neutrophils, monocytes, macrophages), CD19 (B cells), CD36 (platelets, RBCs, and monocytes), and CD38 (T, B, and NK cells) [13]. Alternatively, magnetically labeled monoclonal antibodies and a magnet can be used to isolate CTCs either from whole blood or enriched samples from density gradient isolation. Commercially available immunomagnetic isolation kits use anti-epithelial antibodies for positive selection of CTCs, typically through epithelial cell adhesion molecule (EpCAM) or cytokeratins (CKs) [12]. EpCAM is a cell surface molecule for cell–cell adhesion, and it is highly expressed in epithelial carcinoma [8]. CKs are intermediate filament proteins that are tissue-specific constituents of normal epithelial cells and are well preserved in carcinoma and in the metastatic state [45, 46]. Although immunomagnetic separation can increase enrichment by 4–5 orders of magnitude, antibody isolation can result in false positives due to nonspecific labeling of epithelial specific antibodies [8]. For example, 0–20% CK positive cells have been detected in noncancer patients or controls from normal hematopoietic cells such as leukocytes that may express

Table 16.1 Commonly used markers to isolate CTCs [48].

Epithelial markers	Tumor markers
Epithelial cell adhesion molecule (EpCAM)	α-Fetoprotein (hepatocellular carcinoma)
Cytokeratin (CK) 7, 8, 18, 19, 20	CA-IX (renal cell carcinoma)
Human epithelial antigen (HEA)	Carcinoembryonic antigen (colorectal carcinoma, gastric carcinoma)
	Carcinoembryonic antigen adhesion molecule 2 (colorectal)
	HER2-neu (breast cancer)
	Mammaglobulin (breast cancer)
	Melanoma antigen gene family-A3 (breast carcinoma, melanoma)
	MOC-31 (ovarian carcinoma)
	MUC1/MUC2 (breast cancer)
	N-Acetylgalactosaminyltransferase (breast carcinoma)
	Prostate-specific antigen (prostate cancer)
	Squamous cell carcinoma antigen (squamous cell carcinoma of the esophagus)
	Stanniocalcin-1 (breast carcinoma)

epithelial antigens after cell activation [47, 48]. Cell loss may also occur when tumor cells lose epithelial features such as during EMT in the metastatic process; cell recovery can vary from 50 to 80% [49]. Antibody-based techniques mainly suffer from the fact that no markers are known to be universally present in all tumor cells from a particular tumor type or from the same lesion [50]. Combining multiple markers may be the best alternative to reduce false positive results and cell losses. Table 16.1 shows a list of CTC markers used in clinical studies.

Newer techniques have also successfully isolated CTCs by exploiting differences in tumor and normal cell size and/or deformity [51]. Isolation by size of epithelial tumor cells (ISETs) is by a membrane with 8 μm pore size developed to enrich blood samples by exploiting the general observation that tumor cells are larger than hematological cells. Although ISET can result in low levels of leukocyte contamination, cell loss may result for some cancer cells types, such as small neuroendocrine cancer cells, that are similar in size and shape to lymphocytes [52, 53]. Furthermore, recent studies with microfluidic-based devices to capture CTCs by cell size and deformity also indicate that different membrane pore sizes may be needed to optimize capture of specific types of cancer cells. Specifically, in the microfluidic-based devices, CTCs were captured by flowing blood samples in a chip consisting of rows of columns with channel spacing decreasing from 20 to 5 μm [54]. The larger and less rigid CTCs were found between columns with 5–10 μm spacing. A major drawback of this technique is the long time needed to process the blood; 1 ml of blood takes 1 h to be processed [54]. However, similar to ISET, captured cells remain intact and can be recovered for morphological, immunological, and genetic characterization of CTCs [51, 55, 56]. Although results are promising, further studies are needed to compare such recently developed isolation techniques with the more established density gradient and immunomagnetic isolation.

16.4.2 Detection Techniques

After enrichment, collected cells can be characterized by various methods, some of which combine both enrichment and detection steps on the same platform such as CellSearch, CTC chip, and epithelial ImmunoSPOT (EPISPOT). Detection-only techniques include polymerase chain reaction (PCR), laser flow cytometry, laser scanning cytometry (LSC), and the more recently developed fiber-optic array scanning technology (FAST).

CTCs can be identified by their cancer-specific genetic alterations, such as proto-oncogenes, tumor suppressor genes, and microsatellite instabilities. Reverse transcription polymerase chain reaction (RT-PCR) verifies the presence of CTCs in the sample through the detection of gene expression of mutated or overexpressed genes in tumor cells. After blood withdrawal and sample enrichment, RNA is extracted, cDNA is synthesized, and marker gene cDNA is amplified. RT-PCR is found to be a highly sensitive technique, capable of identifying 1 target cell out of 10^6-10^7 normal nucleated cells [8]. However, the sensitivity depends on the expression of the target gene, which can vary in carcinoma cell lines [55]. Another disadvantage of RT-PCR approaches is low specificity, since no RNA marker is expressed only in tumor cells; thus, gene expression from normal cells may be comparable to those from target tumor cells, making it difficult to determine true positives [8]. Given the heterogeneity of cancer, the use of multiple maker assays targeting tumor-specific and leukocyte-specific markers has been shown to reduce false negatives [57]. The issue of low specificity was also partly addressed by quantitative reverse transcription polymerase chain reaction (qRT-PCR) assays in which the detection of CTCs was assessed by a definitive cut-off value of tumor marker expression level [12]. However, defining a quantitative cutoff may still be problematic, as it is unclear whether general thresholds from studies are adaptable to patients on an individual basis [8]. Moreover, the strict adherence to protocols to minimize artifacts and the fact that RNA is extremely labile may further limit the practical use of PCR in the clinic [47, 57].

Similar to PCR, flow-based cytometry identifies CTCs by positive expression of specific makers detected by antibody labeling. In standard FC, labeled cells are flowed one by one and are enumerated as they cross a laser beam. A flow cytometer system consists of multiple laser sources to simultaneously excite various fluorophores and to provide light scattering information [58]. Since the size and granularity of normal white blood cells are different from those of CTCs, light scattered in the forward and perpendicular directions to the incident laser beam can be used, in conjunction with epithelial or tumor-specific antibody fluorescence labeling, to define CTC signal "gates." Only cells with signals that fall within this defined gate are classified as CTCs. FC can rapidly enumerate cells (10^4–10^5 events/s) with up to 10 fluorescence parameters [14, 59]. Since it has the capacity to simultaneously analyze multiple markers per cell, it has high specificity. Depending on the marker used for detection, general flow cytometers have a sensitivity of $<10^{-4}$–10^{-5}. However, sensitivity can be improved by increasing the blood volume to be analyzed and/or by using multiple markers [14, 60]. For example, combined staining for CK18 and DNA for breast cancer cells was shown to improve sensitivity to 10^{-6} [61]. In a model study, the use of three CK markers, each conjugated to different color dyes, to label cancer cells and a set of markers for normal cells (e.g., leukocytes, platelets, and RBCs) conjugated to a fourth dye showed increased sensitivity of 10^{-7} [62].

LSC combines the quantitative capabilities of FC with the ability to analyze a large number of fluorescently labeled cells on a slide. After enrichment, cells are transferred and fixed on slides, and then stained with epithelial antibodies and CD45 [12]. The system software analyzes the forward scatter and computes the total fluorescence on a per cell basis, correcting for any fluorescence background to better discriminate between CTCs and leukocytes [12]. Unlike standard FC, LSC can confirm true positives. Because LSC is a slide-based technology, additional morphological and subcellar analysis of CTCs is possible. Since the system records the precise location of detected CTCs on the slide, tone can relocate a cell and perform reiterative staining for further cell analysis and characterization [58]. Moreover, LSC allows for the analysis and enumeration of cell aggregates, which are typically excluded from CTC enumeration using FC, as high forward scattering and/or fluorescence from aggregated cells can fall outside of the defined CTC signal gates [59]. However, cell enumeration is slower than in FC since LSC must scan the slide samples. Scan rates can be as slow as 1000 cells/5 min [58]. Since cell images are analyzed and stored, LSC also requires large data storage. As in FC, the final detection rate depends on the enrichment technique. For example, a study showed that 30% of patient samples had detectable number of CTCs if enriched by cell filtration compared with 48% of samples enriched by immunomagnetic isolation [57]. Since LSC and FC are based on similar detection platforms, they share similar issues affecting sensitivity and specificity, such as the enrichment step and the choice of the type as well as the number and expression levels of the chosen markers.

FAST was developed for faster analysis of immunofluorescently labeled cells for CTC detection. The scan rate of automated digital microscopes, typically 80 cells/s, is limited by the camera image acquisition time, exposure time to excite fluorophores, and sample scanning [49]. Exposure time is reduced in FAST by using high-intensity laser sources and sensitive photomultiplier tube (PMT) detectors [49]. The use of specific lasers can reduce background fluorescence and increase the signal-to-noise ratio (SNR) [49]. The scan rate is further enhanced by designing the system with optical fiber bundles to have a large field of view at the expense of lower resolution. Although the scan rate is improved dramatically to 300 000 cells/s, the resolution of the system ($\sim$12 μm) may be too low to visually assess cell morphology, making it difficult to verify true positives [49, 63]. For this reason, FAST may be limited as a preliminary step prior to final verification by standard automated digital microscopes. Indeed, FAST in tandem with automated fluorescence microscopy can reduce processing time by 75% compared with standard automated digital microscopy alone in the detection of CTCs in the peripheral blood of patients with cancer [63].

Combining antibody isolation and automated digital microscopy, the CellSearch system is a combined enrichment and detection platform and is the only technique currently approved by the U.S. Food and Drug Administration (FDA) for CTC monitoring in metastatic breast, prostate, and colorectal cancers. An overview of the CellsSearch system is shown in Figure 16.2a. The system has two components, namely, the CellTracks AutoPrep and the CellTracks analyzer. The CellTracks Autoprep system carries out an automated immunomagnetic enrichment using ferrofluids coated with the epithelial-cell-specific marker for EpCAM. Cells are then permeabilized, fixed, and stained with DAPI nuclear dye, and fluorescently labeled with antibodies for epithelial cells (CK 8919) and leukocytes (CD45). The CellTracks analyzer is a four-color semiautomated fluorescence microscope that identifies CTCs from leukocytes and other contaminants using predetermined criteria, which include cell size, round to oval cell morphology, visible nucleus (positive staining by DAPI), positive staining for CKs, and negative staining for CD45. Image frames meeting the predetermined criteria are captured and automatically presented for interpretation to a trained operator, who makes the final selection of cells as CTCs [60]. An example of an image frame meeting all the criteria and selected as a CTC is shown in Figure 16.2b. The semiautomation

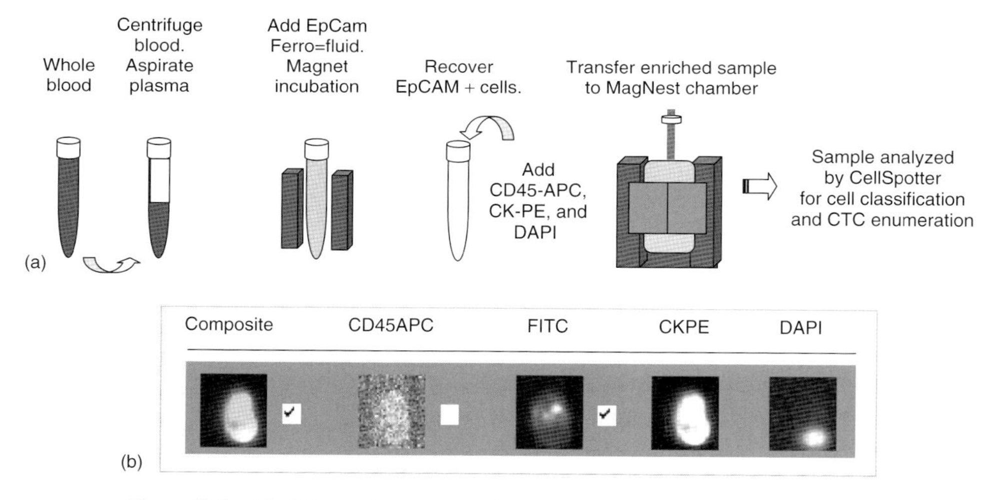

Figure 16.2 (a) CTC enrichment procedure using CellsSearch assay [8] and (b) example of cell characterized as CTC based on corresponding four fluorescence cell images [48].

of the CellSearch system allows samples to be analyzed rapidly and reproducibly. Studies using the CellSearch system have shown high specificity for this technique; only 0.3% of healthy and nonmalignant cases in comparison to $> 60\%$ of patients with metastatic breast cancer showed $\geq$2 CTCs/7.5 ml of blood using CellSearch [33]. In a comparative study, CellSearch detected more blood samples with CTCs and had a higher mean number of CTCs detected when compared to LSC [12]. The average recovery rate from samples spiked with cancer cells is about 80% [64]. However, variability between readers who make the final cell selection as CTCs is one of the main sources of error of this technique [60]. A study found that two independent readers will count the same number of tumor cells in only 50% of the cases [60]. Moreover, CellSearch has limitations similar to antibody-based isolation techniques, that is, the use of specific epithelial markers that can result to nonspecific labeling and cell loss due to tumor cell heterogeneity. These are partly addressed by incorporating avidin–biotin chemistry to increase the magnetic load of EpCAM expressing cells, compensating for possible downregulation of EpCAM in some tumor cells [12]. CellSearch also uses the pan-cytokeratin antibody that recognizes multiple CKs to compensate for downregulation of certain CKs during EMT [12].

Recently, a microfluidic-based device (CTC chip) has been developed for combined CTC isolation and detection. The CTC chip consists of an array of microposts coated with epithelial antibodies for EpCAM [65]. Cell capture was found to be dependent on flow velocity, as it affects attachment of cells to the microposts. Maximum capture rate was $>$60% for CTCs and decreased dramatically to $>$20% when the flow rate was increased from 1 to 2.5 ml h^{-1} [65]. To minimize processing time, the microfluidic-based device analyzed a small volume of blood. One advantage of the CTC chip is that an enrichment procedure is not required, minimizing cell losses. Specifically, it was found that samples preprocessed by RBC lysis resulted in capture rates similar to that in whole-blood samples [65]. Although average purity is only $\sim$50%, the captured cells remain intact and can be further verified as true positives by staining for DNA content and labeling for tumor/epithelial cell-specific antibodies [65]. Moreover, subsequent molecular analysis by PCR is possible by direct lysing of the captured cells to collect mRNA [65]. The major advantages of a microfluidic-based chip for CTC detection are that they are inexpensive and portable, making them appealing for use in personalized treatment of patients with cancer in both developed and underdeveloped countries. Since only a small sample of blood is needed for analysis, it is less invasive than traditional biopsies or most CTC detection techniques (e.g., CellSearch) and allows for more continuous monitoring of patients, potentially enabling treatment to be modified at a much earlier stage. Moreover, the chip-based platform can allow the integration of other techniques to characterize cancer cells, such as genetic analysis, to assess treatment response and to optimize therapy on an individual basis.

Another recently developed combined CTC isolation and detection technique is EPISPOT, an immunological method based on enzyme-linked immunosorbent assay that detects CTCs by specific proteins that are released [66]. After CD45 cell depletion, cells from the resulting enriched sample are seeded into wells with membranes coated with specific antibodies. Cells are incubated and cultured for 48 h. During the incubation period, proteins are released and captured by the antibody-coated membrane. The plates are washed and cells are removed. The fluorescently labeled antibody is then added to reveal the presence of the released protein of interest. Fluorescent spots are then counted by an automatic reader. As each spot corresponds to release by one viable cell, enumeration of CTCs is possible. The main drawback of this technique is the long processing time needed for tissue culturing. Moreover, this technique requires the proteins to be actively secreted or released for detection [66]. Although new approaches such as CTC chip and EPISPOT show potential recovery and enumeration of CTCs, they are less mature with regard to standardization. More studies are needed to determine the clinical utility of these technologies.

16.4.3
Summary

A comparison of the performance of the different CTC detection techniques is summarized in Table 16.2. Although numerous studies have been conducted focusing on CTC detection/assay sensitivity and specificity on various types of cancer and disease stages, no gold standard currently exists for CTC detection.

16.5
In vivo Flow Cytometry (IVFC)

The limitation of sensitivity and specificity of *in vitro*-based CTC detection techniques is, at least in part, a consequence of the invasive extraction of blood cells and the extensive processing that can introduce artifacts and artificially reduce the number of already rare CTCs. Meanwhile, the need to draw blood samples from patients prevents the observation and long-term study of CTCs, since blood samples cannot be drawn rapidly enough. By using *in vivo*-based techniques, such as IVFC, monitoring of CTC depletion kinetics at numerous time points is feasible without disturbing complex biological interactions, which is ideal for studying processes like metastasis. Therefore, by quantifying cells naturally flowing in blood vessels, IVFC can address some of the limitations of *in vitro*-based CTC detection techniques. However, adapting principles of FC from an *in vitro* to an *in vivo* setting must address additional challenges, including light scattering and absorption from blood vessels and surrounding tissue, variations in the blood vessel cross section and flow, and the presence of many normal cells surrounding CTCs.

To address the latter, IVFC systems mainly rely on fluorescent labeling to detect and enumerate CTCs. In order to optimally detect only fluorescence from cells in blood vessels instead of the surrounding tissue, IVFC systems are based on one of two basic design platforms: (i) single-photon IVFC with confocal detection and (ii) multiphoton IVFC. Regardless of the design platform, IVFC systems are limited to interrogation of superficial microvessels due to scattering from tissue and

Table 16.2 Performance, advantages, and disadvantages of *in vitro* CTC detection methods [14].

Method	Estimated sensitivity	Advantages	Disadvantages
RT-PCR methods (including qRT-PCR)	10^{-4}–10^{-6}	• Rapid • Quantitative • High sensitivity • Small sample volume required	• Low specificity • Cell-by-cell analysis is not possible; does not allow for enumeration • Cannot discriminate between viable and nonviable cells • Strict protocol to minimize contamination • Technical issues with mRNA degradation
Flow cytometry	10^{-4}–10^{-5}	• Rapid • Quantitative • Allows for cell-by-cell analysis • Multiparameter • High specificity • Can identify viable and nonviable cells • Cells can be sorted for further characterization	• General systems have limited sensitivity. Higher sensitivity ($>10^{-6}$) has been demonstrated by some research studies • Large sample of volume must be analyzed unless enriched samples are used • Visual confirmation during enumeration is not possible • Aggregated tumor cells can clog the system and can be rejected in the enumeration or be miscounted. • Multiparameter can get technically and analytically challenging

(continued overleaf)

Table 16.2 *(continued)*

Method	Estimated sensitivity	Advantages	Disadvantages
Laser scanning cytometry	10^{-4}–10^{-5}	• Rapid, though not as fast as flow cytometry • Quantitative • Allows for cell-by-cell analysis • Multiparameter • High specificity • Can identify viable and nonviable cells • Allows for morphological analysis • Can recover cells for further characterization	• Limited sensitivity • Technically and analytically challenging
CellSearch	10^{-7}	• High sensitivity and specificity • Semiautomated; highly reproducible • Commercially available • FDA approved • Allows for cell enumeration • Moderate sample volume is needed for CTC enumeration • Can identify viable and nonviable cells	• Use of epithelial specific markers can result in cell loss • Multiple steps prior to enumeration can result in cell loss • Final readout can be partially subjective
CTC chip	10^{-7}+	• High sensitivity and specificity • Quantitative • Allows for cell enumeration • Can recover cells for further characterization	• Flow-dependent capture rate • Not commercially available • Use of epithelial specific marker, EpCam can result in cell loss • Final readout can be partially subjective
EPISPOT	10^{-7}+	• High sensitivity and specificity • Quantitative • Allows for cell enumeration • Only viable cells are detected	• Requires 48 h of cell culture before CTC analysis

absorption by blood. In addition, since very few exogenous labels are approved for humans, IVFC systems are currently limited to measurements in small animals, which can, however, provide valuable insights into the role of CTCs in cancer development, metastasis, and treatment. IVFC sampling of murine microvessels has demonstrated the feasibility and advantage of *in vivo* detection and enumeration of an array of cell types including fluorescently labeled RBCs, leukocytes, stem cells, apoptotic cells, and various cancer cells including those of leukemia, breast, and prostate cancer [67–74].

Experimental investigation of tumor growth and metastasis of human cancers typically involves xenograft transplantation of human tumor cells or tissues in immunocompromised mice. Such models are important in elucidating the role of CTCs during cancer progression and therapeutic interventions. Moreover, experimental studies of CTCs in animal models can be useful in the development of more reliable and sensitive techniques for patients in clinical settings. The study of rare CTCs in small animals using *in vitro* CTC detection techniques is challenging due to the small amount of blood available for analysis. Thus, the kinetics and longitudinal assessment of the metastatic process can be difficult, particularly during early tumor development when the number of CTCs is expected to be even smaller. In mice, $<25-100\ \mu l$ of blood can be drawn for serial measurements [75]. Direct measurement of cancer cells in the circulation, such as by IVFC systems, allows for improved enumeration of CTCs in animal models by alleviating the need for intensive blood withdrawals and sample processing. In the following sections, we review the application of IVFC in cancer metastasis studies after first discussing the main principles of operation and performance characteristics of single-photon and multiphoton IVFC systems.

16.6 Single-Photon IVFC (SPIVFC)

16.6.1 Principles, Advantages, and Limitations

To optimally detect the fluorescence signal from CTCs in the blood vessel instead of the surrounding tissue, single-photon *in vivo* flow cytometry (SPIVFC) systems are designed based on the principle of confocal detection, which is commonly used in imaging optically thick samples. In confocal microscopy, optical sectioning at a given depth is achieved by rejecting scattering and fluorescence above and below the objective focal point through the use of a confocal pinhole. Fluorescence is excited by light, typically from a laser, in the visible region (400–600 nm) and focused onto a point in the specimen by the objective lens. The fluorescence emitted from the same point is re-imaged on to a pinhole placed directly in front of a sensitive detector, such as a PMT. Since the pinhole and the focus of the objective lens are confocal, in-focus fluorescence (and scattering) is passed efficiently to the detector, while out-of-focus fluorescence (and scattering) is largely rejected. Unlike customary point scanning in confocal systems, SPIVFC systems use a line or slit illumination in order to excite fluorescence in the entire blood vessel. This obviates the need for complicated scanning optics and allows for fast data acquisition, needed for detecting rare CTCs particularly in fast-flowing arteries. Although CTCs have been detected by SPIVFC in capillaries and veins, interrogation of faster blood flow in arteries translates to shorter sampling time for a given blood volume [67].

SPIVFC system designs use a number of available standard lasers that can excite a variety of fluorescent dyes/probes, such as those used in standard FC [76]. Since multiple excitation lasers and detectors can easily be added into the system, multiple fluorophore labeling is possible. As in multiparameter FC, the combined use of several markers can increase system specificity and/or allow for simultaneous enumeration and monitoring of more than one cell type. For example, a two-color SPIVFC demonstrated the detection and enumeration of double-labeled hematopoietic stem cells labeled by tail vein injection of PE-conjugated c-kit and APC-conjugated sca-1 antibodies. Double-labeled cells were detected as fluorescence peaks by two separate detectors at the same time point, while cells showing only one fluorescence peak likely corresponded to other cells (i.e., non-hematopoietic stem cells) that the c-kit or sca-1 antibody can label [69]. Representative data traces from double-labeled hematopoietic cells and single-labeled cells are shown in Figure 16.3. In another study, an SPIVFC system was used to excite cells expressing fluorescent proteins incorporated in the DNA such as green fluorescent protein (GFP), which is one of the highly exploited proteins used to visualize metastasis [77, 78]. The use of GFP expressing cells allows for the omission of *in vitro* or *in vivo* fluorescence labeling.

The depth of the blood vessel that SPIVFC can probe such that fluorescence signal from the cells can be detected above background noise is limited by tissue scattering, absorption, and autofluorescence, which are high in the visible region (Figure 16.4a–c). Absorption mainly decreases the number of photons that can reach the focus in the illumination path and the detector in the emission path. The

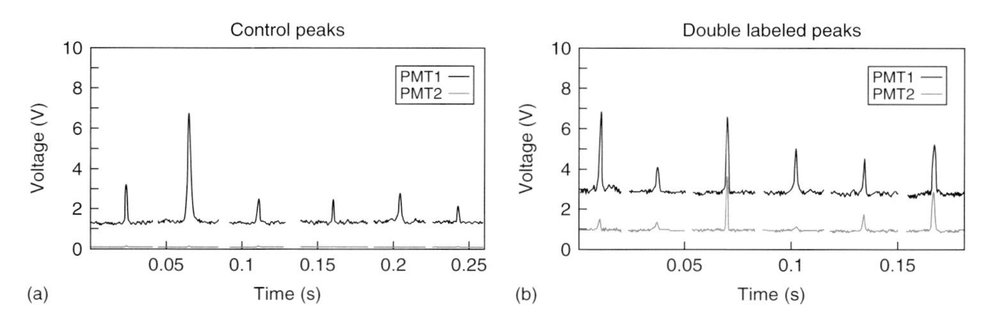

Figure 16.3 Sample time traces of (a) single-labeled and (b) double-labeled cells (black is from PMT1 and gray is from PMT2) [69].

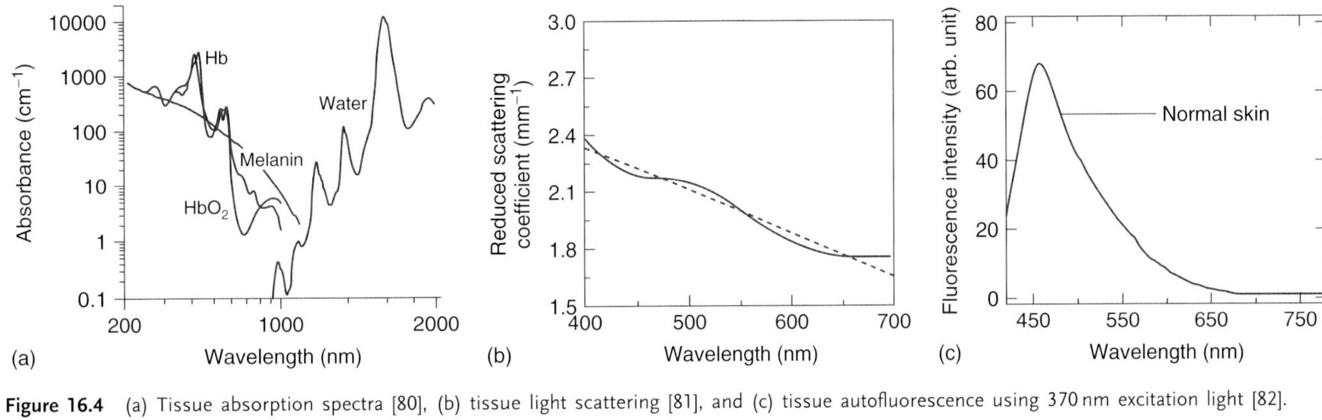

Figure 16.4 (a) Tissue absorption spectra [80], (b) tissue light scattering [81], and (c) tissue autofluorescence using 370 nm excitation light [82].

main tissue absorbers in the visible region are hemoglobin (Hb and HbO_2 present in RBCs) and melanin (Figure 16.4a). Similarly, scattering reduces the number of photons at the focus and out of the tissue by diverting the light off the imaging and detection path. Tissue autofluorescence absorbs photons that would otherwise reach the sample focus and contributes to background fluorescence that degrades the SNR of the detected fluorescence peaks. For these reasons, SPIVFC is limited to imaging depths $<200\,\mu m$, as in confocal microscopy [79]. However, for animal studies these depths are sufficient, since numerous blood vessels 70–100 μm below the skin surface of the mouse ear are available for CTC measurements by SPIVFC [67].

16.6.2 Instrumentation

Since the development of the first SPIVFC demonstrating the potential of enumerating fluorescently labeled cells in murine microcirculation, researchers continued to improve the instrumentation design to allow for multicolor excitation and detection, imaging, and interrogation of multiple blood vessels.

16.6.2.1 Two-Color SPIVFC

A schematic of a typical two-color SPIVFC system is shown in Figure 16.5a. In this particular example, fluorophore excitation is provided by two lasers at 488 and 633 nm, which are combined into a single illumination path by dichroic filters and focused onto a slit by a cylindrical lens. The slit is then re-imaged on the sample at the focus of the objective. To visualize the major veins and arteries of the ear microcirculation and subsequently align the slit across the selected artery, the ear is transilluminated by a light-emitting diode (LED) with a central wavelength of 530 nm, where hemoglobin absorption is strong and produces high contrast between blood vessels and the surrounding tissue. The same objective collects fluorescence from flowing labeled cells that traverse the slit at the focus. The collected light is directed to the detection path and focused into a confocal slit to reject the out-of-focus light. Light that makes it through the slit is separated into distinct emission wavelength bands before being detected by PMTs. Each PMT integrates the collected fluorescence light intensity and converts it to a voltage, which is recorded by the computer for analysis. Detected fluorescence bursts from cells that traverse the excitation slit are displayed as voltage peaks on a recorded time trace. Double-labeled cells are recorded simultaneously by the PMTs, resulting in peaks that are correlated in time as shown in Figure 16.3b. Software-based analysis programs are typically designed to enumerate peaks above a user-defined threshold. Thresholds are typically four to six times the standard deviation above the mean of time traces acquired from control samples or prior to injection of cancer cells in the circulation. Once the program locates a peak above the defined threshold, peak information such as height (maximum signal to background mean), width (typically full-width at half-maximum), and time of occurrence can be stored. The SPIVFC system is capable of acquiring data at a rate of $<100\,kHz$, which is the limit set by the data acquisition board.

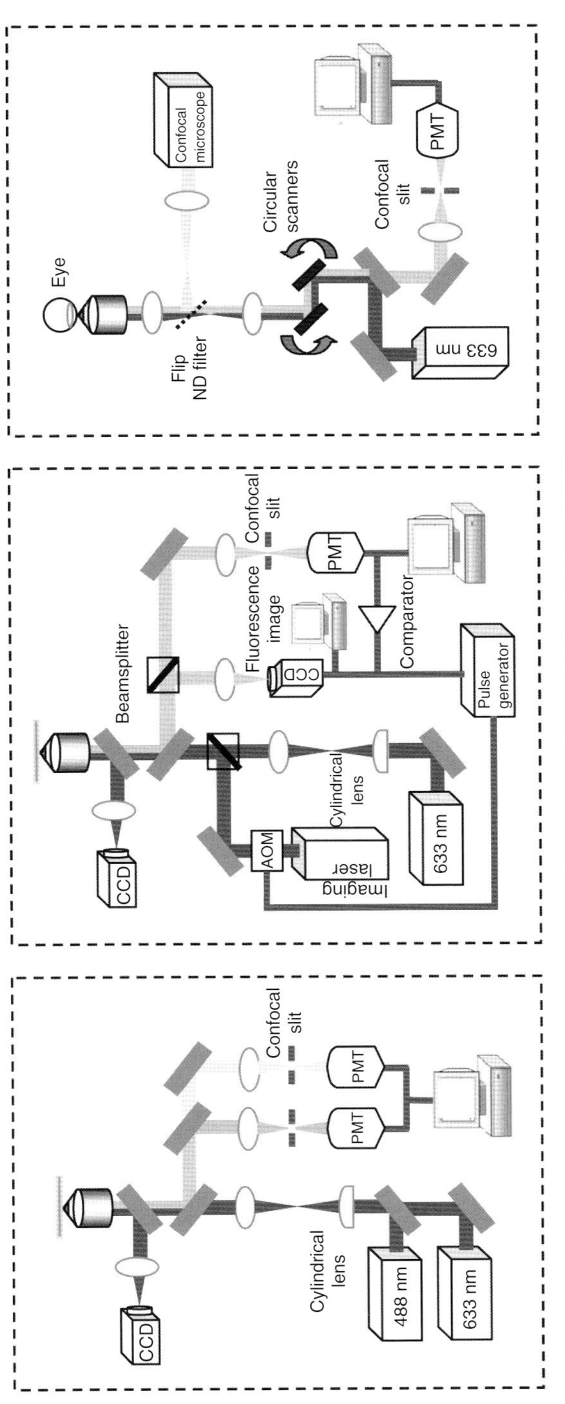

Figure 16.5 Optical schematic of (a) two-color SPIVFC, (b) imaging SPIVFC, and (c) retinal flow SPIVFC.

16.6.2.2 **Imaging IVFC**

The SPIVFC system can be modified to capture the corresponding cell image for each detected PMT peak. A schematic of the system is shown in Figure 16.5b. In the imaging SPIVFC, an acoustooptic modulator (AOM) controlled by a function generator is used to create pulses from a laser used for single or repeated strobe illumination for imaging. This laser is added into the excitation path by a beam splitter, and it illuminates the sample 25 μm laterally displaced from the counting slit [68]. On the detection side, the excited fluorescence signal that passes through the confocal slit is separated by another beam splitter to the PMT fluorescence signal detection path or to the CCD camera fluorescence imaging path [68]. Fluorescence images are captured by a frame grabber and stored in the computer [68]. The PMT signal from a fluorescently labeled cell that traverses the slit and is detected above a certain voltage threshold triggers the function generator that activates the AOM and CCD to capture the image of the cell corresponding to the detected PMT signal [68]. A limitation of the current design is that fluorescence images are captured only from cells with very high or oversaturated PMT peak voltages, likely because the CCD is a less sensitive detector when compared to a PMT [68].

By incorporating an imaging system to SPIVF, cell verification is possible, which allows for more accurate enumeration of the cell population of interest. For example, by capturing images of fluorescently labeled cells, a single cell moving at a slow speed resulting in a wide PMT peak width is distinguishable from a cluster of CTCs that result in a similar peak width. Similarly, the actual number of cells traveling in close proximity to each other resulting in one PMT peak as if the signal originated from a single cell can be confirmed [68]. In addition, using imaging IVFC, researchers were able to ignore 1,1′-dioctadecyl-3,3,3′,3′-tetramethylindodicarbocyanine (DiD)-labeled non-T cells from their enumeration of target cells, that is, DID-labeled T cells, by comparing cell sizes from the captured images. For cells that are labeled *ex vivo*, the captured images and PMT signals can be used to assess uniformity of the labeling procedure and potentially improve fluorescence labeling protocols [68]. In summary, incorporating an imaging system in the SPIVFC allows for morphological cell assessment which can result in more accurate enumeration, improving system specificity and sensitivity.

16.6.2.3 **Retinal Flow IVFC**

To improve the blood volume sampled with the SPIVFC ($<1\ \mu l\ min^{-1}$ for blood sampled from a 40 μm diameter blood vessel with $3–5\ mm\ s^{-1}$ average flow velocity), larger arteries or multiple blood vessels can be sampled [69]. The latter was demonstrated by the retinal flow IVFC, a modified SPIVFC capable of sampling multiple veins and arteries around the optical nerve head by incorporating a scanner that circularly steers the excitation laser beam spot at a rate of 4.8 kHz (Figure 16.5c). The collected fluorescence is descanned and then spatially filtered by a confocal slit in front a PMT. The PMT signal is then transferred to a variable-scan frame grabber that displays each blood vessel around the optical nerve head as a straight line. Image streams are recorded for image analysis to verify enumeration of the labeled cells. A confocal microscope is used verify the focal plane and the region probed by the retinal flow IVFC. By probing smaller but more blood vessels

(10 in total), retinal flow IVFC improved the sampling volume by about five times compared to sampling a single blood vessel in the mouse ear. The excitation scanning system can in principle be incorporated to probe blood vessels in locations other than the eye and improve the detection sensitivity of single-color SPIVFC.

16.6.3
Applications in Enumeration of CTCs

Multiple studies have demonstrated the enumeration of CTCs using SPIVFC and its use in improving the understanding of the role of CTCs in cancer progression and metastasis. Tail vein injections of various carcinoma cell lines in mice have shown different depletion kinetics depending on the cell type and host environment. For instance, after tail vein injection, GFP expressing SUM149 primary breast cancer cells depleted to ~50% from initial counts in less than 15 min, while NALM-6 ALL cells depleted by the same percentage in about 30 min post injection.

In a more extensive study, Georgakoudi *et al.* observed differences in the depletion kinetics of two prostate cancer cell lines, with the more highly metastatic MLL cells decreasing more rapidly compared with the less metastatic LNCaP cells (Figure 16.6a). The observation that the number of CTCs detected remains low after days post injection agrees with the expected survival of a small number of CTCs in the circulation that will ultimately form metastasis in distant organs. The faster depletion rate of the MLL cells was correlated with a higher number of cells arrested in the lungs of these mice compared with that of the LNCaP cells. In addition, MLL cells also appeared more likely to survive in the lungs, once arrested there, in agreement with their higher propensity to form metastasis. Results also show different patterns of CTC depletion depending on the host animal. In Copenhagen rats, initial depletion in CTC counts was followed by an increase, hours post injection (Figure 16.6a), while SCID mice only showed a continuous decrease in CTC counts (Figure 16.6b). Moreover, a relatively high number of circulating MLL cells was persistently observed hours post injection in comparison to LNCaP cells. These results demonstrate the potential of SPIVFC to examine the influence of intrinsic cellular and host environment on CTC kinetics and, ultimately, their metastatic potential.

CTC circulation kinetics was also related to homing of cancer cells, in particular ALL NALM-6 cells, to specialized niches in the bone marrow [71]. Using *in vivo* imaging, NALM-6 cells were shown to be attracted to bone marrow regions that expressed chemokine SDF-1 [71]. Homing to SDF-1 was mediated by a chemokine binding to the cell surface receptor CXCR4, which is found to be highly expressed in NALM-6. Images confirmed that CXCR4 receptor blockage, such as by the small-molecule inhibitor AMD3100, resulted in decreased homing of leukemia cells to SDF-1 hot spots [71]. SPIVFC enumeration of NALM-6 following tail vein injection showed a large depletion (>70%) within the first hour, reflective of the rapid homing to the bone marrow [71]. In contrast, >80% of most of the cells treated with AMD3100 or by SDF-1 induced CXCR4 desensitization, and remained in the circulation as a result of inhibited homing to the bone marrow as shown in

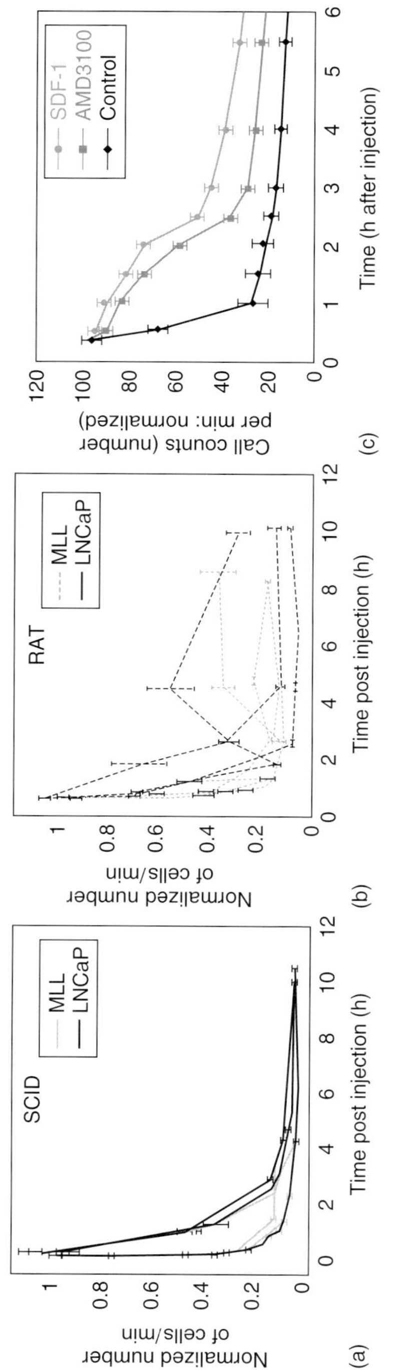

Figure 16.6 (a) Depletion kinetics of circulating two prostate cancer cell lines with different metastatic potential in SCID mice and in (b) Copenhagen rats [74]. Measurements were normalized to number of cells detected immediately after tail vein injection [74]. (c) Change in the circulation kinetics of leukemia cells when cell homing to bone marrow was disrupted by AMD3100 treatment or SDF-1-induced CXCR4 desensitization [71].

Figure 16.6c [71]. Moreover, it was observed that the number of CTCs decreased in correlation to the time when the drug AMD3100 was expected to decline [71]. Thus, SPIVFC in conjunction with *in vivo* imaging can provide a better understanding of the interactions between metastatic cells and host microenvironment, which are important in the development of new tumor treatments.

A more direct use of IVFC to monitor treatment effects was provided by Wei *et al.* [70]. Specifically, prostate cancer cells that were treated with camptothecin chemotherapy, which induced apoptosis to 80% of the cells, were successfully enumerated by SPIVFC following tail vein injection and labeling of fluorescently conjugated annexin-V either *in vitro* (prior to injection) or *in vivo* [70]. By labeling cells directly *in vivo*, cell processing time can be reduced and potential cell loss can be minimized. In a follow-up experiment, untreated cells were injected and their apoptosis in the circulation induced naturally was monitored. Results showed a steady increase of apoptotic cells within the first 3 h [70]. The ability of SPIVFC to monitor apoptosis *in vivo* is significant in the understanding of biological and molecular events contributing to therapeutic responses, which in turn could play an essential role in the development of more effective treatments and in guiding therapeutic interventions.

16.7 Multiphoton IVFC (MPIVFC)

16.7.1 Principles, Advantages, and Limitations

As multiphoton microscopy evolved as an alternative to conventional single-photon confocal microscopy, the application of multiphoton excitation to IVFC after the development of SPIVFC is a natural progression. Similar to confocal microscopy, multiphoton microscopy provides optical sectioning but at depths farther than what can be achieved with confocal microscopy. Penetration depth of multiphoton systems is 500–1000 μm [83, 84]. Multiphoton fluorescence microscopy is based on localized nonlinear excitation of fluorescence in a highly confined volume, which occurs when light of very high intensity is used to illuminate the sample. In two-photon excitation (TPE), two photons with low energy or longer wavelength are absorbed simultaneously, which results in the molecular excitation and subsequent fluorescence emission that would have occurred from single-photon absorption with the photon having approximately double the energy or half the wavelength.

The probability of simultaneous two-photon absorption has a quadratic rather than a linear dependence on light intensity, and since two-photon absorption is significantly less probable than single-photon absorption, multiphoton-based systems require special lasers that can deliver light with high peak power over a short period to maintain a low enough average power and avoid sample photodamage. Ti:sapphire-based lasers are the most commonly used light source for multiphoton excitation. They emit light typically in the 700–1000 nm range in 80–150 fs pulses at a repetition rate of 80–100 MHz [84]. Unlike lasers used in SPIVFC systems, laser sources for multiphoton *in vivo* FC (MPIVFC) are expensive and can require high maintenance, such as cooling and humidity controls, for proper operation [84]. Moreover, dispersion from elements along the optical path, such as mirrors and glasses, can result in pulse broadening, reducing average power, and lowering TPE efficiency. Power can be increased but at the expense of possible sample photodamage or photobleaching. Alternatively, dispersion compensation can be added to the system, making the system more difficult to operate. The cost of special laser sources and optics is one practical disadvantage of MPIVFC when compared to SPIVFC. However, the additional cost and system alignment with MPIVFC is partly offset by the fact that one laser source can excite multiple fluorophores simultaneously. With single-photon-based systems, proper alignment of lasers with respect to each other is critical for ensuring simultaneous excitation of different fluorophores on the same cell.

Alignment of the MPIVFC detection path is also typically simpler than for SPIVFC. In focusing the laser on the sample, TPE probability increases only in the region just around the lens/objective focus where the photon flux is sufficiently high. Away from the focal plane, TPE probability dramatically drops off; therefore, out-of-focus fluorescence is not generated in contrast to single-photon excitation. Therefore, with TPE, optical depth sectioning is achieved naturally without requiring a pinhole in the detection path. Thus, the signal detection path of MPIVFC is simpler and more efficient, requiring only a sensitive detector. Moreover, optical filters along the detection path can easily separate the longer excitation and shorter emission wavelengths associated with TPE fluorescence, in comparison with more overlapping excitation and emission spectra of fluorophores excited using a single photon. However, since the excitation is highly confined, the interrogation volume of MPIVFC can be smaller than that of SPIVFC, requiring a longer sampling duration for an equivalent volume of blood sample. For instance, the SPIVFC system designed by Novak *et al.* had an interrogation slit dimension of 5 μm in width and 100 μm in length with an axial resolution of ±50 μm. In comparison, the MPIVFC setup designed by Tkaczyk *et al.* had an excitation region of 2 μm diameter at the focus and an axial resolution of ±10 μm. Finally, similar to SPIVFC, MPIVFC does not necessarily require beam scanning.

MPIVFC improves the penetration depth limitations of SPIVFC, mainly due to the lower tissue absorption and scattering at the longer excitation wavelengths used (Figure 16.4a,b). Moreover, two-photon absorption in MPIVFC occurs only near the focus; with single-photon excitation; absorption can occur throughout the illumination cone, reducing the number of photons that can reach fluorophores in blood vessels deeper in the tissue. By interrogating deeper blood vessels, which typically are larger and maybe higher in flow rate, the interrogation time can be reduced as the detection rate increases from the larger volume of blood that can be sampled over a period of time.

The same reasons that improve the penetration depth of MPIVFC can also improve the detected SNR. Owing to localized excitation, autofluoresnce outside of the focus is not generated which can decrease SNR. Moreover, light is mainly absorbed by the fluorophores in the focus and not the surrounding tissue, so fluorescence intensity from labeled cells at the focus will be higher. A recent comparison of non-confocal, confocal-based, and multiphoton-based IVFC showed, as expected, that the non-confocal method is inferior to either SPIVFC or MPIVC due to the lack of optical sectioning. MPIVFC measurements yielded an improved detection rate and sensitivity over SPIVFC [72].

16.7.2 Instrumentation

As in SPIVFC, MPIVFC systems continue to evolve from single fluorophore excitation to multilabeling and fiber-based MPIVFC.

16.7.2.1 Two-Color and Extended Laser MPIVFC

A schematic of a two-color MPIVFC setup is shown in Figure 16.7a. Light from a Ti:sapphire laser is focused by an objective into a blood vessel. Dispersion is corrected by precompensating the beam with a prism pair along the laser path [85]. The pulse width is adjusted to increase the excitation light intensity at the focus, maximizing emission signal intensity, as verified by measuring fluorescence from dye solutions [85]. The illumination area at the sample can be a spot or a slit. Fluorescence from the focus is collected by the same objective and split from the excitation wavelength by a dichroic mirror. Fluorescence emission is separated into distinct wavelength regimes and detected by two PMTs. Instead of converting the PMT integrated intensity to a voltage signal, photon counters can be used so the number of photons over a given time bin interval is used for data analysis [73]. As in SPIVFC, a software program is designed to enumerate peaks above a user-defined threshold based on the mean and standard deviation of the detected peak intensities. Information such as peak height, width, and time are stored. *Peak width* can be defined as either full-width half-maximum or width at the threshold. Sampling rate is <25 000 kHz, based on the time resolution of the photon counting scalers, in this case ~41 μs [73].

In an *in vitro* study of GFP cells suspended in phosphate-buffered saline (PBS) compared with cells suspended in blood, results showed lower detection rate, by over 30%, of cancer cells in blood suspension [86]. So although MPIVFC is less susceptible to tissue absorption, scattering, and autofluoresnce than SPIVFC systems, the detected signal can still be degraded by hemoglobin absorption at certain wavelengths and from the amount of overlap between tissue autofluorescence and fluorophore emission spectra [86]. Tkaczyl *et al.* improved SNR from fluorophores by developing a home-built Ti:sapphire laser with a 20 MHz pulse repetition instead of a commercial Ti:sapphire laser with a 76 MHz repetition rate [87]. Because of the quadratic dependence of TPE rate on light intensity, TPE fluorescence intensity increases inversely proportional to the repetition rate for a fixed average power. Improved SNR consequently increased the detection rate of GFP cancer cells in whole blood.

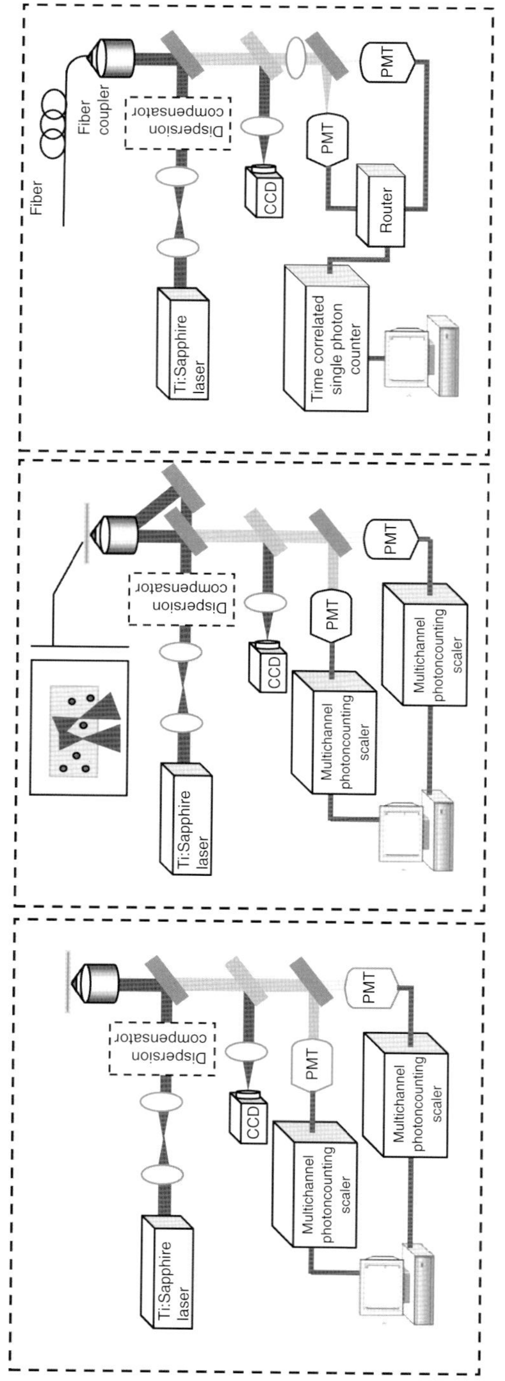

Figure 16.7 Optical schematic of (a) two-color MPIVFC, (b) two-beam MPIVFC, and (c) fiber-based MPIVFC.

16.7.2.2 Two-Beam MPIVFC

In standard FC, cell size information can be obtained from the intensity of the forward scattered light. Since it is not possible to collect forward scattered light directly *in vivo*, cell size can be determined by IVFC through analysis of the peak widths of the detected cells. However, the peak width (or cell travel time in the interrogation region) is also affected by the individual flow of each cell, which can vary naturally as is evident in studies from imaging SPIVFC. By normalizing the peak width to the transient velocity of the cell, size measurements can be made more accurately [85]. To measure cell velocity, the excitation beam can be split and focused into the sample, resulting in two excitation regions laterally separated and lined up along the direction of the flow (Figure 16.7b) [85]. The fluorescence from the two excitation regions is collected by the same objective and directed to two PMTs and photon counting scalers [85]. Initial *in vitro* experiments have been conducted by flowing fluorescent beads of known diameters, two different cell lines (acute T cell leukemia and epidermoid carcinoma), and aggregated cells in glass capillary tubes [85]. The size calculated from the two-beam MPIVFC system was closer to the size distributions of fluorescent beads compared with sizes obtained from peak width analysis using standard single-beam MPIVFC. Results also indicate different size distributions for the two different cell lines [85]. Although it may be possible to differentiate different cells by size, care must be taken in interpreting cell data from two-beam MPIVFC since the size information from labeled CTCs is actually a measure of the volume distribution of the fluorophore in the cell as opposed to the actual size of the cell [85]. However, using this system, signals from cell aggregates are clearly discernible from signals originating from single cells, which is important for accurate enumeration of CTCs [85]. Unfortunately, the application of two-beam MPIVFC *in vivo* is limited to measurements in small capillaries, since in larger blood vessels it would be difficult to differentiate a single cell passing sequentially through the two beams from two different cells passing each beam [85]. Essentially, the cell concentration must be low enough to allow one cell to pass through both beams before the next cell can flow through [85].

16.7.2.3 Fiber-Based MPIVFC

An alternative to increasing the system detection rate is to measure larger blood vessels deeper. To access these deeper blood vessels, fiber-based MPIVFC was developed to bring the excitation and collection optics deep inside the body. In this design, a Ti:sapphire excitation laser is focused by the objective and coupled to a double-clad fiber with a 6 μm diameter core, as shown in Figure 16.7c [86]. Pulses are dispersion-compensated to optimize the intensity out of the fiber [86]. Fluorescence signal is collected by the same fiber and eventually the objective and the same detection path as in two-color MPIVFC. The system's capabilities have been demonstrated both with *in vitro* and *in vivo* circulation models [86]. For *in vivo* measurements, the fiber was inserted into the liver through a needle (fiber-based MPIVFC is minimally invasive compared with other IVFC designs, which are noninvasive) [86]. For fiber-based MPIVFC, it is also necessary that the animal is appropriately restrained to minimize movements that can drag or damage the fiber probe [86]. Moreover, initial measurements *in vivo* indicate that fiber-based IVFC is sensitive to motion artifacts from normal biological conditions, such as respiration, particularly affecting short emission wavelengths, such as those from GFP, more than longer emission wavelengths [86]. Therefore, the choice of fluorophores for measurements is critical for the accuracy of the counts achieved using this approach.

16.7.3
Applications in Enumeration of CTCs

Most animal studies demonstrating the feasibility of IVFC rely on *ex vivo* labeling of isolated cells and injection back into the circulation. However, *ex vivo* labeling of a patient's CTC is not a feasible option if IVFC systems are to be implemented in the clinic. He *et al.* demonstrated that *in vivo* labeling and subsequent *in vivo* detection of CTCs by MPIVFC is possible, similar to *in vivo* labeling and detection of apoptotic CTCs and circulating stem cells by SPIVFC [69, 70]. On the basis of the observation that folate receptors are overexpressed in various carcinomas, He *et al.* designed an assay of fluorescent folate conjugates that were injected in the circulation and directly labeled leukemia cells in Balb/C mice [72]. The clearance kinetics of CTCs labeled by folate, conjugated to rhodamine, was observed [72]. To confirm *in vivo* labeling, leukemia cells that were labeled *ex vivo* with a different fluorophore were injected and then *in vivo* labeled by fluorescently conjugated folate, resulting in correlated peak signals and images of double-labeled cells [72].

The depletion kinetics of breast cancer cells with different metastatic potential were also investigated using MPIVFC. Both cell lines, each of which was labeled with a different fluorophore, were injected into the same mice. The study specifically compared the depletion kinetics of MDA-MB-435 cells with the less metastatic MCF-7 breast cancer cells over a 16 h time course [73]. Results showed that more MCF-7 cells were detected overall compared with the more metastatic MDA-MB-435 [73]. Moreover, MCF-7 cells depleted more rapidly than metastatic MDA-MB-435, which persisted and remained detectable over the 16 h observation period [73]. Forty-eight hours post injection, the lungs and liver were excised and imaged, showing a greater abundance of the highly metastatic MD-MB-435 cells in comparison to less metastatic MCF-7 cells, which are known to go through apoptosis in secondary organs [73]. Results from *ex vivo* imaging and *in vivo* CTC measurements suggest that more MD-MB-435 cells were arrested in the lungs and liver, which accounted for their low CTC count. Meanwhile, the persistence of MD-MB-435 cells in these organs resulted in their slower depletion rate in the circulation [73]. This MPIVFC study demonstrates the feasibility of detecting different cell types in the same host and verifies previous observations from SPIVFC measurements that the depletion rate of CTCs depends on their metastatic potential.

Using fiber-based IVFC, the depletion kinetics of murine sarcoma cells (MCA-207) was investigated by interrogating blood vessels in the mouse liver.

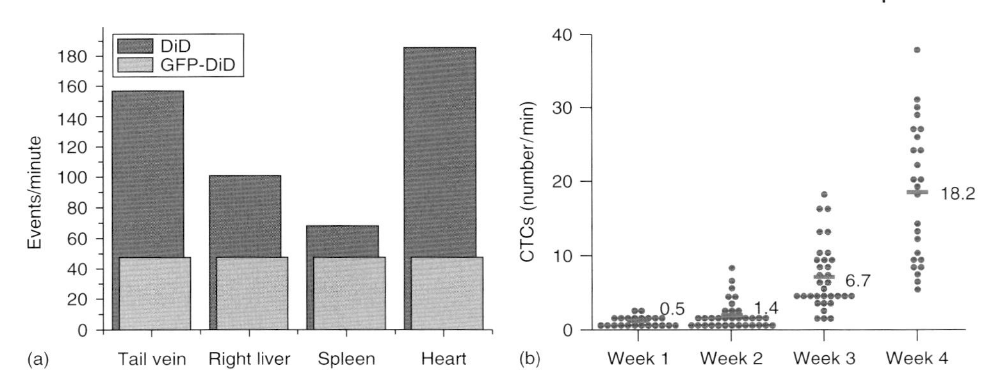

Figure 16.8 (a) Number of cells detected when double-labeled sarcoma cells were injected through different location and (b) change in number of CTCs after tumor induction [72].

Keeping the fiber probe in the liver, cancer cells were introduced into the blood stream via injection in the tail, heart, liver, and spleen [86]. CTC counts decreased over time, as observed by other IVFC studies [86]. The number of detected CTCs was high when cancer cells were introduced directly in the circulation, as by injection in the tail vein and in the heart as shown in Figure 16.8a [86]. CTCs were still observed, though low in numbers, when they were injected in the liver and the spleen, suggesting that cells trapped in these organs can eventually escape into the circulation to form metastases [86]. This study illustrates that by using fiber-based MPIVFC, observations between CTC counts and accessibility of escape into the circulation can be made. Moreover, the ability to access blood vessels of deeper organs in the body has the potential to allow for enumeration of cancer cells as they exit from the primary tumor, potentially allowing for the study of CTCs in relation to tumor vasculature and angiogenesis during tumor development.

In fact, in a tumor induction study, MPIVFC results showed a correlation between the number of detected CTCs and tumor growth. In that study, a tumor model was established by subcutaneous injection of lung cancer cells on the dorsal flanks of mice [72]. Folate conjugates were injected for *in vivo* labeling of circulating CTCs and their number was assessed over a period of four weeks. These measurements demonstrated that CTCs could be observed as early as the first week of implantation even though the number of CTCs was expected to be low during early tumor development [72]. Moreover, the average number of CTCs detected increased exponentially every week in correlation to tumor growth, as shown in Figure 16.8b [72]. Analysis of immunostained and excised tissue samples revealed no metastatic disease at the end of four weeks [72]. However, a few micrometastases were observed in mice one week later (five weeks post injection of tumor cells), suggesting that MPIVFC can quantify CTCs in circulation before metastatic disease is detectable by microscopic examination [72]. This study highlights the application of MPIVFC to sensitive detection of CTCs in relation to primary and early metastatic tumor growth, which is important since CTC enumeration by *in vitro* techniques has been challenging in early stage cancer.

16.8
Summary and Future Directions

The observation of tumor cells circulating in the peripheral blood of patients with malignancies was first described in the mid-nineteenth century [4]. However, only recent technological advances have enabled researchers to characterize and enumerate CTCs. Since metastasis is the leading cause of morbidity in patients with cancer, the number of CTCs is likely to reflect the aggressiveness of the disease, since a portion of CTCs are expected to originate from the primary tumor while the remaining are cells shed from metastatic lesions. The assessment of disease by blood withdrawal is technically simpler and less risky than single or serial biopsies. For these reasons, several researchers have developed *in vitro* CTC assays and detection techniques. The general trend from various such studies indicates a prognostic value for CTCs. Moreover, the enumeration of CTCs can serve as an early marker in assessing a patient's risk for relapse or therapeutic response. However, CTC detection techniques are challenged by the rarity of these cells in circulation and the ability to resolve tumor cells from a large number of host cells. The solution involves extensive blood volume withdrawals and cell processing to enrich rare cells, which ultimately affect system sensitivity and specificity. The variability in results from various experimental designs still questions the reliability of *in vitro* techniques and impedes the support in the use of CTC detection for patient management decisions and disease prognosis in the clinic.

While researchers continue to explore CTCs in clinical settings, animal models are used to further elucidate the complex metastatic process and to validate therapeutic efficacy of novel treatments. In human research studies and animal model systems, long-term assessment of occurrence and enumeration of CTCs are challenging particularly during early stages, when a very small number of disseminated tumor cells are in the peripheral blood. In animal studies, *in vitro* CTC detection is even more complicated by the small volume of blood available for analysis, making kinetic studies particularly difficult. In a study testing CellSearch for CTC enumeration in murine xenograft models, researchers were only able to detect CTCs in tumor-bearing mice by cardiac puncture due to the larger volume of blood that can be obtained with this blood withdrawal technique with little cell contamination from normal epithelial cells [75]. Sensitivity to CTCs and measurement reliability are further hampered by blood sampling and extensive processing on limited blood samples. Therefore, by monitoring CTCs directly in the circulation, serial measurements by SPIVFC or MPIVFC are simpler and allow for repeated or more continuous measurements in the same animal without disturbing the natural biological processes.

By combining fluorescence excitation and optical sectioning, IVFC is sensitive to flowing cancer cells in underlying blood vessels. SPIVFC with confocal detection provides an inexpensive way of achieving depth-sectioning for measurements *in vivo*. MPIVFC improves the depth penetration of SPIVFC, but requires specialized laser sources and other optical equipment. Using either SPIVFC or MPIVFC, circulation kinetics has been correlated with metastatic potential as verified by imaging techniques and histopathological analysis. Variations of IVFC systems have evolved to increase penetration depth and system sensitivity and specificity simply by incorporating additional components such as optical fibers, scanners, and CCD cameras. System modifications enabled verification of non-tumor cells and CTC clusters, improving enumeration accuracy. Both SPIVFC and MPIVFC systems can excite and detect the emission from multiple fluorophores, such as fluorescent proteins and exogenous chromophores, which is important for multilabeling CTCs and for detecting simultaneously two different populations of CTCs. Such capabilities offer opportunities for examining a number of interesting questions related to the role of CTCs in cancer development and metastasis. For example, we can study the circulation kinetics and metastatic potential of two cell types that are derived from the same parent cell line but have slightly different cellular characteristics, such as the expression of a particular chemokine receptor. Alternatively, one can develop a better assessment of patient therapeutic response and prognosis by simultaneously monitoring nonviable/apoptotic and viable CTCs. Such studies could result in the discovery and development of novel therapeutic molecules that disrupt the metastatic process. Similar to results using CellSearch in murine xenograft models, results from IVFC showed a correlation between the number of CTCs detected and tumor development, but without the complications of blood draws and sample processing that can affect the specificity and sensitivity of CTC detection.

Nevertheless, IVFC for the detection and enumeration of CTCs is not without challenges. Because of the nature of metastasis, CTCs are expected to express a variety of markers and with different levels of expression. As in most *in vitro* CTC detection techniques, IVFC relies on CTC labeling, which ideally should be performed *in vivo*. If IVFC systems are to be applied to the clinic, the effect of *in vivo* labeling of CTCs to system performance must be better characterized. To increase sensitivity and specificity, *in vitro* CTC detection techniques employ labeling for multiple markers, an approach that can similarly be utilized in IVFC systems, as already demonstrated by detection of double-labeled CTCs. Since no marker is 100% specific, molecular studies of CTCs and the development of new tumor markers are also essential for CTC characterization to increase specificity of both *in vitro* and *in vivo* CTC detection techniques.

Current SPIVFC and MPIVFC systems are also limited to interrogating shallow peripheral blood vessels, resulting in the sampling of a small volume of blood. System improvements such as scanning of multiple blood vessels and use of fiber probes can increase the volume of blood that can be sampled. However, future research should clearly define the volume of blood that must be sampled to statistically minimize false positives and false negatives. Moreover, system designs must be developed to address motion artifacts and maintain the blood vessel in focus during the required interrogation period. Lastly, IVFC systems must address the toxicity associated with fluorophore conjugated markers. Studies can be performed to clearly assess the concentration of labeling agents that can be tolerated by the patient, yet still allow for suitable detection of CTCs *in vivo*. At the same time, the development of new fluorophores clinically approved for use in humans is critical for IVFC systems to transition into the clinic.

To partly address the dependence of IVFC systems on the use of exogenous labels, a label-free IVFC system was recently developed. Coherent anti-Stokes Raman scattering (CARS) IVFC is based on CARS imaging microscopy, which is a nonlinear technique that is chemically selective to vibrational modes of molecular species, such as lipids. Initial *in vivo* studies with CARS IVFC explored the relationship between excess lipid and cancer aggressiveness [88]. CARS images confirmed from the origin of CARS signal from CTCs that expressed both green and red fluorescent proteins [88]. Results suggested that excess free fatty acids from a high-fat diet resulted in the early appearance and higher number of CTCs, as well as in increased lung metastases after subcutaneous injection of lung carcinoma cells in Balb/C mice [88]. Corresponding *in vitro* studies indicated that free fatty acids induced intracellular lipid accumulation and concentration of cellular content toward one cell pole, which reduced cell-to-cell contact, increased chemotactic motility, and increased surface adhesion capabilities [88]. While this study clearly demonstrated the potential use of an IVFC approach that does not require exogenous labels, CARS IVFC is currently limited to enumeration of CTCs in veins with slow flow rate and small diameter so that the cells flow in a controlled manner across the interrogation volume to reduce signal variability [88]. In addition, a distinct lipid profile is required for the cell population of interest.

As in CARS IVFC, phothermal (PT)/photoacoustic (PA)-based IVFC is capable of detecting unlabeled cells based on chemical selectivity, specifically on the absorption of certain biomolecules in the cells [89]. By exploiting the unique absorption signatures of certain biomolecules that vary in different cell populations, PT/PA-based IVFC can differentiate between cancer cells in the circulation from normal blood cells. PT signal characteristics, such as amplitude, rise time (from cell heating), and decay (from cell cooling) reflect the type of absorbing molecules, its distribution, as well as its concentration in the cells. For example, the RBC's uniform distribution of hemoglobin in contrast to locally distributed cytochromes in leukocytes resulted in larger PT amplitude and faster signal decay from the faster relaxation time [89].

Melanoma, which contain the absorbing molecule melanin in high concentrations as compared to normal blood cells, are ideal cells to study metastasis with PT/PA IVFC. *In vivo* studies with melanoma cells have shown that PT/PA IVFC systems can identify cancer cells from leukocytes and RBCs in 50 and 150 μm blood vessels in the mice (ear and skin) tumor models without the use of exogenous labeling [90]. Within the first week of induction, researchers observed PA signals near the tumor margin corresponding to the intravasation of individual or aggregated melanoma cells in the initial metastatic process [90]. Similar to results in kinetic studies with fluorescence IVFC, researchers observed an increase in the number of circulating cells correlated with tumor size growth over four weeks after tumor induction [90]. Although detection of unlabeled cancer cells has been demonstrated, the differentiation of other types of cancer cells with similar biomolecules as normal blood cells is still challenging with PT/PA-based techniques [91, 92]. In these cases, cancer cells are typically labeled with gold nanopoarticles both to enhance their PA signal and differentiate them from normal leukocytes and RBCs. Details pertaining to the theoretical basis, system designs, and studies based on PT/PA IVFC are described in Chapters 14 and 17.

Before IVFC systems can be implemented in clinical settings, technical improvements such as the ones discussed above will need to be developed. Moreover, further studies are clearly needed to verify *in vivo* results and compare performance to commonly used *in vitro* assays. However, the development of better methods for staging cancer, monitoring patient therapeutic response, and assessing patient risk can improve drug development, clinical management, and, most importantly, increase patients' quality of life. Therefore, the optimization of IVFC systems for use in the clinic is an endeavor worth pursuing.

Acknowledgments

The authors wish to acknowledge Charles Lin, who introduced one of the authors (IG) to the world of IVFC, and Charlotte Kuperwasser, who has had plenty of discussions with the authors on the biological significance of CTCs. In addition, Steven Boutrus and Derrick Hwu have been valuable laboratory members in this area. Finally, funding for the IVFC-related work of the authors has been provided by NIH **R21 CA110103**.

References

1. Pantel, K. and Brekenhoff, R. (2004) Dissecting the metastatic cascade. *Nat. Rev. Cancer*, **4**, 448–456.
2. Chambers, A.F., Groom, A.C., and MacDonald, I.C. (2002) Dissemation and growth of cancer cells in metastatic sites. *Nat. Rev. Cancer*, **2**, 563–572.
3. Fidler, I. (1990) Critical factors in the biology of human cancer metastasis: twenty-eight G.H.A. Clowes memorial award lecture. *Cancer Res.*, **50**, 6130–6138.
4. Ashworth, T.R. (1869) A case of cancer in which cells similar to the tumors were seen in the blood after death. *Med. J. Aust.*, **14**, 146–149.
5. Zieglschmid, V., Hollman, C., and Bocher, O. (2005) Detection of disseminated tumor cells in peripheral blood. *Crit. Rev. Clin Lab. Sci.*, **42**, 155–196.
6. Fidler, I.J. (2003) The pathogensis of cancer metastasis: the 'seed and soil' hypothesis revisited. *Nat. Rev. Cancer*, **3**, 1–6.
7. Gupta, G.P. and Massague, J. (2006) Cancer metastasis: building a framework. *Cell*, **127**, 679–695.
8. Peterlini-Brechot, P. and Benali, N.L. (2007) Circulating tumor cells (CTC)

detection: clinical impact and future directions. *Cancer Lett.*, **253**, 180–204.

9. Lee, J.M., Dedhar, S., Kalluri, S., and Thompson, E.W. (2006) The epithelial–mesenchymal transition: new insights in signaling, development, and disease. *J. Cell Biol.*, **172**, 973–981.
10. Radinsky, D.C., Levy, D.D., Littlepage, L.E. *et al.* (2005) MMP-3-induced Rac1b stimulates formation of ROS, causing EMT and genomic instability. *Nature*, **436**, 123–127.
11. Liotta, L. and Kohn, E.C. (2001) The microenvironment of the tumor-host interface. *Nature*, **411**, 375–379.
12. Mostert, B., Slijfer, S., Foken, J.A., and Gratama, J.W. (2009) Circulating tumor cells (CTCs): detection methods and their clinical relevance in breast cancer. *Cancer Treat. Rev.*, **35**, 463–474.
13. Muller, A., Homey, B., Hortensia, S. *et al.* (2001) Involvement of chemokine receptors in breast cancer metastasis. *Nature*, **410**, 50–56.
14. Allan, A. and Keeney, M. (2010) Circulating tumor cell analysis: technical and statistical considerations for application to the clinic. *J. Oncol.*, **2010**, 1–10.
15. Allard, W.J., Matera, J., Miller, M.C. *et al.* (2004) Tumor cells circulate in the peripheral blood of all major carcinomas but not in healthy subjects or patients with nonmalignant diseases. *Clin. Cancer Res.*, **10**, 6897–6904.
16. Rolle, A., Gunzel, R., Pachmann, U. *et al.* (2005) Increase in the number of circulating disseminated epithelial cells after surgery for non-small cell lung cancer monitored by MAINTRAC is a predictor for relapse: a preliminary report. *World J. Surg. Oncol.*, **3**, 18.
17. Tanaka, F., Yoneda, K., Kondo, N. *et al.* (2009) Circulating tumor cells as diagnostic marker in primary lung cancer. *Clin. Cancer Res.*, **15**, 6980–6986.
18. Naoe, M., Ogawa, Y., Morita, J. *et al.* (2007) Detection of circulating urothelial cancer cells in the blood using the cellsearch system. *Cancer*, **109**, 1439–1445.
19. He, W., Wang, H., Hartman, L.C. *et al.* (2007) In-vivo quantification of rare circulating tumor cells by multiphoton intrvital flow cytometry. *Proc. Natl. Acad. Sci. U.S.A.*, **28**, 11760–11765.
20. Pierga, J., Bidard, F., Mathiot, C. *et al.* (2008) Circulating tumor cell detection predicts early metastatic relapse after neoadjuvant chemotherapy in large operable and locally advanced breast cancer in a phase II randomized trial. *Clin. Cancer Res.*, **14**, 7004–7010.
21. Lobodasch, K., Frohlich, F., Rengsberger, M. *et al.* (2007) Quantification of circulating tumor cells for the monitoring of adjuvanttherapy in breast cancer: an increase in cell number at completion of therapy is a predictor of early relapse. *Breast*, **16**, 211–218.
22. Cristofanilli, M., Budd, G.T., Ellis, M.J. *et al.* (2004) Circulating tumor cells, disease progression and survival in metastatic breast cancer. *N. Engl. J. Med.*, **351**, 781–791.
23. Weitz, J., Kienle, P., Lacroix, J. *et al.* (1998) Dissemination of tumor cells in patients undergoing surgery for colorectal cancer. *Clin. Cancer Res.*, **4**, 343–348.
24. Seiden, M.V., Kantoff, P.W., Krithivas, K. *et al.* (1994) Detection of circulating tumor cells in men with localized prostate cancer. *J. Clin. Oncol.*, **12**, 2634–2639.
25. Camara, O., Rengsberger, M., Egbe, A. *et al.* (2007) The relevance of circulating epithelial tumor cells (CETC) for therapy monitoring during neoadjuvant (primary systematic) chemotherapy in breast cancer. *Ann. Oncol.*, **18**, 1484–1492.
26. Allen-Mersh, T.G., McCullough, T.K., Patel, H. *et al.* (2007) Role of circulating tumor cells in predicting recurrence after excision of primary colorectal carcinoma. *Br. J. Surg.*, **84**, 96–105.
27. Ito, S., Nakanishi, H., Hirai, T. *et al.* (2002) Quantitative detection of CEA expressing free tumor cells in peripheral blood of colorectal cancer patients during surgery with real-time RT-PCR on a LightCycler. *Cancer Lett.*, **183**, 195–203.
28. Pachmann, K., Camara, O., and Kavallaris, A. (2008) Monitoring the

relapse of circulating epithelial tumor cells to adjuvant chemotherapy in breast cancer allow detection of patients at risk of early relapse. *J. Clin. Oncol.*, **26**, 1208–1215.

29. Moreno, J.G., Miller, M.C., Gross, S. *et al.* (2005) Circulating tumor cells predict survival in patients with metastatic prostate cnacer. *Urology*, **65**, 713–718.
30. de Bono, J.S., Scher, H., and Montgomery, R.B. (2008) Circulating tumor cells predict survival benefit from treatment in metastatic castration-resistant prostate cancer. *Clin. Cancer Res.*, **14**, 6302–6309.
31. Cohen, S.J., Punt, C.J.A., Iannotti, N. *et al.* (2009) Prognostic significance of circulating tumor cells in patients with metastatic colorectal cancer. *Ann. Oncol.*, **20**, 1223–1229.
32. Cohen, S.J., Punt, C.J.A., Iannotti, N. *et al.* (2008) Relationship of circulating tumor cells to tumor response progression-free survival, and overall survival patients with metastatic colorectal cancer. *J. Clin. Oncol.*, **26**, 3213–3221.
33. Cristofanilli, M., Hayes, D.F., Budd, G.H. *et al.* (2005) Circulating tumor cells: a novel prognostic factor for newly diagnosed metastatic breast cancer patients. *J. Clin. Oncol.*, **7**, 1420–1430.
34. Budd, G.T., Cristofanilli, M., Ellis, M.J. *et al.* (2006) Circulating tumor cells versus imaging-predicting overall survival in metastatic breast cancer. *Clin. Cancer Res.*, **12**, 6403–6409.
35. Leukemia Classifications (2010) *http://cancer.org* (cited 13 May 2010).
36. Reis, L.A., Smith, M.A., Gurney J.G. *et al.* (1999) *Cancer Incidence and Survival Among Children and Adolescents: United States SEER Program 1975–1995*, National Cancer Institute, SEER Program, Bethesda, MD, NIH Pubication No. 99-4649. *http://seer.cancer.gov/publications/childhood/index.html* (cited 2010).
37. Jemal, A., Siegel, R., Ward, E. *et al.* (2007) Cancer statistics 2007. *CA Cancer J. Clin.*, **57**, 43–66.
38. Gajjer, A., Ribeiro, R., Hancock, M.L. *et al.* (1995) Persistence of circulating blasts after 1 week of multiagent chemotherapy confers a poor prognosis in childhood acute lymphoblastic leukemia. *Blood*, **86**, 1292–1295.
39. Rautonen, J., Hovi, L., and Siimes, M.A. (2010) Slow disappearance of peripheral blasts cells: an independent risk factor indicating poor prognosis in children with acute lymphoblastic leukemia. *Blood*, **71**, 989–991.
40. De Angulo, G., Yuen, C., Palla, S.L., Anderson, P.M., and Zweidler-McKay, P.A. (2008) Absolute lymphocyte count is a novel prognostic indicator in ALL and AML. *Cancer*, **112**, 407–415.
41. Campana, D. and Pui, C.H. (1995) Detection of minimal residual disease in acute leukemia: methodologic advances and clinical significance. *Blood*, **6**, 1416–1433.
42. Ciudad, J., San Miguel, J.F., Lopez-Berges, M.C. *et al.* (1998) Prognostic value of immunophenotypic detection of minimal residual disease in acute lymphoblastic leukemia. *J. Clin. Oncol.*, **16**, 3774–3781.
43. Viehmann, S., Teigler-Schlegel, A., Bruch, J. *et al.* (2003) Monitoring minimal residual disorder by real-time quantitative reverse transcription PCR (RQ-RT-PCR) in childhood acute myeloid leukemia with AML1/ETO rearrangement. *Leukemia*, **17**, 1130–1136.
44. Pachmann, K., Clement, J.H., Schneider, C.P. *et al.* (2005) Standardized quantification of circulating peripheral tumor cells from lung and breast cancer. *Clin. Chem. Lab. Med.*, **43**, 617–627.
45. Egland, K.A., Liu, X.F., Squires, S. *et al.* (2006) High expression of cytokratin-sassociatd proteins in many cancers. *Proc. Natl. Acad. Sci. U.S.A.*, **103**, 5929–5934.
46. Kummar, S., Forgasi, M., Canova, A. *et al.* (2002) Cytokeratin 7 and 20 staining for the diagnosis of lung and colorectal adnocarcinoma. *Br. J. Cancer*, **86**, 1884–1887.

47. Goeminne, J.C., Guillaume, T., and Symann, M. (2000) Pitfalls in the detection of disseminated non-hematological tumor cells. *Ann. Oncol.*, **11**, 785–792.
48. Sleiffr, S., Gratama, J.W., Sieuwert, A.M. *et al.* (2005) Circulating tumor detection on its way to routine diagnostic implementation? *Eur. J. Cancer*, **43**, 2645–2650.
49. Krivacic, R.T., Ladanyi, A., Curry, D.N. *et al.* (2004) A rare-cell detector for cancer. *Proc. Natl. Acad. Sci. U.S.A.*, **101**, 10501–10504.
50. Shin, S.J., Hyjek, E., Early, E., and Knowles, D.M. (2006) Intratumoral heterogeneity of her-2/neu in invasive mammary carcinomas using fluorescence in situ hybridization and tissue microarray. *Int. J. Surg. Pathol.*, **14**, 279–284.
51. Kahn, H., Presta, A., Yang, L. *et al.* (2004) Enumeration of circulating tumor cells in the blood of breast cancer patients after filtration enrichment: correlation with disease stage. *Breast Cancer Res. Treat.*, **86**, 237–247.
52. Bui, M. and Khalbuss, W.E. (2005) Primary small cell neuroendocrin carcinoma of the urinary bladder with coexisting high-grade urothelial carcinoma: a case report and a review of the literature. *Cytojournal*, **2**, 1–5.
53. Marchevsky, A.M., Gal, A.A., Shah, S., and Koss, M.N. (2001) Morphometry confirms the presence of considerable size overlap between 'small cells' and 'large cells' in high-grade pulmonary neuroendocrine neoplasms. *Am. J. Clin. Pathl.*, **116**, 466–472.
54. Mohamed, H., Murray, M., Turner, J.N., and Caggana, M. (2009) Isolation of tumor cells using size and deformation. *J. Chromatogr., A*, **1216**, 8289–8295.
55. Vona, G., Sabile, A., Louha, M. *et al.* (2000) Isolation by size of epithelial tumor cells. *Am. J. Pathol.*, **156**, 57–63.
56. Pinzani, P., Salvadori, B., Simi, L. *et al.* (2006) Isolation by size of epithelial tumor cells in peripheral blood of patients with breast cancer: correlation with real-time reverse transcriptase polymerase chain reaction results and feasibility of molecular analysis by laser microdissection. *Hum. Pathol.*, **37**, 711–718.
57. Ring, A.E., Zabaglo, L., Ormerod, M.G. *et al.* (2005) Detection of circulating pithelial cells in the blood of patients with breast cancer: comparison of three techniques. *Br. J. Cancer*, **92**, 906–912.
58. Harnet, M. (2007) Laser scanning cytometry: understanding the immune system in situ. *Nat. Rev. Immunol.*, **7**, 807–904.
59. Goodale, D., Phay, C., Postenka, C.O. *et al.* (2009) Characterization of tumor cells dissemination patterns in preclinical models of cancer metastasis using flow cytometry and laser scanning cytometry. *Cytometry A*, **75A**, 344–355.
60. Tibbe, A.G., Miller, M.C., and Terstappen, L.W. (2007) Statistical considerations for enumeration of circulating tumor cells. *Cytometry A*, **71**, 154–162.
61. Cruz, I., Ciudad, J., Cruz, J.J. *et al.* (2005) Evaluation of multiparameter flow cytometry for the detection of breast cancer tumors in blood samples. *Am. J. Clin. Pathol.*, **123**, 66–74.
62. Gross, H.J., Verwer, B., Houck, D. *et al.* (1995) Model study detecting breast cancer cells in peripheral blood mononuclear cells at frequencies as low as 10^{-7}. *Proc. Natl. Acad. Sci. U.S.A.*, **92**, 537–541.
63. Hsieh, B., Marriunucci, D., Bethel, K. *et al.* (2006) High speed detection of circulating tumor cells. *Biosens. Bioelectron.*, **21**, 1893–1899.
64. Riethdorf, S., Fritsche, H., Muller, V. *et al.* (2007) Detection of circulating tumor cells in peripheral blood of patients in metastatic breast cancer: a validation study of the CellSearch system. *Clin. Cancer Res.*, **13**, 920–928.
65. Nagrath, S., Sequist, L.V., Maheswaran, S. *et al.* (2007) Isolation of rare circulating tumor cells in cancer patients by microchip technology. *Nature*, **450**, 1235–1239.
66. Alix-Panabieres, C. and Pantel, K. (2008) EPISPOT Assay: detection of proteins secreted by viable disseminating tumor cells in solid tumor patients. *AARC Educ. Book*, **1**, 611–615.

67. Novak, J., Georgakoudi, I., Wei, X. *et al.* (2004) In-vivo flow cytometers for real-time detection and quantification of circulating cells. *Opt. Lett.*, **29**, 77–79.
68. Lee, H., Alt, C., Pitsillides, C. *et al.* (2006) In-vivo imaging flow cytometer. *Opt. Express*, **14**, 7789–7797.
69. Butros, S., Greiner, C., Hwu, D. *et al.* (2007) Portable two-color in-vivo flow cytometers for real-time detection of fluorescently-labeled circulating cells. *JBO Lett.*, **12**, 020507-1–020507-3.
70. Wei, X., Sipkins, D.A., Pitsillides, C.M. *et al.* (2005) Real-time detection of circulating apoptotic cells by in-vivo flow cytometry. *Mol. Imaging*, **4**, 415–416.
71. Sipkins, D.A., Wei, X., Wu, J.W. *et al.* (2005) In-vivo imaging of specialized bone marrow endothelial microdomains for tumor engraftment. *Nat. Lett.*, **16**, 969–973.
72. He, W., Wang, H., Hartman, L.C. *et al.* (2007) In-vivo quantification of rare tumor cells by multiphoton intravital flow cytometry. *Proc. Natl. Acad. Sci. U.S.A.*, **104**, 11760–11765.
73. Tkaczyk, E.R., Zhong, C.F., Ye, J.Y. *et al.* (2008) In-vivo monitoring of multiple circulating cell populations using two-photon flow cytometry. *Opt. Commun.*, **281**, 888–894.
74. Georgakoudi, I., Solban, N., and Novak, J. (2004) In-vivo flow cytometry: a new method for enumerating circulating cancer cells. *Cancer Res.*, **64**, 5044–5047.
75. Eliane, J., Repollet, M., Luker, K.E. *et al.* (2008) Monitoring serial changes in circulating human breast cancer cells in murine xenograft models. *Cancer Res.*, **68**, 5529–5532.
76. Shapiro, H.M. (2004) Excitation and emission spectra of common dyes. *Curr. Protoc. Cytom.*, Unit 1.19.1–1.19.7.
77. Yang, M., Baranov, E., Jiang, P. *et al.* (2000) Whole-body optical imaging of green fluorescent protein-expressing tumors and metastases. *Proc. Natl. Acad. Sci. U.S.A.*, **97**, 1206–1211.
78. Naumov, G.N., Wilson, S.M., MacDonal, I.C. *et al.* (1999) Cellular expression of green fluorescent protein, coupled with high-resolution in-vivo videomicroscopy, to monitor steps in tumor metastasis. *J. Cell Sci.*, **112**, 1835–1842.
79. Collier, T., Follen, M., Malpica, A., and Richards-Kortum, R. (2005) Sources of scattering in cervical tissue: determination of the scattering coefficients by confocal microscopy. *Appl. Opt.*, **44**, 2072–2081.
80. Richards-Kortum, R. (1996) Quantitative optical spectroscopy for tissue diagnosis. *Annu. Rev. Phys. Chem.*, **47**, 555–606.
81. Georgakoudi, I., Jacobson, B.C., Van Dam, J. *et al.* (2001) Fluorescence, reflectance, and light-scattering spectroscopy for evaluating dysplasia in patients with Barrett's esophagus. *Gastroenterology*, **120**, 620–1629.
82. Na, R., Stender, I., and Wulf, H.C. (2001) Can autofluorence demarcate basal cell carcinoma from normal skin? A comparison with protoporphyrin IX fluorescence. *Acta Derm. Venerol.*, **81**, 246–249.
83. Rubart, M. (2004) Two-photon microscopy of cells and tissue. *Circ. Res.*, **95**, 1154–1166.
84. Georgakoudi, I., Rice, W., Hronik-Tupaj, M., and Kaplan, D. (2008) Optical spectroscopy and imaging for noninvasive evaluation of engineered tissues. *Tissue Eng.*, **14**, 321–340.
85. Zhong, C.F., Tkaczyk, E.R., Thomas, T. *et al.* (2008) Quantitative two-photon flow cytometry – in-vitro and in-vivo. *J. Biomed. Opt.*, **13**, 034008-1–034008-19.
86. Chang, Y.C., Ye, J.Y., Thomas, T.P. *et al.* (2010) Fiber-optic multiphoton flow cytometry in whole blood and in-vivo. *J. Biomed. Opt.*, **15**, 047004-1–047004-9.
87. Tkaczyk, E.R., Tkaczyk, A.H., Katnik, S. *et al.* (2008) Extended cavity laser enhanced two-photon flow cytometry. *J. Biomed. Opt.*, **14**, 041319-1–041319-12.
88. Le, T.T., Huff, T.B., and Cheng, J. (2009) Coherent anti-stokes Raman scattering imaging of lipids in cancer metastasis. *BMC Cancer*, **9**, 42.
89. Zharov, V.P., Galanzha, E.I., and Tuchin, V.V. (2005) Integrated

photothermal flow cytometry in vivo. *J. Biomed. Opt.*, **10**, 51502-1–51502-13.

90. Galanzha, E.I., Shankov, E.V., Spring, P.M. *et al.* (2009) In vivo, non-invasive, label-free detection and eradication of circulating metastatic melanoma cells using two-color photoacoustic flow cytometry with a diode laser. *Cancer Res.*, **9**, 7926–7934.

91. Zharov, V.P., Galanzha, E.I., and Tuchin, V.V. (2007) Photothermal flow cytometry in vitro for detecting and imaging of individual moving cells. *Cytometry*, **71A**, 191–206.

92. Galanzha, E.I., Kokoska, M.S., Shashkov, E.V. *et al.* (2009) In vivo fiber photoacoustic detection and photothermal purging of metastasis targeted by nanoparticles in sentinel lymph nodes at single cell level. *J. Biophoton.*, **2**, 528–539.

17
In vivo Photothermal and Photoacoustic Flow Cytometry

Valery V. Tuchin, Ekaterina I. Galanzha, and Vladimir P. Zharov

17.1
Introduction

Flow cytometry (FC) is one of the fruitful and powerful techniques used in cell biology and in the diagnostics of cellular-molecular diseases [1, 2]. There are numerous physical phenomena and techniques based on them, among which the ones that use optical phenomena, such as light absorption, scattering, polarization, fluorescence, Raman scattering, are used in FC. Owing to significant progress in the development of optical measurement technologies and understanding of the interaction of light with cells and tissues, now some of the optical imaging technologies enable not only *in vitro* but also quantitative *in vivo* FC [3].

One of the prospective approaches adopted in *in vivo* FC is based on laser-induced photothermal (PT) and photoacoustic (PA) effects in single cells in flows [4–10]. Most PT and PA methods do not require conventional labeling and are not sensitive to light scattering, and thus represent the advanced alternatives for developing label-free *in vivo* FC [11–21]. In these methods, light absorption by cellular components, most of which naturally have no or weak fluorescence, is measured by monitoring thermal and related effects taking place directly in cells as a result of nonradiative relaxation of the absorbed energy in heat. For nonfluorescent samples, PT methods with laser sources currently offer the highest sensitivity for the absorption coefficient, on the order of 10^{-5}–10^{-6} cm^{-1} [4], which makes it possible to noninvasively (a short-term temperature elevation ≤ 1–$5\,^{\circ}C$) detect a single, unlabeled biomolecule with a sensitivity comparable to that of laser-fluorescence methods (i.e., with labeling) [22]. Because many cellular structures are weakly fluorescent in native state and may create molecular assemblies/clusters, laser energy absorbed by these clustered structures is transformed mostly into heat and then into amplified sound, which makes PA "listening" of individual cells promising for cell biology as an alternative (or supplementary) to fluorescent "seeing" them.

These methods, in particular, PA spectroscopy and time-resolved two-beam (pump-probe) PT spectroscopy, have been used successfully in biochemistry to

Advanced Optical Flow Cytometry: Methods and Disease Diagnoses, First Edition. Edited by Valery V. Tuchin.

study the kinetics of relaxation processes in biomolecules in solution; photochemical and photosynthetic reactions; spectroscopy of hemoproteins such as oxidized and reduced cytochrome (Cyt) *c*, hemoglobin, and other cellular components; monitoring the redox state of the respiratory chain in liquid and thin-layer chromatography; and other related biochemical applications [4–6, 9, 10, 21–27]. Photothermal imaging (PTI) has demonstrated the capability of noninvasively visualizing nonfluorescent absorbing cellular structures, at the living single-cell level in suspension or on a substrate, that are not visible by other techniques [21]. In particular, the thermal-lens microscope on a microchip platform has been successfully used to study the redistribution of cellular Cyt-*c* during apoptosis [24], carcinoembryonic antigen (a marker of colon cancer) in human serum [25], and immunoglobulin molecules labeled with gold nanoparticles (GNPs) on the surface of red blood cells (RBCs) [26]. Another modification of the PT method, via detection of a phase-shift between the two spatially separated beams of an interferometer in frequency-domain mode, has provided high-resolution imaging of 1.7–80 nm GNPs for protein detection in cells [27]. The importance of nanoscale imaging using nanoparticle technologies for many biomedical problems, including *in vivo* cytometry, is underlined in Ref. [28].

In this chapter, we are discussing the basic principles of PT/PA techniques in the context of their application to *in vivo* cytometry, and their integration with other optical methods including transmittance digital microscopy (Chapter 14), and fluorescence (Chapter 16). For illustration of the novel facilities of the technique in comparison with *in vitro* cytometry, we present many examples of *in vivo* FC in preclinical studies of lymph and blood vessels in animal models. It includes cell velocity measurements, real-time monitoring, and enumeration of circulating blood and lymph cells, bacteria, cancer cells, contrast agents, and nanoparticles. Perspectives on early diagnostics and monitoring of diseases, including cancer and infection, as well as feedback control during therapy are also discussed.

17.2
Photothermal and Photoacoustic Effects at Single-Cell Level

17.2.1
Basic Principles

The PT method detects the time-dependent heat generated in a cell via interaction with pulsed or intensity-modulated optical radiation [4–17]. Such an interaction induces a number of thermal and other accompanying effects in a living cell; in particular, it causes the generation of acoustic waves. The detection of acoustic waves is the basis for the PA (termed also optoacoustic, OA) method. The informative features of this method in general enable the estimation of cell thermal, optical, and acoustical properties that depend on the peculiarities of the cell structure.

Typically, two modes are used for excitation of cell thermal response: (i) a pulse of light (usually pulsed laser) excites the cell and the signal is detected in the time

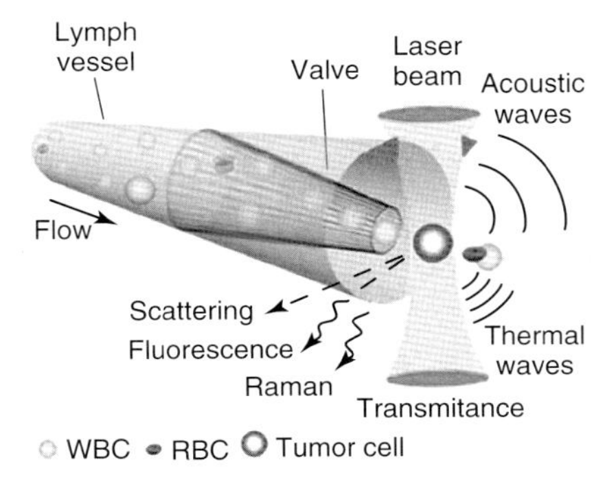

Figure 17.1 Principle of *in vivo* lymph flow cytometry [18].

domain with a fast thermal or acoustic detector attached to a wideband amplifier; signal averaging and gating techniques are used to increase the signal-to-noise ratio; (ii) an intensity-modulated light source (high-intensity lamp or continuous wave (CW) laser) and a wideband or resonance narrow-band transducer are exploited, respectively, to excite and detect PT and PA signals; the measurement is in the frequency-domain; phase-sensitive detection (lock-in-amplification of the signal at the modulation frequency) is used for noise suppression.

In every case, the thermal waves generated by the heat release result in several effects, which have given rise to various techniques [4–17]: photothermal (or optothermal) radiometry (PTR/OTR) (detection of infrared thermal emission); PA or OA methods; photorefractive techniques: thermal blooming, thermal lensing, probe-beam refraction, interferometry, and deflectometry (detection of refractive index gradients above and inside the sample); opto-geometric technique (surface deformation or volume changes); and optical calorimetry or laser calorimetry.

A schematic representation of generation of PT and PA effects applied to cell flow studies is given in Figure 17.1 as components of multimodal optical FC platform, which may also include detection of the intensity of transmitted and scattered light, fluorescence, and Raman signals. Cells move across to the excitation laser beam passing its beam waist; the light wavelength is tuned to an absorption line of the cell chromophore of interest [29] (or exogenous contrast agents); and the optical energy is absorbed by the cell. In most cell chromophores, the collisional quenching rate is significantly higher than the radiative rate; therefore significant part of the absorbed energy transforms to heat. The time-dependent heating leads to all of the above-mentioned thermal and accompanying effects. In PA techniques, a ultrasound (US) transducer that is in acoustic contact with the sample is typically used as a detector to measure the amplitude or phase of the resultant acoustic wave (Figure 17.2). Heating the medium changes its refractive index. The change in the refractive index of the sample can be detected either directly by means of an interferometer, or by a probe laser beam that changes its shape (converging or diverging) (thermal-lens) or is deflected when it passes the region excited by the pump beam [4, 6, 9–15, 20, 21].

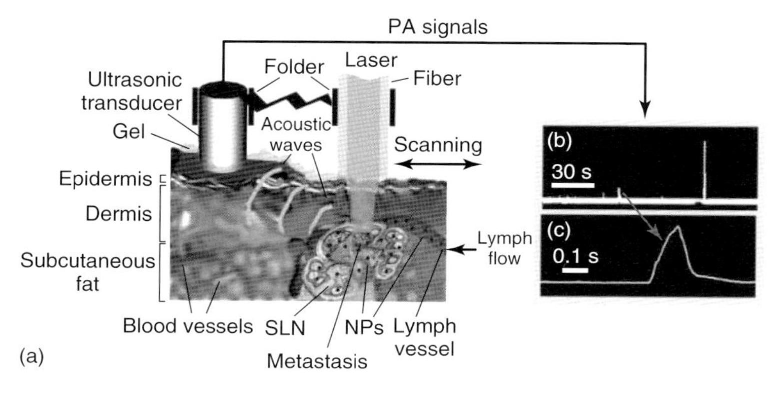

Figure 17.2 *In vivo* fiber-based integrated PA flow lymph cytometry, PA lymphography, and PT therapy. (a) Schematic. (b) Typical trace of digitized signals from flowing melanoma cells in animal ear lymphatics. Each signal is based on the acquisition of ~40 PA signals induced by a high-pulse-rate diode laser. Laser parameters: wavelength, 905 nm; pulse width, 15 ns, pulse rate, 9 kHz; laser fluence, 90 mJ cm^{-2}. (c) Temporal shape of one zoomed digitized signal [19].

The interaction of laser radiation with the absorbing micro- or nanotargets within the cell (which could be intrinsic like a small-scale variations of hemoglobin or melanin respectively within RBC or melanoma cells, and/or exogenous labeling nanoparticles) may be accompanied by many different PT phenomena (Figure 17.3). Laser-beam absorption by a target may produce: (i) heating (temperature rise ΔT); (ii) secondary infrared radiation (PTR technique); (iii) photomechanical stress (ΔF); (iv) thermal expansion (ΔL); (v) OA phenomena – acoustic and shock waves (Δp); (vi) vaporization, bubble formation, and cavitation (at high light energies); and (vii) alterations of the index of refraction (Δn). In general, all of these phenomena may be used to detect cell absorbing inclusions. However, not all of them are suited well to provide fast and sensitive measurements needed at the mode of *in vivo* cytometry.

Two PT-based techniques are mostly suitable for cell examination in blood or lymph streams – they are: PA, PT thermal-lensing, and PTI technique in different modifications. In the PA method, information about an absorbing inhomogeneity size and location can be obtained through measurement of the width of the PA pulse wave and the time delay before it reaches the acoustic transducers. These waves are produced by the thermal expansion of a heated absorber or the thin layer of surrounding liquid. The time-resolved, nonscanning PTI mode can be realized by irradiating moving single cells of interest with a pump pulse, which produces laser-induced temperature-dependent variations in the refractive index in the cells translated into images with a second, collinear probe laser pulse, and with a charge-coupled device (CCD) camera using phase-contrast [30–34], multiplex thermal-lens [36], or thermal-lens [35] schematics. PT images are calculated as the difference between the two probe pulses, the first immediately before the pump pulse and the second together with the pump pulse [30]. In the thermal-lens mode,

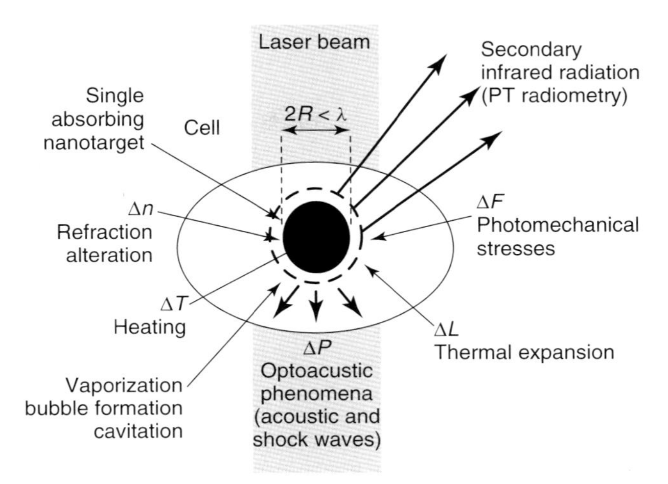

Figure 17.3 Laser-induced PT and related phenomena around an absorbing micro- or nanotarget (endogenous or exogenous inclusions) of radius R in a cell [21].

a pump-laser-induced refractive heterogeneity around the cells causes defocusing of a collinear probe beam, leading to a change in the beam's intensity at its center as detected by a photodiode with a pinhole.

17.2.2 Signal Description

The intensity of the signals obtained with any of the PT or PA techniques depends on the amount of energy absorbed and transformed into heat and on the thermoelastic properties of the sample and its surroundings. Assuming that nonradiative relaxation is the main process in light beam decay, extinction is not very high, and $\mu_a l \ll 1$ (l is the length of a cylinder within the sample occupied by a pulse laser beam), the absorbed energy can be estimated on the basis of Beer's law [4–10, 21]:

$$E_\mathrm{T} \cong E\mu_\mathrm{a} l \tag{17.1}$$

where E is the incident pulse energy and μ_a is the absorption coefficient.

Energy absorption causes an increase in the local temperature ΔT, which is defined by the relation

$$\Delta T = E_\mathrm{T}/c_\mathrm{p} V\rho \cong E\mu_\mathrm{a} l/c_\mathrm{p} V\rho = E\mu_\mathrm{a}/\pi \mathrm{w}^2 c_\mathrm{p}\rho \tag{17.2}$$

where c_p is the specific heat capacity for a constant pressure, $V = \pi w^2 l$ is the illuminated volume, w is the laser-beam radius, and ρ is the medium density. Assuming an adiabatic expansion of an illuminated volume at a constant pressure, one can calculate the change in the volume:

$$\Delta V = \pi(w + \Delta w)^2 l - \pi w^2 l = \beta V \Delta T \cong \beta E\mu_\mathrm{a} l/c_\mathrm{p}\rho \tag{17.3}$$

where Δw is the change in radius of a cylinder illuminated by a laser beam caused by a local temperature increase and β is the coefficient of volumetric expansion.

This expansion induces a wave propagating in a radial direction with the speed of sound. The corresponding change of pressure Δp is proportional to the amplitude of mechanical oscillations $\Delta x_T \sim \Delta w$:

$$\Delta p = 2\pi f_{ac} \upsilon_{ac} \rho \Delta x_T \sim f_{ac} \upsilon_{ac} \rho \Delta w \tag{17.4}$$

where f_{ac} is the frequency of acoustic oscillations and υ_{ac} is the velocity of acoustic waves in a medium.

Using Equation 17.4 and taking into account that $\Delta w \ll w$, we can finally find

$$\Delta p \sim (f_{ac}/w)(\beta \upsilon_{ac}/c_p) E \mu_a \tag{17.5}$$

Information on the absorption coefficient μ_a at a specific wavelength can be obtained from the measurements of the pressure change Δp (PA technique). Using the relation between the focal length of the "thermal lens" f_T, the deflection angle of a probe laser beam φ_d, and the phase-shift in a measuring interferometer $\Delta\Psi$ with the change in a sample temperature ΔT, the approximate expressions describing the photorefractive methods can be written in the form [4–10] for a "thermal lens" technique as

$$(1/f_T) \approx d_p (dn/dT)(\Delta T/w^2) \tag{17.6}$$

and a probe-beam deflection technique as

$$\varphi_d \approx (1/n)(dn/dT)\Delta T \tag{17.7}$$

and a phase shifting (interferometry) technique as

$$\Delta\Psi \approx (2\pi d_p/\lambda_p)(dn/dT)\Delta T \tag{17.8}$$

where dn/dT is the refractive index temperature gradient of the medium (tissue), d_p is the length of the space where the exciting and the probe laser beams are overlapped, and λ_p is the wavelength of the probe beam. These expressions could be easily rewritten to account for the spatial distributions of the temperature-induced refractive index and the focusing–defocusing and deflection parameters.

The time delay (or termed *relaxation time*) between optical, thermal, and acoustic pulses is an important parameter in time-resolved PT (or PA) techniques. For example, the pulse PA method can be characterized by the time delay (or acoustic relaxation time) between short optical and acoustical pulses [4, 5]:

$$t_{ac} \cong R/\upsilon_{ac} \tag{17.9}$$

where R is the radius of the target (at large beam diameters, $w_p > R$) and υ_{ac} is the velocity of acoustic waves in a target.

The thermal relaxation time for the PT technique is defined by the time of the "thermal" wave propagation transverse to the probing laser beam with a radius w_p [4, 5]

$$\tau_T \approx (w_p/2.4)^2/k_T \tag{17.10}$$

where

$$k_T = k/\rho c_p \tag{17.11}$$

is the thermal diffusivity of the medium, k is the heat conductivity, and ρ is the density of the medium.

17.3 PT Technique

17.3.1 PT Spectroscopy

As in conventional absorption spectroscopy, the basic information provided by laser PT spectroscopy is the absorption coefficient wavelength dependence $\mu_a(\lambda)$. However, the difference is that instead of a path-dependent result, which is the product $\mu_a(\lambda) \times d$, where d is the length of the sample, in PT spectroscopy, the measured signals are directly proportional to the absorption coefficient (Equations 17.2 and 17.5). For example, in conventional absorption spectroscopy to measure absorption in a 300 nm long optical path (the typical diameter of an elongated mitochondrion) with an accuracy of ~0.5% in measuring light variation, a minimum detectable coefficient of absorption of $(\mu_a)_{min} \approx 1.5 \times 10^2\ \text{cm}^{-1}$ is necessary [4]. This sensitivity is insufficient to study cellular structures that typically have a lower coefficient of absorption and size. Furthermore, typically the average absorption within a whole cell or a cell layer is measured, which prevents spectral identification of localized cellular components.

PT with tunable laser sources provides sensitivity to absorption at least four to five orders of magnitude higher than that in conventional absorption spectroscopy, and the capability of measuring subcellular localized absorption spectra with a spectral width of 0.1–1 nm [4–6]. In particular, PT spectral identification of just three main cellular components, such as Cyt *c*, P450, and Cyt-*c* oxidase, with distinguishable absorption bands near 450, 550, and 700–900 nm, respectively [29], may provide many potential applications of PT spectroscopy in cell biochemistry, as these components are involved in such important cellular processes as apoptosis, drug kinetics, and cellular energy, respectively. The tunable lasers in visible and near-IR spectral ranges provide the opportunity to identify additional cellular components such as Cyt-*a* (605 nm) and Cyt-*c* oxidase with the copper A center (Cu_A) in reduced (620 nm) and oxidized (820 nm) states, and with the copper B center (Cu_B) in oxidized (680 nm) and reduced (760 nm) states. Measurement of PT responses from oxidized (760 nm) and reduced (900–950 nm) hemoglobin may make it possible to estimate the oxidization of RBCs in flow at the single-cell level. Additionally, using specific absorption bands for RBCs (415–425 nm (Oxy 415, deOxy 425), 542 nm (Oxy), 578 nm (Oxy) and 554 (deOxy), and 760 nm (deOxy); close to maximal difference in absorptions between Oxy and deOxy is at 675 nm, close to minimum at 760 nm, while isosbestic wavelength 805 nm is not to be dependent on blood volume) and white blood cells (WBCs) (525–565 and 750–830 nm, determined by absorption of Cyt-c and Cyt-*c* oxidase, respectively), the presence of rare RBCs among WBCs in lymph flow can be identified [31]. To identify abnormal cells (e.g., cancer cells among normal WBCs) that do not differ significantly in absorption, but differ in the specific PT parameters related to cell shape and structural parameters at specific response times or PT-response amplitude, shape, and duration, the correlation between these parameters and the morphologic cell type can be analyzed.

17.3.2
PT Scanning Image Cytometry

The technical platform for PT cell imaging is microscopy, which enables great flexibility in choosing the signal acquisition algorithms and simplifies cell manipulation. However, to create images, most described techniques require time-consuming scanning of a focused laser beam across the cells. For example, with the PT thermal-lens technique, cell imaging with a resolution of ~1 μm was achieved by laser scanning of one cell for ~1 h [24]. Thus, scanning technology is suitable only for a high-resolution imaging of nonmoving a live cells. To avoid time-consuming conventional scanning with a strongly focused laser beam across the cells and to make possible dynamic cell studies, the adaptation of a wide-field PTI [30, 36] for *in vivo* flow cytometric measurements was performed [31]. In this scheme, the PT image from one cell is obtained in 0.1 s after just one laser shot with an 8 ns pulse width and a broad beam diameter of 15–25 μm covering the entire cells. However, for *in vitro* or *in vivo* conditions when cells are unmovable, PT images from different cells could be captured only one by one. Thus, the laser irradiates one cell first, and then, the position of the laser-beam center must be moved to another cell and the laser then irradiates this cell, and so on. This is a so-called stepwise scanning, which is much faster than the conventional one and allows imaging of cell chains or aggregates. Discrete changes in the position of the laser beam in PT microscopy takes very little time (0.5–5 s) compared to that for conventional continuous laser scanning (0.5–1 min). PT analysis of many cells (≥1000), however, may take from 10–15 min to hours, which is not convenient for routine cell study, and requires the FC mode, that is automatically realized in natural *in vivo* conditions.

17.3.3
PT Flow Cytometery (PTFC)

17.3.3.1 PT Image Flow Cytometry

The high-resolution, high-sensitivity PTI can be realized by recording the laser-induced temperature-dependent variations of the refractive index around absorbing micro- or nanoinclusions within a cell (Figure 17.3). PT-image FC systems are based on irradiation of the absorbing nanotargets flowing alone or as components of flowing cells with a focused short laser pump pulse and recording of the thermal effects with phase-contrast (Figure 17.4) [30–32], thermal-lens

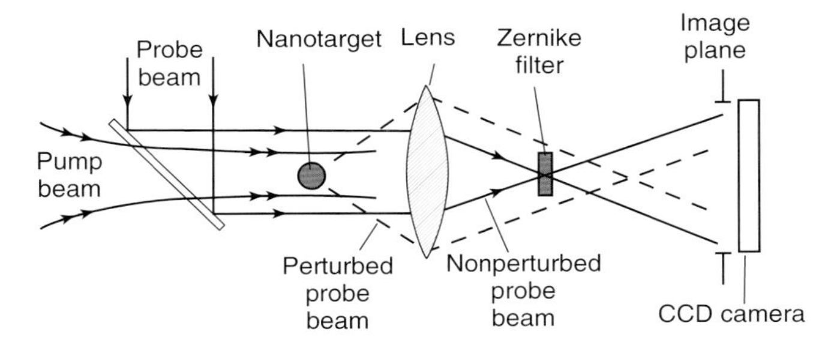

Figure 17.4 Principle of laser-based dual (pump-probe) PT phase-contrast microscopy of micro- or nanoscaled absorbing cell inclusions [12].

[12, 21, 33], and confocal thermal-lens [37] imaging of a second collinear laser pulse that probes the heated cells.

According to Abbe's principle, light cannot create an adequate image of an object whose structural details are smaller than the diffraction limit $R_D \sim 0.61\lambda/\mathrm{NA}$ (i.e., less than the wavelength, λ) of far-field optics [38]. However, image creation based not on the interaction of light with an object itself but on a laser-induced thermal field around the object (Figure 17.5) may overcome Abbe's limitation [12]. This is because the thermal field holds information on the target parameters – size and shape – even during its expansion as a result of heat diffusion. Therefore, analysis of a thermal field's spatial structure when it has grown sufficiently to satisfy Abbe's criterion makes it possible to realize PTI of nanoscaled inclusions and obtain actual information on their sizes. Thus, PT and related effects make it possible to transform nanoscale dimensions invisible with far-field optics into microscale dimensions visible with these techniques, and correlation of microscale with nanoscale events makes it possible to realize nanoscale imaging with far-field optics [12].

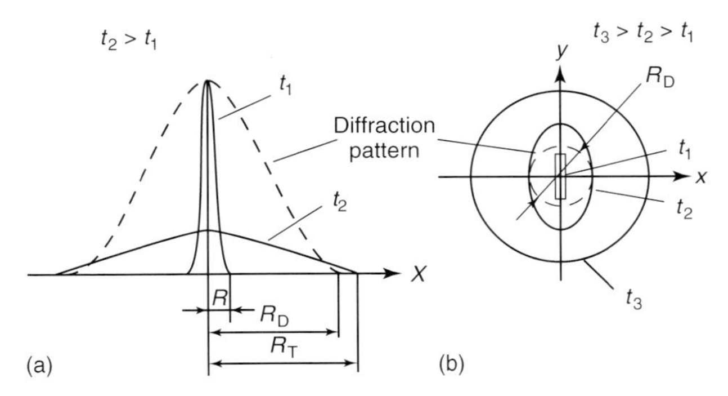

Figure 17.5 Temporal dynamics of thermal fields formed around a nanoobject at a short laser pulse absorption. (a) temperature profiles along the x-axis (solid lines, for times elapsed since laser shot, t_1 and t_2) and diffraction-limited optical image (puncture curve) and (b) margins of temperature distributions for a cylindrical nanotarget at times elapsed since laser shot, $t_3 > t_2 > t_1$ [12].

If the laser pulse width, t_p, is significantly shorter than the thermal relaxation time of the nanoscaled object, τ_T (i.e., $t_p \ll \tau_T$), the initial margin of laser-induced temperature distribution, $\Delta T(r, t)$, will coincide with the target's geometry. For example, for targets with three basic geometries – (i) a sphere with radius R, (ii) a cylinder with radius $R_c \ll \lambda$ and length $l(R_c \ll l)$, and (iii) a planar circle with radius R_p and thickness $d(d \ll R_p)$ – the thermal relaxation time τ_T may be estimated, respectively, as follows [12] (also Equations 17.10 and 17.11):

$$\tau_T = R^2/6.75k_T \tag{17.12a}$$

$$\tau_T = (R_c)^2/4k_T \tag{17.12b}$$

$$\tau_T = d^2/8k_T \tag{17.12c}$$

With Equation 17.12 and the measurement of τ_T by monitoring the target's thermal dynamics, it is possible to obtain information on a target whose size is less than λ when the heat diffusion coefficient k_T is known. For example, for spherical targets with $R = 50$ nm, 500 nm, and 5 μm, estimations of $\tau_T(k_T = 1.44 \times 10^{-3}$ cm^2 s^{-1} for water parameters, Equation 17.11) are approximately 4 ns, 0.4 μs, and 40 μs, respectively. The theoretical limit for this method is characterized by the time of transformation of the absorbed light energy into heat (i.e., at $\tau_T = \tau_{NR}$), which is for typical parameters ~0.1–1 nm minimal size. It is interesting to note that the acoustic relaxation time t_{ac} (Equation 17.9) is always shorter than the thermal relaxation time τ_T (Equations 17.10–17.12), except for very small targets with sizes less than $R = 6.75\ k_T/\upsilon_{ac}$, which is determined from (Equation 17.9) and (Equation 17.12) at $t_{ac} = \tau_T$. Thus, for an approximately 0.1–1 nm target, when both times are equal, this range seems to be the principal limit for different PT methods.

Because a thermal field keeps the memory of the target's spatial shape, analysis of the shape of the thermal ghost can reveal the target's shape (Figure 17.5b). This approach might be valuable in the near zone of relatively large objects with $R > R_D/5(> 50$ nm). In particular, if the cylinder length, l, or the radius of the planar structure, R_p, exceeds λ, such objects can be partly visible through a diffraction-limited microscope. Measurement of their cooling time (Equations 17.12b and 17.12c) yields additional information on invisible nanoscale parameters, such as cylinder diameter or thickness of the planar structure. In particular, this information could be important for the study of platelets, nuclear chromatin, cell or muscle filament diameter, or ellipsoidal mitochondria, especially for monitoring processes such as the growth or modification of spatial structure under different external impacts (e.g., radiation, drug).

The laser-induced thermal effects around a nano-object can be estimated through monitoring of temperature-dependent changes in the refractive index, Δn, by wide-field phase-contrast imaging of a second coaxial probe laser pulse (Figure 17.4). When the probe beam is passed through this heated area of pathway d_p, phase-shift is obtained, which is given by the Equation [11, 32] (basic Equation 17.8)

$$\Delta\Psi = 2\pi d_p(r)\Delta n(r, t)/\lambda = 2\pi (dn/dT) d_p(r)\Delta T(r, t)/\lambda_p \tag{17.13}$$

The first pump beam, after passing through the object, is cut-off with a rejection filter. A Zernike filter has a circular shape that is more suitable for laser measurement (Figure 17.4). After filtering, interference between the unperturbed and the perturbed components of the probe-beam transforms phase changes into images of spatial-intensity distribution that correlate with the thermal fields in the object plane. This image, captured by a CCD camera and referred to as a *PT image*, can be obtained without laser scanning with just one laser shot with a relatively broad beam diameter [21, 30]. A similar scheme may also realize multiplex thermal-lens imaging without scanning with a CCD camera, in which each spatially confined pixel is equivalent to a single photodetector with a pinhole [12, 36].

With a low numerical aperture (NA), the probe beam is diffracted (at $R \sim \lambda$) or spatially redistributed (at $R \ll \lambda$) on a nano-object in broad angles that are greater than the NA angle of the objective, and only a part of the diffracted energy enters the objective. Initially, temperature distribution coincides with the absorbing target's geometry. Then, heat diffusion leads to an increase in the thermal-spot size, R_T (blurring effects), which becomes larger than the target size R. The maximum PTI sensitivity is expected at $R_T = R_D$, when most diffracted energy enters the objective. Using Equation 17.12, target size can be estimated by measuring the temperature evolution outside the diffraction spot at least for two different moments of time at the same spatial point or at different spatial points at the same time. At $R_T \ll R_D$, loss of diffracted light energy might be compensated for by increased energy of the probe beam. Immersed objectives with a high NA make it possible to collect diffracted (redistributed) energy in an angle aperture close to optimal [12] and hence, monitor the initially high temperature level around an object immediately after the pump laser pulse. This PTI mode may yield the highest sensitivity (for objects as small as approximately 1 nm) through the monitoring over time of the average light-intensity changes in the diffraction spot, which keeps the same size during the rapid cooling process.

The time shape of the PT signal represents a high initial peak, because of fast heating of the target, and a much lower exponential tail, corresponding to the cooling time of the heated target (Figure 17.6). The temperature rises practically instantaneously and in the first moment, repeats the spatial distribution of absorbed energy, depending on the spatial distribution of the pump laser beam and the target's coefficient of absorption, respectively. As mentioned above, when τ_T is measured using Equation 17.12 through the cooling time, it is possible, when k_T is known, to obtain information about the target size R.

17.3.3.2 PTFC System and Its Image Resolution

To realize photothermal flow cytometry (PTFC) based on the above-discussed principles of time-resolved and nonscanning PTI, the technical platform of an upright Olympus BX51 microscope (Olympus America Inc., Center Valley, Pennsylvania) was used [18, 21, 31]. Moving single cells of interest were irradiated with a pump pulse of a tunable optical parametric oscillator (OPO) (wavelength range, 410–2300 nm; pulse width, 8 ns; pulse energy, 0.1–1000 µJ; pulse rate 10 Hz; pumped by a Nd:YAG laser; Lotis Ltd., Minsk, Belarus). Laser-induced

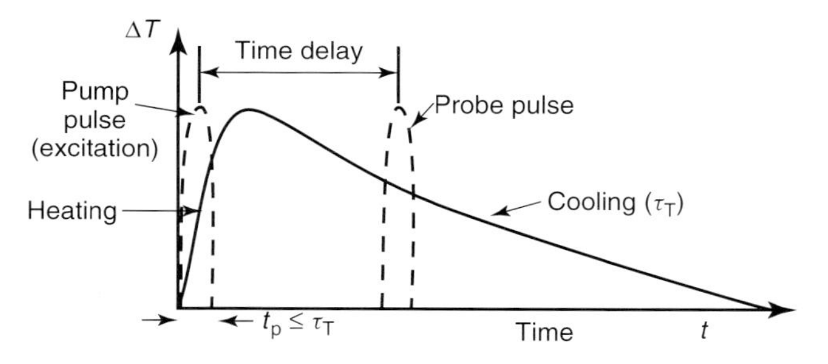

Figure 17.6 Typical time shape of the PT-pulsed response [21].

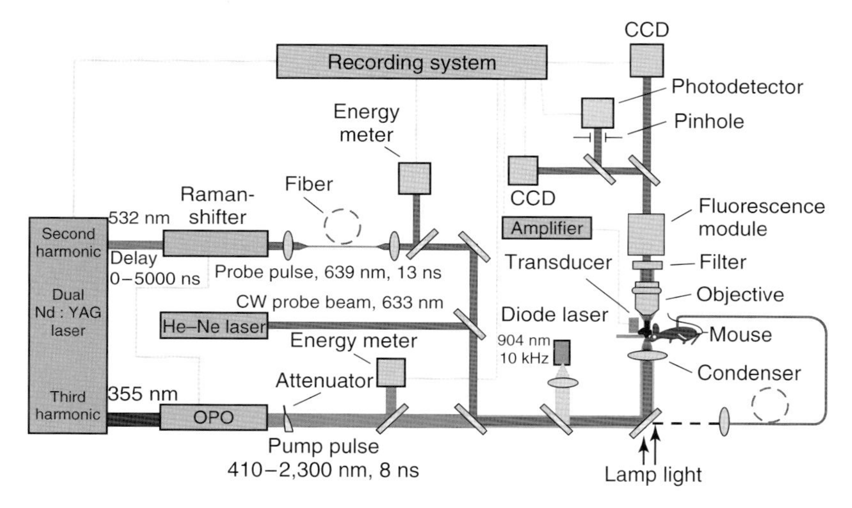

Figure 17.7 *In vivo* flow cytometry integrating photothermal, photoacoustic, fluorescent, and transmission microscopic techniques [18, 31].

temperature-dependent variations in the refractive index in the cells were translated into images with a second, collinear probe laser pulse from a Raman shifter (wavelength, 639 nm; pulse width, 13 ns; pulse energy, 10 nJ) and with a CCD camera (AE-260E, Apogee) having the following parameters: pixel size, 20 μm; area, 512 × 512 pixels; maximum frame rate, 12 fps; and digitization, 14 bits at 1.3 MHz (Figure 17.7).

The diameters of the pump- and probe-beam spots, with smooth intensity profiles controlled by the CCD camera, were adjustable from 15 to 50 μm and from 10 to 40 μm, respectively. Because of the laser beams' relative broadness covering the entire single cells, formation of a PT image required just one pump pulse. The pump pulse was triggered by signals from a photodiode when individual cells in flow crossed the CW helium–neon (He–Ne) laser beam (Figure 17.7) and thus changed its intensity. As a result, PT images of selected individual cells could be captured with each pump laser exposure [17]. PT images were calculated as the difference between the two probe pulses, the first immediately before the pump pulse and the second together with the pump pulse, with the use of customized software [21, 30]. Obtaining an image, which represented a normalization of the differences in the energy of each laser pulse, took 0.1 s.

In particular, in the phase-contrast mode, a customized phase-contrast microobjective (20×) and a Zernike coaxial quarter-wave filter were used to image the probe laser beam [21]. The diameters of the pump and probe-beam spots were varied in the ranges 20–40 and 15–25 μm, respectively. In the presence of gaps between neighboring cells in lymph and blood flow (typical for small lymphatics and blood capillaries), a circular laser-beam geometry (Figure 17.8a,b) was used. At short distances, selected experiments were performed with a linear beam shape (Figure 17.8c). In contrast to transmission digital microscopy (TDM) imaging and velocity measurements (Chapter 14), the PT technique is able to identify low-absorbing cells (e.g., normal and apoptotic WBCs or cancer cells) in blood and lymph flow *in vivo* on the basis of their differences in integral and local absorption associated with specific absorbing chromophores and pigments [21, 30, 39–44].

The PT-image resolution (δ_{PT}) in flow mode depended on the conventional optical resolution of the microscope objective, $\delta_{OPT} = 0.61\lambda/NA$, where λ is the wavelength and NA the numerical aperture value; thermal resolution, $\delta_T = (4\,k_T t_p)^{1/2}$ [12]; motion distortion, $\delta_F = V_F t_p$ due to cell displacement during the exposure or the time between the two probe pulses [44]; and the resolution of the recording system, in particular the CCD camera, $\delta_{CCD} = P/M$ (where P is the pixel size and M the optical magnification) in the plane of analysis:

$$\begin{aligned}\delta_{PT} &= \left\{(\delta_{OPT})^2 + (\delta_T)^2 + (\delta_F)^2 + (\delta_{CCD})^2\right\}^{1/2} \\ &= \left\{(0.61\lambda/NA)^2 + 4k_T t_p + (V_F t_p)^2 + (\delta_{CCD})^2\right\}^{1/2}\end{aligned} \quad (17.14)$$

where k_T is the thermal diffusivity, t_p is the laser pulse duration, and V_F is the flow velocity. For the typical parameters used in experiments: $t_p = 8$ ns, $\lambda \approx 500$ nm, $k_T = 1.44 \times 10^{-3}$ cm^2 s^{-1} (water), NA = 0.4 (20× magnification), $P = 20$ μm, $M = 100$, $\delta_{OPT} \cong 760$ nm, $\delta_{CCD} = 200$ nm (i.e., $\delta_{CCD} < \delta_{OPT}$), $\delta_T \cong 69$ nm, and without

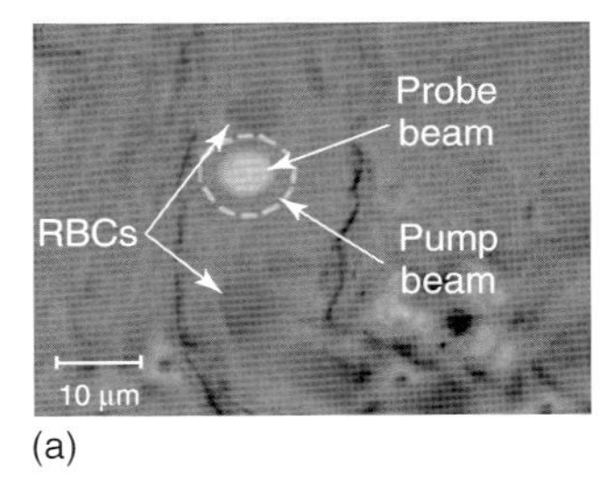

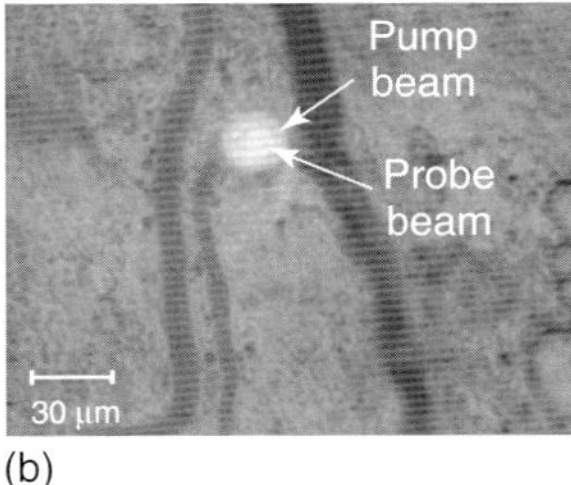

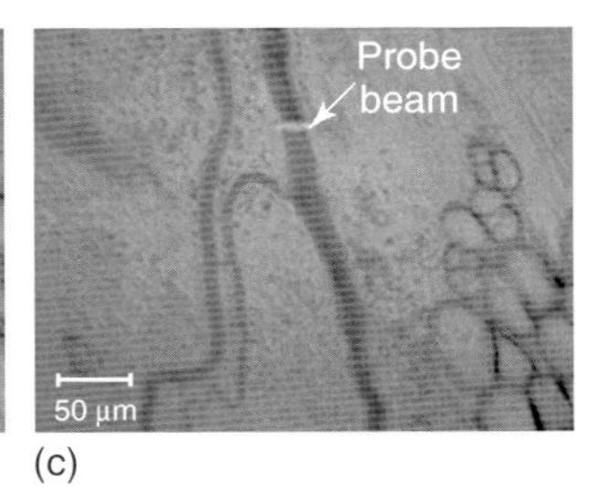

Figure 17.8 Typical positions of probe (red) and pump (green) laser beams during PT imaging. (a) Circular beams in a blood capillary of rat mesentery (cell velocity, 0.5–2 mm s^{-1}; magnification,100×); (b) overlapping pump and probe pulses in an artery of rat mesentery (cell velocity, 2–5 mm s^{-1}; magnification, 10×); and (c) a linear (or ellipsoidal) beam geometry in a blood vessel of rat mesentery (magnification, 10×) [15, 45].

the effect of flow (i.e., at $V_F = 0$), the PT resolution is $\delta_{PT} \cong 763$ nm (i.e., depending mainly on optical resolution). At the highest optical resolution, $\delta_{OPT} \cong 250$ nm (at NA = 1.25), the influence of heat diffusion is still negligible, $\leq 4\%(\delta_{PT} \cong 259$ nm) because of very short laser pulse.

Maximum flow velocity, which does not distort PT resolution during the laser exposure, can be estimated from the component of the lowest resolution (i.e., optical resolution, assuming $\delta_F \leq \delta_{OPT}$ and $\delta_T < \delta_{OPT}$):

$$(V_F)_{max} \leqslant \delta_{OPT}/t_p \quad (17.15)$$

For the above parameters, $(V_F)_{max} \leq 95$ m s^{-1} at a resolution of 763 nm and $(V_F)_{max} \leq 31$ m s^{-1} at a resolution of 259 nm. Thus, theoretically, the short exposure time may allow high-speed high-resolution imaging without reduced resolution at a high flow velocity, comparable to the resolution of conventional FC [1, 2]. This is also valid for a one-probe imaging algorithm suitable for relatively strongly absorbing cells, such as RBCs or melanoma cells. However, use of the differential PT-image algorithm with the currently significant time gap between the two probe pulses may decrease PT resolution in flow. To avoid this problem, a laser with a high pulse repetition rate as well as a high-speed CCD camera operating at 100 kHz and 10^4 fps, respectively, can be used as was demonstrated with TDM for imaging of individual RBCs in fast blood flow [44].

17.3.3.3 PTFC with Thermal-Lens Mode

In PTFC with thermal-lens mode, a pump-laser-induced refractive heterogeneity around cells (a so-called thermal-lens effect) [23, 46] causes defocusing of a collinear, intensity-stabilized, CW laser probe beam. The time-resolved integral PT response of a whole cell is recorded by the low-power beam defocusing detected with a photodiode through a small pinhole. In the PTFC system shown in Figure 17.7, the PT response is measured using a CW He–Ne laser (model 117A, Spectra-Physics, Inc., Mountain View, CA; $\lambda = 633$ nm; power, 1.4 mW), a photodiode C5658 (Hamamatsu Corp., Hamamatsu City, Japan) with a 0.5 mm diameter pinhole, and a Tektronix TDS 3032B oscilloscope, and analyzed with customized software [21].

This mode provides time-resolved (~3 ns) monitoring of the integral PT response, $\Delta T(t)$, from whole single cells, which can be described in the first approximation as follows [12, 21]:

$$\Delta T(t) \approx \Delta T_{max}[\exp(-t/\tau_T) - \exp(-t/\tau_{RT})] \quad (17.16)$$

where ΔT_{max} is the maximum temperature elevation in the irradiated cells, τ_T is the thermal relaxation time (for a spherical cell with radius R, τ_T is defined by Equation 17.12a), and τ_{RT} is the rise time [12]. For small cells (e.g., small-animal RBCs) with radius $R = 2.5$ µm, τ_T is estimated to be 10 µs. The rise time (τ_{RT}) depends mainly on the longest time among the following: nonradiative relaxation time (τ_{NR}), time of response of the photodetector (τ_{PH}), laser pulse width (t_p), and temperature-averaging time within the whole cell (τ_{TA}) determined by heat diffusion from the initially heated local absorbing cellular structures [21]. For the setup (Figure 17.7), in which $\tau_{NR} \approx 10^{-11}$ s, $\tau_{PH} \approx 3$ ns, and $t_p \approx 8$ ns and the cell structure is relatively homogeneous (e.g., RBCs), when $\tau_{TA} \approx 0$, $\tau_{RT} \approx 8.1$ ns; for a heterogeneous cellular structure (e.g., WBCs), $\tau_{TA} \approx 200-400$ ns [21] and hence $\tau_{RT} \approx \tau_{TA}$. Thus, in most cases, $\tau_{RT} < \tau_{T}$ and the PT response demonstrates a high initial peak and a much slower exponential tail corresponding to the cooling time of the heated cell as a whole, with a time range of $\tau_{T} \approx 10-30$ μs, depending on cell size (Figure 17.6). This result corresponds to a theoretical maximum rate of PT cell analysis, f_{cell} ($f_{cell} \approx 1/\tau_{T}$), of 10^4-10^5 cells s^{-1}. To achieve this maximum, lasers with a high pulse repetition rate are required. The presence of a long PT-response tail shape (e.g., for large cells, Figure 17.6) that may overlap the sequent PT response from other cells is not too important in PTFC because cells may flow away from the area of laser exposure after their single exposure.

17.3.3.4 **PTFC Integrating Thermal-Lens and Imaging Modes**

In *in vivo* conditions, cells may travel very quickly through the detected volume, and thus, the concurrent PTI and thermal-lens measurement, which supplement each other and provide additional information on cell types, is a key point in PTFC. Realization of the combined mode involved splitting of the probe beams with semitransparent mirrors (Figure 17.7), with each portion of the beam directed separately to the CCD camera (PTI mode) or the photodetector (PT thermal-lens mode) with the use of a spectral filter (not shown). This filter eliminated the interference of different-wavelength beams (called *cross-talk*), particularly the influence of the CW He–Ne probe beam on the CCD camera and of the pump laser pulse on the photodetector.

With the thermal-lens mode, continuous PT monitoring of moving blood cells can be realized as one of the approaches to PTFC [15, 31]. Figure 17.9 left, (a–c) shows typical PT-signal tracings from cells as a function of the time interval that it takes for the cell to flow through the irradiation area. PT signals from RBCs in flow showing a purely positive component (Figure 17.9 left, a), indicate a standard linear PT response from the cells at a low laser-energy level. The amplitude difference indicates difference in average absorption and shows RBCs' natural heterogeneity. Figure 17.9 left, b shows the detection of single RBC at a low laser-energy level that did not produce notable PT signals from many of the WBCs in lymph flow because of their low absorption. Laser-induced blood vessel injury the nearest lymph vessel led to a fast-growing number of flowing RBCs in lymph vessels (Figure 17.9 left, c) [15]. The next tracing shows that the PT signals from RBCs in the same lymph vessel at a relatively high laser-energy level led to cell damage (Figure 17.9 left, d). The presence of only a negative PT signal indicates the development of strong, dominant nonlinear effects (e.g., microbubble formation around strongly absorbing cellular zones). The presence of both positive and negative components in the PT-signal amplitude tracing (Figure 17.9 left, e) indicates a noninvasive condition for WBCs (which have a high photodamage threshold) and an invasive condition for RBCs (which have a low photodamage threshold) at the same energy level. To demonstrate the capability of PTFC to distinguish cells with different photodamage thresholds, a relatively high laser-energy level and a relatively slow oscilloscope rate were used. The differences in the optical properties of normal and

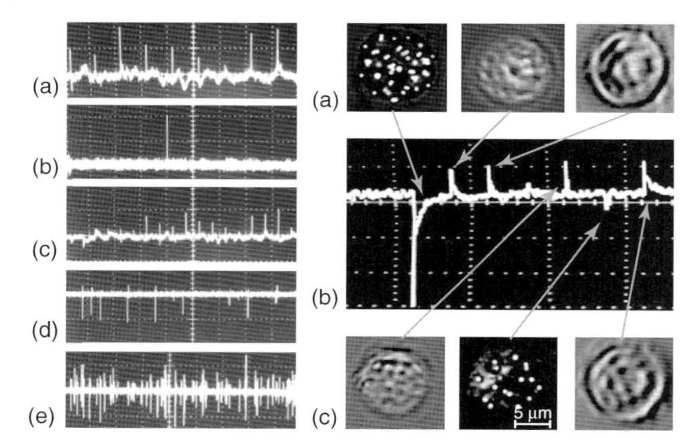

Figure 17.9 Left: Typical tracings of PT signal from blood cells in blood and lymph flows of rat mesentery [15, 31]. (a) RBCs in blood capillary (Figure 17.8a); (b) rare RBC in lymph flow under normal conditions; (c) growing number of RBCs in lymph flow during laser-induced hemorrhage; (d) laser-induced damage of RBCs in lymph flow; and (e) WBCs and RBCs in lymph flow. (a–e) Laser ($\lambda_p = 525$ nm) energy/amplitude/time scale/division: 0.3 μJ/50 mV/100 ms/div, 0.5 μJ/20 mV/1 s/div, 0.6 μJ/100 mV/200 ms/div, 5 μJ/500 mV/4 s/div, and 145 μJ/100 mV/10 s/div, respectively. Right: *In vitro* PT identification of K562 cells labeled with 40 nm gold nanoparticles in artificial flow [39]. (a) and (c) PT images. (b) PT-response amplitude tracings (arrows) (time scale on horizontal axis, 200 ms/div). Laser parameters: $\lambda_p = 525$ nm; pulse width, 8 ns; and laser energy, 35 μJ.

cancerous cells, especially mitochondrial distribution [47], which can be visualized with PTI, allow the use of PTFC to distinguish these cells.

PT identification of cells with similar absorption properties and sizes, such as subpopulations of WBCs, might pose a problem. The high absorption sensitivity of the PT technique offers an opportunity to consider GNPs as a new type of labels for PT, and especially PTFC [13, 21, 25, 27, 28, 39]. These particles have unique properties for PT cell study because they are strong absorbers with adjustable spectral maximum, low toxic, biocompatible, photostable, and can easily be conjugated with proteins and antibodies (details in Section 17.4.4). The PTFC mode may allow high-temporal- and high-spatial-resolution PT monitoring of labeled cells in blood and lymph flow, with quantitation of their time-dependent numbers. In particular, the PT responses from cells with these gold PT labels show specific negative peaks in nonlinear PT responses (associated with nano-and microbubble formation around overheated nanoparticles) compared to the positive amplitude peaks of linear PT responses from cells alone at the same laser-energy level (Figure 17.9 right, c). Simultaneous monitoring of PT images from the same cells confirmed this finding: PT images of cells tagged with gold PT labels (Figure 17.9 right a; left; and c, center) revealed specific, highly localized image contrast, compared with cell PT images showing the relatively smooth contrast of endogenous absorbing structures (e.g., Figure 17.9. right: a, center and right; and c, left and right). The continuous flow speed in this *in vitro* model experiment made it possible to detect up to 10 cells s^{-1}, a rate approximately two orders of magnitude greater than that in conventional PT methods [21]. Besides these two applications, there are many other applications to characterize and differentiate normal and abnormal blood cells as well as to detect cancer cells and pathogens in blood or lymph flows.

In vitro modeling of PTFC was also performed with cell flow realized with a syringe pump-driven system (KD Scientific, Inc., Holliston, MA) and glass tubes with circular and plane geometry ($D_{\text{inside}} = 10–100\ \mu m$ and cross section of $15 \times 20–50\ \mu m^2$). This system provided flow rates in the range of $0.001\ \mu l\ h^{-1}$ to $100\ ml\ min^{-1}$, which corresponds to a minimal linear velocity of $\sim 3.5\ \mu m\ s^{-1}$ and maximal velocity of $200\ m\ s^{-1}$ in a $100\ \mu m$ diameter tube [31, 39]. With this system, no notable change in the classic (heating–cooling phenomena) PT-response amplitude and shape at flow velocities of up to $0.05\ m\ s^{-1}$ was found; with a further increase in the flow velocity to $0.1\ m\ s^{-1}$, a small fluctuation in the PT-response tail appeared, with more profound distortion developing at a flow rate of $2\ m\ s^{-1}$. However, no significant change in the maximum PT-response amplitude was observed even at the high flow velocity, $10–20\ m\ s^{-1}$. The PT images of RBCs of interest at high flow velocities (up to $2–10\ m\ s^{-1}$) without notable motion distortion were also obtained [39].

Such parameters as the amplitude of linear PT-response and nonlinear parameter (δ_{NL}), characterized by the percentage of nonlinear PT responses with a negative component (Figure 17.9), can be used for cell differentiation in flows. Both of these parameters depend on the absorption level. An increase in absorption increases the PT response and parameter δ_{NL}. However, the integral PT response from a cell as a whole depends more on the average absorption, while parameter δ_{NL} depends more on localized absorption, especially the presence of strongly absorbing zones. The PT-response amplitude and parameter δ_{NL} were obtained for every 50 cells and colored calibrated beads ($6.8 \pm 0.5\ \mu m$) (Figure 17.10). These data clearly demonstrate that the use of these two parameters made it possible to more finely distinguish cells with different absorption properties on the basis of their different PT-response amplitudes and different parameters δ_{NL}. Moreover, the use of these two parameters will permit the identification of cells with similar absorption properties but different localized absorption; these cells produce similar PT responses but different values of parameter δ_{NL}. This capability could, for example, be important for identifying RBCs and melanoma cells, which have similar absorption properties in the visible spectral range but different absorbing structures (spatially homogeneous and highly localized, respectively). In particular, the higher parameter δ_{NL} for melanoma cells can be associated with more effective bubble formation around localized zones of strongly absorbing melanin compared with the smooth hemoglobin distribution in RBCs. These features in the formation of linear and nonlinear PT effects may make it possible to detect every single melanoma cell among blood cells *in vitro* and *in vivo*.

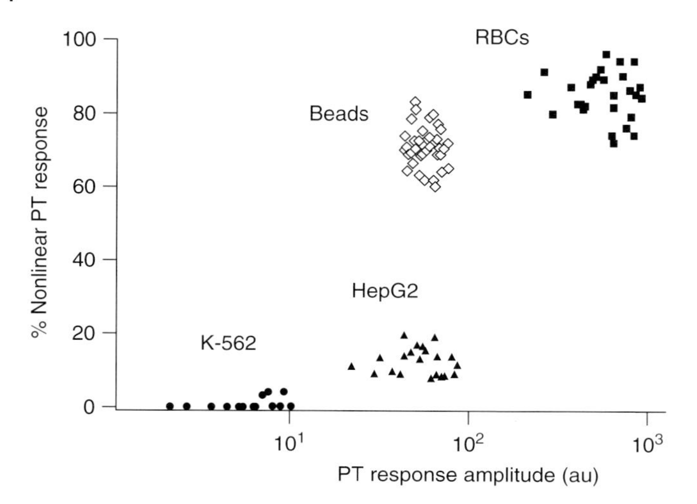

Figure 17.10 Dot-plot display of two types of PT data. PT-response amplitude (horizontal axis) and percentage of nonlinear PT responses (vertical axis) were averaged for every 50 cells for selected samples: rat RBCs, K562 and HepG2 cancer cells, and 6.8 μm diameter colored polystyrene beads [39].

17.4 Integrated PTFC for *In vivo* Studies

17.4.1 General Schematics of the Instrument

The general scheme of integrated multispectral *in vivo* PTFC combines the basic principles of *in vitro* and *in vivo* FC [15, 30, 45, 48–50] (Figure 17.11). See also Chapters 14 and 16.

Cell detection occurs in superficial microvessels in the skin, or in thin, relatively transparent living biostructures (e.g., ear or mesentery), that allow measurement of light transmitted through the vessels. To illuminate selected vessels, different optical sources can be used, ranging from a conventional lamp with filters to lasers in pulsed- or CW mode. Light beams are focused on a small area of the microvessels (the site for analysis). To detect all cells in a cross section of a vessel, the laser-beam diameter must cover the whole vessel. To minimize simultaneous irradiation of several cells along the vessel axis, the beam may have a linear or at least elongated ellipsoidal shape-oriented perpendicular to the vessel axis (Figure 17.8c).

Optical absorption by nonfluorescent cellular structures is measured by PT and PA methods. Specifically, a PT thermal-lens module is incorporated to measure integrated time-resolved PT responses from moving cells, and a PTI module to image moving unlabeled cells or with PT labels (e.g., GNPs) [21, 31]. A PT deflection module with a position-sensitive photodetector is incorporated

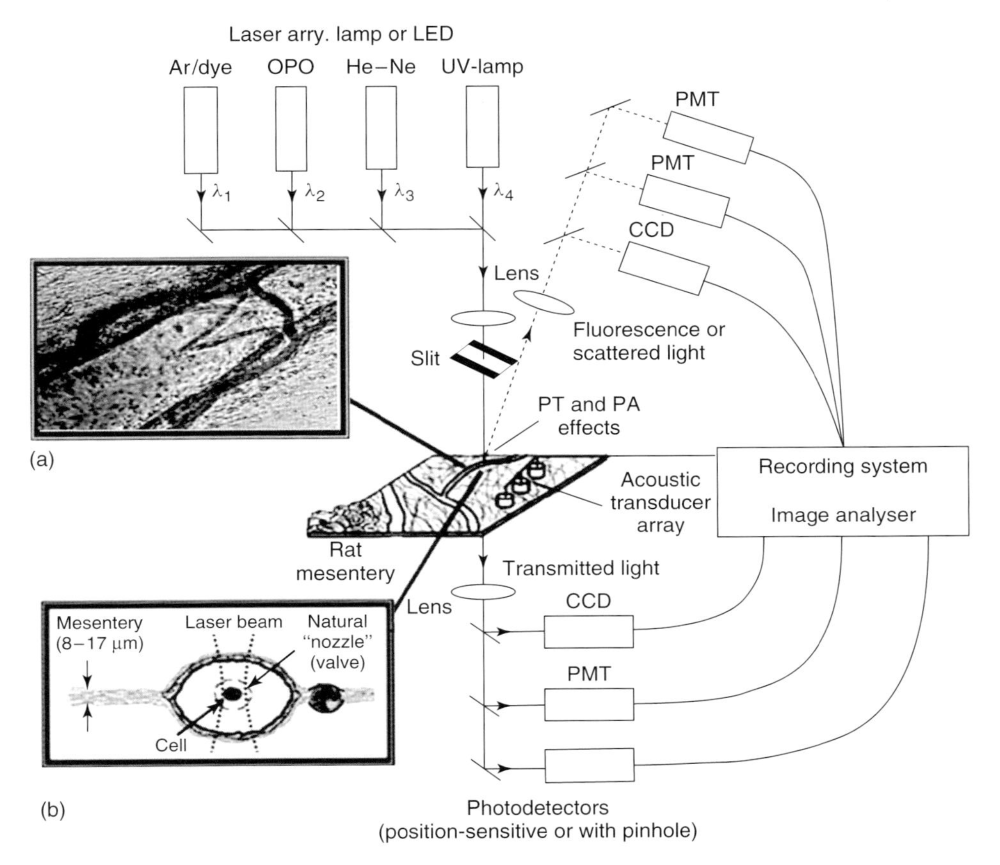

Figure 17.11 Optical scheme of the integrated multispectral flow cytometer for *in vivo* studies, which combines photothermal, optical transmission, and fluorescence techniques; as an animal model for *in vivo* cytometry, rat mesentery with blood and lymph vessels is shown (also two insertions); lymph valve which works as a natural "nozzle" of a cytometer hydrodynamic cell is well seen (also schematic representation in Figures 17.1 and 17.7) [15, 31, 45].

to measure bioflow velocity (PT velocimetry [51]). A PT deflection module with a position-sensitive photodetector is added to measure bioflow velocity using the principle of time-of-flight PT velocimetry [51]). Also, a PA transducer or transducer array allows measurement of PA signals from single cells, either unlabeled or labeled by nanoparticles, as well as measurement of PA signals from cells accumulated at specific sites (vessel wall, surrounding tissue, or specific organ) with a relatively broad laser beam using PA tomography schematics [52–54].

In the PTI mode, a single cell or cells are irradiated with a short, focused pump laser pulse from tunable OPO (Sections 17.3.3.2 and 17.3.3.4). This irradiation leads to an increase in the temperature of local non- or weakly fluorescent cellular absorbing structures. In turn, temperature distribution is transformed into index of refraction distribution. Time-resolved monitoring of temperature-dependent variations in the cell's refractive index is realized with thermal-lens or/and PTI schematics. As mentioned, formation of a PT image required just one pump pulse with a relatively broad beam diameter covering entire single cells. The entire PT image acquisition procedure includes illumination of a cell with three pulses: an initial probe pulse followed by the pump pulse and then a second probe pulse, with a tunable delay with respect to the pump pulse (0–5000 ns). The PT image is calculated as the difference between the two probe images [21] and depends only on absorption contrast rather than on possible phase distortion of the probe beam itself or natural refractive cellular heterogeneities. This technique allows for the capture of PT images of single moving cells with a time delay between laser pulses that is adjustable to cell speed.

The thermal-lens mode made it possible to record a whole cell's time-resolved integral PT response (Sections 17.3.3.3 and 17.3.3.4). The amplitude of PT-response and temporal shape depend on several cellular parameters, such as local absorption coefficient, characteristic relaxation times, and thermodynamic and morphological cell properties. In the pulse mode, the PT response demonstrates a high initial peak because of fast (nanosecond scale) heating of the absorbing structure, and a much longer (microsecond scale) exponential tail corresponding to the cooling of the cell as a whole (Figure 17.6). Because of the laser beam's relatively large breadth, optical resolution is determined by the microscope objective itself (e.g., $\sim$0.7 μm at 20×, NA 0.4; and $\sim$0.3 μm at 60×, NA 1.25), without significant influence of thermal blurring of the diffraction spot during the short pump pulse.

Fluorescent technique, which is widely used in *in vitro* and in *in vivo* FC (Chapter 16), and TDM (Chapter 14) are integrated in the system in order to provide some additional information and sometimes to verify PT data. For that, two additional digital cameras are used: (i) a sensitive CCD camera with a reasonably high frame speed (up to 500 fps; Cascade 650; Photometrics, Roper Scientific, Inc., Trenton, NJ) to image relatively slow-moving WBCs in blood (e.g., rolling leukocytes) and lymph flow and (ii) a high-speed CMOS camera with reasonable sensitivity and a high frame speed (model MVD1024-160-CL8, Photon Focus, Switzerland; 2500 fps for an area 128 × 256 pixels, and 10 000 fps for 128 × 64 pixels) sufficient to image strongly absorbing fast-moving RBCs. A fluorescence module is equipped with colored cameras (Nikon DXM1200 and Olympus DP2-BSW).

The integrated PTFC system provides the measuring threshold for the absorption coefficient on the order of 10^{-2}–10^{-3} cm^{-1}, which is sufficient to detect most intracellular absorbing structures with a typical local absorption coefficient of 10^{-1}–10^{3} cm^{-1} for the visible and near-infrared (NIR) wavelengths. This measuring range corresponds to laser-induced temperature variations of $\sim 10^{-2}$–$10\,^{\circ}$C in a cell. Thus, the temperature fluctuations are small enough not to damage cell structures, especially during the short time of irradiation while cells cross the illuminated area.

17.4.2 High-Resolution Imaging of Flowing Cells

The rat mesentery is one of the most attractive models to prove the concept of integrated PTFC *in vivo* [45]. Rats are narcotized, and the mesentery is exposed and placed on a heated (37.7 °C) stage bathed in warm Ringer's solution (37 °C, pH 7.4). At different magnifications (4–100×), TDM revealed that the rat mesentery provides a unique opportunity to simultaneously study blood and microlymphatic systems, together with the neighboring lymph nodes (Chapter 14) [15, 17, 45]. The rat mesentery contains a blood microvessel network that includes arterioles (diameter, 7–60 μm; velocity, 5–10 mm s^{-1}), venules (10–70 μm; 0.5–3 mm s^{-1}), and capillaries (5–9 μm; 0.1–1.4 mm s^{-1}), as well as well-developed microlymphatic vessels (diameter, 50–250 μm; velocity, 0–4 mm s^{-1}), and sometimes distinguishable initial lymphatics with migrating cells. Some blood microvessels are located very close to lymph microvessels and the parallel geometry of these vessels makes it possible to study cell migration between the two systems. This model is able to provide *in vivo* imaging of lymph nodes close to mesenteric structures with high optical resolution (~300 nm), with a clearly distinguished vascular network and migrating individual cells. This technique may make it possible to monitor the migration of the different cells by their typical pathways: from blood microvessel to tissue interstitium, then to the initial lymphatic, then to the afferent lymphatic, and then to the regional lymph node.

High-resolution TDM (Chapter 14) allows one to determine the morphological types of individual cells through visualization of cell shape and size (for example, parachute-like RBCs, single platelets, and rolling leukocytes). Images of subcutaneous blood microvessels through a dorsal skin-flap window and in animal ear, or images of retinal microvessels of the eye can be also obtained. Despite the relatively low resolution of these images compared to that of images of the mesentery, this achievement potentially allows the simultaneous monitoring of labeled circulating cells in different parts of the rat's body. Image contrast of these microvessels can be improved by the application of so-called optical clearing agents (OCAs), which makes it possible to see microvessels and cells with the enhanced contrast. Optical clearing (less scattering) of the surrounding tissues happens because of the matching of refractive indices of scatterers (tissue and cell components with higher refractive index) and base material (interstitial fluid and cytoplasm) [10, 55, 56]. Hyperosmotic OCAs may have some impact on blood or lymph flow velocity and cell shape; therefore, controlled response of vessels and cells to hyperosmotic OCAs can be used for *in vivo* and *in situ* studies of cell physiology.

In contrast to TDM's low absorption sensitivity (Chapter 14), the PT technique is able to not only monitor small variations in absorption in cells as a whole, but also identify cells on the basis of different PT-image structures associated with specific localized concentrations or the spatial distribution of absorbing biomolecules. As mentioned above, in the presence of gaps between neighboring cells in flow (typical for small blood vessels and lymphatics), a circular laser-beam geometry is used

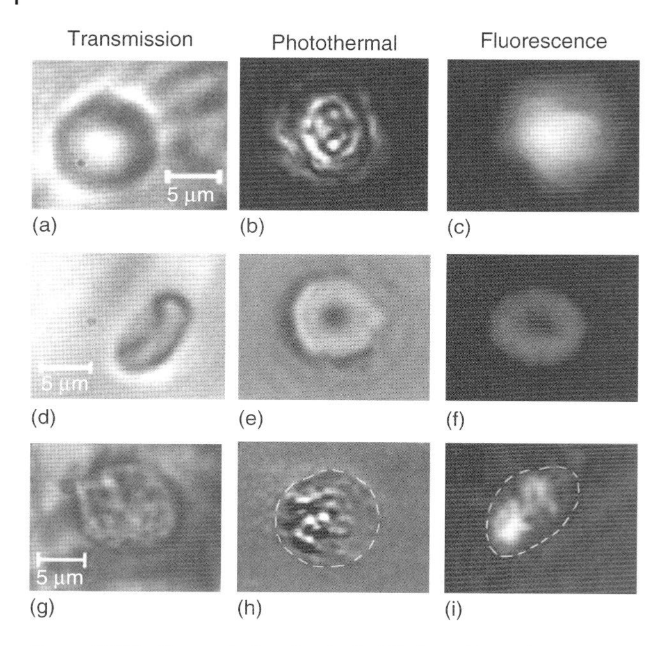

Figure 17.12 Transmission (a, d, g), PT (b, e, h), and fluorescent (c, f, i) images of WBC (a–c), RBC (d–f), and cancer cell, K562 (g–i) in *in vivo* lymph flow in rat mesentery [31].

(Figure 17.8). At short distances between flowing cells and especially, with many overlapping cells, a linear beam shape is more preferable.

To avoid distortion of high-resolution images due to the transverse movement of cells flowing in a cross section of vessels, especially in relatively large lymphatic vessels (up to 50 μm in 150 μm diameter lymph vessels), cells are imaged immediately after they pass through a lymph valve, which provides a natural type of hydrodynamic focusing, thus limiting the fluctuation of cells to within a few micrometers (Figure 17.1 and Chapter 14).

With such focusing, good-quality label-free images (similar to images obtained under static conditions [39]) of single cells in lymph flow, such as WBCs, RBCs, and leukemic cells (K562) (Figure 17.12), could be obtained. PTI revealed specific absorbing cellular structures (Figure 17.12b,e,h) associated with the spatial distribution of cytochromes in WBCs and cancer cells, and hemoglobin in RBCs, that are not quite visible with TDM (Figure 17.12a,d,g, respectively). Fluorescent images obtained with a high-sensitive CCD camera show good image contrast (Figure 17.12c,f,i) similar to that of PT images, but they require additional staining by intravenous injection of fluorescent dyes (e.g., fluorescein isothiocyanate (FITC) for RBCs and MitoTracker Red for K562 cells). It is noteworthy that PT techniques are not sensitive to scattering light.

The rotation of single cells in blood and, especially in lymph flow, can often be observed. This effect opens the way to *in vivo* projection cell tomography, whereby sequent 2D-cell images captured at different angles are reconstructed into 3D-cell images, using well-developed mathematical algorithms.

17.4.3
PT Identification of Cells

First, identification of circulating cells in blood or lymph streams can be based on the differences in cell amplitude of the integral PT response at a different wavelength (PT spectroscopy), which is proportional to the cell average absorption [15]. For example, for RBCs and WBCs, the difference in the amplitude of PT response is ~30–40 folds in visible spectral range. Second, the spatial structure of PT image, which is related to the specific distribution of cellular chromophores, can also be used to determine the cell type (e.g., RBCs with relatively smooth distribution of hemoglobin, or WBCs with localized distribution of cytochromes or other absorbing biomolecules). Third, cells can be roughly identified through differences in PT relaxation (i.e., cooling) time (e.g., 6–10 μs for RBCs with an average size of 5–7 μm, and 20–25 μs for 7–8 μm WBCs) [15].

The PT amplitude from the surrounding tissue (i.e., absorption background) in the linear (i.e., noninvasive) mode is ~30- to 50-fold lower than that from RBCs, which is explained by the low absorption of connective tissue and thin vessel walls. For low-absorbing lymphocytes, the corresponding difference in amplitude (from cells in flow and surrounding tissues) is lower, ~5- to 10-fold only, but still detectable [15].

17.4.4
Cell Targeting with Gold Nanoparticles

The high absorption sensitivity of the PT technique offers an opportunity to consider GNPs as a new type of PT labels [13, 27, 57]. These particles have unique properties for studying cells with a low or nonspecific intrinsic absorption, especially *in vivo*; they are strong absorbers (four to six orders of magnitude greater than any conventional fluorescent label), which makes it possible to detect single particles even as small as ~2 nm. In addition, they are nontoxic, biocompatible, and photostable (i.e., absence of photobleaching or blinking); have a fast thermal response time (up to 10^{-12} s); and can easily be conjugated with proteins and antibodies to label many specific cellular markers. Additionally, they may have adjustable absorption properties in the spectral range of 500–1000 nm, providing high image contrast for their reliable detection against the absorbing cellular background from water, DNA, and lipids [56]. When closely located or during aggregation, GNP-clusters demonstrate red-shift effects; this allows GNPs, in combination with the PT technique, to be used for aggregation immunoassay at the single-cell level [57]. The absorption bandwidths of gold nanoshells (GNSs) and gold nanorods (GNRs) with adjustable spectral properties are 80–150 and 60–100 nm, respectively. These are not as narrow as in fluorescent probes (e.g., quantum dots, QDs), but they should be sufficient for spectral identification.

Gold-labeling protocols could be used as the basis for molecular imaging and/or laser thermal therapy of cancer cells or infections. For example, the WTY-6 subpopulation of the MDA-MB-231 human breast adenocarcinoma cell line was engineered to express seprase, a clustered cell surface protein [57], which was targeted with a primary F19 monoclonal antibody (Ab). Goat anti-mouse immunoglobulin G (H + L) conjugated with 40 nm GNPs was used as a secondary Ab to selectively target the primary Ab. The specific distribution of gold labels attached to seprase and formation of gold nanoclusters were observed in parallel with transmission electron microscopy (TEM) and PTI. This highly specific labeling, which targeted selected cancer markers, was used to demonstrate selective killing of these cells by laser-induced bubble formation around the GNPs.

The high sensitivity of PT responses to positions of closely located nanoparticles opens the way for precise PT monitoring of nanoscale distances between nanoparticles, with expected nanometer-scale accuracy. With PT spectroscopy, the spectral shift of plasmon resonance potentially enables monitoring of the average distances between GNPs in the range of 1–50 nm within individual cells, the particle size ranging from 1.5 to 2 nm (which is the current sensitivity threshold of the PT technique) and higher. Potential applications may include measurement of protein–protein interactions; spatial relocation of GNPs during cell metabolism; cellular dynamics, such as the movement of cell organelles; or response to external stimuli, such as the administration of drugs or radiation.

The conjugation of gold labels with Abs may potentially affect cell–cell interactions or the elimination rate of labeled cells from blood flow because of Ab-enhanced immune response. To avoid (or estimate) these effects, nonspecific gold labeling may be realized with nonconjugated gold particles through conventional endocytosis. It is expected that the 50–100 small GNPs tagged to a cell and detectable with PTFC do not significantly influence cell-adhesive properties, which is crucial to understanding the genuine interaction of abnormal cells with normal cells in flow and in endothelium.

This relatively simple protocol may provide reliable identification of labeled cells in flow. In particular, the PT responses from cells with these gold PT labels show specific negative peaks in nonlinear PT responses compared to the positive amplitude peaks of linear PT responses from nonlabeled cells at the same laser-energy level (Figure 17.9) [30, 57]. Simultaneous monitoring of PT images from the same cells may identify labeled metastatic or apoptotic cells with highly specific localized high-contrast image structure (Figure 17.9).

17.5 Integrated PAFC for *In vivo* Studies

17.5.1 Schematics of the Instrument

In photoacoustic flow cytometry (PAFC) laser-induced PA waves (referred to as *PA signals*) from the cells in the blood (or lymph) flow are detected with an US

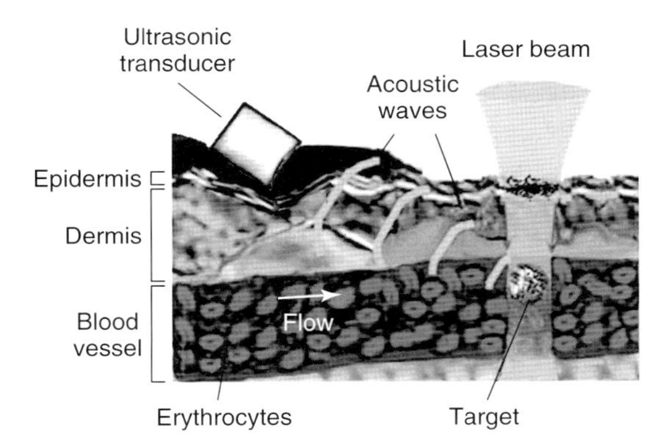

Figure 17.13 Principle of *in vivo* PA detection of single absorbing targets in blood flow [59].

transducer attached to the skin (Figures 17.1 and 17.13) [58, 59]. The PAFC system is built on a platform similar to that of the PTFC system (Figures 17.7 and 17.11). Laser energy is delivered to vessels with conventional lenses or fibers. PA signals from the ultrasonic transducers (unfocused XMS-310, Panametrics, 10 MHz, Olympus NTD, Inc., Waltham, Massachusetts; focused cylindrical V316-SM, Panametrics, 20 MHz, focused lengths of 6.5 and 25 mm; and customized resonance transducers) and amplifier (models 5662 and 5678, Panametrics, bandwidth 50 kHz to 5 and 40 MHz, customized amplifiers with adjustable high and low frequency boundary in the range of 50–200 KHz and 1–20 MHz, respectively; resonance bandwidth of 0.3–1 MHz) are recorded either with a boxcar (Stanford Research Systems, Inc., Sunnyvale, California), a Tektronix TDS 3032B oscilloscope, or a computer with standard and customized software. The boxcar or software provides averaged PA signals from cells in the blood vessels and time-resolved discrimination from background signals from surrounding tissue because of the different time delays. For the purposes of verification, the signals from the oscilloscope screen were also recorded with a digital camera (Sony Corporation of America, NY) and video camera (JVC Company of America, Wayne, NJ).

PAFC is integrated with PTFC (Figures 17.7 and 17.11) [15, 17, 31, 45] in order to validate PAFC data and to provide better reproducibility and objectivity of new data on flowing cell parameters. In particular, in the thermal-lens mode of PTFC, laser-induced temperature-dependent refractive heterogeneities around cellular endogenous absorbing structures (or exogenous contrast agents) cause defocusing of a collinear He–Ne laser probe beam and a subsequent reduction in the beam's intensity at its center as an integrated PT response from a whole cell (also Section 17.4.1).

In analogy to conventional FC and PTFC, to distinguish closely located cells and particles in flow, the minimum laser-beam size should be comparable to or less than the object sizes (i.e., 1–20 μm). The laser beams may have a circular geometry with diameters comparable to vessel diameters for the rare cells in flow (Figure 17.8a). However, for dense cell streams, the laser beam with the use of an additional cylindrical lens and a slit should be adjusted in a linear (elliptical) configuration (e.g., 6 μm wide and 55 μm long, Figure 17.8c). With the use of an elliptical beam shape with a long axis comparable in length to the vessel's diameter, the detection efficiency (i.e., the number of cells detected per the total number of cells passing through the vessels) is typically close to unity in small blood vessels. For the detection of rare cells, the beam diameter should be expanded or laser pulse rate should be increased to preclude any missing cells.

Navigation of the laser beams is controlled with a high-resolution TDM system (Figures 17.7 and 17.11, and Chapter 14). A fluorescence module with colored CCD cameras is used to verify PA/PT data using selected cells tagged with specific fluorescent labels.

A feasibility study was first performed *in vitro* with cells in suspension in conventional microscope slides. To model flow conditions, a flow module fitted with a syringe pump-driven system (KD Scientific, Inc., Holliston, MA) was used with glass microtubes of different diameters in the range from 30–150 μm to 1–4 mm [39]. This dynamic phantom provides flow velocities of 0.1 mm s^{-1} to 50 cm s^{-1}, which are typical of blood flow in animal microvessels and humans [60].

17.5.2 Animal Models

The feasibility studies of *in vivo* PAFC involved mouse (a nude) and rat (White Fisher, F344) models [59]. Most experiments for PA detection of circulating cells were performed in thin (250–300 μm), relatively transparent (i.e., also suitable for PTFC in trans-illumination mode) mouse ear with well-distinguished blood microvessels (Figure 17.14a,b). The ear blood vessels examined were located 30–100 μm deep and had diameters in the range of 30–70 μm and blood velocities of 1–5 mm s^{-1}. After standard anesthesia (ketamine/xylazine, 50/10 mg kg^{-1}), an animal was placed on the customized heated microscope stage, together with a topical application of warm water, which provided acoustic matching between the transducer and the ear. Rat ear (Figure 17.14c) also has distinguished blood microvessels, in spite of its thicker structure compared to mouse ear, and thus is also used as an animal model when thicker tissue (deeper vessels) models should be tested. The rat mesentery (Figure 17.14d) has an almost ideal biostructure for *in vivo* PAFC because it consists of very thin (7–15 μm) transparent connective tissue with a single layer of blood and lymph microvessels (Figure 17.14d). For this procedure, the rat was anesthetized (ketamine/xylazine, 50/10 mg kg^{-1}), and the mesentery, as in PTFC studies (Section 17.3), after exposure was placed on a heated (37.7 °C) stage and bathed in warm Ringer's solution (37 °C, pH 7.4).

17.5.3 Contrast Agents for PAFC

PAFC's capacity could be demonstrated by intravenous injection of exogenous PA targets, such as nanoparticles, contrast dyes, and bacteria, into the animal tail vein. The practical usefulness of the experiments lies not only in showing

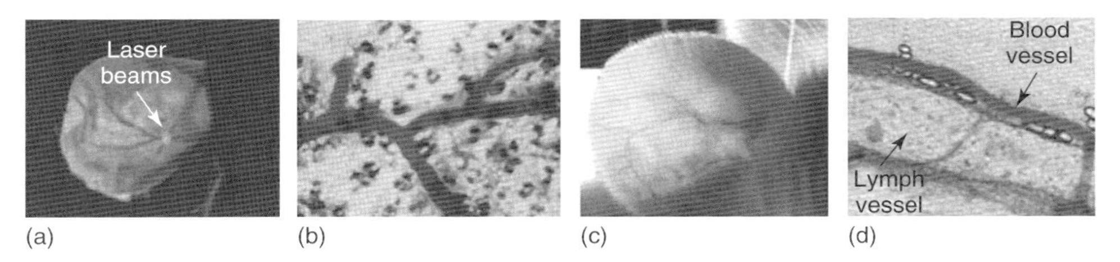

Figure 17.14 (a) Mouse ear photo with the linear-configurated probe laser beam (light red color) and circular pump laser beam (green spot); (b) mouse ear blood microvessels at 10× magnification; (c) rat ear blood microvessels; and (d) rat mesenteric structures; see also Figures 17.8, 17.9, and 17.12, and Chapter 14 [59].

PAFC's sensitivity but also in understanding the mechanisms and kinetics of biodistribution of nanoparticles, dyes, and microorganisms via blood and lymph circulating systems.

GNPs, such as GNRs of an average diameter of 15 nm and a length of 45 nm and GNSs of an average silica core diameter of 145 nm and 5–10 nm gold coating (total diameter around ~155 nm) [61], with concentration in the range of 10^9–10^{11} particles ml^{-1}, were tested. Two types of GNRs and GNSs were used in the experiments: (i) GNPs without surface modification and (ii) GNPs functionalized by thiol-modified polyethylene glycol (PEG) according to the established protocol [62].

Single-walled (Nano-Lab, Inc. Newton, MA) and multiwalled carbon nanotubes (CNTs) (Carbon Nanotechnologies, Inc., Houston, TX) used in the study were processed as described previously [63, 64]. The average length and diameter were 186 and 1.7 nm for the single-walled CNT, and 376 and 19.0 nm for the multiwalled CNT, respectively.

The bacterium *S. aureus* strain designated UAMS-1 was isolated at the McClellan Veterans Hospital in Little Rock, Arkansas, from a patient with osteomyelitis [35]. The strain was deposited with the American Type Culture Collection and is available as strain ATCC 49230. UAMS-1 was cultured in tryptic soy broth and grown aerobically for 16 h at 37 °C. Cells were harvested by centrifugation, resuspended in sterile phosphate-buffered saline (PBS) and incubated with CNTs or Indocyanine green (ICG) (Akorn Inc.) as follows. Before incubation, ICG was filtered through a 0.22 μm pore size filter. An aliquot (150 μl) of bacteria in suspension was incubated with 150 μl (375 μg) of ICG solution for 30 min and 2 h at 37 °C. The labeled bacteria were centrifuged (5000 rpm; 3 min) and the pellet was resuspended in PBS. Before incubation, the CNT solutions were further treated with interrupted US for 10 min: 1.5 min of US (power 3 W) and 0.5 min of interruption. An aliquot (150 μl) of bacteria in suspension was incubated with 150 μl of CNT solution for 30 min and 2 h at room temperature. The labeled bacteria were centrifuged (10 000 rpm; 5 min) and the pellet was resuspended in PBS.

E. coli K12 strain was obtained from the American Type Culture Collection (Rockville, Maryland). *E. coli* was maintained in Luria-Bertani (LB) medium

(1% tryptone; 0.5% yeast extract; 0.5% NaCl; pH 7). An aliquot (100 μl) of *E. coli* in PBS was incubated with 100 μl of the CNT solution for 60 min at room temperature.

As a standard contrast vital dye for testing the capability of PAFC, Lymphazurin™ 1% (Isosulfan Blue; Ben Venue Labs, Inc.) was used.

17.5.4 PAFC Testing and Applications

17.5.4.1 Testing on Noninvasiveness

To prove PAFCs noninvasiveness, the laser-induced damage threshold at the single-cell level should be determined first. The methodology assumes the specific changes in PT images and PT responses from individual cells, associated with cell vital heating and follow-up damage, as a function of the pump-laser energy [21, 65]. For relatively strong absorbing RBCs in the visible spectral range, the *in vitro* photodamage threshold ED_{50} (50% of cells damaged) varied in the range of 1.5–5 J cm^{-2} with a laser wavelength in the range of 417–555 nm (for a 20 μm laser spot). For other cell types with lower absorption, the thresholds were much higher from 12 to 42 J cm^{-2} for lymphocytes and from 36 to 90 J cm^{-2} for K562 blast cells. In the NIR spectral range from 860 to 920 nm, the damage thresholds for RBCs and WBCs were more than one order magnitude less compared to those in the visible spectral range. Within the accuracy of experiments (~15%), the photodamage cell thresholds measured *in vitro* and *in vivo* were similar [15]. The noninvasiveness of PAFC studies was provided by choosing laser fluence much below the estimated cell photodamage threshold due to application of the laser safety standard, which limits fluence to 30–100 mJ cm^{-2} in the NIR spectral range 650–1100 nm in the conditions of PAFC [66].

17.5.4.2 PAFC Testing Using Vital Dyes

The capability of PAFC is typically estimated using vital contrasting dyes, including ICG, Lymphazurin, Evans blue (EB), and Methylene blue [42, 45, 59, 67], which are widely used to study the physiological functions of blood and lymph systems in norm and pathology. For example, study of Lymphazurin kinetics using PAFC is interesting to discuss [59]. Lymphazurin (Isosulfan Blue) is used mostly for the delineation of lymphatic vessels [68, 69]; however its exact pharmacological action and kinetics in the blood circulation and surrounding tissue are still unclear.

In [59] an aqueous solution of Lymphazurin (200 μl; 1% concentration) was injected into mouse and rat blood circulatory systems through the tail vein. The observed animal skin coloration kinetics was very similar in both mice and rats, and low toxicity and safety of the dye at the doses used was proved in line with the early reported data [68]. Lymphazurin has a high permeability through the walls of blood vessels and is distributed into tissue quickly. For PA detection, a wavelength of 650 nm was taken, where absorption of Lymphazurin is close to maximum (broad band from 570 to 660 nm), while the blood absorption drops to a level that is almost one order of magnitude less compared to the maximum blood absorption near 570–580 nm [70]. The PA signals from blood vessels were slightly higher (1.5- to 3-fold) than the PA background signals from surrounding tissue [59]. The continuous monitoring of PA signals from the ear blood microvessels with average diameter 50 μm after intravenous administration of Lymphazurin with the standard (1%) concentration in 100 μl solution revealed fast (within a few minutes) Lymphzurin increase in blood flow, followed by its clearance half-life of 30 min and total disappearance from the blood within approximately 50 min.

Maximum PA signals from the rat ear microvessel with a diameter of 50 μm after intravenous injection of dye in the tail vein at laser fluence of 30 mJ cm^{-2} and beam diameter of 40 μm were approximately three- to fivefold higher than the PA signals from blood vessels before dye administration. The PA signals from tissue around the vessels after dye injections gradually increased 2- to 2.5-fold during 15–20 min and then remained relatively constant for approximately 1–1.5 h. This level was comparable to or even a little higher than the background PA signals from blood vessels when Lymphazurin was cleared from the blood pool. The PA signals at scanning laser beam across ear tissue revealed the readable increase in the PA signals in some local areas that can probably be associated with dye uptake by macrophages. No readable changes in PA signal amplitude and shape as precursors of photodamage phenomena were observed [65] that suggest noninvasive condition for one pulse irradiation at the used laser fluence of 30 mJ cm^{-2}.

The application of contrast dyes is based on their absorption or fluorescent properties, their good permeability, fast excretion, and low toxicity. In particular, after administration EB or Lymphazurin, the vessels can be easily identified visually with the naked eye (below Figure 17.19b). The ability of PAFC to detect conventional contrasting agents is related to their limited quantum yield, typically in the range of 1–20% [67], resulting in transformations of the most absorbed energy into heat, to which the PA technique is very sensitive.

The capability of ICG and EB dyes to enhance the PA image contrast of the blood vessels was demonstrated [71, 72]. However, most PA imaging algorithms currently in use for signal acquisition are not quite suitable for rapid real-time PA monitoring of contrast agent kinetics in fast blood circulation due to their time-consuming signal acquisition process (in most cases seconds, if not minutes). In addition, PA imaging sometimes suffers from relatively poor spatial resolution (in the best case ~50–100 μm) [73], preventing its cytometric application at the single-cell level.

As described, the time-resolved PAFC with its high spatial resolution of 6–20 μm, and up to 0.5–1 μm if integrated with PTI mode (Section 17.4) [31], temporal resolution of 10^{-4}–10^{-1} s (depending on the laser repetition pulse rate), and spectral resolution ≤0.5 nm in the broad spectral range of 420–2300 nm has a great potential to study kinetics of dyes in the blood and lymph circulating systems in different animal models. A quick translation of this technology to humans is expected, because some dyes (e.g., ICG) are already Food and Drug Administration (FDA) approved and laser fluences used meet the safety standard. Detection of both fluorescent light and heat can increase the diagnostic value of integrated *in vivo* FC with PA and fluorescence detection techniques, especially in

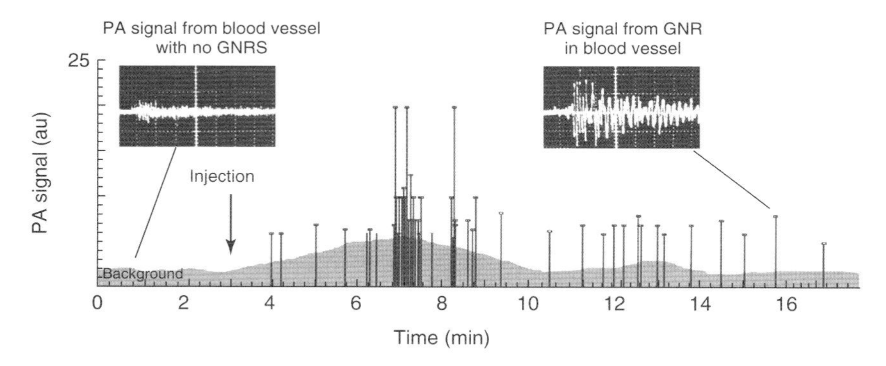

Figure 17.15 PA signals from circulating GNRs in blood flow in the rat mesentery microvessel with a diameter of 50 μm as a function of time after injection. Laser parameters: wavelength 830 nm and laser fluence 100 mJ cm^{-2}. The amplitude/time scale on oscilloscope signals is 100 mV/div/4 μs/div [59].

far-red and NIR spectral ranges corresponding to tissue diagnostic window [10]. The high absorption sensitivity of the PA technique makes it possible to use relatively low concentrations of most dyes, especially dyes with the low fluorescence quantum yields, without notable signs of dye toxicity or photobleaching.

17.5.4.3 PAFC Detection of Circulating Nanoparticles

The 250 μl suspension of GNRs in concentration of 10^{10} ml^{-1} was injected to the rat blood circulation through the vein tail followed by continuous monitoring of PA signals from blood vessels with 50 μm diameter in the rat mesentery [59]. The chosen laser wavelength of 830 nm was close to the maximum absorption of GNRs. Uncoated GNRs were rapidly cleared from the blood circulation within 1–3 min preferentially by the reticuloendothelial system (RES). In contrast, on injection of PEGylated GNRs, the fluctuating strong PA signals appeared with the amplitudes significantly exceeding the PA background signals from blood vessels approximately within the first minute and lasted 14–25 min, depending on the individual animal (Figure 17.15).

Simultaneously, the slightly fluctuating PA background increased approximately 1.5–2 times above the background from blood at 3 min and lasted around 4 min. These measurements were performed in three animals, and the average data of the number of circulating GNRs per minute are shown in Figure 17.16.

Different nanoparticles are intensively being developed for biomedical applications, including diagnostics and therapy [57, 61–63, 74, 75]. Before any clinical application of nanoparticles can be feasible, it is imperative to determine the critical *in vivo* parameters, namely pharmacological profiles, including the clearance half-time from the blood circulation [76, 77]. Data presented on circulation of GNRs in blood stream show promise both as PA and PT labels and as PT therapeutic agents for killing individual cancer cells and bacteria [13, 28, 42, 56, 61, 77–82].

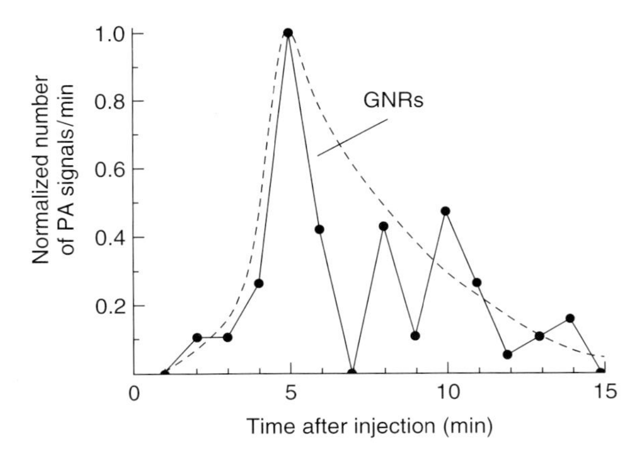

Figure 17.16 Normalized number of circulating GNRs in blood microvessels of the rat mesentery as a function of time after injection. Dashed curve shows an average data with three rats [59].

This technology belongs to the so-called theranostics (therapy/diagnostics) – an approach that provides in the framework of a similar technology – diagnostics (targeting) first and then therapy. In biophotonics, there are two important examples: the fluorescence tumor mapping and the follow-up photodynamic therapy, as well as optical imaging of cancer cells with GNPs (by using their unique absorption and scattering properties) and the follow-up selective photothermolysis of these cells mediated by the GNPs (using their strong absorption) with the irradiation by the same laser but of increased power density [13, 41].

The PEGylated GNRs remained for a longer time in the blood circulation after intravenous administration (Figures 17.15 and 17.16), compared to the non-PEGylated nanoparticles. Fast clearing of uncoated GNRs can be associated with quick GNR aggregation in blood flow immediately after intravenous injection, and the influence of the incompletely purified and partly toxic particle precursors, especially CTAB, which lead to fast responses of RES and the immune system [42]. It is known that PEG coating leads to reduced nanoparticle aggregation, protein binding, and adhesion to blood and endothelial cells [62, 74]. However, the observed clearance rate of around 15 min for the PEGylated GNRs was not as long as expected. It could be related to some defects in PEG coating. The described *in vivo* PAFC technique could also be useful to the evaluation of the influence of the nanoparticle structure and coating technology on nanoparticle clearing.

To estimate the detection threshold sensitivity of PAFC at injection of the 250 μl GNR suspension (10^{10} ml^{-1}) in the blood pool of a rat with a total blood volume of 20–25 ml, the ideal condition with no loss of GNRs during injection and circulation was assumed. Thus, the average GNR concentration in the rat blood pool should be $\sim 10^8$ ml^{-1}. The irradiated (detected) volume in blood vessels is limited by a laser-beam diameter of 25 μm and path length of around 45 μm across the vessel (i.e., comparable with vessel diameter), which gives the volume around

2.2×10^{-8} cm^3. These parameters correspond to the estimated maximum number of GNRs in the detected volume around 2–3. The possible loss of GNRs during injections, such as adhesion in blood vessels and elimination by RES, may decrease this value several times, especially at the end of the clearance process. This strongly implies that the number of GNRs in the small detection volume (i.e., minimum background absorption) should be close to 1. Therefore, each pulse in Figure 17.15 may be associated with the PA signals from individual GNRs or small aggregates. However, immediately after injection, local concentration of GNRs in the blood flow before redistribution in larger volume should be higher (i.e., many GNRs in the detected volume). This was confirmed by the increases in both individual PA signal rates and average PA background above the blood background in approximately 4–9 min (Figure 17.15). In addition, in this time period, the PA signals with larger amplitudes were observed after each laser pulse. Essentially, these fluctuations were not accompanied by decrease of PA signal amplitude to zero, supporting the hypothesis of random appearance of several GNRs in the detected volume at the same time. This is similar to the aforementioned PA monitoring of Lymphazurin, when relatively large amounts of homogeneously distributed absorbing molecules appear in the detected volume, thus providing relatively stable continuous (i.e., not fluctuated) PA signals. On the contrary, only rare PA signals appeared even after many pulses at 10 min, suggesting the observation of the GNR clearance.

Laser fluence used in this study was of 100 mJ cm^{-2}, which is at least two orders of magnitude less than the damage threshold for RBCs at a wavelength of 830 nm [59]. However, high absorption of metal nanoparticles may lead to laser-induced nanoparticle modification, including their melting, evaporation, and explosion, accompanied by enhanced bubble and shock wave formation, even at a relatively low laser fluence up to few milljoules per square centimeter [83–86]. These nonlinear effects can lead to strong enhancement of PA signals from single GNPs due to highly localized bubble formation and shock wave and, which might explain a ultrahigh-sensitivity PAFC observed.

Some additional PA signal enhancement for PEG-coated GNRs in comparison with the uncoated is due to more effective GNR heating and PA signal formation caused by changed thermal properties and average size increase (from 10 to 50 nm depending on chemical procedure). This effect was more profound for GNSs due to probably more effective explosion of thin gold layers around a relatively large silica core (a so-called spherical piston explosion model) [84–86]. In addition, these effects might be important to increase the efficiency of PT nanotherapy with nanoparticles and their clusters used as a PT sensitizer for cancer cells or bacteria [77–87].

The results obtained so far demonstrate high sensitivity of the developed PAFC schematics, enabling *in vivo* single GNP detection in blood flow in small vessels. It also suggests the possibility of using a limited number of GNPs (e.g., 50–100) for cell labeling to achieve reliable PA detection of cells in the blood circulation [42]. In addition, the routine monitoring of the clearance rate of different nanoparticles may be combined with simultaneous real-time PA monitoring of nanoparticle accumulation in different organs in normal and pathological states (e.g., in tumors, lymph nodes, or infected areas).

17.5.4.4 PA Detection of Single Circulating Bacteria

Despite major advances in medicine in the last decade, diseases of microbiological origin continue to present enormous global health problems, especially because of the appearance of multidrug-resistant bacteria strains. The critical steps in the development of bacterial infections include their penetration into the blood and lymph systems, their interactions with blood cells in flow or with endothelial cells, and their toxin translocations in the host organisms [88]. Increasing evidence has implicated the bacterial hematological and lymphatic translocation as the main sources for the induction of sepsis. Unfortunately, little is known about bacteria circulation in the blood pool, its clearance and adherence rates, and other kinetic parameters, which might be very important for understanding the transition from the bacteremic stage to the tissue invasive stage of disease to develop effective therapies of infections.

The capability of PAFC to monitor circulating single *S. aureus* in blood flow was estimated with the mouse ear model [59]. Because the endogenous absorption of bacteria was relatively weak compared to blood absorption even in the NIR spectral range, bacteria were labeled with NIR absorbing contrast agents such as ICG, GNRs, GNSs, and CNTs. The incubation of *S. aureus* with these agents was performed for 30 min and 2 h at 37 °C without antibodies to avoid the potential influence of immunogenicity on bacteria circulation. Except CNTs, the other labeling agents showed slight toxicity at 2 h incubation. Hence, 30 min incubation was used in most experiments to avoid their potential cytotoxic effects. The labeling efficiency was estimated using methodology of the PT aggregation/clusterization assay by comparison of PT signals from control and labeled individual bacteria [45, 57, 59].

At relatively high laser fluences (24 J cm^{-2}), which were nevertheless below the bacteria photodamage in the NIR spectral range (795 nm), the classic linear PT response was observed for the control bacteria without contrasting agents, as indicated by an initial peak due to fast laser-induced heating of the bacteria and a slower exponential tail corresponding to bacteria cooling through heat diffusion into the surrounding solution (Figure 17.6) [21]. Even higher laser fluences are required to provide sufficient PT and PA signal amplitudes from bacteria without labeling for their *in vivo* detection on the background of RBCs. Such fluencies may cause the RBC damage and the influence of background absorption from blood, thus in *in vivo* conditions only labeled bacteria were studied. For example, staining of *S. aureaus* with ICG led to significant PA signal amplitude increase, on average almost two orders of magnitude, which allowed for reduction of the laser fluence to safety level of 34 mJ cm^{-2}.

Labeling *S. aureus* with CNTs and their clusters yielded more dramatic increase of PT-signal amplitude, which was several times higher than that from bacteria with ICG at the same laser energy. The amplitude of the PT response rapidly increased with increase in the pulse energy, until the overheated local effects around CNT clusters led to the bubble formation, which resulted in a PT negative peak due to bubble-dependent changes in local temperature, refractive index, and light scattering phenomena [21, 64]. The degree of increase in PT-signal amplitudes and bubble-associated effects were used to optimize the labeling procedure almost

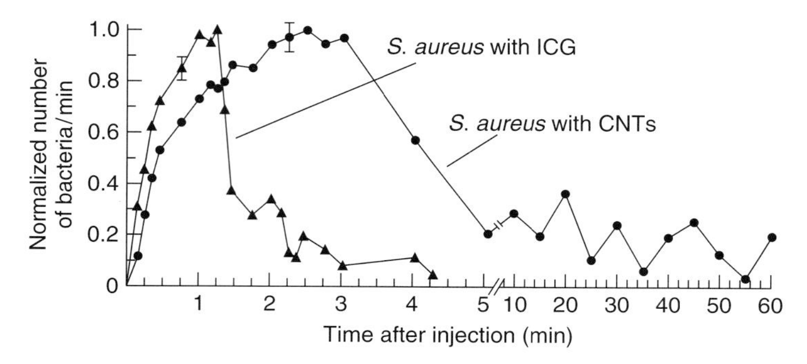

Figure 17.17 Normalized number of circulating *S. aureus* in blood microvessels of mouse ear as a function of time after injection, and label type (ICG and CNTs). Laser parameters: wavelength 795 and 850 nm, respectively. Laser fluence: 50 and 20 mJ cm^{-2}, respectively [59].

in real time with different nanoparticles. The best results were obtained with GNRs, but its toxicity related to the preparation procedure requires further studies [42]. Minimal bacterial binding was obtained both with nonconjugated and PEGylated GNSs.

Owing to relatively high labeling efficiency and low toxicity, ICG and CNTs were chosen for *in vivo* experiments. Labeled bacteria in suspension, with volumes of 100 μl and concentration of 5×10^5 ml^{-1}, were injected into the mouse's circulatory system through the tail vein, followed by clearance monitoring with PAFC in mouse ear blood microvessels with an average diameter of 50 μm (Figure 17.17). The obtained results were similar to both ICG and CNT contrasting agents, such as the rapid appearance of bacteria in the ear blood microvessels after injection within the first minute, followed by their elimination from the blood circulation after 3–5 min.

Laser fluence in these particular experiments was in the range of 20–50 mJ cm^{-2}, which is much less than the damage threshold for RBCs at wavelengths of 790–850 nm [59]. However, multiple laser exposure to bacteria with CNTs at 50 mJ cm^{-2} led to a gradual decrease (several percents) in PA signal amplitude from pulse to pulse that can be associated with strong absorption by CNTs, leading to thermal and bubble-related disintegration of CNT clusters [64].

The similar methodology was also applied for real-time monitoring of *E. coli* labeled with CNTs. Figure 17.18a exemplifies TEM images of an *E.coli* before and after 0.5 h incubation with CNTs, which demonstrates the excellent binding efficiency of CNTs within the bacterial wall structure with formation of local CNT clusters. Furthermore, using the standard viability kits revealed that bacteria viability was not affected by CNTs [64]. The CNT clusters provided more effective heating, and thus generation of PA signals from individual bacteria which exceeded the background PA signals from the blood pool (Figure 17.18b, top and bottom oscilloscope signals, respectively). Bacteria labeled with CNTs have been intravenously administrated to mice through the tail vein and monitored through

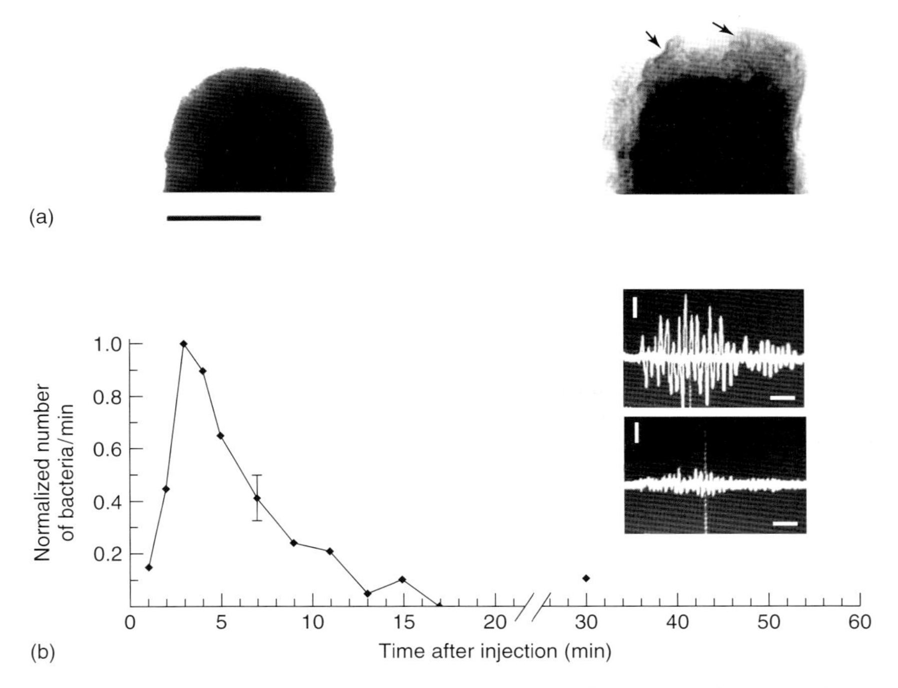

Figure 17.18 (a) TEM images of an *E. coli* fragment before (left) and after (right) incubation with CNTs. Arrows indicate CNT clusters within the bacteria wall structure. Scale bars represent 500 nm. Normalized number of circulating *E. coli* in blood microvessels of mouse ear as a function of time after injection. Oscilloscope signals: PA signals from labeled *E.coli* in blood (top) and from blood alone (bottom). Amplitude/time scale: 200 mV/div/2 μs/div. Scale bar represents amplitude/time scale: 200 mV/div/2 μs/div, respectively. Laser parameters: wavelength 850 nm and laser fluence 100 mJ/cm^2 [59].

the CNT's intrinsic NIR absorption at the wavelength of 850 nm with PAFC. After fast appearance of the bacteria, their concentrations in the blood decreased almost exponentially within a half-life of 6 min (Figure 17.18b).

The absence of bi- or multiexponential kinetics indicates that there was no significant temporary accumulation of bacteria in a specific area (e.g., vessel wall or liver) that could act as reversible reservoirs. After a few days, the experimental animals displayed normal behavior and there was no evidence of adverse effects from bacteria injections.

CNTs demonstrate great potential for labeling bacteria such as *S. aureus* and *E. coli* for FC diagnostics. High CNT binding efficiency may eliminate the bioconjugation steps with biological recognition components, such as antibodies. This alleviates the potential influence of the additional biocomponents on clearance rates. Besides binding efficiency, the advantage of CNTs is their intrinsic strong absorption in the NIR, thus they may serve as excellent NIR high-contrasting agents for PT and PA diagnostics, in addition to the previous reports showing their potential as high-contrast PT sensitizer for cancer [89] and antimicrobial therapy [64].

The clearance of bacteria with CNTs was a little slower than that with ICG (Figure 17.17), which can be explained by a slight ICG toxicity and more effective RES response, on the one hand, or by more effective bacteria labeling with CNTs (i.e., the presence of more labeled bacteria in the blood) on the other hand. The observed advantage of CNTs as a PT label compared to ICG can be explained by the stronger absorption of individual CNTs, and especially by CNT clusters, compared to ICG absorption [21, 74, 79]. However, the absorption band of ICG is more narrow and specific compared to the not quite selective absorption of CNTs. Thus, besides binding efficiency, the choice of nanoparticles and dyes may depend also on requirements to selectivity. Encapsulation and stabilization of ICG within poly(styrene-alt-maleic anhydride) blockpoly(styrene) micelles or other biocompatible materials as micro- or nano-sized particles could be a way to provide spectral selectivity and targeting. Another prospective way is to integrate narrow-band absorptive dyes with different kinds of nanoparticles [90–94]. This approach is rather new, however, some positive results especially in bacteria killing are already achieved [91–96].

The ultrasensitive, rapid PA detection of infection on a single bacterium level may supplement or replace conventional assays, for example, the time-consuming polymerase chain reaction (PCR) and others that are available currently *in vitro* only. The PA technique described is capable of detecting approximately 10^2 bacteria every $\sim$15 min in the peripheral circulation of small animals, such as mice and rats, and has the potential to improve this parameter to ten bacteria (or 1 bacterium ml^{-1}). Theoretically, this technique enables detection of just a single circulating cell or bacterium in the whole mouse blood pool during continuous PA monitoring in the relatively large vessels approximately every hour. The obtained clearance rate for *S. aureus* (3–5 min) is in line with data obtained with fluorescent-labeled bacteria monitored with the dorsal skinfold chamber technique [88]. However, the time-resolved PAFC can observe the temporal fluctuation of clearance rate (Figure 17.17). There are two possible explanations for the nature of these observations. First, this may be related to the bacterial adherence to endothelial cells, followed by bacteria extravasations in different organs, which are considered to play a central role in the invasive disease stage [88]. Indeed, the scanning of the laser beam along vessels revealed the appearance of strong PA signals at different locations, indicating the adherence of bacteria with CNTs along the vessel wall. The adherence might be a dynamic process and can lead to periodical detachment of bacteria from blood vessel walls. The adherence could be influenced by bacteria labeling, especially with CNTs, which may prevent effective interaction of bacteria with other cells because of their high-density clustering within the bacterial wall (Figure 17.18). Additionally, the possibility of partial CNT detachment cannot be excluded, especially in the relatively large CNT clusters, from bacteria surfaces during their interactions with other cells in fast blood flow. The detached CNTs could circulate longer than bacteria themselves, at least a few hours.

The developed PAFC system demonstrates, similar to the GNRs, extremely high sensitivity in detecting individual CNTs and especially their small clusters.

Therefore at scanning of the focused laser beam across tissue surrounding the examined blood vessels (PA mapping) the strong local PA signals that can be associated with the presence of labeled bacteria in interstitium were detected. This technique applied *in vivo* and *in vitro* also revealed strong PA signals in liver (e.g., 24 h after administration), indicating the accumulation of bacteria in this organ as well.

Evidently, additional studies are desired to elucidate the behavior of CNT-coated bacterial cells, as well as to optimize the processes of labeling bacteria with CNTs, however these results demonstrate well that time-resolved PA method with continuous *in vivo* monitoring is better suited for studying clearance rates of bacteria and contrast agents, especially when clearance time is relatively short (a few minutes).

17.5.4.5 PAFC Benefits and Potentialities

The innovative combination of PA and FC techniques, each of which has been used for many years separately, provides new facilities for studying absorbing nanoparticles in the blood and lymph flows. PAFC is a prospective technique used for real-time *in vivo* monitoring of the depletion kinetics of circulating contrasting agents, blood and cancer cells, bacteria, and different nanoparticles in native biological environments in animal models [42, 58, 59].

Considering that PAFC is a new tool in biological research, requirements for different applications and further possible improvements of the technique should be discussed on the basis of the results obtained. The detection of fast-moving absorbing micro- or nanoobjects of different origins on the background of the absorption of blood cells and surrounding tissue presents a new challenge for the PA technique. For noninvasive real-time detection of living cells or bacteria, most laser parameters are important, including wavelength λ, pulse duration t_p, pulse repetition rate f, and laser fluence Φ.

As CW lasers are widely used in conventional *in vitro* FC to provide fluorescent or scattering detection methods [1–3, 97], this mode is not quite suited to PA methods in FC applications, because it does not provide an effective generation of PA signals, especially from small single objects [4]. To enhance the PA signals, intensity modulation in the CW mode or pulse regime is required [17]. To generate the maximum PA signal in the pulse mode with more efficient conversion of light energy into thermal, and then acoustic energy, the laser pulse duration t_p must meet the thermal and, especially more strict, acoustic (stress) confinement criteria [59, 73], $t_p \leq \tau_A = R/\upsilon_{ac}$, Equation 17.9, where τ_A is the transit time of the acoustic wave traveling through an absorbing object, R is the radius of the absorbing target, and υ_{ac} is the speed of sound in the object. For water, $\upsilon_{ac} = 1.5 \times 10^5$ cm s^{-1}. Thus, the pulse durations required for an object with a diameter of $2R = 5$–15 µm (the typical mammalian cell sizes) and 0.5–1 µm (typical bacteria or cell organelle sizes) are $t_p \leq 1.6$–5 ns and $t_p \leq 0.16$–0.32 ns, respectively.

Thus, the 8 ns pulse width used in PAFC is close to the optimum for detecting cells and bacteria, although smaller objects such as single nanoparticles in linear mode require shorter laser pulses (e.g., for 100 nm particles $\tau_A \cong 33$ ps). With nanosecond laser pulses, however, PA signals from nanoparticles can be enhanced through nonlinear phenomena (e.g., bubble formation, or thermal explosion nanoparticles or special coating) as well as by clustering nanoparticles into cells or bacteria [35, 57, 64]. Nanoparticle clusters of larger size and dense nanoparticle distribution may better meet the acoustic confinements. Therefore, the dependence of PA signals on nanocluster sizes may be used for real-time PA monitoring of nanoparticle clustering or aggregations.

To expose each fast-flowing object, the laser must have a pulse repetition rate of $f \geq V_F/D$, where V_F is the flow velocity and $D = 2R$ [44]. For example, in small-animal blood microvessels, typical flow velocity ranges from 1 mm s^{-1} (capillary) to 10 mm s^{-1} (arterioles) [60], and for $2R = 20\ \mu m$, $f \geq 50$–500 Hz, respectively. To detect smaller objects, or at faster flow, the pulse repetition rate should be higher. The high rate is also essential to increase PAFC sensitivity during multipulse object exposure, because the signal-to-noise ratio is proportional to the square root of the number of laser pulses.

A maximum data acquisition rate is presently limited by the 50 Hz laser pulse repetition rate, which may lead to missing some fast-circulating objects. To avoid this problem, lasers with higher pulse repetition rates such as diode and diode-pumped solid-state lasers as well as CW lasers with high-frequency intensity modulation must be used.

The laser fluence must meet the American National Standards Institute (ANSI) safety standard [66], which is 30–100 mJ cm^{-2} for NIR spectral range from 650 to 1100 nm radiation. The high sensitivity of the acoustic detection system with a customized transducer and an optimized signal acquisition provides reliable detection of PA signals from single labeled bacteria in the blood flow *in vivo* in the presence of electronic noise and background signals from RBCs and surrounding tissue. The high PA sensitivity, together with the high RBC damage thresholds (17–80 J cm^{-2}) at the NIR, provide PAFC noninvasiveness in the *in vivo* study.

The wavelength, which renders the maximum absorption contrast between labeled bacteria (or cancer cells) [42] and RBCs, lies in the NIR spectral range of 650–940 nm, depending on the optical properties of the PA labels (e.g., CNTs, GNSs, GNRs, and ICG). Possible spectral shifts of PA label absorption bands due to the interaction with blood components, labeling, aggregation, or clustering (e.g., as in the case of ICG or GNPs) [10, 57, 59, 98] should be taken into account. Thus, further verification is required to determine the optimal wavelength in each specific case in the presence of the background signals from blood and surrounding tissues and possible red and blue shifts [57]. For example, the PA signals from blood microvessels in mouse ear were approximately 1.5–2.5 times higher than the PA background signals from surrounding tissue in the spectral range of 650–850 nm [59]. PA signals from thin unpigmented tissue, such as mesentery, were three to four times less compared to PA signals from thicker and more melanin pigmented ear tissue with bulk blood perfusion (i.e., capillary network). The real-time multispectral *in vivo* detection and identification of cells and bacteria in the blood circulation (as in modern *in vitro* multicolor FC) on the background of absorption by RBCs and surrounding tissue using multicolor PA labels should now be accomplished by the developed PAFC with extended spectral regions (415–2300 nm) and multibeam schematics [23, 30, 42, 45, 57, 59].

In the studies [15, 42, 45, 59], with focused laser-beam schematics and spatial resolution of 10–30 μm, peripheral (dermal) vein, capillaries, arteries, and lymphatics with diameters of 10–150 μm were assessed using rat and mouse tissue at depths of 50–3000 μm. The attenuation of laser radiation in tissue is not critical in PAFC, because even after significant attenuation, the laser-energy level is still sufficient to generate PA signals in deep vessels, which are detectable with highly sensitive PA schematics taking into account a low attenuation of acoustic waves in tissue [73].

Besides, laser energy could be increased without skin photodamage caused by skin surface cooling broadly used in laser medicine. Nevertheless, laser-beam blooming due to the scattering light may cause a decrease in PAFC spatial resolution. It should not be too important in cases of detection of rare objects with large distances between them (100 μm or higher). However, to assess closely located moving objects in deep vessels, somewhat invasive approaches should be adopted, such as bypasses, and delivering light to vessels using fiber incorporated in tiny needles or catheters into the vessels [15]. An alternative noninvasive approach is the time-resolved detection of PA signals from local vessel areas in combination with focused cylindrical high-frequency US transducers near the fiber or with the fiber built into the transducer with the central hole.

Another alternative is the tissue optical clearing technology based on surrounding tissue impregnation by biocompatible OCAs which considerably and reversibly reduce light scattering for 10–20 min after OCA topical application to animal model [10, 19, 55, 56].

According to preliminary theoretical and experimental estimations, the expected resolutions are around 10–200 μm at depth up to a few centimeters, especially using strongly absorbing PT/PA labels) [19, 42, 58, 59].

In conclusion, the results of this study strongly suggest the excellent potential of PAFC as a new promising tool in biological research. This technology provides an unprecedented capability for real-time detection of circulating bacteria (and cells) with ultrahigh sensitivity as one bacterium (or cancer cell) in the background of billion normal blood cells. This is unachievable with existing techniques, including advanced PCR, *in vitro* FC, and immunochemistry assays, even *in vitro* with cell enrichment.

17.6 *In vivo* Lymph Flow Cytometery

17.6.1 Principles and Main Applications of Lymph FC

17.6.1.1 Schematics of Integrated Lymph FC

Compared with the blood test, the lymphatic test has been paid much less attention, although recent discoveries in the biology of lymphatics (e.g., new receptors,

LYVE-1, VEGFR-3, and PROX1) emphasized the key role of lymphatic function in the development of many specific diseases such as tumor metastasis, infections, or immunity disorders [99–105]. To date, many questions of lymphatic function are still not quite answered. For example, despite dramatically growing evidence of the crucial role of the lymphatic route in the spread of metastatic cells such as melanoma or breast cancer to other organs, researchers knew almost nothing about transport of these cells through lymph vessels before their invasion into lymph nodes [99, 100].

Another very important issue is transportation of immune-related cells in lymphatics. According to a few *in vitro* animal studies, afferent lymph is composed mainly of immune-related cells: 85% T lymphocytes, 5% B-lymphocytes, and 10% other cell types (dendritic, and macrophages, among others) [103, 104]. Besides differences in composition, the total concentration of cells in lymphatics is lower and varies significantly (10^3–10^5 cells μl^{-1}) than in the blood flow ($\sim 5 \times 10^6$ cells μl^{-1}). Since, the sampling of lymph is not well established, it is still a low-accurate procedure, the exact composition of cells in lymphatics (e.g., amount RBCs), as well as cell rate, especially, at pathologies, is not still clear. Despite significant progresses in diagnostic techniques (e.g., magnetic resonance imaging (MRI), lymphoscintography, or fluorescence lymphography), to date, none has demonstrated the ability to real-time detection of individual cells in the lymph flow either without labeling or use nanoparticles as advanced low-toxic PA/PT high-contrast agents.

The discovery of natural, hydrodynamic focusing of cells led us to develop an advanced lymph PAFC (Figure 17.1), in which individual cells are irradiated near a valve nozzle when they form single-file flow. To verify this concept, we integrated PAFC with well-established TDM (Chapter 14), fluorescence, and PT techniques, which also made PAFC more universal for analyzing cells with different absorption, fluorescence, or scattering properties (Figures 17.7 and 17.11). Navigation of the laser beam on colorless lymphatic vessels in noninvasive skin models (Figure 17.14d) was based on lymphography with a contrast agent whose spectral range of absorption, for example, in the visible range, did not interfere with cell absorption in the NIR range (e.g., with absorption from unlabeled melanoma cells or cells labeled with NIR nanoparticles).

As results, PAFC provided detection of cells and simultaneous visualization of lymph with a standard tags, such as FITC (Figure 17.19a,c), EB dye (Figure 17.19b), and ICG (Figure 17.19d). In particular, fast scanning of a focused laser beam across the ear tissue that was stained with EB within the field of interest provided PA mapping/visualization of lymph vessels at 639 nm, where absorption of EB was still significant. In addition, interference with blood at 639 nm was minimal because blood absorption dropped to a level that was almost one order magnitude less compared with the maximum absorption near 580 nm. This may make it possible to determine the lymph vessels' margins at a low concentration of EB (or when a vessel has a deep location), which is invisible with the naked eye.

The strong focusing of the laser beam into the vessels on areas near a valve nozzle provided selective detection of signals from cells preferentially in this area

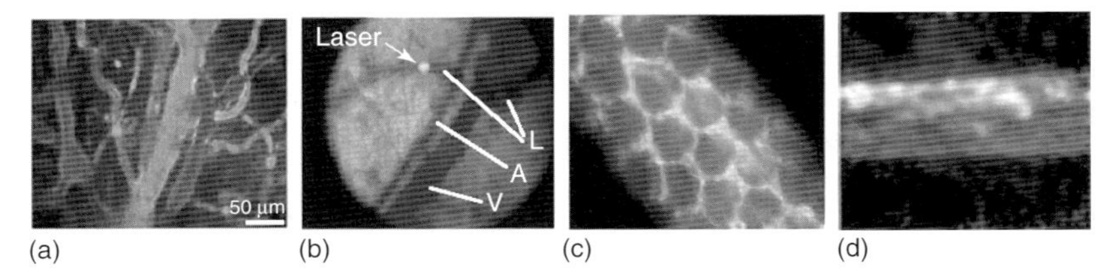

Figure 17.19 (a) Fluorescent image of blood (green) and lymph (red) microvessels in mouse ear stained *in vivo*. (b) TDM image of a mouse ear fragment: A, arteriole; V, vein; L, EB-stained lymph vessel; linear (red) and the circular (green) laser beams on a lymph vessel. (c,d) Fluorescent images of a mesenteric lymphatic after intravenous injection of FITC and ICG, respectively [18].

because a defocused relatively large laser beam generates smaller PA signals than a focused small-diameter beam from the same targets.

17.6.1.2 Label-Free Counting of Metastatic Melanoma Cells

PA detection of melanoma cells is based on laser-induced PT and accompanied PA effects in melanin as an intrinsic melanoma cell marker. To optimize the detection parameters, *in vitro* studies were undertaken first [18]. Signal-to-noise ratios in the range of 10–50 depend on cell "absorption" heterogeneity, and specific PT-image signatures associated with cellular chromophore (pigment) distribution, such as cytochromes in WBCs, hemoglobin in RBCs, and melanin in melanoma cells. PT signals from single cells in a linear mode without cell photodamage [15, 18, 59] demonstrated the standard fast-rising unipolar peak associated with rapid (nanosecond scale) cell heating and a slower (microsecond scale) tail corresponding to cell cooling (Figure 17.6).

The PA signal from the same cells had a bipolar shape that was transformed into a pulse train due to reflection and resonance effects (Figure 17.20). The signals were compressed either at a slow (ms scale) oscilloscope rate or after signal acquisition and presented as vertical unipolar positive lines similar to Figure 17.15.

Because melanin has a broad absorption spectrum with some absorption decrease in the NIR spectral range, melanoma cells were detected at a wavelength of 850 nm, with simultaneous laser navigation on a lymph vessel using EB dye, which absorbs in the visible-light range (Figure 17.19b). Specifically, right position of laser beam on lymph vessel was adjusted when the PA signal amplitude reached its maximum at the laser wavelength of 639 nm. In mice, early melanoma metastatic cells with a rate of 0.26 ± 0.05 cells min^{-1} were observed to appear in a lymphatic vessel of the mouse's ear on the first week after inoculation, with further gradual increase a rate over the course of the next three weeks (2.13 ± 0.3 cells min^{-1} at second week). The population of melanoma cells in a lymph flow was heterogeneous and contained live and necrotic (partly destroyed) cells. Routine examination of the sentinel lymph node (SLN) (the first location of metastasis) by histology and immunohistochemistry demonstrated no metastasis at the first week and a few micrometastasis at the second week. However, the high-sensitive PA mapping technique was able to detect single micrometastases and clusters of metastatic cells

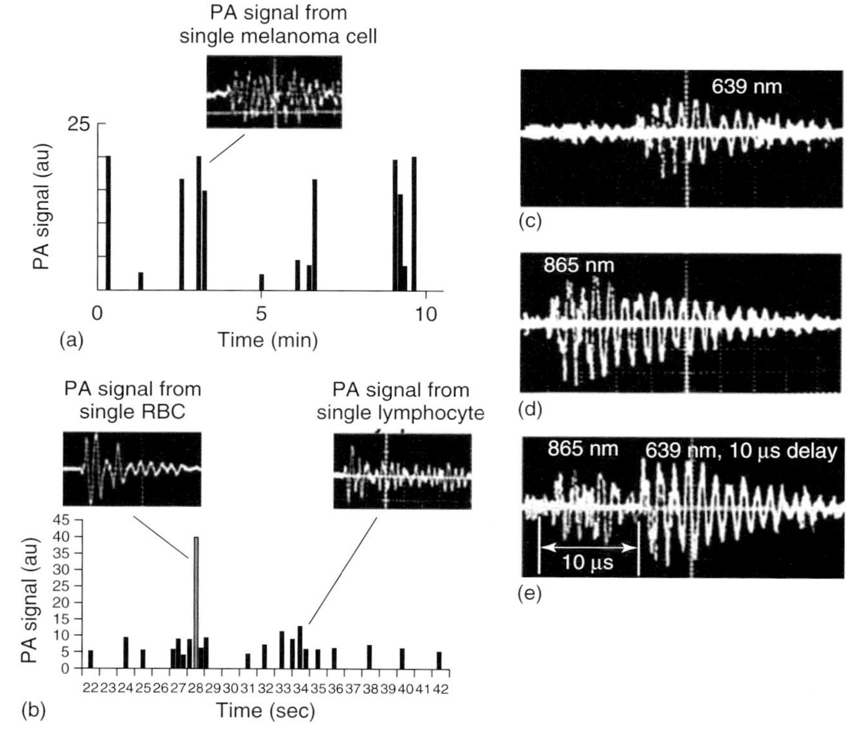

Figure 17.20 PA signals, real (top) and compressed with the positive part only (bottom) from (a) melanoma cells in an ear lymphatic in an ear tumor model at second week after inoculation, and (b) a single RBC among lymphocytes (multiplied ×5). PA signal from (c) rare necrotic lymphocytes labeled with GNRs absorbing at 639 nm, (d) apoptotic lymphocytes labeled with GNSs absorbing at 865 nm, and (e) live neutrophils labeled with CNTs absorbing at both wavelengths. Laser parameters (wavelength, fluence) (a) 850 nm, 30 mJ cm^{-2}; (b) 550 nm, 100 mJ cm^{-2}; and (c–e) 639 and 865 nm, 25 and 30 mJ cm^{-2}, respectively [18].

in SLNs *in vivo* on the first week after inoculation (Section 17.7). In mice with a late melanoma tumor in the ear and distant metastasis in the lungs (fourth week after tumor inoculation), metastatic cells appeared in distant (mesenteric) lymphatics at a rate 8.2 ± 1.1 cells min^{-1} [18, 19].

17.6.1.3 **Label-Free PA Detection of Lymphocytes**

In vitro spectroscopic studies revealed that PA signals from lymphocytes reached maximal amplitude in the visible-spectral range near 530–550 nm associated with absorption by cytochromes mainly Cyt *c* as the intrinsic absorption marker [18]. Background PA signals from vessels and surrounding tissues were approximately three- to fivefold less than from single lymphocytes at this wavelength due to the low level of background absorption and laser focusing effects. The obtained cell rate in the intact lymph vessel was 60 ± 12 cells min^{-1}; cell heterogeneity resulted in 2- to

2.5-fold fluctuations in PA signal amplitude from cell to cell. We also observed rare signals (~1%) that had strong amplitudes exceeding those of lymphocyte signals by 6–15 times; spectral and imaging multiparameter analysis identified rare single RBCs as the sources of these signals.

17.6.1.4 Real-Time Two-Wavelength Lymph FC with Multicolor Probes

Because of the relatively short (10^{-2}–10^{-3} s) time during which cells in the lymph flow appear in the detected volume after the valve nozzle, existing time-consuming imaging and spectral scanning methods are not quite adequate for multispectral *in vivo* PAFC with pulse lasers. Therefore, a novel approach for real-time multispectral exposure of fast-moving cells labeled with multicolor probes with experimental proof for two-wavelength modes and three nanoparticle types was introduced [18].

Necrotic and apoptotic lymphocytes, and live neutrophils were labeled with GNRs, GNSs, and CNTs, respectively. The GNSs and, especially GNRs had relatively narrow absorption bands, 180 and 75 nm (at a 50% amplitude level) with maximum absorption at 860 and 640 nm, respectively; in contrast, the absorption spectrum of CNTs was relatively broad covering visible and NIR range [89]. These labeled cells, mixed in equal proportions, were injected into a rat tail vein. After 6 h, in mesenteric lymphatics irradiated with two laser pulses at wavelengths of 865 and 639 nm with a 10 μs delay between the pulses, we observed rare PA signals associated with cells labeled by different nanoparticles. In particular, very rare PA signals from necrotic lymphocytes were generated by a laser pulse at 639 nm only after a 10 μs delay (Figure 17.20c), while PA signals from apoptotic lymphocytes were generated by laser pulse at a wavelength of 865 nm with no delay between the pulses (Figure 17.20d). Live neutrophils yielded two PA signals with a 10 μs delay because of CNT absorption at both 639 and 865 nm wavelengths (Figure 17.20e).

17.6.2 High-Resolution Single-Cell Imaging

17.6.2.1 High-Speed Imaging

Integrated technique allows simultaneous use of PTI, TDM, and fluorescent imaging. Here we illustrate the capability of last two imaging modalities. The high-speed TDM provided *in vivo* label-free visualization of natural transport of single cells from tissue to regional (e.g., sentinel) lymph nodes (Figures 17.2 and 14.24), including monitoring of their passage through the endothelial wall to an initial lymphatic vessel (Figure 14.23b,c) [18]. To count lymphocyte's subsets (e.g., B-lymphocytes), the fluorescence-labeled antibodies specific to CD45 markers of B-lymphocytes were injected into rat peritoneum *in vivo* followed by time-resolved fluorescence monitoring of mesenteric lymphatics (Figure 14.22i). The mean velocity of labeled B-lymphocytes in lymph flow was determined to be 1720 ± 18 μm s^{-1}, and the proportion of these B-lymphocytes in prenodal lymph was 4.5%. To capture an image of apoptotic cells, apoptosis was induced by intraperitoneal injection of dexamethasone followed 6 h later by the injection of Annexin V-FITC and then by fluorescence monitoring of apoptotic cells. The majority of apoptotic lymphocytes were found mainly in mesenteric tissues. Single lymphocytes adhered to a vessel wall were also observed. Only a few apoptotic cells were detected in lymph flow during the 1 h observation (Figure 14.22j).

17.6.2.2 **Capability of Multimodal Cell Imaging**

In the past, PA effects have been used mainly for biomedical imaging of tissue consisting only of many cells, however, with a depth of a few centimeters and potential micron-sized resolution [54, 106, 107]. As a fundamentally related technique, PAFC using a cell focusing phenomenon and PA effect within single cells has potential to address to single-cell lymph testing *in vivo* [18]. As demonstrated above, that endogenous cellular absorption can serve as intrinsic cell-specific markers having distinguished spectral and spatial signatures and its level is enough to produce sound in single cells readable with conventional US technique. This is noninvasive procedure because required a laser fluence is in the range of below 34–100 mJ cm^{-2}, which is the ANSI laser safety standard at the spectral range of 700–1100 nm, respectively [66]. The absorption-sensitivity threshold of PAFC at the single-cell level was estimated to be 10^{-2} cm^{-1}, which is several orders of magnitude better than that of absorption spectroscopy for the same condition.

This high sensitivity can be further enhanced during the interaction of pulse laser with absorbing nanoclusters associated with self-assembly of endogenous chromophores (e.g., cytochromes and melanin) or exogenous nanoparticles. Laser-induced optical, thermal, acoustic, and accompanied phenomena from individual closely located chromophore molecules or nanoparticles into clusters may overlap and provide synergistic amplification of PT/PA signals [57]. The high absorption sensitivity of PAFC together with low background absorption in colorless lymphatics (compared with blood vessels) provides the opportunity for label-free identification and quantification of RBCs, melanoma cells, and different types of WBCs, including immune-related cells in normal, apoptotic, and necrotic stages. In particular, PAFC revealed the presence of low concentration of RBCs in lymphatics, which may have diagnostic value for evaluation some diseases, in particular, venous dysfunctions, or diabetes accompanied by increasing blood vessel permeability [105, 108].

The label-free identification of a subpopulation of WBCs as well as other cancer cells (e.g., breast or prostate) has not been realized yet. However, it should be possible, on the basis of spectral fingerprints of the different cytochromes, for which concentration and mutual proportion may vary during cell metabolism, apoptosis, or carcinogenesis [47].

The most significant achievement is a real-time label-free detection of metastatic melanoma cells in lymph flow (Figure 17.20a) [18]. PAFC at a WBC rate of 10^2 cells s^{-1} in vessels with a diameter of 200 μm is able to detect just one melanoma cell during 3 h of observation; this rate is approximately equivalent to one melanoma cell in the background of 10^6 WBCs. This unprecedented sensitivity threshold in *in vivo* lymphatics is unachievable with any other existing technique. In larger lymph vessels, the sensitivity threshold and the observation time should be improved because of the higher WBC rate.

To prove the concept, superficial lymphatic vessels at a depth of 100–200 μm were taken; however, recent advances with PA imaging of blood vessels 1–3 cm deep [54, 106] support the potential of the new tool to sense lymph vessels and lymph nodes at depths of a few centimeters with an expected resolution of 0.03–0.1 mm with NIR laser and focused cylindrical US transducers [59].

In general, for the first time [18], *in vivo* evidence was obtained that (i) lymph vessels transport live metastatic cells and remove destructed melanoma cells; (ii) melanoma cells travel in lymph flow preferentially as small aggregates; (iii) the lymphatic route involves the earliest stage of metastatic disease, even before metastasis in SLNs are detected by conventional histological assays; and (iv) the lymphatic system can promote dissemination of tumor cells at the late stage of metastatic disease by passing cells between distant organs. In particular, from the primary tumor in mouse ear the metastatic melanoma cells may be spread by mesenteric lymphatics within the abdominal cavity [18].

The PA visualization of "colorless" lymphatics and, especially, SLNs using conventional blue dyes (e.g., EB, Methylene Blue, or Lymphazurin) that absorb in the visible-spectral range without interference with cell detection in the NIR range may help resolve complex and relatively unexplored issues, including integration *in vivo* noninvasive SLN mapping, single-cell detection, and metastasis assessment. This includes also the diagnostic value of lymphogenously disseminated rare circulating tumor cells (CTCs) for metastasis development [99, 109].

The specificity of PAFC can be enhanced by the use of multicolor nanoparticles with different absorption spectra as was demonstrated for selective detection of neutrophils, and necrotic and apoptotic lymphocytes in lymphatics. These nanoparticles and their clusters, and especially gold-based probes, have shown recently excellent potential for optical labeling because they do not photobleach or blink, are photostable, strong absorbers, can easily be conjugated with many proteins and antibodies [57]. The GNPs were recently approved for clinical trials, including PT therapy of cancer. (The safety of GNPs is subject of ongoing clinical trials, e.g., [110].) The *in vivo* lymph test concept may be further developed to extend its utility beyond diagnosis to theranostics, and to include the possibility of *in vivo* lymph purging by elimination, at the appropriate laser-energy level, of cancer cells labeled with nanoparticles in flow without harmful effects on surrounding normal cells and tissue.

To verify the concept of the *in vivo* rapid multispectral lymph test, a two-wavelength mode was applied [18]. The number of spectral channels can be easily increased by the use of low-cost, compact diode lasers with selected wavelengths and high pulse repetition rates (up to 10–50 kHz). Assuming the appearance of cells in the irradiated volume approximately during 1 ms (for a flow velocity of 5 mm s^{-1} and linear beam width of 5 μm) and a minimum delay of 10 μs between laser pulses, approximately $10^{-3}/10^{-5}$ s = 100 laser pulses with 100 different wavelengths can be used for spectral cell identification (in modern FC only 8–17 wavelengths) [111].

The integrated PA and fluorescent lymph FC schematics provide real-time *in vivo* detection of individual lymphocytes in norm and apoptosis by their labeling with conventional fluorescent probes (Figures 14.22i,j). Moreover, conventional well-established vital fluorescent molecules can be used as PA/PT labels because the limited quantum yield of most of them results in the transformation of most absorbed energy into heat [59, 112–115]. The high sensitivity of the PA/PT technique enables labeling at low concentration of dye molecules. Furthermore, the replacement of fluorescent dyes on GNPs could solve the potential cytotoxicity problem.

The phasic contractions of lymph vessels acting as a natural concentrator of flowing cells near a vessel's center in combination with lymphatic valves acting as natural nozzles provided cell positioning with minimum radial fluctuation at 5–10 μm in the lymph vessel with a diameter of 150–200 μm. It was sufficient for counting each cell near the valve exit using a linear laser-beam shape with a length of 40–70 μm and a width of 5–15 μm (Figures 17.1 and 17.8c, and Chapter 14).

To avoid counting the same cells during their backward motion, we applied the monitoring cell motion with TDM with further exclusion of data obtained for backward cell motion. Besides, the contractions in some lymphangions and cycles of valve activity (opening-closing) were found to be relatively stable, with an average rate of (12 ± 1) and (9 ± 1) min^{-1}, correspondingly [45]. So laser exposure was adjusted to these particular rhythms. In addition, detection of forward moving cells only can be achieved by synchronization of laser exposure with the vessel wall or valve motion coinciding to forward cell motion, controlled with a pilot laser.

The majority of examined lymphangions had well-developed bicuspid funnel-shaped valves and operated mainly under viscous fluid forces. This anatomic structure was not quite optimal for creation of the natural small nozzles with clear margins. In addition, some external factors that are not well controlled, such as muscle contractions, respiratory movements, and intestinal peristalsis, can influence valve function and phasic constrictions. Nevertheless, most selected active valves provided excellent natural hydrodynamic focusing of cells into single-file flow in analogy with artificial hydrodynamic focusing in conventional *in vitro* FC.

Reversible laser-induced short-term lymph vessel squeezing (Figure 14.15) [18] is also very attractive for the purpose of controllable forming single-file cell flow, although its mechanism (e.g., laser activation of pacemakers, or smooth muscle), possible invasive nature, side effects, and control degree require further study.

17.6.2.3 Conclusion Remarks

A novel combination of PA microscopy/spectroscopy and FC may be considered as a new, powerful tool in lymphatic research with a broad spectrum of potential applications. The integrated PAFC has many innovations, including *in vivo* natural cell focusing, PA effects in single cells, endogenous absorption as an intrinsic cell signature, real-time multispectral identification of cells with multicolor nanoparticles, combined navigation/detection schematics using contrast agents for visualization of "colorless" lymphatics without interference with cell detection, an almost ideal preclinical animal models to test the concept of lymph FC *in vivo*, and integration of PA, PT, fluorescent, and TDM techniques, which makes the *in vivo* lymph test more universal. Potential applications are related to real-time detection of various types of cells (e.g., RBCs, lymphocytes, dendritic, metastatic, infected, inflamed, and stem cells) at different states (e.g., norm, apoptosis, or necrosis), pathogens such as bacteria and viruses, contrast agents, nanoparticles, liposomes, and drugs in a variety of lymphatic structures (e.g., initial, afferent, and efferent lymphatics, and lymph nodes) under normal and pathological conditions (e.g., tumors, infections, lymphedema).

A quick translation of this technology to use in humans is anticipated with the development of a portable device that can be attached to lymph vessels and nodes for single-cell diagnostics, metastasis detection, or assessment of therapy (e.g., *in vivo* radiosensitivity and chemosensitivity testing) or different impacts (e.g., nicotine [116]) through counting metastatic and apoptotic cells.

A similar approach using vein valves for cell focusing could also be applied to FC in vein blood stream.

17.7 *In vivo* Mapping of Sentinel Lymph Nodes (SLNs)

17.7.1 Motivation for Cancer Prognoses

Cancer metastases, which originate from detached tumor cells, are major causes of mortality [109, 117, 118]. These cells can disseminate through afferent (prenodal) lymph vessels to SLNs, which have dual roles as the first identifiable metastatic site and immune barrier for CTCs. The presence of metastatic cancer cells in SLNs (i.e., positive SLNs) is one of the most important prognostic factors for survival in cancer patients [109, 118]. Traditionally, SLN assessment involves mapping of SLNs and surgical excision for histological and immunohistochemical detection of intranodal metastasis [119]. This procedure is invasive and time-consuming, and can lead to complications related to surgical procedure.

The progress in *in vivo* mapping of lymph nodes with reasonable precision has been made with many advanced technologies such as computed tomography, positron emission tomography, multiphoton microscopy, MRI with magnetic nanoparticles (MNPs), radio-lymphoscintigraphy with ^{99m}Tc-sulfur-colloid-, albumin-, or gadolinium-labeled contrast agents, and optical lymphography with blue dyes or multicolor QDs [119–127]. However, most of these techniques are limited in their ability to assess SLN status because of their still poor sensitivity and depth resolution (e.g., a few millimeters for optical technique), require high concentration of contrast agents because of the influence of background signals, and provide only anatomic and nonfunctional maps, which cannot image lymphatic tracers together with real-time counting of tumor cells in lymph flow and lymph nodes. As a result of these challenges, SLNs with micrometastases and, especially cancer cells in lymphatics, may remain undetected until there is sufficient tumor burden to be detected by currently available techniques. This may delay the initiation of systemic therapy for a patient or result in erroneous clinical decision-making and treatment.

The PA techniques using endogenous chromophores (e.g., melanin or hemoglobin) or synthetic nanoparticles as PA contrast agents have demonstrated tremendous potential for *in vivo* imaging of tumors with higher resolution in deeper tissues (up to 3–5 cm) compared to existing optical modalities [18, 59, 72, 106, 128, 129]. In particular, we performed pioneered applications of CNTs [59], QDs [130], and golden carbon nanotubes (GNTs) [131] and their clusters with red-shift effects [57] as PA contrast agents using *in vivo* FC PT [17, 51] and PA [42] detection schematics for real-time counting of CTCs targeted by advanced nanoparticles. We also first suggested the possibility of using the PA method for assessing SLNs [18, 19, 59, 87, 131, 132] and demonstrated multicolor PA detection of disseminated tumor cells (DTCs) in lymphatics using tumor bearing animal models [18].

Recently, PA imaging technique with microscopic schematic was successfully applied to mapping of lymph nodes with depths up to ~3 cm in animal models using blue dyes and nanoparticles as contrast agents [133–135]. The diagnostic capability of PA technique for detecting metastasis in SLNs and counting metastatic cells transported by lymph to SLNs has been also recently demonstrated [19]. It was shown that PA detection of metastasis in SLNs can be integrated with their PT purging using a fiber-based multimodal diagnostic-therapeutic platform with time-resolved multicolor PA lymphography, PA lymph FC, and PT therapy.

17.7.2
Fiber-Based Multimodal Diagnostic-Therapeutic Platform

17.7.2.1 Experimental Arrangement and Methodology

The experimental setup was based on an upright Olympus BX51 microscope (Olympus America, Inc., USA), with incorporated PA fiber-based, PT, fluorescent and TDM, and fiber modules as previously described [18, 19, 31, 126, 127, 131, 132, 136] (also Figures 17.2, 17.7, and 17.11 and Chapter 14). Briefly, a tunable pulse OPO (Lotis Ltd., Minsk, Belarus) was used to irradiate samples at the following parameters: wavelength, 420–2300 nm; pulse width, 8 ns; beam diameter, 10–100 μm; fluence range, $1–10^4$ mJ cm^{-2}; and pulse repetition rate, 10–50 Hz. In selected experiments, a diode laser (905 nm, 15 ns, 10 kHz, 0.01–0.7 J cm^{-2}, model 905-FD1S3J08S, Frankfurt Laser Company) with driver IL30C (Power Technology Inc., Little Rock, AR, USA) was used for the detection of melanoma cells. PA waves were detected by US transducers (model 6528101, 3.5 MHz, 5.5 mm in diameter; Imasonic Inc., Besançon, France), then amplified (amplifier 5660B, band 2 MHz, gain 60 dB; Panametrics) and recorded with a Boxcar (SR250, Stanford Research Systems Inc., USA), Tektronix TDS 3032B oscilloscope, and a computer using standard and customized software. The US gel or warm water was used to provide acoustic and optical coupling between tissue and transducer.

Two-color mapping of lymphatics and identification of nanoparticles was realized with two laser pulses at wavelengths of 850 nm (OPO) and 639 nm (Raman shifter) with a 10 μs delay, followed by time-resolved detection of corresponding PA waves [18]. Delivery of laser radiation to samples was performed either with a microscope schematic or with a 400 μm diameter fiber with focusing tip fixed in a customized folder (Figure 17.2). PA microscope was integrated with a PT module [31]. In PT thermal-lens mode, defocusing of a collinear He–Ne laser probe beam was measured using a pinhole. A fluorescence module with colored CCD camera (Nikon DXM1200) was added to verify the PA/PT data.

Navigation of the laser beams as well as anatomic microstructure of SLNs was controlled with high-resolution (300 nm) TDM (Chapter 14). An *in vitro* study was performed with cells in suspension on conventional 120 μm thick microscope slides. PA mapping was achieved *in vivo* and *in vitro* by spatial scanning of focused beam or fiber across samples using a computer controlled microscopic stage (Conix Research, Inc., USA) and visual basic software with a total scanning time of 1–10 min, depending on the sample size.

17.7.2.2 Contrasting Agents, Cells under Study, and Animal Model

The magnetic nanoparticles (MNPs) as well recognized lymphographic tracer [123] and novel golden carbon nanotubes (GNTs) [131] were used as PA contrast agents [19]. The 30 nm spherical PEG-coated MNPs with a Fe_2O_3 core were provided by Ocean NanoTech (Springdale, AR, USA). Customized GNTs were synthesized as reported previously [131] and consisted of hollow single-walled CNTs surrounded by thin gold layers with an average dimension of 12.8 nm × 91.7 nm. They exhibited high water solubility and biocompatibility, low cytotoxicity (due to the protective layer of inert gold [137] around the CNTs [131]), and high plasmon absorption in the NIR range of 850 nm [131]. The GNTs were conjugated with folate, which are highly expressed in selected human breast tumors and are not expressed in normal endothelial cells in lymphatics.

B16F10 mouse melanoma cells and MDA-MB-231 human breast cancer cells (American Type Culture Collection, Manassas, VA, USA) were cultured according to the vendor's specifications [19]. Viable cells were resuspended in PBS for all tests. For verification purpose, the melanoma cells (1×10^4 cells) were stained with FITC using standard procedure (15 min at 37 °C), while the bulk of breast cancer cells (1×10^4 cells) were double labeled with GNT-folate conjugates (overnight at 37 °C) and, then, with FITC. The labeling efficiency was estimated *in vitro* using fluorescent technique (e.g., ∼99% efficiency for fluorescein isothiocyanate (FITS)) and by comparing PA/PT signals from labeled and control cells at selected wavelengths.

Nude *nu/nu* mice weighing 20–25 g, purchased from Harlan Sprague–Dawley (Harlan Sprague-Dawley Indianapolis, IN, USA), were used to verify the efficacy of PA/PT technique in accordance with the protocols approved by the UAMS Institutional Animal Care and Use Committee [19]. The mice were anesthetized with intraperitoneal injection of ketamine/xylazine of 50 mg/10 mg/kg and then placed on a warmed microscopic stage. Melanoma tumors were created in the mouse ear by inoculating 10^6 B16F10 melanoma cells in 50 μl of PBS with a 30 gauge needle. The progression to metastatic disease was estimated by measurements of primary tumor size and subsequently verified with PA evaluation and histology. Also the breast cancer cells were injected with a microsyringe (Beckton Dickson and Co., Franklin Lakes, NJ, USA) directly in the lymph nodes to mimic metastasis.

17.7.3 *In vivo* and *In vitro* SLN Studies

17.7.3.1 SLN Examination

In vivo and *in vitro* PA mapping and microscopic and H&E pathological examination of lymph nodes were performed at one and two weeks after melanoma tumor cell inoculation [19]. *In vivo* assessment of SLNs and lymphatic vessels was performed noninvasively through intact skin and intraoperatively through a small skin incision overlying the SLN and direct attachment of the folder with a fiber to skin and the lymph node, respectively. For *in vitro* study, lymph nodes were excised without fat and surrounding tissues from the mouse and placed in a well (8 mm in diameter and 2 mm in depth; Molecular Probes) of a microscope slide to obtain optical images at different magnifications (4×, 10×, 20×, 40×, 100× with oil immersion).

To decrease beam blurring due to light scattering in lymph nodes, the nodes were embedded for 5 min in 40% glucose, 100% DMSO, and 80% glycerol as hyperosmotic OCAs, which reduce tissue scattering [10, 55, 56, 138]. Comparisons of optical images with different agents and PBS (control) revealed that glycerol had a maximal optical clearing (Figure 14.23). The treatment of SLNs with 80%-glycerol permitted the label-free imaging of a fresh node at the cellular level including localization of immune-related, metastatic, and other cells (e.g., lymphocytes, macrophages, dendritic cells, and melanoma cells) as well as surrounding microstructures (e.g., afferent lymph vessel, subcapsular and transverse sinuses, the medulla, a reticular meshwork, follicles, and the venous vessels) (Figure 14.24).

The combination of PA mapping of SLN metastases with high-resolution TDM mode (up to 300 nm) enables assessment of the exact distribution and number of melanoma metastasis and laser-induced damage area [19]. To confirm that the PA signals and following PT purging are related to the tumor cells (evaluation of false-positive signals), the cells were additionally labeled with FITC and imaged with fluorescence techniques. For histological and immunohistochemistry analysis, the examined SLNs were fixed in phosphate-buffered 10%-formalin (pH 7.2) and then embedded in paraffin. Multiple 5 μm sections at 200 μm intervals were cut through the entire node and studied with conventional histology with H&E according to the manufacturer's instructions.

17.7.3.2 Two-Wavelength PA Lymphography with Multicolor Nanoparticles

PA technique was employed for identification of lymph vessels and node location using MNPs and GNTs as PA lymphographic contrast agents [19]. To evaluate two lymphatic basins, MNPs and GNTs (each $\sim10^{11}$ NPs ml^{-1} in 5 μl of PBS) were intradermally injected in the tip of the left and right mouse ears of intact healthy animals ($n = 4$). The prior measurement of PA spectra of NPs revealed that 639 nm is the optimal wavelength providing the best PA contrast of MNPs in tissues and minimal interference with GNTs that have a maximum absorption at 850 nm [131]. The dual-color PA mapping using two laser pulses at 639 and 850 nm was performed immediately after injection and then periodically every 10 min for 5 h. Then, co-injection of GNTs with conventional EB dye (5 μl) provided additional visualization of lymphatic vessels and SLNs. This dye with absorption in the visible region (below 660 nm) was chosen to avoid interference from GNTs with maximum absorption in the NIR region. This study revealed that nanoparticles quickly entered into lymphatics and migrated in less than 3–5 min to left and right cervical lymph nodes at a depth of 1–3 mm. This is in line with rapid uptakes of small NPs by lymphatics [139, 140].

The PA signal amplitude at 850 nm on the site over SLNs before and after injection of GNTs increased from 1.5 ± 0.2 to 30.1 ± 3.2 a.u. (10 min after injection, $p = 0.05$). The PA signal amplitude at 639 nm on the site over SLNs before and after injection of MNPs increased from 2.2 ± 0.3 to 8.7 ± 0.9 a.u. ($p = 0.01$). This led to an increase in PA contrast (as ratio of signals before and after injection at the skin site overlying SLNs) of ~20 for GNTs and ~4 for MNPs at 850 and 639 nm, respectively. The PA signals from a SLN with MNPs were stable for three hours with slight decreasing amplitude at 5 h after injection; however, the clearance of GNTs was faster, with decreasing amplitude to almost background 1–1.5 h after injection. Co-injection of GNTs with EB dye yielded co-localization of PA signals from GNTs (850 nm) and EB (639 nm) verifying GNTs as a novel PA high-contrast NIR lymphographic tracer.

To confirm accumulation of MNPs and GNTs in SLNs, SLNs were also inspected directly by incision and retraction the overlying skin. PA mapping by scanning of the fiber across the samples demonstrated that the high-amplitude PA signals were exclusively from the SLNs with PA contrast of 55 for GNTs and 14 for MNPs at 850 and 639 nm, with signal averaging of 1 s.

17.7.3.3 Melanoma Model: Detection of Tumor Cells in Lymphatics and SLNs

The feasibility of the integrated PA platform for assessment of lymphatic DTCs and SLN status was first tested in a melanoma tumor model [19]. Melanoma is a malignancy that is increasing epidemically. It can quickly progress to incurable metastasis [141]. Conventional diagnostic tools have a high false-negativity (~25%) in detecting positive SLNs [142]. From a clinical perspective, melanin is a very promising endogenous PA contrast agent and provides PA signals from individual pigmented melanoma cells above blood background in the NIR range [18, 136]. Monitoring of the PT signals from individual B16F10 cells (total 300) in suspension demonstrated a high level of heterogeneity: the 5–10% of cells with increased pigmentation produced PT signals with amplitudes much greater than PT-signal amplitudes from cells with less pigmentation [136]. Approximately $82 \pm 3.8\%$ of tumor cells were detectable with the PT/PA technique.

The entire diagnostic procedure takes 10–20 min and involves injection of PA lymphographic agent (5 μl of MNPs or EB) near the ear tumor and PA mapping of lymphatic vessels with counting of melanoma cells in the collecting prenodal lymph vessel and mapping of the SLN for detection of metastasis. Specifically, after a primary melanoma tumor had developed in the mouse ear, the injection of MNPs provided stronger PA signals from lymphatics and SLNs at 639 nm. However, similar NIR absorption spectra of MNPs and melanin made spectral identification of melanoma metastasis in the presence of MNPs difficult. To resolve this problem, EB dye was used, which absorbed at 639 nm only, while the presence of melanoma cells in lymphatics and in a node was identified through the appearance PA signals at 850 nm. Because absorption of melanin [18, 136] also occurs at 639 nm above the background from EB dye, for additional identification of melanoma cells, specific ratio of PA signals at 639 and 850 nm was used [136].

In addition, EB injection was used to visually differentiate the afferent lymph vessel that was exclusively collecting lymph from the tumor (Figure 17.21a, top, callout). Many (20–50) diode laser induced PA signals from the same flowing melanoma cells were acquired using the Boxcar system and transformed into signals with width determined by the transit time of the cells through the laser beam (Figure 17.2b,c; details of acquisition algorithms in [18, 136]). Real-time PA monitoring of prenodal lymph vessels and the SLN revealed that the number of metastatic cells producing rare flash PA signals, which increased from 0.26 ± 0.05 to 2.13 ± 0.3 cells/ min at one and two weeks after melanoma inoculation. At one week after inoculation, the primary tumor size was 1.0 ± 0.2 mm^2 (Figure 17.21, top row, left) and 493 PA signals associated with individual melanoma cells (single spots covering 6% of the examined area) and three SLN micrometastases were identified with PA contrast of 10.6 ± 1.2 (Figure 17.21, middle row, left). At two weeks, the primary tumor size was increased to 3.6 ± 0.5 mm^2 (Figure 17.21, top row, right) and much larger numbers of PA signals (3188 spots covering 39% of the examined area) with a mean contrast of 22.5 ± 0.9 were detected, including 7–10 high-amplitude signals, which were indicative of metastases (Figure 17.21, middle row, right). The histology of these nodes showed an absence of metastases at one week (Figure 17.21, bottom row, left), but single metastases were present on histology at two weeks post-inoculation (Figure 17.21, bottom row, right). This result is consistent with the ability of PA technique to detect single-cell metastases with an improved sensitivity over conventional histology [18, 19, 136].

Thus, an increase in the quantity of lymphatic DTCs by ~10 times is accompanied by a ~6.5-fold increase in the number of PA signals and a 2-fold increase in their mean contrast from the SLNs that were associated with an increased number and size of melanoma micrometastases in the SLNs. This demonstrates the capability of PA technique to not only detect early micrometastasis including individual metastatic cells (at week 1), which cannot be achieved by conventional histological assays, but also confirm a correlation between lymphatic DTCs, primary tumor size, and SLN metastasis status.

17.7.3.4 **PA Detection of Breast Cancer Metastases in SLNs with Functionalized Nanoparticles**

To explore the feasibility of the PA technology to detect nonpigmented metastasis [19], low-absorbing MDA-MB-231 breast cancer cells were labeled with GNT-folate conjugates [131, 143]. The SLN was identified by the PA technique at 639 nm with

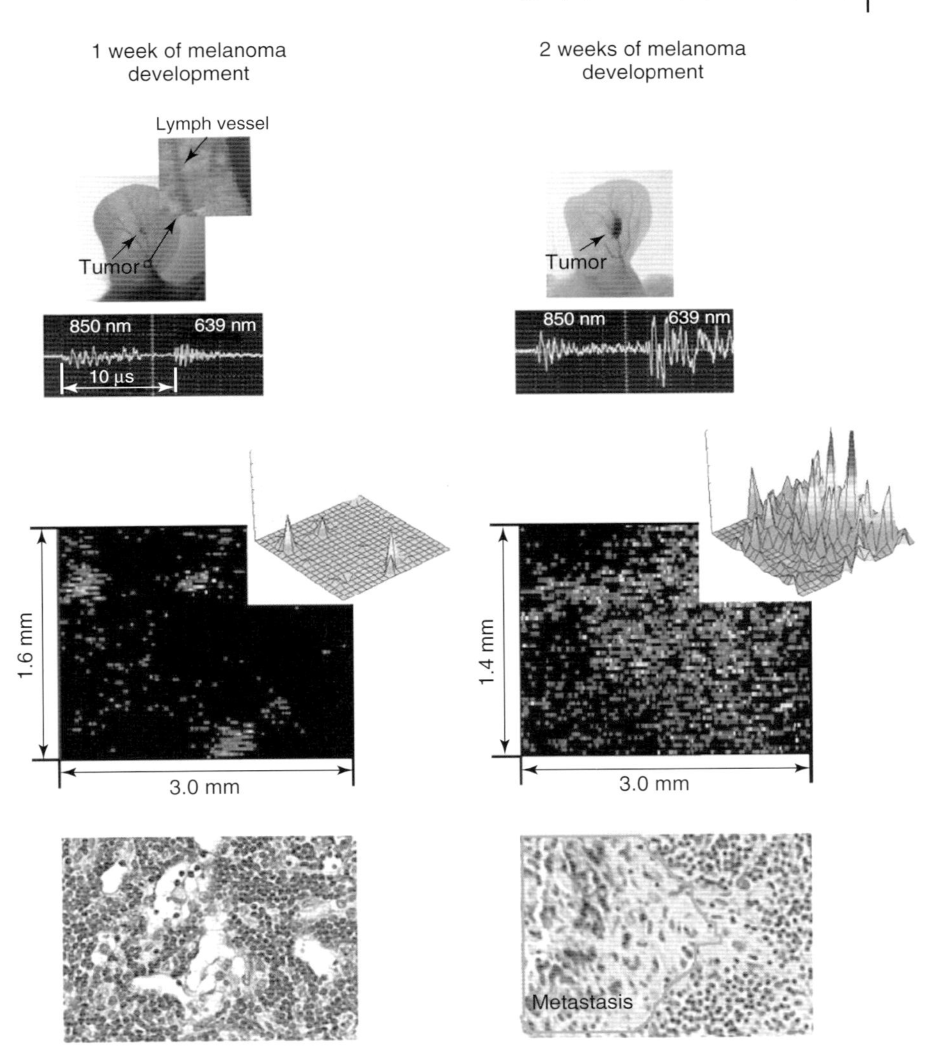

Figure 17.21 PA detection and counting of melanoma metastasis during tumor development. (top row) Photo of tumor with visualization of the lymph vessel (using EB dye) collecting lymph from primary tumor area (top, callout) and *in vivo* two-wavelength PA detection (bottom oscillogram) of melanoma metastasis with tumor progression in the SLN at first (left) and second (right) week. (middle row) *Ex vivo* PA mapping of the SLN with melanoma metastasis at single cell level at 1 (left) and 2 (right) week(s) of primary tumor development. The data are presented as 2-D high-resolution (bottom) and 3-D low-resolution (top, callout) simulation. Each single spot on bottom is associated with single metastatic cells showed on right panel (top, left callout). Red pseudo-color peaks indicate the photoacoustic signals with maximum amplitudes (bottom row). Histological images of the investigated SLNs demonstrating no histological changes at week 1 (left) and the detectable metastases, contoured by green line, at week 2 after tumor inoculation (right) [19].

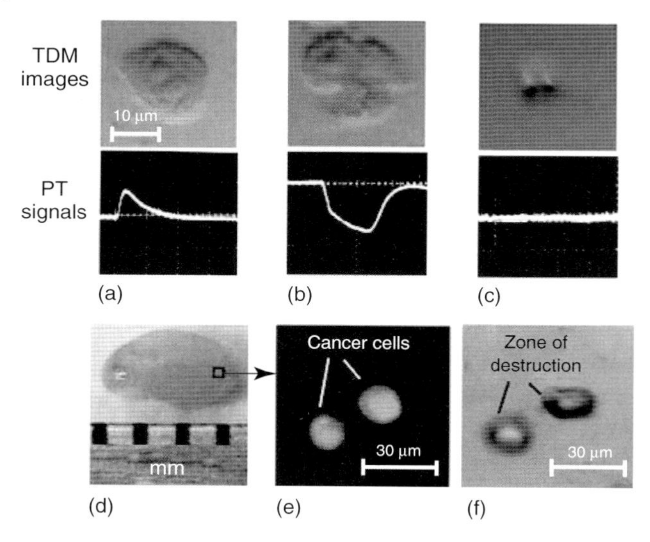

Figure 17.22 Targeted laser purging of SLN metastasis. (a–c) Laser ablation of MDA-MB-231 breast cancer cells labeled by GNT-folate conjugates: linear PT response from live cell (a), nonlinear PT response from damaged cell (b), and the disappearance of PT signal after cell disintegration by high energy laser pulse as feedback control of therapeutic efficacy (c). (d–f) Laser-targeted ablation of single breast cancer cells double labeled by GNT-folate and FITC injected in the lymph node *ex vivo* (d) was verified at single cell level by fluorescent microscopy (e) with identifying the zone of the high local destruction (30–40 μm) within the lymph node after cancer cell elimination (f) [19].

injection of 5 μl of MNPs as described above. Then, 10 μl of suspension with a limited number (~100) of labeled breast cancer cells was injected directly into target lymph nodes. Immediately after this injection, PA and PT signals at 850 nm were observed. To ensure that these signals were from cancer cells, additional dual labeling of cells with folate-GNTs and fluorescent dye (FITC) were performed. The *in vitro* mapping of SLNs with injected cells using PA and fluorescent techniques shows that PA signals are exclusively from fluorescent cells (Figure 17.22d,e).

17.7.3.5 **Targeted Laser PT Purging Metastases in a SLN**

To examine the application of this novel tool for targeted therapy of metastases, PA diagnosis was integrated with PT therapy based on laser-induced microbubbles around overheated GNTs [131] or melanin clusters [136] as a main mechanism of cell damage [19]. A low laser pulse energy (20 mJ cm^{-2}) at 850 nm produced a linear PT response *in vitro* from single breast cancer cells labeled with GNTs with a characteristic unipolar positive pattern associated with fast cell heating and slower cooling (Figure 17.22a, bottom). However, at an increased pulse energy from 20 to 100 mJ cm^{-2}, a negative peak was observed (Figure 17.22b, bottom), which is indicative of microbubble formation, with subsequent cell disintegration (Figure 17.22b and c). When the double labeled (GNTs–folate–FITC) breast cancer cells were injected into lymph nodes and exposed once with low-energy laser pulse to image them and then with increased energy pulse, there was an *in situ* disappearance of both fluorescence and PA/PT signals at the second laser exposure, which correlated with cell disintegration (Figure 17.22d–f). The strong localization of damaged areas in the lymph node tissue was confirmed by transmission images (Figure 17.22f).

A similar protocol was used for purging melanoma metastasis in SLNs *in vivo* at two weeks after tumor inoculation in the mouse ear (Figure 17.23a). [19]. Briefly, the application of 1 ns laser pulse at 639 nm with different fluences of 80, 200, and 400 mJ cm^{-2} led to a decrease of PA signals during secondary laser pulses (e.g., Figure 17.23c, right). Specifically, the PA signal levels detected using low fluence laser pulses after exposure with high fluence laser pulses were comparable to the background signals in an intact SLN without melanoma, indicating that there was a complete laser ablation of melanoma metastasis. The skin above the SLNs was intact at low energy (like in Figure 17.23a) and was partly damaged at high laser energy (Figure 17.23b).

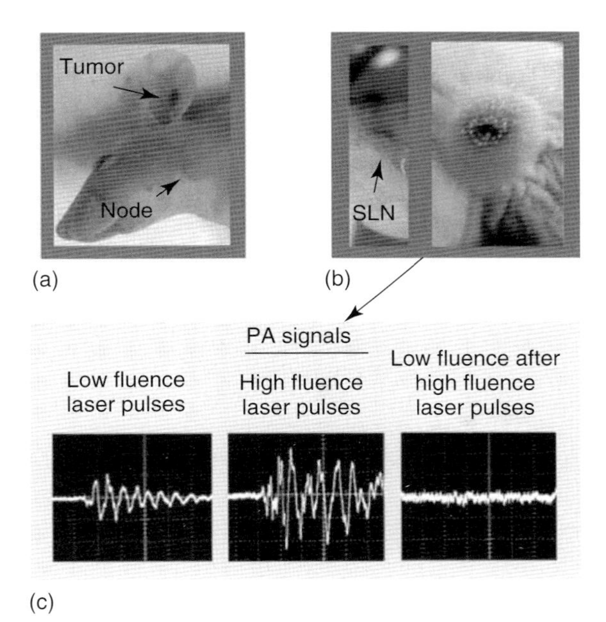

Figure 17.23 (a) Melanoma in the mouse ear with micrometastasis in the SLN at week 2 after tumor inoculation. (b) Laser-induced localized damage of the SLN containing melanoma micrometastasis at different laser energy. Laser parameters: wavelength, 639 nm; 10 laser pulses with overlapping beams with diameter on skin of 1 mm; laser fluence, 200 mJ cm^2 (b, left) and 750 mJ cm^2 (b, right). (c) PA signals from the SLN using a low-fluence laser pulse, 50 mJ cm^2 (left); at high fluence laser pulse, 400 mJ cm^2 (middle); and at low-fluence laser pulse again, 50 mJ cm^2 after high-fluence laser pulse (right). Amplitude (vertical axis), time scale (horizontal axis): 100 mV/div, 2 μs/div (left); 50 mV/div, 2 μs/div (middle); and 20 mV/div, 4 μs/div (right) [19].

17.7.4
Discussion of PA/PT Platform Benefits and Perspectives

To the best of our knowledge, Ref. [19] is the first report of a robust multifunctional PA/PT platform, which by integrating time-resolved multicolor PA lymphography, PA lymph FC, and PT therapy provides real-time mapping of lymph vessels and SLNs, counting of DTCs in prenodal lymphatics, and diagnosis of early metastasis in SLNs and its purging.

The integrated PA diagnostic and PT therapy can be performed at single cancer cell level as verified in the studies [18, 19, 59, 131, 136]. Thus, this technology overcomes the limitation of existing techniques and permits simultaneously distinguishing lymphatic flow and SLNs with flowing and static tumor cells. Using this platform, the diagnostic value of SLN status could be significantly improved by quantitative molecular detection of DTCs in prenodal lymph before they colonize in SLNs. As a result, prevention or at least inhibition of metastatic progression can be potentially achieved by noninvasive targeted PT therapy of early DTCs and micrometastasis precisely in SLNs. The applicability of this approach was recently demonstrated for circulating melanoma cells in blood [136].

A challenge of this approach will be to select cancer metastasis biomarkers and nanoparticle parameters and to determine how to distinguish molecular-based targeting of metastasis compared to nonspecific uptake of dyes and nanoparticles by SLNs (e.g., due to phagocytosis of nanoparticles by macrophages or natural killer cell action) that leads to increased false-positivity [123, 140].

As we discovered, there is a shorter retention period for GNTs in SLNs (1–1.5 h) compared to other nanoparticles and dyes (from many hours to several days) [124, 131]. This finding confirms other observations that phagocytosis can be minimized by the "worm"-like shape of nanopartilces such as GNTs [144]. More data can be collected with the PA technique on the drainage dynamics of functionalized nanoparticles, such as GNTs, with different sizes and shapes by real-time counting of nanoparticles before, during, and after transit in SLNs. If this is successful, metastasis in SLNs can be potentially identified by the prolonged retention of optimized nanoparticles in SLNs, whereas in SLNs without metastasis, the nanoparticles clear quickly. The identification of micrometastasis can be based also on the synergistic enhancement of PA signals from nanoparticles within the metastatic lesions due to a high local concentration of nanoparticles in targeted tumor cells, and, especially, nanoparticle clustering around dense or clustered cancer markers [57, 59, 145] compared to random nonspecific distribution of nanoparticles across SLNs.

Actually, we have already demonstrated that (i) clustered GNPs or melanin aggregates enhance both the PA and PT signals from 5 to 100 times depending on the cluster sizes [57, 131, 136] and (ii) laser-induced nanobubble generation around overheated nanoparticles is an effective amplifier of PA signals [130, 136].

The high-resolution optical imaging developed with optical clearing [10, 19, 55, 56, 138] with its ability to differentiate main SLN microstructures (Figure 14.24) should be important for the verification of metastasis and nanoparticle (NP) location in SLNs by precise guidance of diagnostic laser beams at least in *in vitro* experiments.

Currently, there are some uncertainties with regard to the clinical significance of micrometastasis detection because of the limitation of currently available diagnostic modalities [109, 117, 118]. The highly sensitive PA technology can adequately address this and also clarify how SLN micrometastasis and even small dormant clusters of metastatic cells are related to tumor progression in their natural environment. The PA technique could also further define the correlation between the presence of lymphatic DTCs and SLN status. In addition, simultaneous PA counting of tumor cells in the lymph [18, 19] and blood [136] systems could reveal the main pathways for tumor spread including its entrance into blood vessels from SLNs and *vice versa*.

Therefore, it remains to be determined if CTCs enter into blood vessels during SLN metastasis progression. The results of this study suggest that PA technology is an ideal tool to answer this problem because it permits molecular detection/counting of CTCs *in vivo* with potential sensitivity in humans of one CTC in 100 billion of RBCs [136].

In the future, PA/PT technology can be extended by combining it with other promising and established methods, for example, MRI using MNPs as multimodal (PA-PT-MRI) contrast agents or QD-based lymphography [119, 122]. As we have demonstrated the utility of QDs as PA/PT contrast agents [130], QDs could serve as triple lymphographic tracers for advanced fluorescent imaging, highly sensitive and deeply penetrating PA cytometry, and PT elimination of metastatic cells, if the concern over potential QD toxicity is resolved. The integrated PA/PT diagnostic-therapeutic platform needs to be compared with other recently developed approaches, like SLN purging using adoptive transfer of normal T cells with oncolytic virus that kills metastatic cells [144]. We have already demonstrated that viruses with covalently incorporated gold NPs have the potential to be used for integrated PT-gene cancer therapy [57, 146]. Thus, a combination of both technologies is quite feasible.

The label-free nature of PA/PT technique using melanin as an intrinsic cell marker helps avoid potential problems related to the labeling of tumor cells *in vivo*. This may provide the precedent for noninvasive, *in vivo* staging and purging of SLNs in melanoma, which is a highly aggressive epidemic malignancy and is often widely metastatic at an early stage. Approximately 18% (and potentially more) of melanoma cells with low melanin expression can be missed with label-free PA techniques [136]. However, this is less than or comparable to conventional assays, which can miss up to 30–70% of tumor cells because of the limited expression of cancer biomarkers [109, 117, 118]. Further studies are necessary to determine whether label-free PT therapy of melanoma metastasis is effective enough to be used alone or if it should be used in conjunction with chemo and radiation therapy. Further improvement could be achieved by targeting melanoma with strongly absorbing nanoparticles as we demonstrated for breast tumor cells [131].

Taking into account the successful clinical applications of PA techniques with target (blood, lymph, or tumor) depth up to 3 cm (maximal depth for breast SLNs) [72, 148–150] using the safe range of laser fluence [66], nontoxic (e.g., melanoma) or low-toxic (e.g., GNTs) contrast agents [131], and sterile disposable fibers (at least for interstitial therapy of SLNs), we expect that PA/PT technologies can be rapidly translated to the bedside through the use of a portable multicolor device with flexible multiple-strand fibers and miniature focused US transducers. For diagnostic purpose a high pulse repetition rate (10–100 kHz) laser array as demonstrated recently for the detection of circulating melanoma cells in blood [136] should be used.

In clinical applications, a high-resolution PA imaging for mapping SLNs is useful but not absolutely necessary. Instead, rapid (minute scale) identification of lymph vessel and SLN and margins with reasonable resolution can be achieved with simpler robust multifiber schematics. Moreover, these schematics with multicolor NPs [18] injected in different sites can provide assessment of several lymph vessels that drain into shared SLNs, allowing the differentiation of nonmetastatic lymph nodes from metastatic ones and the noninvasive treatment of the metastatic ones.

The expected clinical benefits include rapid, ultrasensitive, and noninvasive SLN assessment integrated with *in vivo* high-targeted localized PT therapy. This may preserve SLNs as an immunological barrier against cancer cells and allow avoiding surgical complications (e.g., bleeding, lymphedema, nerve damage, and abnormal scarring). In addition, the PA technique might rapidly assess the therapeutic efficacy through *in vivo* counting of DTCs and micrometastasis.

17.8 Concluding Remarks and Discussion

A crucial advantage of PTFC/PAFC conventional PT and PA technique, called *PT/PA scanning cytometry* (e.g., photothermal scanning cytometry, PTSC) [15, 39], is the significant increase in the speed of cell analysis in flows with velocities up to $2\,m\,s^{-1}$. In particular, in conventional technique, the PT image, as well as the integral PT response from individual cells, is formed 5–30 μs after just one pump pulse with a relatively large beam diameter of 15–25 μm covering one entire cell [21]. Additional time is required for realization of the probe image subtraction algorithm. Currently, this time is ~0.1 s owing to a low-speed CCD camera operating at <12 fps and a low laser pulse repetition rate of 10 Hz.

In general, to create an image of one cell, PTI mode does not need time-consuming scanning with a strongly focused laser beam across a cell, which currently takes at least a few minutes [24, 27]. This time could be shorter with the use of a laser having a high pulse repetition rate. Nevertheless, to obtain PT images and PT responses from different cells, the center of a laser beam must be moved from cell to cell. Discrete changes in the position of the laser beam in PT microscopy take a short time (0.5–1 s for one mode). Nevertheless, the analysis of ~1000 cells with PTSC may take from 10–15 min to hours.

Furthermore, obtaining in sequence both the PT image and the integral PT response from one cell, if necessary, takes at least 5–10 s (if not much longer due to switching from one mode to another). Currently, with combined modes (PTI and PT thermal-lens response), PTFC's rate of approximately 10 cells s^{-1} is limited by both the pump laser repetition rate and the time for PT-image signal acquisition. Nevertheless, PTFC is approximately 10^2 times faster than existing PTSC and can be used in many traditional applications specific for PTSC [21].

These applications include (i) label-free PT detection of cells with different endogenous absorption properties [21]; (ii) detection of cells labeled with GNPs [21, 30, 57]; (iii) optimization of selective nanophotothermolysis of cancer cells, bacteria, and viruses [13, 36, 150]; (iv) real-time monitoring of the cellular response to low doses of radiation or drugs including nicotine, alone or in different combinations [151, 152]; and (v) monitoring of nanoparticles (e.g., GNRs, GNSs, CNTs, MNPs) penetration into cells and cell tracking during endocytosis, phagocytosis, optoporation, or particle aggregation (PT aggregation immunoassay) at the single-cell level [57]. Many of these applications are specific to the PT technique and cannot be realized with conventional FC.

In contrast to conventional FC, in which detection is based on fluorescence and scattered light, the main diagnostic parameter in PTFC and PAFC is the absorption coefficient, which is similar to that used in conventional absorption spectroscopy. Using specific absorption signatures (so-called fingerprints) of biomolecules, PTFC/PAFC has the potential for label-free identification of many cellular components, including flavoproteins (e.g., NADH-dehydrogenase), cytochromes (P450, a, b, c), Cyt-*c* oxidase, and many others [31]. Most of these components have distinctive, relatively narrow (10–30 nm), isolated spectral bands in the visible and NIR spectral range of 400–900 nm, without the significant influence of DNA, RNA, lipids, and water, which absorb mainly in the UV (<350 nm) and the IR (>1000 nm) ranges.

The lowest absorption background is determined by water; in the spectral range of 400–900 nm, the absorption coefficient of water is 5×10^{-4} to 5×10^{-3} cm^{-1}, with a further increase to $\sim 2 \times 10^{-1}$ cm^{-1} in the NIR range (1 μm) [31]. In contrast, most cellular chromophores have much stronger localized absorption (10^{-1}–10^{3} cm^{-1}) in the visible and the NIR ranges [21, 153]. From a diagnostic perspective, cellular chromophores are involved in cellular signaling pathways that are extremely sensitive to cell metabolism and to various therapeutic interventions (e.g., drugs, radiation) [30, 31]. In particular, PT spectral identification of just three main cellular components, Cyts *c*, P450, and Cyt-*c* oxidase, may provide many potential applications of PT spectroscopy, as these components are involved in cellular apoptosis, pharmacokinetics, and cellular energy, respectively.

Cells themselves can be identified through the monitoring of linear and nonlinear PT/PA image and PT response parameters related to localized and average cellular absorption associated with the concentration and spatial distribution of endogenous chromophores. Thus, absorption might be considered as an additional parameter, together with fluorescence and light scattering, for detecting and identifying cells in flow with a PT/PA technique, especially if the cells are weakly fluorescent or low-scattering.

The absorption-sensitivity threshold of PTFC at the single-cell level is 10^{-2}–10^{-3} cm^{-1} [21]. This level of sensitivity makes possible label-free detection of most individual cells, even those with relatively low absorption in the visible and especially NIR spectral range (e.g., WBCs or cancer cells) [21, 42]. PTFC's high absorption sensitivity and dynamic range also enable selective detection of individual cells on the basis of their different endogenous absorption properties, especially relatively strongly absorbing cells among low-absorbing cells. For example, PTFC operating in the broad spectral range of 420–920 nm can detect single RBCs or pigmented melanoma cells among many WBCs in the visible-spectral range of 420–680 nm. Operating in the NIR range of 750–800 nm, PTFC can detect single melanoma cells among many RBCs [42] because the absorption coefficient of the melanoma cells is several times greater (depending on pigmentation) than that of RBCs. The use of labeling with GNPs significantly increases PT-signal amplitude from labeled cells [17, 21, 31, 42, 57].

Detection efficiency parameter, expressed as the ratio of the number of cells detected to the total number of cells passing the analysis point, is expected to be close to one at a relatively low flow velocity confined by the laser pulse repetition rate and the beam diameter/shape. Because cells appear randomly in the area of detection even at the same flow velocity, in pulse mode to expose every cell or selected cells, the pump pulse must be synchronized with cell motion. This regime was realized in the PTI mode for the capturing of images of selected cells of interest because the pump pulse is triggered with electrical signals from a photodetector generated as the cells cross the CW He–Ne probe laser beam.

However, the feasibility of PTFC was demonstrated without such synchronization using a simpler regime with a fixed maximum pulse rate in combination with a broader laser-beam diameter of 30–40 μm. Depending on cell velocity and number, this mode might lead to some cells being missed out or, in contrast, to simultaneous exposure of several closely located cells. In linear, noninvasive mode, multiple exposures pose no significant problem, especially when rare, relatively strongly absorbing cells are being examined among less strongly absorbing cells (e.g., pigmented metastatic melanoma cells among WBCs). Nevertheless, to avoid multiple exposures and beam-overlap effects, the pulse repetition rate should be increased and confined to the cell velocity.

The precision of PT measurements in flow may be expressed as the sum of the relative deviations resulting from the effects of turbulence, probe-beam intensity fluctuation (including beam-pointing noise), variations in the pump pulse laser-energy level, alignment errors (e.g., as a result of vibration or acoustic noise), photodetector or CCD camera errors, and instrumentation errors resulting from signal digitization. The first three factors, turbulence effects and probe and pump-beam intensity fluctuations are typically predominant but are significantly minimized by the use of pulsed mode and an intensity-stabilized He–Ne probe laser. With these features, the accuracy of the PT-response amplitude measurement in PT thermal-lens mode is relatively high, approximately 3% [21]. The reproducibility of PT responses depends on the pulse-to-pulse energy stability and, for typical laser parameters, reaches 8% [65].

In PTI mode using optical immersion and small-diameter tubes in the flow module, PTFC may provide the highest optical resolution up to 250–300 nm in an ideal laminar flow. However, in nonlaminar flow, changes in the positions of the cells in a cross section are crucial to obtaining high-resolution PT images. On using a low-magnification (20×) micro-objective with a depth of focus of approximately 2.6 μm to minimize defocusing effects, a resolution of 0.7–1 μm as in conventional FC is provided [1, 2, 154–156]. Nevertheless, for high-resolution imaging [44], especially with a confocal schematic [37], spatial fluctuations of flowing cells might be a problem, thus hydrodynamic focusing of cells in flow should be used [1, 2, 154–156].

The significant advantage of modern FC, with its long history of development compared with that of the PTFC/PAFC prototypes described here, is FC's ability to automatically measure simultaneously multiple multispectral parameters of fast-moving cells. To provide PTFC label-free, simultaneous measurements of at least two parameters at different wavelengths, the PT-response amplitude and the percentage of nonlinear PT responses (parameter δ_{NL}), could be used. This capability makes it possible to easily identify strongly absorbing cells (e.g., RBCs or melanoma cells) in the presence of low-absorbing cells (e.g., WBCs).

Analysis of PT images provides additional opportunities for identifying cells having a specific spatial distribution of absorbing endogenous cellular chromophores; RBCs, for example, can be recognized by their smooth, nonlocalized distribution of hemoglobin; WBCs by their localized distribution of cytochromes in mitochondria [151, 152], and tumor cells by the localized, multipoint-like distribution of GNPs (or melanin) [13, 17, 36, 42].

We expect that highly sensitive PTFC will make it possible to identify metastatic cancer cells at early stages of cancer development on the basis of changes in their absorbing structures, without the need for cell labeling. In particular, because the specific shape, distribution, and aggregation of mitochondria can be markers for identifying cancer cells, as has already been demonstrated with fluorescence labeling technique [47], a label-free PT assay that is extremely sensitivity to mitochondrial "nanomorphology" [142] has great potential for identifying cancer cells.

From a technical standpoint, the further development of PTFC may include the following: (i) increasing the speed of cell analysis; (ii) extending the spectral range to the UV and the IR regions; (iii) extending the spectral modes with a fast spectrum-scanning laser [157], multiplex spectral detection [158], or laser-frequency modulation [160]; and (iv) combining PA and scattering techniques [30, 45, 42, 43].

From the point of view of applications, PTFC can be further developed through (i) study of the cellular responses to the impact of different factors (e.g., γ-radiation, laser radiation, glucose) [151], (ii) the monitoring of cell viability and especially cell death (apoptosis and necrosis) [151, 152], (iii) molecular imaging with the use of PT contrast agents [42], and (iv) integration of PT diagnostics with PT therapeutics [36, 42].

In particular, increased speed can be achieved by the use of hydrodynamic focusing of flow as in conventional FC, in combination with the use of available lasers with a high pulse repetition rate (e.g., a diode-pumped solid laser with a pulse repetition rate up to 50–500 kHz or a diode laser with a pulse repetition rate up to 10–50 kHz). With such lasers, PTI mode can be realized with commercially available digital cameras with high frame speeds of up to 10^4 fps [44], which could be crucial for morphologic analysis of cells in flow. For the purpose of detecting rare cells, a simpler solution would be the simultaneous irradiation of many cells with a broad-diameter pump laser beam measuring ≥ 50 μm. This approach was demonstrated by the PT identification of GNP-labeled cancer cells among nonlabeled cells under static conditions [161].

Two-photon luminescence (TPL) from GNPs shows considerable potential in biological imaging [162] and can be used in *in vivo* PAFC/PTFC as an additional channel for recognition of nanoparticle-labeled cancer cells: one particle – two signals (PT/PA) and fluorescence. In addition fluorescence lifetime imaging microscopy (FLIM) with GNPs provides images with better contrast and sensitivity than intensity imaging. The characteristic fluorescence lifetime of GNPs with a longitudinal plasmon mode centered at around 750 nm, and a weak transverse plasmon mode at 550 nm is found to be less than 100 ps [162]. This characteristic lifetime may provide a significant selectivity and can be used to distinguish nanopartilce labeled cancer cells from other fluorescent labels and endogenous fluorophores in lifetime imaging.

In analogy to standard *in vitro* FC, the PTFC/PAFC may have a broad spectrum of similar *in vivo* applications, as well as provide many new applications that may include detection of various targets with intrinsic absorption or with PA/PT labels targeting specific antigens and receptors (e.g., normal and abnormal cells, bacteria [36, 64, 59], viruses [57] labeled by bioconjugated advanced nanoparticles) in a variety of vessels (e.g., arterioles, capillaries, veins, and afferent and efferent lymphatics) in assorted locations (e.g., skin, ear, retina, conjunctiva, liver, and lymph nodes) for many disease models (e.g., cancer, infections, inflammation, and immune system disorders) and their combinations.

Further development of the PTFC/PAFC may solve very complex and largely unexplored areas of medicine related to detection of pathogens and stem, dendritic, and metastatic cells in different functional states (e.g., apoptosis) in lymph and blood flow at the single-cell level. Especially, it may be used for early diagnostics of infection during hematogenous spread with bacteria into different organs, vascular grafts, and stents. In particular, these infections commonly result in death at sepsis, despite aggressive treatment at the developed disease stage.

A quick development of this technology for application in humans is anticipated with the advancement of portable noninvasive devices attached to the skin over vessels or lymph nodes for alarm control of bacterial infection dissemination, cancer recurrence, metastasis development, therapy assessment (through controlling the number of circulating bacteria or metastatic cells), or monitoring of circulating drugs using nanoparticle-based drug carries (e.g., liposomes). PTFC/PAFC technology may allow one to develop portable "personal" flow cytometers for blood testing without a needle stick using compact robust, low-cost, laser diode arrays with different wavelengths.

Acknowledgments

This work was supported by: The National Institute of Health; grants **R01 EB000873, R01 CA131164, R01 EB009230**, and **R21 CA139373** The National Science Foundation; grant **DBI 0852737**; Arkansas Biosciences Institute; Grant **224014 PHOTONICS4LIFE** of **FP7-ICT-2007-2**; Projects 2.1.1/4989 and 2.2.1.1/2950 of RF Program on the Development of High School Potential; The Special Program of RF "Scientific and Pedagogical Personnel of Innovative Russia"; and Governmental contracts 02.740.11.0484, 02.740.11.0770, and 02.740.11.0879.

References

1. Shapiro, H.M. (2003) *Practical Flow Cytometry*, 4th edn, Wiley-Liss, New York.
2. Givan, A.L. (2001) Principles of flow cytometry: an overview. *Methods Cell Biol.*, **63**, 19–50.
3. Tuchin, V.V., Tárnok, A., and Zharov, V.P. (Guest Editors) (2009) Towards *in vivo* flow cytometry. *J. Biophoton.*, **2** (8–9), 457–547.
4. Zharov, V.P. and Letokhov, V.S. (1989) *Laser Opto-Acoustic Spectroscopy*, Springer-Verlag, New York.
5. Braslavsky, S.E. and Heihoff, K. (1989) Photothermal methods, in *Handbook of Organic Photochemistry* (ed. J.C. Scaiano), CRC Press, Boca Raton, pp. 327–355.
6. Priezzhev, V., Tuchin, V.V., and Shubochkin, L.P. (1989) *Laser Diagnostics in Biology and Medicine*, Nauka, Moscow.
7. Gusev, V.E. and Karabutov, A.A. (1993) *Laser Optoacoustics*, AIP Press, New York.
8. Mandelis, A. and Michaelian, K.H. (eds) (1997) Special section on photoacoustic and photothermal science and engineering. *Opt. Eng.*, **36** (2), 301–534.
9. Tuchin, V.V. (1998) *Lasers and Fiber Optics in Biomedical Science*, Saratov University Press, Saratov.
10. Tuchin, V.V. (2007) *Tissue Optics: Light Scattering Methods and Instruments for Medical Diagnosis*, 2nd edn, PM 166, SPIE Press, Bellingham, WA.
11. Lapotko, D., Romanovskaya, T., and Zharov, V. (2002) Photothermal images of live cells in presence of drug. *J. Biomed. Opt.*, **7** (3), 425–434.
12. Zharov, V. (2003) Far-field photothermal microscopy beyond the diffraction limit. *Opt. Lett.*, **28**, 1314–1316.
13. Zharov, V., Galitovsky, V., and Viegas, M. (2003) Photothermal detection of local thermal effects during selective nanophotothermolysis. *Appl. Phys. Lett.*, **83** (24), 4897–4899.
14. Galitovsky, V., Chowdhury, P., and Zharov, V.P. (2004) Photothermal detection of nicotine-induced apoptotic effects in a pancreatic cancer cells. *Life Sci.*, **75**, 2677–2687.
15. Zharov, V.P., Galanzha, E.I., and Tuchin, V.V. (2005) Integrated photothermal flow cytometry *in vivo*. *J. Biomed. Opt.*, **10**, 51502-1–51502-13.
16. Maslov, K., Stoica, G., and Wang, L.V. (2005) *In vivo* dark-field reflection-mode photoacoustic microscopy. *Opt. Lett.*, **30** (6), 625–627.
17. Zharov, V.P., Galanzha, E.I., and Tuchin, V.V. (2005) Photothermal image flow cytometry *in vivo*. *Opt. Lett.*, **30** (6), 628–630.
18. Galanzha, E.I., Shashkov, E.V., Tuchin, V.V., and Zharov, V.P. (2008) In vivo multispectral, multiparameter, photoacoustic lymph flow cytometry with natural cell focusing, label-free detection and multicolor nanoparticle. *Cytometry*, **73**, 884–894.
19. Galanzha, E.I., Kokoska, M.S., Shashkov, E.V., Kim, J.-W., Tuchin, V.V., and Zharov, V.P. (2009) *In vivo*

fiber photoacoustic detection and photothermal purging of metastasis targeted by nanoparticles in sentinel lymph nodes at single cell level. *J. Biophoton.*, **2**, 528–539.

20. Douglas-Hamilton, D.H. and Conia, J. (2001) Thermal effects in laser-assisted pre-embryo zona drilling. *J. Biomed. Opt.*, **6** (2), 205–213.
21. Zharov, V.P. and Lapotko, D.O. (2005) Photothermal imaging of nanoparticles and cells. *IEEE JSTQE*, **11** (4), 733–751.
22. Tokeshi, M., Uchida, M., Hibara, A., Sawada, T., and Kitamori, T. (2001) Determination of suboctomole amounts of nonfluorescent molecules using a thermal lens microscope: subsingle-molecule determination. *Anal. Chem.*, **73**, 2112–2116.
23. Zharov, V.P. (1986) in *Laser Analytical Spectrochemistry* (ed. V.S. Letokhov), Bristol, Boston, MA, pp. 229–271.
24. Tamaki, E., Sato, K., Tokeshi, M., Aihara, M., and Kitamori, T. (2002) Single-cell analysis via scanning thermal lens microscope with a microchip: direct monitoring of cytochrome c distribution during apoptosis process. *Anal. Chem.*, **74**, 1560–1564.
25. Sato, K., Tokeshi, M., Kimura, H., and Kitamori, T. (2001) Determination of carcinoembryonic antigen in human sera by integrated bead-bed immunoassay in a microchip for cancer diagnosis. *Anal. Chem.*, **73**, 1213–1218.
26. Kimura, H., Sekiguchi, K., Kitamori, T., Sawada, T., and Mukaida, M. (2001) Assay of spherical cell surface molecules by thermal lens microscopy and its application to blood cell substances. *Anal. Chem.*, **73**, 4333–4337.
27. Cognet, L., Tardin, C., Boyer, D., Choquet, D., Tamarat, P., and Lounis, B. (2003) Single metallic nanoparticles imaging for protein detection in cells. *Proc. Natl. Acad. Sci.*, **100**, 11350–11355.
28. Tuchin, V.V., Drezek, R., Nie, S., and Zharov, V.P. (Guest Editors) (2009) Special section on nanophotonics for diagnostics, protection and treatment of cancer and inflammatory diseases. *J. Biomed. Opt.*, **14** (2), 020901; 021001–021017.
29. Jacobs, A. and Worwood, M. (1974) *Iron in Biochemistry and Medicine*, Academic Press, London, New York.
30. Lapotko, D., Kuchinsky, G., Potapnev, M., and Pechkovsky, D. (1996) Photothermal image cytometry of human neutrophils. *Cytometry*, **24**, 198–203.
31. Zharov, V.P., Galanzha, E.I., and Tuchin, V.V. (2006) *In vivo* photothermal flow cytometry: imaging and detection of individual cells in blood and lymph flow. *J. Cell Biochem.*, **97** (5), 916–930.
32. Lapotko, D.O., Romanovskaya, T.R., Shnip, A., and Zharov, V.P. (2002) Photothermal time-resolved imaging of living cells. *Lasers Surg. Med.*, **31**, 53–63.
33. Lapotko, D., Romanovskaya, T., Kutchinsky, G., and Zharov, V. (1999) Photothermal studies of modulating effect of photoactivated chlorin on interaction of blood cells with bacteria. *Cytometry*, **37**, 320–326.
34. Lapotko, D.O., Zharov, V.P., Romanovskaya, T.R., and Kuchinskii, G.S. (1999) Investigation of the influence of the photodynamic effect on microorganisms using the laser photothermal cytometry method. *Quant. Electron.*, **29**, 1060–1065.
35. Zharov, V.P., Mercer, K.E., Galitovskaya, E.N., and Smeltzer, M.S. (2006) Photothermal nanotherapeutics and nanodiagnostics for selective killing of bacteria targeted with gold nanoparticles. *Biophys. J.*, **90**, 619–627.
36. Zharov, V.P., Galitovskiy, V., Lyle, C.S., and Chambers, T.C. (2006) Super high-sensitive photothermal monitoring of individual cell response to antitumor drug. *J. Biomed. Opt.*, **11**, 064034.
37. Zharov, V.P., Galanzha, E.I., Ferguson, S., and Tuchin, V.V. (2005) Confocal photothermal flow cytometry *in vivo*. *Proc. SPIE*, **5697**, 167–176.
38. Born, M. and Wolf, E. (1964) *Principles of Optics: Electromagnetic Theory of Propagation, Interference and Diffraction of Light*, 2nd edn, Macmillan, New York.

39. Zharov, V.P., Galanzha, E.I., and Tuchin, V.V. (2007) Photothermal flow cytometry *in vitro* for detection and imaging of individual moving cells. *Cytometry A*, **71A**, 191–206.
40. Galanzha, E.I., Tuchin, V.V., and Zharov, V.P. (2005) *In vivo* integrated flow image cytometry and lymph/blood vessels dynamic microscopy. *J. Biomed. Opt.*, **10**, 54018-1–54018-8.
41. Zharov, V.P., Galanzha, E.I., Shashkov, E.V., Khlebtsov, N.G., and Tuchin, V.V. (2006) *In vivo* integrated photoacoustic flow cytometry: Application for monitoring circulating cancer cells labeled with gold nanorods. Fifth Workshop on Optical Imaging from Bench to Bedside at the National Institutes of Health, p. 126.
42. Zharov, V.P., Galanzha, E.I., Shashkov, E.V., Khlebtsov, N.G., and Tuchin, V.V. (2006) *In vivo* photoacoustic flow cytometry for monitoring of circulating single cancer cells and contrast agents. *Opt. Lett.*, **31**, 3623–3625.
43. Zharov, V.P., Menyaev, Yu., Shashkov, E.V., Galanzha, E.I., Khlebtsov, B.N., Scheludko, A., Zimnyakov, D.A., and Tuchin, V.V. (2006) Fluctuations of probe beam in thermolens schematics as potential indicator of cell metabolism, apoptosis, necrosis and laser impact. *Proc. SPIE*, **6085**, 10–21.
44. Zharov, V.P., Galanzha, E.I., Menyaev, Yu., and Tuchin, V.V. (2006) *In vivo* high-speed imaging of individual cells in fast blood flow. *J. Biomed. Opt.*, **11** (5), 054034-1–054034-4.
45. Galanzha, E.I., Tuchin, V.V., and Zharov, V.P. (2007) Advances in small animal mesentery models for *in vivo* flow cytometry, dynamic microscopy, and drug screening (invited review). *World J. Gastroenterol.*, **13** (2), 198–224.
46. Bialkowsky, E. (1996) *Photothermal Spectroscopy Methods for Chemical Analysis*, Wiley-Interscience Publication, New York.
47. Gourley, P.L., Hendricks, J.K., McDonald, A.E., Copeland, R.G., Barrett, K.E., Gourley, C.R., Singh, K.K., and Naviaux, R.K. (2005) Mitochondrial correlation microscopy and nanolaser spectroscopy – New tools for biophotonic detection of cancer in single cells, *Technol. Cancer Res. Treat.*, **4**, 585–592.
48. Novak, J., Georgakoudi, I., Wei, X., Prossin, A., and Lin, C.P. (2004) *In vivo* flow cytometer for real-time detection and quantification of circulating cells. *Opt. Lett.*, **29**, 77–79.
49. Lee, H., Alt, C., Pitsillides, C. *et al.* (2006) *In vivo* imaging flow cytometer. *Opt. Express*, **14**, 7789–7797.
50. Butros, S., Greiner, C., Hwu, D. *et al.* (2007) Portable two-color *in vivo* flow cytometers for real-time detection of fluorescently-labeled circulating cells. *J. Biomed. Opt.*, **12**, 020507-1–020507-3.
51. Zharov, V.P., Galanzha, E.I., and Tuchin, V.V. (2004) Photothermal imaging of moving cells in lymph and blood flow *in vivo* animal model. *Proc. SPIE*, **5320**, 256–263.
52. Karabutov, A.A. and Oraevsky, A.A. (2002) in *Handbook of Optical Biomedical Diagnostics*, Chapter 10, PM107 (ed. V.V. Tuchin), SPIE Press, Bellingham, WA, pp. 585–674.
53. Oraevsky, A.A. and Karabutov, A.A. (2003) in *Biomedical Photonics Handbook*, Chapter 34 (ed. T. Vo-Dinh), CRC Press, Boca Raton, pp. 34-1–34-34.
54. Wang, L. (ed.) (2009) *Photoacoustic Imaging and Spectroscopy*, CRC Press, Taylor & Francis Group, London.
55. Tuchin, V.V. (2006) *Optical Clearing of Tissues and Blood*, PM 154, SPIE Press, Bellingham, WA.
56. Tuchin, V.V. (2007) A clear vision for laser diagnostics. *IEEE J. Select. Topics Quantum Electron.*, **13** (6), 1621–1628.
57. Zharov, V.P., Kim, J.-W., Everts, M., and Curiel, D.T. (2005) Self-assembling nanoclusters in living systems: application for integrated photothermal nanodiagnostics and nanotherapy (review). *Nanomedicine*, **1**, 326–345.
58. Zharov, V.P., Galanzha, E.I., Shashkov, E.V., Khlebtsov, N.G., and Tuchin, V.V. (2006) *In vivo* photoacoustic flow cytometry for real-time monitoring of circulating cells and nanoparticles. *SPIE News Room*, 1–3. doi: 10.1117/2.1200609.0391

59. Zharov, V.P., Galanzha, E.I., Shashkov, E.V., Kim, J.-W., Khlebtsov, N.G., and Tuchin, V.V. (2007) Photoacoustic flow cytometry: principle and application for real-time detection of circulating single nanoparticles, pathogens, and contrast dyes *in vivo*. *J. Biomed. Opt.*, **12** (5), 051503-1–051503-14.
60. Charm, S.E. and Kurland, G.S. (1974) *Blood Flow and Microcirculation*, John Wiley & Sons, Inc., New York.
61. Khlebtsov, N.G. and Dykman, L.A. (2010) in *Handbook of Photonics for Medical Science*, Chapter 2 (ed. V.V. Tuchin), CRC Press, Taylor & Francis Group, London, pp. 37–85.
62. Liao, H. and Hafner, J.H. (2005) Gold nanorod bioconjugates. *Chem. Mater.*, **17**, 4636–4641.
63. Kim, J.-W., Kotagiri, N., Kim, J.H., and Deaton, R. (2006) *In situ* fluorescence microscopy visualization and characterization of nanometer-scale carbon nanotubes labeled with 1-pyrenebutanoic acid, succinimidyl ester. *Appl. Phys. Lett.*, **88**, 213110.
64. Kim, J.-W., Shashkov, E.V., Galanzha, E.I., Kotagiri, N., and Zharov, V.P. (2007) Photothermal antimicrobial nanotherapy and nanodiagnostics with self-assembling carbon nanotube clusters. *Lasers Surg. Med.*, **39**, 622–634.
65. Lapotko, D.O. and Zharov, V.P. (2005) Spectral evaluation of laser-induced cell damage with photothermal microscopy. *Lasers Surg. Med.*, **36**, 22–33.
66. ANSI (2000) Z136.1. *American National Standard for Safe Use of Lasers*, American National Standards Institute.
67. Bornhop, D.J., Contag, C.H., Licha, K., and Murphy, C.J. (2001) Advances in contrast agents, reporters, and detection. *J. Biomed. Opt.*, **6** (2), 106–110.
68. Hirsch, J.I., Tisnado, J., Cho, S.R., and Beachley, M.C. (1982) Use of Isosulfan blue for identification of lymphatic vessels: experimental and clinical evaluation. *Am. J. Roentgenol.*, **139**, 1061–1064.
69. Burgoyne, L.L., Jay, D.W., Bikhazi, G.B., and De Armendi, A.J. (2005) Isosulfan blue causes factitious methemoglobinemia in an infant. *Paediatr. Anaesth.*, **15**, 1116–1119.
70. Roggan, M.F., Dorschel, K., Andreas, H., and Müller, G. (1999) Optical properties of circulating human blood in the wavelength range 400–2500 nm. *J. Biomed. Opt.*, **4** (1), 36–46.
71. Pilatou, M.C., Marani, E., de Mul, F.F., and Steenbergen, W. (2003) Photoacoustic imaging of brain perfusion on albino rats by using Evans Blue as contrast agent. *Arch. Physiol. Biochem.*, **111**, 389–397.
72. Wang, X., Ku, G., Wegiel, M.A., Bornhop, D.J., Stoica, G., and Wang, L.V. (2004) Noninvasive photoacoustic angiography of animal brains *in vivo* with near-infrared light and an optical contrast agent. *Opt. Lett.*, **29**, 730–732.
73. Xu, M. and Wang, L.V. (2006) Photoacoustic imaging in biomedicine. *Rev. Sci. Instrum.*, **77**, 041101.
74. Hirsch, L.R., Gobin, A.M., Lowery, A.R., Tam, F., Drezek, R.A., Halas, N.J., and West, J.L. (2006) Metal nanoshells. *Ann. Biomed. Eng.*, **34**, 15–22.
75. Khlebtsov, B.N., Bogatyrev, V.A., Dykman, L.A., and Khlebtsov, N.G. (2007) Spectra of resonance light scattering of gold nanoshells: effects of polydispersity and limited electron free path. *Opt. Spectrosc.*, **102**, 233–241.
76. Cherukuri, P., Gannon, C.J., Leeuw, T.K., Schmidt, H.K., Smalley, R.E., Curley, S.A., and Weisman, R.B. (2006) Mammalian pharmacokinetics of carbon nanotubes using intrinsic near-infrared fluorescence. *Proc. Natl. Acad. Sci. U.S.A.*, **103**, 18882–18886.
77. Terentyuk, G.S., Maslyakova, G.N., Suleymanova, L.V., Khlebtsov, B.N., Kogan, B.Ya., Akchurin, G.G., Shantrocha, A.V., Maksimova, I.L., Khlebtsov, N.G., and Tuchin, V.V. (2009) Circulation and distribution of gold nanoparticles and induced alterations of tissue morphology at intravenous particle delivery. *J. Biophoton.*, **2** (5), 292–302.
78. Terentyuk, G., Akchurin, G., Maksimova, I., Shantrokha, A., Tuchin, V., Maslyakova, G., Suleymanova, L., Kogan, B., Khlebtsov, N., and Khlebtsov, B. (2009) Tracking gold

nanoparticles in the body. *SPIE News Room*. doi: 10.1117/2.1200907.1619

79. Huang, X., El-Sayed, I.H., Qian, W., and El-Sayed, M.A. (2006) Cancer cell imaging and photothermal therapy in the near-infrared region by using gold nanorods. *J. Am. Chem. Soc.*, **128**, 2115–2120.
80. Terentyuk, G.S., Maslyakova, G.N., Suleymanova, L.V., Khlebtsov, N.G., Khlebtsov, B.N., Akchurin, G.G., Maksimova, I.L., and Tuchin, V.V. (2009) Laser-induced tissue hyperthermia mediated by gold nanoparticles: toward cancer phototherapy. *J. Biomed. Opt.*, **14** (2), 021016-1–021016-8.
81. Chen, W.R., Li, X., Naylor, M.F., Liu, H., and Nordquist, R.E. (2010) in *Handbook of Photonics for Medical Science* (ed. V.V. Tuchin), CRC Press, Taylor & Francis Group, London, pp. 739–761.
82. Terentyuk, G.S., Akchurin, G.G., Maksimova, I.L., Maslyakova, G.N., Khlebtsov, N.G., and Tuchin, V.V. (2010) in *Handbook of Photonics for Medical Science*, Chapter 29 (ed. V.V. Tuchin), CRC Press, Taylor & Francis Group, London, pp. 763–797.
83. Letfullin, R.R., Joenathan, C., George, T.F., and Zharov, V.P. (2006) Laser induced explosion of gold nanoparticles: potential role for nanophotothermolysis of cancer. *Nanomedicine*, **1**, 473–480.
84. Khlebtsov, B., Zharov, V., Melnikov, A., Tuchin, V., and Khlebtsov, N. (2006) Optical amplification of photothermal therapy with gold nanoparticles and nanoclusters. *Nanotechnology*, **17**, 5167–5179.
85. Akchurin, G., Khlebtsov, B., Akchurin, G., Tuchin, V., Zharov, V., and Khlebtsov, N. (2008) Gold nanoshell photomodification under single nanosecond laser pulse accompanied by color-shifting and bubble formation phenomena. *Nanotechnology*, **19**, 015701-1–015701-8.
86. Aguirre, C.M., Moran, C.E., Young, J.F., and Halas, N.J. (2004) Laser-induced reshaping of metallodielectric nanoshells under femtosecond and nanosecond plasmon resonant illumination. *J. Phys. Chem.*, **B108**, 7040–7045.
87. Zharov, V., Galanzha, E., Shashkov, E., Maksimova, I., Khlebtsov, B., Khlebsov, N., Terentuk, G., Akchurin, G., and Tuchin, V. (2007) *In vivo* detection and killing of individual circulating metastatic cells. *Lasers Surg. Med. (Suppl.)*, **19**, 47–48.
88. Laschke, M.W., Kerdudou, S., Herrmann, M., and Menger, M.D. (2005) Intravital fluorescence microscopy: a novel tool for the study of the interaction of *Staphylococcus aureus* with the microvascular endothelium *in vivo*. *J. Infect. Dis.*, **191**, 435–443.
89. Kam, N.W., O'Connell, M., Wisdom, J.A., and Dai, H. (2005) Carbon nanotubes as multifunctional biological transporters and near-infrared agents for selective cancer cell destruction. *Proc. Natl. Acad. Sci. U.S.A.*, **102**, 11600–11605.
90. Rodriguez, V.B., Henry, S.M., Hoffman, A.S., Stayton, P.S., Li, X., and Pun, S.H. (2008) Encapsulation and stabilization of indocyanine green within poly(styrene-alt-maleic anhydride) blockpoly(styrene) micelles for near-infrared imaging. *J. Biomed. Opt.*, **13** (1), 014025-1–014025-10.
91. Geddes, C.D., Cao, H., and Lakowicz, J.R. (2003) Enhanced photostability of ICG in close proximity to gold colloids. *Spectrochim. Acta*, **A59**, 2611–2617.
92. Gil-Toma's, J., Tubby, S., Parkin, I.P., Narband, N., Dekker, L., Nair, S.P., Wilson, M., and Street, C. (2007) Lethal photosensitisation of *Staphylococcus aureus* using a toluidine blue O–tiopronin–gold nanoparticle conjugate. *J. Mater. Chem.*, **17**, 3739–3746.
93. Kuo, W.-S., Chang, C.-N., Chang, Yi.-T., and Yeh, C.-S. (2009) Antimicrobial gold nanorods with dual-modality photodynamic inactivation and hyperthermia. *Chem. Commun.*, 4853–4855.
94. Tuchina, E.S. and Tuchin, V.V. (2010) Photodynamic/photocatalytic effects on microorganisms processed by nanodyes. *Proc. SPIE*, **7576**, doi: 10.1117/12.840532

95. Tuchina, E.S. and Tuchin, V.V. (2010) TiO_2 nanoparticle enhanced photodynamic inhibition of pathogens. *Laser Phys. Lett.*, **7**, 1–6. doi: .
96. Ratto, F., Matteini, P., Centi, S., Rossi, F., Fusi, F., Pini, R., Khlebtsov, B.N., Khlebtsov, N.G., Tuchina, E.S., Tuchin, V.V., Dunsby, C., Neil, M., French, P., Manohar, S., Subramaniam, V., Bendsoe, N., and Svanberg, K. (2010) The P4L project, NIR-laser-activated gold nanoparticles: perspectives in minimally invasive diagnosis and therapy. SPIE Photonics Europe, Biophotonics: Photonic Solutions for Better Health Care, 12-16 April, Brussels, Belgium, paper # [7715-120].
97. Hogan, H. (2006) Flow cytometry moves ahead. *Biophotonics Int.*, **13**, 38–42.
98. Wei, X., Runnels, J.M., and Lin, C.P. (2003) Selective uptake of indocyanine green by reticulocytes in circulation. *Invest. Ophthalmol. Visual Sci.*, **44**, 4489–4496.
99. Brown, P. (2005) Lymphatic system: unlocking the drains. *Nature*, **436**, 456–458.
100. Alitalo, K., Tammela, T., and Petrova, T.V. (2005) Lymphangiogenesis in development and human disease. *Nature*, **438**, 946–953.
101. Schmid-Schonbein, G.W. (1990) Microlymphatics and lymph flow. *Physiol. Rev.*, **70**, 987–1028.
102. Aukland, K. and Reed, R.K. (1993) Interstitial-lymphatic mechanisms in the control of extracellular fluid volume. *Physiol. Rev.*, **73**, 1–78.
103. Haig, D.M., Hopkins, J., and Miller, H.R. (1999) Local immune responses in afferent and efferent lymph. *Immunology*, **96**, 155–163.
104. Olszewski, W.L. (2003) The lymphatic system in body homeostasis: physiological conditions. *Lymphat. Res. Biol.*, **1**, 11–21.
105. Browse, N., Burnand, K.G., and Mortimer, P.S. (2003) *Diseases of the Lymphatics*, Arnold, London.
106. Zhang, H.F., Maslov, K., Stoica, G., and Wang, L.V. (2006) Functional photoacoustic microscopy for high-resolution and noninvasive *in vivo* imaging. *Nat. Biotechnol.*, **24**, 848–851.
107. Zhang, E.Z., Laufer, J.G., Pedley, R.B., and Beard, P.C. (2009) *In vivo* high-resolution 3D photoacoustic imaging of superficial vascular anatomy. *Phys. Med. Biol.*, **54**, 1035–1046.
108. Galanzha, E.I., Tuchin, V.V., and Zharov, V.P. (2007) Optical monitoring of microlympatic disturbances at experimental lymphedema. *Lymphat. Res. Biol.*, **5**, 11–28.
109. Christofori, G. (2006) New signals from the invasive front. *Nature*, **441**, 444–450.
110. (a) Nanospectra Biosciences, Inc., *www.nanospectra.com*, (accessed 17 January, 2011); (b) National Cancer Institute (NCI), TNF-Bound colloidal gold in treating patients with advanced solid tumors, *http://clinicaltrials.gov/ct/show/NCT00356980?order=2*, (accessed 21 April, 2009).
111. De Rosa, S.C., Brenchley, J.M., and Roederer, M. (2003) Beyond six colors: a new era in flow cytometry. *Nat. Med.*, **9**, 112–117.
112. He, W., Wang, H., Hartmann, L.C., Cheng, J.X., and Low, P.S. (2007) *In vivo* quantitation of rare circulating tumor cells by multiphoton intravital flow cytometry. *Proc. Natl. Acad. Sci. U.S.A*, **104**, 11760–11765.
113. Georgakoudi, I., Solban, N., Novak, J., Rice, W.L., Wei, X., Hasan, T., and Lin, C.P. (2004) *In vivo* flow cytometry: A new method for enumerating circulating cancer cells. *Cancer Res.*, **64**, 5044–5047.
114. Wei, X., Sipkins, D.A., Pitsillides, C.M., Novak, J., Georgakoudi, I., and Lin, C.P. (2005) Real-time detection of circulating apoptotic cells by *in vivo* flow cytometry. *Mol. Imaging*, **4**, 415–416.
115. Slavik, J. (1996) *Fluorescence Microscopy and Fluorescent Probes*, Plenum, New York, London.
116. Galanzha, E.I., Chowdhury, P., Tuchin, V.V., and Zharov, V.P. (2005) Monitoring of nicotine impact on microlymphatics of rat mesentery with time-resolved microscopy. *Lymphology*, **38**, 181–192.

117. Van Trappen, P.O. and Pepper, M.S. (2002) Lymphatic dissemination of tumour cells and the formation of micrometastases. *Lancet Oncol.*, **3**, 44–52.
118. Leong, S.P., Cady, B., Jablons, D.M., Garcia-Aguilar, J., Reintgen, D., Jakub, J., Pendas, S., Duhaime, L., Cassell, R., Gardner, M., Giuliano, R., Archie, V., Calvin, D., Mensha, L., Shivers, S., Cox, C., Werner, J.A., Kitagawa, Y., and Kitajima, M. (2006) Clinical patterns of metastasis. *Cancer Metast. Rev.*, **25**, 221–232.
119. Kobayashi, H., Hama, Y., Koyama, Y., Barrett, T., Regino, C.A., Urano, Y., and Choyke, P.L. (2007) Simultaneous multicolor imaging of five different lymphatic basins using quantum dots. *Nano Lett.*, **7**, 1711–1716.
120. Barrett, T., Choyke, P.L., and Kobayashi, H. (2006) Imaging of the lymphatic system: new horizons. *Contrast Media Mol. Imaging*, **1**, 230–245.
121. Ogata, F., Azuma, R., Kikuchi, M., Koshima, I., and Morimoto, Y. (2007) Novel lymphography using indocyanine green dye for near-infrared fluorescence labeling. *Ann. Plast. Surg.*, **58**, 652–655.
122. Kim, S., Lim, Y.T., Soltesz, E.G. *et al.* (2004) Near-infrared fluorescent type II quantum dots for sentinel lymph node mapping. *Nat. Biotechnol.*, **22**, 93–97.
123. Guimaraes, R., Cle'ment, O., Bittoun, J., Carnot, F., and Frija, G. (1994) MR lymphography with superparamagnetic iron nanoparticles in rats: pathologic basis for contrast enhancement. *Am. J. Roentgenol.*, **162**, 201–207.
124. Kobayashi, H., Ogawa, M., Kosaka, N., Choyke, P.L., and Urano, Y. (2009) Multicolor imaging of lymphatic function with two nanomaterials: quantum dot-labeled cancer cells and dendrimer-based optical agents. *Nanomedicine*, **4**, 411–419.
125. Makale, M., McElroy, M., O'Brien, P., Hoffman, R.M., Guo, S., Bouvet, M., Barnes, L., Ingulli, E., and Cheresh, D. (2009) Extended-working-distance multiphoton micromanipulation microscope for deep-penetration imaging in live mice and tissue. *J. Biomed. Opt.*, **14**, 024032.
126. Wessels, J.T., Busse, A.C., Mahrt, J., Dullin, C., Grabbe, E., and Mueller, G.A. (2007) In vivo imaging in experimental preclinical tumor research--a review. *Cytometry A*, **71**, 542–549.
127. Fujiki, Y., Tao, K., Bianchi, D.W., Giel-Moloney, M., Leiter, A.B., and Johnson, K.L. (2008) Quantification of green fluorescent protein by in vivo imaging, PCR, and flow cytometry: comparison of transgenic strains and relevance for fetal cell microchimerism. *Cytometry A*, **73**, 11–118.
128. Eghtedari, M., Oraevsky, A., Copland, J.A., Kotov, N.A., Conjusteau, A., and Motamedi, M. (2007) High sensitivity of in vivo detection of gold nanorods using a laser optoacoustic imaging system. *Nano Lett.*, **7**, 1914–1918.
129. De la Zerda, A., Zavaleta, C., Keren, S., Vaithilingam, S., Bodapati, S., Liu, Z., Levi, J., Smith, B.R., Ma, T.J., Oralkan, O., Cheng, Z., Chen, X., Dai, H., Khuri-Yakub, B.T., and Gambhir, S.S. (2008) Carbon nanotubes as photoacoustic molecular imaging agents in living mice. *Nat. Nanotechnol.*, **3**, 557–562.
130. Shashkov, E.V., Everts, M., Galanzha, E.I., and Zharov, V.P. (2008) Quantum dots as multimodal photoacoustic and photothermal contrast agents. *Nano Lett.*, **8**, 3953–3958.
131. Kim, J.-W., Galanzha, E.I., Shashkov, E.V., Moon, H.-M., and Zharov, V.P. (2009) Golden carbon nanotubes as multimodal photoacoustic and photothermal high-contrast molecular agents. *Nat. Nanotechnol.*, **4**, 855–860.
132. Galanzha, E., Tuchin, V., and Zharov, V. (2007) Optical clearing in small animals: an application for in vivo lymph flow cytometry. *Lasers Surg. Med. (Suppl.)*, **19**, 10.
133. Song, K.H., Stein, E.W., Margenthaler, J.A., and Wang, L.V. (2008) Noninvasive photoacoustic identification of sentinel lymph nodes containing methylene blue in vivo in a rat model. *J. Biomed. Opt.*, **13**, 054033.
134. Song, K.H., Kim, C., Cobley, C.M., Xia, Y., and Wang, L.V. (2009) Near-infrared

gold nanocages as a new class of tracers for photoacoustic sentinel lymph node mapping on a rat model. *Nano Lett.*, **9**, 183–188.

135. Song, K.H., Kim, C., Maslov, K., and Wang, L.V. (2009) Noninvasive in vivo spectroscopic nanorod-contrast photoacoustic mapping of sentinel lymph nodes. *Eur. J. Radiol.*, **70**, 227–231.

136. Galanzha, E.I., Shashkov, E.V., Spring, P., Suen, J.Y., and Zharov, V.P. (2009) *In vivo*, noninvasive, label-free detection and eradication of circulating metastatic melanoma cells using two-color photoacoustic flow cytometry with a diode laser. *Cancer Res.*, **69**, 7926–7934.

137. Schipper, M.L., Nakayama-Ratchford, N., Davis, C.R., Kam, N.W., Chu, P., Liu, Z., Sun, X., Dai, H., and Gambhir, S.S. (2008) A pilot toxicology study of single-walled carbon nanotubes in a small sample of mice. *Nat. Nanotechnol.*, **3**, 216–221.

138. Tuchin, V.V. (2005) Optical clearing of tissue and blood using immersion method. *J. Phys. D: Appl. Phys.*, **38**, 2497–2518.

139. Sage, H.H., Sinha, B.K., Kizilay, D., and Toulon, R. (1964) Radioactive colloidal gold measurements of lymph flow an functional patterns of lymphatics and lymph nodes. *J. Nucl. Med.*, **5**, 626–642.

140. Nopajaroonsri, C. and Simon, G.T. (1971) Phagocytosis of colloidal carbon in a lymph node. *Am. J. Pathol.*, **65**, 25–42.

141. Jemal, R.S., Ward, E., Murray, T., Xu, J., and Thun, M.J. (2007) Cancer statistics, 2007. *CA Cancer J. Clin.*, **57**, 43–66.

142. Karim, R.Z., Scolyer, R.A., Li, W., Yee, V.S., McKinnon, J.G., Li, L.X., Uren, R.F., Lam, S., Beavis, A., Dawson, M., Doble, P., Hoon, D.S., and Thompson, J.F. (2008) False negative sentinel lymph node biopsies in melanoma may result from deficiencies in nuclear medicine, surgery, or pathology. *Ann. Surg.*, **247**, 1003–1010.

143. Sega, E.I. and Low, P.S. (2008) Tumor detection using folate receptor-targeted imaging agents. *Cancer Metast. Rev.*, **27**, 655–664.

144. Qiao, J., Kottke, T., Willmon, C., Galivo, F., Wongthida, P., Diaz, R.M., Thompson, J., Ryno, P., Barber, G.N., Chester, J., Selby, P., Harrington, K., Melcher, A., and Vile, R.G. (2008) Purging metastases in lymphoid organs using a combination of antigen-nonspecific adoptive T cell therapy, oncolytic virotherapy and immunotherapy. *Nat. Med.*, **14**, 37–44.

145. Ross, J.S., Linette, G.P., Stec, J., Clark, E., Ayers, M., Leschly, N., Symmans, W.F., Hortobagyi, G.N., and Pusztai, L. (2003) Breast cancer biomarkers and molecular medicine. *Expert. Rev. Mol. Diagn.*, **3**, 573–585.

146. Everts, M., Saini, V., Leddon, J.L., Kok, R.J., Stoff-Khalili, M., Preuss, M.A., Millican, C.L., Perkins, G., Brown, J.M., Bagaria, H., Nikles, D.E., Johnson, D.T., Zharov, V.P., and Curiel, D.T. (2006) Covalently linked Au nanoparticles to a viral vector: potential for combined photothermal and gene cancer therapy. *Nano Lett.*, **6**, 587–591.

147. Champion, J.A. and Mitragotri, S. (2009) Shape induced inhibition of phagocytosis of polymer particles. *Pharm. Res.*, **26**, 244–249.

148. Petrov, Y.Y., Petrova, I.Y., Patrikeev, I.A., Esenaliev, R.O., and Prough, D.S. (2006) Multiwavelength optoacoustic system for noninvasive monitoring of cerebral venous oxygenation: a pilot clinical test in the internal jugular vein. *Opt. Lett.*, **31**, 1827–1829.

149. Ermilov, S.A., Khamapirad, T., Conjusteau, A., Leonard, M.H., Lacewell, R., Mehta, K., Miller, T., and Oraevsky, A.A. (2009) Laser optoacoustic imaging system for detection of breast cancer. *J. Biomed. Opt.*, **14**, 024007.

150. Zharov, V.P., Galitovskaya, E.N., Jonson, C., and Kelly, T. (2005) Synergistic enhancement of selective nanophotothermolysis with gold nanoclusters: potential for cancer therapy. *Laser Surg. Med.*, **37**, 219–226.

151. Zharov, V.P., Galitovsky, V., and Chowdhury, P. (2005) Nanocluster model of photothermal assay: application for high-sensitive monitoring of nicotine induced changes in metabolism, apoptosis and necrosis at a cellular level. *J. Biomed. Opt.*, **10**, 044011.
152. Zharov, V.P., Galitovsky, V., Lyle, C.S., and Chambers, T.C. (2006) Superhigh-sensitivity photothermal monitoring of individual cell response to antitumor drug. *J. Biomed. Opt.*, **11**, 064034.
153. Letokhov, V.S. (1991) Effects of transient local heating of spatially and spectrally heterogeneous biotissue by short laser pulses. *Nuovo Cimento*, **D13**, 939–948.
154. Kubota, F., Kusuzawa, H., Kosaka, T., and Nakamoto, H. (1995) Flow cytometer and imaging device used in combination. *Cytometry*, **21**, 129–132.
155. Kubota, F. (2003) Analysis of red cell and platelet morphology using an imaging-combined flow cytometer. *Clin. Lab. Haematol.*, **25**, 71–76.
156. Tibbe, A.G., de Grooth, B.G., Greve, J., Dolan, G.J., and Terstappen, L.W. (2002) Imaging technique implemented in cell tracks system. *Cytometry*, **47**, 248–255.
157. Zharov, V.P. (1986) in *Laser Analytical Spectrochemistry* (ed. V.S. Letokhov), Adam Hilger, Bristol & Boston, MA, pp. 229–271.
158. Zharov, V.P. and Montanary, S.G. (1983–84) Multiplex detection for photothermal identification of absorbing compounds in flow. *J. Photoacoustics*, **1**, 355–364.
159. Willard-Mack, C.L. (2006) Normal structure, function, and histology of lymph nodes. *Toxicol. Pathol.*, **34**, 409–424.
160. Zharov, V.P. and Montanary, S.G. (1983–84) Discrete frequency modulation in optoacoustic flow measurements. *J. Photoacoustics*, **1**, 445–461.
161. Zharov, V.P., Galitovskaya, E., and Viegas, M. (2004) Photothermal guidance for selective photothermolysis with nanoparticles. *Proc. SPIE*, **5319**, 291–300.
162. Zhang, Y., Yu, J., Birch, D.J.S., and Chen, Yu. (2010) Gold nanorods for fluorescence lifetime imaging in biology. *J. Biomed. Opt.*, **15** (2), 020504-1–020504-3.

18
Optical Instrumentation for the Measurement of Blood Perfusion, Concentration, and Oxygenation in Living Microcirculation

Martin J. Leahy and Jim O'Doherty

18.1
Introduction

Perhaps the earliest and best-known automated flow cytometer was the Coulter counter [1], which facilitated the "complete cell count" from a blood sample and stimulated a whole industry of hematology. The concentration and flow of blood in living microcirculation has seen three periods of extraordinary growth. Galileo's compound microscopes were not as successful as his telescope because of blurred images and it was not until the second half of the seventeenth century that "high-power" simple lenses, made most notably by van Leeuwenhoek, allowed observation of blood corpuscles moving in the capillaries of transparent organisms such as the louse by Malpighi, van Leeuwenhoek, and Swammerdam. Malpighi also used this technology to confirm Harvey's theory of the passage of blood from arteries to veins; secondly, in the 1960s, after the establishment of the microcirculatory societies from 1954 and the spreading use of video capillaroscopy; thirdly, from the 1980s, after the development in the mid-1970s of automated, continuous, noninvasive measurement of perfusion (laser Doppler) and oxygenation (pulse oximetry). This latest explosion of interest in the microcirculation has continued to this day with continual development of new technologies illuminating this complex world of the microcirculation and its vital role in the health of both organ and organism [2].

Continuous, noninvasive monitoring of microvascular blood flow and arterial oxygen saturation were made possible by the development of laser Doppler flowmetry and pulse oximetry. The importance of the latter is demonstrated by the rapid uptake of this technique in anesthesiology and intensive care medicine. Routine use led to criticism of the technique in certain clinical situations. The technique was shown to be unreliable in conditions of low perfusion at the site of measurement, normally the finger, the presence of abnormal hemoglobins (e.g., hemoglobin bound to molecules other than oxygen), motion at the site of measurement, and interfering contaminants in the light path such as nail polish. However, despite these criticisms, the use of a pulse oximeter has been firmly established as a "standard of care" in every operating room and intensive care unit in the developed

Advanced Optical Flow Cytometry: Methods and Disease Diagnoses, First Edition. Edited by Valery V. Tuchin.

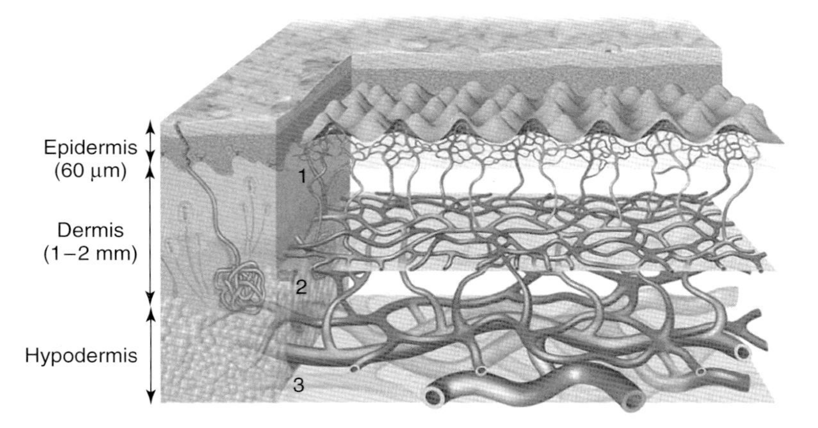

Figure 18.1 Microcirculation in the skin. Three specific planes of vessels can be observed; the superficial plexus, in the papillary dermis approximately 1–2 mm below the epidermis (layer **2**), branches to form the dermal papillary loops (layer **1**), whose tips comprise capillaries. Vessels in the deep plexus lying between the dermal and cutaneous layer (layer **3**) run parallel to the surface of the skin and this constitutes a large surface area for blood flowing through this bed to exchange heat with the external environment.

world [3]. Pulse oximetry is a combination of the techniques of spectrophotometry and photoplethysmography, both well-known techniques with basic theoretical, instrumentational, and operational principles, the synergy of which allowed rapid acceptance in clinical and surgical settings.

The laser Doppler technique, on the other hand, has followed a more gradual acceptance due to a number of issues associated with measurement and interpretation. More fundamentally, it is not possible to assign absolute units of blood flow using this technique due to the stochastic nature of the skin microcirculation architecture (Figure 18.1), creating many scattering events for photons to undergo. System calibration is manufacturer dependent, which makes device cross comparison difficult. Reproducibility of results via various manufacturers' systems is compromised, and thus represents a stumbling block in clinical acceptance of Doppler technology. Owing to the lack of a fundamental measuring unit, clinicians remain unable to correctly interpret the results of laser Doppler experiments. The applicability of the laser Doppler technique to the measurement of blood "flow" at any tissue site has brought about widespread research use in evaluating microcirculation function under certain pathological states, but only modest clinical acceptance.

Visible and near-infrared light, particularly in the wavelength region of 600–1100 nm, offers a window into human and animal tissues due to reduced light scattering and absorption. Nevertheless, the remaining scattering obscures images of the microcirculation everywhere but the eye and nailfold plexus (where blood vessels are parallel to the skin and sufficiently close to the surface).

Low-frequency ($f < 10$ MHz) ultrasonic propagation has many advantages due to its low scattering and ability to provide satisfactory images of larger and deeper pathology and blood vessels. Ultrasound, however, requires higher frequencies for the detection of smaller vessels causing difficulties close to the surface. Although high-frequency Doppler ultrasound ($f > 20$ MHz) has been applied for imaging the microcirculation, it has significant limitations in imaging microvessels close to the surface [4] and may cause heating of the tissue. These effects include spectral broadening due to scanning tissue and difficulty in detecting small Doppler shifts due to slowly moving blood. Difficulties also arise in detecting the scattered signal from small volumes of blood and significant tradeoffs must be made between frame rate, lateral resolution, and the minimum detectable blood velocity [5]. Array transducers are not routine for Doppler ultrasound, and thus imaging is time consuming. The axial component of velocity is the only component estimated, and thus not representative of the complete nutritional flow. Resolution, however, remains good with vessels 30–100 μm in diameter being visible using the technique [6]. Research into Doppler ultrasound unfortunately revolves around small animal models at thin tissue sites, or carefully controlled phantoms [5, 7]. Recent promising results for the prediction of diabetic complications have been obtained using contrast transit velocities as a surrogate for blood velocity in a similar manner to Xe clearance [8].

The use of magnetic resonance imaging (MRI) is well documented in literature concerning microcirculation imaging, although its use is not widespread. The use of phase-contrast magnetic resonance velocity imaging estimates the phase shift of hydrogen nuclei as they pass through a magnetic field, which in turn is proportional to velocity [9]. The acquisition time of MRI images is considerable, with the patient usually supine in the bore, thus limiting the field of observation. Owing to the acquisition time, MRI is more suited to steady-state assessment rather than transient circulatory responses. Indirect measurements may be made; indeed T1-weighted contrast MRI has been used to assess an increase in signal to noise ratio (SNR) when a microvascular bed is irradiated with a 780 nm laser due to an increase in contrast, as a result of vasodilation by the laser [10, 11]. The lack of a clinical use for MRI assessment of skin microcirculation, and the difficult logistics of performing trials in a strong magnetic field also contribute to a lack of application in the area [12].

Direct (optical) observation of skin and other tissues is of course the oldest method of biomedical imaging. Ancient Greeks understood that the pallor of the skin was a significant discriminator between health and disease. However, the decomposition of white light, the Doppler effect, and the identification of hemoglobin as the substance that changes its color and that of the skin when carrying oxygen provided a scientific basis for reliable assessment of the microcirculation developed over the past 150 years [13]. Over the past 30 years, in particular, tremendous technological advancements have been made in single-point biophotonic measurements of microvascular state/activity. Two-dimensional imaging of microvascular structure now represents the minimum standard for research into microcirculatory evaluation.

The vital role of microcirculatory blood supply, and the oxygen it carries, in the health of the individual has ensured the development of many different techniques for its assessment. The principles upon which the methods are based vary widely, as does the suitability (i.e., environmental conditions and physical size), cost, accessibility, ease of use, and result interpretation. The depth of interdisciplinary knowledge required to specify the appropriate technique often means that the wrong technique is used, or the correct one is used inappropriately. This often leads to unfair criticism of techniques used in situations for which they were never designed. A small number of the methods are noninvasive and even fewer are operated remotely without the need to be in physical contact with the skin. Noninvasive *in vivo* techniques have obvious advantages in providing information without disturbing the normal environment. In the same way, it poses considerable difficulties for developers through the need to make accurate measurements in complex environments subject to enormous (biological) variability. For routine clinical use, the technologies also need to conform to the Medical Device Directive (in Europe) and the corresponding Food and Drug Administration (FDA) regulations (in the United States).

The imaging technologies discussed here can be separated by physical resolution and sampling depth (Figure 18.2) with confocal microscopy and ultrasound included for completeness. Pixel resolution is improved by 2 orders of magnitude in commercially available units by going from laser Doppler perfusion imaging (LDPI) to laser speckle perfusion imaging (LSPI) and again from LSPI to tissue viability imaging (TiVi). There is a clinical need for simple-to-use, reliable, user-independent, and inexpensive technologies for bed-side monitoring and for

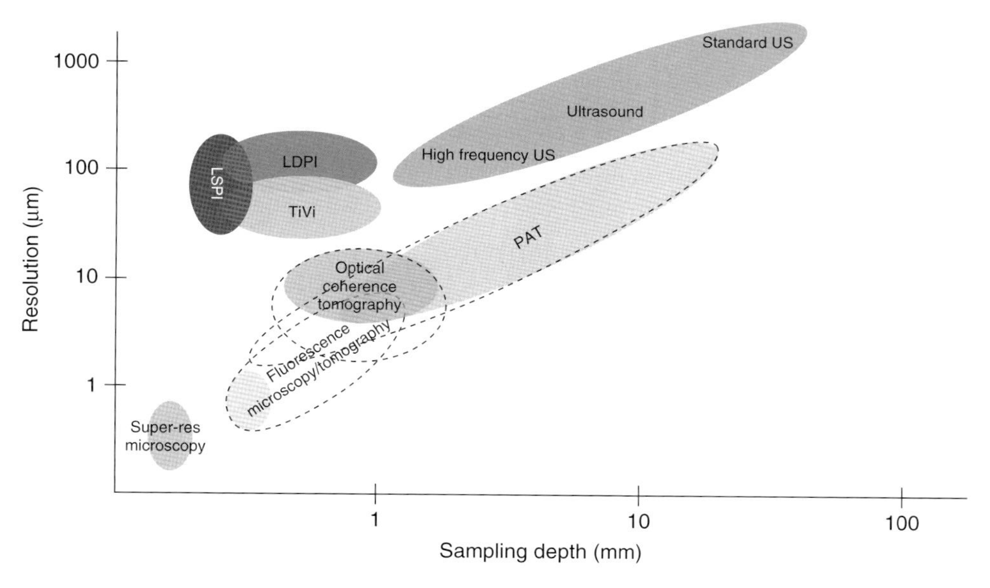

Figure 18.2 Nonionizing microcirculation imaging domains. Resolution is expressed as a "best" range of typical spatial resolutions achieved by regular biophotonic use of the instrument. Dashed lines represent recent efforts to extend the technologies to provide 3D, micron-resolution imaging at depths of more than 1 mm.

diagnosis that facilitate comparison of data recorded at different laboratories and clinical wards throughout the world. This allows the exclusion of exam repetition and points out the need for interlaboratory comparison.

To understand why optical imaging of the skin and microcirculation is not universally applied, it is necessary to consider how light interacts with tissue and which situations allow for direct visualization of the skin and its constituents. Light microscopy and direct visual inspection are hampered by scattering, regular reflection from the surface, and absorption by superficial chromophores. Bloodless melanin-free skin provides little absorption, but significant scattering by cell constituents across all colors of visible light, and the fibrous collagen structure makes the skin appear white. Regular, mirror-like reflections in the skin surface or tissue boundaries account only for a small percentage of the incident photons. However, absorption is the main contrast mechanism and backscattering is required to return photons to the detector/eye.

These techniques represent a mere fraction of the devices developed to perform noninvasive optical measurements on blood in the human microcirculation. The principle of each technology and the technical specifications of each commercial imager are briefly described here, and the interested reader is invited to examine the references to accompanying literature.

18.2 Xe Clearance

The clearance of xenon from the microcirculation has long been used as a reference standard for the measurement of blood flow [14]. It is used as a radiopharmaceutical, which is inhaled to assess pulmonary function via imaging of the lungs. It is also often used to image blood flow, particularly in the brain. A tracer of radioactive 133Xe is injected intradermally and its elimination from the microcirculation of the skin provides a measurement of flow in absolute units (ml/min/100 g of tissue).

Other invasive methods have been shown to correlate well to this "gold standard" of measurement methods, such as the heat clearance method, where the skin is heated to between 2 and 10° above resting levels. When a plateau is reached, the heat is discontinued and the temperature decay is monitored over time. Flow is calculated using a mathematical equation [15]. When blood flow is high, the heat is dissipated at a greater rate.

18.3 Nailfold Capillaroscopy

One of the most useful methods of assessing blood flow in a small number of blood vessels in the skin is by direct observation using capillary microscopy [16]. Capillaries are the smallest of the blood vessels and their purpose is to link the arterial and venous systems together, allowing the exchange of carbon and oxygen as well as removal of waste metabolites. Only in certain places in the body do the capillaries come close enough to the surface of the skin to be naturally visible, that is, without the use of specialized optical equipment or optical clearing agents. This is one of the reasons that the study of the skin overlapping the base of the fingernail and toenail is so important, although the capillaries of this region may not be representative of other microvascular networks of the body. In the nailfold, the capillaries come within 200 μm of the surface of the skin, implying that the methods of examining them are simple. Another reason is that the fingernail is easily fixed in position, free from any movement, due to arterial pulsations or respirations, and the capillaries run parallel to the skin surface. Initial microscopes (biomicroscope or stereomicroscope) had a magnification range between $\times 15$ and $\times 100$ and a wide depth of field to account for the differing depths of capillaries. Optics were required to be within centimeters of the examination site, and could be recorded by connecting a camera or a video camera [17].

Using a microscope with a magnification of between $\times 200$ and $\times 600$, it is possible to clearly see the erythrocytes moving within the blood vessel. As vessels at this level are transparent, the red blood cell (RBC) motion in a single capillary can be measured. They appear as hairpin-like loops (6–15 μm in diameter) appearing parallel to one another and pointing toward the tips of the fingers or toes. On other parts of the skin, only the tips of the loops can be seen due to the increasing depth of the vessels, where the effects of light scattering and absorption overcome visibility.

If a video camera is attached to the microscope [18, 19], the motion can be examined in a frame-by-frame procedure yielding accurate velocity information for RBC flow in the capillary. The flow may be altered by the fixing procedure and/or the heating effect of the light source. Examining the capillary condition and density can aid the diagnosis of certain diseases. Capillary density and diameter are dependent on age, with younger children having a lower density and capillary thickness than adults [20]. The presence of abnormal vessels, in addition to these factors, has been attributed to various diseases, and the vessels are normally assessed as measured by changes in the vessel diameter following agonist or antagonist stimulation. Some conditions can be detected very early as capillary deformations can be observed before other symptoms occur [21].

Figure 18.3 is an image of a nailfold capillary arrangement of a healthy adult [22]. The capillaries are hairpin-like loops, arranged into regular rows. Each loop consists of two parallel limbs; a thinner arterial limb, and a thicker venous limb. The wall of each blood vessel is transparent and the RBCs are seen passing through the capillaries [23].

A typical experimental setup would consist of a microscope, a monitor, and a video-recording system [21, 23, 24]. Fiber-optic illumination is often used [20, 25]. The finger can be immobilized easily by use of a clamp [24]. To increase the transparency of the skin, a clearing agent such as immersion oil is usually applied to the nailfold [20, 21, 23]. A fluorescent solution such as fluorescein can be used to distinguish the capillaries from the surrounding skin [26]. The software used for analyzing the images and the preparation of the subjects is dependent on what is being measured.

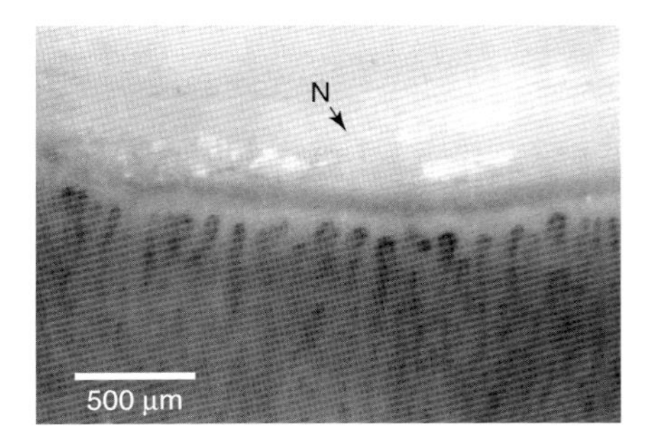

Figure 18.3 Nailfold capillary pattern of a healthy adult [22]. The finger nail is marked as *N*. A typical capillary density for a healthy adult is approximately 30–40 mm^{-2} [23]. Reproduced with permission from Wiley-Blackwell Publishing Ltd.

18.3.1 Data Analysis

A disease-specific "pattern" can be identified by analyzing the geometry and density of the capillaries and the presence of abnormal or very large blood vessels [23]. Other relevant factors include avascular areas (absence of capillaries) [22] and RBC velocity (RBV) through the capillaries by tracking individual RBCs in the vessels [24]. By tracking the pattern over two consecutive frames, the velocity can be approximated. A drawback of this method is that these patterns are difficult to detect in larger vessels. A typical value of RBV would be $\sim$0.3–1 mm s^{-1} [24]. Examples of avascular areas, and giant and dilated capillaries (more than 50 μm in diameter or greater than five times their normal caliber), can be seen in Figure 18.4.

The geometry of each capillary is defined by taking measurements at specific points. An example of this is shown in Figure 18.5.

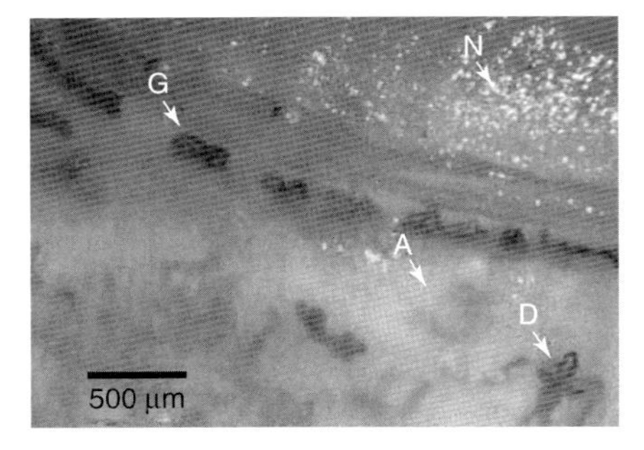

Figure 18.4 Nailfold capillaries in a patient with dermatomyositis. Examples of avascular areas, A; giant capillaries, G; and dilated loops, D are shown [22]. Reproduced with permission from Wiley-Blackwell Publishing Ltd.

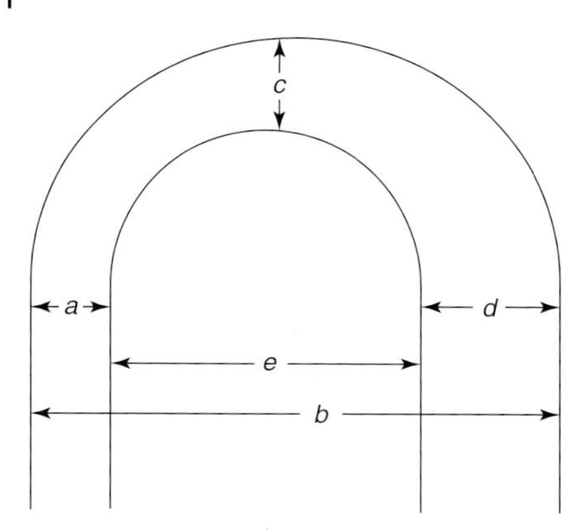

Figure 18.5 Measurements of dimensions of a capillary [23]. The arterial capillary diameter, *a*. The venous capillary diameter, *d*. The loop diameter, *c*. The width of the capillary loop, *b*. The distance between the limbs, *e*.

The most difficult calculation in nailfold capillaroscopy is RBV, and there are many different methods of determining microvascular blood flow. There are two main advantages of nailfold capillaroscopy over other wide-area microcirculation imaging methods. The first is that wide-area methods are not limited to a specific area, meaning that the measurements are not taken on a specific capillary. As a result, the measurements are made less accurate by RBCs with various velocities [26]. The second drawback is that the signals produced by other technologies are directly related to the concentration and velocity of blood cells, but do not allow absolute measurements of velocity or flow in standard SI units. It is impossible to measure the absolute units of blood flow because of spatially varying and *in vivo* unknown tissue optical properties [27].

Other methods of analysis include cross-correlation methods, which can be further subdivided into two methods. The first works by measuring the average intensity at two independent regions of the capillary. Using a temporal cross correlation and the distance between the regions, the RBV can be calculated [28, 29]. The second method works by creating a series of spatial intensity distributions on a particular section. Using a temporal spatial correlation and the time separation between consecutive frames, the velocity can be calculated [30]. A spatiotemporal autocorrelation function has been used to develop software to perform calculations of the RBV [31]. When calculating RBV in a single capillary, it must be borne in mind that this RBV is generally not representative of neighboring capillaries. Ensembles of capillaries must be used to accurately describe the functionality of a specific microvascular bed.

18.3.2 Applications

Capillaroscopy is an old technique, and has fundamentally shown its use in detecting and diagnosing impaired microvascular function, and is an established procedure for differentiating primary Reynaud's from the secondary type; in the primary type, the vessels appear normal, whereas in the secondary the vessels appear disordered and abnormal [32, 33]. It has historically been used in quantification of vascular effects of food [34]. Measurement of transit times of RBCs through such vessels has been very useful in diagnosing microvascular diseases, such as Raynaud's phenomenon and systemic scleroderma, the latter of which is associated with the presence of giant capillaries, and with a heterogeneous decrease in capillary density and disorganization of capillary loops [17].

Research has showed evidence that, for patients with primary Raynaud's phenomenon, the loop diameter, capillary width, and capillary length (inner length + outer length/2) were greater in patients who went on to develop undifferentiated connective-tissue disease than in those who did not [21]. Bukhari *et al.* [25] studied the dimensions and density of capillaries and showed that there was a significant increase in all dimensions except the distance between the limbs in patients with Raynaud's phenomenon and systemic sclerosis (SSc) compared to a control group.

Also, capillary density is reduced in patients with SSc. A decrease in capillary density in the nailfold of patients with psoriasis compared to a control group has been observed [34], and a characteristic pattern of capillary abnormalities has been shown to occur in patients with scleroderma [35]. The RBV functional capillary density has been studied [24] and capillary morphology differences in patients with primary Sjögren's syndrome and a healthy control group noted. It was found that the affected patients had a higher functional capillary density and lower RBV than the control group.

18.3.3 Limitations and Improved Analysis

The depth of tissue penetration represents the major limitation to the technique and, so, this approach is restricted to the nutritive portions of the skin where the vessels rise close to the surface [19]. The measurements required for analyzing the capillary dimensions and patterns are more time consuming than the examination itself. Computerized systems have been developed [36–38] for analyzing images of capillary networks and improving their quality by excluding disturbances caused by hair, liquid, reflections, and so on. Combining videocapillaroscopy and various mathematical methods, the statistical properties of the capillaries can be analyzed automatically by extracting capillary count, position, density, size, and so on, from the images. Triangulation methods to allow statistical analysis of distances between nearest neighboring capillaries have been employed [36, 37], which is useful for looking at areas that are susceptible to development of necrosis. The Markov-chain-based detection algorithm for updating capillaroscopy images to reduce them to a grayscale image containing only the capillaries is under development currently. More fundamentally, a 55.8% coefficient of variation in blood flow velocity was noted when examining the same patient via deep inspiration and ischemia after a time period of seven days. This may show either inherent variation in the microcirculatory architecture or dynamic technique, or may also be due to low, long-term reproducibility of the technology.

While nailfold capillaroscopy is important in the diagnosis of certain connective-tissue diseases, some of the methods described above cannot be applied to skin sites other than the nailfold for the simple reason that the capillaries are not clearly visible. This necessitates other methods of measuring the condition and density of the capillaries, and the velocity of the RBCs.

18.4 LDPM/LDPI

Laser Doppler perfusion monitor (LDPM) and LDPI are well-established technologies, mentioned in over 8000 research articles. Long-running companies such as Oxford Optronix, Moor Instruments, and Perimed are involved with commercial sales of perfusion monitors and imagers. LDPI has been developed into a high-resolution device; using a specialized scanner head, the beam can be focused to obtain a lateral resolution of 40 μm. Combining this with a scanning step size of 25 μm allowed *in vitro* measurements of microvessels with an average difference of just 11 μm (half a pixel) from independent microscopic measurements [39]. In its basic form, a perfusion image is a collection of LDPM points placed together on a color-coded map showing areas of low and high perfusion. LDPI has found use in many aspects of dermatology, including melanomas [40, 41], drug development [42], ulcers [43], burns and skin grafts [44], Raynaud's syndrome [45], and cerebral measurements [46].

18.4.1 Principles of Operation of LDPI

Coherent laser light is incident on the tissue surface from a distance of approximately 15 cm. A portion of the light is scattered by static tissue and retains its incident frequency, and another portion is scattered by moving RBCs, which causes a frequency shift in the light returning to the detector (due to the Doppler effect). The resultant mixing (interference) of these dissimilar frequencies at a sensitive detector results in a beat frequency, which is equal to the Doppler shift between the two portions of light. As the scattering that occurs in tissue is of a completely random nature, not every Doppler-shifted photon has a similar Doppler shift. A wide range of velocities is present, thus creating a Doppler frequency spectrum. The power spectrum $P(f)$ of the fluctuations of the photodetector current contains information about the blood perfusion. Through the use of signal-processing techniques [47] taking the first moment of the power spectrum, an RBC flux is described, which is the product of the concentration and the root mean square of the RBC speed.

Conventional LDPI achieves the perfusion image by rastering of a laser beam horizontally and vertically across the tissue surface, momentarily stopping to perform calculations and moving to the next point. Each point represents a point on the perfusion map. New generations of commercial imagers raster a line of pixels in parallel at a single time step, thereby substantially reducing the acquisition time of single images. A later development of LDPI method utilizes CMOS technology, and is devoid of any scanning components as it acquires data from all points at the same time through the use of a high-powered expanded laser beam. Temporal resolution is drastically improved, requiring between 2 and 10 s for acquisition and display of a 256 × 256 pixel perfusion image, rather than 4 min using conventional (point scanning) LDPI.

18.5 Laser Speckle Perfusion Imaging (LSPI)

It is interesting to note that a similar result can be achieved [48, 49] by considering the scattering particle to be a reflector of light, like a mirror in a Michelson interferometer, moving with velocity v. The reflected light will interfere with the reference beam and the resultant intensity will vary with the difference in optical path, (optical path difference, OPD), between the two beams. The number of interference fringes to pass the detector in time, t, will be given by OPD/λ or $2d/\lambda$, where d is the actual distance moved by the particle. Speckle results from the pattern produced due to interference of coherent light randomly traveling minutely different optical path distances from a scattering surface. Upon interaction with a scattering medium, coherent light, which has been scattered from various positions within the medium, produces both constructive and destructive interference on the surface of a photodetector. If a 2D array such as a CCD camera is used to image the pattern, it is seen to vary in space and thus produces a randomly varying intensity pattern, commonly termed "*speckle*." Movement of the particle causes fluctuations in the interference pattern, appearing as intensity variations at the CCD surface. The number of interference fringes to pass the detector in time, t, is given by

$$\frac{\mathrm{OPD}}{\lambda t} = \frac{2d}{\lambda t} \tag{18.1}$$

where d is the actual distance moved by the particle and the factor 2 refers to the fact that the light must travel to and from. The number of fringes per second is then:

$$f = \frac{2d}{\lambda t} = \frac{2v}{\lambda} \tag{18.2}$$

Thus, the relationship of the resultant frequency to velocity is the same in each case.

The speckle-imaging technique was originally developed by Fercher and Briers [50] in the early 1980s for retinal blood flow imaging. Digital laser speckle imaging is a recent technology, with the first commercial speckle imager released in 2007 by Moor Instruments fast laser Doppler perfusion imager (FLPI). In recent times, it has found use in many areas, such as blood flow in the optic nerve [51], flow in animal knee joint [52], cerebrum [53], and rodent skin folds [54], and has even been adapted to fit an endoscopic setup [55]. A comprehensive text on the application of laser speckle to blood flow cytometry is given in the next chapter.

18.6 TiVi

TiVi imaging is a technology that allows high-resolution (~54 μm) imaging of the RBC concentration in the upper human dermal tissue [56]. This technique was developed to allow online real-time video imaging of the microcirculation at a frame size of 400 × 400 pixels at 8 fps. Offline, the temporal resolution can be increased to 25 fps at 720 × 576 pixels [57]. Using the video mode, spatial resolution is decreased to 75 μm. The system has been evaluated with fluid models of skin tissues in order to validate its image-processing algorithm. Using the system, it is possible to capture an area of skin measuring 15 cm × 10 cm with 5 million measurement pixels at a device–tissue distance of 5 cm (the most recent version has 10 million pixels). Using the 400 × 400 pixel acquisition frame for video studies, a minimum area of 4 cm × 3 cm can be imaged. The technique has shown many uses in drug development, burn investigations, pressure studies, and general research maneuvres due to the ease of use, portability, and low cost [58].

18.6.1 Principles of Operation of TiVi

White incoherent light is linearly polarized by a filter and is incident on the surface of the skin. A small fraction of the light is reflected from the epidermal layers and retains its polarization state. Approximately half of the remaining light is absorbed by the tissue, while the other half is diffusely backscattered by the dermal tissue. This backscattered portion is exponentially depolarized mainly due to scattering events by chromophores present in the tissue [59]. Upon reemerging from the tissue structure as diffusely reflected light, approximately one-fifth is parallel and one-fifth perpendicular polarized with respect to the direction of the original filter. This light contains information about the main chromophores in the epidermis (melanin) and dermis (hemoglobin), while the surface reflections contain information about the surface topography, such as texture and wrinkles.

The surface reflections and a fraction of the diffusely backscattered light can be differentiated by placing another polarization filter over the detector parallel or perpendicular with respect to the direction of the transmission axis of the filter over the light source. With the transmission axes of both filters oriented parallel or perpendicular with respect to each other, copolarized (CO) or cross-polarized (CR) data respectively is obtained. For the TiVi-imaging system, the CR arrangement is used, allowing gating of photons from the subepidermal region, optically removing the effects of the superficial layers of the skin. Implicit in its use is the assumption that weakly scattered light (near-surface reflections) retains its polarization state, whereas strongly scattered light will successively depolarize with each scattering event.

CR images are color-separated by a standard camera CCD, and an algorithm sensitive to the concentration of RBCs only is adapted to provide a linear relationship over a physiological range of 0–4% RBC volume fraction. To process the images, an algorithm is used that relates the higher absorption values of blood in the green (500–600 nm) to lower reference values in the red (600–700 nm) wavelength regions, while the absorption of light in both wavelength regions in surrounding tissue is of a similar and low degree. Image acquisition for the high-resolution mode is instantaneous (1/60 s), and TiVi-processed images are presented in a similar color map to LDPI and LSPI images, with red and blue representing high and low RBC concentration respectively.

Further, a real-time system is currently under development, whereby RBC concentration can be displayed online [60]. Current technology can capture CR frames at 25 fps at a frame size of 400 × 400 pixels (also capable of 15 fps at 1024 × 768 pixels). Video studies will lead to increased temporal resolution, with 0.04 s between each image. Thus, theoretically, the human pulse can be resolved from the time average (duplex) mode of the video imager, and also spatial data over time can be observed. Occlusion of the cephalic vein for the upper arm, with 80 mmHg leads to higher RBC concentration due to the RBCs being trapped by the occlusion, as the brachial artery pumps RBCs into the area. An example of the duplex mode is shown in Figure 18.6, detailing the average TiVi index of a 50 × 50 pixel region of interest (ROI) on the volar forearm for an acquisition of 25 fps for cephalic vein occlusion. Clear vasomotion of the vessels can be observed at a rate of approximately 5 cycles per min during the postocclusion period. Future work

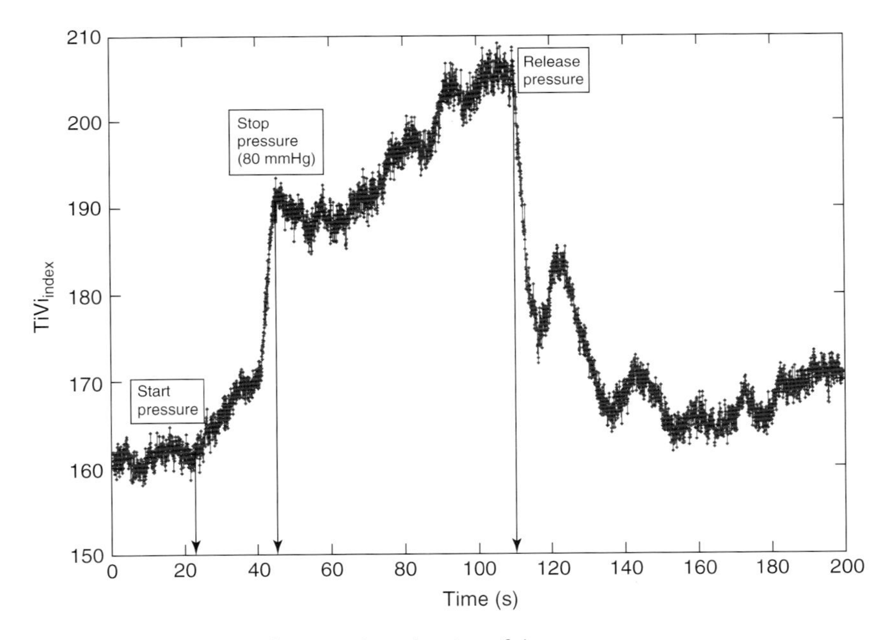

Figure 18.6 Time trace of a 50 × 50 pixel region of the volar forearm of a subject. A pressure cuff on the upper forearm causes an occlusion of the cephalic vein. This demonstration shows clear signs of vasomotion following the release of cuff pressure.

will investigate pulsations of the vascular bed; pulse models have already shown promising results.

Limitations of this new technique include the lack of SI units in measurements (as is also the case with other microcirculation imaging technologies) and inaccurate measurements on persons with dark skin due to high absorption of melanin. The effect of melanin in the epidermis causes an increase in the TiVi index. Image-subtraction algorithms have been developed so as to display only the reaction on the skin site, while subtracting out the static components including melanin-rich spots in the CR images. This is also helpful in removing the unwanted effect of deep veins in the volar forearm, a site which is used commonly in routine skin tests. Owing to the availability of measuring tissue hematocrit only, low cost, ease-of-use, and portability, the TiVi imaging system is an attractive alternative to more expensive and complex equipment such as LDPI and LSPI for investigation of tissue RBC concentration.

18.7 Comparison of TiVi, LSPI, and LDPI

In order to accurately validate the results of any microcirculation technique, the opinion of a clinical expert is usually required. Limitations/precautions of skin assessment visual scoring systems such as the skin blanching assay include the requirement of (i) the application of the test substance at equal amounts in each patch in random order, (ii) the need to make all observations under standardized lighting conditions, and (iii) the participation of a well trained and skilfull observer. If these requirements are not met, interlaboratory results may vary substantially and be difficult to compare. Updating of protocol for UVB skin-sensitivity testing involves the use of a radially divergent beam, which provides a range of dose to the skin tissue. Both LDPI and TiVi have been used successfully in the analysis of results from the new protocol, each with its own merits and limitations [61, 62]. As yet there is no modern technology that has supplanted the fundamental visual observation technique in the clinical environment [63], even though clinicians' interpretations of images can be bypassed entirely by a computerized analysis, such as neural networks [64].

The following section represents a brief introduction to the instrumentation chosen to assist in basic skin testing. A summary of principles and operation of each instrument is shown in Table 18.1. The actual instruments used are briefly described here.

18.7.1 Laser Doppler Imaging – Moor Instruments – MoorLDLS

The Moor instruments, laser Doppler line scanner (LDLS), is a line scanning laser Doppler imager with a 780-nm laser diode. Owing to the capture of 64 points simultaneously, the temporal resolution is markedly improved from conventional LDPI. The minimum image acquisition time is dependent on the number of data

Table 18.1 Comparison table outlining mechanical and operational principles of the different devices evaluated.

Property	LDLS	FLPI	TiVi
Principle	Doppler effect	Reduced speckle contrast	Polarization spectroscopy
Variable recorded	Perfusion (speed × concentration)	Perfusion (speed × concentration)	Concentration
Variable units	Perfusion units (A.U.)	Perfusion units (A.U.)	$TiVi_{index}$ (A.U.)
Measurement sites	16 384	431 680	Up to 12 million
Best resolution (μm)	≈500	≈100	≈50
Measurement depth (mm)	≈1	≈300	≈500
Repetition rate	6 s (50 × 64)	4 s (568 × 760)	5 s (4000 × 3000)
Best capture time	6 s	4 s	1/60th s
Distance dependence	Yes	Yes	No
Zoom function	No	Yes	Yes
Laser driver	Required	Not required	Not required
Movement artifact sensitivity	Substantial	Dependent on temporal averaging	None
Video mode	No	Yes	Yes
Video rate	–	25 fps	25 fps
Video size	–	113 × 152	Up to 720 × 576
Pixel resolution (per cm^2)	100	40 000	600 000
View angle	Top only	Top only	User selectable
Imager dimensions	19 cm × 30 cm × 18 cm	22 cm × 23 cm × 8 cm	15 cm × 7 cm × 5 cm
Weight (kg)	Scanner 3.2	2	1
Calibration	Motility standard	Motility standard	None
Ambient light sensitivity	Intermediate	Substantial	Negligible

points per Fast Fourier Transforms (FFTs): at maximum resolution (256 × 64 pixels), the best image acquisition time is 26 s for 256 FFTs and 51 s for 1024 FFTs. The 26- and 51-s image scans require 100 and 200 ms respectively for a single scan of a line of pixels. The maximum scan area is 20 cm × 15 cm, and the optimal distance to the tissue surface (given by the manufacturer) is 15 cm. The device is also capable of single-point multichannel acquisition at 1, 2, 5, 7, and 10 Hz for the entire line of detectors simultaneously.

18.7.2
Laser Speckle Imaging – Moor Instruments – MoorFLPI

For laser speckle studies, the Moor instruments FLPI system was chosen. The 780-nm laser source is expanded over an area, rather than a line as in LDLS.

There are two modes available: high resolution/low speed or high speed/low resolution. For low resolution/high speed, the integration time is = 0.3 s (check these), providing 25 fps at a size of 152 × 113 pixels. High Resolution/low speed allows an adjustable integration time, constant for temporal averaging (1 s per frame – 25 frames, 4 s per frame – 100 frames, 10 s per frame – 250 frames, and 60 s per frame – 1500 frames) at 568 × 760 pixels for each image. The imager allows a repeat image function, video mode (25 fps), and single image acquisitions, all simultaneously if required. The camera gain is adjustable in order to increase the area of image acquisition, although care should be taken so as not to saturate the CCD.

18.7.3 Polarization Spectroscopy – Wheels Bridge – Tissue Viability TiVi Imager

The tissue viability imager (TiVi600) is a small and portable device for high-resolution imaging of the concentration of RBCs in superficial skin tissue. A white xenon flash lamp provides illumination over the 450–750 nm wavelength region, and a wide-area CCD detects the backscattered light from the tissue. The available commercial system utilizes a high-resolution CCD offering instantaneous image acquisition (1/60th of a second) with a refresh rate of one image every 5 s at a resolution of 4272 × 2848 pixels. An enabled macromode allows focusing at a closest distance of 4 cm from the tissue surface. A research prototype high-speed mode has also been developed, allowing 25 fps at 400 × 400 pixels. This prototype employs a constant broadband light source with the same CCD as that used in the high-resolution version of TiVi. The spectra of the light sources are equivalent for both the high-resolution and high-speed TiVi systems.

18.7.4 Comparison of Imaging Systems

Tests comparing the use of three microcirculation imagers, TiVi, LDPI, and LSPI show inherent differences in technology, operational performance, and measurement method.

Physical size prevented all three technologies from being used at the same time on the same test area, meaning that tests with the FLPI had to be carried out separately from TiVi and LDLS studies. The white flash light source on the TiVi did not interfere with the LDLS system, as there are filters present to remove much of the extraneous light. Finally, the TiVi readings were not affected by the scanning beam of the LDLS system as they were outside the sensitive region of the detector.

The effects of the skin blanching steroid provided visual whitening due to vasoconstriction of the underlying blood vessels in each subject. Images of the microvasculature from each device can be seen in Figure 18.7. From the figure, it can be seen that the LDLS and FLPI do not show any difference between the areas

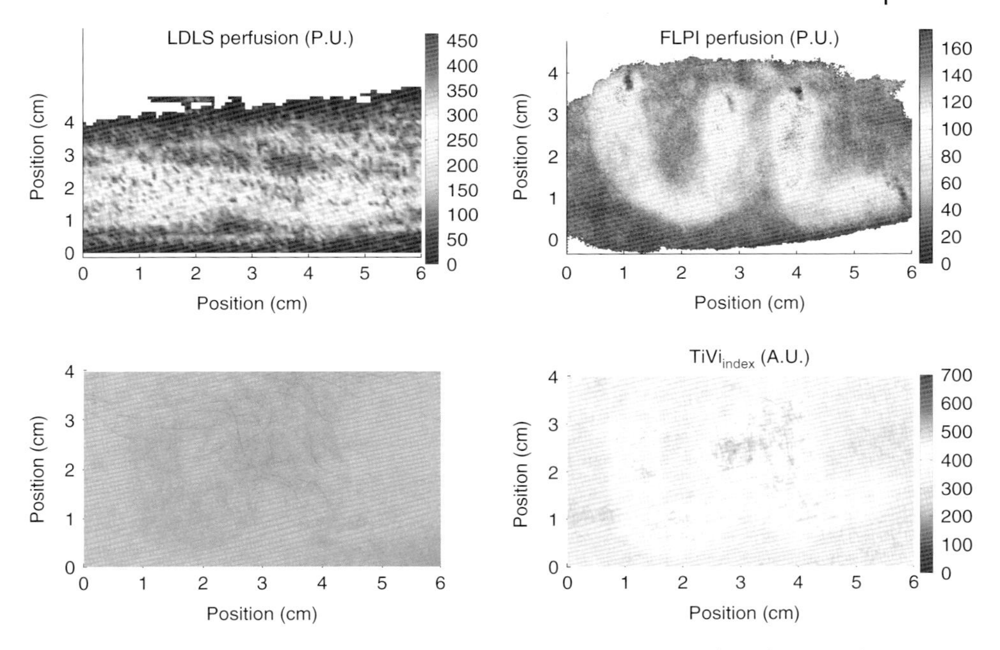

Figure 18.7 The imaging of skin blanching due to a vasoconstricting corticosteroid by three different microcirculation imaging techniques. The black dots in the FLPI and LDLS images show the area affected by the drug. The FLPI image is shown over a smaller scale as the entire arm could not be imaged due to the background level. Thus, the border between affected and unaffected tissue is shown, to the left of the dots is where the drug was administered; to the right of the dots is unaffected tissue.

of baseline and the area where the corticosteroid was administered. Only the TiVi system shows the marked vasoconstriction caused by the steroid.

Endothelial function of the skin microcirculation can also be assessed by the use of electrical stimulation of a vasoactive drug through the permeable cutaneous tissue, known as *iontophoresis* [65, 66]. LDPI has principally been used to quantify the reaction by the creation of dose-response curves [67], which facilitates the comparison of the effect of different drugs, or the response of different groups of patients. TiVi has been used to compare against LDPI during the process of iontophoresis [68]. Their results show that, because TiVi can measure below the biological resting baseline level, it can be directly applied to measurement of vasoconstriction without the need for heating the tissue first, as is common in LDPI studies. This technique of local warming increases the number of confounding factors because it has been noted that local heating may reduce vasoconstrictor response [69]. Results further seem to suggest that when the blood perfusion increases during vasodilatation in skin the initial phase relies mainly on an increase in RBC concentration, whereas the further perfusion increase is due to an increase in RBV. Output comparison of LDPI and TiVi values simultaneously

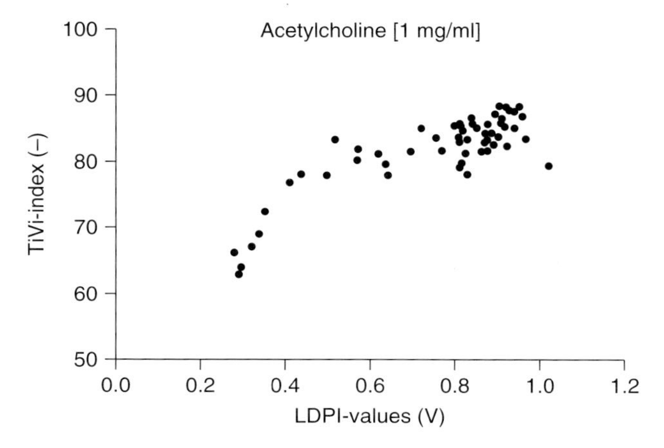

Figure 18.8 Simultaneously recorded average TiVi and average LDPI response values. A correlation coefficient of 0.85 was found between the systems [68].

measured from the iontophoresis of a vasodilating agent (acetylcholine) is shown in Figure 18.8 [68]; a correlation coefficient of 0.85 was determined between the technologies.

As the concentration of RBCs in iontophoresis skin testing is of more relevance than the actual blood flow, TiVi seems more suited to the application. The higher-resolution imaging allows spatial statistics inside the electrode (25 mm diameter) to be determined, and also comparison of regions of differing spatial heterogeneity.

18.7.5
Discussion

Even with the massive technological advancements of the past 30 years, current "gold-standard" methods of examination of skin lesions, erythema, and tumors still involve simple microscope techniques, that is, a closer look at the affected area, carried out by a skilled clinician, be it a pathologist or dermatologist. Similarly, no modern imaging technique has been proven to consistently outperform the opinion of a clinical expert in relation to patch or prick testing [70]. Oftentimes, the capital cost of a technology without an extremely high success rate is not justified, even if the workload of the clinician is drastically reduced.

The ideal blood flow/concentration monitor should have the characteristics of observational measurements, be able to provide reproducible results, have little or no dependency on the tissue optical properties, have instantaneous image acquisition (to help against patient motion), and provide a variable in standard SI units. The commercial devices examined in this paper provide relative changes in flow/RBC concentration due to the random nature of the microvascular architecture. Since LDPI is the oldest technology used in this work (20 years), more research publications and commercial products exist on LDPI than in the other technologies. A full-field laser Doppler imager has been developed and is close to commercial production, providing 256 × 256 points every 2–10 s [71].

The most important feature of the technologies presented is that both the LDLS and FLPI measure perfusion, albeit at different depths in dermal tissue, while TiVi technology measures relative concentration of RBCs in the tissue. It is important to remember that the new TiVi imaging technology does not provide measurements of blood flow. Indeed, it is similar to processing of the unweighted moment of the laser Doppler power spectrum to provide measurement of concentration, except that the Doppler version is valid only with a low amount of scattering events at low concentrations, and is dependent on the total light intensity.

With the larger measurement depth of LDPI compared to the other techniques, it proves to be inefficient for very superficial measurements, such as the erythema produced by the topical analgesic. Acquisition on the temporal scale is also hindered with the LDPI due to rastering of the laser line. As the microcirculation is a highly dynamic environment, its perfusion values can change rapidly. Cardiac pulse (circa 1 Hz), breathing (circa 0.3 Hz), and vasomotion (circa 0.1 Hz) in highly perfused tissues are evidences of the rapid temporal nature of microvascular activity. Thus, in laser Doppler studies and the relatively long acquisition time of a single image, the temporal changes in perfusion may often be misinterpreted as spatial heterogeneity [72]. This occurs because it has to be assumed that the perfusion of the site under investigation is stationary and constant, thus the study of dynamic changes is severely compromised. It remains to be seen if the recent CMOS LDPI technology can bridge the gap in temporal resolution in LDPI.

The laser Doppler and speckle signals are complex functions of many parameters and are also dependent on source wavelength, coherence properties of the laser, tissue optical properties, and distance between the scanner and tissue surface.

Much variation is reported in the literature as to the depth of penetration of a light beam in tissue, owing to the technological variations such as light power and wavelength, and biological factors such as pigmentation level, age, skin transparency, and epidermal thickness. However, the important parameter is the measurement depth rather than penetration or even sampling depth. Indeed, laser Doppler imaging measurement depth in the forearm is in the range of 300–600 μm. This measurement depth is highly dependent on the wavelength and optical arrangement, and thus depends on which device has been employed.

Physiologically, in the vessel compartments the total cross-sectional area (A) increases from the precapillary vessels to the superficial dermal plexus and then decreases again in the veins because the blood flux is constant in a closed system and the speed must be proportional to $1/A$. Thus, the speed is lower in the narrower vessels than in the larger (and deeper) vessels as well as in shunting vessels (anastomosis). It is for this reason that high frequencies are attributed to flux originating in the deeper vessels. Low Doppler frequencies originate, however, from the more superficial vessels, but to some extent also from the larger and deeper vessels, because narrow angle scattering and scattering in low speed RBCs close to the vessel wall result in comparatively low Doppler shifts. Indeed, approximately

15% of skin blood flow originates from the nutritional flow, while 85% results from the deeper thermoregulatory flow. LSPI has neither the sensitivity to high frequencies (5 ms integration time leaves a maximum frequency of 200 Hz) nor the frequency weighting to be sensitive to the faster moving blood, which is mostly in the deeper vessels. This partly explains the observation that many speckle studies are carried out on surgically exposed tissue and the appearance of microvessels in the processed images.

Although the TiVi system is not responsive to blood flow, it does return a signal from the deeper veins in the forearm due to the high concentration of blood. Because of the appearance of the blue veins, the TiVi returns an almost zero value in its variable, $\text{TiVi}_{\text{index}}$, even though a nearby skin site in the upper microcirculation can have a high $\text{TiVi}_{\text{index}}$ value. Again, it should be remembered that the TiVi imager is designed to investigate the upper microcirculation and that RBC concentration values from anywhere other than this region may not be accurate. The measurement volume (the volume from which the bulk of the signal is acquired) is well inside the reticular dermis of the microcirculation. For laser Doppler measurements, the measurement volume has been determined as approximately 1 mm^3 or smaller in the imaging variant, and in laser speckle systems the volume is highly dependent on the type of system being used.

18.8 Pulse Oximetry

Scientists have been attempting to measure the relative quantities of oxygenated and deoxygenated hemoglobin in human blood-bearing tissue for more than 100 years. Even before this, the degree of cyanosis, as determined by clinical observation, was considered to be a fundamental diagnostic parameter for many centuries. The clinicians' eye, like many instruments developed in the past century, was used to assess the relative absorption of red light. The role of much higher scattering toward the blue end of the spectrum, including Rayleigh scattering from the granular melanin, mitochondria, and cell nuclei, play a large (but unrecognized) role, as even in the deoxygenated state hemoglobin absorbs far more blue light than red. This section reviews these developments and the improvements in clinical observation that have been achieved.

18.8.1 Historical and Literature Review

Woodcock [73] reported that Plethysmography was originally developed by Glisson in 1622 and Swammerdam in 1737. In 1672, Sir Isaac Newton presented his discovery of the composition of white light [74]. "I placed my prism at this light entrance, that it might be refracted to the opposite wall. It was at first a pleasing divertissement to view the vivid and intense colors produced thereby . . . ." After 6 years of experimental work and thought, he proposed a thesis which was to become the basis of spectrophotometric analysis and gave his conclusion thus: "Colors are not *Qualifications of Light*, derived from Refractions, or Reflections of natural Bodies (as 'tis generally believed), but *Original and connate properties*, which in divers Rays are divers."

Later [75], the relationship between light absorption and the concentration of absorbent was identified and formulated this relationship now known as the *Beer–Lambert law* [76]. This law states that the light transmitted through a solution is logarithmically related to path length, concentration of absorbent, and its absorption coefficient at the wavelength of the incident light.

The substance that gives blood its color (hemoglobin) was first recognized as an oxygen carrier in 1864 by Irish scientist Stokes [13]. These discoveries provided the basis for the development of the theory of *in vivo* oximetry over 100 years ago.

Von Veirordt first described the application of the spectrometer to measure Hb/O_2Hb on a human subject in 1876. The main problem encountered was the absorption by tissue, which varied from site to site and among individuals. This principle was applied in the development of the first instrument for continuous monitoring of blood oxygen saturation by transilluminating tissue [77]. Although this instrument could reliably follow trends in oxygen saturation, it could not make absolute measurements. During World War 2, a compact and lightweight device was produced to investigate the loss of consciousness of fighter pilots during battle, and the term "*oximeter*" was coined [78]. Further progress was made by researchers who compressed the tissue in order to zero the instrument in the absence of blood [79, 80]. An instrument was developed "for clinical conditions," called the "*Cyclops*" because the probe was fixed to the forehead [81]. Soon after, it was finally discovered that the fundamental principle (Beer–Lambert law) on which the theory of dual wavelength oximetry is based, inadequately described light transmission in whole blood, where the hemoglobin molecules are contained in little packets (RBCs), instead of a homogenous solution. Scattering of light is a function of particle size and wavelength, so even if one assumes a constant size of RBC the difference in wavelength (typically 660 and 940 nm) accounts for significant differences in the optical path length traveled in the absorbent. This conflict between theory and practice was exposed where considerable deviations from the Beer–Lambert law for high concentrations of RBCs were found [82].

This discovery dampened enthusiasm for this technique and was to plague *in vivo* oximetry up to the present day. Thus, the emphasis shifted to other methods of oxygen measurement and the advent of the Clark electrode [83] focused attention on oxygen tension rather than saturation.

A new approach to oxygen saturation (according to Severinghaus and Astrup [84]) employed eight wavelengths facilitating self-calibration and clinically acceptable accuracy. In the form of the Hewlett Packard ear oximeter, this device became very popular in the 1970s particularly in research, where difficulties with application and heating the site were more acceptable. More rigorous tests in the late 1970s demonstrated inaccuracies at low SaO_2 ($<70\%$) and interference from abnormal hemoglobins (e.g., Wukitsch *et al.* [85]).

An exciting new idea first formulated by Aoyagi *et al.* [86] in a patent filed by the Nikon Kohden Company of Japan was to become the basis for continuous *in vivo* arterial oxygen saturation. This novel concept, which used the fact that light transmission in human tissue varies with arterial pulsations, to isolate the relative absorption of red and infrared light by arterial blood was later published [87]. They postulated that the cardiac synchronous variations in the light intensity reaching the detector were entirely attributable to blood volume changes in the small arteries and arterioles "seen" by the probe. Therefore, comparing only the pulsating portion of the red and infrared signals should yield absorption data for arterial blood alone, the absorption of venous blood and other tissues being constant or slowly varying. This filtering mechanism was of tremendous benefit in the area of critical care medicine because the oxygen status of the blood supply could be measured rather than that of the mixed vascular bed. This is a parameter which is subject to reasonably precise regulation in the normal healthy individual at rest (approx. 95–97% saturation), but may reduce to dangerous levels suddenly in the critically ill. In smokers, the maximal value can be only 84% due to a COHb of 15% [88].

A modification of this device using microprocessor control and human volunteer calibration data became the first instrument with clinically acceptable accuracy. This instrument (Biox 2) was successfully applied to research in pulmonary function laboratories. The clinical significance of this instrument, particularly for anesthesiology, was realized by William New, himself an anesthesiologist at Stanford University. New developed and marketed a pulse oximeter (Nellcor N100) primarily for monitoring oxygen saturation in patients under anesthesia. Because of the tremendous importance of the technique, more than 20 manufacturers now produce pulse oximeters. Widespread use has, however, led to significant criticism of the technique in certain clinical situations. The instrument was shown to be unreliable in conditions of low perfusion (at the site of measurement, normally the finger), subject movement, or shivering, the presence of abnormal hemoglobins (hemoglobin bound to molecules other than oxygen), and even by the presence of interfering contaminants such as nail polish in the light path. Despite these criticisms, however, the use of a pulse oximeter has been firmly established as a "standard of care" in every operating room and intensive care unit in the developed world [89].

The criticisms of the technology have led to improvements in the algorithms and the way in which the information is displayed. For example, it is regularly recommended to use a pulse oximeter with a plethysmographic and signal strength display in order to judge the reliability of the reading. "Unreliable signals" are often filtered out and the last reliable value may be held for 24 s before discontinuing the display (Ohmeda Technical Support, personal communication). Some manufacturers have employed signal weighting methods where the signals are classified during acquisition according to their signal-to-noise ratio and deviation from the previous values, so that their contribution to the final SpO_2 signal can be manipulated. This further processing certainly leads to readings being available more of the time, but one must be careful as to how much of the signal is of physiological or microprocessor origin. For example, one study of the more up-to-date instruments concluded that connecting pulse oximetry (PO) probes during motion has significantly higher total failure rates compared to those instituting motion after the PO probes are connected [90].

All commercially available pulse oximeters have been subject to a fundamental limitation of approximately 0.2% signal-to-noise ratio, with erroneous readings being frequent at lower signal levels. One technologist claimed to have overcome this limitation by filtering out noise within the physiological bandwidth by "generating a signal that represents the noise," (Masimo Corp. – personal communication). Indeed, a prototype based on this technology has performed well on some US trials, with a specified minimum S/N ratio of 0.02% and much improved performance during motion [91–93]. A comprehensive review of the principles and practice of pulse oximetry gives invaluable information for any existing or prospective user [94]. The history of all aspects of blood gas analysis is comprehensively reviewed by Severinghaus and Astrup [84].

18.8.2 Alternative Methods for Measurement of Oxygenation

A large number of methods are available for the measurement of oxygenation due to its fundamental importance to almost all animal life. This section is confined to only those most relevant to the subject of this thesis.

18.8.3 Transcutaneous Oxygen Pressure ($TCpO_2$)

$TCpO_2$ is most commonly measured using the Clark arrangement of the polarographic technique. A voltage is generated between a platinum cathode and silver anode in KCl solution by an electrochemical reaction in the presence of O_2. A PTFE membrane can be used to separate the blood or tissue from the anode, cathode, and electrolyte [95]. Research showed that when a finger was immersed in an electrolyte at 45 °C, the pO_2 of the solution approximated that of the arterial blood [96]. The arterial vasodilation induced is a necessary condition of measurement, as skin surface pO_2 is near zero at normal skin temperatures (<30 °C), although some success has been achieved at 37 °C, when neural, hormonal, and local metabolic influences remain in active control [97]. The increased temperature has three effects [98]:

1) Vasodilation of the dermal capillaries, thereby "arterializing" the capillary blood.
2) A rightwards shift of the oxyhemoglobin dissociation curve, that is, pO_2 increases at the electrode site. At a constant SaO_2 of 90%, the pO_2 is 45 mmHg at 32 °C but increases to approximately 73 mmHg at 40 °C.
3) An increase in oxygen diffusion through the stratum corneum of the skin.

A further effect of heating is an increased metabolism, which at the right probe temperature (approx. 43.5 °C), may cancel effect (ii) and produce a reliable measurement of arterial pO_2.

The usefulness of this method in assessment of skin microcirculation is therefore rather dubious because one is always measuring some proportion of central arterial pO_2 if the temperature is insufficient to completely arterialize the blood. However, it has been found useful in the assessment of arterial insufficiency, especially when the skin and systemic PaO_2 are normal [99]. They showed that $TCpO_2$ is very sensitive to arteriovenous pressure differences, especially when these differences are small. Electronic drifting remains a problem and frequent calibration is necessary.

18.8.4
Near Infrared Spectroscopy (NIRS)

The improvements in optical detectors and emitters has perhaps nowhere in medicine had such an impact as in the near infrared region. This is largely due to the optical window between 600 (Hb) and 1300 nm (H_2O) and the distinct variations in absorption by important chromophores, especially those involved in the respiratory chain – hemoglobin and cytochrome "*a*" – in this region depending on whether they are bound to oxygen. Because both contain a haem moiety, they have similar absorption spectra, but they can be resolved in the near infrared (NIR). Furthermore, the oxidized form of cytochrome *a* has a weak absorption peak in this region, which changes depending on its redox state. The use of this phenomenon in the measurement of cat brain oxygenation by transillumination has been demonstrated [100]. Other researchers, notably Delpy at University College London and Chance at University of Pennsylvania, soon realized the potential of extending this technique to neonatal monitoring where the head is only a factor of two greater. Delpy and coworkers recognized that absolute measurements in this highly scattering and complex medium would only be possible where the actual path length of the laser light was known. Subsequent "time of flight" studies [101] using picosecond lasers and a synchroscan streak camera demonstrated that this pathlength was a simple multiple of the geometric distance between source and detector, 5.9 for brain and 3.6 for forearm muscle.

Commercial instruments are available from Hamamatsu and Johnson & Johnson in which the optical path length is determined from the phase delay of light intensity modulated at several hundred megahertz. The oxygenation values are computed by microprocessor by applying a modified Beer–Lambert law. Cytochrome "*a*" has the distinction of being fixed in the membrane of the mitochondria, and such measurements can be considered to be of tissue oxygenation. A lucid and readable account of near infrared spectroscopy (NIRS) is given in the literature, also describing the perinatal application of the Hamamatsu device registering the rapid change in hemoglobin oxygenation after the first breath of life [102].

18.8.5
Luminescence Quenching (Fluorescence and Phosphorescence)

The absorption of high energy (short wavelength) photon by a molecule raises an electron to a higher energy level, where it will remain until it relaxes to the lowest vibrational level of the excited state by loss of thermal energy. It remains here for a short "lifetime" due to the large band gap, but must return to the ground state by either direct photon emission (fluorescence) or via the "spin-forbidden" triplet state and subsequent emission at a longer wavelength. The "forbidden" route has a longer lifetime (approx. 10^{-5} s) due to its lower probability. These lifetimes can be shortened by interaction with other molecules, especially with oxygen which has one of the lowest triplet energies occurring in the ground state and a low-energy singlet state. The diffusion-limited process of collision with the luminescing molecules thus permits a measurement of oxygen tension.

For both processes, the probe consists of a quantity dye dissolved and encapsulated or immobilized in a silicone rubber membrane, the latter being preferred for invasive measurements. The luminescing dye is excited by a xenon flash lamp or blue silicon carbide LEDs (450 nm, where an appropriate absorption peak permits), and the longer wavelength luminescence is detected by a photomultiplier tube (PMT). Earlier systems which simply measured this intensity are now being replaced by high-frequency excitation and phase-sensitive detection systems, avoiding any problems of instability of light source, probe, or instrumentation. The pO_2 can be extracted using the Stern–Volmer relation:

$$\frac{\tau_0}{\tau} = 1 + k_Q\,[pO_2] \tag{18.3}$$

where τ_0 and τ are the lifetimes in the presence and absence of oxygen respectively, and k_Q is the quenching constant for which each probe must be calibrated. The immobilization of the dyes can cause the detected lifetimes to exhibit multiexponential decay characteristics complicating analysis. However, phase fluorometric and phosphorometric instruments have been described [103]. These techniques have great potential for measurement of other analytes such as CO_2 [104]. Furthermore, if the naturally occurring luminescing compounds can be exploited, then noninvasive imaging of NADH (the reduced form of nicotinamide adenosine diphosphate) at the top of the respiratory chain [102] and metalloporphyrins might be possible, although in this case it is difficult to envisage how k_Q can be determined. Indeed, an instrument for phosphorescent imaging of the retinal and choroidal vessels by injection of a metalloporphyrin derivative has been developed [105].

18.9
Conclusions

We have compared the operation of an established microcirculation imaging technique, laser Doppler (LDPI), with laser speckle and TiVi in human skin tissue using the occlusion and reactive hyperemia response. However, TiVi differs from the other techniques in its response to occlusion of the brachial artery. As this is a global occlusion of the forearm, all instruments were sensitive to the effect. LDPI and LSPI showed a large decrease as they are both sensitive to red cell concentration and speed. However, TiVi, being sensitive to concentration only, showed stasis when a full arterial occlusion (130 mmHg) was applied quickly and an increased RBC concentration when the arterial inflow remained open with the venous outflow occluded (90 mmHg). It also shows improved performance, reliability, and analysis over LDPI data in the iontophoresis of vasoactive agents. LSPI and TiVi are both welcome tools in the study of the microcirculation, but care must be taken in the interpretation of the images as blood flow and blood concentration in tissue are essentially different parameters. However, to date LDPI is the only one of these imaging techniques that has been approved by regulatory bodies including the FDA. Laser Doppler imaging has been shown to accurately assess burn depth and predict wound outcome with a large weight of evidence [106].

We have discussed the *modus operandi* and applications of capillaroscopy, pulse oximetry, LDPI, LSPI, and TiVi in the context of alternative technologies. Each has its own niche applications owing to user experience. Proper application of each is dependent on user knowledge of light interaction with tissue and the basic workings of the device used. Choice of technology for *in vivo* imaging of the microcirculation is dependent on the sampling depth, resolution (physical and pixel), field-of-view, and whether structure or function is to be investigated. None of the techniques have come close to the elegance of the pulse oximeter in simplicity, clinical relevance, or commercial success. Key elements of this success include the singular importance of the measurement of oxygen saturation, the discovery by Aoyagi of the cardiac pulse as a biological filter, the recognition by William New that arterial oxygen saturation was of life-saving importance in general anesthesia, the establishment of pulse oximetry as a standard of care (legal requirement) by the FDA, and all similar authorities in the developed world and luck. Luck plays an important role in the commercial success of many inventions and it is often related to timing. Too early and the market is not ready, too late and the need is already fulfilled. In this case, all of that and the ethical approval gained to insert radial artery canulas in Olympic athletes and feed them oxygen air mixtures down to 11%, until their arterial oxygen saturation reached 70% in order to develop a calibration curve for the Nellcor pulse oximeter. Most of all, however, any entrepreneur will tell you what is required is perseverance and this is well illustrated in the timescale of development of the pulse oximeter.

Acknowledgments

The authors would like to acknowledge the support of the IRCSET (Irish Research Council for Science, Engineering & Technology) for funding this project. This research was also supported by the National Biophotonics Imaging Platform (NBIP) Ireland funded under the Higher Education Authority PRTLI Cycle 4, cofunded by the Irish Government and the European Union – Investing in your future.

References

1. Coulter, W.H. (1953) Means for counting particles suspended in a fluid. US Patent 2, 656, 508, filed Aug. 27, 1949 and issued Oct. 20, 1953.
2. Leahy, M.J. (2011) *Microcirculation Imaging*, Wiley-VCH Verlag GmbH, Heidelberg.
3. Kelleher, J.C. (1989) Pulse oximetry. *J. Clin. Monit.*, **5**, 37–62.
4. Yang, V.X.D., Needles, A., Vray, D., Lo, S., Wilson, B.C., Vitkin, I.A., and Foster, F.S. (2004) High frequency ultrasound speckle flow imaging – comparison with Doppler optical coherence tomography (DOCT). *Ultrason. Symp. IEEE*, **1**, 453–456.
5. Goertze, D., Yu, J.L., Kerbel, R.S., Burns, P.L., and Foster, F.S. (2003) High-frequency 3-D color-flow imaging of the microcirculation. *Ultrasound Med. Biol.*, **29**, 39–51.
6. Marion, A., Aouida, W., Basarabb, A., Delachartrea, P., and Vray, D. (2010) Blood flow evaluation in high-frequency, 40 MHz imaging: a comparative study of four vector velocity estimation methods. *Ultrasonics*, **50** (7), 683–690.
7. Christopher, D.A., Burns, P.A., Starkoski, B.G., and Foster, F.S. (1997) A high-frequency pulsed-wave Doppler ultrasound system for the detection and imaging of blood flow in the microcirculation. *Ultrasound Med. Biol.*, **23**, 997–1015.
8. Duerschmied, D., Maletzki, P., Freund, G., Olschewski, M., Seufert, J., Bode, C., and Hehrlein, C. (2008) Analysis of muscle microcirculation in advanced diabetes mellitus by contrast enhanced ultrasound. *Diabetes Res. Clin. Pr.*, **81**, 88–92.
9. Yucel, E.K., Anderson, C., Edelman, R.R., Grist, T., Baum, R.A., and Manning, W.J. (1999) Magnetic resonance angiography: update on applications for extracranial arteries. *Circulation*, **100**, 2284–2301.
10. Schaffer, M., Bonel, H., Sroka, R., Schaffer, P.M., Busch, M., Reiser, M., and Duhmke, E. (2000) Effects of 780 nm diode laser irradiation on blood microcirculation: preliminary findings on time-dependent T1-weighted contrast-enhanced magnetic resonance imaging (MRI). *J. Photoch. Photobio.*, **B54**, 55–60.
11. Schaffer, M., Bonel, H., Sroka, R., Schaffer, P.M., Busch, M., Reiser, M., and Duhmke, E. (2000) Magnetic resonance imaging (MRI) controlled outcome of side effects caused by ionizing radiation, treated with 780 nm-diode laser -preliminary results. *J. Photoch. Photobio. B*, **59**, 1–8.
12. Oliver, J.J. and Webb, D.J. (2003) Noninvasive assessment of arterial stiffness and risk of atherosclerotic events. *Arterioscler. Thromb. Vasc.*, **23**, 554–566.
13. Stokes, G.G. (1864) On the reduction and oxygenation of the colouring matter of the blood. *Philos. Mag.*, **28**, 391.
14. Monteiro, A.A., Svensson, H., Bornmyr, S., Arborelius, M., and Kopp, S. (1989) Comparison of ^{133}Xe clearance and laser Doppler flowmetry in assessment of blood flow changes in human masseter muscle induced by isometric contraction. *Arch. Oral Biol.*, **34**, 779–786.
15. Midttun, M., Sejrsen, P., and Colding-Jørgensen, M. (1966) Heat-washout: a new method for measuring cutaneous blood flow in areas with and without arteriovenous anastomoses. *Clin. Physiol.*, **16**, 259–274.
16. Ryan, T.J. (1973) in *The Physiology and Pathophysiology of The Skin*, vol. **1** (ed. A Jarrett), Academic Press, London, pp. 653–679.
17. Carpentier, P.H. (1999) New techniques for clinical assessment of the peripheral microcirculation. *Drugs*, **58**, 17–22.
18. Butti, P., Intaglietta, M., Reimann, H., Holliger, C., Bollinger, A., and Anliker, M. (1975) Capillary red blood cell velocity measurements in human nailfold by video densitometric method. *Microvasc. Res.*, **10**, 220–227.

19. Fagrell, B., Fronek, A., and Intaglietta, M. (1977) A microscope television system for studying flow velocity in human skin capillaries. *Am. J. Physiol.*, **233**, 318–321.
20. Dolezalova, P., Young, S.P., Bacon, P.A., and Southwood, T.R. (2003) Nailfold capillary microscopy in healthy children and in childhood rheumatic diseases: a prospective single blind observational study. *Ann. Rheum. Dis.*, **62**, 444–449.
21. Ohtsuka, T., Tamura, T., Yamakage, A., and Yamazaki, S. (1998) The predictive value of quantitative nailfold capillary microscopy in patients with undifferentiated connective tissue disease. *Br. J. Dermatol.*, **139**, 622–629.
22. Nagy, Z. and Czirják, L. (2004) Nailfold digital capillaroscopy in 447 patients with connective tissue disease and Raynaud's disease. *J. Eur. Acad. Dermatol. Venereol.*, **18**, 62–68.
23. Allen, P.D., Taylor, C.J., Herrick, A.L., and Moore, T. (1999) Image analysis of nailfold capillary patterns from video sequences, in *Medical Image Computing and Computer-Assisted Intervention–MICCAI'99*, vol. 1679 (eds C. Taylor and A. Colchester), Springer Berlin/Heidelberg, 698–705. *http://dx.doi.org/10.1007/10704282_76.*
24. Aguiar, T., Furtado, E., Dorigo, D., Bottino, D., and Bouskela, E. (2006) Nailfold videocapillaroscopy in primary Sjögren's syndrome. *Angiology*, **57**, 593–599.
25. Bukhari, M., Herrick, A.L., Moore, T., Manning, J., and Jayson, M.I. (1996) Increased nailfold capillary dimensions in primary Raynaud's phenomenon and systemic sclerosis. *Br. J. Rheumatol.*, **35**, 1127–1131.
26. Nilsson, G.E., Jakobsson, A., and Wårdell, K. (1991) Tissue perfusion monitoring and imaging by coherent light scattering. *Biooptics: Opt. Biomed. Environ. Sci.*, **15**, 90–109.
27. Stern, M.D. (1975) In vivo evaluation of microcirculation by coherent light scattering. *Nature*, **254**, 56–58.
28. Bollinger, A., Butti, P., Barras, J.P., Trachsler, H., and Siegenthaler, W. (1974) Red blood cell velocity in nailfold capillaries of man measured by a television microscopy technique. *Microvasc. Res.*, **7**, 62–72.
29. Mawson, D.M. and Shore, A.C. (1998) Comparison of capiflow and frame by frame analysis for the assessment of capillary red blood cell velocity. *J. Med. Eng. Technol.*, **22**, 53–63.
30. Rosen, B. and Paffhausen, W. (1993) On-line measurement of microvascular diameter and red blood cell velocity by a line-scan CCD image sensor. *Microvasc. Res.*, **45**, 107–121.
31. Sourice, A., Plantier, G., and Saumet, J.L. (2005) Red blood cell velocity estimation in microvessels using the spatiotemporal autocorrelation. *Meas. Sci. Technol.*, **16**, 2229–2239.
32. LeRoy, E.C. and Medsger, T.A. (1992) Raynaud's phenomenon: a proposal for classification. *Clin. Exp. Rheumatol.*, **10**, 485–488.
33. Bhushan, M., Moore, T., Herrick, A.L., and Griffiths, C.E.M. (2000) Nailfold video capillaroscopy in psoriasis. *Br. J. Dermatol.*, **142**, 1171–1176.
34. Wright, C.I., Kroner, C.I., and Draijer, R. (2005) Raynaud's phenomenon and the possible use of foods. *J. Food Sci.*, **70**, R67–R75.
35. Wong, M.L., Highton, J., and Palmer, D.G. (1988) Sequential nailfold capillary microscopy in scleroderma and related disorders. *Ann. Rheum. Dis.*, **47**, 53–61.
36. Sainthillier, J.M., Degouy, A., Gharbi, T., Pieralli, C., and Humbert, P. (2003) Geometrical capillary network analysis. *Skin Res. Technol.*, **9**, 312–320.
37. Zhong, J., Asker, C.L., and Salerud, G. (2000) Imaging, image processing and pattern analysis of skin capillary ensembles. *Skin Res. Technol.*, **6**, 45–57.
38. Hamar, G., Horváth, G., Tarján, Z., and Virág, T. (2007) Markov chain based edge detection algorithm for evaluation of capillary microscopic images. *IFMBE Proc.*, **16**, 18–21.
39. Linden, M., Golster, H., Bertuglia, S., Colanfouni, A., Sjöberg, F., and

Nilsson, G. (1998) Evaluation of enhanced high resolution laser Doppler imaging (EHR-LDI) in an *in vitro* tube model with the aim of assessing blood flow in separate microvessels. *Microvasc. Res.*, **56**, 261–270.

40. Ilias, M., Wårdell, K., Stücker, M., Anderson, C., and Salerud, E.G. (2004) Assessment of pigmented skin lesions in terms of blood perfusion estimates. *Skin Res. Technol.*, **10**, 43–49.
41. Stücker, M., Springer, C., Paech, V., Hermes, N., Hoffman, M., and Altmeyer, P. (2006) Increased laser Doppler flow in skin tumors corresponds to elevated vessel density and reactive hyperemia. *Skin Res. Technol.*, **12**, 1–6.
42. Kragh, M., Quistorff, B., and Kristjansen, P.E.G. (2001) Quantitative estimates of angiogenic and anti-angiogenic activity by laser Doppler flowmetry (LDF) and near infra-red spectroscopy (NIRS). *Eur. J. Cancer*, **37**, 924–929.
43. Malanin, K., Havu, H.K., and Kolari, P.J. (2004) Dynamics of cutaneous laser Doppler flux with concentration of moving blood cells and blood cell velocity in legs with venous ulcers and in healthy legs. *Angiology*, **55**, 37–42.
44. Jeng, J.C., Bridgeman, A., Shivnan, L., Thornton, P.M., Alam, H., Clarke, T.J., Jablonski, K.A., and Jordan, M.H. (2003) Laser Doppler imaging determines need for excision and grafting in advance of clinical judgement: a prospective blinded trial. *Bol. Inst. Estud. Med. Biol. Univ. Nac. Auton. Mex.*, **29**, 665–670.
45. Kanetaka, T., Komiyama, T., Onozuka, A., Miyata, T., and Shigematsu, H. (2004) Laser Doppler skin perfusion pressure in the assessment of Raynaud's phenomenon. *Eur. J. Vasc. Endovasc. Surg.*, **27**, 414–416.
46. Tonnesen, J., Pryds, A., Larsen, E.H., Paulson, O.B., Hauerberg, J., and Knudsen, G.M. (2005) Laser Doppler flowmetry is valid for measurement of cerebral blood flow autoregulation lower limit in rats. *Exp. Physiol.*, **90**, 349–355.
47. Arildsson, M.L., Wårdell, K., and Nilsson, G.E. (1997) Higher order moment processing of laser Doppler perfusion signals. *J. Biomed. Opt.*, **2**, 358–363.
48. Briers, J.D. (2001) Laser Doppler, speckle and related techniques for blood perfusion mapping and imaging. *Physiol. Meas.*, **22**, R35–R66.
49. Leahy, M.J. (1996) Biomedical instrumentation for monitoring microvascular blood perfusion and oxygen saturation. DPhil thesis, University of Oxford.
50. Fercher, F. and Briers, J.D. (1981) Flow visualization by means of single exposure speckle photography. *Opt. Commun.*, **37**, 326–329.
51. Yaoeda, K., Shirakashi, M., Funaki, S., Funaki, H., Nakatsue, T., Fukushima, A., and Abe, H. (2000) Measurement of microcirculation in optic nerve head by laser speckle flowgraphy in normal volunteers. *Am. J. Opthalmol.*, **130**, 606–610.
52. Forrester, K., Stewart, C., Tulip, J., Leonard, C., and Bray, R. (2002) Comparison of laser speckle and laser Doppler perfusion imaging: measurement in human skin and rabbit articular tissue. *Med. Biol. Eng. Comput.*, **40**, 687–697.
53. Li, P., Ni, S., Zhang, L., Zeng, S., and Luo, Q. (2006) Imaging cerebral blood flow through the intact rat skull with temporal laser speckle imaging. *Opt. Lett.*, **31**, 1824–1826.
54. Choi, B., Kang, N.M., and Nelson, J.S. (2004) Laser speckle imaging for monitoring blood flow dynamics in the in vivo rodent dorsal skin fold model. *Microvasc. Res.*, **68** (2), 143–146.
55. Bray, R.C., Forrester, K.R., Reed, J., Leonard, C., and Tulip, J. (2006) Endoscopic laser speckle imaging of tissue blood flow: applications in the human knee. *J. Orthop. Res.*, **24**, 1650–1659.
56. O'Doherty, J., Henricson, J., Anderson, C., Leahy, M.J., Nilsson, G.E., and Sjöberg, F. (2007) Sub-epidermal imaging using polarized light spectroscopy for assessment of skin microcirculation. *Skin Res. Technol.*, **13**, 472–484.

57. O'Doherty, J., Henricson, J., Nilsson, G.E., Anderson, C., and Leahy, M.J. (2007) Real time diffuse reflectance polarisation spectroscopy imaging to evaluate skin microcirculation. *Proc. SPIE*, **6631**, 66310.
58. Nilsson, G.E., Anderson, C., Henricson, J., Leahy, M.J., O'Doherty, J., and Sjöberg, F. (2008) Assessment of tissue viability by polarization spectroscopy. *Opto-Electron. Rev.*, **16**, 59–63.
59. Ishimaru, A. (1989) Diffusion of light in turbid media. *Appl. Opt.*, **28**, 2210–2215.
60. McNamara, P., O'Doherty, J., O'Connell, M.L., Fitzgerald, B.W., Anderson, C., Nilsson, G.E., Toll, R.J., and Leahy, M.J. (2009) Tissue viability (TiVi) imaging: temporal effects of local occlusion studies in the volar forearm. *J. Biophotonics*, **3**, 66–74.
61. O'Doherty, J., McNamara, P., Clancy, N.T., Enfield, J.G., and Leahy, M.J. (2009) Comparison of instruments of microcirculatory blood flow and red blood cell concentration. *J. Biomed. Opt.*, **14**, 034025.
62. Leahy, M.J., Enfield, J.G., Clancy, N.T., O'Doherty, J., McNamara, P., and Nilsson, G.E. (2007) Biophotonic methods in microcirculation imaging. *Med. Laser Appl.*, **22**, 105–126.
63. Rallan, A. and Harland, C.C. (2004) Skin imaging: is it clinically useful? *Clin. Exp. Dermatol.*, **29**, 453–459.
64. Tomatis, S., Bono, A., Bartoli, C., Carrara, M., Lualdi, M., Tragni, G., and Marchesini, R. (2003) Automated melanoma detection: multispectral imaging and neural network approach for classification. *Med. Phys.*, **20**, 212–221.
65. Droog, E.J., Henricson, J., Nilsson, G.E., and Sjoberg, F. (2004) A protocol for iontophoresis of acetylcholine and sodium nitroprusside that minimises nonspecific vasodilatory effects. *Microvasc. Res.*, **67**, 197–202.
66. Curdy, C., Kalia, Y.N., and Guy, R.H. (2001) Non-invasive assessment of the effects of iontophoresis on human skin in-vivo. *J. Pharm. Pharmacol.*, **53**, 769–777.
67. Henricson, J., Tesselaar, E., Persson, K., Nilsson, G.E., and Sjoberg, F. (2007) Assessment of microvascular function by study of the dose–response effects of iontophoretically applied drugs (acetylcholine and sodium nitroprusside) – methods and comparison with in vitro studies. *Microvasc. Res.*, **73**, 143–149.
68. Henricson, J., Nilsson, A., Tesselaar, E., Nilsson, G., and Sjöberg, F. (2009) Tissue viability imaging: microvascular response to vasoactive drugs induced by iontophoresis. *Microvasc. Res.*, **78**, 199–205.
69. Wilson, T.E., Cui, J., and Crandall, C.G. (2002) Effect of whole-body and local heating on cutaneous vasoconstrictor responses in humans. *Auton. Neurosci.*, **97**, 122–128.
70. Lamminen, H. and Voipio, L. (2008) Computer-aided skin prick test. *Exp. Dermatol.*, **17**, 975–976.
71. Serov, A., Steinacher, B., and Lasser, T. (2005) Full-field laser Doppler perfusion imaging and monitoring with an intelligent CMOS camera. *Opt. Express*, **13**, 3681–3689.
72. Leahy, M.J., de Mul, F.F., Nilsson, G.E., and Maniewski, R. (1999) Principles and practice of the laser-Doppler perfusion technique. *Technol. Health Care*, **7**, 143–162.
73. Woodcock, J.P. (1974) Plethysmography. *Biomed. Eng.*, **9**, 409–417.
74. Newton, I. (1672) A letter from Mr. Isaac Newton, proffessor of the mathematicks in the University of Cambridge containing his new theory about light and colours. *Philos. Trans. R. Soc.*, **80**, 3075–3087.
75. Doppler, J.C. (1842) Ueber das farbige Licht der doppelsterne und einiger anderer gestirne des Himmels. [On the coloured light from double stars and some other Heavenly Bodies]. *Abh. Konigl. Bohmischen Ges. Wiss.*, **2**, 465 [Treatises of the Royal Bohemian Society of Science].
76. Ingle, J.D.J. and Crouch, S.R. (1988) *Spectrochemical Analysis*, Prentice Hall.
77. Matthes, K. (1935) Untersuchungen ober die sauerstoffsattingungen des

menechlichen arterienblutes. *Arch. Exp. Pathol. Pharmacol.*, **179**, 698–711.
78. Millikan, G.A. (1942) Oximeter, an instrument for measuring continuously oxygen saturation of arterial blood. *Rev. Sci. Instrum.*, **13**, 434.
79. Squire, J.R. (1940) Instrument for measuring the quantity of blood and its degree of oxygenation in the web of the hand. *Clin. Sci.*, **4**, 331–339.
80. Goldie, E.A.G. (1942) Device for continuous indication of oxygen saturation of circulating blood in man. *J. Sci. Instrum.*, **19**, 23.
81. Brinkman, R. and Siijlstra, W.G. (1949) Determination and continuous registration of the percentage oxygen saturation in clinical conditions. *Arch. Chir. Neerl.*, **1**, 177–183.
82. Kramer, K., Elam, J.O., Saxton, G.A., and Elam, W.N. (1951) Influence of oxygen saturation, erythrocyte concentration and optical depth upon the red and near-infrared light transmittance of whole blood. *Am. J. Physiol.*, **165**, 229–246.
83. Clark, L.C., Wolf, L.C., Granger, D., and Taylor, Z. (1953) Continuous recording of blood oxygen tension by polarography. *J. Appl. Physiol.*, **6**, 189.
84. Severinghaus, J.W. and Astrup, P.B. (1987) History of blood gas analysis. *Int. Anesthesiol. Clin.*, **25**, 1–224.
85. Wukitsch, M.W., Petterson, M.T., Tobler, D.R., and Pologe, J.A. (1988) Pulse oximetry; analysis of theory technology and practice. *J. Clin. Monit.*, **4**, 290–301.
86. Aoyagi, T., Kishi, M., and Yamaguchi, K. (1974) Improvement of the earpiece oximeter. Poster presented at the 13th Annual Meeting of the Japan Society of Medical Electronics and Biomedical Engineering, pp. 90–91.
87. Nakajima, S., Hirai, Y., Takase, H., and Kuse, A. (1975) New pulse-type earpiece oximeter. *Resp. Circ.*, **23**, 709–713.
88. Zander, R. (1991) in *The Oxygen Status of Arterial Human Blood* (eds R. Zander and F. Mertzlufft), Karger, Basel, pp. 1–13.
89. Kelleher, J.F. (1989) Pulse oximetry. *J. Clin. Monit.*, **5**, 37–62.
90. Shah, N., Barker, S.J., Hyatt, J., Shah, E.S.M., Clack, S., and Trivedi, N. (1995) Impact of type and timing of motions on the performance of new pulse oximeters in human volunteers at normoxia and hypoxia. *Anesthesiology*, **83**, A452.
91. Barker, S.J. and Tremper, K.K. (1987) The effect of carbon monoxide inhalationon on pulse oximetry and transcutaneous pO2. *Anesthesiology*, **66**, 677–679.
92. Shah, N., Barker, S.J., Hyatt, J., Shah, E.S.M., Clack, S., and Trivedi, N. (1995) Comparison of alarm conditions between new pulse oximeters during motion at normal & lower oxygen saturations. *Anesthesiology*, **83**, A1067.
93. Elfadel, I.M., Weber, W., and Barker, S.J. (1995) Motion-resistant pulse oximetry. *J. Clin. Monit.*, **11** (4), 262.
94. Moyle, J.T.B. (1994) in *Pulse Oximetry. Principles and Practice Series* (eds C.E.W. Hahn and A.P. Adams), BMJ Publishing Group, London.
95. Clark, L.C. (1956) Monitor and control of blood and tissue oxygen tension. *T. Am. Soc. Art. Int. Org.*, **2**, 41–48.
96. Baumberger, J.P. and Goodfriend, R.B. (1951) Determination of arterial oxygen tension in man by equilibration through intact skin. *Fed. Proc. Fed. Am. Soc. Exp. Biol.*, **10**, 10–11.
97. Webster, M.W. and Steed, D.L. (1993) in *Vascular Medicine* (ed. H. Boccalon), Elsevier Science Publishers B.V., pp. 501–505.
98. Parker, D. (1987) Sensors for monitoring blood gases in intensive care. *J. Phys. E: Sci. Instrum.*, **20**, 1103–1112.
99. Agache, P.G., Agache, A., and Lucas, A. (1993) in *Vascular Medicine* (ed. H. Boccalon), Excerpta Medica.
100. Jobsis, F. (1977) Noninvasive, infrared monitoring of cerebral and myocardial oxygen sufficiency and circulatory parameters. *Science*, **198**, 1264–1267.
101. Delpy, D.T., Cope, M., Van der Zee, P., Arridge, S., Wray, S., and Wyatt, J. (1988) Estimation of optical pathlength through tissue from direct time of flight measurement. *Phys. Med. Biol.*, **33**, 1433–1442.

102. Delpy, D.T. (1994) Optical spectroscopy for diagnosis. *Phys. World*, **7**, 34–39.
103. Shambot, S.B., Holavanahali, R., Lakowicz, J.R., Carter, G.M., and Rao, G. (1994) Phase fluorometric sterilisable optical oxygen sensor. *Biotechnol. Bioeng.*, **43**, 1139–1145.
104. Gewehr, P.M. and Delpy, D.T. (1993) Optical sensor based on phosphorescence lifetime quenching and employing a polymer immobilised metalloporphyrin probe. *Med. Biol. Eng. Comput.*, **31**, 2–10.
105. Shonat, R.D., Wilson, D.F., Riva, C.E., and Pawlowski, M. (1992) Oxygen distribution in the retinal and choroidal vessels of the cat as measured by a new phosphorescence imaging method. *Appl. Opt.*, **31**, 3711–3718.
106. Monstrey, S., Hoeksema, H., Verbelen, J., Pirayesh, A., and Blondeel, P. (2008) Assessment of burn depth and burn wound healing potential. *Burns*, **34**, 761–769.

19
Blood Flow Cytometry and Cell Aggregation Study with Laser Speckle

Qingming Luo, Jianjun Qiu, and Pengcheng Li

19.1
Introduction

Laser speckle is produced from random interference of the laser light scattered from an optically rough surface or a diffuse medium such as biological tissue [1]. Different motion speeds of the scattering particles lead to different levels of intensity fluctuations of the speckle image, so analyzing the speckle pattern may yield the speed information of the object. Laser speckle contrast imaging (LSCI) obtains a two-dimensional blood flow map by calculating the speckle contrast values of the speckle images recorded from biological tissue [2]. It has been demonstrated to be a useful tool of *in vivo* blood flow cytometry with the advantages of high spatial and temporal resolution [3–10]. An integration of speckle technique with other cytometry methods enables us to assess physiologic and pathologic processes *in vivo* in detail [11, 12]. Besides LSCI, speckle phenomenon has also been utilized to investigate the evolution of cell parameters, such as aggregation of red blood cells (RBCs) [13, 14]. In this chapter, we introduce some fundamentals and the instrumentation of LSCI, discuss some important imaging parameters for its optimum imaging conditions, and review some of its recent advancements, including the spatio-temporal scheme for laser speckle contrast analysis and fast blood flow visualization using graphics processing unit (GPU). We also discuss the application of laser speckle phenomenon in investigating aggregation of RBCs.

19.2
Laser Speckle Contrast Imaging

19.2.1
Fundamentals

LSCI is a noninvasive optical technique used for obtaining two-dimensional maps of blood flow without scanning. Laser speckle is generated from random interference of the laser light scattered from an optically rough surface or a diffuse medium [1].

Advanced Optical Flow Cytometry: Methods and Disease Diagnoses, First Edition. Edited by Valery V. Tuchin.

It is a granularly structured pattern with bright and dark areas due to different path lengths [15]. Such granularly structured pattern is modulated by the scattering particles. Stationary scattering particles cause a static speckle pattern that does not change over time, and moving scattering particles lead to a dynamic speckle pattern that changes in time. Using a charged coupled device (CCD) camera with a limited integration time to record such pattern, a spatially blurred time-integrated speckle image is obtained.

To quantify such spatial blurring, the concept of speckle contrast K is introduced, which is defined as the ratio of the standard deviation σ_I of the intensity I to the mean intensity $\langle I \rangle$ of the speckle pattern [1]:

$$K = \frac{\sigma_I}{\langle I \rangle} \tag{19.1}$$

For slow motion of the scattering particles, K is large and vice versa. Goodman deduced that K could be related to the correlation time τ_c of the scattering particles by [1]

$$K = \frac{2}{T} \int_0^T \left(1 - \frac{t}{T}\right) \left|g_1(t)\right|^2 \mathrm{d}t \tag{19.2}$$

where T is the exposure duration, and the electric field autocorrelation function $g_1(t)$ has correlation with τ_c. Assuming that the speed of the scattering particles follows Lorentzian or Gaussian distribution, $g_1(t)$ can be expressed using Equation 19.3 or 19.4:

$$g_1(t) = \exp\left(-\frac{t}{\tau_c}\right) \tag{19.3}$$

$$g_1(t) = \exp\left(-\left(\frac{t}{\tau_c}\right)^2\right) \tag{19.4}$$

Therefore, the relation between K and τ_c can be given by Equations 19.5 and 19.6, respectively [16]:

$$K = \sqrt{\frac{\exp(-2x) - 1 + 2x}{2x^2}} \tag{19.5}$$

$$K = \sqrt{\frac{\exp\left(-2x^2\right) - 1 + \sqrt{2\pi}x\operatorname{erf}\left(\sqrt{2}x\right)}{2x^2}} \tag{19.6}$$

where $x = T/\tau_c$, and τ_c is usually assumed to be inverse proportional to speed [17]. In practical applications, it has been discovered that the difference between the relative flow changes measured by Equations 19.5 and 19.6 is small [18]. Though LSCI can hardly provide absolute speed information till now, in many situations, most attention is paid to the relative speed rather than the absolute speed.

Basically, there are mainly two methods for calculating K, which are laser speckle spatial contrast analysis (LSSCA) [2] and laser speckle temporal contrast analysis

(LSTCA) [19, 20]. For LSSCA, a spatial window of 7×7 pixels is usually used to compute the speckle contrast image, which is defined as [2]

$$K_s = \frac{\sqrt{\frac{1}{N_S^2 - 1} \sum_{i=1}^{N_S^2} \left(I_i - \frac{1}{N_S^2} \sum_{j=1}^{N_S^2} I_j \right)^2}}{\frac{1}{N_S^2} \sum_{j=1}^{N_S^2} I_j} \tag{19.7}$$

where N_s is the length of the square matrix and I_i is the CCD counts of an arbitrary pixel in the square matrix. Because LSSCA is based on spatial statistics, it preserves high temporal resolution with compromised spatial resolution. For LSTCA, images acquired along a few time points are involved for computing speckle contrast image, which is defined as [20]

$$K_t = \frac{\sqrt{\frac{1}{N-1} \sum_{n=1}^{N} \left(I_{x,y}(n) - \frac{1}{N} \sum_{j=1}^{N} I_{x,y}(j) \right)^2}}{\frac{1}{N} \sum_{j=1}^{N} I_{x,y}(j)} \tag{19.8}$$

where $I_{x,y}(n)$ is the CCD counts at pixel (x, y) in the nth laser speckle image and N is the number of images acquired. Because LSTCA is based on temporal statistics, it preserves high spatial resolution with compromised temporal resolution. Depending on practical applications, one should select the appropriate method for computing speckle contrast images.

19.2.2
Instrumentation

Figure 19.1 shows a schematic setup of *in vivo* laser speckle blood flow imaging of rat cortex. An adult male Wistar rat weighing about 250 g is anesthetized and fixed in a stereotaxic frame. For better viewing of the vessels, a craniotomy is made over the left parietal cortex to form an approximately 6.0 mm $\times$ 8.0 mm cranial window. A laser beam is expanded and collimated to illuminate the cranial window with a spot size of approximately 1 cm diameter. Laser speckle images generated from random interference of the laser light scattered from the surface of the rat cortex are recorded by a 12 bit digital CCD camera attached to a microscope. The system magnification and diaphragm are carefully adjusted to make the minimum speckle size equaling approximately 2 camera pixels to guarantee optimum imaging results [21]. A variable attenuator is used in the light path to make sure that the light intensity is within the dynamic range of the CCD camera. The whole setup is fixed on a vibration-isolator table. To reduce statistical noise, a ground glass is also introduced in the light path to rotate slowly in some investigations to ensure statistical independence from adjacent speckle frames [22]. The setup illustrated in Figure 19.1 is also appropriate for other applications of LSCI such as in investigating mesenterial [23, 24] and skin blood flow [6].

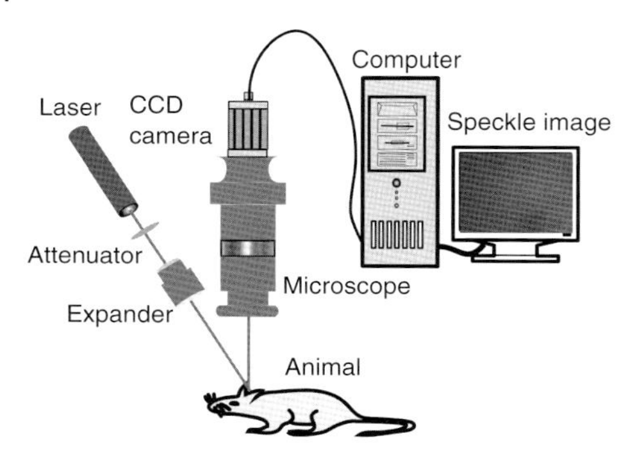

Figure 19.1 Schematic setup of laser speckle contrast imaging.

Data analysis of the recorded laser speckle images can be performed either off-line or on-line, depending on practical needs. Offline data processing of laser speckle images is conventionally adopted, in which speckle contrast images are computed either in spatial domain or temporal domain using a custom-made software based on central processing unit (CPU). Online data processing of speckle images introduces the GPU into speckle contrast image computation and achieves real-time visualization of blood flow maps [25]. Detailed description of fast blood flow visualization using GPU is given in Section 19.5.

19.3 Investigation of Optimum Imaging Conditions with Numerical Simulation

It is of great importance for obtaining optimum results for LSCI by carefully controlling imaging parameters. Numerical simulation of speckle patterns is a useful tool for investigating optimum imaging conditions of LSCI. By simulating static and dynamic speckle fields, the relation between the minimum speckle size and speckle contrast, and the relation between specificities of CCD camera and speckle contrast are investigated quantitatively. It is worth noting that the simulated speckle images are objective (nonimaged) rather than subjective (imaged) ones. However, objective speckle images are easier to simulate, and in most cases, objective speckle images have statistical properties similar to subjective speckle images, as long as the number of scattering particles is large enough [1].

19.3.1 Static Speckle Field Simulation

When the scatterers are stationary, the speckle pattern in the receiving screen will not change with time. Such a speckle image is called *static speckle image*. Assuming that the distance between the scattering surface and the receiving screen is long enough, the field in an arbitrary point (x, y) of the receiving screen can be expressed as [1]

$$\mathbf{U}(x, y) = \sum_{j=1}^{n} a_j \exp(i\phi_j) \tag{19.9}$$

where n is the number of the scattering particles and ϕ_j is the phase of the jth scatterer with a uniform probability density on the interval $(-\pi, \pi)$, which is easy to be simulated using MATLAB. Equation 19.9 can be operated using fast Fourier transformation (FFT). The intensity value in point (x, y) is

$$I(x, y) = |\mathbf{U}(x, y)|^2 \tag{19.10}$$

Assuming that we have simulated a phase matrix R with the size of 256×256 pixels following uniform distribution on the interval of $(-\pi, \pi)$, then the minimum speckle size r_s of the speckle image to be simulated is considered. Following Duncan's method [26], static speckle images with various r_s value can be considered as follows: If we want to simulate a speckle image with r_s equal to or larger than 2 pixels, we can first generate a matrix A with a size of $(256 \times r_s/2) \times (256 \times r_s/2)$ pixels, in which the values of all the elements are set to be 0. Then, we set $A(1:256, 1:256) = R$. By performing FFT on A, we obtain a matrix U with the size of $(256 \times r_s/2) \times (256 \times r_s/2)$. Then we get a matrix $U_2 = U(1:256, 1:256)$. U_2 is then utilized to compute the intensity matrix I using Equation 19.10. Subsequently, an objective speckle image with a size of 256×256 pixels and a minimum speckle size of r_s is obtained. If we want to simulate a speckle image with r_s less than 2 pixels, we can first simulate a speckle image with r_s equaling 2 pixels. Then, the image is low pass filtered with a spatial window with a size of $r_s/2 \times r_s/2$ pixels to produce the final speckle image. Figure 19.2a shows a simulated speckle image with r_s equaling 4 pixels. Figure 19.2b shows the corresponding histogram of intensity values. Theoretically, the intensity values of a fully developed speckle image are negative-exponential distributed [1]. Our simulation result agrees with the expectance very well.

The r_s value of the simulated speckle image in Figure 19.2 can be estimated by calculating the normalized autocovariance function of the intensity [27]. To do this, we reshape the two-dimensional speckle intensity matrix into a one-dimensional

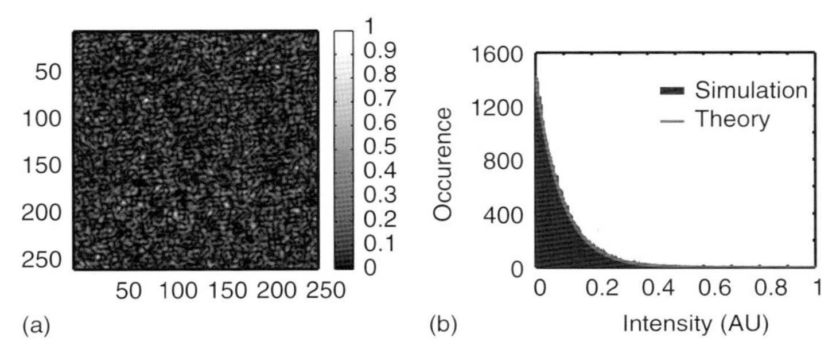

Figure 19.2 (a) Numerical simulated static speckle image and (b) its intensity histogram.

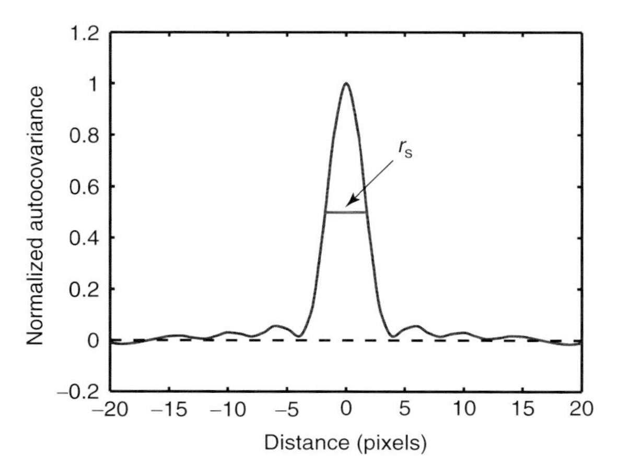

Figure 19.3 Estimation of the minimum speckle size r_s through normalized autocovariance of the intensity.

array and calculate its normalized autocovariance function. Subsequently, r_s is estimated as the full wide half magnitude (FWHM) of the normalized autocovariance function curve [27]. Figure 19.3 shows that the estimated r_s is approximately 3.7 pixels, which is very close to the theoretical 4 pixels.

19.3.2
Dynamic Speckle Field Simulation

On the basis of Duncan's copula method [28], we simulate a sequence of statistically independent time-integrated dynamic speckle images. The copula method for simulating a dynamic speckle sequence is as follows: a sequence of statistically independent Z values following Gaussian distribution are generated by [28]

$$Z(k) = \sqrt{-2\ln X_1}\cos\left(2\pi X_2 + \frac{\pi}{2} \times \frac{k-1}{N-1}\right) \quad (19.11)$$

where N is the length of the sequence and $k = 1, 2, \ldots, N$, X_1, and X_2 are two statistically independent random variables following uniform distribution on the unit interval. By performing percentile transformation for Z values, N frames of uniformly distributed T values on the unit interval that are statistically correlated are obtained. Subsequently, N frames of correlated fully developed speckle images are obtained by performing FFT of the random phasor, which is as follows [28]:

$$\phi(k) = \exp(2\pi mT(k)) \quad (19.12)$$

where m is a multiplicative factor that influences the decorrelation time τ_c of the simulated speckle images. τ_c is large when m is small and vice versa.

On the basis of the method described above, a time-integrated dynamic speckle image can be obtained by averaging N_0 frames of the generated fully developed speckle images, where N_0 is smaller than or equal to N. The consideration of r_s is

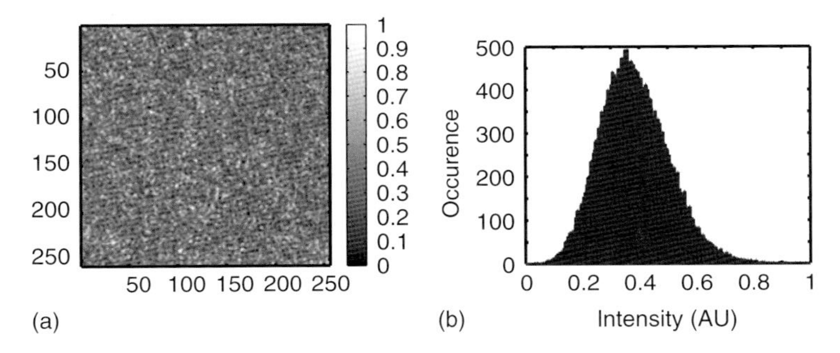

Figure 19.4 (a) Numerical simulated time-integrated dynamic speckle image and (b) its histogram.

the same as the static speckle image simulation. Figure 19.4a shows a simulated time-integrated dynamic speckle image, in which the size of X_1 and X_2 is set to be 256×256 pixels, m is 40, N is 50, and N_0 is 30. Similar to Figure 19.2a, r_s is also set to be 4. As is shown in Figure 19.4b, the intensity distribution is nearly Gaussian shaped rather than a negative exponential one owing to spatial blurring of the speckle intensity.

19.3.3
Speckle Size and Speckle Contrast

There exists a longstanding misconception that the minimum speckle size r_s should be equal to the size of 1 pixel to optimize the results of LSCI [29]. Kirkpatrick *et al.* investigated the relation between speckle contrast and r_s through simulation and phantom experiment and pointed out that the minimum speckle size should be at least 2 pixels to maximize speckle contrast [21]. In their study, only the condition of static speckle image was considered for both simulation and experiment. Here, the condition of time-integrated dynamic speckle images with different τ_c values was also taken into consideration. Using the method described previously in this section, we simulated three types of speckle images: static speckle images, time-integrated dynamic speckle images with shorter τ_c (faster speed), and time-integrated dynamic speckle images with longer τ_c (slower speed). For each type, five frames of statistically independent speckle images with a size of 256×256 pixels are simulated. For simulating dynamic speckle images, N is 50, and N_0 is 30, m is set to be 2 for images with longer τ_c and 40 for images with shorter τ_c. Figure 19.5 shows the relation between global speckle contrast and r_s for the three types of simulated speckle images. Global speckle contrast is defined as the ratio of the standard deviation of intensity in the whole speckle image to the mean intensity in the whole speckle image [26]. As is shown, for each type of speckle image, there is approximately 30% reduction of speckle contrast to the maximum one when r_s is equal to 1 pixel. The results in Figure 19.5 demonstrate again that r_s should be at least 2 pixels to maximize speckle contrast, not only for static but also for dynamic speckle images.

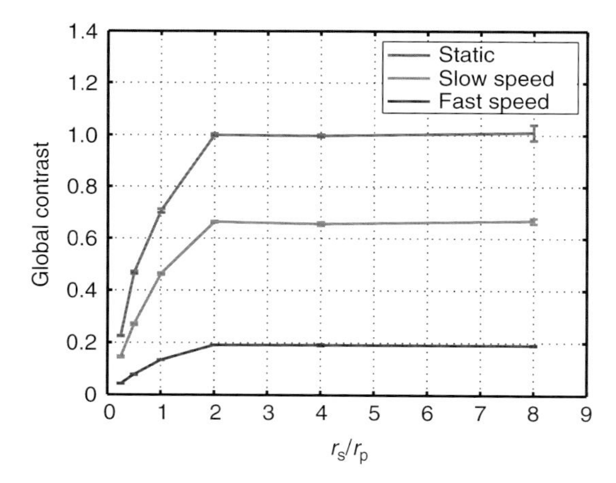

Figure 19.5 Global speckle contrast of the simulated speckle images as a function of r_s/r_p, where r_s is the minimum speckle size and r_p is the pixel size.

19.3.4 Specificities of CCD Camera and Speckle Contrast

Specificities of CCD camera, such as dynamic range, dark counts, and readout noise, may have impact on speckle contrast [16, 29]. Through numerical simulation of static speckle images, we quantitatively investigate the impact of these specificities on speckle contrast.

According to the parameters of the PixelFly VGA CCD camera, the dynamic range of the CCD camera is 12 bit; the full well capacity is 35 000 e^- and the A/D conversion factor is 7.5 e^-. Because $7.5 \times 2^{12} = 30\ 720$ e^-, which is smaller than 35 000 e^-, so there exists a baseline value of 4280 e^- for A/D conversion. First, we assume that there is no other noise during A/D conversion. In such a case, the impact of digitization on speckle contrast is shown in Figure 19.6, in which three frames of statistically independent speckle images are investigated. As is shown, when the mean intensity is high, some of the pixels in the images are in saturation state; the statistical error will increase with increasing mean intensity. When the mean intensity is weak, the statistical error increases with decreasing mean intensity. However, the statistical error is not significant for weak intensities. For example, the statistical error is within 1% when the mean intensity is no less than 50 counts.

In practice, there exist noises other than digitization noise that may influence the value of speckle contrast, such as the dark counts and the readout noise. Again, taking the PixelFly VGA CCD camera, for example, the average dark charge in 1 pixel is only 3 e^-/s, which means that the dark counts are not more than 1 count. Therefore, the impact of dark counts on speckle contrast is extremely small and can be ignored. The readout noise is not shown in the operating instructions,

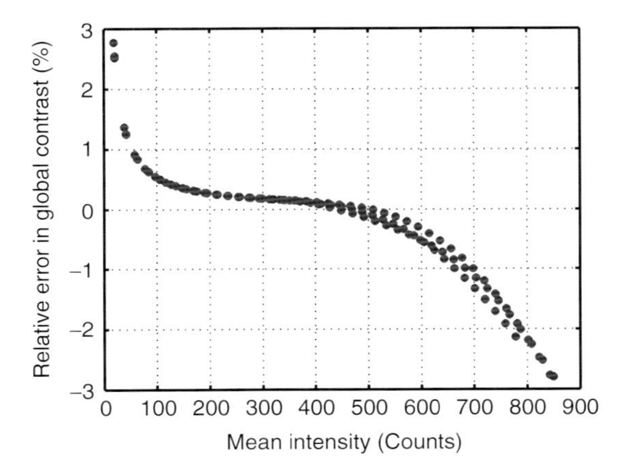

Figure 19.6 Impact of digitization on speckle contrast.

but through practical measurement, it is 34 ± 3 counts and follows Gaussian distribution [30]. As shown in Figure 19.7, the impact of readout noise on speckle contrast is significant. When the mean intensity is weak, the statistical accuracy can be more than 40%. When the mean intensity is high, the statistical accuracy is also very high. In addition, the maximum value of speckle contrast is less than 1 when the mean intensity is about 700 counts. There exists about 5% statistical error in speckle contrast compared with the theoretical value. By subtracting a DC (discocyte) noise of 34 counts, such statistical error is significantly reduced. As long as the pixels are not saturated, such statistical error of speckle contrast can be controlled within 3%. Actually, we have discovered previously that subtracting a DC count is also valid for dynamic speckle images for obtaining optimum speckle contrast values [30].

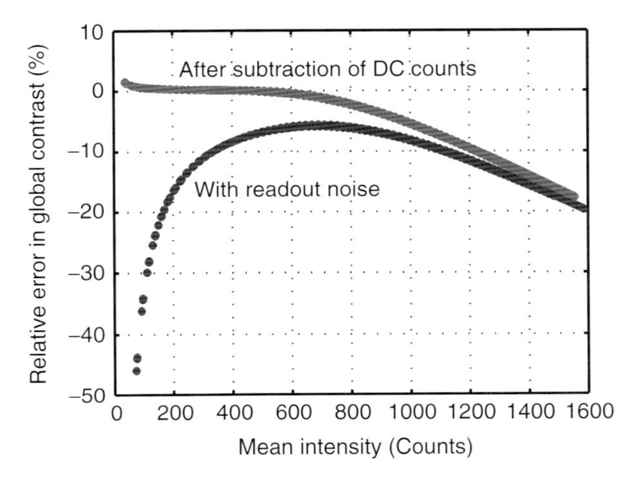

Figure 19.7 Impact of readout noise on speckle contrast.

19.4
Spatio-Temporal Laser Speckle Contrast Analysis

As has been mentioned in Section 19.2, speckle contrast analysis can be performed in either spatial or temporal domain, each of which has its advantages and limitations. LSSCA achieves high temporal resolution with the loss of spatial resolution, impeding its application in monitoring blood flow changes in small vessels. LSTCA preserves the original spatial resolution by sacrificing the temporal resolution, making it inappropriate in applications where video frame rate visualization of blood flow is required. To balance spatial and temporal resolutions, several methods that attempted to combine spatial and temporal statistics have been presented [3, 31, 32]. Figure 19.8 shows a schematic diagram of laser speckle contrast analysis methods based on spatial statistics, temporal statistics, and a combination of both. In this section, we review the methods that are based on a combination of spatial and temporal statistics and compare the statistical accuracy of these methods through numerical simulation and phantom experiment. Related work has been published in the *Journal of Biomedical Optics* [33].

19.4.1
Spatial Based Method

As has been presented by Völker *et al.*, the statistical noise is $N_p^{-0.5}$, where N_p is the total number of pixels involved for statistics [22]. For LSSCA, the statistical noise is typically around $(N_s \times N_s)^{-0.5}$, where N_s is the length of the selected spatial window. If a 7×7 spatial window is utilized, then the statistical noise is about $49^{-0.5} \approx 14.3\%$, which is very high. The spatial based laser speckle contrast analysis (sLASCA) method is an improvement over LSSCA [3]. It averages the derived spatial speckle contrast images along a few of time points. If $K_{s1}, K_{s2}, \ldots, K_{sN}$ denote the consecutive contrast values in frames $1, 2, \ldots, N$, then the contrast value K_{sLASCA}

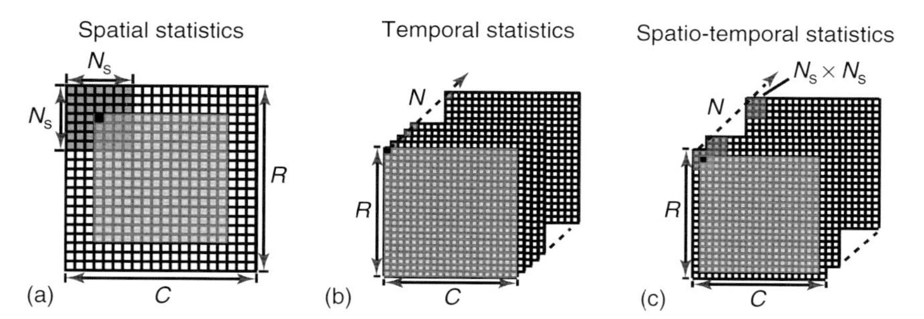

Figure 19.8 Schematic diagrams of speckle contrast analysis methods based on (a) spatial statistics, (b) temporal statistics, and (c) a combination of both. The $R \times C$ arrays represent the original speckle images. The elements with dark grey represent the pixels involved in calculation of the local speckle contrast of the pixel with black color. The squares with light grey represent the derived speckle contrast images.

calculated by sLASCA can be defined as follows:

$$K_{\text{sLASCA}} = \frac{K_{\text{s}1} + K_{\text{s}2} + \cdots + K_{\text{s}N}}{N} \tag{19.13}$$

Evidently, by performing temporal average, sLASCA yields lower statistical noise than LSSCA does. For example, if the number of frames N is 10, then the statistical noise can be reduced to about 4.5%. Therefore, sLASCA is often used in practical applications instead of LSSCA to reduce noise.

19.4.2
Temporal Based Method

The temporal based laser speckle contrast analysis (tLASCA) method averages the derived temporal speckle contrast image using a small spatial window such as 3×3 pixels. tLASCA is defined as [31]

$$K_{\text{tLASCA}(x,y)} = \frac{1}{N_{\text{s}}^2} \sum_{r=x-(N_{\text{S}}-1)/2}^{x+(N_{\text{S}}-1)/2} \sum_{c=y-(N_{\text{S}}-1)/2}^{y+(N_{\text{S}}-1)/2} \frac{\sigma_{r,c,N}}{\langle I_{r,c,N} \rangle} \tag{19.14}$$

where N_{s} is the length of the selected spatial window, $\sigma_{r,c,N}$ is the standard deviation, and $\langle I_{r,c,N} \rangle$ is the mean intensity of all pixels at (r, c) in N frames along the temporal dimension. By performing spatial average, tLASCA yields speckle contrast with reduced statistical noise than LSTCA does. For instance, 25 or 49 frames of speckle images used for computing a temporal speckle contrast image with LSTCA lead to about 20 and 14.3% statistical noise, respectively. Using tLASCA, the statistical noise can be reduced to one-third of the value compared to LSTCA. In other words, tLASCA computed speckle contrast image achieves acceptable signal to noise ratio with less number of frames than that by LSTCA.

19.4.3
Spatio-Temporal Based Method

Recently a spatio-temporal laser speckle contrast analysis (stLASCA) method has been presented, in which speckle contrast is calculated within a $N_{\text{s}} \times N_{\text{s}} \times N$ pixels stack, where N is the number of frames and N_{s} is the length of the spatial window. stLASCA is defined as [32]

$$K_t = \frac{\sqrt{\frac{1}{N\times N_{\text{S}}^2-1} \sum_{i=1}^{N\times N_{\text{S}}^2} \left(I_i - \frac{1}{N\times N_{\text{S}}^2} \sum_{j=1}^{N\times N_{\text{S}}^2} I_j \right)^2}}{\frac{1}{N\times N_{\text{S}}^2} \sum_{j=1}^{N\times N_{\text{S}}^2} I_j} \tag{19.15}$$

where I_i is the CCD counts of the pixels within the $N_{\text{s}} \times N_{\text{s}} \times N$ pixels stack. In contrast to sLASCA and tLASCA, stLASCA performs spatial and temporal statistics simultaneously. There also exist differences in the correlation of speckle contrast values to the size of spatial window and number of frames computed by sLASCA, tLASCA, and stLASCA. For sLASCA, the mean speckle contrast is dependent only on the size of the selected spatial window. For tLASCA, the mean speckle contrast is dependent only on the number of frames. But for stLASCA, the mean speckle contrast is correlated with both the size of the selected spatial window and the number of frames. Therefore, if the same size of spatial window and the same number of frames are used for sLASCA, tLASCA, and stLASCA methods, then stLASCA will achieve the maximum mean speckle contrast value.

19.4.4 Theoretical and Experimental Comparisons

The statistical accuracy of a method for computing speckle contrast is very important owing to the close correlation of speckle contrast value with speed of motion. In this subsection, we investigate the correlations of mean speckle contrast values and relative noises of the speckle contrast images computed by sLASCA, tLASCA, and stLASCA to the length of the spatial window (N_s) and the number of frames (N) theoretically and experimentally.

First, 20 frames of statistically independent time-integrated dynamic speckles with the size of 256×256 pixels and the minimum speckle size r_s of 4 pixels are simulated using the method mentioned in Section 19.3. In the simulation procedure, the length of the Gaussian distribution sequence Z in Equation 19.11 is set to be 50 and m is set to be 40. For simulating each single dynamic speckle image, 50 frames of statistically dependent but fully developed speckle images are obtained using Equation 19.12 and then 30 frames of these images are averaged to obtain a time-integrated dynamic speckle image. With certain N_s and N values, each of the three methods, sLASCA, tLASCA, and stLASCA, obtain a speckle contrast image by processing these simulated speckle images. For each speckle contrast image, the mean speckle contrast μ_K and the contrast of contrast σ_K/μ_K are calculated. Figure 19.9 shows (a–c) the μ_K values and (d–f) the σ_K/μ_K values computed by sLASCA, tLASCA, and stLASCA methods for the simulated speckle images as functions of N and N_s. As is shown, the μ_K value computed by sLASCA increases with increasing spatial window but keeps unchanged with increasing number of frames. The μ_K value computed by tLASCA increases with increasing number of frames but remains unchanged with increasing spatial window. The μ_K value computed by stLASCA increases with either increasing number of frames or increasing spatial window. When $N \geq 5$ and $N_s \geq 5$, the μ_K value computed by stLASCA remains almost constant, and the maximum relative error is within 1.5%. But for sLASCA and tLASCA, such relative error is about 7 and 7.5%, respectively. As is shown in Figure 19.9d–f, the σ_K/μ_K values computed by sLASCA, tLASCA, and stLASCA decrease with either increasing N or N_s values. If N and N_s are large enough, there exist little difference in the σ_K/μ_K values computed by these methods, and the σ_K/μ_K values are approximately $(N_s \times N_s \times N)^{-0.5}$. However, when N and N_s are small, the difference in the σ_K/μ_K values computed by these methods is obvious. The maximum σ_K/μ_K values for sLASCA, tLASCA, and stLASCA are approximately 16.7, 20.7, and 17.3%, respectively, when $N = 3$ and

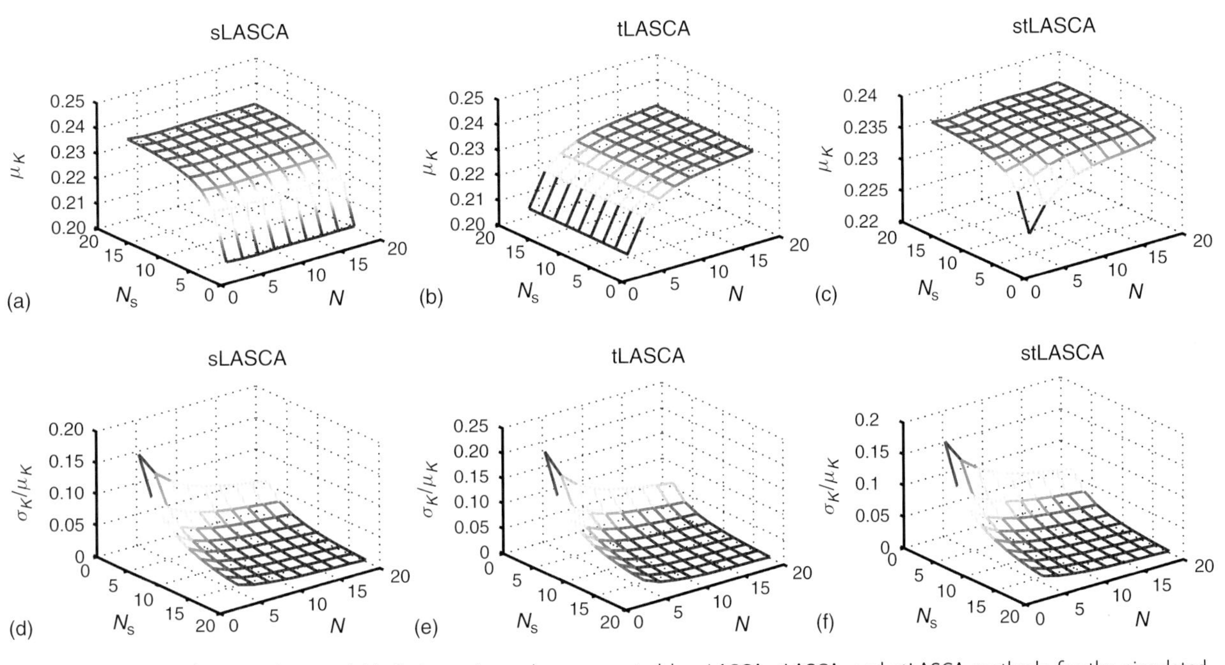

Figure 19.9 (a–c) The μ_K values and (d–f) the σ_K/μ_K values computed by sLASCA, tLASCA, and stLASCA methods for the simulated time-integrated dynamic speckle sequence as a function of the number of frames N and the length of spatial window N_s.

$N_s = 3$. The σ_K/μ_K value computed by tLASCA is larger than those computed by sLASCA and stLASCA because of significant statistical error in μ_K when N is small.

We also compare the results obtained by sLASCA, tLASCA, and stLASCA through phantom experiment. We use a white porcelain plate with rough surface as the object. To obtain a sequence of dynamic speckle images, the porcelain plate is motivated by a stepping motor to move horizontally with a speed of 1.5 mm s^{-1}. Twenty frames of speckle images recorded from the surface of the porcelain plate are obtained for data processing. The r_s values of these speckle images are adjusted to be around 4 pixels, which is comparable to the simulated speckle images. Figure 19.10 shows the μ_K values (a–c) and the σ_K/μ_K values (d–f) computed by sLASCA, tLASCA, and stLASCA methods for the simulated speckle images as functions of N and N_s. Obviously, there exist similar changes in μ_K and σ_K/μ_K values with increasing N and N_s values as the simulation results. Combining the simulational and experimental results, we conclude that stLASCA most effectively utilizes the statistical number of pixels and achieves the highest statistical accuracy in μ_K than the other methods.

19.5 Fast Blood Flow Visualization Using GPU

19.5.1 CPU-Based Solutions for LSCI Data Processing

As has been mentioned in Section 19.2, data processing of the recorded laser speckle images can be performed either offline or online. The conventional data processing of the speckle images using the CPU of an ordinary personal computer is time consuming because of the heavy computation required for speckle contrast analysis. In practice, video frame rate visualization of blood flow can hardly be achieved for the high-resolution image data with such conventional computing solution, and data processing of the image data is usually performed offline. Typically, for the commercial LSCI system, moorFLPI released by Moor Instruments Ltd in the United Kingdom, the online video frame rate visualization of blood flow can only be achieved for 113×152 pixels and 8 bit data, which does not match the spatial resolution required for some potential applications of LSCI, such as microsurgery. Several attempts have been made to reduce the processing time of speckle contrast analysis. Cheng *et al.* found that the relation between speckle contrast and the decorrelation time τ_c can be reduced simply as $\tau_c = 1/(TK^2)$, where T is the exposure duration and K is the speckle contrast [34]. Though some study has pointed out a serious misconception in Cheng's paper [35], the value of τ_c is practically small for most applications of LSCI and such simple relation is found to be consistent with the result computed by Newton iteration quite well [34]. This method avoids the time-consuming Newton iteration for computing τ_c and saves a lot of computational time. Other attempts have been made to reduce computational time of speckle contrast. Tom *et al.* investigated the expansion

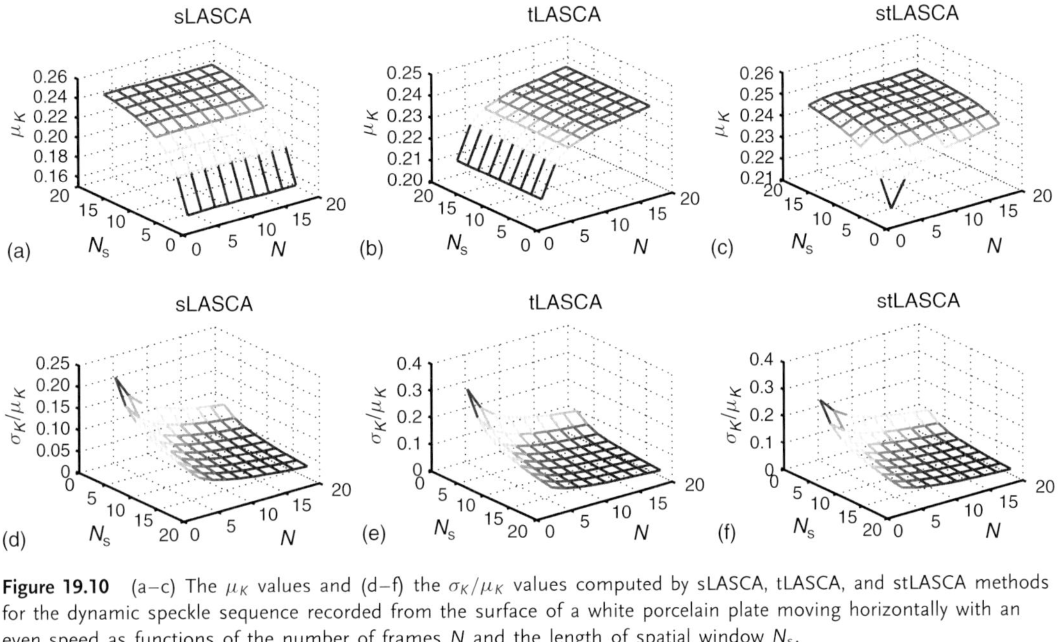

Figure 19.10 (a–c) The μ_K values and (d–f) the σ_K/μ_K values computed by sLASCA, tLASCA, and stLASCA methods for the dynamic speckle sequence recorded from the surface of a white porcelain plate moving horizontally with an even speed as functions of the number of frames N and the length of spatial window N_s.

formula for calculating speckle contrast and presented a roll algorithm to reduce processing time [36]. With the roll method, the results of prior steps can be applied into the subsequent steps for computing the contrast pixels. This method avoids many redundant computations and about 291 raw speckle images with the resolution of 640×480 pixels and dynamic range of 8 bit can be processed per second [36]. Although the roll method is very efficient, it is still difficult to achieve real-time processing of high-resolution LSCI data.

19.5.2
GPU-Based Solution for LSCI Data Processing

GPU for general computation has been widely applied for processing data in the field of biomedical optics because of its powerful parallel processing ability. At present, the most popular and simple language for operating GPU is the Compute Unified Device Architecture (CUDA) of Nvidia company. A great number of threads can be launched simultaneously on the GPU by programming with the CUDA. And these threads are organized in the form of the CUDA bank. Therefore, it is quite appropriate for parallel processing of LSCI data since each thread can calculate 1 pixel of the blood flow map. Liu *et al.* [25] introduced GPU into the data processing framework of LSCI to achieve fast and high-resolution blood flow visualization on PCs by exploiting the high floating-point processing power of commodity graphics hardware and achieved a 12- to 60-fold performance enhancement in comparison to the optimized CPU implementations [25]. This solution enhances the performance greatly and makes it possible to be used in clinical applications.

The implementation on the host is responsible for capturing raw speckle image from CCD and then delivers the data into the video memory. After that, the kernel is started to run on the GPU by the host implementation. The raw speckle image is divided into many small tiles and each CUDA bank owns a part of the tiles with some other extra pixels that are used for calculating the border pixels of the respective tile. Each thread of the CUDA bank is responsible for calculating 1 pixel of the contrast image from a sliding window independently and stores the result into video memory. After all the threads complete the calculation, the kernel quits. The host reads the results out of the video memory and displays the resultant image in the screen or stores in the disk.

Figure 19.11a [25] illustrates a comparison of GPU-based implementation (Geforce 8600GT) and the CPU-based ones. The CPU-based implementation includes SSE (streaming SIMD extensions) optimized version and non-SSE optimized version. The SSE is an SIMD technology that allows for "single instruction multiple data" and can make the processing much quicker. It can be seen that GPU-based implementation runs much faster than the two CPU-based ones. Perhaps, 12- to 17-folds of performance enhancement can be obtained. The time for processing a single frame is much less than the video frame rate threshold time. It means that a fast blood flow visualization of LSCI is possible. Figure 19.11b [25] indicates that more performance enhancement can be obtained using the shared memory of the GPU. By introducing GPU to the architecture of LSCI data

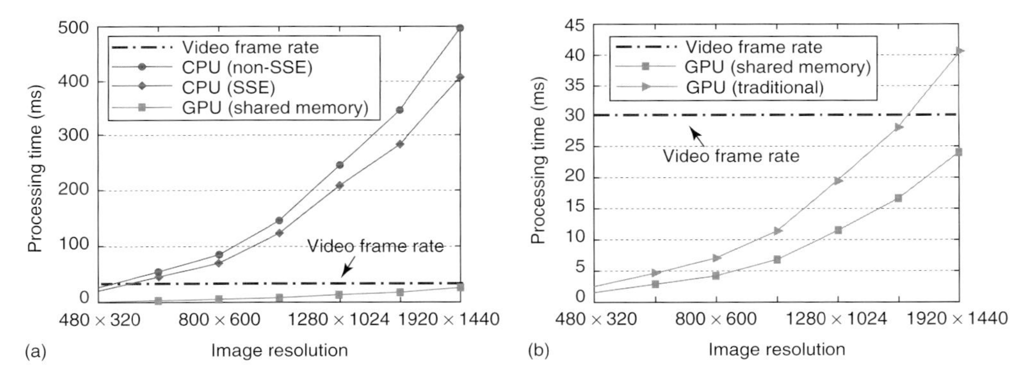

Figure 19.11 (a) Comparison among two CPU codes with different compiler options and the GPU-based version. (b) Performance difference between the shared memory version GPU-based data processing and non-shared-memory (traditional) one. Video frame rate threshold is shown in both figures. Reproduced with permission from Liu *et al.* [25].

processing, a fast blood flow visualization of LSCI can be made, which is quite significant in medical usage. Evidently, further work can be done to develop an optimized data processing scheme based on GPU by introducing some efficient data processing schemes, such as the roll scheme, into computation.

19.6 Detecting Aggregation of Red Blood Cells or Platelets Using Laser Speckle

Laser speckle phenomenon can also be used in investigating evolution of cell parameters, besides being applied in *in vivo* blood flow cytometry. A typical application of laser speckle is on the investigation of aggregation process of RBCs or platelets.

The RBCs are an important component of blood. They are biconcave disc shaped and have a tendency to form aggregates [14, 37], even in healthy organisms. These aggregates have a significant impact on blood viscosity and blood flow *in vivo* [38, 39]. Laser speckle can be used as a label-free method to monitor the aggregation process *in vivo* or *in vitro*. Aggregation of RBCs changes the absorption and the reduced scattering coefficients of the scattered light and leads to speckle patterns that vary in time when a beam of laser light is used for illumination. The aggregation process of RBCs has been investigated by Pop *et al.* [14] with a light-scattering technique at small angles. In their study, the RBCs were separated from venous blood samples of healthy adult donors and then resuspended in autologous plasma with hematocrit values in the range of $3.5 \times 10^{-2} \leq \text{H} \leq 10^{-1}$ and used for light-scattering measurements. A 632.8 nm linearly polarized HeNe laser with 0.81 mm diameter at $1/e^2$, 1 mrad beam divergence, and power of 5 mW was directed onto a 200 μm thick optical glass cuvette (20 μl volume) containing the

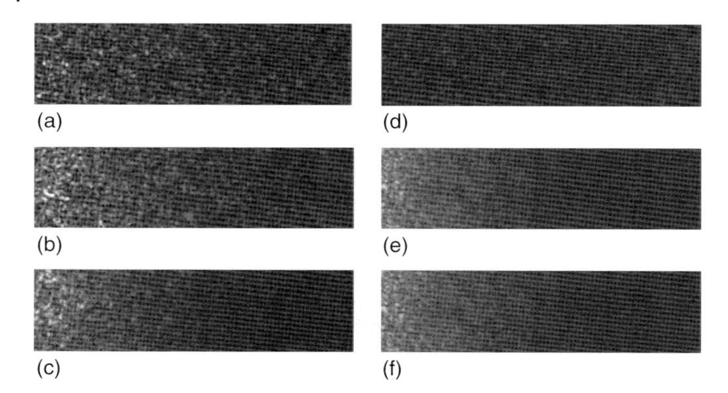

Figure 19.12 Speckle images during aggregation process of RBCs in suspension at $H = 3.5 \times 10^{-2}$ for (a) 0 min, (b) 4 min, and (c) 8 min, and at $H = 10^{-1}$ for same moments of time in (d–f). H is the hematocrit (volumic fraction of RBCs). Reproduced with permission from Pop *et al.* [14].

sample of RBCs in suspension at rest. To monitor the RBC aggregation process, a CMOS (complementary metal oxide semiconductor) camera with a resolution of 1280×1024 pixels at 27 fps was used to acquire images at each 10 s during 10 min for each sample. Figure 19.12 [14] shows speckle images recorded from the RBC suspensions during the RBC aggregation process for 0 min (a and d), 4 min (b and e), and 8 min (c and f). Figure 19.12a–c corresponds to H (hematocrit) of 3.5×10^{-2} and Figure 19.12d–f corresponds to H of 10^{-1}. As is shown, for each H value of RBCs in suspension, the speckle intensity becomes higher and higher during the RBC aggregation process owing to changes in absorption and reduced scattering coefficient values. Besides, the minimum speckle size seems to become larger after the aggregation, which tends to be consistent with the result discovered by Piederrière *et al.* [13].

Platelets are usually smaller than RBCs and have a major role in hemostasis. The aggregation process of platelets was investigated by Piederrières *et al.* They studied the evolution of the minimum speckle size during aggregation of platelets [13]. In their study, the platelet suspensions in plasma were prepared from whole blood by removing the larger red and white cells. The ADP (adenosine diphosphate) was added to the samples to induce aggregation. The imaging setup is similar to the one used by Pop *et al.* [14] but with a slight difference. In Piederrières' study, the detector was a CCD camera with resolution of 788×268 pixels and was positioned at a small angle with the optical axis to avoid the directly transmitted laser light.

Figure 19.13 [13] shows the minimum speckle size dy as a function of the time t at two ADP concentrations. As is shown, higher ADP concentration corresponds to a faster aggregation and thereby causes a strong increase in the speckle size 1 min after ADP addition and a subsequent slow increase of the speckle size when the aggregation process slows down. Weaker ADP concentration corresponds to smaller aggregation and thereby causes smaller increase in speckle size. So it can

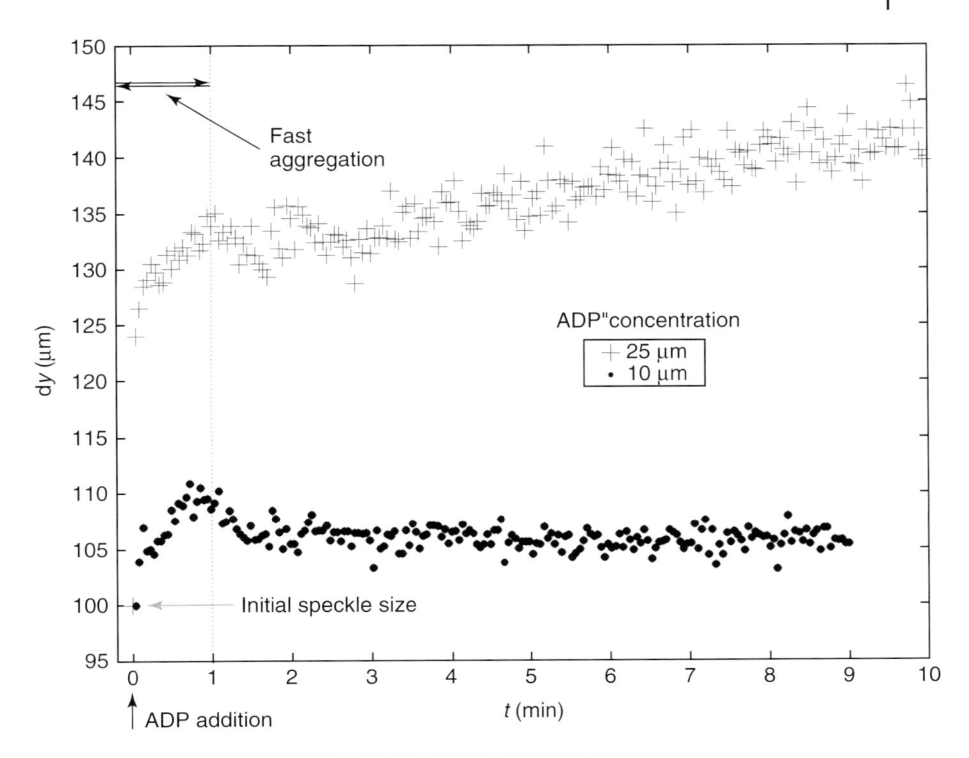

Figure 19.13 The minimum speckle size dy as a function of the time *t* during platelet aggregation process obtained for two ADP concentrations. Reproduced with permission from Piederrière *et al.* [13].

be concluded that the minimum speckle size increases during the aggregation process.

Actually, other values such as speckle contrast, cross correlation function, and degree of polarization of the speckle images can also be measured simultaneously with speckle size estimation to provide more information on the properties of cells during the aggregation process of RBCs or platelets. In conclusion, laser speckle technique is noninvasive and cost-effective, which provides an additional choice for investigating cell aggregation, fragmentation of aggregates, or hemolysis induced by drugs [13] or other stimulus [38].

19.7 Conclusion

Laser speckle is sensitive to the speed and morphological changes of the scattering particles, and is thus very suitable for *in vivo* blood flow cytometry and investigation of evolutions of cell parameters. This chapter has discussed the optimum imaging conditions for LSCI and reviewed some of its recent advancements. The application of laser speckle on cell aggregation study has also been reviewed.

Acknowledgments

This work is supported by the National High Technology Research and Development Program of China (Grant No.2007AA02Z303), the PhD Programs Foundation of the Ministry of Education of China (Grant No.20070487058), the NSFC–RFBR International Joint Research Project (Grant No.30711120171), and the Program for Changjiang Scholars and Innovative Research Team in University.

References

1. Goodman, J.W. (2006) *Speckle Phenomena in Optics: Theory and Applications*, Roberts and Company, Berlin.
2. Briers, J.D. and Webster, S. (1996) Laser speckle contrast analysis (LASCA): a nonscanning, full-field technique for monitoring capillary blood flow. *J. Biomed. Opt.*, **1**, 174–179.
3. Dunn, A.K., Bolay, H., Moskowitz, M.A., and Boas, D.A. (2001) Dynamic imaging of cerebral blood flow using laser speckle. *J. Cereb. Blood Flow Metab.*, **21** (3), 195–201.
4. Durduran, T., Burnett, M.G., Yu, G., Zhou, C., Furuya, D., Yodh, A.G., Detre, J.A., and Greenberg, J.H. (2004) Spatiotemporal quantification of cerebral blood flow during functional activation in rat somatosensory cortex using laser-speckle flowmetry. *J. Cereb. Blood Flow Metab.*, **24**, 518–525.
5. Ayata, C., Dunn, A.K., Gursoy-Ozdemir, Y., Huang, Z., Boas, D.A., and Moskowitz, M.A. (2004) Laser speckle flowmetry for the study of cerebrovascular physiology in normal and ischemic mouse cortex. *J. Cereb. Blood Flow Metab.*, **24** (7), 744–755.
6. Choi, B., Kang, N.M., and Nelson, J.S. (2004) Laser speckle imaging for monitoring blood flow dynamics in the in vivo rodent dorsal skin fold model. *Microvasc. Res.*, **68** (2), 143–146.
7. Strong, A.J., Anderson, P.J., Watts, H.R., Virley, D.J., Lloyd, A., Irving, E.A., Nagafuji, T., Ninomiya, M., Nakamura, H., and Dunn, A.K. (2007) Peri-infarct depolarizations lead to loss of perfusion in ischaemic gyrencephalic cerebral cortex. *J. Cereb. Blood Flow Metab.*, **130** (4), 995.
8. Luo, W., Wang, Z., Li, P., Zeng, S., and Luo, Q. (2008) A modified mini-stroke model with region-directed reperfusion in rat cortex. *J. Cereb. Blood Flow Metab.*, **28** (5), 973–983.
9. Cheng, H., Yan, Y., and Duong, T.Q. (2008) Temporal statistical analysis of laser speckle images and its application to retinal blood-flow imaging. *Opt. Express*, **16** (14), 10214–10219.
10. Wang, Z., Luo, W., Li, P., Qiu, J., and Luo, Q. (2008) Acute hyperglycemia compromises cerebral blood flow following cortical spreading depression in rats monitored by laser speckle imaging. *J. Biomed. Opt.*, **13**, 064023.
11. Galanzha, E.I., Tuchin, V.V., and Zharov, V.P. (2007) Advances in small animal mesentery models for in vivo flow cytometry, dynamic microscopy, and drug screening. *World. J. Gastroenterol.*, **13** (2), 192.
12. Galanzha, E.I., Tuchin, V.V., and Zharov, V.P. (2005) In vivo integrated flow image cytometry and lymph/blood vessels dynamic microscopy. *J. Biomed. Opt.*, **10** (5), 054018.
13. Piederrière, Y., Le Meur, J., Cariou, J., Abgrall, J., and Blouch, M. (2004) Particle aggregation monitoring by speckle size measurement; application to blood platelets aggregation. *Opt. Express*, **12** (19), 4596–4601.

14. Pop, C.V.L. and Neamtu, S. (2008) Aggregation of red blood cells in suspension: study by light-scattering technique at small angles. *J. Biomed. Opt.*, **13** (4), 041308.
15. Rigden, J.D. and Gordon, E.I. (1962) The granularity of scattered optical maser light. *Proc. IRE*, **50** (11), 2367–2368.
16. Bandyopadhyay, R., Gittings, A.S., Suh, S.S., Dixon, P.K., and Durian, D.J. (2005) Speckle-visibility spectroscopy: a tool to study time-varying dynamics. *Rev. Sci. Instrum.*, **76**, 093110.
17. Bonner, R. and Nossal, R. (1981) Model for laser Doppler measurements of blood flow in tissue. *Appl. Opt.*, **20** (12), 2097.
18. Wang, Z., Hughes, S., Dayasundara, S., and Menon, R.S. (2006) Theoretical and experimental optimization of laser speckle contrast imaging for high specificity to brain microcirculation. *J. Cereb. Blood Flow Metab.*, **27** (2), 258–269.
19. Cheng, H., Luo, Q., Zeng, S., Chen, S., Cen, J., and Gong, H. (2003) Modified laser speckle imaging method with improved spatial resolution. *J. Biomed. Opt.*, **8** (3), 559–564.
20. Li, P., Ni, S., Zhang, L., Zeng, S., and Luo, Q. (2006) Imaging cerebral blood flow through the intact rat skull with temporal laser speckle imaging. *Opt. Lett.*, **31** (12), 1824–1826.
21. Kirkpatrick, S.J., Duncan, D.D., and Wells-Gray, E.M. (2008) Detrimental effects of speckle-pixel size matching in laser speckle contrast imaging. *Opt. Lett.*, **33** (24), 2886–2888.
22. Völker, A., Zakharov, P., Weber, B., Buck, F., and Scheffold, F. (2005) Laser speckle imaging with an active noise reduction scheme. *Opt. Express*, **13** (24), 9782–9787.
23. Cheng, H., Luo, Q., Liu, Q., Lu, Q., Gong, H., and Zeng, S. (2004) Laser speckle imaging of blood flow in microcirculation. *Phys. Med. Biol.*, **49** (7), 1347–1357.
24. Ulyanov, S.S. (2001) High-resolution speckle-microscopy: study of the spatial structure of a bioflow. *Physiol. Meas.*, **22** (4), 681–692.
25. Liu, S., Li, P., and luo, Q. (2008) Fast blood flow visualization of high-resolution laser speckle imaging data using graphics processing unit. *Opt. Express*, **16**, 14321–14329.
26. Duncan, D.D., Kirkpatrick, S.J., and Wang, R.K. (2008) Statistics of local speckle contrast. *J. Opt. Soc. Am. A*, **25** (1), 9–15.
27. Piederrière, Y., Cariou, J., Guern, Y., Le Jeune, B., Le Brun, G., and Lortrian, J. (2004) Scattering through fluids: speckle size measurement and Monte Carlo simulations close to and into the multiple scattering. *Opt. Express*, **12** (1), 176–188.
28. Duncan, D.D. and Kirkpatrick, S.J. (2008) The copula: a tool for simulating speckle dynamics. *J. Opt. Soc. Am. A*, **25**, 231–237.
29. Briers, J.D., Richards, G.J., and He, X.W. (1999) Capillary blood flow monitoring using laser speckle contrast analysis (LASCA). *J. Biomed. Opt.*, **4** (01), 164–175.
30. Qiu, J., Li, P., Ul'yanov, S.S., Zeng, S., and Luo, Q. (2008) Enlarging the linear response range of velocity with optimum imaging parameters modified data processing in laser speckle imaging. *Proc. SPIE*, **6863**, 68630T.
31. Le, T.M., Paul, J.S., Al-Nashash, H., Tan, A., Luft, A.R., Sheu, F.S., and Ong, S.H. (2007) New insights into image processing of cortical blood flow monitors using laser speckle imaging. *IEEE Trans. Med. Imaging*, **26** (6), 833–842.
32. Duncan, D.D. and Kirkpatrick, S.J. (2008) Spatio-temporal algorithms for processing laser speckle imaging data. *Proc. SPIE*, **6858**, 685802.
33. Jianjun Qiu, P.L., Luo, W., Wang, J., Zhang, H., and Luo, Q. (2010) Spatio-temporal laser speckle contrast analysis for blood flow imaging with maximized speckle contrast. *J. Biomed. Opt.*, **15** (1), 016003.
34. Cheng, H. and Duong, T.Q. (2007) Simplified laser-speckle-imaging analysis method and its application to retinal blood flow imaging. *Opt. Lett.*, **32** (15), 2188–2190.
35. Duncan, D.D. and Kirkpatrick, S.J. (2008) Can laser speckle flowmetry

be made a quantitative tool? *J. Opt. Soc. Am. A*, **25** (8), 2088–2094.

36. Tom, W.J., Ponticorvo, A., and Dunn, A.K. (2008) Efficient processing of laser speckle contrast images. *IEEE Trans. Med. Imaging*, **27** (12), 1728–1738.
37. Enejder, A.M.K., Swartling, J., Aruna, P., and Andersson-Engels, S. (2003) Influence of cell shape and aggregate formation on the optical properties of flowing whole blood. *J. Biomed. Opt.*, **42** (7), 1384–1394.
38. Bishop, J.J., Nance, P.R., Popel, A.S., Intaglietta, M., and Johnson, P.C. (2001) Effect of erythrocyte aggregation on velocity profiles in venules. *Am. J. Physiol.*, **280** (1), 222–236.
39. Kalchenko, V., Brill, A., Bayewitch, M., Fine, I., Zharov, V., Galanzha, E., Tuchin, V., and Harmelin, A. (2007) In vivo dynamic light scattering imaging of blood coagulation. *J. Biomed. Opt.*, **12** (5), 052002.

20
Modifications of Optical Properties of Blood during Photodynamic Reactions *In vitro* and *In vivo*

Alexandre Douplik, Alexander Stratonnikov, Olga Zhernovaya, and Viktor Loshchenov

20.1
Introduction

Blood tissue is particularly important for photodynamic therapy (PDT) for several reasons. First, blood is the main oxygen supplier of biotissues and oxygen is one of the key components of PDT. Second, blood is a main carrier of photosensitizers (PSs) administrated systemically – orally or intravenously (IV). Third, blood is one of the most significant absorbers of visible light because of hemoglobin (Hb) and water (H_2O). This review is focused on photodynamic modalities exploiting the currently used PSs and red light radiation. In this chapter, we analyze and estimate the influence of blood and its properties upon delivery of light. Modifications of the optical properties of PSs caused by blood as a solvent are also considered. We review PS uptake by blood cells, and photodynamic reactions directly involving blood and blood components. Modifications of the optical properties of blood resulting from *in vivo* and *in vitro* PDT reactions are also rigorously analyzed. We examined a blood sample *in vitro* as a test model to assess and benchmark the photodynamic activity of various PS.

20.2
Description and Brief History of PDT

PDT is a nonsurgical treatment modality based on photochemical reactions. PDT is extremely precise and controllable targeting malignant lesions. PDT is time effective as a procedure (e.g., a large malignancy can be exposed to laser light at once), and it does not develop a resistance, demonstrating a very low mutagenic potential. The limitations of PDT include a delay in the onset of treatment effect for hours or days after the session and shallow depth of treatment ($\sim$1 cm). Nevertheless, the latter can be interpreted also as an advantage avoiding complications resulting from damage due to overexpansion. PDT is particularly important for certain superficial clinical applications such as destroying a large lesion of cancer *in situ* in hollow organs such as esophagus or lungs. In the biophotonics market

Advanced Optical Flow Cytometry: Methods and Disease Diagnoses, First Edition. Edited by Valery V. Tuchin.

sector, PDT demonstrates the highest growth, almost 40% annually (Dennis L. Matthews, private communication). Pharmaceutical companies, such as Axcan, Pharmaceuticals Inc, Nigma, Glaxo-Welcome Inc, DUSA, Photocure, Galdderma, and others have become important players in the PDT field.

The history of modern clinical PDT begins from very topical applications, particularly ultraviolet (UV) irradiation of skin where endogenous amino acids were exploited as PSs. In 1903, Niels Finsen was awarded the Nobel Prize for his work exploiting UV light from a carbon arc lamp for the treatment of skin tuberculosis [1]. In 1974, a combination of Psoralen® (external dye furocoumarin) and UV light began to be used for the treatment of psoriasis (PUVA therapy) in 1974 [2]. Psoralen absorbs light at 300–380 nm, which penetrates into tissue only to a depth of 100 μm. The demand for the treatment of malignant *in situ* or invasive tumors with relatively large volume led to the usage of hematoporphyrin derivatives (HPDs) [3] exploiting red light (∼630 nm), which penetrates into biotissue much deeper than the UV-blue-yellow part of the spectrum. The next step was the usage of phthalocyanines with strong absorption at 670 nm [4], thus providing deeper light penetration into biological tissue than at 630 nm. The recent trend of searching for new PSs is prescribed by their ability to absorb in near-IR range toward so-called "transparency window" for biotissues (650–950 nm) [5]. Higher specificity of PS uptake, intracellular and intratissue localization, and quantum yield of photodestruction and photoinactivation are the next highest priorities for further clinical applications.

20.3
PDT Mechanisms

The interaction between laser radiation and material depends on various factors such as power density, wavelength, interaction time, and material properties (e.g., absorption coefficient). The physical processes involved in the interaction between laser beams and material are divided into three steps: (i) absorption of some of the light energy; (ii) transformation of this energy into chemical energy and/or into heat; and (iii) eventually, chemical reaction and/or phase transformation. The PDT domain is within the circle on the graph representing the photobiological effects depending on the fluence rate and the interaction time duration proposed by Boulnois in 1986 [6].

As it can be seen from Figure 20.1, PDT does not require large doses of light energy and usually coagulation or hyperthermal effects are insignificant. It has been recently suggested that PDT is a photochemical reaction involving light, photosensitive molecules absorbing light or PS, and ambient molecular oxygen (O_2) to generate reactive oxygen species (ROS), which in turn destroy biotissue [7]. These include Type I (sensitizer-substrate) and Type II (sensitizer-oxygen) reactions. Type I photochemical reactions result in the formation of superoxide anions by transfer of an electron from the PS to the molecular oxygen. Superoxide anions can react to produce hydrogen peroxide (H_2O_2), which can easily pass

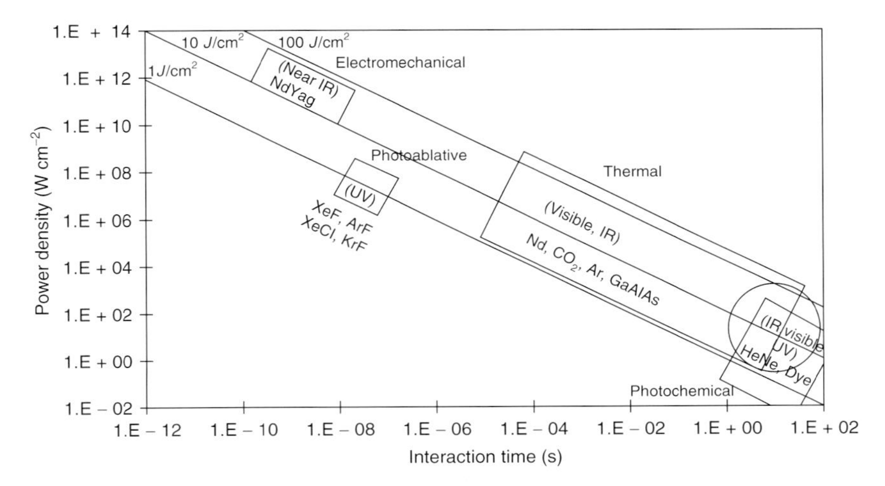

Figure 20.1 Laser tissue interaction effects depending on power density and interaction time. The circler shows the PDT domain. (Modified from Ref. [6].)

through biological membranes and produce cellular damage. Type II photochemical reactions represent the transfer of energy to molecular oxygen. During type II photochemical reaction, singlet oxygen ($^1O_2$) is formed. In this chapter, we consider only the second type of oxygen-dependent photochemical reaction, which comes with release of singlet oxygen ($^1O_2$), as according to the contemporary PDT concept, the main type of the clinical photochemical mechanism [8]. In terms of light, we consider only the red and infrared range as mostly exploited in contemporary PDT. We do not consider the photoreaction developing without PSs, as such a contribution to cell and tissue destruction and inactivation is considered rather insignificant [8].

Jablonsky diagram of PDT reaction is presented in Figure 20.2. PDT via singlet oxygen mechanism may be termed a *photo catalytic oxidation* with the PS as the catalyst. The main advantages of PDT over chemotherapy are that PS works as a catalyst in the reaction of photocatalytic oxidation. It means that one PS molecule, being not toxic itself, should be capable of generating a lot of reactive species. Indeed, during PDT, the PS itself is not consumed. In most cases, light absorption by organic compounds relies on the presence of a π-system (chromophore) to promote an electron from the singlet electronic ground state (S^0, electron spins paired) to the excited singlet state (S^{1*}) of the chromophore ($\pi \rightarrow \pi^*$ or $n \rightarrow \pi^*$) [9]. The lifetime of 4 μs is $^1O_2$ in pure water [10] and less than 0.05 μs in cells, which provides its ability to travel about 125 and 10–20 nm respectively from the site of generation [11]. This must be the reason why PDT has a very low mutagenic potential: most PSs are localized outside the nuclei [12]. Intranuclear DNA is generally not a primary target for $^1O_2$ as lipophilic PSs tend to accumulate at the nuclear membrane rather than inside the nucleus [13]. It is

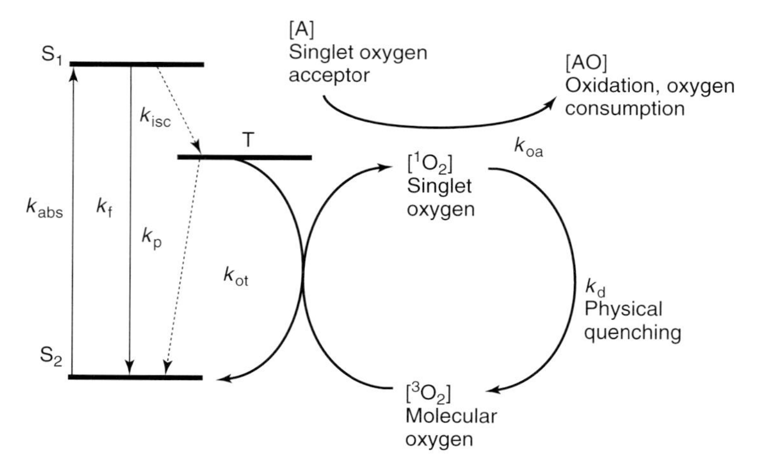

Figure 20.2 Jablonsky diagram of main processes during PDT via singlet oxygen mechanism. (All the constants can be found in Glossary.)

still not clear whether the main portion of reactions with singlet oxygen occurs with solvents (water) or biomolecules [14]. A typical lifetime of ^{3}PS is on the order of 200 μs [15]. Oxygen can exist in two singlet excited states. The longer lived form, $^1\Delta_g$, with an excess energy of 23 kcal mol^{-1} (corresponding to a photon energy of about 1 eV $=$ 1270 nm) is the principal species involved in photodynamic reactions [16].

The PDT initiates lipid peroxidation, a chain degenerative process that affects cell and intracellular membranes and the lipid-containing structures under conditions of oxidative stress [17]. In general, this phenomenon is initiated by the attack of any chemical species that has sufficient reactivity to abstract a hydrogen atom from a methylene carbon in the side chain. The hydrogen atom is a free radical and its removal leaves behind an unpaired electron on the carbon to which it was originally attached. That new radical can give a peroxyl radical after the reaction with oxygen. Peroxyl radicals can combine with each other or can attack membrane proteins, but they are also capable of abstracting hydrogen from adjacent fatty acid side chains in a membrane, and thus propagating the chain reaction of lipid peroxidation. Membrane lipids may be a central site of photodamage if sensitizing agents localize in the membrane bilayer [18]. Such a mechanism is confirmed by the observation that the presence of antioxidants, primarily reducing agents prone to scavenge reactive species in one way or another, usually reduces the efficacy of PDT [19]. However, it is also noted that at high concentrations, antioxidants may act as pro-oxidants and ultimately enhance PDT efficacy that may be related to extra turning on the first type of photochemical reaction forming additional ROS [20]. Cell membranes are characterized by their ability to self-repair. The lipid peroxidation shifts the repair–damage balance, leading to the cell death developing necrosis or apoptosis [21]. Another mechanism by which cells might be damaged during the PDT of tumors is via the covalent crosslinking of proteins to proteins or to other molecules in the cell. It has been suggested that photodynamically generated singlet oxygen interacts with photo-oxidizable amino acid residues such as His, Cys, Trp, and Tyr in one protein molecule to generate reactive species, which in turn interact nonphotochemically with residues of these types or with free amino groups in another protein molecule to form a crosslink [22]. The lipid peroxidation and protein cross-linking may ultimately lead to necrosis and apoptosis. Necrosis is characterized by swelling of cytoplasm and mitochondria, loss of membrane integrity, total cell lysis, and release of cellular content. Apoptosis includes nuclear and cytoplasmic condensation, membrane blebbing without loss of integrity, and aggregation of chromatin at the nuclear membrane, and ends by formation of membrane-bound vesicles (apoptotic bodies). The pathophysiology of necrosis and that of apoptosis are different. Apoptosis is part of the normal cell recycle process proceeding without inflammation, while necrosis is a response for abnormal cell damage and controlled by inflammation and immunological mechanisms. The dominant mode of cell death depends on the particular PS used for treatment, the localization of the PS, and the treatment protocol [23]. Luo *et al.* [24] demonstrated that the mode of cell death depends on the photodynamic dose: high doses result in a necrotic cell death, while lower doses result in an apoptotic mode of death. Another study of these authors demonstrated that PSs located in mitochondria result in apoptotic cell death, in contrast to PSs localized in lysosomes or cell membrane. Dellinger [25] demonstrated that PS incubation time before light treatment determines the cell death mechanism. Long incubation time (24 h) leads to apoptosis, while 1 h of incubation induces necrosis of cells. The PDT-induced apoptosis is triggered mainly as a result of mitochondria damage [26]. Microtubule damages cause a strong effect on cell division and proliferation [27]. Proteins also undergo significant transformation of secondary and tertiary structures under PDT affecting their architectural and metabolic functionality, ultimately resulting in respiratory failure. Disruption of protein synthesis/routing and lysosomes leads harmful acidic hydrolases being released into the cytosol [28]. The diagram of cell fate pathways after PDT depending on the severity of damage [29] is presented in Figure 20.3.

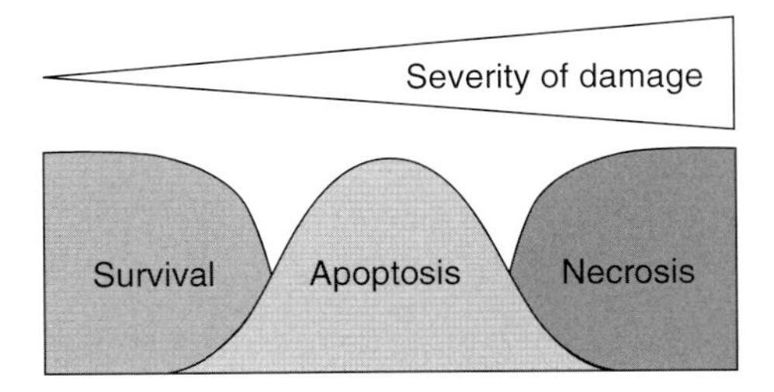

Figure 20.3 Diagram of cell fate pathways after PDT depending on the severance of damage. (Modified from Ref. [29].)

Kessel *et al.* proposed a concept that lipophilic and liposome or lipoprotein-bound dyes sensitize the tumor cells *in vivo* directly, whereas water-soluble dyes mainly sensitized the vascular system in the tumor to shut this down [30]. The immunomodulation effects of PDT are also significant for tumor destruction and prevention of tumor recurrence [31] and for anticancer vaccination [32]. The main tissue effects are determined by the drug-light time interval and PS pharmacokinetics.

20.4 Blood and PDT

Blood tissue is particularly important for PDT as already mentioned in Section 20.1. Blood plays an important role in the development of the after-PDT reactions such as necrosis and apoptosis. Blood plasma proteins and white blood cells (WBCs, leucocytes) mediate facilitate and control these processes. Blood delivers immunocompetent cells and antibodies mediating immune response for PDT treatment and its consequences [33, 34]. PDT has recently begun to be applied as photodynamic antimicrobial chemotherapy (PACT) to viral and infectious inactivation for blood clearing [35]. Limitations of PDT are also closely linked with blood. Blood plays a key role in supplying oxygen to biotissues, which may undergo the photodynamic reaction. In normal physiology, the oxygen content of any biotissue is prescribed by a balance between oxygen consumption due to cell metabolism and oxygen supply by blood circulation (Figure 20.4), that is, erythrocytes filled out by hemoglobin. The photodynamic reaction causes extra oxygen consumption. On top of such extra oxygen usage, PDT may lead to destruction of red blood cells (RBCs) and blood vessels. Once all the molecular oxygen in the vicinity of 100 μm (molecular oxygen diffusion length) is over, the PDT stops.

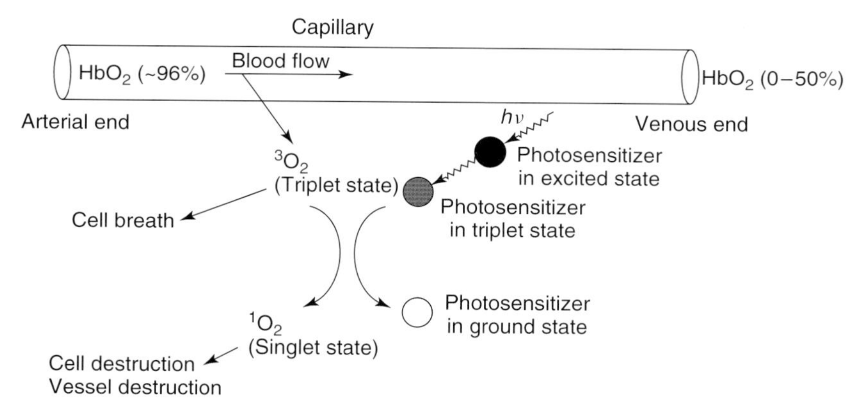

Figure 20.4 The oxygen transport and its consumption in tissues during photodynamic therapy. (Modified from Ref. [36].)

Blood contributes 3–10% of total tissue volume, depending on the type of tissue; however, blood determines 5–50% of the light penetration into biotissues via its absorption and scattering in the visible and near-infrared ranges. The hemoglobin profile can be recognized through observation of any spectra acquired *in vivo* from mucosa (inner epithelium lining hollow organs such as GI and lungs) or interstitially. A certain contribution to tissue scattering comes from blood cells particularly from the cell and cell organelle membranes.

20.5 Properties of Blood, Blood Cells, and Photosensitizers: Before Photodynamic Reaction

20.5.1 Main Physiological Properties of Blood (Hematocrit, Hemoglobin, Oxygenation, Share Rate) Coupled with Its Absorption, Scattering, and Autofluorescence

Blood plasma is a watery extracellular matrix that contains dissolved substances. Blood plasma is about 91.5% water and 8.5% solutes, most of which (7% by weight) are proteins. Some of the proteins in blood plasma are also found elsewhere in the body, but those confined to blood are called *plasma proteins*. The plasma proteins include albumins (54% of plasma proteins), globulins (38%), and fibrinogen (7%). Besides proteins, other solutes in plasma include electrolytes; nutrients; regulatory substances such as enzymes and hormones; gases; and waste product such as urea, uric acid, creatinine, ammonia, and bilirubin, which makes blood plasma yellowish [37]. The optical properties of blood plasma are characterized by strong absorption in UV range because of the proteins and relative transparency in visible range.

Erythrocytes (RBC) are the oxygen carriers for all oxygen-dependent metabolic reactions in an organism. They are the only blood cells without nuclei, which allows them to bind and exchange great numbers of O_2 molecules. Their physiological biconcave disk shape with a thick rim provides optimal plasticity. RBCs live only about 120 days because of the wear and tear their plasma membranes under go as they squeeze through blood capillaries. Without a nucleus and other organelles, RBCs cannot synthesize a new component to replace damaged ones. Each RBC contains about 280 million hemoglobin molecules. A hemoglobin molecule consist of a protein called *globin* composed of four polypeptide chains (two alpha and two beta chains); a ring like nonprotein pigment called a *heme* is bound to each of the four chains. At the center of the hem ring is an iron ion (Fe^{2+}) that can combine reversibly with one oxygen molecule, allowing each hemoglobin molecule to bind four oxygen molecules. Each oxygen molecule picked up from the lungs is bound to an iron ion. As blood flows through tissue capillaries, the iron–oxygen reaction reverses. Hemoglobin releases oxygen, which diffuses first into the interstitial fluid and then in to cells. Hemoglobin also transports about 23% of the total carbon dioxide, a waste product of metabolism. Blood flowing through tissue capillaries picks up carbon dioxide, some of which combines with amino acids in the globin part of hemoglobin. As blood flows through the lungs, the carbon dioxide is released from hemoglobin and then exhaled.

Unlike RBCs, WBCs or leukocytes have nuclei and do not contain hemoglobin. WBC are classified as either granular or agranular, depending on whether they contain conspicuous chemical-filled cytoplasmic granules (vesicles) that are made visible by staining. Granular leukocytes include neutrophils, eosinophils, and basophils; agranular leukocytes include lymphocytes and monocytes. In a healthy body, some WBCs especially lymphocytes, can live for several months or years, but most live only a few days. During a period of infection, phagocytic WBCs may live only a few hours. WBCs are far less numerous than RBCs; at about 5000–10 000 cells/μl of blood, they are outnumbered by RBCs by a ratio of about 700 : 1 [37].

Platelets (thrombocytes) are a type of blood cells that are derived from megakaryocytes. They are anuclear irregularly shaped cells with an average normal diameter of 2–3 μm and a lifespan of 8–12 days. Platelets are involved in hemostasis, leading to the formation of blood clots. In the activated state, a platelet is a spicular spheroid and in a nonactivated state, it is a discoid cell with a diameter of 2–4 μm and a thickness of 0.5–2 μm. Activation of platelets starts when oxygen appears in the blood plasma. The process of platelet activation includes transformation of the cell to aggregate into a thrombus – pseudopodia emerge from a discoid platelet which becomes spicular [38]. Hence, a homogeneous oblate spheroid with 14 rectilinear parallelepipeds with refractive index ranging from 1.36 to 1.44 (with a wavelength of 0.66 μm and refractive index of the medium of 1.333) can be suggested as an optical model of the thrombocyte [39].

Examples of blood cells are shown in Figure 20.5. Blood is denser and more viscous than water. The temperature of blood is 38 °C, and it has a pH ranging from 7.35 to 7.45. Blood constitutes about 20% of extracellular fluid, amounting to 8% of the total body mass. The total volume of blood is 4–6 l in an average-sized adult. Several hormones, regulated by negative feedback, ensure that blood volume

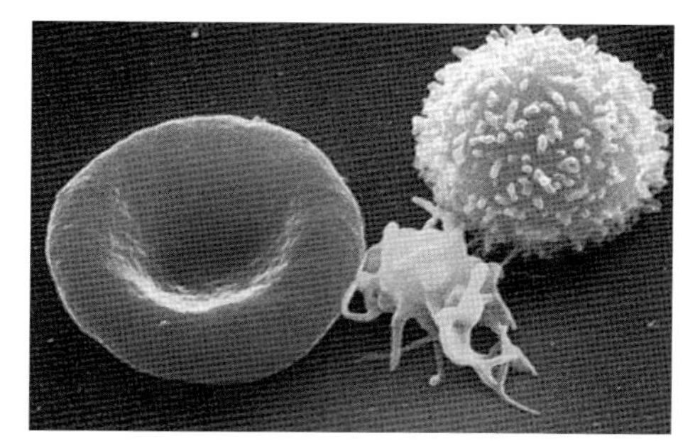

Figure 20.5 A three-dimensional ultrastructural image analysis of a T-lymphocyte (right), an activated platelet (center), and a red blood cell (left), using a Hitachi S-570 scanning electron microscope (SEM) equipped with a GW Backscatter Detector. (Modified from the open access picture from Electron Microscopy Facility at The National Cancer Institute at Frederick (NCI-Frederick).)

and osmotic pressure remain almost constant [40]. The oxygen supply capability of blood is determined by its oxygen capacity BO_2, that is, the maximum amount of hemoglobin-bound oxygen per unit volume of blood (mmol l^{-1}). Oxygen is dissolved in blood water and bound with hemoglobin. The capacity of hemoglobin to carry oxygen is two orders of magnitude higher than that of the water, so the main part of oxygen in blood is bound with hemoglobin. The oxygen capacity of tissue depends mostly on tissue blood volume fraction, hemoglobin concentration, and hemoglobin oxygenation (fraction of oxygenated hemoglobin over total hemoglobin). It also depends on pH, temperature, and so on.

Hemoglobin concentration in turn is proportional to the hematocrit (volume fraction of blood cells in blood) and hemoglobin index (average amount of normal hemoglobin per single erythrocyte). As the normal concentration of RBCs is higher by two orders of magnitude compared to the other types of blood cells all along (leucocytes and thrombocytes), hematocrit is mostly prescribed by the volume fraction of erythrocytes. Every 100 ml of blood contains 14–16 g of hemoglobin. It was proved by C. G. von Hüfner that every 1 g of crystalline hemoglobin can be combined with 1.34 ml of oxygen, but nowadays this value is determined as 1.39. On practice this value is a bit smaller because some amount of hemoglobin can be in an inactive state. The oxygen capacity of arterial blood can be experimentally estimated by the equation [41]:

$$BO_2 = 0.0138 \cdot [\mathrm{Hb}] \cdot SO_2 + 0.0031 \cdot pO_2 \qquad (20.1)$$

where [Hb] is hemoglobin concentration, SO_2 is oxygen saturation of blood, and pO_2 is the partial pressure of oxygen. Under normal physiological conditions, the amount of oxygen in human arterial blood is about 17–20/100 ml.

Hemoglobin can also exist in several forms – fetal and adult hemoglobin, depending on the age of the patient – that may differ both in terms of its the oxygen bounding and spectral absorption properties. Only reduced and oxygenated hemoglobin are considered as normal types of hemoglobin responsible for oxygen supply. The other forms such as sulpha-, met, CO hemoglobin complexes are altered forms not capable of carrying and supplying oxygen. The absorption spectra of hemoglobins are depicted in Figure 20.6 [42]. Usually, normal hemoglobin does not undergo transformation into pathological forms during PDT. Hemoglobin concentration, its oxygenation, hematocrit, and tissue blood volume fraction determine, to a very significant extent, the ability of blood to allow light to pass through it. The difference between bloodless and blood containing tissue in terms of visible light transmission may differ significantly depending on wavelength.

Figure 20.7 shows the penetration depth spectra (at transmission of 1/e, i.e., 37%) simulated from a 2-layer tissue model including epithelium and blood containing derma obtained via the single back scattering approximation [43]. The total volume fraction of blood is 0% (dashed curve), 3% of fully oxygenated ($SO_2 = 100\%$) (solid curve), and completely deoxygenated blood ($SO_2 = 0\%$) (dotted curve).

Depending on the velocity gradient profile inside the blood vessels, the aggregation of blood cells (formation of multi cell complexes starting immediately after

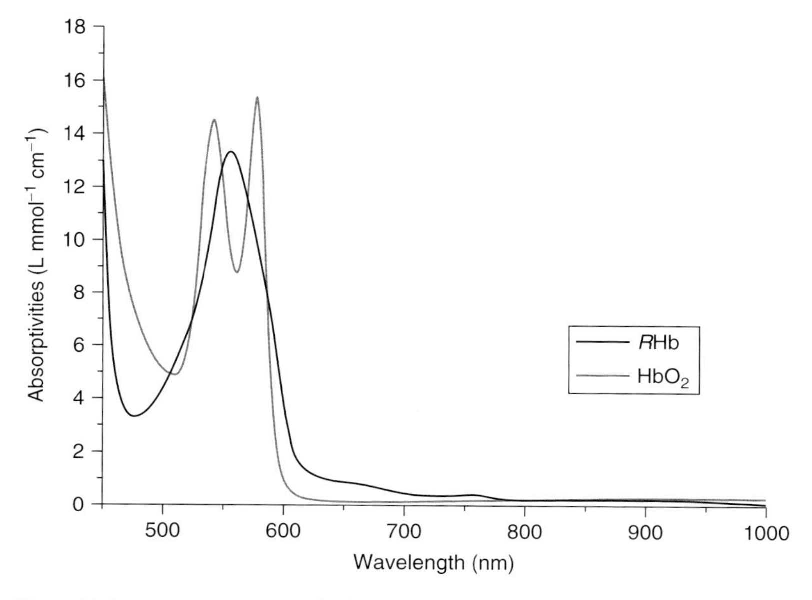

Figure 20.6 Extinction spectra of adult reduced (*R*Hb, dark grey curve) and oxygenated (HbO_2, light grey curve) hemoglobin. (This graph is created based on data from Ref. [42].)

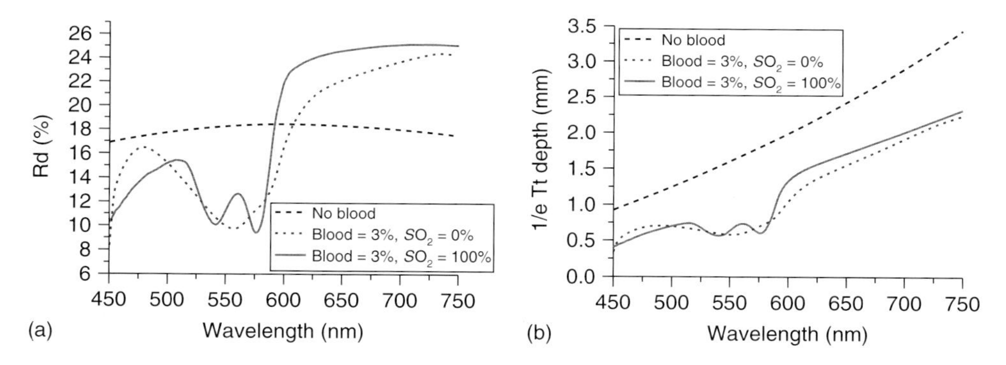

Figure 20.7 Diffuse reflectance (a) and penetration depth (b) spectra at transmission of 1/e (37%). The total volume fraction of blood is 0% (dashed curve), 3% of fully oxygenated ($SO_2 = 100\%$) (solid curve), and completely deoxygenated blood ($SO_2 = 0\%$) (dotted curve).

blood stops moving) may vary depending on the blood flow share rate. The size of blood aggregates determines the ability of blood to scatter light. In general, the higher the share rate, the smaller the aggregates (down to a single-cell scale), the higher the reflectance, and consequently, the lower the transmission. The anisotropy factor drops to 0.96–0.97 and reduced scattering coefficient grows from

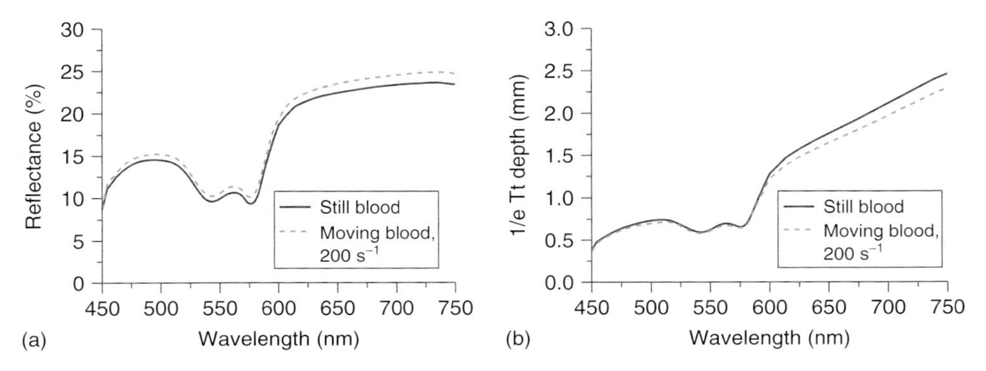

Figure 20.8 Diffuse reflectance (a) and penetration depth (b) simulated spectra using the same model as for Figure 20.7. The total volume fraction of partially oxygenated ($SO_2 = 60\%$) blood is 3%. The solid curve presents still blood which is highly aggregated. The dashed curve presents circulating blood at shear rate 200 s^{-1}.

the highly aggregated blood (still) to low aggregated (circulating at a shear rate of 200 s^{-1}) by 20–25% consequently for visible and near-infrared range [44, 45]. Based on these data, Figure 20.8 shows diffuse reflectance and penetration depth simulated (using the same model as for Figure 20.7) spectra with high (blue) and low (red dashed curves) indices of RBC aggregation. It is also noteworthy that absorption coefficient also depends on the shear rate because of the flow orientation of the erythrocytes, RBC axial migration, and deformation. From the theoretical point of view, it is possible to assume that the depth of absorption coefficient modulation depending on the blood flow shear rate will be determined by the ratio between the geometrical cross sections of the frontal and the lateral surfaces of erythrocytes [44].

The common notion is that oxygenation of hemoglobin is not directly connected with the properties of light scattering. However, there are published data that report that a number of blood parameters related to the scattering depend on oxygenation of hemoglobin, including erythrocyte sedimentation [46], RBC aggregation index (AI), and erythrocyte deformability [47]. Formation of aggregation branched rouleaux networks proceeds faster for deoxygenated blood. In general, these studies report that at higher oxygenation the scattering is also higher.

20.5.2
Blood and Blood Component Autofluorescence

Autofluorescence is an intrinsic parameter of biological tissues and cells arising from endogenous fluorophores under excitation of UV or visible radiation. The autofluorescence of blood takes place predominantly in UV range. Alterations of morphological and physiological state of cells as well as their chemical and physical properties result in modification of autofluorescence spectra of cells and their components. The intrinsic fluorescence of blood proteins is determined by fluorescent amino acids – tryptophans, tyrosines, and phenylalanines [48]. Human blood serum has two characteristic regions of fluorescence: a highly fluorescent UV region (300–370 nm) and a near-UV /visible region (390–580 nm) [49]. The first region is characterized by three to four evident peaks; the last has one dominant single peak, resulting from the strong fluorescence of protein-bound tryptophans, with the maximum at 287 nm (excitation) and 337 nm (emission). Experimental studies revealed that blood has fluorescence peaks in the range 300–700 nm [50]. It was shown that intensity and the position of fluorescence peaks of human blood under action of UV light with various wavelengths depend on the excitation wavelength. There is a similar behavior of fluorescence spectra in the range 420–680 nm at 350 and 400-nm excitation, when only two slight wide peaks appear near 490 and 613 nm. An excitation wavelength of 300 nm induces more evident peaks at 327 and 396 nm and a sharp and intense peak at 485 nm. Different fluorescence spectra for various wavelengths in the range 200–250 nm with the most evident peaks at 314, 332, and 435 nm were observed for excitation wavelengths of 200, 210, and 250 nm, respectively. Fluorescence spectrum of blood plasma from a normal human shows a prominent emission peak at 461 nm under excitation at 405 nm, while at 420 nm excitation fluorescence spectrum of blood plasma from normal subjects has an emission peak at 499 nm [51]. Fluorescence spectra of mouse blood under various irradiation powers of Ar^+ laser excitation have evident sharp spectral peaks in a range 600–860 nm, which are located at 616, 666, 708, 739, 752, 766, 800, 812, and 844 nm [52]. The autofluorescence peak at 666 is much more intense and evident than others, and the intensity of this peak varies almost linearly with the laser irradiation power. The fluorescence properties of human WBCs depend on the cell type [53]. Differences in the intrinsic properties seem to be responsible for the autofluorescence difference observed for several types of cells, such as Lymphocytes, Monocytes, Neutrophils, and Eosinophils. They can be distinguished according to the intensity and spectral shape of the autofluorescence emission in the visible range from 440 to 580 nm, which may provide a discriminating parameter for counting human WBCs.

20.5.3
Overview of Optical Properties of Contemporary Photosensitizers Used for Systemic Administration

An ideal PS should be accumulated with a high contrast ratio by the malignant entity, demonstrating high singlet oxygen quantum yield, nontoxic, and completely removable from tissue or solutions (once the photodynamic session is completed), photostable, broadband absorbing in near-infrared, and cost effective. Preparation and functionalization of PSs for therapy include biocompatible coating such as PEGulation, MF BSA, and MF DEXTRAN to reduce the toxicity of nanoparticles. Antibodies, peptides, liposomes or other functional molecules targeting, and/or

Table 20.1 Some contemporary and prospect photosensitizers.

Photosensitizers	Absorption maxima in buffer solution at wavelength range used in PDT, $\Delta\lambda$, nm	Extinction at $\Delta\lambda$	Quantum efficiency of singlet oxygen generation
Hematoporphyrin derivates (HPD)	**600–650**	**3500–4400 cm^{-1} mol^{-1} l^{-1}**	**Monomers – 0.64** [54] **Dimers – 0.11**
Al sulphonated phthalocyanines	**670–700**	**90 000–1 10 000 cm^{-1} mol^{-1} l^{-1}**	**0.3–0.5**
Chlorines e6	**665–695**	**<41 000 cm^{-1} mol^{-1} l^{-1}**	**0.4–0.6** [55]
PPIX (via ALA-5)	**630–640**	**<5 000 cm^{-1} mol^{-1} l^{-1}**	**0.65–0.7** [56]
Methylene blue	**600–635**	**75 000 cm^{-1} mol^{-1} l^{-1}**	**0.5–0-55**
TOOKAD (d-acteriopheophorbide)	**760–770**	**88 500–1 05 000 cm^{-1} mol^{-1} l^{-1}**	**0.5–0.6** [57]
Conjugated Porphyrin dimmers	700–920 (2 *photon absorption*)	5 000–25 000 GM (10^{-50} cm^4 s $photon^{-1}$)	>0.15
Si nanoparticles	265–400		0.06
TiO_2 nanoparticles	220–430		0.21–0.23 [58]

Clinically exploited contemporary PSs are in bold; prospective PSs are italic.

angio-contrast are used to enhance the normal/abnormal contrast. Some of recently used and prospective PSs are presented in Table 20.1.

The contemporary clinically used PSs are usually aromatic molecules that are efficient in the formation of long-lived triplet excited states under irradiation by 630–670-nm waves. PDT is used clinically for various neoplastic and nonneoplastic conditions. For example, porfimer sodium (Photofrin®) is currently licensed in the UK for the treatment of non-small cell lung cancer and obstructing oesophageal cancer. 5-Aminolevulinic acid (ALA) (Levulan®) and its methyl ester (Metvix®) are licensed in the USA and Europe for treatment of nonmelanoma skin cancers. Verteporfin (Visudyne®) is used worldwide for treatment of "wet" age-related macular degeneration [59]. The PSs that are currently used in clinical or medical research practice are summarized in Table 20.1. Phthalocyanines, chlorines, porphyrins, PPIX (ALA-5), methylene blue, and Si nanoparticle [60] PSs fluoresce, while TOOKAD® and TiO_2 nanoparticles do not. Some PSs can be excited in a 2-photon regime such as HPDs. Conjugated porphyrin dimmers demonstrate a high 2-photon absorption cross section [61]. The advantages of 2-photon PDT include better light penetration due to doubling of the excitation/absorbing wavelength and high spatial confinement of the PDT effect due to the quadratic dependence of the 2-photon absorption probability [7]. Some nanoparticles are themselves photoactive and can generate singlet oxygen. For instance, porous silicon NPs disposing a high surface-to-volume ratio generate singlet oxygen at light absorption through direct energy transfer [62, 63]. One of the biologically most relevant features of fullerenes (carbon Bucky balls C60, C70) is the ability to form a long-lived triplet excited state upon photosensitization, acquiring the potential to generate singlet oxygen ($^1O_2$) with a high quantum yield [64]. The efficiency of $^1O_2$ formation (unity for C60 at 532 nm) for C60 is the highest among all PS molecules investigated to date [65]. The semiconductor nanoparticles such as TiO_2 adn ZnO are known as *effective photocatalysts*. Recently, these materials were shown to be capable of generating singlet oxygen by monitoring the photooxidation of methyl oleate and 2,2,6,6-tetramethyl-4-piperidone [66]. The generation of singlet oxygen in TiO_2 was attributed to an energy transfer and electron transfer mechanism, that is, electron transfer from the excited TiO_2 chelate to the conduction band of TiO_2 and energy transfer between the excited triplet state of the chelate and ground state oxygen [67]. The modern approaches to deliver, specifically activate and regulate singlet oxygen generation of the PS include PDT molecular beacons, DNA aptamer-PS-single-walled carbon nanotubes (SWNTs) complexes. The novel PDT molecular beacons comprise a PS and a quencher, linked by a target-labile linker that keeps the PS and quencher in close proximity so that Forster resonant energy transfer (FRET) prevents activation of the PS. Upon interacting with the target, the linker is either broken or opened, so that the PS and quencher are separated, allowing the PDT action to occur [68]. The DNA aptamer-PS-SWNT complexes combine the significant features of SWNT, aptamer, and PS to form a simple, but efficient and elegant PDT agent (AP-SWNT). The ability to use infrared light to initiate PS excitation in PDT could expand its applications by increasing tissue penetration and the efficiency of tumour-specific cell death through the combined usage of the PS and the up-converting phosphor agent. Up-converting phosphors are ceramic materials containing inert, rare earth oxides embedded in a crystalline matrix (e.g., Sunstones™). These rare earth up-converting gadolinium oxide (80–200 nm diameter) nanocrystals up-convert infrared energy to specific visible wavelengths. The up-conversion mechanism can be described either as sequential excitation of the same atom or as excitation of two centers and subsequent energy transfer. The wavelength distribution of the emission can be tailored to a region appropriate for excitation of PSs commonly used in PDT [69].

Most PSs have also substantial fluorescence quantum yield. Despite the fact that fluorescence is competing with PDT, this property is very useful as it allows making additional tumor demarcation to specify light irradiation zone and to control PS concentration in tumor before and during light irradiation. Moreover, the PS fluorescence quantum yields as low as several percents, which may be sufficient to detect PS fluorescence in tissues during PDT.

The results of the fluorescence measurements of the PSs in phosphate buffered saline (PBS) solution are presented in Figure 20.9a,b [70]. Concentrations of PSs have been selected in such a way as to ensure similar fluorescence intensity for each system, allowing the presentation of all curves in one scale.

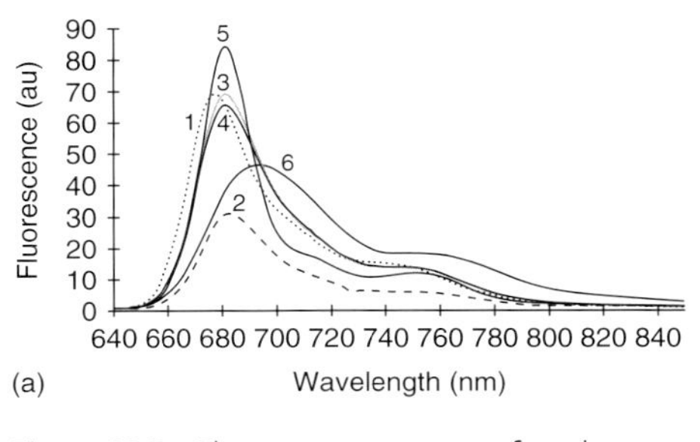

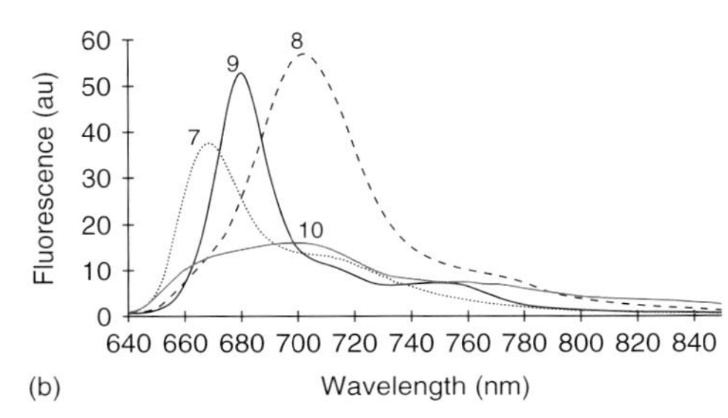

Figure 20.9 Fluorescence spectra of studied photosensitizers in PBS solution. Excitation $\lambda = 633$ nm. (1) Zinc phthalocyanine I – 7 mg l^{-1}; (2) zinc phthalocyanine II – 13 mg l^{-1}; (3) zinc phthalocyanine III – 12 mg l^{-1}; (4) zinc phthalocyanine IV – 13 mg l^{-1}; (5) aluminum phthalocyanine (Photosens®) 1 mg l^{-1}; (6) methylene blue 10 mg l^{-1}; (7) chlorine p6 I 10 mg l^{-1}; (8) chlorine p6 II –10 mg l^{-1}; (9) uroporphyrin III – 23 mg l^{-1}; and (10) hematoporphyrin derivative (Photogem®) – 50 mg l^{-1}. (Modified from Ref. [70].)

20.5.4
Interaction of Photosensitizers with Blood Cells: Uptake Locations and Pharmacokinetics

The potential cellular targets of PDT depend on the specific pharmacokinetic characteristics and localization of the PS [23, 71]. Cellular PS distribution can be influenced by such parameters as length of exposure and PS concentration [72]. PSs may even relocalize after photodamage to an initial site of accumulation such as from lysosomes to other, possibly more sensitive, cellular locations, where they will then be available for activation [73]. The major targets of PSs are mitochondria, lysosomes, nucleus, and plasma membrane.

It was found that temoporfin *meta*-tetra(hydroxylphenyl) chlorine (mTHPC) localized mainly at the mitochondria and the perinuclear region [74]. Photodynamic treatment of murine leukemia cells under the action of mTHPC caused rapid appearance of the apoptogenic protein cytochrome c in the cytosol. The loss of integrity of the mitochondrial membrane and the appearance of chromatin condensation were observed 1 h after light irradiation. Luo *et al.* [24] reported that chloroaluminum phthalocyanine is localized throughout the cytoplasm and catalyzed both lysosomal and mitochondrial photodamage. The membrane photodamage progressively depends on light doses resulting in apoptosis of cells. Georgiou *et al.* [75] observed that meso-tetra(4-*N*-methylpyridyl) porphine (TMPP) is principally located in the nucleus, while another porphyrin, meso-tetra(4-*N*-hexylpyridyl) porphin (THPP), exhibited a particulate distribution in the cytoplasm. Studies of Photofrin and Photochlor® pharmacokinetics showed that both PSs have long circulating half-lives and can be found more than three months after intravenous infusion [76]. Photofrin and Photochlor are cleared slowly, but the pharmacokinetic parameter estimates indicate considerably longer retention of Photofrin (25.8 ml h^{-1}) than Photochlor (84.2 ml h^{-1}) in patients. Photofrin binds to several plasma lipoproteins with the following distribution: 40% albumins (ALB), 37% high-density lipoprotein (HDL), 16% low-density lipoprotein (LDL), and 7% very low-density lipoprotein (VLDL). Amphiphilic agents (for example, benzoporphyrin derivative) may be administered either in an aqueous medium or via a carrier. Aqueous vehicles such as Hyskon (Dextran 70 in dextrose solution) are employed for topical PDT with ALA. Triesscheijn *et al.* [77] investigated the pharmacokinetics of mTHPC, which is a hydrophobic PS. It was observed *in vitro* that it binds to plasma lipoproteins after intravenous injection to some unknown protein and subsequently redistributes to lipoprotein fractions. Pharmacokinetic profiles of mTHPC in a group of human subjects showed that constant mTHPC levels were kept for at least 10 h. Detailed examination of individual profiles showed that the initial (5 min) plasma drug levels were, on average, 86% of the maximal plasma concentration, which occurred at about 5 h after injection.

20.5.5
Alteration of the Optical Properties of the Photosensitizers Dissolved in Blood

For most PSs, an absorption maximum in the red part of the spectrum is used for the treatment. To achieve optimal treatment efficacy, it is important to choose a wavelength corresponding to the absorption spectrum *in vivo*, as the absorption peak is often shifted because of the bioenvironmental influence such as bounding with biochemical components, tissue optical properties, and blood perfusion. We can see an example of the absorption peak shift of Photosens® (Pc) in Figure 20.10. The absorption peak in blood-Intralipid® phantom is shifted from 675 nm in water to 671 nm in the phantom. With broadband light sources, such as filtered lamps, it is easy to cover the absorption peak. However, much of the treatment light will not be efficiently absorbed by the PS but rather by tissue chromophores, leading to a temperature increase, which in turn may change the tissue state with stimulated

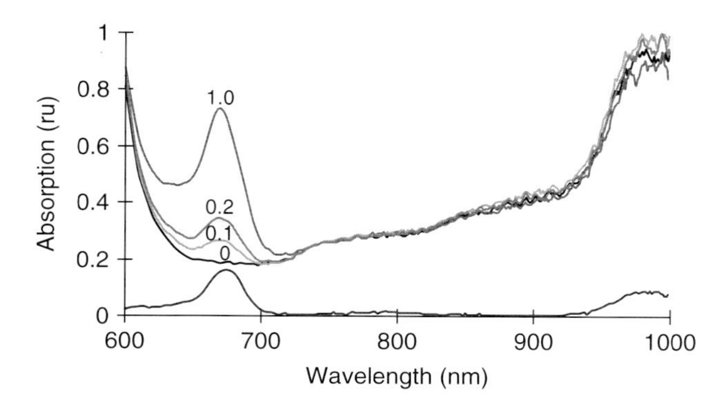

Figure 20.10 Absorption spectra of tissue phantom containing 1.6% of Intralipid®, 2% of blood, and different photosensitizer Photosens® concentrations (0, 0.1, 0.2, and 1.0 mg l^{-1}). The absorption spectrum of photosensitizer at concentration 2 mg l^{-1} in water (pathlength – 0.5 cm) is also shown in the figure as the bottom curve. (Modified from Ref. [78].)

blood perfusion, and, as an ultimate consequence, may lead to overall growth of the light absorption.

In general, if we compare the absorption of PS in water with interstitial distribution *in vivo*, the absorption shift is caused by (i) the tissue background absorption (other than blood), the gradual function increasing with wavelength [79] and (ii) the blood absorption. Put simply, both factors act as a long pass color correction filter suppressing the blue component of the transmission and reflectance spectra from the tissue. The contribution of the latter factor is dominant in blood containing tissues. We can suggest the following formalism for the light absorption spectrum of PS (A^{t}_{PS}) in the biological tissue at moment t:

$$A^{t}_{PS}(t) \sim A_{t} A_{RHb} A_{HbO_2} A^{w}_{PS} K_{c}(t) \quad (20.2)$$

where A^{w}_{PS} is the absorption of PS in water, A_{t} is the absorption of tissue taken into consideration without blood content, A_{RHb} and A_{HbO_2} are the absorption of reduced and oxygenated hemoglobin respectively, and K_c is the coefficient taking into account the bounding effects of the environment. We assume that tissue optical properties do not undergo considerable changes as a result of PDT after Δt period, and can be simplified (Equation 20.2) as

$$A^{t}_{PS}(t + \Delta t) \sim A_{RHb} A_{HbO_2} A^{w}_{PS} K_{c}(t + \Delta t) \quad (20.3)$$

Taking into account the known hemoglobin spectra, it is possible that the absorption peak of most of the PSs in red range will be blue shifted. This effect is in contrary to the diffuse reflectance and transmission as the absorption is presented in negative logarithm form according to Equation 20.4.

$$A(\lambda) = \lg\left(\frac{I_{ref} - I_{dark}}{R_{d} - I_{dark}}\right) \quad (20.4)$$

where I_{ref} is the measured reflectance signal from the reference sample ($BaSO_4$), which has the uniform diffuse reflectance near unity in the spectral range under consideration (400–1000 nm), I_{dark} is the signal in the absence of any light (dark current of CCD), and R_d is the diffuse reflectance signal from the tissue. Equation 20.4 accounts for the spectral nonuniformity of the light source, the fiber transmission, and the detector sensitivity. It should be noted that $A(\lambda)$ as given by Equation 20.4 is defined to be within some constant value. However, the exact value of the constant in $A(\lambda)$ is not important as only its wavelength dependence does matter. The spectra taken from tissue phantom containing 1.6% Intralipid, 2% blood, and various Photosens concentrations are given in Figure 20.10. The contribution of PS to attenuation is clearly observed as a peak with a maximum of about 675 nm. The absorption spectrum of Pc in physiological solution measured with the same spectrometer in a cuvette with a path length of 5 mm is also given in Figure 20.10. The contribution of PS to absorption was calculated as an area between $\lambda \pm 15$ nm after subtracting the linear background. The 30 nm wide spectral range for quantifying absorption contribution quantification corresponds to the PS absorption peak width (FWHM). The background approximated by linear

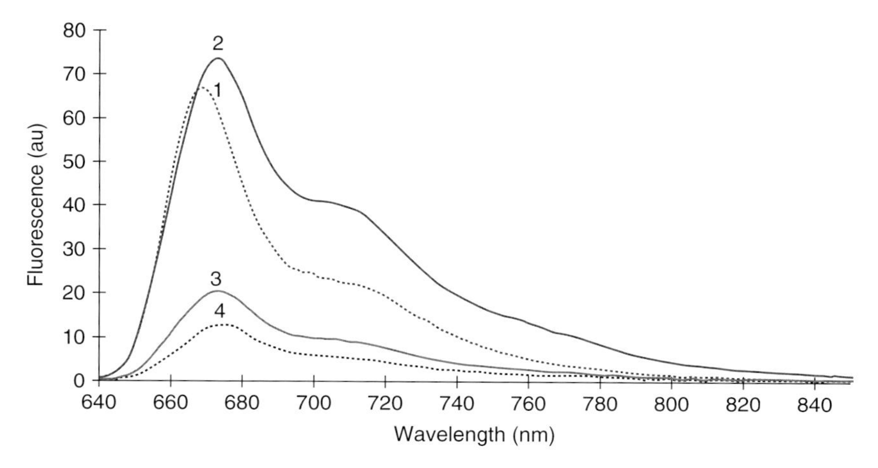

Figure 20.11 Fluorescence spectra of chlorine p6 I ($10\ mg\,l^{-1}$) in (1) PBS buffer; (2) plasma; (3) whole blood, and (4) erythrocytes. Excitation $\lambda = 633$ nm. (Reproduced by permission from Ref. [70].)

function takes into account the scattering contribution as well as varying absorption due to different blood content and different blood oxygenation. The absorption peak shift was confirmed by Stratonnikov *et al.* [78] with Al phthalocyanine (Photosens). For a HpD Photofrin, the absorption spectrum [80] and the fluorescence excitation spectrum [81] all have a maximum at 625 nm *in vivo*, as against 630 nm for *in vitro*. A redshift of the absorption peak of Photosens in tumor versus normal tissue [82] was also reported. A similar case is the fluorescence properties of PS in blood. Fluorescence spectra presented in Figure 20.11 were measured for Chlorine p6 I in whole blood before (curve 3) and after (curves 2 and 4) sedimentation [70]. This provides an observation of PS distribution between blood plasma and erythrocytes compartments. We supposed that fluorescence amplitude indicates PS concentration. The majority appears localized in blood plasma; hence, the concentration of Chlorine p6 I in whole blood and plasma must be similar. However, the fluorescence amplitude in blood plasma is considerably higher than in whole blood because of strong attenuation of exciting and fluorescent light by erythrocytes scattering and hemoglobin absorption [83]. Apparently, determination of relative PS concentration from fluorescence measurements can be quantified only in samples with similar types of optical properties. This imposes certain limitations on the *in vivo*, *ex vivo*, or *in vitro* usage of fiberoptic fluorescence spectroscopy in real biological tissues. It is seen that the spectral shape and peak positions of PSs in whole blood and erythrocytes are the same. PS spectra in the whole blood and blood components, compared with PBS buffer spectra, exhibit a red shift of the fluorescence peak by ~7 nm, which is most probably due to hemoglobin absorbance and also due to binding of the PSs with the blood components (e.g., plasma proteins or erythrocytes).

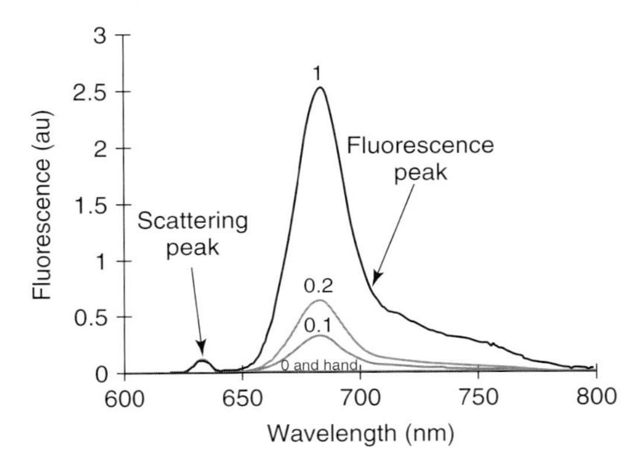

Figure 20.12 Excitation backscattering and fluorescent spectra of tissue phantom containing 1.6% of Intralipid®, 2% of blood and various photosensitizer (Photosens®) concentrations (0, 0.1, 0.2, and 1.0 mg L^{-1}). (Reproduced by permission from Ref. [78].)

The scattering of blood affects the absorption and fluorescence of PS collected by backward signal collection somewhat differently. In order to estimate this, a phantom containing Intralipid and hemolyzed blood to separate absorption and scattering of blood environment, was exploited containing 1.6% Intralipid, 2% of hemolyzed blood, and different Photosens concentrations [78]. Figure 20.12 presents the spectra measured from the blood–Intralipid phantom.

The peak at 632.8 nm is due to backscattering of the excitation light, while the broad peak with maximum at 685 nm is due to PS fluorescence. One can observe that fluorescence contribution of Intralipid itself and tissue (hand) is negligible compared with Photosens contribution at typical concentrations applied for PDT. At an Intralipid concentration of 1.6%, the backscattered contribution from the phantom (peak at 632.8 nm) is approximately equal to that of from the tissue [84]. The Intralipid absorption is neglected at this wavelength. The fluorescence and scattering contribution calculated as mentioned above at different PS concentrations in tissue phantom are given in Figure 20.13. This diagram demonstrates the linear dependence between fluorescence contribution and PS concentration with small deviations at higher PS concentrations (up to 5 mg l^{-1}) when the deviation of the fluorescence from linear dependence became substantial with backscattering contribution decreasing due to PS absorption (data are not shown). Calculating the slope from Figure 20.13, one can retrieve the coefficients between fluorescence contribution and PS concentration in the tissue. However, as it was shown, the more convenient and accurate approach is to use the ratio of fluorescence to scattering contribution to quantify PS concentration in tissues by fluorescence spectroscopy [85].

Figure 20.14 presents the dependence of PS absorption contribution calculated from the spectra shown in Figure 20.10. There is a substantial deviation from linear

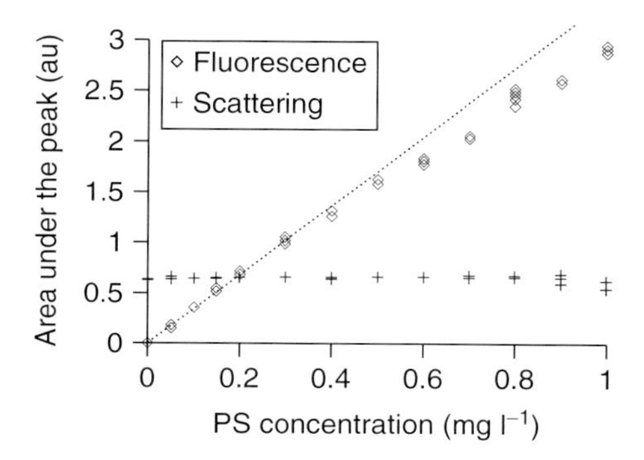

Figure 20.13 Dependence of the fluorescence and excitation scattering (computed as areas under the peaks in Figure 20.12) contribution on photosensitizer (Photosens®) concentration. Fluorescence dependence may be approximated by linear function in the range of PS concentrations used for PDT. (Reproduced by permission from Ref. [78].)

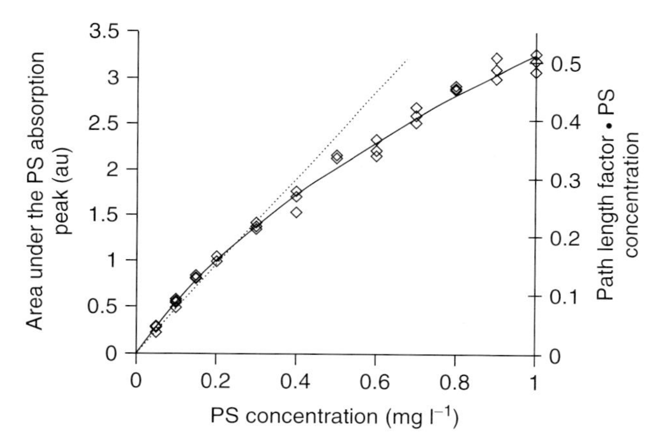

Figure 20.14 Dependence of photosensitizer (Photosens®) absorption contribution from its concentration in tissue phantom calculated from the spectra shown in Figure 20.10. At high PS concentration the data deviate from linear relation (compare with Figure 20.13) due to decrease of differential path length with increase of absorption coefficient. The line shows the calculation that takes into account the correction for differential path length factor with diffusion theory. (Reproduced by permission from Ref. [78].)

dependence at higher PS concentrations compared to the fluorescence method [86] shown in Figure 20.13. This deviation may be attributed to the decrease of path length factor with increasing PS concentration. Indeed, the evaluated PS absorption contribution is proportional to the product of PS concentration and the path length factor, β. The path length factor depends on scattering and absorption; this dependence in diffusion approximation is given by Equation 20.5 [87].

$$\beta = \frac{\sqrt{3}}{2}\frac{d}{d+\delta}\sqrt{\frac{\mu_s'}{\mu_a}} \qquad (20.5)$$

where $\delta \approx \frac{1}{3}\sqrt{\mu_a \mu_s'}$ is the light penetration depth (diffusion length) and μ_a and μ_s are the absorption and reduced scattering coefficients respectively, and d is the space between delivery and receiving fibers. The increase of PS concentration results in the increase of total tissue absorption coefficient and consequently the decrease of the path length factor. It is evident that at high tissue absorption the photon trajectories with shorter path lengths will be more contributing to the measured diffuse reflected signal, resulting in the decrease of the averaged path length. The product of PS concentration and the path length factor calculated according to Equation 20.5 with reduced scattering coefficient $\mu_s' = 19\ \text{cm}^{-1}$ is plotted in Figure 20.14 with solid line. In the range of PS concentrations typical of PDT (up to $0.4\ \text{mg}\ \text{l}^{-1}$), this dependence may be approximated by a linear function, the slope (dotted curve) of this line may be used as a coefficient for calculating PS concentration in tissue from its measurements.

It is evident that variability in tissue scattering should affect the measured fluorescence signal due to changes in excitation light distribution in tissue and catching volume of the fluorescence signal. The variability in tissue absorption also modifies the excitation as well as fluorescence reabsorption. It is difficult to predict a priori whether the fluorescence signal will reduce or grow with increase of scattering, although it is evident that the fluorescence signal should be diminished with increase in the tissue absorption. Figure 20.15 presents the

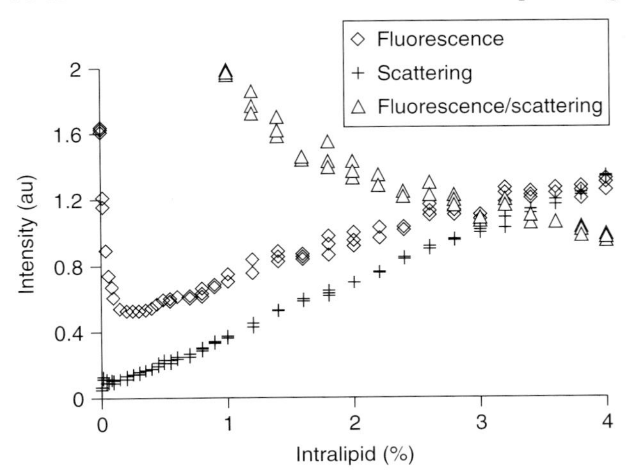

Figure 20.15 Dependence of fluorescence and excitation peaks in solution containing photosensitizer (Photosens®) ($0.2\ \text{mg}\ \text{l}^{-1}$) on Intralipid® concentration simulating the variation of scattering in tissues. The ratio of the fluorescence over the excitation back scattering peak is also given. (Reproduced by permission from Ref. [78].)

relationship between fluorescence and backscattered signal at various Intralipid concentrations modeling variability in tissue scattering. The Photosens concentration was 0.2 mg l^{-1}, which is typical of tissue during PDT. In the absence of scatterer the backscattered signal is close to zero (the small signal is due to the specular reflection from the bottom of the sample), while the fluorescence signal is rather high due to isotropic fluorescence emission. When the scattering is increased, the backscattered signal uniformly increases while fluorescence sharply goes down and then slowly increases again with the increase of scattering. This behavior is explained by the fact that in transparent media the catching depth of fluorescence is rather great and in fact reaches the bottom of the glass (2 cm) resulting in high fluorescence signal. With addition of scatterers, the catching depth is sharply diminished resulting in a sharp fluorescence drop. This slow fluorescence increase is due to the increase in the excitation in the vicinity of the detection fiber as confirmed by the backscattering signal growth. In the range of scattering variation typical of tissues the fluorescence signal is not constant but varies within 15% resulting in corresponding errors in quantifying PS concentration through fluorescence signal. The method of using the ratio of the fluorescence over the excitation back scattering does not diminish the errors in variation of signal due to tissue scattering variation as shown in Figure 20.15.

Though both fluorescence and backscattered signal are increased in the range of tissue scattering variation, the ratio of fluorescence to scattering is also not constant and varies within 25% in this scattering range. So, the ratio method even increases the errors resulting from tissue scattering variations. However, the total error up to 25% is quite acceptable for our purpose. Figure 20.16 demonstrates the influence of tissue absorption variation on the fluorescence measurements modeled by changing the blood content. As can be observed from Figure 20.16 both the fluorescence and backscattered signal are diminished as the blood volume increases, which is quite evident due to absorption of laser irradiation and fluorescence reabsorption. The concentration of Intralipid was 1.6% to match tissue scattering and Photosens concentration was again 0.2 mg l^{-1}. The variation of fluorescence signal in the range of physiological variation of blood volume (0–5%) is up to 65%. However, the ratio of the fluorescence to scattering varies just within 30% in this blood volume range. It justifies the use of fluorescence to scattering ratio when quantifying PS concentration by fluorescence method. It will be even more justified when one applies green or blue light for PS fluorescence excitation as it is more strongly absorbed by blood. Thus, for quantifying PS concentration in patients, one can apply the ratio of the fluorescence to the excitation scattering signal, which partly diminishes the errors due to tissue absorption variations, the excitation laser power drift and variation of angle between fiber probe and normal to the tissue being taken into account.

Now, we discuss the influence of tissue scattering and absorption variation on PS quantification by diffuse reflectance method. It should be mentioned again that PS absorption contribution measured by diffuse reflectance is proportional to the product of PS concentration to the averaged photon path length

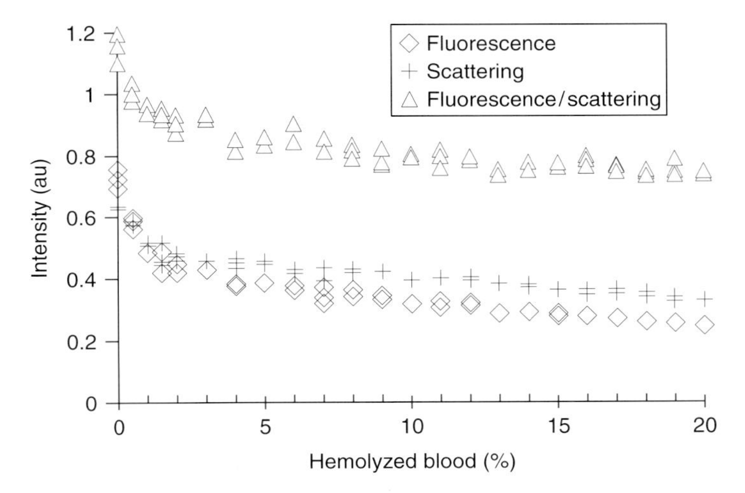

Figure 20.16 Dependence of fluorescence and laser scattered peaks in tissue phantom containing Intralipid® (1.6%) and photosensitizer (Photosens®) (0.2 $mg\,l^{-1}$) from hemolyzed blood concentration simulating the variation of absorption in tissues. The hemoglobin oxygenation was 75%. The ratio of the fluorescence over the excitation back scattering peak quantifying PS concentration is also given. (Reproduced by permission from Ref. [78].)

through the tissue. It is evident that the increase of scattering results to the increase of averaged photon path length thus overestimating the PS concentration. The increase of tissue absorption results in decrease of the averaged photon path length underestimating the measured PS concentration by diffuse reflectance method.

Figure 20.17 demonstrates the measured PS absorption contribution as a function of Intralipid concentration (scattering).

The measurements were done without blood and with 4% of hemolyzed blood. The dye (Photosens) concentration was 0.2 $mg\,l^{-1}$ in both cases. The path length factor dependence calculated with Equation 20.5 is shown in Figure 20.18 by solid lines, confirming that variations in PS absorption contribution are due to changes in the averaged photon path length. In the range of typical tissue scattering variations the corresponding error in PS concentration is within 30%. Figure 20.18 presents the dependence of the PS absorption contribution on the blood absorption. The Intralipid concentration was 1.6% to match the biotissue scattering ($\mu_s' = 19\,cm^{-1}$). The path length factor dependence is shown again with solid lines. As in the previous case, the influence of absorption may be totally attributed to the changes in differential path length. In the region of typical tissue absorption variations the changes in PS absorption contribution are within 50%.

Summing up the above-mentioned observations, we can expect that error in PS concentration due to tissue variation is within 30% for the fluorescence method and within 50% for the diffuse reflectance method.

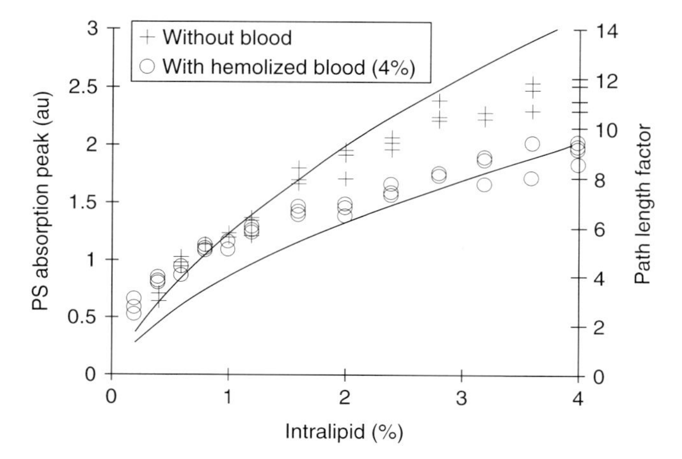

Figure 20.17 Dependence of PS absorption contribution in solution of photosensitizer (Photosens®) (0.2 mg l^{-1}) without blood and with hemolyzed blood (4%) on Intralipid® concentration simulating the variation of scattering in tissues. The growth of the signal with the increase of scattering may be attributed to the increase of differential pathlength factor. The pathlength factor dependence calculated with diffusion theory is also shown by the thin lines. (Reproduced by permission from Ref. [78].)

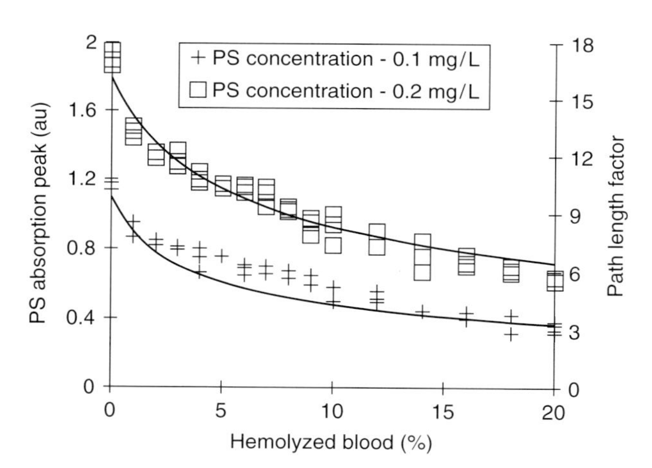

Figure 20.18 Dependence of PS absorption contribution in solution containing Intralipid® (1.6%), photosensitizer (Photosens®) (0.1 and 0.2 mg l^{-1}) from blood concentration simulating the variation of absorption in tissues. The decrease of the signal with the increase of background absorption (blood) may be attributed to the decrease of differential path length factor. The path length factor dependence calculated with diffusion theory is also shown by the thin lines (the upper curve is multiplied by 2 to scale the double PS concentration). (Reproduced by permission from Ref. [78].)

20.6
Photodynamic Reactions in Blood and Blood Cells, Blood Components, and Cells

20.6.1
Photodynamic Modifications and Alterations of Blood Plasma and Plasma Proteins

The interactions of sulphonated chloroaluminum phthalocyanine ($AlPcS_n$) with human LDL were studied *in vitro* in human plasma and in an isolated LDL fraction, in order to understand the potential effects of the sensitizer against LDL. The $AlPcS_n$ added to plasma distributes in all lipoproteins as observed by the drastic color changes of the separated fractions by ultracentrifugation. In isolated LDL, incubation with $AlPcS_n$ causes fluorescence quenching of the apoprotein tryptophan residues. Furthermore, $AlPcS_n$ incorporates in liposomes, with a lipid composition similar to that of the external monolayer of human LDL, as indicated by absorbance spectroscopy. The photosensitizing properties of $AlPcS_n$ in LDL particles were studied on the basis of the fluorescence quenching of previously incorporated *cis*-parinaric acid (PnA), used as an oxidation probe, and of O_2 consumption [88]. Photodynamic virus inactivation was studied in whole human fresh plasma mediated by visible light in the presence of the phenothiazine dyes, methylene blue, or toluidine blue. The activities of the coagulation factors I, VIII, IX, X, and XI were affected to a certain degree, while most of other plasma proteins were not. There was no indication that the photodynamic virus inactivation procedure applied considerably influences the properties of plasma proteins [89]. Blood plasma proteins are protected by antioxidant activity of carotenoids during photosensitized oxidation of human plasma, and thus carotenoids can protect tocopherols from the oxidative loss by $^1O_2$ [90].

20.6.2
Photodynamic Modifications and Alterations of Red Blood Cells (Erythrocytes)

The PDT effects reported on RBC include alteration of erythrocytes aggregation, deformability, photohemolysis, and cell shape. Photohemolysis of human or animal erythrocytes sensitized by PS is very often used to study possible chemical modifications of PDT. Photohemolysis is observed by exposing an RBC suspension to light and then spinning down the intact cells leaving hemoglobin in the supernatant. A convenient assay procedure utilizes light scattering because the transmission of far-red light by an RBC suspension is proportional to the concentration of intact cells. The photohemolysis rate may be quantified by either of two techniques. In continuous photohemolysis (CPH) the fraction of hemolyzed cells is measured during the irradiation. In postirradiation or delayed photohemolysis (DPH) the cells are exposed to a brief, nonhemolysing irradiation, and hemolysis is measured during incubation in the dark. DPH and CPH curves are sigmoidal frequently with an initial delay period. The time required to achieve 50% hemolysis (t_{50}) is a useful rate parameter. t_{50} is measured from the start of irradiation for

CPH and from the start of dark incubation for DPH. Photohemolysis kinetics is characterized by a nonlinear dependence of the rate on the incident fluence and PS concentration [91]. The DPH rate for many different PSs correlates with the empirical relation:

$$1/t_{50}\,(\mathrm{DPH}) \sim f^{k'} C_t^{j'} \tag{20.6}$$

where t_{50} is the dark incubation time required for 50% hemolysis, C_t is the total dye concentration, f is fluence, and $j'$and $k'$are the "as measured" exponents. The functional form of Equation 20.6 is assumed in the detailed colloid-osmotic kinetics model of Pooler [92]. The empirical rate expression for CPH is

$$1/t_{50}\,(\mathrm{CPH}) \sim C_t^{j'} \tag{20.7}$$

Some sensitizers led to $k' \sim 2$ and $j' \sim 2$, which implies that the photohemolysis rate varies with the square of the absorbed fluence. Other sensitizers led to $k' \sim 2$ and $j' \sim 1.2$ 1.4 according to Grossweiner *et al.* The lower values of j' have been attributed to various factors, including partitioning of the sensitizer between the cells and the aqueous phase and sensitizer photobleaching [91]. Priezzhev found out concentrate-dependent and dose-dependent increase of the level of hemolysis with Russian PSs Photogem® (Ph) (29 mg l^{-1}) and Photosens (Pc) (33 mg l^{-1}), and was noticed that Ph- and Pc-photosensitized hemolysis was inhibited by a-tocopherol at 1.3–130 mM concentrations with more effective inhibition for Pc-photosensitized hemolysis [93]. Cell culture observations in the colon carcinoma HT29 cell line showed that 0.33–1 mM of a-tocopherol can enhance PDT activity of *m*THPC in cell culture, while lower concentrations of the antioxidant (0.001–0.1 mM) had no significant effect in the same system [94]. The method of scanning electron microscopy allowed observing changes of the erythrocytes shape and size under photosensitization. It was found that the normal discoid erythrocytes become echinocytes in case of sensitization with the HPD [95]. The laser irradiation of sample leads to spherocytic transformations of RBCs while after the long tune irradiation the hemolysis of erythrocytes takes place. In case of sensitization with phthalocyanines and following irradiation of sample, a spherocytic transformation accompanied with an increase of the cell volume was observed [96]. It is not clear whether the alteration of mechanical properties of erythrocytes and their interaction ability (aggregation) at sensitization with platelet concentrates (PCs) under the conditions of dark incubation and laser irradiation take place. Priezzhev *et al.* presented that Pc administration into blood (29 mg l^{-1}, 670 nm, 200 J cm^{-2}) causes a significant increase in the strength of erythrocyte aggregates (by more than two times). After the laser irradiation, this parameter also increased but not to the extent mentioned earlier. The study of lipid peroxidation judged by the accumulation of products reacting with Thiobarbituric acid (TEA) [97], which was expressed as an equivalent amount of malone dialdehyde (MDA) has shown that the accumulation MDA in RBC suspension proportionally grows at increase of irradiation dose. At the fluence, equal to 20.9 J cm^{-2}, the amount of products of oxidation has reached $0.135 \pm 0.018\ \mu mol\ l^{-1}$ of RBC suspension [98]. Protoporphyrin-sensitized photo-oxidation in human RBC membranes leads to severe deterioration of membrane structure and function. Dubbelman *et al.* yet in 1978 have shown that the membrane damage is also caused by direct oxidation of amino acid residues, with subsequent cross-linking of membrane proteins. The chemical nature of these cross-links was studied in model systems, isolated spectrin, and red cell ghosts, indicated that a secondary reaction between free amino groups and a photo-oxidation product of histidine, tyrosine, or tryptophan is involved in photodynamic cross-linking. Accordingly, succinylation of free amino groups of membrane proteins or addition of compounds with free amino groups protects against cross-linking. Quantitative data and consideration of the reaction mechanisms of photodynamic oxidation of amino acids make it highly probable that an oxidation product of histidine rather than of tyrosine or tryptophan is involved in the cross-linking reaction, via a nucleophilic addition by free amino groups [99]. PDT does not affect, or slightly reduces, RBC adherence to endothelial cells (ECs) [100]. Photodynamic treatment of blood for virus inactivation with sulphonated aluminum phthalocyanine ($AlPcS_4$) and the silicon phthalocyanine Pc4 has previously been shown to enhance RBC aggregability measured by cell flow properties analyzer (CFA); it induces the formation of large irregular aggregates which require higher than normal shear stress for their dispersion and the clusters of cells were huge and abnormally shaped, unlike the rouleaux formed by untreated RBC. This aggregation was prevented when a mixture of antioxidants was included during PDT. Addition of the antioxidants after PDT reduces aggregation only partially [101]. Adding antioxidant ascorbat at large concentrations (more than 1 mM) generates photodynamic damage according to Type I photochemical, ascorbat-assisted mechanism. The direct correlation was found between the lysis rate and ascorbat concentration over the range 0.1–1 mM, assumed that this additive is a reactant and not a catalyst. In the presence of quercetin (plant flavonoid), the phtalocyanine PS induced photohemolysis was inhibited [102]. In HPD sensitized photohemolysis the antioxidative function of quercetin was enhanced by ascorbat even under conditions of which ascorbat functioned as a pro-oxidant when it was added alone [103]. However, origination of Type I reaction in blood may lead to hemoglobin oxidation from Fe^{2+} to Fe^{3+}, which alters the oxygen supplying function of blood. Using method of real time confocal scanning microscopy [104] to monitor immediate effects of photodynamic activation of photosensitized erythrocytes by Rollan *et al.* has enabled to distinguish between the differential susceptibility of age-density resolved subpopulations of human erythrocytes to photodynamic activation using HPD and Rose Bengal as PSs [105]. This study demonstrates that younger (low age-density) subpopulations of photosensitized erythrocytes are less susceptible than older (high age-density) subpopulations to photodynamic activation with no significant difference in uptake of PS by both populations observed using absorbance spectrophotometry. It has also been shown that lipid peroxidation increases on the membranes of ageing erythrocytes [106].

20.6.3
Photodynamic Modifications and Alterations of White Blood Cells (Leucocytes)

PDT often causes an inflammatory reaction in irradiated tissues. Dellian *et al.* [107] investigated the effects of PDT on leukocyte-endothelium interaction in the microvasculature of tumors and normal tissue. The maximum increase in adhering leucocytes was observed in postcapillary venules of tissue 1 h after PDT. In contrast, enhanced leukocyte–endothelium interaction was missing in tumor vessels and in control groups. Thus, induction of leukocyte adhesion in the PDT-treated normal tissue was suggested as a contribution to the inflammatory response of PDT. The study of lipid peroxidation performed via monitoring of MDA has shown that at 0.8 J cm^{-2}, with Photosens (Al sulphated phthalocyanine), with increase of the PS concentration, the level of MDA was also increased [98]. Addition of Fe^{2+} allowed judging whether accumulation of lipid hydroperoxides in the system has occurred. It is noteworthy that on irradiation, the leucocytes suspension, not containing Photosens, also causes the accumulation of MDA, which probably is, in turn, the effect of the oxidation of human blood leucocytes caused by endogenous PSs [98]. Chen, Cui, Liu, 2002 presented the study suggesting that a photodynamic cell therapy process (PDT) can selectively deplete host alloantigen-specific T cells to prevent graft-versus-host disease (GvHD), while preserving immune and antileukemia functions Rhodamine derivative 123 (TH9402) selectively retained in the mitochondria of activated host-reactive cells but not tumour- or third-party-specific resting cells. The treated cells were subsequently exposed to visible light (514 nm) to deplete the TH9402-enriched activated host-reactive cells. Treatment with photodynamic cell purging process (PDP) inhibited antihost responses measured by cytotoxic T-lymphocytes (CTLs) by 93%, and interferon production by 66%. By contrast, anti-BCL1 (BALB/c-origin leukemia/lymphoma), and anti-third-party C3H/HeJ (H-2k) responses were preserved [108].

20.6.4
Photodynamic Modifications and Alterations of Blood Platelets (Thrombocytes)

PDT may cause damage to platelets that come from photoactivation of PS bound to platelet or from activation of PS in plasma. Morphological changes of mouse platelets during PDT were described by Zhou *et al.* [109]. In their experiments, mice received HPD in a dose of 10 mg/b.w.kg 24 or 48 h prior to red light irradiation (600–700 nm, 100 J cm^{-2}). The platelet-rich plasma obtained by centrifugation of whole blood was irradiated by red light. The platelets in control group (without irradiation and PS) were in normal condition, while platelets of experimental groups showed structural changes. Immediately after irradiation, a small number of platelets were necrotized, having a swollen irregular shape. Considerable deformations were observed in many other platelets. Some of them were swollen and enlarged, had irregular shape; their cytoplasm contained alpha-granules, mitochondria; some platelets showed a decrease number of alpha-granules, and only a small number of platelets kept their normal shape. In samples studied 8 h after irradiation, about one-fourth or more of platelets had already become necrotic. Other platelets experienced considerable deformations: they were swollen and enlarged; there were a decrease of granule number and dilatation of canalicular membrane system. The platelets after action of PS but without exposure of light showed much less severe changes. All platelets were necrotized 16 h after the PDT treatment, while platelets which weren't irradiated by light but were under action of PS, necrotized only one-fourth of all cells and the majority of platelets were intact. Under action of HPD and irradiation one-fourth of platelets were necrotized after 16 h after photodynamic treatment and necrosis of all platelets occurred after 16 h after irradiation. Most likely the similar photodamage during PDT experience human platelets. The release of variety of vasoconstriction, coagulating, and anticoagulation agents results in process of vascular damage induced by PDT. The role of circulating platelets on blood flow stasis and vascular damage during and after PDT using Photofrin was assessed in an intravital animal model [110]. Animals were injected with 25 mg/b.w.kg of the PS intravenously 24 h before light irradiation (630 nm, 135 J cm^{-2}). Thrombocytopenia was induced in animals by administration of 3.75 mg kg^{-1} of rabbit antirat platelet antibody intravenously 30 min before the initiation of the light treatment. A chondrosarcoma tumor model was used in the study. Action of Photofrin and light to the tumor caused significant increases in the intravascular release of thromboxane. This could be improved by induction of thrombocytopenia before the light treatment. Treatment with both Photofrin and light led to fourfold increases in plasma thromboxane compared to untreated controls. Only slight rises in thromboxane were observed when light treatment was given for thrombocytopenic animals, and observed plasma levels of thromboxane in that animals given the combination of Photofrin and light did not exceed that of control animals given no treatment or Photofrin alone. Complete blood flow stasis occurred within the first half of light treatment (15 min) and no significant difference in the timing or magnitude of this event was seen between vessels in normal tissues or tumor. Focal areas of platelet deposition and thrombus formation could be observed at specific locations within vessels. Also, it was shown that thromboxane release is a necessary component of vessel response after PDT. Leakage to albumin in tumor and normal vessels occurred shortly after initiation of light treatment of photosensitized animals with normal platelet amount. Induction of thrombocytopenia in animals given Photofrin completely inhibited vascular leakage after light treatment in both normal and tumor tissues. Photodynamic treatment also influence on aggregation of platelets. Exposure to Photofrin either at light or at high concentrations of this PS inhibits the aggregation ability of platelets [111]. Platelets that may have become adherent to areas of microvasculature will receive significantly greater light doses compared to platelets in blood, and these platelets may be responsible for the observed thromboxane release and platelet aggregation. *In vivo* platelet damage may be possibly limited by the short time of light irradiation of individual cells.

20.7 Types of Photodynamic Reactions in Blood: *In vitro* versus *In vivo*

In a way, the photodynamic reaction in blood involves all above considered photodynamic reactions with blood components including RBC, WBC, blood platelets, and blood plasma and heavily depends on the PS type applied. The most common features of PDT reactions in blood throughout all types of currently used PS are associated mainly with erythrocytes and include reduction of blood oxygenation [112], RBC hemolysis [113], hemoglobin destruction [70], increase of RBC aggregation [70] and, assumingly, blood viscosity. Blood disoxygenation is easily observable via blood color change from scarlet to dark red. Hemolysis can be detected as a presence of hemoglobin in plasma and RBC aggregation growth may be linked to blood thickening. Once hemolysis occurred, the PS can bind directly to the hemoglobin molecule and inactivate it. The attempts of the reoxygenation of the PDT deoxygenated (with Photosens and methylene blue PSs) blood samples were reported successful restoring pO_2 in the samples with PSs after light irradiation back to 100–170 mmHg and the blood oxygen saturation level to almost 97% observing both pH and pCO_2 virtually stable [70]. This implies that the use of Photosens and methylene blue does not cause oxygen transport dysfunction of blood itself during PDT. Chlorine p6 I in the cuvette after laser irradiation (0.025 mg ml^{-1}, 675 nm, 35 mW cm^{-2}) leads to significant photohemolysis (Figure 20.19a) and hemoglobin photodestruction (Figure 20.19b) evaluated with fully oxygenated whole (for photohemolysis) or hemolyzed (for hemoglobin photodestruction) blood evaluating via absorption spectra immediately after irradiation, relative to the spectra of the samples without PS [70].

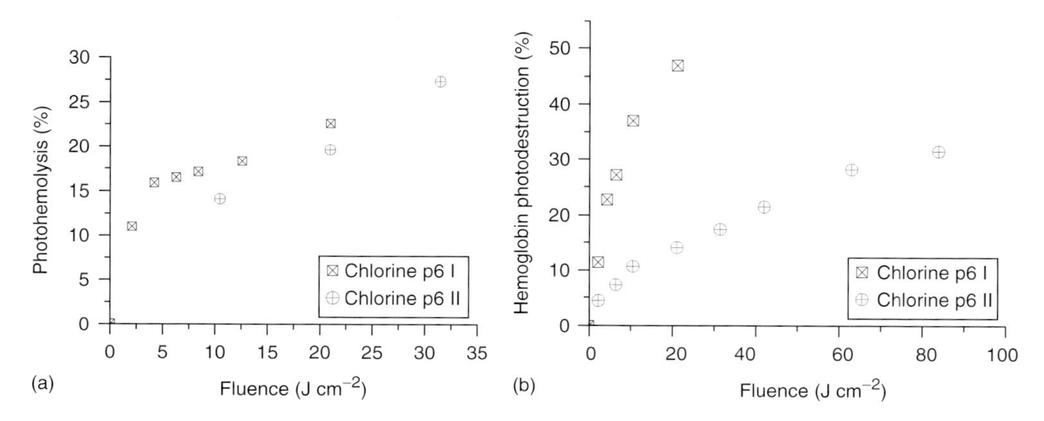

Figure 20.19 Monitoring of photohemolysis in whole blood (a) and hemoglobin photodestruction in hemolyzed blood (b) during irradiation *in vitro*. Both the photohemolysis and photodestruction of hemoglobin are measured in percent of the control (non-irradiated) sample, which is the reference. $\lambda = 675$ nm, $P = 35$ mW cm^{-2}. Concentrations: chlorine p6 I – 25 mg l^{-1}; chlorine p6 II – 25 mg l^{-1}. (Updated from Ref. [70].)

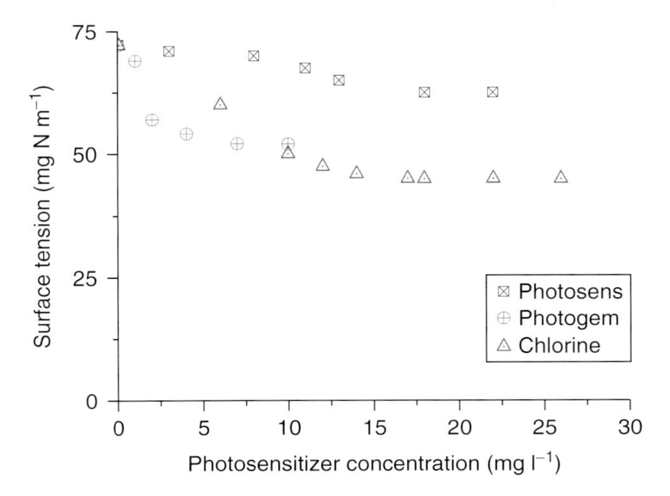

Figure 20.20 Surface tension of whole blood depending on PS concentration. Photosens®, Pthotogem®, and chlorine p6 I were studied.

The surface tension of blood drops with a growth of PS concentration (Figure 20.20).

Douplik *et al.* observed a substantial increase in viscosity of whole blood photosensitized with Chlorine p6 in cuvette after laser irradiation (25 mg l^{-1}, 675 nm, 120 J cm^{-2}). This phenomenon can be also related to protein denaturation in the Chlorine p6 I system, due to its expected high ability for binding with erythrocyte membrane and hemoglobin [70]. It is not clear whether viscosity within a single blood cell is increasing during PDT. A recent report by Kuimova *et al.* stated that there is a dramatic increase in the viscosity inside the cancer cells leading to the cancer cell's further deterioration by slowing down vital communication and transport processes inside the cell, and the toxic oxygen molecule's mission to kill the cell is slowed down too [114]. The difference between PDT reactions in blood *in vitro* versus *in vivo* is mainly prescribed by the presence of blood circulation including renewable oxygen supply, blood flow shear rate and shear stress conditions, homeostasis, and blot clotting (usually blocked *in vitro* by anticoagulants). Therefore, photodynamic reactions in blood *in vivo* are always a balance between local effects and blood supply, while PDT *in vitro* is usually nonreversible and noncompensated process. *In vitro*, oxygen consumption is purported to be an irreversible process because of limited content of oxygen in the sample. In living tissue, the blood oxygen level is a balance between arterial input and venous output, which are the parts of blood microcirculation supply system of tissue metabolism (Figure 20.4). The photodynamic process may cause damages of microvasculature and causes additional consumption of oxygen in the tissue and directly in the blood, thus shifting this balance. The decrease of blood oxygen saturation following PDT depends on the fluence, fluence rate, PS type, its concentration in blood and surrounding tissue, and extends blood microcirculation within the volume of light irradiation. An attempt has been reported to use a blood circulating circuit including membrane oxygenator, peristaltic pump, and mixing chamber [46] for study PDT reactions in blood at conditions closer to *in vivo* [83]. PSs during PDT undergo photobleaching mediated either by singlet oxygen or by the sensitizer triplet state, and that the relative importance of each depends on the ambient oxygen concentration [115]. The fluorescence reduction is assumed to be proportional to loss of photodynamic activity of the PS accompanied by changes of its phototoxicity [116, 117]. In turn, the PS phototoxicity depends on oxygen concentration in surrounding media [118]. On the other side, photobleaching itself can lead to decrease of oxygen consumption [119]. Rotomskis *et al.* observed photobleaching *in vitro* at bleaching rates [120]. The molecular oxygen consumed in PDT reaction, is a convenient parameter to quantify PDT. Actually, the consumed oxygen concentration during PDT is equal to the concentration of oxidized centers (defects) in the tissue. Once the concentration of the defects overbalances a threshold value, the tissue is irreversibly damaged. However, this threshold value depends on PS localization within the cells. As singlet oxygen diffusion length in tissue is as low as 10–15 nm [11], the tissue damage takes place in the vicinity of PS binding site. If PS is bound to the vitally important sites such as cell nuclei or mitochondria, the fewer amounts of defects can be generated to kill the cell. Thus, different PSs in case of different localization require different threshold dose of consumed oxygen for tissue damage. In studies of Douplik *et al.* (2000), photobleaching was not observed in PBS solution and blood plasma and most expressed photobleaching effect took place on precipitated RBCs. Figure 20.21 summarizes the results of photobleaching monitoring in precipitated erythrocytes.

The fluorescence recovery processes for Chlorine p6 I, HPD (Photogem®), Aluminum phthalocyanine (Photosens), and methylene blue suggests that absence of light irradiation may lead to the recovery of the PS fluorescence in biological objects *in vitro* up to the very initial level [121]. Due to the observed photobleaching of PSs, *PDT* can be defined as a self-inhibiting process unless it is externally supported. Below we present a model for PDT monitoring *in vitro*.

20.8
Blood Sample *In vitro* as a Model Studying Photodynamic Reaction

Whole blood is a convenient medium for investigation of these effects as the oxygen consumption during PDT could be easily measured *in situ* by evaluating hemoglobin oxygen saturation from its absorption spectra. For the first time the phenomenon of hemoglobin disoxygenation in blood samples incubated with sulphonated aluminum phthalocyanines was observed in [122]. In [112] this effect has been confirmed for blood samples incubated with various PSs. The detailed studies of oxygen consumption and fluorescence intensity in blood samples incubated with different PS concentrations light fluence rates were carried out in

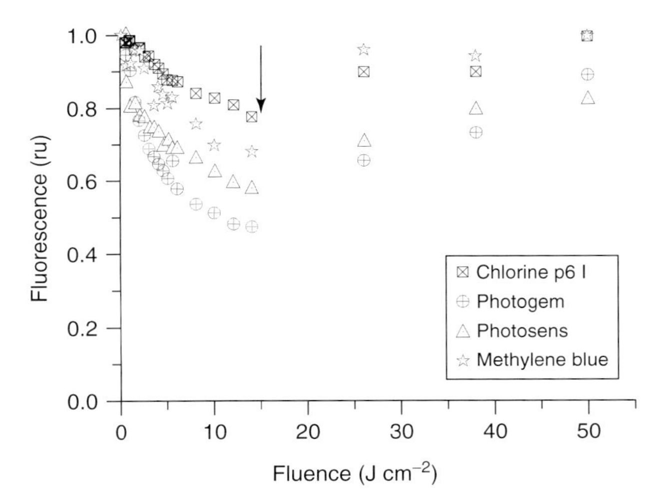

Figure 20.21 Photobleaching of photosensitizers in precipitated erythrocytes. Fluorescence presented is normalized by the initial measurements within the first 1–3 s of excitation, which is assumed to have 100% of its initial fluorescence level. Black arrow indicates the moment of the excitation laser off. Excitation $\lambda = 633$ nm, $P = 100$ mW cm^{-2}. Photosens® (triangles) – 1 mg l^{-1}; chlorine p6 I (squares) – 10 mg l^{-1}; Photogem® (circles) – 50 mg l^{-1}; and methylene blue – 10 mg l^{-1}. (Updated from Ref. [70].)

[123]. Apart from blood samples the effect of hemoglobin disoxygenation during PDT was also observed *in vivo* in mice models [36]. Two effects might be responsible for hemoglobin disoxygenation – extra oxygen consumption due to PDT quantum reaction and vessel destruction or vasoconstriction [36, 124] as a consequence of PDT. The last mechanism should be especially pronounced for PSs acting through "vessel mechanism." The predominating capillary destruction should also be observed when light irradiation is applied during a short time interval after PS injection, when significant amount of the PS still circulates in the blood. Capillary thrombosis and capillary hemorrhage will also alter the tissue oxygen supply. The blood sample model can be considered a convenient experimental object to study PDT reactions in tissues *in vivo*. An experimental setup for monitoring of blood samples during PDT proposed by Stratonnikov, Douplik *et al.* facilitated simultaneous detection of transmitted light in the range of 450–630 nm, excitation backscattered light at 670 nm and fluorescence in the range of 690–850 nm by means of a single CCD array as shown in Figure 20.22.

The transmission spectra were used to evaluate hemoglobin oxygen saturation and relative hemoglobin concentration. The algorithm applied is described in details in [86]. The decrease of light intensity in the sample depth due to blood absorption is also non substantial as the absorption length at 670 nm $l_a = 1/\mu_a$ is greater than the sample thickness. Taking into account that extinction coefficients of hemoglobin in deoxygenated and oxygenated form are ε (*R*Hb, 670 nm) = 3045 cm^{-1} M^{-1} and

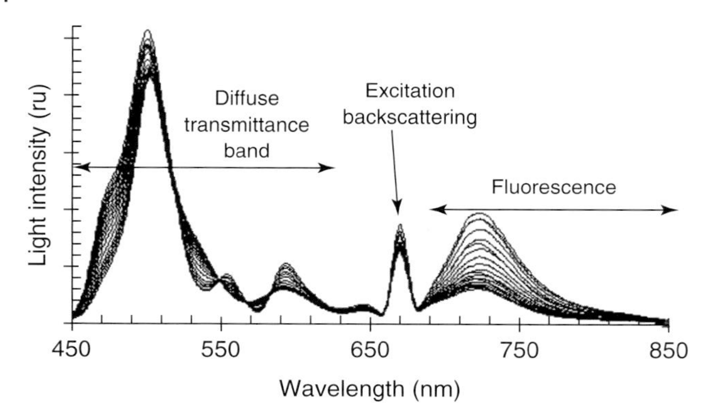

Figure 20.22 The raw spectra detected on a single CCD detector combining the diffuse transmittance, laser backscattering, and fluorescence bands during blood PDT monitoring. The spectra were obtained on blood sample incubated with methylene blue at 50 $\mathrm{mg\,l^{-1}}$ during laser irradiation at fluence rate 200 $\mathrm{mW\,cm^{-2}}$.

$\varepsilon\,(\mathrm{HbO_2}, 670\,\mathrm{nm}) = 321\,\mathrm{cm^{-1}\,M^{-1}}$ and concentration of hemoglobin in whole blood is $C_{\mathrm{Hb}} = 145\,\mathrm{g\,l^{-1}}$ one can easily calculate that corresponding contribution of hemoglobin to blood absorption coefficients. The calculation gives that $\mu_{\mathrm{a}}(R\mathrm{Hb}, 670\,\mathrm{nm}) = 15.2\,\mathrm{cm^{-1}}$, $\mu_{\mathrm{a}}(\mathrm{HbO_2}, 670\,\mathrm{nm}) = 1.6\,\mathrm{cm^{-1}}$, and the decrease of light intensity (without correction for scattering) at the far surface of 125 μm blood layer is 83 and 98% correspondingly for deoxygenated and oxygenated blood. So the average changes of light intensity through the blood sample volume due to deoxygenating are only 7.5% allowing one to consider the spatial light distribution inside the blood sample to be uniform and equal to that of incident on the sample surface. Thus, the problems of tissue optics are not substantial in this model.

20.8.1
Blood Heating Effects during PDT *In vitro*

It is necessary to evaluate also the possibility of blood heating due to light absorption as we applied sufficiently high light irradiancies up to $2\,\mathrm{W\,cm^{-2}}$. In our experiment hemoglobin absorption was mainly responsible for blood heating. The heat generated by light absorption in blood layer is dissipated by thermo conductivity through surrounding glass into the ambient medium. At equilibrium that should be attained in some time interval after starting irradiation the amount of generated heat is equal to the heat flow due to the thermo conductivity. By equating these two values one can easily get the temperature difference between blood and ambient media at equilibrium:

$$\Delta T = \frac{P\mu_{\mathrm{a}}\mathrm{d}L}{2k} \tag{20.8}$$

where P is light fluence rate, $d = 125$ μm – blood layer thickness, $L = 1.2$ mm – glass thickness, and $k = 0.01$ W (cm K)$^{-1}$ – glass thermo conductivity. Substituting numerical values in this equation one can easily obtain that temperature difference between blood and ambient medium does not exceed 1 and 0.1°C for deoxygenated and oxygenated case correspondingly at light irradiance of 1 W cm^{-2}. We also made an experiment to evaluate heating by irradiating the blood sample without PS for 3 h at fluence rate 1.2 W cm^{-2} and observed no visible changes induced by heating. All these facts allow one to conclude that blood heating is not significant here.

20.8.2
Monitoring of Oxygen Consumption, Photosensitizer Concentration, and Fluence Rate during Photodynamic Therapy in Whole Blood and Individual Blood Cells

Direct measurement of ROS in tissues is rather a difficult problem. A more attractive way for PDT control is implicit dosimetry based on the control of parameters directly correlating with final biological effects [125]. The efficiency of oxygen consumption during PDT when deficiency of oxygen itself is not a limiting factor, and at low light fluence rates when depletion of PS ground state may be neglected, is determined by the following expression:

$$\Gamma_{\mathrm{PDT}} = k_{\mathrm{PDT}}\,[C] \cdot P \tag{20.9}$$

where Γ_{PDT} is oxygen consumption rate measured in molar per second, k_{PDT} is oxygen consumption coefficient measured in joules per square centimeter, $[C]$ is PS concentration measured in molar, and P is light fluence rate measured in milliwatts per square centimeter. The oxygen consumption coefficient has the sense of consumed oxygen molecules per one PS molecule at the delivered fluence of 1 J cm^{-2}. The product of delivered fluence during PDT session to oxygen consumption coefficient gives the number of consumed oxygen molecules per one PS molecule (PS photobleaching is neglected). The inverse value to oxygen consumption coefficient may be called the *oxygen consumption fluence*: $f_{\mathrm{PDT}} = 1/k_{\mathrm{PDT}}$, as it has the dimension of fluence and has the sense of the delivered fluence, at which for every PS molecule one molecule of oxygen is consumed in PDT reactions. The values of f_{PDT} and k_{PDT} are very convenient for estimation of PDT efficiency of PSs working via singlet oxygen formation. As per mentioned above, the PDT is a catalytic process and its advantage is that small amount of PS, which works as a catalyst, may generate many oxidation acts within the region exposed to light. Let us estimate the value of oxygen consumption coefficient, k_{PDT}, so that during light session one PS molecule could generate at least several oxidation acts. Typical light fluences f during PDT are of orders of 100 J cm^{-2}. Consequently, the oxygen consumption coefficient should be much greater than $1/f$: $k_{\mathrm{PDT}} \gg 10^{-2}$ (cm^2 J^{-1}). In this case, the advantage of PDT as a catalytic process will be implemented. Now, let us estimate the oxygen consumption coefficient for PS from its photochemical properties and assure that one PS molecule really produce a lot oxidation acts during light session. The oxygen consumption coefficient is defined mainly by PS extinction and singlet oxygen quantum yield and may be written as

$$k_{\mathrm{PDT}} \equiv \frac{1}{f_{\mathrm{PDT}}} = \frac{\sigma}{E_{h\nu}} \varphi_{\Delta} \cdot \varphi_{\mathrm{ch}} = \frac{\ln 10 \times 10^3}{N_A} \cdot \frac{\lambda}{\hbar c} \cdot \varepsilon \cdot \varphi_{\Delta} \cdot \varphi_{\mathrm{ch}} = K \cdot \varepsilon \cdot \varphi_{\Delta} \cdot \varphi_{\mathrm{ch}} \tag{20.10}$$

where σ is the PS absorption cross section (cm^2), ε is the PS extinction coefficient ($M^{-1}\ cm^{-1}$), φ_{Δ} is the quantum yield of singlet oxygen formation, φ_{ch} is the quantum yield of chemical singlet oxygen quenching, $E_{h\nu}$ is the photon energy (J), $\hbar$ is the Plank constant, N_A is the Avogadro constant, c is the speed of light, and λ is the photon wavelength. For $\lambda = 670$ nmthe numerical constant K in the last part of Equation 20.10 is 1.3×10^{-2} ((M cm) $\cdot$ ($cm^2\ J^{-1}$)). For typical PSs which has extinction coefficient in the range of $10^5\ M^{-1}\ cm^{-1}$ and $\varphi_{\Delta} \sim 0.1{-}0.3$ we can expect that oxygen consumption coefficient is in the range of $100\ cm^2\ J^{-1}$ (assuming that chemical quenching of singlet oxygen is prevailing ($\varphi_{\mathrm{ch}} \sim 1$)). The corresponding oxygen consumption fluence is expected $f_{\mathrm{PDT}} = 1/k_{\mathrm{PDT}} \sim 10\ mJ\ cm^{-2}$ or lower. So, for typical PSs we estimate that one PS molecule is capable to generate up to 10^4 oxidation acts during light session at typical fluence of $100\ J\ cm^{-2}$, if photobleaching and oxygen deficiency are not substantial. Let us show that for whole blood the oxygen consumption coefficient may be evaluated from the slope of hemoglobin oxygen saturation curve. In whole blood some oxygen is bound to hemoglobin and some oxygen is dissolved. The value of hemoglobin oxygen saturation in blood is defined by the following relation:

$$SO_2 = \frac{[\mathrm{HbO_2}]}{[\mathrm{Hb}] + [\mathrm{HbO_2}]} \tag{20.11}$$

where $[R\mathrm{Hb}]$ and $[\mathrm{HbO_2}]$ are molar concentrations of heme complexes free and bound to oxygen molecule respectively. It should be reminded that one hemoglobin molecule contains four structural units of hem. In fact, hemoglobin oxygen saturation is defined by relative occupation of oxygen bound sites. The total molar concentration of oxygen bound to hemoglobin in blood is determined by

$$[\mathrm{O_2}]_{\mathrm{bound}} = 4\,[\mathrm{Hb}T] \cdot SO_2 = 4 \frac{C_{\mathrm{Hb}}}{M_{\mathrm{Hb}}} SO_2 \tag{20.12}$$

where $[\mathrm{Hb}T] = ([R\mathrm{Hb}] + [\mathrm{HbO_2}])/4$ – total molar concentration of hemoglobin in blood, C_{Hb} – hemoglobin concentration in blood measured in $g\,l^{-1}$, and $M_{\mathrm{Hb}} = 66\,500\ g\,mol^{-1}$ – molecular weight of hemoglobin molecule (four structural units). In our blood samples the hemoglobin concentration was $C_{\mathrm{Hb}} = 145\ g\,l^{-1}$ giving the value of molar concentration of bound oxygen at 100% saturation to be $[\mathrm{O_2}]_{\mathrm{bound}} = 8.7 \times 10^{-3}$ M. Apart from bound to hemoglobin some oxygen is dissolved in blood. The hemoglobin oxygen saturation is in dynamic equilibrium with concentration of dissolved oxygen $-[\mathrm{O_2}]$, its dependence defined by empiric Hill equation:

$$SO_2 = \frac{[\mathrm{O_2}]^n}{[\mathrm{O_2}]^n + [\mathrm{O_2}]_{50}^n} \tag{20.13}$$

with $n \approx 2.34$ and $[\mathrm{O_2}]_{50} \approx 35\ \mu M$ – the concentration of dissolved oxygen when hemoglobin is half saturated [126]. The molar concentration of dissolved oxygen according to Henry–Dalton law is proportional to partial oxygen pressure pO_2:

$$[O_2] = \frac{\alpha}{22.4(\mathrm{l\ mol^{-1}})} \cdot \frac{pO_2(\mathrm{mmHg})}{760(\mathrm{mmHg})} \qquad (20.14)$$

where the dimensionless value α is Bunzen solubility coefficient. For blood and water at 37 °C this value is 0.024 [127], resulting in the following simple relation between molar oxygen concentration and partial oxygen pressure in blood: $[O_2]\,(\mu M) \approx 1.4 \cdot pO_2\,(\mathrm{mmHg})$. It can be easily seen that in the range of SO_2 values from several percent up to 95% concentration of bound oxygen is almost two orders of magnitude higher than that of dissolved oxygen. Thus, the main surplus of oxygen in blood is bound to hemoglobin and dissolved fraction may be neglected in evaluating the oxygen consumption rate during PDT. The dose of consumed oxygen D_{PDT} and oxygen consumption rate Γ_{PDT} during PDT in blood samples may be estimated with good accuracy from the changes in hemoglobin oxygen saturation:

$$D_{PDT} = 4\,[HbT]\,\Delta SO_2 \qquad (20.15)$$

$$\Gamma_{PDT} = -4\,[HbT]\,\frac{dSO_2}{dt} \qquad (20.16)$$

Combining the Equations 20.9 and 20.16 one can easily evaluate the oxygen consumption coefficient from the slope of oxygenation curve at the initial moment after starting light irradiation:

$$k_{PDT} = \frac{-4\,[HbT]\,\dfrac{dSO_2(0)}{dt}}{[C_0]\cdot P} \qquad (20.17)$$

This equation will be applied for estimation of oxygen consumption coefficient from experimental data and comparing it with the theoretical value given above by Equations 20.10. The effects of the oxygen deficiency at PDT are revealed when monomolecular PS decay rate of excited triplet to ground state defined by k_p value prevails over the quenching of PS triplet state by molecular oxygen defined by k_{ot} (Figure 20.2 and Glossary). It should be noted that the effects of oxygen deficiency in PDT are revealed when monomolecular PS decay rate of excited triplet to ground state defined by k_p value prevails over the quenching of PS triplet state by molecular oxygen defined by k_{ot}. It is achieved when local concentrations of molecular oxygen drop below its critical value defined by the following relation [128]:

$$\left[{}^3O_2\right]_{crit} = \frac{k_p}{k_{ot}} \qquad (20.18)$$

Substituting the available values for sulphonated aluminum phthalocyanines [118, 129]: $k_p \approx 2 \times 10^3\ s^{-1}$ and $k_{ot} = 2.3 \times 10^9\ M^{-1}\ s^{-1}$ we obtain that $[{}^3O_2]_{crit} = 0.86\ \mu M$ for this PS. For HPDs (Photofrin) and Nile blue selenium (EtNBSe) there were obtained [117] substantially higher values for $[{}^3O_2]_{crit}$ – 11.9 and 222 μM respectively. It means that sulphonated aluminum phthalocyanines may be more efficient in the deficiency of oxygen. Assuming that concentration of dissolved oxygen is in equilibrium with hemoglobin oxygen saturation defined by Hill equation (Equation 20.13), we obtain that the deficiency of oxygen became appreciable when SO_2 value drops below $1.7 \times 10^{-2}\%$. However, during irradiation the local concentration of oxygen in space between erythrocytes may be somewhat lower than the equilibrium value due to limited diffusion. This decrease may be evaluated by solving diffusion equation with consumption in one dimensional space:

$$\Delta\left[{}^3O_2\right] = -\frac{1}{2}\frac{\Gamma_{PDT}}{D}R^2 \quad (20.19)$$

where D is diffusion coefficient of oxygen $D \approx 2 \times 10^{-5}\,\text{cm}^2\,\text{s}^{-1}$, and R is an average half distance between erythrocyte borders. Taking this distance to be several microns in whole blood and substituting the value for oxygen consumption rate $\Gamma_{PDT} = 250\,\mu\text{M}\,\text{s}^{-1}$ achieved in our experiments at the highest fluence rate ($1200\,\text{mW}\,\text{cm}^{-2}$) and PS concentration (20 μM) we obtain that the reduction in oxygen concentration is $\Delta[{}^3O_2] = 1\,\mu\text{M}$. Owing to erythrocyte aggregation the size of volumes free from erythrocytes may be greater than the average distance between erythrocytes resulting in higher value of oxygen decrease in these regions. However, even at O_2 drop 10 times greater than estimated above – (10 μM) the corresponding SO_2 value calculated from Hill equation (Equation 20.13) will be as low as 5%. Thus, neglecting the range of very low SO_2 values (less than several percent) the oxygen deficiency during light irradiation is not substantial in our model of blood samples *in vitro*. We shall further assume that strong variations in the oxygen consumption rate dynamics observed experimentally with the blood samples are due to PS photobleaching or rather PS photo modification. In terms of catalytic reaction PS photobleaching during PDT is the catalyst poisoning. In this case the oxygen consumption rate will be not constant during light session. It may be taken into account by substituting in Equation 20.8 the time-dependent PS concentration. For limiting cases of first- and second-order photobleaching, this dependence is given by

$$\frac{[C](t)}{[C_0]} = \exp(-t/\tau_1) \quad \text{first order bleaching} \quad (20.20)$$

$$\frac{[C](t)}{[C_0]} = \frac{1}{1 + t/\tau_2} \quad \text{second order bleaching} \quad (20.21)$$

where τ_1, τ_2 are the photobleaching time constants for the first- and second-order correspondingly.

One of the ways to control photobleaching during PDT is monitoring PS fluorescence. If we assume that fluorescence signal is proportional to PS concentration multiplied by light fluence rate: $I_{flu} \sim [C] \cdot P$, then using Equation 20.9 we obtain that oxygen consumption rate should be proportional to PS fluorescence signal: $\Gamma_{PDT}(t) \sim I_{flu}(t)$. On the other hand, according to Equation 20.16 the oxygen consumption rate is proportional to the slope of oxygen saturation curve and consequently, this slope should be proportional to the fluorescence dynamics:

$$\frac{dSO_2}{dt}(t) \sim I_{flu}(t) \quad (20.22)$$

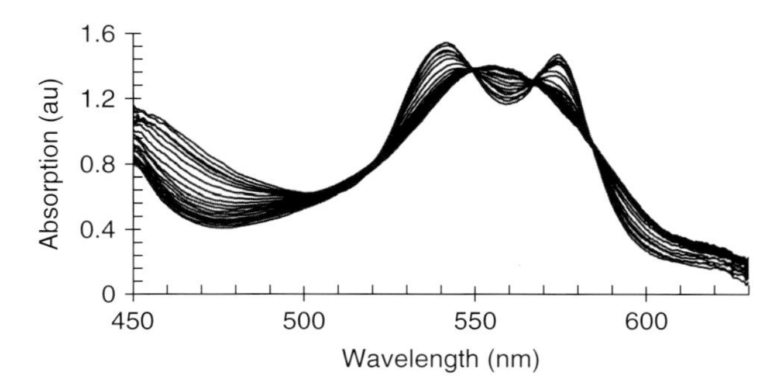

Figure 20.23 The absorption spectra variation in blood sample incubated with methylene blue during laser irradiation. The initial two-hump spectrum of HbO_2 transformed into a single peak spectrum of *R*Hb as a result of PDT reaction.

By simultaneous measuring the PS fluorescence and blood oxygen saturation one can evaluate whether it is possible to apply fluorescence monitoring for real time control of PDT efficiency (oxygen consumption rate). For example, if during PDT, due to the photo modification, the PS lost its photodynamic activity, but retained its fluorescence, it would not be an adequate PDT dosimetry by monitoring the fluorescence alone. Further, it will be shown that the fluorescence dynamics of methylene blue adequately represents its temporal PDT efficiency variations. However for Photosens the PDT efficiency drops down sharply after starting irradiation, but the fluorescence decreases much slower. The resulting absorption spectra calculated from the diffuse transmission spectra (Figure 20.22) are shown in Figure 20.23. The observed changes in the absorption spectra are due to hemoglobin deoxygenating during laser irradiation.

The spectral fluorescence band with excitation backscattered peak is shown in Figure 20.24. One can observe the strong increase in fluorescence (outburning) induced by laser irradiation followed by photobleaching. There are also small changes in the fluorescence spectral line shape (5 nm red shift) in the course of laser irradiation observed for methylene blue. For Photosens the spectral line shape changes are even less (data are not shown). In the following analysis we calculated the complete area under fluorescence curve (685–850 nm) to quantify the fluorescence signal. The excitation backscattered signal was also calculated as area under the laser peak (660–680 nm).

The dynamics of oxygenation, fluorescence, and laser backscattering normalized on theirs initial values observed in blood samples incubated with methylene blue and Photosens during laser irradiation at 400 mW cm^{-2} are shown in Figure 20.25. There is striking difference between these two PSs. First, for methylene blue (a), a strong fluorescence outburning induced by laser irradiation followed by fluorescence decrease due to photobleaching is observed. For Photosens the fluorescence outburning is negligible and for other PS concentrations and fluence

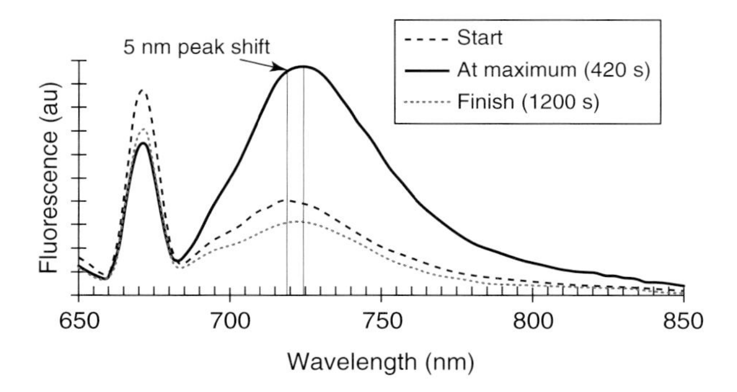

Figure 20.24 Fluorescence variation in blood sample incubated with methylene blue during laser irradiation at fluence rate $P = 400$ mW cm^{-2}. Dashed – initial spectrum; solid – 420 s after starting irradiation, when fluorescence outburning is maximal; dotted gray – 1200 s after starting irradiation.

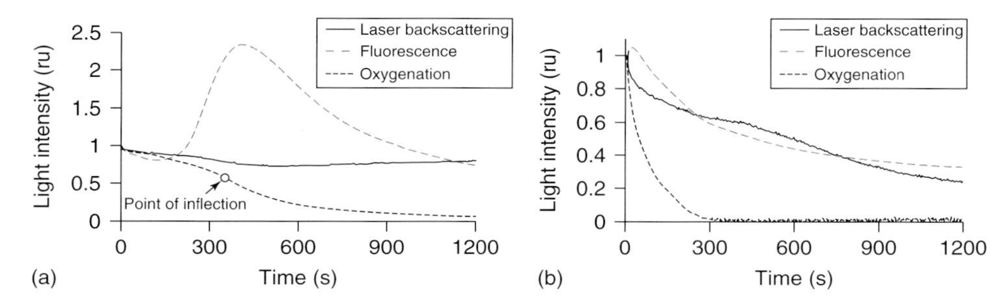

Figure 20.25 Dynamics of excitation backscattering (solid), fluorescence (dashed), and oxygenation (dotted) for blood sample incubated with methylene blue (a) at concentration of 50 mg l^{-1} and Photosens® (b) at concentration of 20 mg l^{-1} during laser irradiation at fluence rate $P = 400$ mW cm^{-2}. The inflection point on the oxygenation curve matches the maximum of the fluorescence outburning for methylene blue.

rates (see below) is not observed at all. We observe just a smooth fluorescence decrease during laser irradiation. Second, the oxygenation curve for methylene blue has evident sigmoid decrease with distinct point of inflection attained in time close to the fluorescence maximum. For Photosens the oxygenation curve has sharp drop down with the decrease rate (curve slope) strongly diminishing in the course of irradiation. And, third, the laser backscattering is strongly decreased for Photosens while for methylene blue the small initial decrease of the laser backscattering is followed by the increase of this value. It has been quantitatively estimated above that the light intensity at passing the 125 μm whole blood layer is decreased from 98 to 83% after deoxygenating. Approximately the same drop (15%) is observed with the excitation backscattered signal, as its average path length through the blood sample in this case is about the sample thickness. The value measured is 20% (see the initial drop of curve 1 in Figure 20.25) higher than that of estimated. This difference may be due to modification of scattering during the disoxygenation as discussed before in Section 20.5.1. During light irradiation of sensitized blood the erythrocytes are hemolyzed. As it has been observed experimentally that at the first stage of photohemolysis the erythrocytes swell and then burst out [130]. Apparently, the blood scattering properties are altered during PDT. Erythrocyte's swelling should result in increase of backscattering, while their destruction should sharply reduce the excitation backscattering. Hence, the backscattering signal may serve as a quantitative value of erythrocyte hemolysis rate. At low PS concentrations and low light fluence rates, CPH was not observed, but the distinct DPH was registered at 24–48 h after irradiation. At high light fluence rates and PS concentrations, CPH was discovered. After initial reduction of backscattering coefficient, which is attributed to the increase of blood absorption during deoxygenating as has been mentioned previously; a slight growth of backscattering is observed due to erythrocyte swelling followed by a sharp decrease due to erythrocyte destruction. It should be noted here that decreasing rate at the last stage is determined rather by the total fluence delivered to the sample than by the current fluence rate. The drop of backscattering after switching off the laser at the final stages of hemolysis does not continue.

The phenomenon of fluorescence outburning during PDT has been widely studied for aluminum phthalocyanines [131] and other PSs [132]. The outburning of PSs such as phthalocyanines and methylene blue during PDT is usually attributed to the initial dye localization in lysosomes at high concentration with low fluorescence quantum yield. The laser irradiation destroys lysosomes as a result of PDT damage releasing PS into cytoplasm where its fluorescence quantum yield goes up. At the same time the irreversible destruction of PS by laser irradiation leads to the fluorescence decrease (photobleaching). As a result, the initial fluorescence increase after passing through the maximum is followed by the fluorescence reduction. However, in the case of whole blood where the erythrocytes are prevailing over the other cells, it is improbable that substantial amount of PS is localized in lysosomes. As erythrocytes have no lysosomes it is possible that there are some other binding sites where PS is localized initially at high local concentration. High fluorescence outburning during light irradiation of cells incubated with methylene blue was observed in [133]. The oxygenation curve slopes in Figure 20.25a for methylene blue is $\mathrm{d}SO_2/\mathrm{d}t = 1.4 \times 10^{-3}\ \mathrm{s}^{-1}$ at the inflection point while the corresponding slope for Photosens at the initial time ($t = 0$) is (Figure 20.25b) $\mathrm{d}SO_2/\mathrm{d}t = 2.2 \times 10^{-2}\ \mathrm{s}^{-1}$. Substituting these data into Equation 20.17 we obtain the following values for oxygen consumption coefficient (k_{PDT}) of methylene blue and Photosens correspondingly 0.19 and 24 $\mathrm{cm}^2\ \mathrm{J}^{-1}$. It should be reminded that we calculated the oxygen consumption coefficient at the time when it attains its maximum value (inflection point for methylene blue and initial moment for Photosens). For Photosens oxygen consumption efficiency drops down dramatically in the first few minutes after starting irradiation while for methylene blue the oxygen consumption is low both at the beginning and end of irradiation. Assuming that chemical quenching of singlet oxygen is prevailing ($\varphi_{\mathrm{ch}} \sim 1$) and substituting the values of extinction coefficients 75 000 and 100 000 $l\,mol^{-1}\,cm^{-1}$ and singlet oxygen generation quantum yield 0.52 and 0.3 (Table 20.1) we get the values of oxygen consumption coefficient 500 and 390 $cm^2\,J^{-1}$ for methylene blue and Photosens correspondingly. We see that these values are substantially higher than those obtained above experimentally from blood deoxygenating. Possible reasons for this disagreement can be either domination of physical quenching of singlet oxygen without oxygen consumption ($\varphi_{ch} \ll 1$) or aggregation of the PS in blood, that is, fluence rate effects. It should be noted that typical light fluences delivered in PDT sessions are in the range of 50–200 $J\,cm^{-2}$. PS may be considered perspective for II-type PDT (via singlet oxygen) if its oxygen consumption coefficient is substantially higher than 0.01 ($cm^2\,J^{-1}$). In the opposite case one PS molecule will destroy just one target.

Figure 20.26 presents the dynamics of oxygenation and fluorescence during fractionated light irradiation (the fluorescence monitoring is possible only during light interval). For Photosens (Figure 20.26a) we observe that there are no changes both in fluorescence and oxygen saturation during dark time interval. Once laser is off, the oxygen saturation drop stops immediately (the time response in this experiment is 0.5 s). For methylene blue (Figure 20.26b) the strong outburning of fluorescence is observed during first and second dark intervals with weak

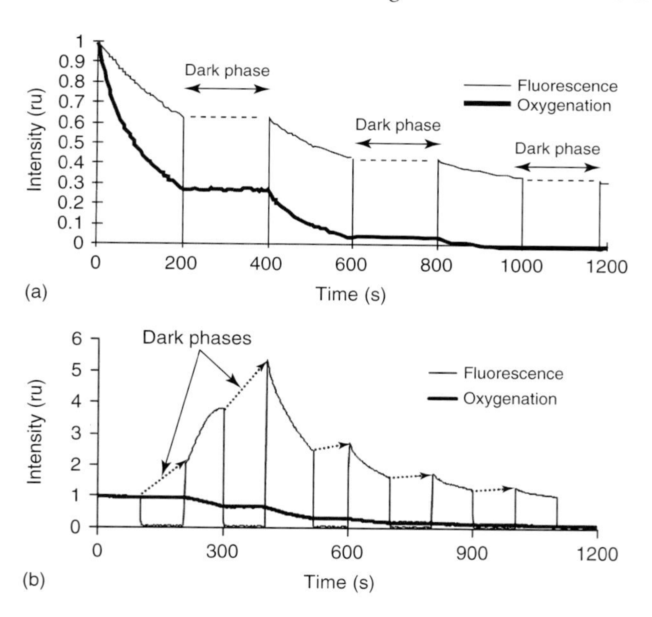

Figure 20.26 Fluorescence (thin curve) and oxygen saturation (thick curve) dynamics for Photosens® (a) (10 $mg\,l^{-1}$, $P = 400$ $mW\,cm^{-2}$) and methylene blue (b) (50 $mg\,l^{-1}$, $P = 800$ $mW\,cm^{-2}$) at fractionated mode of laser irradiation. During dark intervals fluorescence could not be measured and its extrapolated behavior is given by dashed lines.

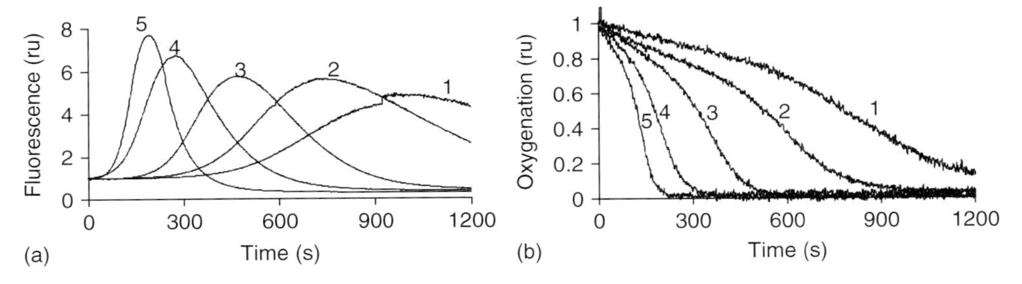

Figure 20.27 Fluorescence (a) and oxygenation (b) dynamics for blood samples incubated with methylene blue (100 mg l^{-1}) during laser irradiation at different fluence rates: (1) 200, (2) 400, (3) 800, (4) 1200, and (5) 2000 mW cm^{-2}.

fluorescence increase in the following dark intervals. The oxygen saturation is not changed during dark intervals; however, the slope to oxygen saturation curve increases sharply after the first dark interval. A detailed consideration of the fluorescence and oxygenation dynamics for methylene blue at various light fluence rates is shown in Figure 20.27a,b correspondently. A strong fluorescence outburning for all fluence rates is presented. As can be seen from the figure, the fluorescence outburning depends on the light fluence rate applied.

Considering Figures 20.26b and 20.27a, one can conclude that the outburning phenomenon of the fluorescence of methylene blue takes place within the dark interval and also depends on the light fluence rate applied. Photobleaching and disoxygenation in whole blood with Al phthalocyanine (Photosens) at different fluence rates and PS concentrations are shown in Figure 20.28. The following two features should be noted about the PS photobleaching – a sharp reduction of fluorescence intensity at the very initial moment of the irradiation observed only for the low PS concentrations (1.2 μM) (data not shown), and the second is a slight

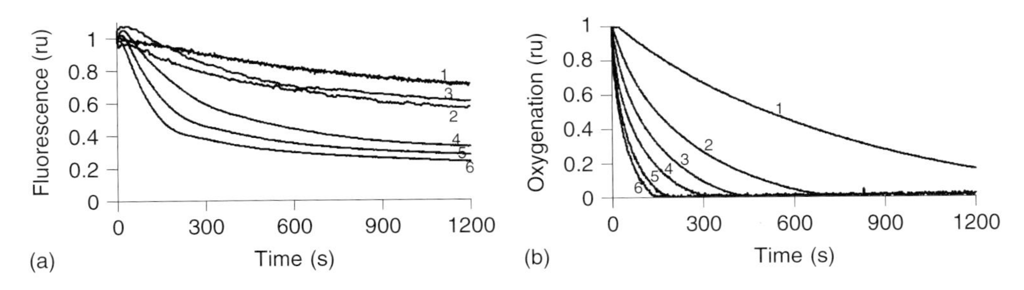

Figure 20.28 Fluorescence (a) and oxygenation (b) dynamics for blood samples incubated with Photosens® (20 mg l^{-1}) during laser irradiation at different fluence rates: (1) 10, (2) 50, (3) 100, (4) 400, (5) 800, and (6) 1200 mW cm^{-2}. (Modified from Ref. [86].)

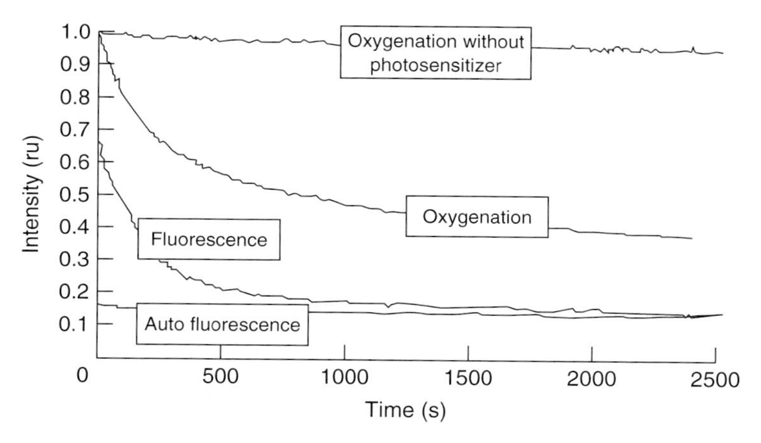

Figure 20.29 Dynamics of oxygenation and fluorescence in the limit of low photosensitizer concentration for blood samples incubated with Photosens® (1 μM), $P = 1200$ mW cm^{-2}. The PS fluorescence drops down to the autofluorescence level. (Reproduced by permission from Ref. [86].)

initial increase of the fluorescence intensity (outburning) observed for the high PS concentrations (Figure 20.28a). For the lower concentrations of PS the sharp drop mentioned above is disguised by the fluorescence outburning resulting in appearance of an inflection point at the fluorescence dynamic curve.

In conclusion to this section Figures 20.29 and 20.30 demonstrate the results for fluorescence and oxygenation monitoring in two limiting cases of low and high PS concentrations (Photosens). A special interest to these instances is that at low PS concentrations and high fluences all PS may be photo destroyed, while sufficient amount of oxygen is still preserved in blood.

It is possible to say that at the final stage we observe the oxygen consumption after total photobleaching of PS (Figure 20.29).

In the second limiting case of high PS concentration all the oxygen reserve will be utilized while high amount of PS is saved. In this case at the final stage the photobleaching kinetics is observed in the absence of oxygen as shown in Figure 20.30.

20.8.3
Theoretical Model of Oxygen Consumption and Photobleaching in Blood during PDT *In vitro*

A simple singlet oxygen-mediated PDT model taking into account photobleaching can be described by the following set of kinetic equations [128, 134, 135]. The values of $\Gamma_{PDT}(0)$ calculated from expression (20.16) as functions of PS concentration are linear while dependencies on fluence rate have relatively slower growth at all PS concentrations. To interpret these relations we apply the following analytical

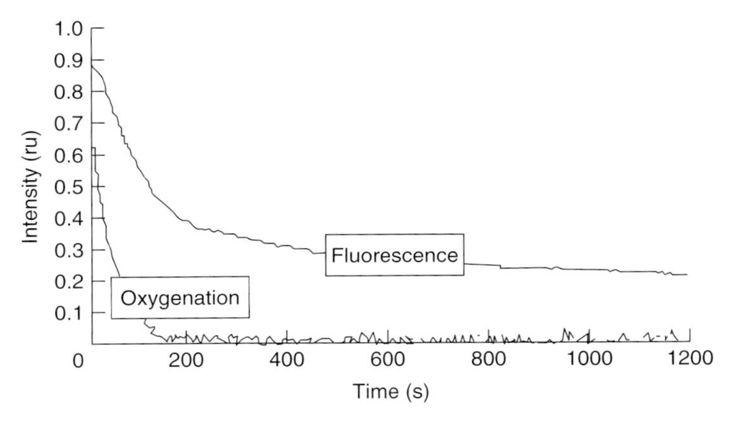

Figure 20.30 Dynamics of oxygenation and fluorescence in the limit of high photosensitizer concentration for blood samples incubated with Photosens® (20 μM), $P =$ 1200 mW cm^{-2}. (Reproduced by permission from Ref. [86].)

expression resulting from the theoretical considerations [86]:

$$\Gamma_{\mathrm{PDT}}\,(0) = \frac{1}{f_{\mathrm{PDT}}}\,[C_0]\,\frac{P}{1 + P/P_{\mathrm{sat}}} \tag{20.23}$$

where f_{PDT} is the oxygen consumption fluence, $[C_0]$ the initial molar PS concentration, P the fluence rate, and P_{sat} the saturation fluence rate. Our interpretation is based on the assumption that the PS ground state concentration is depleted during light irradiation. In a simple two-level model the following dependence of P_{sat} from the lifetime in excited state (τ) can be written as $P_{\mathrm{sat}} = \frac{E_{h\nu}}{\sigma\tau}$. Thirty six experimental points shown in Figure 20.31 at the fluence rates and PS

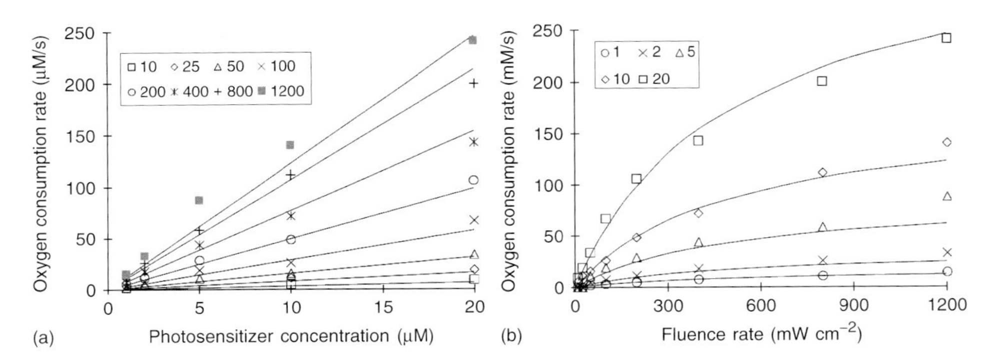

Figure 20.31 Oxygen consumption rate at initial time (Γ_{PDT} (0)) in blood samples as a function of photosensitizer concentration at various light fluence rates (a) and a function of fluence rate at various photosensitizer Photosens® concentrations (b). (Reproduced by permission from Ref. [86].)

concentrations varying about two orders of magnitude were fitted with only two parameters – oxygen consumption fluence f_{PDT} and saturation fluence rate P_{sat}. The best fit of these data with expression (20.23) was obtained at the following values of oxygen consumption fluence $-f_{PDT} = 26 \pm 6$ mJ cm^{-2} and saturation fluence rate $-P_{sat} = 500 \pm 120$ mW cm^{-2}. The corresponding theoretical curves according to Equation 20.23 with fitted values of oxygen consumption fluence and saturation fluence rate are plotted in Figure 20.31 with solid lines. The oxygen consumption fluence f_{PDT} is a convenient value for estimation of PDT efficiency. It has the sense of the delivered fluence, at which for every PS molecule one molecule of oxygen is consumed in PDT reactions when fluence rate effects are not substantial ($P \ll P_{sat}$). Since the tissue destruction during PDT is defined eventually by a dose of generated singlet oxygen that took part in irreversible chemical reactions with tissue (dose of consumed molecular oxygen), then, in the absence of photobleaching, we may write the following simple relation defining the total light fluence f, which should be delivered to the tissue site to achieve such dose:

$$D_{PDT} = \frac{f}{f_{PDT}} [C_0] \frac{1}{1 + P/P_{sat}} \tag{20.24}$$

Analyzing the dependence of the oxygen consumption rate from PC concentration we obtained a good linear relation in all range from 1 to 20 μM as predicted by Equation 20.23 (Figure 20.31a). The dependence of oxygen consumption from fluence rate (Figure 20.31b) revealed an obvious deviation from linear relationship, which may be described by the factor of $1/(1 + P/P_{sat})$ with $P_{sat} \approx 500$ mW cm^{-2} (Equation 20.23).

Assuming that the maximum value for the lifetime in excited triplet state τ is 5×10^{-4} s [129], we can easily obtain from relation (20.23) that the effects of depleting ground states are achieved at much more higher fluence rates ($P_{sat} = 2.8$ W cm^{-2}) compared to that of obtained above from our experimental data ($P_{sat} = 0.5$ W cm^{-2}). In the presence of molecular oxygen being the main quencher of triplet states the lifetime will be substantially less. For example, at oxygen concentration as low as $[^3 O_2] = 10$ μM and taking the bimolecular rate of triplet PS quenching by ground state oxygen to be $k_{ot} = 2.3 \times 10^9$ M^{-1} s^{-1} [26] we obtain that the lifetime in triplet state is $\tau = 1/(k_{ot}[^3O_2]) = 4.3 \times 10^{-5}$ s and the corresponding value for $P_{sat} = 33$ W cm^{-2}. To interpret the actual low values of P_{sat} observed in the experiments, we assume that the PS is capable of forming in blood long living complexes in excited states resulting in effective decreasing PS concentration taking part in photo catalytic reactions. The formation of such complexes of sulphonated aluminum phthalocyanines with human serum albumin and the corresponding increase in lifetime of triplet state was observed in [129].

The effect of decreasing PDT efficiency with higher fluence rate at the same fluence has been observed by many researchers for *in vivo* [136, 137] and *in vitro* [117, 138] models including blood samples [139]. In *in vivo* models, a fast depletion of molecular oxygen in tissues at high fluence rates is resulted due to the high oxygen consumption rate during PDT. To overcome this and increase PDT efficiency, the fractionated mode of irradiation with light and dark time intervals required to restore oxygenation in irradiation area has been proposed [117, 140, 141]. The average distance between blood capillaries, the main oxygen suppliers *in vivo*, is in the range of 50–200 μm that matches the molecular oxygen diffusion path length. If the oxygen consumption prevails over the oxygen supply, then oxygen can be completely depleted in the peripheral regions [136]. As it was mentioned the deficiency of molecular oxygen in tissues may result also from the capillary destruction or vasoconstriction during irradiation as a consequence of PDT effects [36, 124] such as destruction of capillaries itself. The PS photobleaching important for PDT dosimetry [142–147] can be described by the mixed-order bleaching model based on Equations 20.20 and 20.21 [148]

$$\frac{[C](t)}{[C_0]} = \frac{\exp(-t/\tau_1)}{1 + \frac{\tau_1}{\tau_2}\left(1 - \exp(-t/\tau_1)\right)} \qquad (20.25)$$

The values τ_1 and τ_2 are called *bleaching times* for the first- and second-order bleaching respectively. The photobleaching curves obtained in our experiment (Figure 20.28a) cannot be described by Equation 20.25. The behavior of fluorescence decrease is too complicated to be fit by such a simple one-component mixed-order bleaching model. First, a slight outburning of fluorescence for high PS concentrations was observed, and second, a sharp drop in fluorescence intensity was obtained for low PS concentrations. The possible involvement of fluorescent photoproducts may also complicate the interpretation of the experimental data. A sharp drop of fluorescence at the initial moment implies that a fraction of PS has a very low bleaching time. We assume that in whole blood there exist binding sites at which PS photobleaching is highly increased. It should be reminded that ambient medium is responsible for photodecomposition, as no PS destruction was observed in physiological solution (Figure 20.32). It is also possible that some active acceptor of PS triplet state resulting in its destruction is present in blood in small concentrations. Our estimation for the concentration of this acceptor is of the order of 1 μM. When the initial PS concentration is much higher than the concentration of such active acceptor, only a small fraction of PS can be destroyed resulting in smearing the initial drop of the fluorescence during irradiation.

From Figures 20.27b and 20.28b it is clearly seen that oxygen consumption rate, which is proportional to the slope of the curves plotted is substantially reduced in the process of light irradiation. If there had been no change in oxygen consumption rate, then the decrease of hemoglobin oxygen saturation SO_2 during PDT would have been linearly described except for the range of the low SO_2 values (few percent) when oxygen deficiency becomes substantial. We assume that the main reason for the decrease of oxygen consumption rate during the course of light irradiation is the PS photodecomposition. Since the oxygen consumption rate is proportional to PS concentration, its destruction during irradiation should inevitably result in the decrease of Γ_{PDT}. Presuming that PS fluorescence intensity is proportional to its concentration we can write the following relation between time derivative s'_t of the normalized SO_2 curves shown in Figures 20.27b and 20.28b and relative

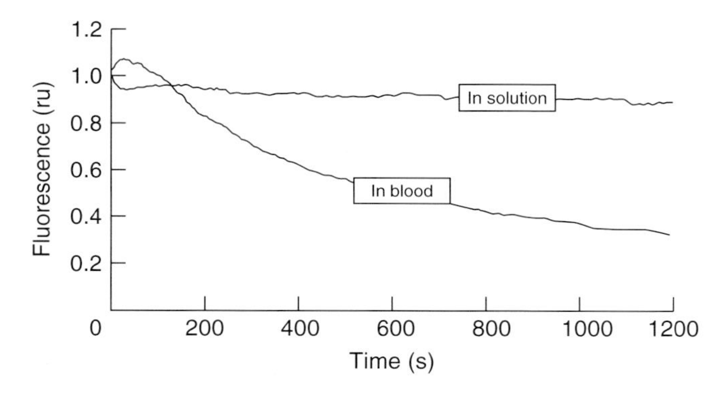

Figure 20.32 Comparison of Photosens® photobleaching dissolved in physiological solution and incubated in whole blood at the same concentration (5 μM) and irradiated at the same fluence rate (200 mW cm^{-2}). (Reproduced by permission from Ref. [86].)

fluorescence intensities presented in Figures 20.27a and 20.28a.

$$s'_{\mathrm{t}} = -\frac{1}{4\,[\mathrm{Hb}T]\,SO_2\,(0)}\,\frac{1}{f_{\mathrm{PDT}}}\,\frac{P\,[C_0]}{1 + P/P_{\mathrm{sat}}}\,\frac{I_{\mathrm{fl}}\,(t)}{I_{\mathrm{fl}}\,(0)} \qquad (20.26)$$

where $I_{\mathrm{fl}} \sim k_{\mathrm{f}}\,[S_1]$ (Figure 20.2) describes the fluence rate effect upon the oxygen consumption rate. However, the analysis of experimental data showed that the time derivative of oxygen saturation is not proportional to the relative fluorescence intensity as predicted by Equation 20.26. A simple visual comparison reveals the fact that the fluorescence intensity is decreased much slower than the oxygen consumption. Even the initial drops and outburning of fluorescence mentioned above are not apparent in the corresponding curves for hemoglobin oxygen saturation. It means that PS fluorescence intensity does not reflect its efficient concentration taking part in PDT reactions. The efficient PS concentration and, consequently, the oxygen consumption are reduced faster during light irradiation than the corresponding value of fluorescence intensity. It seems it has been the case that in this experimental model only a small fraction of total PS is active for oxygen consumption, while the entire amount of PS contributes into the fluorescence signal. This active fraction is responsible for oxygen consumption that subjected to intensive photobleaching, while the rest part of PS is destroyed at slower rate. This assumption is confirmed also by the fact that we obtained an increased quantity for f_{PDT} value. If one assumes that only 20% of total amount of PS is active for oxygen consumption, than the value for f'_{PDT} will be in correspondence with Equation 20.24 under assumption that chemical quenching of singlet oxygen is predominating. Taking into account the photobleaching of the first and second order, the following relation may be obtained for hemoglobin oxygen saturation changes during irradiation of blood samples incubated with PS

$$\frac{SO_2(t)}{SO_2(0)} = 1 - \frac{1}{4[HbT]\cdot SO_2(0)}\cdot\frac{[C_0]}{1+P/P_{sat}}\frac{f_2}{f_{PDT}}\cdot\ln\left(1+\frac{f_1}{f_2}\left(1-\exp(-Pt/f_1)\right)\right) \tag{20.27}$$

where f_1 and f_2 are bleaching fluences for the first and second orders, respectively, which can be written in the following form

$$\frac{1}{f_1} = \varphi_t k_{at}[A_t]\frac{1}{k_p + k_{ot}[^3O_2]}\cdot\frac{1}{f_a} \tag{20.28}$$

where the PS absorption fluence $f_a = \frac{E_{h\nu}}{\sigma}$ having the sense of the delivered fluence at which every PS molecule would absorb, on average, one photon; φ_t is the quantum yield of PS triplet state $\varphi_t = k_{isc}/(k_{isc} + k_m)$, k_p is the monomolecular PS decay rate of excited triplet to ground state; k_{ot} is the bimolecular rate of triplet PS quenching by ground state oxygen; k_{isc} is the PS intersystem crossing rate; k_m is the monomolecular PS decay rate of excited to ground state (radiative and nonradiative) (Figure 20.2 and Glossary).

$$\frac{1}{f_2} = \varphi_\Delta\frac{k_{os}}{k_d + k_{oa}[A_o]}\cdot\frac{k_{ot}\left[^3O_2\right]}{k_p + k_{ot}\left[^3O_2\right]}\cdot\frac{[C_0]}{1+P/P_{sat}}\frac{1}{f_a} \tag{20.29}$$

where φ_Δ is the quantum yield of singlet oxygen formation $\varphi_\Delta = \varphi_t \cdot S_\Delta$; k_{os} is the bimolecular rate of reaction between singlet oxygen and PS ground state; k_d is the monomolecular decay rate of singlet oxygen to ground state; k_{oa} is the bimolecular rate of reaction between singlet oxygen and acceptor (A_o); k_p is the monomolecular PS decay rate of excited triplet to ground state (see Figure 20.2 and Glossary). The physical meaning of the bleaching fluence value is the fluence required to decrease the PS concentration by 2.718 times for first-order bleaching and two times for second-order bleaching. The inverse of the bleaching fluence value is also called the *bleaching coefficient*. In the two limiting cases of the first-order bleaching ($f_2 \to \infty$) and second-order bleaching ($f_1 \to \infty$), the relations for PS concentration written in fluence form are similar to Equations 20.20 and 20.21, respectively. The modeling curves of hemoglobin oxygen saturation for different bleaching modes are plotted in Figure 20.33. Without photobleaching the oxygen consumption would remain constant, and the corresponding hemoglobin oxygen saturation dependence would be a linear function (curve 1 in Figure 20.33) except for the low SO_2 values when oxygen deficiency became predominating prescribing the SO_2 level approaching zero.

Without photobleaching, the oxygen depletion rate would be linear (curve 1 in Figure 20.33). Curves 2 and 4 in Figure 20.33 describe the SO_2 behavior, which would have been observed in the limiting cases of either the second or first-order photobleaching correspondingly. These curves were calculated via Equation 20.27 assuming the same values for the first and second-order bleaching fluence: $f_1 = f_2 = 25\ \mathrm{J\,cm^{-2}}$. The SO_2 at the first-order bleaching (curve 4) has more sharp deviation from the initial linear dependence approximating the SO_2 value to some nonzero constant at $t \to \infty$. It reflects the scenario when all PS present in the blood sample is destroyed and remained oxygen is not consumed with irradiation continuing any more. The maximum dose of oxygen, which may be consumed

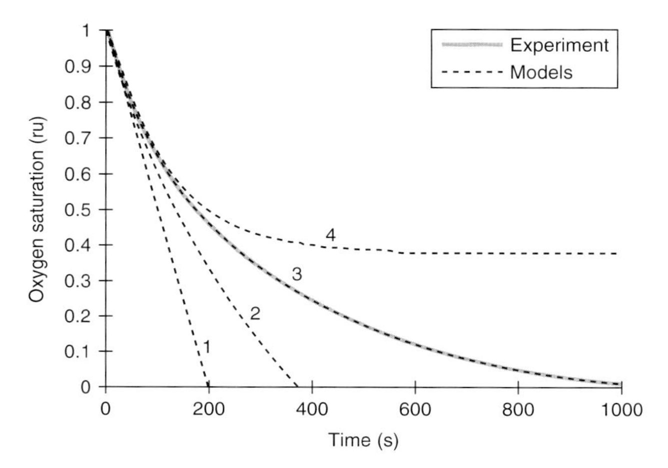

Figure 20.33 Hemoglobin oxygen saturation dynamics during light irradiation of blood samples for different models of PS photobleaching: (1) without photobleaching, (2) second-order photobleaching with bleaching fluence $f_2 = 25\ \mathrm{J\,cm^{-2}}$, (3) best fit to experimental data with mixed-order bleaching ($f_1 = 120\ \mathrm{J\,cm^{-2}}$, $f_2 = 17\ \mathrm{J\,cm^{-2}}$), and (4) first-order bleaching with bleaching fluence $f_1 = 25\ \mathrm{J\,cm^{-2}}$. The thick gray curve refers to experimental data obtained for the blood sample with a PC concentration of 5 μM and light fluence rate 200 mW cm^{-2}. (Modified from Ref. [86].)

in the first-order bleaching model at high fluence rates [149], is defined by the following relation:

$$D^{\max}_{\mathrm{PDT1}} = \frac{f_1}{f_{\mathrm{PDT}}} \frac{[C_0]}{1 + P/P_{\mathrm{sat}}} \tag{20.30}$$

In the case plotted in Figure 20.33, the initial reserve of oxygen in the blood sample ($4SO_2(0) \cdot [\mathrm{Hb}T]$) was higher than this maximum dose, resulting in a nonzero SO_2 value at high fluence equal to $SO_2(t \to \infty) = SO_2(0)(1 - \mathrm{D}^{\max}/4SO_2(0)[\mathrm{Hb}T])$. The SO_2deviation from the no photobleaching linear dependence (curve 1) versus the second-order bleaching model (curve 2 in Figure 20.33) is not as sharp as that of for the first order. PS destruction rate during irradiation in this case is attenuated as second power of PS concentration [86], and, consequently, its decline will be slower than in the case of the first-order bleaching. It should be also noted that for the second-order photobleaching there is no maximum dose in consumed oxygen in the limit of high fluence. However, it demonstrates a slow logarithm growth depending on the delivered fluence [146]:

$$D_{\mathrm{PDT2}} = \frac{f_2}{f_{\mathrm{PDT}}} \frac{[C_0]}{1 + P/P_{\mathrm{sat}}} \cdot \ln\left(1 + f/f_2\right) \tag{20.31}$$

The best fit of the experimental SO_2 data (thick gray curve in Figure 20.33, $[C_0] = 5\ \mu\mathrm{M}$, $P = 200\ \mathrm{mW\,cm^{-2}}$) gives the fitting taking into account both the first- and second-order PS photobleaching simultaneously with the following values of bleaching doses $f_1 = 120\ \mathrm{J\,cm^{-2}}$, $f_2 = 17\ \mathrm{J\,cm^{-2}}$. This dependence is given by curve 3 in Equation 20.33. The corresponding fitted curves with the use of relation (20.27) and their comparison with experimental data for other samples with different PS concentrations and fluence rates are given in Figure 20.31. The fitted variables for these plots were f_{PDT}, f_1, and $f_{20} = f_2[C_0]/(1 + P/P_{sat})$, while the variable P_{sat} was fixed at the magnitude of 500 mW cm^{-2}. The following values for fluence constants were obtained from this fit: $f_{PDT} = 24 \pm 6$ mJ cm^{-2}, $f_1 = 120 \pm 30$ J cm^{-2}, and $f_{20} = 90 \pm 20$ M μJ cm^{-2}. We see that for the low PS concentrations (~1 μM) the bleaching fluences for the first and second order are comparable, while for higher PS concentrations the second-order bleaching will be predominating at initial moment ($f_2 < f_1$). It is interesting to note that the dimensionless ratio (oxygen consumption number) $N_{PDT} = f_1/f_{PDT}$ or f_2/f_{PDT} has the sense of a number of consumed oxygen molecules in PDT reactions catalyzed by one PS molecule before the last will be destroyed due to photodecomposition of the first or second order correspondingly. For efficient PS this number must be much more than unity. It was mentioned above that PS photodecomposition is due to its interaction with plasma proteins and erythrocyte membrane as no photobleaching was observed in physiological solution. Consequently, the oxygen consumption rate obtained in physiological solution is expected to be constant.

20.9
Monitoring of Oxygen Consumption and Photobleaching in Blood during PDT *In vivo*

The *in vivo* measurement data are more difficult to interpret as in addition to the blood oxygen saturation decrease due to the enhanced oxygen consumption by the PDT reaction there is another factor related to vessel destruction as a result of PDT. Figure 20.34 demonstrates SO_2 dynamics in mouse tumor *in vivo* during first (a) and second (b) PDT sessions with multiple laser irradiations. The direct correlation between laser switching on/off and SO_2 is clearly observed. The difference in SO_2 recovery rate after switching off the laser is significant between the first and second sessions. It may be due to the fact that during the first session performed 6 h after IV injection of the PS, its concentration in blood is relatively high resulting in capillary destruction during laser irradiation. After the third laser activation SO_2 level drops down to 25% and is not restored to its initial value (Figure 20.34a).

During the second session carried out 28 h post PS injection, a different SO_2 response to laser irradiation can be seen. At first, SO_2 does not fall to zero during laser irradiation but it is stabilized at 25% not depending on laser irradiation time. Next, a quick recovery of SO_2 to its initial values is observed after switching off the laser. Assumingly, there is no irreversible capillary destruction during the second session and the SO_2 decrease in this case results mainly from the extra oxygen consumption due to PDT. The characteristic saw-like-curve shown in the range of 1200–1800 s in the Figure 20.34b is not changed after switching the laser on and off 20 times (not shown).

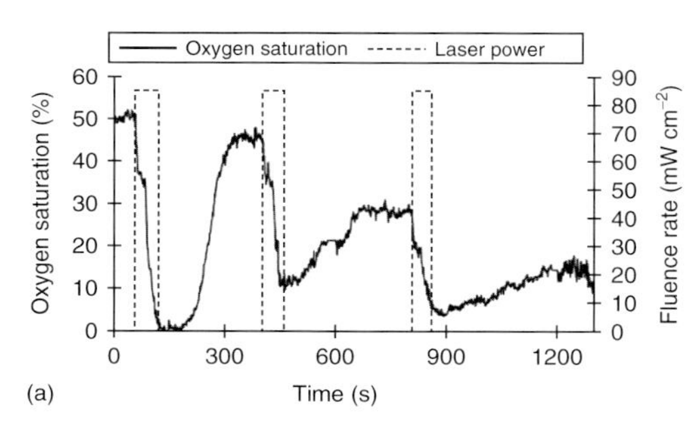

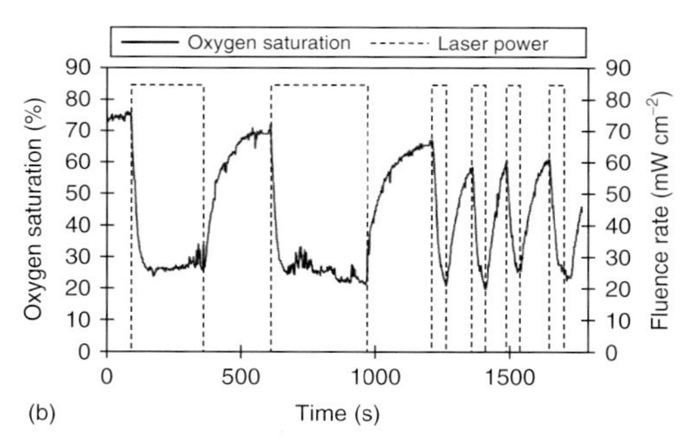

Figure 20.34 The influence of laser irradiation (85 mW cm^{-2}, 670 nm) on hemoglobin oxygen saturation in tumor microcircular blood vessels of mice during photodynamic therapy with Photosens® (sulphanated Al phthalocyanine) at 4 mg/b.w.kg. The laser power is shown by dotted lines. (a) First PDT session (PS was IV injected 6 h prior to the PDT session). The irreversible vessel damage is observed. (b) Second PDT session (PS was IV injected 28 h prior to the PDT session). The oxygen saturation is reversibly restored after switching the laser on and off. (Reproduced by permission from Ref. [36].)

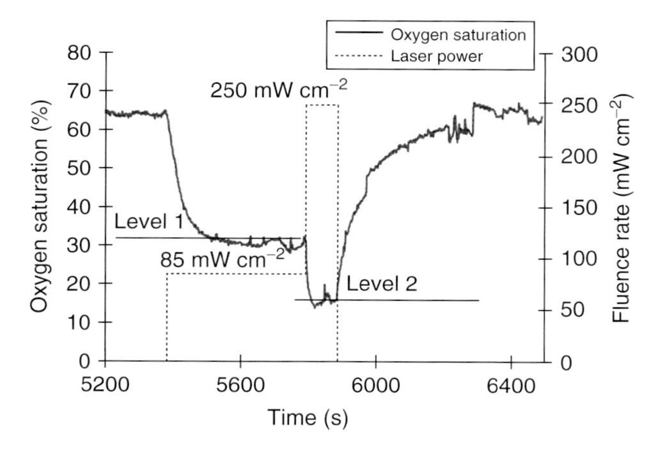

Figure 20.35 The influence of fluence rate (laser power density) on oxygen saturation stabilization level during photodynamic therapy with Photosens® at 4 mg/b.w.kg. The laser power is shown by dotted lines. PS was IV injected 28 h prior to the PDT session. The dynamic equilibrium between oxygen saturation and laser power density is achieved. (Reproduced by permission from Ref. [36].)

Figure 20.35 demonstrates the dependence of SO_2 level stabilized depending on laser power density as $SO_2 = 32\%$ at 85 mW cm^{-2}. The threefold increase of laser power density up to 250 mW cm^{-2} does not induce fall of SO_2 to zero but rather stabilizes it at the level of 15%. After switching the laser off, the SO_2 recovers to its previous values. It proves that a certain dynamic equilibrium is achieved between the oxygen supply by blood and oxygen consumption due to PDT.

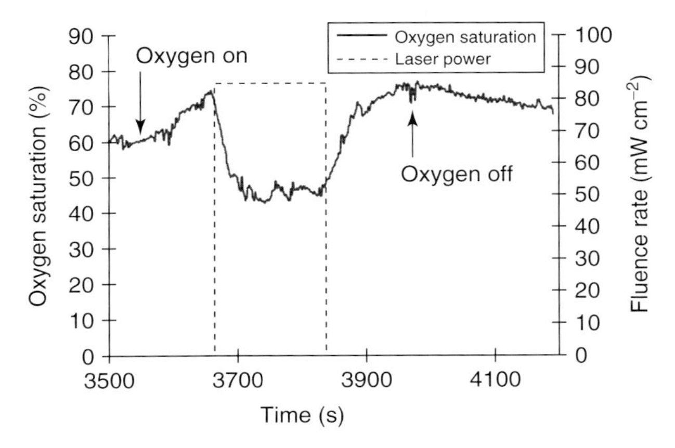

Figure 20.36 The influence of pure oxygen breathing upon hemoglobin oxygen saturation in tissue blood vessels of mouse during photodynamic therapy with Photosens® at 4 mg/b.w.kg. The laser power is shown by dotted lines. PS was IV injected 28 h prior to the PDT session. The arrows show the oxygen flow on and off. As a result of extra oxygen breathing, the dynamic equilibrium is achieved at higher oxygen saturation values. (Reproduced by permission from Ref. [36].)

Figure 20.36 shows the influence of pure oxygen breathing on SO_2 values during PDT. The pure oxygen was delivered to the mouse for breathing. The time of switching oxygen flow on and off is shown on the figure by arrows. We see that both the initial and recovering SO_2 values during laser irradiation are increased by 12–15%.

20.10 Photodynamic Disinfection of Blood

Transfusion of red blood cell concentrates (RBCCs) and PC involves a certain risk of transmission of pathogenic viruses despite serological screening and donor self-exclusion protocols. The risk of hepatitis B, hepatitis C (HCV), and human immunodeficiency (HIV) viral transmission in the United States with a single blood unit has been estimated at 0.0005, 0.03, and 0.0005%, respectively [149]. Suppression of pathogenic flora is one of the strategic goals in contemporary patient care. The wide use of antibiotics in the second half of the twentieth century led to the crisis of antibiotic and chemical resistance among pathogenic bacteria and viruses. Intensive efforts are now placed on the advancement of alternative antimicrobial therapeutics, to which bacteria are not able to easily develop resistance. Conventional treatments such as using X-rays or UV light, or chemical treatment based on solvents and surfactants cause damage to therapeutic components in the blood, so more targeted approaches are being sought. One of these may be a modality recently termed *photodynamic antimicrobial chemotherapy* [35]. PACT, as a treatment of infection, will only be useful if it can be tolerated by the patient. It has been shown that human cells (keratinocytes and fibroblasts) can survive PACT conditions that are lethal to microorganisms [150]. The differences in susceptibility between human body cells and infections are likely to be due to differences in cell size, and structure. Human cells are between 25 and 50 times larger than bacterial cells and thus require more damage to induce cell death [151]. Methylene blue mediated PACT has been an established means of blood product disinfection in some parts of Europe for over five years [152]. The benefit of using PACT over conventional methods is the nonspecific nature of the antimicrobial target of PACT: fungi, viruses, and bacteria can be targeted simultaneously, as well as nucleate WBCs, which have the potential to cause an immune response in the recipient [153]. A potential problem with PACT of blood products could be host toxicity, if, for example, the photosensitizing agent was not completely removed prior to transfusion [154]. In blood transfusion, a whole blood can be used or blood fractions separated by centrifugation. The major difference between the fractions is cellular in nature. Plasma is an aqueous suspension of proteins and infecting microorganisms are also in suspension, whereas platelets, erythrocytes, and leukocytes are different cell types and the pathogens may be intra-, extracellular, or both. A further difference lies in the fact that WBCs contain nucleic acid where the other cell types, and plasma, do not. WBCs (leukocytes) are involved in defence against disease. In blood for transfusion they can cause pathogenic disease since they may be associated with the infectious agents. White cell reduction (leukodepletion) is therefore an important process. Since, in general, the requisite blood fractions (plasma, platelets, and RBCs) do not contain nucleic acids, but the infecting microorganisms and WBCs do, the targeting strategy for most approaches to the problem have been aimed at denaturing or otherwise inactivating nucleic acid [153]. It is established that nucleic acids react to photodynamic treatment mainly through guanosine residues [155].

Regardless of the type of activating energy, it should be remembered that the practical aspect of pathogen inactivation in blood products requires the treatment of large volumes of liquid, usually contained in plastic bags of the order of hundreds of milliliters. Consequently, the use of pinpoint accuracy in light delivery, as in the related photodynamic treatment of tumors is of little use. Photoactivated compounds for use in RBCs should exhibit significant absorption outside the hemoglobin absorption bands in the range 300–630 nm as platelets and plasma are substantially transparent above 300 nm. The phthalocyanines and phenothiazinium dyes can be exploited in RBCC while HPDs, widely used in cancer PDT, are not the optimal compounds to be employed in this instance [153]. Light-mediated killing of many pathogens *in vitro* has been extensively investigated. A wide range of organisms from the gram-positive *Staphylococcus aureus* [156] to the gram-negative *Pseudomonas aeruginosa* [157] in addition to fungi and protozoa [158, 159] have been proven as susceptible to PACT with a number of different PSs *in vitro*. All viruses contain nucleic acid, either DNA or RNA (but not both), and a protein coat, which encases the nucleic acid. Some viruses are also enclosed by an envelope of fat and protein molecules. In its infective form, outside the cell, a virus particle is called a *virion*. Many types of virus have a glycoprotein envelope surrounding the nucleocapsid. The envelope is composed of two lipid layers interspersed with protein molecules (lipoprotein bilayer) and may contain material from the membrane of a host cell as well as that of viral origin. The virus obtains the lipid molecules from the cell membrane during the viral budding process. However, the virus replaces the proteins in the cell membrane with its own proteins, creating a hybrid structure of cell-derived lipids and virus-derived proteins [160]. Photodynamic treatment has been used mainly for inactivation of enveloped viruses. Although in some studies inactivation of nonenveloped viruses has been reported, the inactivation of nonenveloped viruses like adenovirus, poliovirus, or parvovirus, has never been very efficient [161–163]. Most of these PSs are amphiphilic and localize in membranes, suggesting a preference to inactivate lipid enveloped, rather than nonenveloped viruses [164]. HPD has proved to be efficient in the inactivation of different enveloped viruses in tissue culture medium and whole blood [165]. Nonenveloped viruses were not inactivated by the treatment with these sensitizers and light, indicating that the major target for the photodynamic effect is the virus envelope [113]. The benzoporphyrin derivative monoacid A (BPD-MA) has a strong absorption peak at 692 nm, thereby making it a more suitable sensitizer for use in RBCC than HPD, which absorbs weakly at 630 nm. BPD-MA has been shown to inactivate the enveloped virus vesicular stomatitis virus (VSV) in whole blood (maximal $> -7\log_{10}$ kill). The RBC damage was determined (hemolysis) and was minimal 48 h after treatment [166, 167]. Recently it has been shown that free and cell-associated HIV-1 could be inactivated in blood by treatment with BPD-MA and red light. It was suggested that this approach could be a beneficial treatment modality for the management of blood-borne viral infections, such as HIV-1, either alone or in combination with other treatments [168]. MC540 is thought to exert its effects at the viral envelope. However, photodamage to the erythrocyte membrane has also been reported [169]. Methylene blue (phenothiazinium compound, 660–665 nm) is used for the sterilization of plasma in Germany and Switzerland [170]. It is effective in the inactivation of different enveloped viruses in RBC. At a dose where efficient virus-kill is achieved, RBC surface is altered by the photodynamic treatment, resulting in specific IgG binding and K^+ leakage [171]. Lower light fluences cause less RBC damage but also reduced virus-kill. In addition, methylene blue has a genotoxic potential, increasing the mutation frequency in mouse lymphoma cells (at the thymidine kinase locus) approximately fourfold [172]. Both toluidine blue O and methylene violet are known to cause nucleic acid photodamage [173]. methylene blue has been widely used by several European transfusion services in the photodecontamination of blood plasma and has been shown to be particularly effective in the inactivation of enveloped viruses. Both the commercial *Pathinact-MB*® (Baxter) and *Blueflex*® (MacoPharma) systems employ this PS [174]. To allow intracellular activity and decrease collateral photodamage, the drive for improved MB derivatives has been based on increased lipophilicity [175]. As PSs, phthalocyanines are generally efficient in the production of singlet oxygen (Table 20.1). Aluminum phthalocyanines (AlPcs) and its di- and tetra-sulphonated derivatives $AlPcS_2$ and $AlPcS_4$ are effective in sensitizing the killing of VSV [162] added to whole blood or RBCC as well as cell-free and cell-associated HIV-1. However, the nonenveloped encephalomyocarditis virus was not inactivated [176]. RBC damage is minimal, as indicated by the low hemolysis after treatment. In further experiments with $AlPcS_4$, it was shown that following virucidal treatment hemolysis was lower than 2%, the osmotic fragility of the RBC was only slightly increased, and autologous RBC had a normal recovery and circulation survival in baboon [163].

The mechanisms of virus inactivation appear to be different than those leading to RBC damage. From studies using different quenchers, which are specific for radicals generated by either electron transfer reactions (type I) or energy transfer reactions (type II), it was possible to conclude that virus killing is caused mainly by a mechanism involving singlet oxygen generation (type II). Another way to reduce RBC damage selectively without affecting virus inactivation is the use of a high light fluence rate (irradiance). A high fluence rate (80 mW cm^{-2}) is less damaging for RBC than a lower one (5 mW cm^{-2}), while virus-kill is not affected by a change in fluence rate. The explanation for this phenomenon is not known, but the effect could be used to increase the selectivity of the treatment [139]. The silicon phthalocyanine Pc4 proved to be capable of inactivating HIV in all its forms (cell-free, cell-associated, and latent) at doses that were even lower than those required for complete VSV kill. Thus, RBC treated for inactivation of human pathogenic viruses will also lose the ability to transmit malaria and Chagas disease. In conclusion, the most promising method for the sterilization of RBC appears to be the use of phthalocyanines in combination with different quenchers of type I reactions [113].

The main obstruction to be overcome is likely to be the ethical approval of DNA-targeting agents, or obtaining convincing proof that these can be completely removed before transfusion of the treated product. Therefore, a particular interest is in development of completely removable PSs. Inactivation of influenza virus the suspension of crystal fullerene C60 was reported [177]. Its marked advantage is that there are no products of destruction and that the compound can be easily removed from solution by centrifugation. Another promising attempt with photodynamic singlet oxygen generation was demonstrated with silicon nanoparticles that may potentiate also targeting delivery of the PS to the object and its complete clearing from blood [63].

20.11 Photodynamic Therapy of Blood Cell Cancer

Blood cancers arise from cells of the hematopoietic lineage [178]. All cells in this lineage originate from stem cells which are located in the bone marrow. At any type of cancer, blood cells become malignant due to malfunction of regulatory processes responsible for controlling of blood cell's amount and lifespan. Hematological cancers can be classified by their derivation from erythroid, myeloid, or lymphoid cells at specific stages of development as determined by their morphology and by protein markers. The use of hematopoietic stem cell transplantation in cancer treatment has increased dramatically in recent years [179]. To reduce the possible risk of relapse from the infused autologous stem cells, purging of these cells before infusion or storage is employed. Purging methods can be distinguished as positive or negative selection. In the former, stem cells are selected by biophysical or immunological techniques or are preferentially expanded. Negative selection targets the cancer cells to eradicate them and leave the stem cells intact. This includes antibody methods as well as biochemical, biophysical, and biological methods. All of these methods usually achieve 3–5 logs of cancer cell depletion. However, 6–8 log reductions are needed to achieve complete elimination of tumor cells in the graft. This implies that more than two purging methods will likely be needed.

Photodynamic purging of cancer cells in bone marrow was pioneered by Sieber *et al.* using merocyanine 540 as a PS [180]. Despite a decade of research involving this dye, including a clinical trial, it has not been approved, presumably because it is not an effective sensitizer. More recently, other PSs have been studied, such as aluminum phthalocyanine [181] and BPD [182]. The latter has been shown to lower tumor burden by 4 logs with virtually no loss of essential hemopoietic progenitors at extremely low concentration ($25\,\mu g\,l^{-1}$). Even more important is the observation that even multiple-drug-resistant (MDR) cells are responsive to photodynamic treatment [183].

Molecular response of mouse lymphoma cells L5178Y (strains LY-R and LY-S) to PDT was studied by Agarwal *et al.* [184] Strains LY-R and LY-S were treated with 1 μM of AIPcCI and then exposed to graded fluences of red light (600 nm) to produce 10, 50, 90, or 95% cell killing. Extensive DNA fragmentation was observed after 2 h after LD_{90} and LD_{95} PDT doses of light treatment (10 and 15 $kJ\,m^{-2}$ for strain LY-R and 1.5 and 1.8 $J\,cm^{-2}$). Cells of both strains LY-R and LY-S respond to PDT by inducing apoptosis: the chromatin becomes condensed around the periphery of the nucleus and the DNA is degraded into nucleosome-size fragments apparently by activation of a constitutive endonuclease. In contrast to many other agents and treatments (e.g., glucocorticoids and antibodies to CD3/T-cell receptor complex) which induce apoptosis over a period of several hours or days, the induction by PDT in LSI78Y cells is rapid: DNA fragmentation was visible 30 min after LD_{90} dose. Increased fragmentation of DNA was observed at 1 and 2 h after photodynamic treatment. The increase of fragmentation with time indicates that this response is an enzymic process related to cell killing activated by PDT and not a direct photochemical damage to DNA. Two structurally related PS, tetrahydroxyhelianthrone (DTHe) and hypericin (HY) which cause death of human leukemia cells were investigated by Lavie *et al.* [185] The phototoxicities of DTHe and HY to HL-60 cells were compared after exposures to two doses of light irradiation: 4.8 or 14.4 $J\,cm^{-2}$, obtained by irradiation for 20 or 60 min respectively. The results suggested that DTHe exhibits a more potent phototoxic activity in comparison with HY. It was found that cell death with DTHe occurred with an LD_{50} of 1 mM at 4.8 $J\,cm^{-2}$, while required dose of HY was threefold bigger (3 mM). A more potent phototoxic activity of DTHe was also seen at the higher light dose. Cells did not lose their viability under treatment with DTHe or HY conducted in the absence of light, or when the cells were exposed to light in the absence of the compounds. It was reported by the authors that cell death resulted from the photodynamic effects of the compounds, not from nonphotodynamic, intrinsic toxicities. Merocyanine 540 (MC540) and temoporfin (mTHPC) were compared to find the most appropriate agent for photodynamic purging of leukemia and lymphoma cells from autologous bone marrow transplants [186]. mTHPC was found as more potent: 1.3 μM of mTHPC and $0.42\,\mathrm{J\,cm^{-2}}$ were required for killing of 90% of murine myeloid leukemia M1 and WEHI 3B (JCS) cells, while required concentration of MC450 was 20-fold bigger and light dose was 11-fold larger for achieving the same results. Three times less, or 15%, of the coincubated erythrocytes were destroyed by mTHPC than by MC540. Confocal microscopy was used for determination of localization of both PSs in the cells. It was shown that MC540 and mTHPC accumulated diffusely inside the cytoplasm in a similar way, but mTHPC induced a more extensive apoptosis in photosensitized JCS cells. Gulliya *et al.* [187] also reported the use of merocyanine 540 for leukemia treatment. Acute promyelotic leukemia (HL-60) cells were incubated with $20\,\mathrm{mg\,l^{-1}}$ MC540 and exposed to light irradiation (514 nm, $93.6\,\mathrm{J\,cm^{-2}}$). This dose was enough to kill 99.99% of leukemic cells. It was found that addition of salicylic acid during *in vitro* MC540-mediated PDT causes a marked increase in leukemic cell killing [188]. Salicylate enhancement was dose dependent with significant effects observable at concentrations >0.1 mM. Improvement of leukemia cells killing was considered as an enhancement of the effects of PDT treatment in a presence of salicylic acid, since neither preincubation with SA followed by washing prior to illumination nor addition of salicylate following irradiation altered MC540-mediated cell killing. Chloroaluminum phthalocyanine was described as a highly effective PS for treatment of murine leukemia [110]. The PS doze of 0.3 μM and red light dose of $0.045\,\mathrm{J\,cm^{-2}}$ was reported to be enough for obtaining 90% of apoptotic cells within 60 min. The sensitizer catalyzed both lysosomal and mitochondrial photodamage at this light dose. The same PS was applied for acute myeloid leukemia cells and normal peripheral blood leukocytes [189]. Significant and preferential photokilling of leukemia cells in comparison to normal cells were observed upon light irradiation, while lower phototoxicity was noticed for normal cells (peripheral blood leukocytes). This type of PSs also may be considered as efficient PDT agent for photodynamic treatment of leukemia.

PDT can be used in hematopoietic stem cell transplantation for purging remission marrow of residual leukemic cells. Chemotherapy combined with autologous transplantation using bone marrow or peripheral blood-derived stem cells is now widely used in the treatment of hematologic malignancies. Purging procedures are needed to eliminate the risk of recurrence of disease caused by presence of malignant cells in autografts. An amphipathic PS, $ZnPcS_2P_2$ was probed on murine erythroblastic leukemia EL9611 cells in mouse model by Huang *et al.* [190]. $5 \times 10^6\,\mathrm{ml^{-1}}$ mouse bone marrow cells or (BMCs) mixed with various number of EL9611 cells (ranging from 50 to $5 \times 10^4\,\mathrm{ml^{-1}}$) were untreated or treated with the purging procedure. The cell suspensions were incubated with $4\,\mathrm{mg\,ml^{-1}}$ $ZnPcS_2P_2$

for 5 h and then irradiated by red light at 670 nm (53 mW cm^{-2}, 2.1 J cm^{-2}). Cell suspensions in 0.2 ml final volume were injected into the lateral tail vein of mice 4 h post irradiation. Complete cure was achieved in mice infused with ZnPcS2P2-PDT-purged bone marrow cells/EL9611 cells whereas 100% incidence of leukemia occurred in all the control receiving nonpurged cell mixtures. The findings imply that ZnPcS2P2-PDT retained sufficient hematopoietic progenitors to allow complete hematopoietic reconstitution to overcome aplasia in the early phase after lethal irradiation, and killed sufficient EL9611 cells to restrain leukemia incidence. For the purposes of purging of autografts in multiple myeloma, the PS 4,5-dibromorhodamine methyl ester (TH9402) was investigated and found as potential for destroying multiple myeloma cell lines [191]. An eradication rate of more than 5 logs is obtained when using a 40 min incubation with 5–10 μM dye followed by a 90 min washout period and a light dose of 5–10 J cm^{-2} (2.8 mW cm^{-2}) in all cell lines. Absence of significant toxicity toward normal hematopoietic stem cells makes TH9402 an appropriate purging agent for autologous transplantation in multiple myeloma treatment. The silicon phthalocyanine Pc4 is highly effective in killing breast cancer cells and HL-60 leukemia cells. At a dose that eliminated 6 log10 of cancer cells (20 nM Pc 4 and 20 J cm^{-2} red light) there was 100% recovery of bone marrow progenitors (CFU-GM).70 If Pc 4-PDT will prove as effective against cancer cells expressing the MDR phenotype this could be a highly efficient purging method [176]. Extracorporeal irradiation technique is one of the possible methods which may be used in practice for application of PDT for blood cancer [192]. In this method, a portion of the patient's peripheral blood is continuously circulated through an extracorporeal circuit in which the blood is exposed to the irradiation for a predetermined period and returned to the patient.

20.12 Summary

In this chapter we reviewed photodynamic reactions where blood, blood cells, and blood pigments are involved. We briefly reviewed a history of PDT, PDT mechanisms on a quantum, biophysical, biochemical, cell, and physiological levels. We also reviewed contribution of blood into tissue optical properties including fluorescence properties of blood component. We considered optical properties of contemporary and prospect PSs (including quantum yield of singlet oxygen generation) and their modification due to dissolving in blood, and interaction with blood cells and plasma. Photodynamic reactions with blood and blood cells were reviewed in details including modification and alterations of properties of blood components and cells due to photodynamic reactions and also types of photodynamic reactions in blood *in vitro* versus *in vivo* including oxygen consumption, photobleaching, and heating effects with both experimental data and theoretical analysis. Examples of direct clinical applications of PDT with blood such as photodynamic disinfection of blood and PDT of blood cell cancer were also briefly considered.

Acknowledgments

The authors gratefully acknowledge the great help and support from Biospec (Russia) and Medspeclab (Canada); thank Anastasia Ryabova (Moscow) and Gennadiy Sayko (Toronto, Canada) for their valuable suggestions and discussions and Julia Davydova (Toronto, Canada) for assistance with the figures. We also acknowledge funding of the Erlangen Graduate School in Advanced Optical Technologies (SAOT) by the German National Science Foundation (DFG) in the framework of the excellence initiative for support of this review.

Glossary

Notation	Units	Description
$[S_0]$	M	Photosensitizer (PS) concentration in ground (singlet) state
$[S_1]$	M	PS concentration in the first excited singlet state
$[T]$	M	PS concentration in triplet state
$[C]$	M	Total PS concentration ($[C] = [S_0] + [S_1] + [T]$)
$[C_0]$	M	Total PS concentration at initial time ($t = 0$)
$[^3O_2]$	M	Concentration of oxygen in ground (triplet) state
$[^1O_2]$	M	Concentration of oxygen in singlet state
$[A_o]$	M	Concentration of acceptor reacting with singlet oxygen
$[A_t]$	M	Concentration of acceptor reacting with PS triplet state
k_m	s^{-1}	Monomolecular PS decay rate of excited to ground state (radiative and nonradiative)
k_f	s^{-1}	Monomolecular PS decay rate of excited to ground state with fluorescence emission
k_{isc}	s^{-1}	PS intersystem crossing rate
k_p	s^{-1}	Monomolecular PS decay rate of excited triplet to ground state
k_{ot}	$M^{-1}\,s^{-1}$	Bimolecular rate of triplet PS quenching by ground state oxygen
k_{at}	$M^{-1}\,s^{-1}$	Bimolecular rate of reaction between PS triplet and acceptor (A_t)
k_{oa}	$M^{-1}\,s^{-1}$	Bimolecular rate of reaction between singlet oxygen and acceptor (A_o)
k_{os}	$M^{-1}\,s^{-1}$	Bimolecular rate of reaction between singlet oxygen and PS ground state
k_d	s^{-1}	Monomolecular decay rate of singlet oxygen to ground state
φ_f	–	Fluorescence quantum yield $\varphi_f = k_f/(k_m + k_{isc})$
S_Δ	–	Fraction of quenching PS triplets by oxygen resulting in the formation of singlet oxygen
φ_t	–	Quantum yield of PS triplet state $\varphi_t = k_{isc}/(k_{isc} + k_m)$
φ_Δ	–	Quantum yield of singlet oxygen formation $\varphi_\Delta = \varphi_t S_\Delta$
φ_{os}	–	Probability of singlet oxygen to destroy the same PS molecule at which it was generated
$E_{h\nu}$	J	Photon energy $E_{h\nu} = hc/\lambda$
Φ	$cm^{-2}\,s^{-1}$	Photon flux
σ	cm^2	PS absorption cross section

(*continued*)

Notation	Units	Description
ε	$cm^{-1}\ M^{-1}$	PS extinction coefficient $\varepsilon = \sigma N_A \cdot 10^{-3}/\ln 10$
P	$W\ cm^{-2}$	Fluence rate $P = \Phi E_{h\nu}$
P_{sat}	$W\ cm^{-2}$	Saturation fluence rate
f	$J\ cm^{-2}$	Fluence
f_a	$J\ cm^{-2}$	PS absorption fluence $f_a = E_{h\nu}/\sigma$
f_{PDT}	$J\ cm^{-2}$	Molecular oxygen consumption fluence $f_{PDT} = f_a(k_d + k_{oa}[A_o])/\varphi_\Delta k_{oa}[A_o]$
f_1, f_2	$J\ cm^{-2}$	PS bleaching fluences for first- and second-order bleaching respectively
τ_1, τ_2	s	PS bleaching decay times for first- and second-order bleaching, respectively, $\tau_{1,2} = f_{1,2}/P$
N_{PDT}	–	Oxygen consumption number $N_{PDT} = f_{1,2}/f_{PDT}$
D_{PDT}	M	Dose of consumed oxygen due to PDT
Γ_{PDT}	M/s	Oxygen consumption rate due to PDT $\Gamma_{PDT} = dD_{PDT}/dt$
$[RHb]$	M	Concentration of hemoglobin structural units free from oxygen molecules
$[HbO_2]$	M	Concentration of hemoglobin structural units bound to oxygen molecules
SO_2	%	Hemoglobin oxygen saturation in blood $SO_2 = [HbO_2]/([RHb] + [HbO_2])$
pO_2	mm Hg	Partial oxygen pressure
$s(t)$	–	Relative hemoglobin oxygen saturation $s(t) = SO_2(t)/SO_2(0)$
CPH	–	Continuous photohemolysis
DPH	–	Post-irradiation or delayed photohemolysis
$[HbT]$	M	Total molar concentration of hemoglobin (4 units) in blood $[HbT] = ([RHb] + [HbO_2])/4$
HPD	–	Hematoporphyrin derivatives
α	–	Bunzen solubility coefficient
D	$cm^2\ s^{-1}$	Oxygen diffusion coefficient
μ_a	cm^{-1}	Absorption coefficient
μ_s'	cm^{-1}	Reduced scattering coefficient

References

1. Moan, J. and Peng, Q. (2003) An outline of the hundred-year history of PDT. *Anticancer Res.*, **23** (5A), 3591–3600.
2. Parrish, J.A., Fitzpatrick, T.B., Taneubaum, L., and Pathac, M.A. (1974) Photochemotherapy of psoriasis with oral methoxalen and longwave ultraviolet light. *N. Engl. J. Med.*, **291**, 1207–1211.
3. Dougherty, T.J. (1987) Photosensitizers: therapy and detection of malignant tumors. *Photochem. Photobiol.*, **45**, 879–889.
4. Ben-Hur, E. and Rosenthal, I. (1985) Photosensitized inactivation of Chinese hamster cells by phthalocyanines. *Photochem. Photobiol.*, **42**, 129–133.
5. Zimnyakov, D.A. and Tuchin, V.V. (2002) Optical tomography of tissues. *Quantum Electron.*, **32** (10), 849–867.

6. Boulnois, J.L. (1986) Photophysical processes in recent medical laser developments: a review. *Lasers Med. Sci.*, **1**, 47–66.
7. Wilson, B.C. and Patterson, M.S. (2008) The physics, biophysics and technology of photodynamic therapy. *Phys. Med. Biol.*, **53**, R61–R109.
8. Dougherty, T.J., Gomer, C.J., Henderson, B.W., Jori, G., Kessel, D., Korbelik, M., Moan, J., and Peng, Q. (1998) Photodynamic therapy. *J. Nutl. Cancer Znst.*, **90**, 889–905.
9. Schweitzer, C. and Schmidt, R. (2003) Physical mechanisms of generation and deactivation of singlet oxygen. *Chem. Rev.*, **103**, 1685–1757.
10. Rodgers, M.A.J. and Snowden, P.T. (1982) Lifetime of O_2(1delta-G) in liquid water as determined by time-resolved infrared luminescence measurements. *J. Am. Chem. Soc.*, **104**, 5541–5543.
11. Moan, J. and Peng, Q. (2003) in *Photodynamic Therapy. Comprehensive Series in Photochemistry and Photobiology*, vol. 2 (ed. T. Patrice), The Royal Society of Chemistry, pp. 1–18.
12. Moan, J. and Sommer, S. (1983) Uptake of the components of hematoporphyrin derivative by cells and tumors. *Cancer Lett.*, **21**, 167–174.
13. Bisland, S.K., Lilge, L., Lin, A., Rusnov, R., Bogaards, A., and Wilson, B.C. (2004) Erratum: metronomic photodynamic therapy as a new paradigm for photodynamic therapy: rationale and preclinical evaluation of technical feasibility for treating malignant brain tumors. *Photochem. Photobiol.*, **80** (2), 374–374.
14. Niedre, M., Patterson, M.S., and Wilson, B.C. (2002) Direct nearinfrared luminescence detection of singlet oxygen generated by photodynamic therapy in cells in vitro and tissues in vivo. *Photochem. Photobiol.*, **75**, 382–391.
15. Redmond, R.W. and Kochevar, I.E. (2006) Spatially resolved cellular responses to singlet oxygen. *Photochem. Photobiol.*, **82**, 1178–1186.
16. Rodgers, M.A.J. (1985) in *Primary Photo-Processes in Biology and Medicine* (eds R.V. Bensasson, G. Jori, E.L. Land *et al.*), Plenum, New York, pp. 181–195.
17. Girotti, A.W. (2001) Photosensitized oxidation of membrane lipids: reaction pathways, cytotoxic effects, and cytoprotective mechanisms. *J. Photochem. Photobiol. B: Biol.*, **63** (1–3), 103–113.
18. Girotti, A.W. (1990) Photodynamic lipid peroxidation in biological systems. *Photochem. Photobiol.*, **53**, 497–509.
19. Perotti, C., Casas, A., Battle, A.M., and Del, C. (2002) Scavengers protection of cells against ALA-based photodynamic therapy-induced damage. *Lasers Med. Sci.*, **17**, 222–229.
20. Schwartz, J.L. (1996) The dual roles of nutrients as antioxidants and prooxidants: their effects on tumor cell growth. *J. Nutr.*, **126**, S1221–S1227.
21. Darzynkiewicz, Z., Juan, G., Li, X., Gorczyca, W., Murakami, T., and Traganos, F. (1997) Cytometry in cell necrobiology: analysis of apoptosis and accidental cell death (necrosis). *Cytometry*, **2**, 1–20.
22. Shen, H.-R., Spikes, J.D., Kopecekovci, P., and Kopecek, J. (1996) Photodynamic crosslinking of proteins. I. Model studies using histidine- and lysine-containing N- (2-hydroxypropyl) methacrylamide copolymers. *J. Photochem. Photobiol. B: Biol.*, **34**, 203–210.
23. MacDonald, I.J. and Dougherty, T.J. (2001) Basic principles of photodynamic therapy. *J. Porphyr. Phthalocyanines*, **5**, 105–129.
24. Luo, Y. and Kessel, D. (1997) Initiation of apoptosis versus necrosis by photodynamic therapy with chloroaluminum phthalocyanine. *Photochem. Photobiol.*, **66** (4), 479–483.
25. Dellinger, M. (1996) Apoptosis or necrosis following photofrin® photosensitization: influence of the incubation protocol. *Photochem. Photobiol.*, **64** (1), 182–187.
26. Kessel, D. and Luo, Y. (1998) Mitochondrial photodamage and PDT-induced apoptosis. *J. Photochem. Photobiol. B: Biol.*, **42**, 89–95.
27. Berg, K., Moan, J., Bommer, J.C., and Winkelman, J.W. (1990) Cellular

inhibition of microtubule assembly by photoactivated sulphonated meso-tetraphenylporphines. *Int. J. Radiat.Biol.*, **58**, 475–487.

28. Verweij, H., Dubbelman, T.M., and Van Steveninck, J. (1981) Photodynamic protein cross-linking. *Biochim. Biophys. Acta*, **647** (1), 87–94.
29. Plaetzer, K., Kiesslich, T., Verwanger, T., and Krammer, B. (2003) The modes of cell death induced by PDT: an overview. *Med. Laser Appl.*, **18**, 7–19.
30. Kessel, D., Thompson, P., Saatio, K., and Nantwi, K.D. (1987) Tumor localization and photosensitization by sulfonated derivatives of tetraphenylporphine. *Photochem. Photobiol.*, **45**, 787–790.
31. Canti, G., Marelli, O., Ricci, L., and Nicolin, A. (1981) Hematoporphyrin treated murine lymphocytes: in vitro inhibition of DNA synthesis and light-mediated inactivation of cells responsible for GVHR. *Photochem. Photobiol.*, **34**, 589–594.
32. Korbelik, M. and Sun, J. (2006) Photodynamic therapy-generated vaccine for cancer therapy. *Cancer Immunol. Immunother.*, **55** (8), 900–909.
33. Douplik, A.Y., Loshchenov, V.B., Voroztsov, G., Kogan, E., Kusin, M., Ablitsov, Y., and Illyina, O. (1997) Phenomenon of PDT-induced post-irradiadion apoptosis in biological liquids and cancer cells using sulphonated phthalocyanine aluminum photosensitizer. *Proc. SPIE*, **3191**, 124–129.
34. Korbelik, M. (1996) Induction of tumor immunity by photodynamic therapy. *J. Clin. Laser Med. Surg.*, **14**, 329–334.
35. Wainwright, D. (1998) Photodynamic antimicrobial chemotherapy (PACT). *J. Antimicrob. Chemother.*, **42** (1), 13–28.
36. Stratonnikov, A.A., Douplik, A.Y., Klimov, D.V., Loschenov, V.B., Meerovich, G.A., Mizin, S.V., Fomina, G.I., Kazachkina, N.I., Yakubovskaya, R.I., and Budenok, Y.V. (1997) Absorption spectroscopy as a tool to control blood oxygen saturation during photodynamic therapy. *Proc. SPIE*, **3191**, 58–66.
37. Tortora, G. and Derrickson, B. (2006) *Principles of Anatomy and Physiology*, John Willey & Sons, Inc.
38. Michelson, A.D. (2002) *Platelets*, Academic Press/Elsevier Science, New York.
39. Kolesnikova, I.V., Potapov, S.V., Yurkin, M.A., Hoekstra, A.G., Maltsev, V.P., and Semyanov, K.A. (2006) Determination of volume, shape and refractive index of individual blood platelets. *J. Quant. Spectrosc. Radiat. Transf.*, **102** (1), 37–45.
40. Traxler, P. (2002) *Blood*, Piatkus.
41. Tryphonov, E.V. (2009) *Human Psychophysiology*, Saint-Petersburg.
42. Zijlstra, W.G., Buursma, A., and Assendelft, O.W. (2000) *Visible and Near Infrared Absorption Spectra of Human and Animal Hemoglobin: Determination and Application*, Brill Academic Publisher.
43. Stürmer, M., Sayko, G., Prahl, S., and Douplik, A. Evaluation of optical transparency window of biotissues depending on blood perfusion parameters, in preparation, to be submitted to APPL PHYS LETT.
44. Douplik, A.Y., Loschenov, V.B., Malinin, V., Proscurnev, S., and Klimov, D. (1996) Changes of optical parameters of moving whole blood depending on shear rate. *Proc. SPIE*, **2678**, 35–43.
45. Roggan, A., Friebel, M., Dorschel, K., Hahn, A., and Muller, G. (1999) Optical Properties of circulating human blood in the wavelength range 400–2500 nm. *J. Biomed. Opt.*, **4** (1), 36–46.
46. Douplik, A., Yaroslavsky, I.V., Loschenov, V.B., Alexandrov, M.T., and Sirkin, A.L. (1994) Identification of optical parameters of the whole blood depending on its oxygenation level. *Proc. SPIE*, **2326**, 319–325.
47. Uyuklu, M., Meiselman, H., and Baskurt, O. (2009) Effect of hemoglobin oxygenation level on red blood cell deformability and aggregation parameters. *Clin. Hemorheol. Microcirc.*, **41**, 179–188.

48. Nagel, R.E. (ed.) (2003) *Methods in Molecular Medicine. Hemoglobin Disorders: Molecular Methods and Protocols*, Humana Press.
49. Leiner, M.J.P., Schaur, R.J., Desoye, G., and Wolfbels, O.S. (1986) Fluorescence topography in biology. III: characteristic deviations of tryptophan fluorescence in sera of patients with gynecological tumors. *Clin. Chem.*, **32** (10), 1974–1978.
50. Gao, S., Peng, C., and Liu, Y. (2005) Study on ultraviolet light-excitated blood fluorescence spectra characteristics. *Proc. SPIE*, **5630**, 553–557.
51. Madhuri, S., Vengadesan, N., Aruna, P., Koteeswaran, D., Venkatesan, P., and Ganesan, S. (2003) Native fluorescence spectroscopy of blood plasma in the characterization of oral malignancy. *Photochem. Photobiol.*, **78** (2), 197–204.
52. Gao, S., Liu, Y., Peng, C., Ni, X., and Lu, J. (2003) Investigation of Ar^+ laser-induced blood fluorescence spectra. *Proc. SPIE*, **5255**, 111–114.
53. Monici, M., Pratesi, R., Bernabei, P.A., Caporale, R., Ferrini, P.R., Croce, A.C., Balzarini, P., and Bottiroli, G. (1995) Natural fluorescence of white blood cells: spectroscopic and imaging study. *J. Photochem. Photobiol. B: Biol.*, **30** (1), 29–37.
54. Tamielian, C., Scweitzer, C., Mechin, R., and Wolff, C. (2001) Quantum yield of singlet oxygen production by monomeric and aggregated forms of hematoporphyrin derivative. *Free Radic. Biol. Med.*, **30** (2), 208–212.
55. Bachor, R., Shea, C.R., Gillies, R., and Hasan, T. (1991) Photosensitized destruction of human bladder carcinoma cells treated with chlorin e6-conjugated microspheres. *Proc. Natl. Acad. Sci. U.S.A.*, **88**, 1580–1584.
56. Detty, M., Gibson, S., and Wagner, S. (2004) Current clinical and preclinical photosensitizers for use in photodynamic therapy. *J. Med. Chem.*, **47** (16), 3897–3915.
57. Brandis, A.S., Salomon, Y., and Scherz, A. (2007) *Bacteriochlorophyll Sensitizers in Photodynamic Therapy in Advances in Photosynthesis and Respiration in Chlorophylls and Bacteriochlorophylls*, Springer, Netherlands.
58. Li, W., Gandra, N., Courtney, S.N., and Gao, R. (2009) Singlet oxygen production upon two-photon excitation of TiO_2 in chloroform. *ChemPhysChem*, **10** (11), 1789–1793.
59. Martin, J., Mehta, D.K., Jordan, B., MacFarlane, C.R., Ryan, R.S., and Wagle, S.M. (2007) *British National Formulary*, 53rd edn, BMJ Publishing Group Ltd.
60. Sudeep, P.K., Page, Z., and Emrick, T. (2008) PEGylated~silicon~nanoparticles: synthesis and characterization. *Chem. Commun.*, **46**, 6126–6127.
61. Dahlstedt, E., Collins, H.A., Balaz, M., Kuimova, M.K., Khurana, M., Wilson, B.C., Phillips, D., and Anderson, H.L. (2009) One- and two-photon activated phototoxicity of conjugated porphyrin dimmers with high two-photon absorption cross sections. *Org. Biomol. Chem.*, **7**, 897–904.
62. Kovalev, D., Gross, E., Künzner, N., Koch, F., Timoshenko, V.Y., and Fujii M. (2002) Resonant electronic energy transfer from excitons confined in silicon nanocrystals to oxygen molecules. *Phys. Rev. Lett.*, **89** (13), 137401.1–137401.4.
63. Rioux, D., Laferrière, M., Douplik, A., Shah, D., Lilge, L., Kabashin, A.V., and Meunier, M. (2009) Silicon nanoparticles produced by femtosecond laser ablation in water as novel contamination-free photosensitizers. *J. Biomed. Opt.*, **14** (2), 021010.1–021010.5.
64. Prat, F., Stackow, R., Bernstein, R., Qian, W., Rubin, Y., and Foote, C.S. (1999) Triplet-state properties and singlet oxygen generation in a homologous series of functionalized fullerene derivatives. *J. Phys. Chem. A*, **103**, 7230–7235.
65. Aborgast, J.W., Darmanyan, A.P., Foote, C.S., Diederich, F.N., Rubin, Y., Diederich, F., Alvarez, M., Anz, S.J., and Whetten, R.L. (1991) Photophysical properties of sixty atom carbon molecule (C60). *J. Phys. Chem.*, **95** (1), 11–12.

66. Yamamoto, Y., Imai, N., Mashima, R., Konaka, R., Inoue, M., and Dunlap, W.C. (2000) Singlet oxygen from irradiated titanium dioxide and zinc oxide. *Methods Enzymol.*, **319**, 29–37.
67. Wang, S., Gao, R., Zhou, F., and Selke, M. (2004) Nanomaterials and singlet oxygen photosensitizers: potential applications in photodynamic therapy. *J. Mater. Chem.*, **14**, 487–493.
68. Zheng, G., Chen, J., Stefflova, K., Jarvi, M., Li, H., and Wilson, B.C. (2007) Photodynamic molecular beacon as an activatable photosensitizer based on protease-controlled singlet oxygen quenching and activation. *Proc. Natl. Acad. Sci. U.S.A.*, **104** (21), 8989–8994.
69. Collins, J.E., Lakshman, T.V., Finlay, J.E., Kumar, A., Bell, H., Nguyen, B.T., Belov, V., Luo, J., and Friedberg, J.S. (2007) Infrared light utilized for photodynamic therapy by activation of rare earth phosphors for visible light generation. *Proc. SPIE*, **6427**, 642717.1–642717.12.
70. Douplik, A.Y., Stratonnikov, A.A., Loshchenov, V.B., Lebedeva, V.S., Derkacheva, V.M., Vitkin, A., Rumyanceva, V.D., Kusmin, S.G., Mironov, A.F., and Luk'anets, E.A. (2000) Study of photodynamic reactions in human blood. *J. Biomed. Opt.*, **5** (3), 338–349.
71. Vo-Dinh, T. (ed.) (2003) *Biomedical Photonics Handbook*, CRC Press.
72. Kessel, D. (1986) Sites of photosensitization by derivatives of hematoporphyrin. *Photochem. Photobiol.*, **44** (4), 489–493.
73. Berg, K., Madslien, K., Bommer, J.C., Oftebro, R., Winkelman, J.W., and Moan, J. (1991) Light induced relocalization of sulfonated meso-tetraphenylporphines in NHIK 3025 cells and effects of dose fractionation. *Photochem. Photobiol.*, **53** (2), 203–209.
74. Chen, J.Y., Cheung, N.H., Fung, M.C., Wen, J.M., Leung, W.N., and Mak, N.K. (2000) Subcellular localization of merocyanine 540 (MC540) and induction of apoptosis in murine myeloid leukemia cells. *Photochem. Photobiol.*, **72** (1), 114–120.
75. Georgiou, G.N., Ahmet, M.T., Houlton, A., Silver, J., and Cherry, R.J. (1993) Measurement of the rate of uptake and subcellular localization of porphyrins in cells using fluorescence digital imaging microscopy. *Photochem. Photobiol.*, **59** (4), 419–422.
76. Bellnier, D.A., Greco, W.R., Loewen, G.M., Nava, H., Oseroff, A.R., and Dougherty, T.J. (2006) Clinical Pharmacokinetics of the PDT photosensitizers porfimer sodium (Photofrin), 2-[1-Hexyloxyethyl]-2-Devinyl Pyropheophorbide-a (Photochlor) and 5-ALA-induced Protoporphyrin IX. *Las. Surg. Med.*, **38**, 439–444.
77. Triesscheijn, M., Ruevekamp, M., Out, R., Berkel, T., Schellens, J., Baas, P., and Stewart, F. (2007) The pharmacokinetic behavior of the photosensitizer meso-tetra-hydroxyphenyl-chlorin in mice and men. *Cancer Chemother. Pharmacol.*, **60** (1), 113–122.
78. Stratonnikov, A.A., Ermishova, N.V., Meerovich, G.A., Kudashev, B.V., Vakulovskaya, E.G., and Loschenov, V.B. (2002) Simultaneous measurement of photosensitizer absorption and fluorescence in patients undergoing photodynamic therapy. *Proc. SPIE*, **4613**, 162–173.
79. Stratonnikov, A. and Loschenov, V. (2001) Evaluation of blood oxygen saturation in vivo from diffuse reflectance spectra. *J. Biomed. Opt.*, **6** (4), 457–467.
80. Cubeddu, R., Canti, G., Musolino, M., Pifferi, A., Taroni, P., and Valentini, G. (1994) Absorption-spectrum of hematoporphyrin derivative in-vivo in a murine tumor-model. *Photochem. Photobiol.*, **60**, 582–585.
81. Gijsbers, G.H.M., Breederveld, D., Gemert, M.J.C., Boon, T.A., Langelaar, J., and Rettschnick, R.P.H. (1986) *In vivo* fluorescence excitation and emission spectra of hematoporphyrinderivative. *Lasers Life Sci.*, **1**, 29–48.
82. Stratonnikov, A.A., Meerovich, G.A., Ryabova, A.V., Savel'eva, T.A., and Loshchenov, V.B. (2006) Application of backward diffuse refection spectroscopy for monitoring the state of tissues

in photodynamic therapy. *Quantum Electron.*, **36** (12), 1003–1010.

83. Douplik, A. (1996) Investigation of influence of physiological parameters of human blood upon optical properties of the blood in visible and nearinfrared range. PhD Dissertation, Department of Biophysics, Institute of Physical-Chemical Medicine, Moscow.
84. Flock, S.T., Jacques, S.L., Wilson, B.C., Star, W.M., and Gemert, M.J.C. (1992) Optical properties of intralipid: a phantom medium for light propagation studies. *Las. Surg. Med.*, **12** (5), 510–519.
85. Loschenov, V.B. and Steiner, R.W. (1994) Spectra investigation methods of biological tissues: technique, experiment, andclinics. *Proc. SPIE*, **2081**, 96–108.
86. Stratonnikov, A., Douplik, A., and Loschenov, V. (2003) Oxygen consumption and photobleaching in whole blood incubated with photosensitizer induced by laser irradiation. *Laser Phys.*, **13** (1), 1–21.
87. Stratonnikov, A.A., Edinak, N.E., Klimov, D.V., Linkov, K.G., Loschenov, V.B., Luckjanets, E.A., Meerovich, G.A., and Vakulovskaya, E.G. (1996) The control of photosensitizer in tissue during photodynamic therapy by means of absorption spectroscopy. *Proc. SPIE*, **2924**, 49–60.
88. Santos, A.E., Laranjinha, J.A.N., and Almeida, L.M. (1998) Sulfonated chloroaluminum phthalocyanine incorporates into human plasma lipoproteins: photooxidation of low-density lipoproteins. *Photochem. Photobiol.*, **67** (4), 378–385.
89. Mohr, H., Knüver-Hopf, J., Lambrecht, B., Scheidecker, H., and Schmitt, H. (1992) No evidence for neoantigens in human plasma after photochemical virus inactivation. *Ann. Hematol.*, **65**, 224–228.
90. Ojima, F., Sakamoto, H., Ishiguro, Y., and Terao, J. (1993) Consumption of carotenoids in photosensitized oxidation of human plasma and plasma low-density lipoprotein. *Biol. Med.*, **15** (4), 377–384.
91. Grossweiner, L.I., Fernandez, J.M., and Bilgin, M.D. (1998) Photosensitisation of red blood cell haemolysis by photodynamic agents. *Laser Med. Sci.*, **13**, 42–54.
92. Pooler, J.P. (1985) The kinetics of colloid osmotic haemolysis. II. Photohaemolysis. *Biochim. Biophys. Acta*, **812**, 199–205.
93. Priezzhev, A.V., Firsov, N.N., Stranadko, E.F., Ryaboshapka, O.M., Chichuk, T.V., and Suschinskaya, O.V. (1996) Aggregation and hemolysis of human erythrocytes at photodynamic therapy. *Proc. SPIE*, **2924**, 215–221.
94. Melnikova, V., Bezdetnaya, L., Belitchenko, I., Potapenko, A., Merlin, J.L., and Guillemin, F. (1999) *Meta*-tetra (hydroxyphenyl) chlorin-sensitized photodynamic damage of cultured tumor and normal cells in the presence of high concentrations of a-tocopherol. *Cancer Lett.*, **139**, 89–95.
95. Lev, B., Hanania, J., and Malik, Z. (1993) Morphological deformations of erythrocytes induced by hematoporphyrin and light. *Lasers Life Sci.*, **5** (3), 219–230.
96. Ben-Hur, E., Malik, Z., Dubbelman, T., Margaron, P., Ali, H., and Lier, J.E. (1993) Phthalocyanine induced photohemolysis: structure-activity relationship and the effect of fluoride. *Photochem. Photobiol.*, **58** (3), 351–355.
97. Asakava, T. and Matsushita, S. (1980) Coloring conditions of tiobarbituric acid test for detecting lipid hydroperoxides. *Lipids*, **15** (3), 137–140.
98. Chichuk, T.V., Stranadko, E.F., Lubchenko, G.N., Podgornaya, E.V., Pozdnyakova, E.E., and Klebanov, G. (2000) Study of the sensitized action of low level laser radiation mechanism. *Proc. SPIE*, **4059**, 112–118.
99. Dubbelman, T.M., Goeij, A.F., and Steveninck, J. (1978) Photodynamic effects of protoporphyrin on human erythrocytes. Nature of the cross-linking of membrane proteins. *Biochim. Biophys. Acta*, **511** (2), 141–151.
100. Yedgat, S., Hovav, T., and Barshtein, G. (1999) Red blood cell intercellular interactions in oxidative stress states.

Clin. Hemorheol. Microcirc., **21** (3–4), 189–193.

101. Ben-Hur, E., Barschtein, G., Chen, S., and Yedgar, S. (1997) Photodynamic treatment of red blood concentrates for virus inactivation enhances red blood cell aggregation: protextion with antioxydants. *Photochem. Photobiol.*, **66**, 509–512.
102. Ben-Hur, E. and Rosenthal, I. (1993) Effect of antioxidants on phthalocyanine photosensitization of human erythrocytes. *Proc. SPIE*, **1881**, 334–342.
103. Sorata, Y., Takahama, U., and Kimura, M. (2008) Cooperation of quercetin with ascorbate in the protection of photosensitized lysis of human erythrocytes in the presence of hematoporphyrin. *Photochem. Photobiol.*, **48** (2), 195–199.
104. Rollan, A., Ward, T., Flynn, G., McKerr, G., McHale, L., and McHale, A.P. (1996) Use of real time confocal laser scanning microscopy to study immediate effects of photodynamic activation on photosensitized erythrocytes. *Cancer Lett.*, **101**, 165–169.
105. Rollan, A. and McHale, L. (1997) Differential response of photosensitized young and old human erythrocytes to photodynamic activation. *Cancer Lett.*, **111**, 207–213.
106. Ando, K., Beppu, M., and Kikugdwa, K. (1995) Evidence for accumulation of lipid hydroperoxides during the aging of human red blood cells in circulation. *Biol. Pharm. Bull.*, **18**, 659–663.
107. Dellian, M., Abels, C., Kuhnle, G.E.N., and Goetz, A.E. (1995) Effects of photodynamic therapy on leucocyte-endothelium interaction: differences between normal and tumour tissue. *Br. J. Cancer*, **72** (5), 1125–1130.
108. Chen, B.J., Cui, X., Liu, C., and Chao, N.J. (2002) Prevention of graft-versus-host disease while preserving graft-versus-leukemia effect after selective depletion of host-reactive T cells by photodynamic cell purging process. *Blood*, **99** (9), 3083–3088.
109. Zhou, C., Chi, S., Deng, J., Zhang, H., Liang, J., and Ha, X. (1993) Effect of photodynamic therapy on mouse platelets. *Proc. SPIE*, **1881**, 60–66.
110. Fingar, V.H., Wieman, T.J., and Haydon, P.S. (1997) The effects of thrombocytopenia on vessel stasis and macromolecular leakage after photodynamic therapy using photofrin. *Photochem. Photobiol.*, **66** (4), 513–517.
111. Henderson, B.W., Owczarczak, B., Sweeney, J., and Gessner, T. (1992) Effect of photodynamic treatment of platelets or endothelial cells in vitro on platelet aggregation. *Photochem. Photobiol.*, **56** (4), 513–521.
112. Douplik, A.Y., Loschenov, V.B., Lebedeva, V.S., Derkacheva, V.M., Rumyanceva, V.D., Chasovnikova, L.V., Kusmin, S.G., Mironov, A.F., Lukyanets, E.A., and Sergienko, V.I. (1998) Oxygen consumption during photodynamic reactions in human blood. *Proc. SPIE*, **3254**, 461–468.
113. Ben-Hur, E., Moor, A.C.E., Margolis-Nunno, H., Gottlieb, R., Zuk, M.M., Lustigman, S., Horowitz, B., Brand, A., Steveninck, J., and Dubbelman, T.M.A.R. (1996) The photodecontamination of cellular blood components: mechanisms and use of photosensitization in transfusion medicine. *Transfus. Med. Rev.*, **10** (1), 15–22.
114. Kuimova, M.K., Botchway, S.W., Parker, A.W., Balaz, M., Collins, H.A., Anderson, H.L., Suhling, K., and Ogilby, P.R. (2009) Imaging intracellular viscosity of a single cell during photoinduced cell death. *Nat. Chem.*, **1**, 69–73.
115. Finlay, J.C., Mitra, S., Patterson, M.S., and Foster, T.H. (2004) Photobleaching kinetics of Photofrin *in vivo* and in multicell tumour spheroids indicate two simultaneous bleaching mechanisms. *Phys. Med. Biol.*, **49**, 4837–4860.
116. Juzenas, P., Didziapetriene, J., Staciokiene, L., Gudelis, V., Slavenas, J., and Rotomskis, R. (1996) Photomodifications of photodrugs in tumor. *Proc. SPIE*, **2625**, 499–506.
117. Georgakoudi, I., Nichols, M., and Foster, T.H. (1997) The mechanism of

photofrin photobleaching and its consequences for photodynamic therapy. *J. Photochem. Photobiol.*, **65** (1), 135–144.

118. Rosenthal, I. and Ben-Hur, E. (1995) Role of oxygen in the phototoxicity of phthalocyanines. *Int. J. Radiat. Biol.*, **67** (1), 85–91.
119. Nichols, M.G. and Foster, T.H. (1994) Oxygen diffusion and reaction kinetics in the photodynamic therapy of multicell tumour spheroids. *Phys. Med. Biol.*, **39**, 2161–2181.
120. Rotomskis, R., Streckyte, G., and Bagdonas, S. (1997) Phototransformation of sensitizers. 1. Significance of nature of the sensitizer in the photobleaching process and photoproduct formation in aqueous solution. *J. Photochem. Photobiol. B*, **39**, 167–171.
121. Kim, S., Huang, H., Pudavar, H.E., Cui, Y., and Prasad, P.N. (2007) Intraparticle energy transfer and fluorescence photoconversion in nanoparticles: an optical highlighter nanoprobe for two-photon bioimaging. *Chem. Mater.*, **19**, 5650–5656.
122. Douplik, A., Loschenov, V.B., and Stratonnikov, A.A. (1997) Spectra changes of whole blood with sulphonated aluminum phthalocyanine photosensitizer during photodynamic therapy in vitro. *Proc. SPIE*, **2923**, 44–48.
123. Stratonnikov, A.A., Douplik, A.Y., Klimov, D.V., Loschenov, V.B., and Mizin, S.V. (1998) Influence of light irradiation on blood oxygen saturation level in vitro and in vivo during photodynamic therapy. *Proc. SPIE*, **3247**, 128–136.
124. Silva, F.A.M. and Newman, E.L. (1994) Time-dependent photodynamic damage to blood vessels: correlation with serum photosensitizer levels. *Photochem. Photobiol.*, **61**, 414–416.
125. Wilson, B.C., Patterson, M.S., and Lilge, L. (1998) Implicit and explicit dosimetry in photodynamic therapy: New Paradigm. *Laser Med. Sci.*, **12**, 182–199.
126. Fournier, R.L. (1995) *Basic Concepts of Biomedical Engineering*, Greyden Press, Columbus, OH.
127. Schimidt, R.F. and Thews, G. (1989) *Human Physiology*, Springer-Verlag, Berlin.
128. Georgakoudi, I. and Foster, T.H. (1998) Singlet oxygen- versus nonsinglet oxygen-mediated mechanisms of sensitizer photobleaching and their effects on photodynamic dosimetry. *Photochem. Photobiol.*, **67**, 612–625.
129. Foley, M.S.C., Beeby, A., Parker, A.W., Bishop, S.M., and Phillips, D. (1997) Excited triplet state photophysics of the sulphonated aluminium phthalocyanines bound to human serum albumin. *J. Photochem. Photobiol. B*, **38**, 10–17.
130. Ben-Hur, E., Malik, Z., Dubbelman, T.M.A.R., Margaron, P., Ali, H., and Lier, J.E. (1993) Phthalocyanine-induced photohemolysis: structure-activity relationship and the effects of fluoride. *Photochem. Photobiol.*, **58**, 351–355.
131. Moan, J., Iani, V., and Ma, L.W. (1998) In vivo fluorescence of phthalocyanines during light exposure. *J. Photochem. Photobiol.*, **B 42**, 100–103.
132. Moan, J., Iani, V., Ma, L.W., and Peng, Q. (1996) Photodegradation of sensitizers in mouse skin during PCT. *Proc. SPIE*, **2625**, 187–193.
133. Rück, A., Beck, G., Bachor, R., Akguen, N., Gschwend, M., and Steiner, R. (1996) Dynamic fluorescence changes during photodynamic therapy in vivo and in vitro of hydrophilic Al(III) phthalocyanine tetrasulphonate and lipophilic Zn(II) phthalocyanine administered in liposomes. *J. Photochem. Photobiol. B*, **36**, 127–133.
134. Foster, T.H., Murant, R.S., Bryant, R.G., Knox, R.S., Gibson, S.L., and Hilf, R. (1991) Oxygen consumption and diffusion effects in photodynamic therapy. *Radiat. Res.*, **126**, 296–303.
135. Yuan, J., Mahama-Relue, P.A., Fournier, R.L., and Hampton, J.A. (1997) Predictions of mathematical models of tissue oxygenation and generation of singlet oxygen during photodynamic therapy. *Radiat. Res.*, **148**, 386–394.
136. Foster, T.H., Scott, L., Gibson, L., Gao, L., and Hilf, R. (1992) Analysis of photochemical oxygen consumption effects

in photodynamic therapy. *Proc. SPIE*, **1645**, 104–114.

137. Sitnik, T.M., Hampton, J.A., and Henderson, B.W. (1998) Reduction of tumor oxygenation during and after photodynamic therapy in vivo: effects of fluence rate. *Br. J. Cancer*, **77**, 1386–1394.

138. Moor, A.C.E., Lagerberg, J.W.M., Tijssen, K., Foley, S., Truscott, T.G., Kochevar, I.E., Brand, A., Dubbelman, T.M.A.R., and VanSteveninck, J. (1997) In vitro fluence rate effects in photodynamic reactions with AlPcS4 as sensitizer. *Photochem. Photobiol.*, **66**, 860–865.

139. Ben-Hur, E., Geacintov, N.E., Studamire, B., Kenney, M.E., and Horowitz, B. (1995) The effect of irradiance on virus sterilization and photodynamic damage in red blood cells sensitized by phtjalocyanines. *Photochem. Photobiol.*, **61**, 190–195.

140. Loschenov, V., Konov, V., and Prokhorov, A. (2000) Photodynamic therapy and fluorescence diagnostics. *Laser Phys.*, **10** (6), 1188–1207.

141. Bisland, S.K., Lilge, L., Lin, A., Rusnov, R., and Wilson, B.C. (2004) Metronomic photodynamic therapy as a new paradigm for photodynamic therapy: rationale and preclinical evaluation of technical feasibility for treating malignant brain tumors. *Photochem. Photobiol.*, **80** (1), 22–30.

142. Potter, W.R. and Mang, T.J.D. (1987) The theory of photodynamic therapy dosimetry: consequences of photodestruction of sensitizers. *Photochem. Photobiol.*, **46**, 853–858.

143. Grossweiner, L.I. (1997) PDT light dosimetry revisited. *J. Photochem. Photobiol. B*, **38**, 258–268.

144. Ma, L.W., Moan, J., Grahn, M.F., and Iani, V. (1996) A comparison of meso-tetrahydroxyphenyl-chlorin and meso-tetrahydroxyphenyl-bacteriochlorin with respect to photobleaching and PCT efficiency *in vivo*. *Proc. SPIE*, **2924**, 219–225.

145. Robinson, D.J., Stringer, M.R., and Crum, W.R. (1996) A rate-equation analysis of protoporphyrin IX photo-oxidation. *Proc. SPIE*, **2625**, 413–418.

146. Forrer, M., Glanzmann, T., Braichotte, D., Wagnieres, G., Bergh, H., Savary, J.F., and Monnier, P. (1996) *In vivo* measurement of fluorescence bleaching of meso-tetra hydroxy phenyl chlorin (mTHPC) in the esophagus and the oral cavity. *Proc. SPIE*, **2627**, 33–39.

147. Hawkes, R.P., Farrell, T.J., Patterson, M.S., and Weersink, R.A. (1997) Changes in fluorescence emission during PDT due to photobleaching: potential usefulness as marker of tissue damage. *Proc. SPIE*, **2975**, 208–221.

148. Stratonnikov, A.A., Meerovich, G.A., and Loschenov, V.B. (2000) Photobleaching of photosensitizers applied for photodynamic therapy. *Proc. SPIE*, **3909**, 81–91.

149. Dodd, R.Y. (1992) The risk of transfusion-transmitted infection. *N. Engl. J. Med.*, **327**, 419–421.

150. Soukos, N.S., Wilson, M., Burns, T., and Speight, P.M. (1996) Photodynamic effects of toluidine blue on human oral keratinocytes and fibroblasts and Streptococcus sanguis evaluated in vitro. *Lasers Surg. Med.*, **18**, 253–259.

151. Zeina, B., Greenman, J., Corry, D., and Purcell, W.M. (2003) Antimicrobial photodynamic therapy: assessment of genotoxic effects on keratinocytes in vitro. *Br. J. Dermatol.*, **148**, 229–232.

152. Politis, C., Kavallierou, L., Hantziara, S., Katsea, P., Triantaphylou, V., Richardson, C., Tsoutsos, D., Anagnostopoulos, N., Gorgolidis, G., and Ziroyannis, P. (2007) Quality and safety of fresh-frozen plasma inactivated and leucoreduced with the Theraflex methylene blue system including the Blueflex filter: 5 years' experience. *Vox Sang*, **92**, 319–326.

153. Wainwright, M. (2002) The emerging chemistry of blood product disinfection. *Chem. Soc. Rev.*, **31**, 128–136.

154. Wainwright, M. (2002) Pathogen inactivation in blood products. *Curr. Med. Chem.*, **9**, 127–143.

155. Burrows, C.J. and Muller, J.G. (1998) Oxidative nucleobase modifications

leading to strand scission. *Chem. Rev.*, **98** (3), 1109–1152.

156. Griffiths, M.A., Wren, B.W., and Wilson, M. (1997) Killing of methicillin-resistant Staphylococcus aureus in vitro using aluminium disulphonated phthalocyanine, a light-activated antimicrobial agent. *J. Antimicrob. Chemother.*, **40**, 873–876.

157. Donnelly, R.F., McCarron, P.A., Cassidy, C.M., Elborn, J.S., and Tunney, M.M. (2007) Delivery of photosensitizers and light through mucus: investigations into the potential use of photodynamic therapy for treatment of Pseudomonas aeruginosa cystic fibrosis pulmonary infection. *J. Control. Release*, **117**, 217–226.

158. Smetana, Z., Ben-Hur, E., Mendelson, E., Salzberg, S., Wagner, P., and Malik, Z. (1998) Herpes simplex virus proteins are damaged following photodynamic inactivation with phthalocyanines. *J. Photochem. Photobiol. B*, **44**, 77–83.

159. Schagen, F.H., Moor, A.C., Cheong, S.C., Cramer, S.J., Ormondt, H., Eb, A.J., Dubbelman, T.M.A.R., and Hoeben, R.C. (1999) Photodynamic treatment of adenoviral vectors with visible light: an easy and convenient method for viral inactivation. *Gene Ther.*, **6**, 873–881.

160. Mahy, B.W.J. (1985) *Virology – A Practical Approach*, IRL Oxford Press, Washington, DC.

161. Lambrecht, B., Mohr, H., Knuver-Hopf, J., and Schmitt, H. (1991) Photoinactivation of viruses in human fresh plasma by phenothiazine dyes in combination with visible light. *Vox Sang*, **60**, 207–213.

162. Horowitz, B., Williams, B., Rywkin, S., Prince, A.M., Pascual, D., Geacintov, N., and Valinsky, J. (1991) Inactivation of viruses in blood with aluminium phthalocynine derivatives. *Transfusion*, **31**, 102–108.

163. Horowitz, B., Rywkin, S., Margolis-Nunno, H., Williams, B., Geacintov, N., Prince, A.M., Pascual, D., Ragno, G., Valeri, C.R., and Huima-Byron, T. (1992) Inactivation of viruses in red cell and platelet concentrates with aluminum phthalocyanine (AlPc) sulfonates. *Blood Cells*, **18**, 141–150.

164. Matthews, J.L., Sogandares-Bernal, F., Judy, M., Gulliya, K., Newman, J., Chanh, T., and Marengo-Rowe, A. (1992) Inactivation of viruses with photoactive compounds. *Blood Cells*, **18**, 75–89.

165. Lewin, A.A., Schnipper, L.E., and Crumpacker, C.S. (1980) Photodynamic inactivation of herpes simplex virus by hematoporphyrin derivative and light. *Proc. Soc. Exp. Med.*, **163**, 181–190.

166. North, J., Neyndorff, H., and Levy, J.G. (1993) Photosensitizers as virucidal agents. *J. Photochem. Photobiol. B*, **17** (2), 99–108.

167. Neyndorff, H.C., Bartel, D.L., Tufaro, F., and Levy, J.G. (1990) Development of a model to demonstrate photosensitizer-mediated viral inactivation in blood. *Transfusion*, **30**, 485–490.

168. North, J., Coombs, R., and Levy, J. (1994) Photodynamic inactivation of free and cell-associated HIV-1 using the photosensitizer, benzoporphyrin derivative. *J. Acquir. Immune Defic. Syndr.*, **7**, 891–898.

169. Lagerberg, J.W.M., Uberriegler, K.P., Krammer, B., Steveninck, J., and Dubbelman, T.M.A.R. (2000) Plasma membrane properties involved in the photodynamic efficacy of merocyanine 540 and tetrasulfonated aluminum phthalocyanine. *Photochem. Photobiol.*, **71**, 341–346.

170. Zeiler, T., Riess, H., Wittmann, G., Hintz, G., Zimmermann, R., Müller, C., Heuft, H.G., and Huhn, D. (1994) The effect of methylene blue phototreatment on plasma proteins and in vitro coagulation capability of single-donor fresh-frozen plasma. *Transfusion*, **34**, 685–689.

171. Wagner, S.J., Storry, J., Mallory, D.A., Stromberg, R.R., Benade, L.E., and Friedman, L.I. (1993) Red cell alterations associated with virucidal methylene blue phototreatment. *Transfusion*, **33**, 30–36.

172. Wagner, S.J., Cifone, M.A., Murli, H., Dodd, R.Y., and Myhr, B. (1995)

Mammalian genotoxicity assessment of methylene blue in plasma: Implications for virus inactivation. *Transfusion*, **35**, 407–413.
173. Morrison, H., Mohammad, T., and Kurukulasuriya, R. (1997) Photobiological properties of methylene violet. *Photochem. Photobiol.*, **66**, 245.
174. Mohr, H., Bachmann, B., Klein-Struckmeier, A., and Lambrecht, B. (1997) Virus inactivation of blood products by phenothiazine dyes and light. *Photochem. Photobiol.*, **65**, 441–445.
175. Skripchenko, A., Robinette, D., and Wagner, S.J. (1997) Comparison of methylene blue and methylene violet for photoinactivation of intracellular and extracellular virus in red cell suspensions. *Photochem. Photobiol.*, **65**, 451–455.
176. Ben-Hur, E., Hoeben, R.C., Ormondt, H., Dubbelman, T.M.A.R., and Steveninck, J.V. (1992) Photodynamic inactivation of retroviruses by phthalocyanines: the effect of sulphonation, metal ligand and fluoride. *J. Photochem. Photobiol. B.*, **13**, 145–152.
177. Kasermann, F. and Kempf, C. (1998) Buckminsterfullerene and photodynamic inactivation of viruses. *Rev. Med. Virol.*, **8**, 143–151.
178. Schulz, W.A. (2005) *Molecular Biology of Human Cancers. An Advanced Student's Textbook*, Springer.
179. Gulati, S.C. and Bennett, C. (1992) Role of granulocytemacrophage colony stimulating factor (GM-CSF) after autologous BMT for Hodgkin's disease. *Ann. Intern. Med.*, **116**, 177–182.
180. Sieber, F., Spivak, J.L., and Sutcliffe, A.M. (1984) Selective killing of leukemic cells by merocyanine 540-mediated photosensitization. *Proc. Natl. Acad. Sci. U.S.A.*, **81**, 7584–7587.
181. Singer, C.R.J., Linch, D.C., Bown, S.G., Huehns, E.R., and Goldstone, A.H. (1988) Differential phthalocyanine photosensitization of acute myeloblastic leukaemia progenitor cells: A potential purging technique for autologous bone marrow transplantation. *Br. J. Haematol.*, **68**, 417–422.
182. Jamieson, C., Richter, A., and Levy, J.G. (1993) Efficacy of benzoporphyrin derivative, a photosensitizer, in selective destruction of leukemia cells using a murine tumor model. *Exp. Hematol.*, **21**, 629–634.
183. Lemoli, R.M., Igaroshi, T., Knizewski, M., Acaba, L., Richter, A., Jain, A., Mitchell, D., Levy, J., and Gulati, S.C. (1993) Dye-mediated photolysis is capable of eliminating drug-resistant (MDR+) tumor cells. *Blood*, **81**, 793–800.
184. Agarwal, M.L., Clay, M.E., Harvey, E.J., Evans, H.H., Antunez, A.R., and Oleinick, N.L. (1991) Photodynamic therapy induces rapid cell death by apoptosis in LS178Y mouse lymphoma cells. *Cancer Res.*, **51**, 5993–5996.
185. Lavie, G., Kaplinsky, C., Toren, A., Aizman, I., Meruelo, D., Mazur, Y., and Mandel, M. (1999) A photodynamic pathway to apoptosis and necrosis induced by dimethyl tetrahydroxyhelianthrone and hypericin in leukaemic cells: possible relevance to photodynamic therapy. *Br. J. Cancer*, **79** (3/4), 423–432.
186. Chen, J.Y., Mak, N.K., Wen, J.M., Leung, W.N., Chen, S.C., Fung, M.C., and Cheung, N.H. (1998) A comparison of the photodynamic effects of temoporfin (mTHPC) and MC540 on leukemia cells: efficacy and apoptosis. *Photochem. Photobiol.*, **68** (4), 545–554.
187. Gulliya, K.S. and Pervaiz, S. (1989) Elimination of clonogenic tumor cells from HL-60, Daudi and U-937 cell lines: implication for autologous marrow transplantation. *Blood*, **73**, 1059–1066.
188. Anderson, M.S., Kalyanaraman, B., and Feix, J.B. (1993) Enhancement of merocyanine 540-mediated phototherapy by salicylate. *Cancer Res.*, **53**, 806–809.
189. Daziano, J.-P., Humeau, L., Henry, M., Mannoni, P., Chanona, M., Chabannon, C., and Julliard, M. (1998) Preferential photoinactivation of leukemia cells by aluminum phthalocyanine. *J. Photochem. Photobiol. B: Biol.*, **43** (2), 128–135.
190. Huang, H.F., Chen, Y.Z., Wu, Y., and Chen, P. (2006) Purging of

murine erythroblastic leukemia by ZnPcS2P2-based-photodynamic therapy. *Bone Marrow Transplant.*, **37**, 213–217.

191. Brasseur, N., Ménard, A., Forget, R., el Jastimi, R., Hamel, R., Molfino, N.A., and Lier, J.E. (2000) Eradication of multiple myeloma and breast cancer cells by TH9402-mediated photodynamic therapy: implication for clinical *ex vivo* purging of autologous stem cell transplants. *Photochem. Photobiol.*, **72** (6), 780–787.

192. Lee, K.H. and Garro, J. (1989) Engineering aspects of extracorporeal photochemotherapy. *Yale J. Biol. Med.*, **62**, 621–628.

Index

Advanced Optical Flow Cytometry: Methods and Disease Diagnoses, First Edition. Edited by Valery V. Tuchin.

o

p

q

r

s

t

v

w